石油管材及装备材料服役行为与结构安全国家重点实验室科研成果汇编

(2020年)

中国石油集团石油管工程技术研究院
石油管材及装备材料服役行为与结构安全国家重点实验室
中国石油集团石油管工程重点实验室
陕西省石油管材及装备材料服役行为与结构安全重点实验室
编

石油工业出版社

内 容 提 要

本书汇编了中国石油集团石油管工程技术研究院、石油管材及装备材料服役行为与结构安全国家重点实验室、中国石油集团石油管工程重点实验室和陕西省石油管材及装备材料服役行为与结构安全重点实验室在2020年正式发表在国际国内刊物上的论文和2020年授权专利及省部级以上获奖成果，反映了近几年石油管工程的科研成果及进展。内容涉及输送管与管线安全评价、油井管与管柱失效预防、腐蚀防护与非金属材料等方面。

本书可供从事石油管工程的技术人员和石油院校相关专业师生参考。

图书在版编目（CIP）数据

石油管材及装备材料服役行为与结构安全国家重点实验室科研成果汇编. 2020年／中国石油集团石油管工程技术研究院等编. —北京：石油工业出版社，2020.8
ISBN 978-7-5183-4791-9

Ⅰ.①石⋯ Ⅱ.①中⋯ Ⅲ.①石油管道—管道工程—文集 Ⅳ.①TE973-53

中国版本图书馆CIP数据核字（2021）第157404号

出版发行：石油工业出版社
　　　　　（北京安定门外安华里2区1号　100011）
　　　　　网　　址：www.petropub.com
　　　　　编辑部：（010）64523687　图书营销中心：（010）64523633
经　　销：全国新华书店
印　　刷：北京中石油彩色印刷有限责任公司

2021年8月第1版　2021年8月第1次印刷
787×1092毫米　开本：1/16　印张：34.5
字数：859千字

定价：200.00元
（如出现印装质量问题，我社图书营销中心负责调换）
版权所有，翻印必究

《石油管材及装备材料服役行为与结构安全国家重点实验室科研成果汇编（2020年）》编辑委员会

顾　问：黄维和　李鹤林　高德利　赵怀斌　张建军

主　任：刘亚旭

副主任：冯耀荣　霍春勇

委　员：（按姓氏笔画排序）

陈　平　崔红升　高惠临　郭文奇　韩恩厚
乐　宏　李国顺　李贺军　李建军　刘汝山
罗　超　马秋荣　苗长贵　闵希华　孙　军
汤晓勇　王铁军　王香增　魏志平　胥志雄
闫相祯　张士诚　赵新伟　郑新权　周建良
周　敏

主　编：马秋荣

副主编：罗金恒　尹成先　韩礼红　戚东涛　池　强
宫少涛

编辑组：林　凯　林元华　姜　放　房　军　黄桂柏
陈宏远　冯　春　付安庆　李厚补　马卫锋
王　鹏　刘心可

前　言

　　石油管材及装备材料服役行为与结构安全国家重点实验室成立于2015年10月，与中国石油天然气集团有限公司石油管工程重点实验室和陕西省石油管材及装备材料服役行为与结构安全重点实验室两个省部级重点实验室并行运行（以下简称重点实验室），是我国在石油管及装备材料服役安全研究领域的科技创新基地、人才培养基地和学术交流基地。

　　重点实验室依托中国石油集团石油管工程技术研究院，设置输送管与管线安全评价研究、油井管与管柱失效预防研究、石油管材及装备腐蚀与防护研究和先进材料及应用技术研究4个研究方向，围绕我国油气工业发展战略需求，特别是大口径高压输气管道建设和复杂工况油气田勘探开发的技术需求，以管材及装备服役过程中的断裂、变形、泄漏、腐蚀、磨损、老化等突出失效行为为对象，深入开展石油管及装备材料服役领域的基础和应用基础研究，积累服役性能数据，创新研究试验方法，突破国外的技术封锁和壁垒，形成自主创新的技术体系和知识体系，为我国油气重大工程的选材、安全评估与寿命预测提供科学技术支撑。2020年，获省部级、中国石油集团及各类学会等社会力量科技奖励二等奖以上8项，其中特等奖1项，一等奖5项，二等奖2项；SCI/EI收录论文32篇，出版著作和研究文集5部；授权发明专利37件；参与制修定国际标准1项，制修定国家标准1项，行业标准4项。

　　在输送管与管线安全评价研究领域，完成高应变海洋管线管及相关产品试制，为中俄东线、西气东输四线等管道建设工程制定了相关管材标准；针对长输管道环焊缝断裂与变形问题，论证了离散断裂韧性的管道环焊缝断裂评估方法，形成了药芯半自动焊工艺的管道环焊缝强度预测技术；开展了冻土地区管道断裂控制模型的研究，结合冻土性质，提出了适合冻土的土壤约束修正系数，完善了现有的基于巴特尔双曲线的延性断裂控制模型；将现有基于应变设计管道拉伸应变容量模型的适用范围由 1219mm×25.4mm 拓展到了 1422mm×38.5mm，解决了我国未来新建西四线等大口径大壁厚管道的变形能力评估预测难题；形成了电弧增材制造三通的打印方案和工艺，实现了低温高韧性三通的增材制造；开发了用于油气领域低碳微合金钢增材打印焊丝，为增材制造在油气领域的应用奠定了基础。

在油井管与管柱失效预防研究领域，针对页岩油气井套管大量变形限制井下作业的突出瓶颈，完善了非常规油气井筒全尺寸模拟试验装备，实现全井筒模拟试验、非均匀载荷及位移控制下模拟试验评价新能力，提出了预防西南页岩气井套变的井筒协调变形技术，并进行了3口井的现场试验；研发了超深井钻机用轻型675大钩、750吊环和TMCP轧制H型钢新材料新工艺，材料性能提升了15%以上；针对超深井、短半径井等高效安全钻井需求，研制了105ksi级钛合金钻杆，形成了含钛合金钻杆的钻柱结构设计校核技术、钛合金钻杆材料热处理优化技术、适用性评价、使用维护技术等关键技术体系；国产化ϕ88.9mm钛合金钻杆完成了国内首次7000m超深短半径井的现场下井，为管材及装备的轻量化提供了技术支撑。

在石油管材及装备腐蚀与防护技术研究领域，针对塔里木油田高温高压气井超级13Cr油管断裂问题，在实验室模拟还原了现场管柱的断裂过程，明确了超级13Cr油管在受污染环空保护液体系中的断裂机理；发明了基于"主成分分析+支持向量机"机器学习理论的冶金双金属复合管逆向材料设计技术，形成了冶金双金属复合管界面结合力定量预测与调控；基于树脂分子链三维互穿技术和链端基团互补设计，自主研发了常温固化（经济性）无溶剂（环保）有机硅改性环氧涂层；针对加氢高压换热器铵盐腐蚀穿孔与结晶的突出问题，开发了加氢装置铵盐结晶模型；针对典型承压设备随机耦合环境下多损伤交互作用导致的安全评价不准确的瓶颈，建立了加氢反应器、换热器、深海立管、长输管线和LNG储罐多损伤失效模式下的概率失效评定方法。

在先进材料及应用技术研究领域，建立了聚乙烯复合管复合加载试验技术，并确定了管材应用关键技术指标；建立了聚乙烯复合管的温度折减系数，开发了两种类型聚乙烯复合管服役寿命预测技术；完善了聚乙烯复合管线工作压力（设计压力）计算方法，充分考虑管材性能、服役环境以及管道安全等级，提高了管材的应用可靠性；研发形成了热塑性塑料管多层共挤复合制备技术，采用新技术制备了2种复合内衬（PA/PE、ETFE/PE）高温高压抗硫非金属管，并在塔里木油田得到现场应用4.7km，内衬共挤复合制备的非金属管比原产品降低成本约50%；攻关研究了复合材料增强层超声相控阵缺陷无损检测技术，具有分辨率高、检测效率高、结果直观的特点；探索了光纤光栅（FBG）传感器技术，通过预埋FBG的方式，研究了光纤技术对复合材料增强层应变、温度监测的可行性。

本成果汇编收集了重点实验室于 2020 年在国内外重要刊物和学术会议上发表的 51 篇论文，并介绍了 2020 年授权专利及省部级以上获奖成果，这些资料从一个侧面反映了重点实验室近期所取得的研究成果。可以为从事油气管道工程、油气井工程、石油工程材料、安全工程等方面的工程技术人员、研究人员和管理人员提供参考。

由于编者水平有限，经验不足，错误和不妥之处在所难免，敬请广大读者批评指正。

编 者

2021 年 4 月

目 录

第一篇 论文篇

一、输送管与管线安全评价

输送管产品开发与工程应用技术支撑体系 …………… 李鹤林 霍春勇 池 强 等（4）

中国 X80 管线钢和钢管研发应用进展及展望 …………… 冯耀荣 吉玲康 李为卫 等（14）

大应变管线钢和钢管的关键技术进展及展望 …………… 冯耀荣 吉玲康 陈宏远 等（24）

316L 衬里复合管道主要失效形式及其完整性检测技术研究
……………………………………………………………… 赵新伟 魏 斌 杨专钊 等（34）

Constitutive Equation for Describing True Stress-Strain Curves over a Large
Range of Strains …………………………… Cao Jun Li Fuguo Ma Weifeng et al（43）

A Review of Dynamic Multiaxial Experimental Techniques
…………………………………………… Nie Hailiang Ma Weifeng Wang Ke et al（52）

Fatigue Failure Analysis of Dented Pipeline and Simulation Calculation
………………………………………… Luo Jinheng Zhang Yani Li Lifeng et al（66）

Failure Analysis on Tee Pipe of Duplex Stainless in an Oilfield ………… Zhang Shuxin（81）

Cracking Analysis of a Newly Built Gas Transmission Steel Pipe
………………………………………… Ding Han Qi Dongtao Qi Guoquan et al（95）

Microstructure and Impact Toughness of X70-L245 Butt Girth Weld in Natural Gas Station
………………………………………… Ren Junjie Ma Weifeng Hui Wenying et al（105）

Dynamic Impact Damage of Oil and Gas Pipelines
………………………………………… Nie Hailiang Ma Weifeng Sha Shengyi et al（112）

X70 管道自保护药芯焊丝环焊接头力学性能及影响因素
……………………………………………………………… 何小东 薛 如 李为卫 等（119）

埋地管道泄漏数值模拟分析 ………………………………… 王 俊 封 辉 高 琦 等（125）

二、油井管与管柱失效预防

复杂工况油套管柱失效控制与完整性技术研究进展及展望
……………………………………………………………… 冯耀荣 付安庆 王建东 等（138）

·1·

复杂压裂页岩气井套管变形机制及控制方法 .... 韩礼红　杨尚谕　魏风奇　等（148）
Investigation on Impact Absorbed Energy Index of Drill Pipe .... Li Fangpo（159）
Aging Treatment Effect on Microstructure and Mechanical Properties of Ti-5Al-3V-1.5Mo-2Zr
　　Titanium Alloy Drill Pipe .... Feng Chun　Li Ruizhe　Liu Yonggang et al（168）
Tribological Properties of Ni/Cu/Ni Coating on the Ti-6Al-4V Alloy after Annealing
　　at Various Temperatures .... Luo Jinheng　Wang Nan　Zhu Lixia et al（187）
Application Analysis of Epoxy-Coated Tubing in an Oilfield
　　............ Zhu Lijuan　Feng Chun　Han Lihong　et al（200）
Failure Analysis of Casing Dropping in Shale Oil Well during Large Scale Volume Fracturing
　　............ Wang Hang　Zhao Wenlong　Shu Zhenhui et al（206）
Numerical Analysis of Casing Deformation under Cluster Well Spatial Fracturing
　　............ Wang Jianjun　Jia Feipeng　Yang Shangyu et al（218）
Fracture Failure Analysis of C110 Oil Tube in a Western Oil Field
　　............ Zhu Lixia　Kuang Xianren　Xiong Maoxian et al（225）
石油管用 Ti-6Al-4V-0.1Ru 钛合金高温流变行为及预测模型研究
　　............ 刘　强　白　强　田　峰　等（232）
油气开采用钛合金石油管材料耐腐蚀性能研究 .... 刘　强　惠松骁　汪鹏勃　等（242）
钛合金油套管抗挤毁性能计算与实验 .... 刘　强　申照熙　李东风　等（254）

三、腐蚀防护与非金属材料

Facile Fabrication of SnO$_2$ Modified TiO$_2$ Nanorods Film for Efficient Photocathodic Protection
　　of 304 Stainless Steel under Simulated Solar Light
　　............ Zhanga Juantao　Yangb Hualong　Wanga Yuan et al（266）
Effect of CO$_2$/H$_2$S and Applied Stress on Corrosion Behavior of 15Cr Tubing in Oil
　　Field Environment .... Zhao Xuehui　Huang Wei　Li Guoping et al（281）
Corrosion Behavior of Cr-Bearing Steels in CO$_2$-O$_2$-H$_2$O Multi-Thermal-Fluid Environment
　　............ Yuan Juntao　Zhu Kaifeng　Jiang Jingjing et al（293）
Corrosion Behavior of Reduced-Graphene-Oxide-Modified Epoxy Coatings on N80 Steel in
　　10.0wt% NaCl Solution .... Feng Chun　Cao Yaqiong　Zhu Lijuan et al（306）
Experimental and Simulation Investigation on Failure Mechanism of a Polyethylene Elbow Liner
　　Used in an Oilfield Environment .... Kong Lushi　Fan Xin　Ding Nan et al（317）
Failure Analysis on the Oxygen Corrosion of the Perforated Screens Used in a Gas Injection
　　Huff and Puff Well .... Fan Lei　Gao Yuan　Yuan Juntao et al（327）
Investigation on Leakage Cause of Oil Pipeline in the West Oilfield of China
　　............ Liu Qiang　Yu Haoyu　Zhu Guochuan et al（337）

Comparative Study on Hydrogen Embrittlement Susceptibility in Heat-Affected Zone of
　　TP321 Stainless Steel ·················· Xu Xiuqing　Niu Jing　Li Chengzheng et al（349）
External Stress Corrosion Cracking Risk Factors of High Grade Pipeline Steel
　　································· Zhu Lixia　Luo Jinheng　Wu Gang et al（357）
Analysis of Corrosion Behavior on External Surface of 110S Tubing
　　································ Han Yan　Li Chengzheng　Zhang Huali et al（365）
Failure Analysis and Solution to Bimetallic Lined Pipe
　　································· Li Fagen　Li Xunji　Li Weiwei et al（376）
外加电位对 X80 管线钢在轮南土壤模拟溶液中应力腐蚀行为的影响
　　······································ 朱丽霞　贾海东　罗金恒　等（382）
页岩气输送用转角弯头内腐蚀减薄原因分析 ··········· 朱丽霞　罗金恒　李丽锋　等（391）
16Mn 管线钢的焊缝表面冲蚀机理研究 ············· 武　刚　李德君　罗金恒　等（399）
酸洗钝化对 316L/L415 双金属机械复合管环焊缝耐蚀性的影响
　　······································ 宋成立　王福善　冯　泉　等（409）
干湿交替砂土环境下 X80 管线钢的腐蚀行为研究 ······ 李丽锋　李　超　罗金恒　等（415）
一种乙烯裂解炉管高温损伤评估的新方法 ············ 徐秀清　吕运容　尹成先　等（425）
炼化企业常压塔顶露点温度计算及缓蚀剂性能研究
　　······································ 范　磊　刘宏铭　杜笑怡　等（435）
连续纤维复合材料环形试样蠕变行为研究 ············ 张冬娜　邵晓东　蔡雪华　等（439）
非金属智能连续管拉伸层力学特性研究 ············· 丁　楠　李厚补　古兴隆　等（445）

四、其他

A Strain Rate Dependent Fracture Model of 7050 Aluminum Alloy
　　································· Cao Jun　Li Fuguo　Ma Weifeng et al（456）
Effects of Pre-Oxidation on the Corrosion Behavior of Pure Ti under Coexistence of
　　Solid NaCl Deposit and Humid Oxygen at 600℃：the Diffusion of Chlorine
　　································· Fan Lei　Liu Li　Lv Yuhai et al（470）
Failure Analysis of Crankshaft of Fracturing Pump
　　······························ Wang Hang　Yang Shangyu　Han Lihong et al（487）
Failure Analysis of a Sucker Rod Fracture in an Oilfield
　　································ Ding Han　Zhang Aibo　Qi Dongtao et al（500）
Stress Analysis of Large Crude Oil Storage Tank Subjected to Harmonic Settlement
　　··························· Zhang Shuxin　Liu Xiaolong　Luo Jinheng et al（509）
Study on Remaining Oil Distribution of Single Sand Body in Yan 10 Reservoir in
　　Zhenbei Area, Ordos Basin ············ Wang Shuai　Luo Jinheng　Li Sen et al（515）

第二篇 成果篇

一、省部级科技奖励

1. OD1422mm X80 管线钢管研制及应用技术 …………………………………………（525）
2. 复杂工况油气井管柱腐蚀控制技术及工程应用 ………………………………………（526）
3. 西部油气田集输管线内腐蚀控制技术及工程应用 ……………………………………（530）
4. 石油工业用高性能膨胀管及其性能评价技术 …………………………………………（534）

二、授权专利目录

第一篇 论文篇

一、输送管与管线安全评价

输送管产品开发与工程应用技术支撑体系

李鹤林　霍春勇　池　强　杜　伟

(中国石油集团石油管工程技术研究院·石油管材及装备材料服役行为与结构安全国家重点实验室)

摘　要：随着天然气需求的持续增长，中国高钢级油气输送管道建设快速发展，已进入国际领跑者行列。在高钢级油气输送管材应用过程中，产品开发与工程应用技术支撑体系发挥了重要作用。该技术支撑体系由质量监督与评价、标准化、科学研究、失效分析4个部分组成，分别阐述了各组成部分的内涵、发展历程及现状。该技术支撑体系在中国油气输送管产业发展与工程应用及中俄东线天然气管道工程X80输送管产品开发与工程建设中发挥了重要作用，并将继续为本质安全的管道工程建设提供技术支撑。

关键词：输送管；技术支撑体系；质量监督与评价；标准化；科学研究；失效分析

自1959年建成新疆克拉玛依—独山子输油管道以来，中国油气管道建设已经过60余年的发展历程。2012年建成的西气东输二线，主干线管径1219mm、钢级X80、输气压力12MPa/10MPa，干线全长4895km，标志着中国油气管道建设技术水平已跻身世界先进行列。2019年12月，中俄东线天然气管道工程投产通气，其主干线管径1422mm、钢级X80，输气压力12MPa，代表了目前国际上X80天然气管道建设的最高水平[1-12]。截至2018年，中国已建成油气管道总里程13.6×10^{4}km，其中天然气管道7.9×10^{4}km，形成了"北油南运""西油东进""西气东输""海气登陆"的油气输送格局，初步实现了横跨东西、纵贯南北、联通海外、覆盖全国的油气管道运输体系。

20世纪50—70年代，中国管线钢主要采用鞍钢等厂家生产的A3、16Mn；20世纪70年代后期及80年代，采用从日本进口的TS52K（相当于X52）[13]。"六五"到"七五"10年间，武钢牵头开展了X系列管线钢科技攻关。"八五"期间，进一步开发了高性能X52—X70API系列管线钢。进入21世纪，在西气东输及西气东输二线等重大管道工程的推动下，石油管工程技术研究院（以下简称管研院）联合国内多家钢厂、管厂、科研院所开展多项科研攻关，制定了X70/X80管线钢及钢管标准，解决了X70/X80钢管应用中的诸多问题，推动了X70/X80钢管产品的国产化与规模化应用，使中国管线钢管从跟仿国外技术发展至加入全球领跑者行列[14-29]。目前，中国已启动开展X90、X100管线钢管的开发与应用技术研究[30-33]。中国在高钢级管线钢及钢管研发应用方面起步较晚，但研发与应用速度较快，近20年走完了

基金项目：中国石油科学研究与技术开发项目"公司发展战略与科技基础工作决策支持研究"专题"石油管工程学科体系建设研究"（2017D-5001-10）。

作者简介：李鹤林，男，1937年生，教授级高工，中国工程院院士，1961年毕业于西安交通大学金属材料及热处理专业，现主要从事石油管及装备材料工程技术方向的研究工作。地址：陕西省西安市锦业二路89号(710077)，电话：029-81887866，E-mail：lihelin@cnpc.com.cn。

发达国家管线钢管50多年的研发进程。究其原因，油气输送管产品开发与工程应用技术支撑体系的建立与完善起到了举足轻重的作用。

1 技术支撑体系的形成与发展

管研院及其前身石油管材研究所、石油管材研究中心是推动油气输送管开发与工程应用的重要机构。建所伊始，便提出了适合石油管研制与工程应用的失效分析、科研攻关、技术监督"三位一体"的工作模式，也是最初石油管材开发与工程应用的技术支撑体系。在该体系中，通过失效分析发现问题，通过科研攻关找出答案，通过技术监督解决问题。充分利用该体系，不仅提高了技术监督与失效分析水平，而且从根本上实现了科研成果向生产力的转化。几十年的实践证明，该体系为油气输送管产业发展及管道安全运行提供了保障。标准化工作最初被划分在技术监督体系内，但随着技术监督工作的不断扩展与升级，标准化工作范围愈加广泛，发挥的作用也更大，有必要将其独立出来。基于此，将油气输送管产品开发与工程应用技术支撑体系的内涵调整为质量监督与评价、标准化、科学研究、失效分析四个方面(图1)。

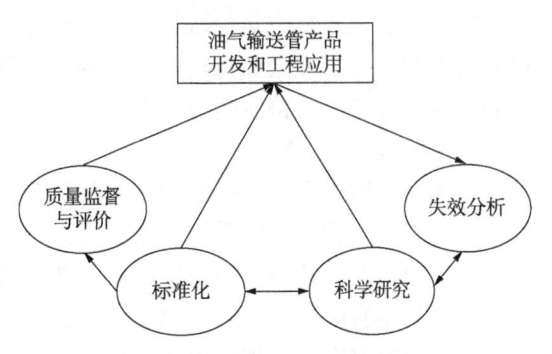

图1 油气输送管产品开发与工程应用技术支撑体系示意图

2 技术支撑体系的内涵

2.1 质量监督与评价

油气输送管的质量监督与评价包括三个方面：(1)生产许可证相关工作，即压力管道元件制造行政许可评审，由国家质检总局授权的机构负责；(2)产品的第三方检测与评价，由国家或行业的石油管材质量监督检验中心负责；(3)驻厂监造，对产品生产及厂内检验过程进行全程跟踪监督，由业主委托第三方监理机构执行。经过多年发展，油气输送管材质量监督机构已较为完善，质量监督与评价在管材开发与工程应用中的作用凸显。

2.1.1 生产许可证相关工作

为了提高压力管道的整体质量水平，从根本上降低安全风险，国家质量监督检验检疫总局特种设备局对压力管道实行制造许可制度。压力管道元件制造许可包括申请、受理、产品试制、型式试验、鉴定评审、审批、发证共七个环节，其中型式试验与鉴定评审是企业获得制造许可的两个关键环节。型式试验是对压力管道元件的设计、制造工艺、产品功能进行验证，目的是审查被设计、制造的产品是否存在不满足安全性能的缺陷，验证制造企业生产符合安全性能产品的能力，属于产品的可靠性试验范畴。鉴定评审是对申请单位的法定资格、资源条件、体系建立及实施情况、型式试验结果、产品质量等进行审查，判定其是否具备行政许可条件。国家质检总局核准授权行业的权威技术机构来承担型式试验及鉴定评审任务。管研院以及依托于管研院的国家石油管材质量监督检验中心分别是国家质量监督检验检疫总局授权的压力管道元件制造行政许可鉴定评审机构及国家压力管道元件制造行政许可型式试验机构。

20世纪90年代初，管研院承担焊接钢管生产许可评审工作，2006年之后，又作为首批特种设备行政许可鉴定评审机构之一，开展压力管道元件制造许可评审工作。经过多年发展，管研院建立并完善了型式试验、鉴定评审实施细则与作业程序，培养了一支业务精通、技术过硬、原则性强的专家队伍，积累了丰富的经验。此项工作对于促进企业加强质量管理、提高生产技术水平发挥了重要作用，整体上提升了产品质量层次，保障了油气管道安全运行。

2.1.2 产品的第三方检测与评价

在油气输送管产业迅速发展的背景下，成立了与管材相关的国家及行业的第三方检测与评价机构，最主要的机构是国家石油管材质量监督检验中心（以下简称国家质检中心）。国家质检中心的依托单位是管研院，其拥有先进仪器设备300余台套，已形成完备的石油管材检验与检测能力。在金属材料方面，具备化学成分、金相分析、无损检测、力学性能、防腐检测、实物性能六类检测能力，满足油气输送管产业发展及管道工程建设的需要。

2.1.3 驻厂监造

中国油气输送管材生产的大规模驻厂监造始于20世纪90年代，管研院借鉴国外经验，最早建立了输送管监理体系与实施细则，经过20余年的发展，已形成完善的油气输送管材第三方监理体系。从事此项工作较早、具有代表性的单位是由管研院出资成立的北京隆盛泰科石油管科技有限公司（以下简称隆盛公司）。驻厂监造是甲方委托具有资质的第三方监理机构，按照国家有关法规、规章、技术标准及合同规定，对乙方承包商产品的制造过程实施质量监督。输送管监造工作范围包括检查制造单位质量体系运行的有效性；抽查确认钢管产品质量的符合性；从原材料入厂，到生产、检验完成，直至钢管发运的全程监督。常见监造方式包括文件审查、现场监控、现场见证、停止点检查、随机抽查检验等。驻厂监造工作在保证发运钢管质量的同时，也促进了承包商质量控制水平的提升，确保油气管道项目建设进度与运行安全。

2.1.4 中俄东线管材产品质量监督

中俄东线天然气管道工程采用外径1422mm、X80钢管，此规格产品为国内首次研发生产，钢管制造厂家需扩大制造许可范围，重新进行型式试验与鉴定评审。管研院承担了型式试验与鉴定评审工作，严格执行程序，客观评价厂家制造能力，为中俄东线天然气管道工程建设保驾护航。

在钢管生产过程中，管研院制定了完备的产品第三方检测评价方案，具体试验由国家质检中心承担。中俄东线天然气管道工程涉及的管材产品主要包括直缝钢管、螺缝钢管、弯管、管件等，通过严格把关，提升了产品开发水平，保障了产品质量的稳定性。隆盛公司承担了驻厂监造工作，截至2020年，监造钢管超过$80×10^4$t，弯管管件与绝缘接头4000余个，涉及宝鸡、宝世顺、巨龙、华油、宝钢等九家焊管厂商，中油管道机械、恒通、隆泰迪、西安泵阀等十家弯管管件及绝厂商，天钢、衡钢等三家站场无缝管厂商。

2.2 标准化

2.2.1 中国石油管材标准化组织及其工作

中国石油管材标准体系由国家标准（GB）、石油天然气行业标准（SY）、中国石油天然气集团有限公司企业标准（Q/SY）共同组成。中国的石油管材标准主要等同或等效采用API/ISO标准。标准制修定等工作的组织机构由全国石油天然气标准化技术委员会石油专用管材分技术委员会、石油工业标准化技术委员会石油管材专业标准化技术委员会、中国石油天然气集团公司标准化委员会石油石化设备与材料专业标准化技术委员会石油管材分技术委员会组成。这些机构的秘书处均设立在管研院。

近几年，针对重大管道工程，结合科研成果，制定了油气输送管道用管材通用系列标准，包括钢管(含板、卷)及弯管、管件等系列标准，对规范管道建设用管材订货技术要求，保证管材质量与经济性发挥了重要作用。针对中俄东线天然气管道工程的具体情况，总结近几年X80高钢级管材关键技术指标最新研究成果，编制了中俄东线天然气管道工程用板材(外径为1422mm、X80)、螺旋缝埋弧焊管、直缝埋弧焊管、弯管、管件等11项管材产品标准，极大地支持了工程建设。

2.2.2 国际标准化组织及其工作

管研院是国际标准化组织ISO/TC67/SC2石油、石化、天然气工业用材料设备与海上结构——管道输送系统分委会的国内技术归口单位，同时也是ISO/TC67/SC2并行秘书处单位。管研院始终与ISO/TC67/SC2保持密切关系，积极参加国际标准化项目，及时投票，并每年派人参加国际标准化会议，承担SC2专项标准化工作组的工作，成立专家专业组并细分专业，面向石油管道规划、设计、建设、运营管理各领域，广泛、有效传播ISO/TC67/SC2标准化动态及标准技术。近几年，针对油气输送管方向，提出6项API提案(表1)，组织提交ISO标准提案并成功立项6项(表2)。

表1 油气输送管API国际标准提案项目统计表

从属领域	项目名称	进展情况	提交时间
API Spec 5L	异常断口评定	API成立了新项目组，负责API 5L3修订	2010.06
API Spec 5L	SAW钢管焊缝焊偏测量	已纳入ISO/API标准	2010.06
API Spec 5L	焊偏时埋弧焊管焊接热影响区冲击试样缺口位置	焊偏时HAZ冲击取样：根据要求提交了新的取样图，采纳了理想状态下取样图	2010.06
API Spec 5L	钢管横向屈服强度拉伸试样的选择	横向拉伸试样：超出标准范围，未采用	2010.06
API Spec 5L	管线钢管拉伸试验伸长率	API成功立项WI 4238，经过2年工作，由于绝大多数成员认为相关指标的修订影响不大，决定暂不做更改	2015.01
API 5L3	试样断口"三角区"评定方法	API 5L3修订考虑	2012.01

表2 油气输送管ISO/TC67/SC2国际标准提案项目统计表

从属领域	项目名称	进展情况	提交时间
ISO 3183	基于应变设计地区使用的PSL2级钢管	2013年SC2年会上进行了汇报与讨论，拟作为ISO 3183一个新附录。后API成立相关工作组，相关成果纳入API新增附录。ISO 3183：2019规范性引用API全文	2012.11
新工作项目	管道完整性管理规范	2013年SC2年会上做了报告，通过会议议决纳入工作组讨论，2014年通过立项投票。经过5年工作，2019年正式发布ISO 19345-1：2019、ISO 19345-2：2019	2012.11
新工作项目	油气管道地质灾害风险管理技术	2014年经SC2年会讨论确认可以立项，2014年通过立项投票。经过5年工作，2019年正式发布ISO 20074：2019	2013.11
新工作项目	油气管道直流杂散电流防护技术	2016年通过立项投票，2017年SC2年会决定由英国代表出任召集人，中国专家参与制定，已完成DIS稿(ISO/DIS 21857)	2016.01

续表

从属领域	项目名称	进展情况	提交时间
新工作项目	管道完整性评价	2017年经SC2年会讨论确认可以立项，2017年通过立项投票，正在编制工作组草案(ISO/AWI 22974)	2017.05
新工作项目	管道输送系统用耐蚀合金内覆复合弯管及管件	2019年通过立项投票，正在编制工作组草案(ISO/AWI 24139-1、ISO/AWI 24139-2)	2018.12

2.3 失效分析

2.3.1 油气输送管失效分析案例

20世纪50年代以来，随着油气管道的大量敷设，管道事故屡有发生，并造成灾难性后果。迄今为止，破裂裂缝最长的管道失效事故是在1960年美国Trans-Western公司的输气管道脆性破裂事故，该管道直径为30in(1in=25.4mm)，钢级为X56，裂缝长度达13km[34]。损失最惨重的是1989年苏联乌拉尔山隧道附近的输气管道爆炸事故，烧毁两列列车，伤亡1024人，其中约800人死亡[35]。

据美国管道与危险物资安全管理局数据，1999—2010年，美国共发生2840起重大天然气管道失效事故，包括992起致死、致伤事故，323人死亡，1327人受伤。近20年，加拿大油气管道干线平均每年发生30~40起失效事故。1971—2000年，欧洲油气管道干线平均每年发生13.8起失效事故。

中国油气管道建设起步较晚，但失效事故也屡见不鲜。1966年，威远气田内部集输管道通气试压时，4天时间内连续爆裂3次。经失效分析及再现性试验，确认爆裂是天然气所含H_2S在含水条件下引起的应力腐蚀开裂所致[36]，这是中国油气管道的第一起重大失效事故。1971—1976年，东北曾发生3次输油管道破裂事故，其中一次发生在1974年冬季大庆—铁岭输油管道复线的气压试验过程中。当时气温为-30~-25℃，裂缝长度为2km，断口几乎全部为脆性断口。四川气田在1970—1990年共发生108次输气管道爆裂事故[37]。1992年，轮库输油管道试压时发生14次爆裂事故。1999年，采石输油管道发生12次试压爆裂事故。

管研院承担了大部分国内油气输送管道失效分析项目，近10年来，较为典型的事故有以下几起：(1)2010年7月20日—9月22日，西气东输二线东段18标段桩间管道试压后排水作业过程中，发生管道破裂事件，其原因是管道高度起伏变化导致试压过程中产生弥合水击作用。(2)2012年2月，中俄原油管道漠河—大庆段工程发生环焊缝开裂，其原因是管道承受轴向弯曲应力，在该事故调查中发现的管道质量情况也引起中国石油天然气集团有限公司的重视。(3)2013年11月22日，青岛东黄输油复线发生管道破裂，其原因是管道腐蚀减薄，泄漏原油进入市政涵道，导致市政涵道内发生爆炸，造成严重后果。(4)2017年7月2日，中缅管道晴隆段发生环焊缝断裂，国家应急管理部特别重视这次失效分析工作，多次组织专家进行研讨。失效分析有助于管理人员与技术人员总结经验，提升管理与技术水平，并对新建管道工程提供重要借鉴。

2.3.2 失效分析机构

中国的石油管材失效分析开始于1966年对四川油田天然气输送管道严重爆破事故的分析，1979年，华北油田某井接连发生两起G105钻杆断裂事故，经失效分析进一步认识到石油管材研究及失效分析的重要性。1981年，石油部成立石油管材试验研究中心，石油管材的失效分析工作普遍开展起来。石油部要求"凡失效，皆分析"，对失效分析工作起到了极大的促进作用。1986年，石油管材试验研究中心成立失效分析与预防研究室，专门从事石油管失效分析工作。

1996年初，石油管材研究所建立的失效分析质量管理体系通过ISO 9002认证，这是中国技术服务领域首次通过该认证，标志着石油管材失效分析质量管理与国际接轨，走向科学化与规范化。2006年，中国科学技术协会工程学会联合会失效分析与预防中心同意在石油管材研究所成立中国科协工程联失效分析和预防中心石油管材与装备分中心。2010年，根据中国石油学会石油管材专业委员会油管专字〔2010〕02号通知，决定成立中国石油学会石油管材与装备失效分析及预防中心，秘书处设在管研院。

2.3.3 失效分析与失效控制

机械产品的零件或部件处于下列3种状态之一时，定义为失效：(1)完全不能工作；(2)仍可工作，但不能满意地实现预期功能；(3)受到严重损伤不能安全可靠地工作，必须进行修理与更换。对于失效事件，按一定的思路与方法判断失效性质、分析失效原因、研究失效事故处理方法与预防措施的技术活动及管理活动，统称失效分析。失效分析预测预防是产品或装备安全可靠运行的保证，是提高产品质量的重要途径，目的在于不断降低装备的失效率，提高可靠性，防止发生重大失效事故，促进经济持续稳定发展。

基于失效分析，提出防止失效的措施，对失效分析结果进行反馈，即失效控制[38]。油气管道失效控制(图2)的主要内容包括：(1)大量搜集国内外失效案例，建立油气管道失效信息案例库；(2)对失效案例进行综合统计分析，确定油气管道的主要失效模式；(3)研究各种失效模式的发生原因、机理及影响因素；(4)研究并提出各种失效模式的控制措施与方法。失效信息案例库是失效控制的基础，应具有较强的数据处理与统计分析功能，并拥有尽可能多的案例，一方面广泛搜集国内外油气管道已发生的重大失效案例，另一方面加强对新发生的油气管道失效事故的搜集分析。在大量失效分析的基础上，总结一些重大共性科学问题进行较深入的研究。随着失效案例的不断增多，失效信息案例库不断充实，失效模式及其原因、机理、影响因素随之动态变化与调整，失效控制措施和方法也不断完善。

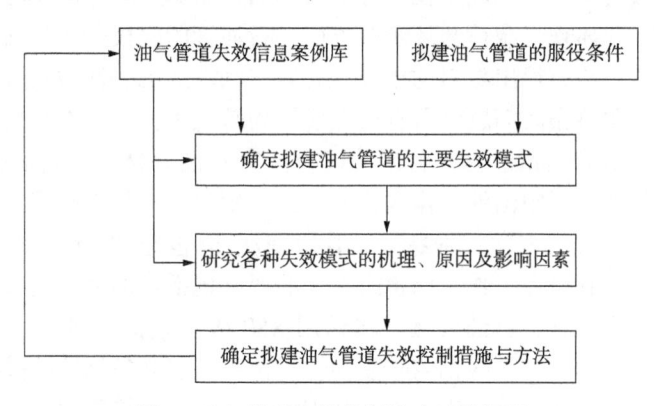

图2 油气管道失效控制基本思路框图

2.4 科学研究

油气输送管的科学研究包括两个方面：(1)油气输送管工程应用与应用基础研究；(2)新产品开发与制管工艺研究。工程应用与应用基础研究由用户的研究机构及一些独立的研究机构承担，油气输送管新产品开发与制管工艺研究的主体是生产企业，同时联合研究机构、高校等。

2.4.1 油气输送管工程应用与应用基础研究

20世纪90年代,管研院开始了石油管工程应用与应用基础研究,承担了国家及中国石油的许多重大科研课题,在取得一系列重要研究成果的同时,总结、梳理形成了石油管工程学(图3)。

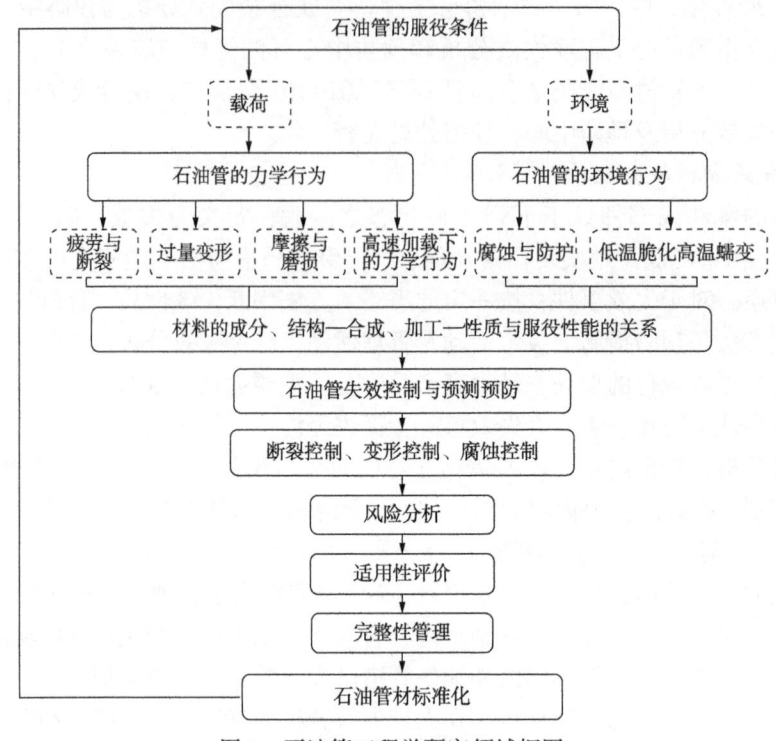

图3 石油管工程学研究领域框图

石油管工程学致力于研究不同服役条件下石油管的失效规律、机理及克服失效的途径。石油管的服役条件主要体现在载荷与环境两方面,石油管的服役行为包括力学行为、环境行为及两者的复合。通过研究得出材料的成分/结构、合成/加工、性质与服役性能的关系,解决失效控制与失效的预测预防问题。所有研究成果均转化为技术标准与规范。

2015年,国家科技部批准,挂靠管研院建设国家石油管材与装备服役行为与结构安全重点实验室,进一步推动了中国油气输送管的工程应用与应用基础研究,为油气管道安全运行保驾护航。基于石油管工程学技术脉络,管研院从20世纪90年代开始,针对油气输送管,开展了大量应用基础研究工作,承担或参加的课题包括:油气输送管止裂韧性试验方法的研究、石油天然气用焊管、石油天然气输送用X80级管线钢管研制、长距离输气管道材质和管型选用技术研究、油气管线钢的止裂韧性以及与动态断裂韧性的关系、油气输送用焊管包申格效应的研究、高性能管线钢的重大基础研究、西气东输管道使用国产螺旋缝埋弧焊管的可行性研究、X80管线钢管的开发、西气东输二线工程关键技术研究等,在管材组织成分、关键技术指标、管材失效控制与预测预防等方面,取得了系列研究成果,支撑了陕京管道、西气东输管道、中亚管道等重大管道工程的建设。2012年,中国石油天然气集团有限公司设立重大专项"第三代大输量天然气管道工程关键技术",管研院系统开展了外径1422mm、X80钢管应用技术研究,综合考虑管材加工性能及焊接性,提出了产品化学成分范围;通过理论计算与全尺寸试验,提出并验证了管道延性止裂韧性;基于试验研究与分析，提出管材产品技术指标，制定产品检测方法。以上科研成果均应用于中俄东线天然气管道工程管材技术标准编制与产品生产。

2.4.2 新产品开发与制管工艺研究

油气输送钢管(特别是长距离输送钢管)的主体是焊管。中国油气输送焊管的主要开发模式是：由业主组织，研究机构技术牵头并对产品进行评价，钢铁企业研究开发管线钢(钢板、板卷)，制管企业从钢铁企业购置钢板(板卷)，加工生产焊管产品。目前，仅上海宝钢既开发管线钢钢板(板卷)，又生产油气输送焊管。油气输送焊管新产品开发包括管线钢的研究开发及制管工艺研究，前者主要由宝钢、武钢、鞍钢、首钢、沙钢、太钢等大型钢铁企业的科研机构承担，后者主要由宝钢、渤海装备等焊管企业承担。

中俄东线天然气管道工程用管材产品研发过程中，基于工程需求与产品技术标准，由管道公司、管研院、钢铁与制管企业联合协作，按照管研院提出的油气输送管材试制过程质量控制体系(图4)，开展产品试制工作。第一阶段主要是考察钢铁与制管企业的工艺可行性，为单炉试制；第二阶段是验证产品生产工艺的可靠性，通常为500~1000t的批量试制，在该阶段，严格按正式产品生产流程进行试制，实施全程驻厂监造。基于研究成果及以上工作，成功开发了中俄东线天然气管道工程用系列管材产品，包括外径1422mm，壁厚21.4mm、25.7mm、30.8mm、32.1mm焊管，以及外径1422mm，壁厚25.7mm、30.8mm、33.8mm、35.2mm弯管等产品，保障了中俄东线天然气管道工程的顺利实施。

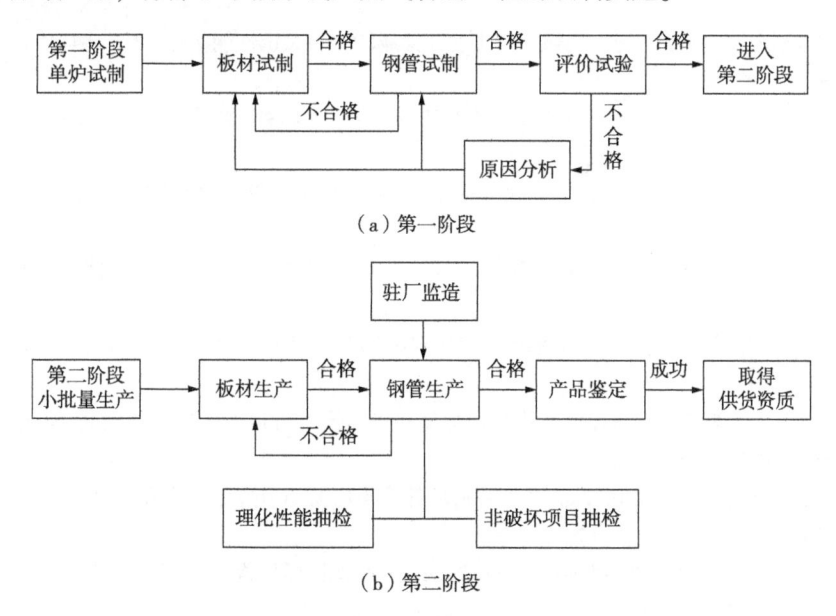

（a）第一阶段

（b）第二阶段

图4 油气输送管材试制过程质量控制体系框图

3 结束语

油气输送管产品开发与工程应用技术支撑体系涵盖质量监督与评价、标准化、科学研究、失效分析四部分内容，四位一体，协调配合，在油气输送管材国产化与工程应用全周期中发挥了重要作用，其成功经验可供其他行业或领域参考。随着长输油气管道工程建设的高速发展，油气输送管产品开发与工程应用技术支撑体系也将不断完善，持续促进油气输送管产品研发与应用技术的进步，进而为油气管道工程建设提供技术支撑。

参 考 文 献

[1] 姜昌亮．中俄东线天然气管道工程管理与技术创新[J]．油气储运，2020，39(2)：121-129．
[2] 程玉峰．保障中俄东线天然气管道长期安全运行的若干技术思考[J]．油气储运，2020，39(1)：1-8．
[3] 蒲明，李育天，孙骥姝．中俄东线天然气管道工程前期工作关键点及创新成果[J]．油气储运，2020，39(4)：371-378．
[4] 张振永．中俄东线X80钢级φ1422mm管道工程设计关键技术应用[J]．焊管，2019，42(7)：64-71．
[5] 陈小伟，嵇峰，白学伟．中俄东线X80钢级φ1422mm×30.8mm钢管理化性能研究[J]．焊管，2019，42(5)：10-17．
[6] 赵新伟，池强，张伟卫，杨峰平，许春江．管径1422mm的X80焊管断裂韧性指标[J]．油气储运，2017，36(1)：37-43．
[7] 张振永，周亚薇，张金源．现行设计系数对中俄东线OD1422mm管道的适用性[J]．油气储运，2017，36(3)：319-324．
[8] 张振永，张文伟，周亚薇，等．中俄东线OD1422mm埋地管道的断裂控制设计[J]．油气储运，2017，36(9)：1059-1064．
[9] 刘迎来，许彦，王高峰，等．中俄东线-45℃低温环境油气管道工程用X80钢级φ1422mm×33.8mm感应加热弯管研发[J]．焊管，2019，42(7)：48-54．
[10] 尤泽广，王成，傅伟庆，等．中俄东线站场用直径1422mm×1219mm三通设计[J]．油气储运，2020，39(3)：347-353．
[11] 蒋庆梅，张小强，钟桂香，等．中俄东线黑龙江穿越段管材关键性能指标对比与确定[J]．油气储运，2020，39(1)：92-98．
[12] 张小强，侯宇，蒋庆梅，等．中俄东线直径1422mm X80钢级冷弯管设计参数的确定[J]．油气储运，2020，39(2)：222-225．
[13] 李鹤林，冯耀荣，霍春勇，等．关于西气东输管线和钢管的若干问题[J]．中国冶金，2003(4)：36-40．
[14] 张圣柱，程玉峰，冯晓东，等．X80管线钢性能特征及技术挑战[J]．油气储运，2019，38(5)：481-491．
[15] 任俊杰，马卫锋，惠文颖，等．高钢级管道环焊缝断裂行为研究现状及探讨[J]．石油工程建设，2019，45(1)：1-5．
[16] 庄传晶，李云龙，冯耀荣．高强度管线钢环焊缝强度匹配对管道性能的影响[J]．理化检验(物理分册)，2004，40(8)：383-386．
[17] 张宏，吴锴，刘啸奔．直径1422mm X80管道环焊接头应变能力数值模拟方法[J]．油气储运，2020，39(2)：162-168．
[18] 毕宗岳，黄晓辉，牛辉．X80级φ1422mm×21.4mm大直径厚壁焊管的研发及性能研究[J]．焊管，2017，40(4)：1-7．
[19] 冯耀荣，霍春勇，吉玲康，等．我国高钢级管线钢和钢管应用基础研究进展及展望[J]．石油科学通报，2016，1(1)：143-153．
[20] 熊庆人，杨扬，许晓锋，等．切环法和盲孔法测试大口径厚壁X80钢级埋弧焊管的残余应力[J]．机械工程材料，2018，42(12)：27-30，67．
[21] 杨坤，池强，李鹤，等．高钢级天然气输送管道止裂预测模型研究进展[J]．石油管材与仪器，2019，5(4)：9-14．
[22] 霍春勇，李鹤，张伟卫，等．X80钢级1422mm大口径管道断裂控制技术[J]．天然气工业，2016，36(6)：78-83．
[23] 崔天燮，张彦睿．太钢X80级管线钢热轧卷板技术开发与应用[J]．焊管，2009，32(12)：32-39．

[24] 杨忠文, 毕宗岳, 牛辉. 高钢级管线钢焊管研制[J]. 焊管, 2011, 34(4): 5-11.

[25] 李少坡, 姜中行, 李永东, 等. 超低碳贝氏体厚壁X80宽厚板的研发[J]. 钢铁, 2012, 47(4): 55-59.

[26] 陈小伟, 王旭, 李国鹏, 等. 我国高钢级、大直径油气输送直缝埋弧焊管研究进展[J]. 钢管, 2016, 45(5): 1-8.

[27] 谢仕强, 桂光正, 郑磊, 等. X80钢级大直径UOE直缝埋弧焊管的开发及应用[J]. 钢管, 2011, 40(4): 29-36.

[28] 张伟卫, 吉玲康, 陈宏远, 等. 一种X70或X80抗大变形钢管生产方法: 201010550557.4[P]. 2013-07-31.

[29] 刘文月, 任毅, 张禄林, 等. 一种X80抗大变形管线钢及制造方法: 201510336908.4[P]. 2018-08-31.

[30] 史立强, 牛辉, 杨军, 等. 大口径JCOE工艺生产X90管线钢组织与性能的研究[J]. 热加工工艺, 2015, 44(3): 226-229.

[31] 刘刚伟, 毕宗岳, 牛辉, 等. X90高强度螺旋埋弧焊管组织性能研究[J]. 焊管, 2015, 38(10): 9-13.

[32] 王红伟, 吉玲康, 张晓勇, 等. 批量试制X90管线钢管及板材强度特性研究[J]. 石油管材与仪器, 2015, 1(6): 44-51.

[33] 李鹤, 封辉, 杨坤, 等. 断口分离对X90焊管断裂阻力影响试验[J]. 油气储运, 2019, 38(10): 1104-1108.

[34] 潘家华. 油气管道断裂力学分析[M]. 北京: 石油工业出版社, 1989: 2-10.

[35] Starostin V. Pipeline disaster in the USSR: It had to happen, yet it could have been averted[J]. Pipes Pipelines Int, 1990, 35(2): 7-8.

[36] 李鹤林. 某管线试压爆破原因分析[M]//李鹤林文集——石油管工程专辑. 北京: 石油工业出版社, 2017: 601-607.

[37] MAO H G. Failure analysis of the natural gas pipelines in Sichuan[C]. Beijing: International Symposium on Structural Technique of Pipeline Engineering, 1992: 20-30.

[38] 李鹤林. 油气管道失效控制技术[J]. 油气储运, 2011, 30(6): 401-411.

本论文原发表于《油气储运》2020年第39卷第7期。

中国 X80 管线钢和钢管研发应用进展及展望

冯耀荣　吉玲康　李为卫　刘迎来　霍春勇

(中国石油集团石油管工程技术研究院·石油管材及装备材料服役行为与结构安全国家重点实验室)

摘　要：随着油气管道建设的快速发展，X80 管线钢及钢管得到了大规模应用。回顾了中国 X80 高钢级管线钢及钢管的研发应用历程与主要进展，指出了面临的挑战，提出了发展建议。经过近 20 年的发展，建成 X80 油气输送管道约 17000km，单管输气量达到 380×10^{8}m^3/a，X80 管材生产及管道建设技术进入国际领跑者行列。形成了 X80 管线钢及钢管组织分析鉴别与评定、强度试验与屈强比控制、断裂与变形控制等关键技术，以及 X80 系列热轧板卷与大口径厚壁螺旋埋弧焊管制造技术、宽厚板与大口径厚壁直缝埋弧焊管制造技术、大应变管线钢及钢管制造技术、感应加热弯管及管件设计与制造技术。有力支撑了西气东输二线、中俄东线等重大管道工程建设。对于未来发展，提出如下建议：加强对制管用板卷或钢板的可焊性评价与控制，深化现场焊接技术及质量性能控制研究，推进高钢级管线钢及钢管在大输量管道建设中的应用，强化油气管道失效控制、完整性理论与技术的研究及应用。

关键词：X80 管线钢；X80 埋弧焊管；X80 管道；研究进展；发展展望

为了满足石油天然气特别是天然气大输量长距离输送需求，提高管道输送效率与经济性，大口径、高压力成为油气管道的重要发展方向，因而促进了高钢级管线钢及钢管的研发与应用。自从 1985 年德国瓦卢瑞克·曼内斯曼钢管公司（今欧洲钢管公司）研发生产的 X80 管线钢及钢管在 Ruhr Gas/Megal Ⅱ 项目首次应用至今，X80 管线钢及钢管的发展已有 35 年的历史。Ruhr Gas 在 1992—1993 年采用欧洲钢管公司生产的 X80 直缝埋弧焊钢管分别建成了两条管径 1220mm、壁厚 18.3mm 与 19.4mm、输送压力为 10MPa 的天然气输送管道，长度约 250km，标志着 X80 管线钢及其直缝埋弧焊钢管技术基本成熟[1]。加拿大 TransCanada 公司是高钢级管线钢管应用的推进者，建成多条 X80 高压输气管道，甚至将 X80 列为新建管道的首选钢级[2,3]。美国 2004 年建成的 Cheyenne Plains 管道，是国际上 X80 管线钢及其螺旋埋弧焊管成功应用的范例[4]，其规格为管径 914mm，壁厚 11.74mm 与 16.94mm，长度 608km。俄罗斯巴甫年科沃—乌恰管道[5] 长度 1106km，包括两条并行敷设的管径 1420mm 管道，设计压力 11.8MPa，管材钢级 K65（相当于 X80），壁厚 23.0mm、27.7mm 及 33.4mm，

基金项目：中国石油天然气集团有限公司科学研究与技术开发项目"高钢级管道可靠性设计及失效控制技术研究"（2019B-3008）。

作者简介：冯耀荣，男，1960 年生，教授级高工，孙越崎能源大奖获得者，1982 年博士毕业于西安交通大学金属材料及热处理专业，现主要从事石油管材及装备材料服役安全研究及重大工程技术支持工作。地址：陕西省西安市锦业二路 89 号（710077），电话：029-81887699，E-mail：Fengyr@cnpc.com.cn。

分别于2013年和2017年建成投产，标志着X80管线钢及钢管研发与应用进入新的发展阶段。迄今为止，国外已建成X80管道长度约8000km。

中国从2000年开始进行X80管线钢及钢管的应用基础研究与技术开发工作，在中国石油天然气集团公司的组织下，石油行业联合冶金行业、专业院所、高等院校协同攻关，先后开展了高钢级管线钢应用基础研究、X80热轧板卷与宽厚板研制、焊管与管件开发、焊材研制、现场焊接技术研究，以及X80管线钢在大口径、高压力、大输量长输天然气管道的应用研究。基于上述研究工作，2005年在西气东输—陕京二线联络线（冀宁联络线）进行了X80管线钢管首次工程应用段的敷设[6,7]，随后又建设了以X80管线钢管为主的西气东输二线（2008—2012年）、西气东输三线（2012—2014年）、中亚C线（2013—2014年）、陕京四线（2016—2017年）、中缅输气管道（2010—2014年）及中俄东线（2017—2020年，在建）等重要管道工程。目前，中国建设的X80管道管径1016~1422mm，壁厚为15.3~30.8mm，总里程约17000km，约是国外X80管道总长度的2倍。中国是世界上生产、应用X80管材最多的国家，X80管材也成为中国大输量、高压力输气管道的首选钢级，X80管材生产及管道建设技术进入国际领跑者行列[8,9]。回顾中国X80管材研发及管道建设历程，总结取得的技术成果与先进经验，对于进一步提升X80管材性能质量与管道建设水平、保障管道长期安全运行具有重要意义。

1 研发与应用历程

1.1 在冀宁联络线的首次应用

在对国际上X80管材生产与应用情况进行系统调研、交流、分析的基础上，中国石油天然气集团公司组织开展了X80管材关键技术指标与技术标准研究、X80管线钢及钢管研发、国内外X80管材性能质量综合分析评价、X80管道安全性分析等工作，论证了建设X80管道工程应用段的可行性，研究开发了X80管道现场焊接工艺与配套技术，制定了X80管道线路焊接施工及验收规范。基于此，在西气东输—陕京二线冀宁联络线进行X80工程应用段的敷设，采用管径1016mm、壁厚18.4mm的X80直缝埋弧焊管与管径1016mm、壁厚15.3mm的X80螺旋埋弧焊管，分别替代管径1016mm、壁厚21.0mm的X70直缝埋弧焊管与管径1016mm、壁厚17.5mm的X70螺旋埋弧焊管，通过采用标准化设计、严于国际上X80管道的建设标准、性能等同于国际上同类X80焊管的产品、适用的焊接工艺与标准，成功完成了7.71km冀宁联络线X80工程应用段的敷设[6,7]，为此后X80管线钢及钢管在中国的大批量规模化应用奠定了基础。

1.2 在西气东输二线的应用

西气东输二线是中国继西气东输管道之后的又一伟大工程，长度约8700km，承担中国每年从中亚进口的$300×10^8 m^3$天然气的输送任务，若沿用西气东输管道10MPa输送压力的X70焊管技术，需要双管敷设，投资巨大、占地翻倍、输送效率低，而采用12MPa输送压力的X80焊管，则有很多关键技术需要突破。中国石油天然气集团公司果断决策，启动西气东输二线工程关键技术重大专项研究，其核心是新一代X80焊管应用关键技术与产品研发。经过石油企业与冶金企业的联合攻关，研究提出了西气东输二线采用X80板材及管材的关键技术指标与检测评价方法，突破了X80管线钢及钢管在西气东输二线应用的10多项关键技术与瓶颈技术，制定了先进适用的西气东输二线管材系列标准，为在西气东输二线成功应用奠定了基础；研究制定了兼顾安全性与经济性的西气东输二线断裂控制方案，在国际上首次开展模拟实际工况条件的X80螺旋埋弧焊管全尺寸气体爆破试验,验证了以X80管线钢及钢管为主的西气东输二线的安全可靠性;研究提出了基于应变设计地区使用的直缝埋弧焊管关键技术指标要求及标准,推动了基于应变的管道设计方法及大应变钢管在西气东输二线工程中的批量应用;在系统研究弯管与管件成分、组织、性能、工艺相关性的基础上,提出了弯管与管件的成分设计、强韧性控制指标及配套的热加工工艺,成功开发了西气东输二线工程急需的厚壁弯管与管件;成功研发了西气东输二线X80板材与管材系列新产品(螺旋埋弧焊管壁厚18.4mm,直缝埋弧焊管壁厚达26.4mm,弯管壁厚达32mm,管件壁厚达55mm),满足了西气东输二线工程建设的急需,推动了X80高钢级管线钢及钢管在中国乃至世界上的大规模应用[8-11]。

1.3 在中俄东线的应用

中俄东线是迄今为止中国输气量最大的管道工程。新建干支线全长3169km,利用已建管道1802km。干线管道包括黑河—长岭、长岭—沈阳、沈阳—永清、安平—泰兴、南通—甪直5段。其中干线管道黑河—长岭段起自黑龙江省黑河首站,止于吉林省长岭末站,管径1422mm,设计压力12MPa,钢级为X80,全长737km,设计输量380×10^{8}m^3/a。管道沿线地处中国东北寒冷地区,冬季最冷月平均气温约-14~-24℃,极端最低温度-48℃[12-14]。针对中俄东线建设的实际需求,中国石油天然气集团公司适时启动第三代管线钢及钢管应用关键技术研究重大专项,攻关内容之一是管径1422mm的X80管线钢管的应用研究,通过石油系统与冶金系统的联合攻关,形成了X80钢级、1422mm、12MPa管道断裂控制技术,建成了中国第一个全尺寸管道爆破试验场,在国际上首次开展了X80钢级、1422mm、12MPa、使用天然气介质的全尺寸管道爆破试验;制定了X80钢级、管径1422mm管材及管件系列技术条件,研发了X80钢级、管径1422mm、壁厚21.4mm的螺旋埋弧焊管,X80钢级、管径1422mm、壁厚25.7mm与30.8mm的直缝埋弧焊管,X80钢级感应加热弯管,管径1422mm与1219mm、厚度57mm三通等管件;形成了X80钢级、管径1422mm焊管现场焊接工艺及冷弯工艺;研发了适应于X80钢级、管径1422mm管道施工的配套对口器、坡口机、内焊机、外焊机、机械化补口机等装备;制定了《1422管道线路工程设计及施工技术规定》等13项标准规范,在国内首次实现了大口径高压输气管道的全自动化焊接。基于上述研究成果,全长1067km的中俄东线北段(黑河—长岭)"一干三支"已于2019年12月顺利建成投产[12-25]。

2 关键技术的重要突破

2.1 应用基础研究

2.1.1 X80管线钢及钢管组织分析与性能控制

揭示了X80管线钢成分、组织、性能、工艺之间的相关性,阐明了X80管线钢的微观组织特征,出版了《管线钢显微组织的分析与鉴别》图谱,提出了针状铁素体管线钢带状组织与晶粒度评定方法。揭示了X80板卷/螺旋钢管、钢板/直缝钢管纵向与横向条形试样、圆棒试样屈服强度的对应关系,确定了采用圆棒试样进行X80横向拉伸强度测试的方法。基于先屈服后断裂(Yield Before Break,YBB)准则,研究得到了焊管屈强比与焊管缺陷尺寸、宽板试验试样几何尺寸之间的关系;采用失效评价图技术,系统研究了屈强比对焊管使用安全性的影响;提出了X80焊管屈强比控制指标。系统研究了螺旋埋弧焊管残余应力的大小、分布规律及影响因素,阐明了环切试样切口张开量与残余应力之间的对应关系,提出了X80螺旋埋弧焊管切口张开量控制指标[26-30]。

2.1.2 X80高压输气管道断裂与变形控制

基于起裂判据及防止低温脆断原则，提出X80钢管焊缝与环焊缝的起裂韧性要求、母材落锤撕裂试验(Drop-Weight Tear Test, DWTT)要求。针对西气东输二线、中俄东线等管道工程的具体特点，基于国内外全尺寸钢管气体爆破试验数据库、Battelle双曲线模型及相关软件对管道的止裂韧性要求进行了系统研究，综合考虑输气管道安全可靠性与经济可行性，提出了西气东输二线与中俄东线管道管材止裂韧性要求。开展了模拟西气东输二线与中俄东线实际运行工况的X80管道实物气体爆破试验，验证了管道止裂韧性指标的合理性及管道的服役安全性。研究提出了基于高钢级管道能量释放率及流变应力修正的止裂预测模型TGRC-1、基于落锤撕裂能量的止裂预测模型TGRC-2，并开发了相应的预测软件。研制了钢管内压+弯曲大变形实物试验装置，研发形成了钢管实物模拟变形试验技术，提出了使用多个应力比、屈强比、均匀伸长率等多参量联合表征评价及控制钢管变形行为的方法，揭示了钢管材料应力比等关键技术指标与钢管临界屈曲应变的关系，提出了钢管临界屈曲应变能力的预测方法，建立了X80大应变管线钢及钢管产品技术指标体系与标准[31-40]。

2.1.3 提高X80管道设计系数的可行性

系统研究了提高强度设计系数对管道安全性及失效风险的影响，确定了0.8设计系数管材关键性能指标及质量控制要求，制定了西气东输三线0.8设计系数管道用X80螺旋缝埋弧焊管技术条件。系统研究了X80厚壁三通的断裂抗力及极限承载能力，在确保三通极限承载能力不小于3.5倍管道设计压力的情况下，设计系数由0.4提高至0.5，管件壁厚减薄20%~30%，因而减小了厚壁管件设计的过度保守性，降低了制造难度[41-43]。

2.1.4 X80大输量管道应用基础研究

结合中俄东线单管$380×10^8m^3/a$大输量管道建设需求，开展了管径1422mm、压力12MPa的X80管道的可行性研究，内容包括：优化管材钢级、管径、输送压力组合，综合分析生产制造可行性、运行安全性、经济合理性。通过多方案比选，确定采用1422mm管径、X80钢级、12MPa设计压力、单管输送的方案，在年输量相同的条件下，可比双管1219mm管径、X80钢级、12MPa方案节约投资$46×10^8$元，节省土地约$1500×10^4m^2$。开展了管径1422mm、压力12MPa和13.3MPa的X80直缝埋弧焊管及螺旋埋弧焊管使用天然气介质管道的全尺寸爆破试验，提出了管材止裂韧性要求，制定了管径1422mm的X80管材系列技术标准[13,19,44-45]。

2.2 产品开发

2.2.1 X80系列热轧板卷及大口径厚壁螺旋埋弧焊管

采用低碳、超低碳，加Mn，低S、P，Nb、V、Ti微合金化，Mo、Cr、Ni、Cu多元少量合金化，成分综合调控，以及纯净钢冶炼、控轧控冷、低温卷取等技术，成功研发了具有高强度、高韧性及良好焊接性能的针状铁素体型X80热轧板卷。在综合研究材料形变强化、包申格效应、残余应力的基础上，研发形成了螺旋埋弧焊管低应力成型与力学性能控制技术。研制了X80高强度、高韧性、高洁净度埋弧焊接材料，研发了X80螺旋埋弧焊管高速焊接工艺，解决了高强度螺旋埋弧焊管焊缝强度、韧性匹配、焊速矛盾，在焊缝质量及性能满足要求的前提下，焊接速度显著提高。建成10余条高水平螺旋埋弧焊管生产线，形成大批量规模化生产供货能力，管径达到1422mm，壁厚达到23mm[46-49]。

2.2.2 X80系列宽厚板及大口径厚壁直缝埋弧焊管

采用低碳、超低碳，加 Mn，低 S、P，Nb、V、Ti 微合金化，Mo、Cr、Ni、Cu 多元少量合金化，以及纯净钢冶炼、控轧控冷、成分/工艺综合调控等技术，成功研发了综合力学性能及焊接性能良好的针状铁素体型 X80 宽厚板。基于金属塑性成形理论，揭示了板材力学性能、钢管成型工艺参数与钢管力学性能、外观尺寸之间的关系，研发了 X80 钢管 JCOE 和 UOE 成型工艺及力学性能控制技术。研制了 X80 厚壁直缝焊管埋弧焊接专用材料，形成了焊缝质量满足技术条件要求的焊接工艺，实现了高强度、高韧性、良好尺寸精度直缝埋弧焊管的稳定生产。建成多条 JCOE 和 UOE 直缝埋弧焊管生产线，形成大批量规模化生产供货能力，管径达到 1422mm，壁厚达到 33.4mm[50-52]。

2.2.3 X80HD 大应变管线钢及钢管

采用低碳或超低碳，加 Mn，Nb、Ti 微合金化，少量加 Cr、Mo、Ni、Cu，控制 N，严格控制 P、S 等有害元素及氢、氧等有害气体含量，Ca 处理控制夹杂物形状；采用具有合适尺寸及比例的"多边形铁素体+贝氏体"双相组织设计，硬相(贝氏体)为管线钢提供必要的强度，软相(铁素体)保证足够的塑性；发明了用热轧方法制备 X80HD 大应变管线钢的工艺技术，成功开发多项新型控轧规程，通过低温控轧、弛豫—快冷相变、新型控轧控冷模型、低温矫直等多项技术集成，实现了"多边形铁素体+贝氏体"双相组织与性能的精准调控，以及 X80HD 大应变管线钢板的稳定生产[53-55]。

揭示了铁素体+贝氏体钢板制管成型、扩径过程关键力学性能的变化规律及其相关性，确定了钢板强度、塑性、韧性、应力比等指标要求；研发了大应变钢管 JCOE 和 UOE 专有成型、扩径等关键工艺技术，优化成型步长及扩径率，实现了制管过程材料强度、塑性、韧性、应力比等关键性能指标的精准控制；发明了高强韧匹配焊接材料，自主研发了集焊接坡口设计、电流、电压及焊材匹配的大应变钢管多丝埋弧焊接技术，焊缝及热影响区性能达标率 100%；揭示了热涂覆温度对钢管应变时效的影响规律，优选涂覆材料，涂覆温度控制在 200℃左右，有效控制了热涂覆过程钢管屈服强度升高及均匀伸长率下降[53-57]。

2.2.4 感应加热弯管及管件

兼顾材料淬透性与可焊性，采用低碳 Mn-Nb-Ni-Mo 合金体系，UOE 或 JCOE 成型、埋弧焊接、弯曲成型+感应加热淬火并回火(弯管)或多次热加工成型淬火并回火(管件)，发明了具备高强度、高韧性及良好焊接性能的 X80 钢级弯管与管件的制备方法。系统研究了材料化学成分、加热温度、冷却速度以及循环加热对弯管与管件材料强韧性的影响规律，提出了 X80 弯管与管件的化学成分及热加工工艺参数。采用全管体在线淬火+整体回火工艺技术替代传统的局部淬火+整体回火工艺，解决了热煨弯管过渡区性能波动大的问题；研发了 X80 厚壁三通埋弧自动焊接技术，焊缝质量显著提高，相比之前的手工多道焊，生产效率提高 2 倍以上。在掌握 X80 弯管与管件成分、组织、性能、工艺相关性及工艺参数的基础上，成功制造了管径 1016~1422mm、壁厚 22~60mm 的 X80 弯管与管件，在保证强度及焊接性的同时，-45℃夏比冲击功达到 60J 以上[23-24,58]。

3 挑战与展望

石油天然气在中国能源消费结构中占有十分重要的地位，中国国民经济与社会发展对石油天然气的需求以及油气在一次能源消费结构中的占比呈增长趋势。近年来，中国油气对外依存度已达到 70%，因而提出加大油气勘探开发力度的要求，石油天然气特别是天然气清洁能源呈加快发展态势。中国天然气、页岩气、煤制气资源主要集中在西部地区，如塔里木天然气、新疆煤制气、长庆天然气、四川天然气与页岩气，以及海洋天然气；进口天然气包括中亚天然气、俄罗斯天然气、缅甸天然气以及沿海进口的LNG。上述天然气与煤制气资源均需要通过管道实现西气东输、北气南送、海气登陆。根据国家发展改革委、国家能源局发布的《中长期油气管网规划》，伴随国家油气管网公司的成立，中国油气管道建设以及对高钢级管线钢与钢管的需求将呈现波动式增长。

中国天然气长输管道工程的发展趋势是进一步增大输量、确保安全并尽可能降低建设运行成本，大口径、高压力、高钢级管材成为必然选择，X80将继续成为今后大输量管道的首选。总体来说，中国X80管材性能与质量已达到国际先进水平，已建及在建管道钢管本身基本未出现大的质量问题。但是，近年来油气管道环焊缝失效事故时有发生[59,60]，其中涉及部分已建X80管道。原因是多方面的，涉及地层移动产生的附加载荷、管道环焊缝性能与质量、施工因素引起的附加载荷等，多数属于脆性起裂、弹塑性断裂或塑性失稳失效，个别引发严重后果。为此，国家应急管理部及国家能源局要求对油气管道进行隐患排查治理，相关管道公司积极主动作为，隐患排查治理工作有序进行且卓有成效。在对管道隐患排查过程中，发现管道环焊缝存在一定比例的裂纹或超标缺陷。含有环焊缝裂纹的管道必须修复才能保障管道的完整性与安全运行。管道失效分析或缺陷分析表明，面型缺陷或裂纹缺陷对管道安全性影响最大，绝大多数产生于直管段与弯管或管件的不等壁厚连接焊缝部位，处于返修口及最后焊接的连头口。这些部位大多存在面型焊接缺陷、错边、截面变化、组对应力、应力集中、地形地貌变化等情况，加之焊缝韧性偏低或分散，焊缝强度或承载能力又低于直管段，在工作应力、组对应力、焊接残余应力、应力集中、地层位移产生的附加载荷等共同作用下，环焊缝焊接缺陷处首先发生脆性起裂、裂纹持续扩展，导致管道发生弹塑性断裂或塑性失稳失效。为了确保X80及以上钢级管道建设质量，保障管道长期安全可靠运行，应持续加强以下几个方面的工作。

（1）进一步加强对制管用板卷或钢板的可焊性评价与控制。除板卷或钢板常规的成分、组织、性能等指标应满足要求外，现场焊接性能的评价与控制十分重要且必不可少。不同冶金厂家供应板卷或钢板化学成分、组织、性能、工艺的差异，特别是化学成分的变化，将导致现场焊接性能的变化。建议选用焊接工艺窗口较宽的原材料，避免使用因焊接工艺窗口较窄导致现场可焊性较差的原材料。一旦通过现场焊接工艺评定，板卷或钢板的化学成分就应基本固定，如果化学成分发生实质性变化，必须重新进行现场焊接工艺评定。只有通过现场焊接工艺评定的板卷或钢板，才能用于焊接钢管的生产。现场焊接工艺评定应由业主指定的相关专业机构进行。

（2）进一步深化现场焊接技术与质量性能控制研究。包括焊接材料、焊接方法与工艺、焊接匹配、缺陷容限与无损检测要求、断裂韧性要求等。根据断裂力学分析，焊缝处的应力、缺陷容限、韧性要求存在内在联系。焊缝高韧性可允许较大的缺陷尺寸，反之亦然。对于容易产生裂纹或发生断裂的直管段与弯管/管件的连接焊缝部位或管道焊接的连头口，高附加应力难以完全避免，而这些部位的焊接难度又很大，加严无损检测要求受到很大限制，使得提高焊缝断裂韧性要求成为必然选择，而要提高环焊缝的断裂韧性要求，则需要焊接材料、焊接方法与工艺的变革。中俄东线全面采用自动焊接技术，提高了管道焊接质量性能的稳定性[61]，宜继续推广。建议持续深化现场焊接技术研究，选用或研发能获得高韧性的新型焊接材料，优化焊接工艺，合理控制焊接热输入、预热温度、层间温度，采用预备回火焊道处理、补充回火焊道处理、焊后热处理等措施,确保焊缝韧性。对于基于应变设计地区管道环焊缝、直管段与弯管/管件的连接焊缝部位或管道焊接的连头口焊缝应采用高匹配原则,使焊缝轴向强度水平高于两侧管段。若材料强度不能满足高匹配要求,可在结构上采取措施,如管段连接部分局部适当加厚、适当增加焊缝余高,确保焊缝承载能力不低于两侧管段。同时应完善管道环焊缝性能质量指标体系,开展管道环焊缝性能质量及服役性能评价,以实现对复杂工况下油气管道失效的有效控制。

（3）持续有序推进高钢级管线钢与钢管在大输量管道建设中的应用。在继续完善X80管道系列技术的基础上,适时建设X90工程应用段。X90管线钢与钢管的研发及应用配套技术,是第三代管线钢与钢管应用关键技术重大科技专项成果之一,研究形成了X90管线钢与钢管显微组织鉴定图谱,提出了X90管材的关键技术指标、检测评价方法及配套系列标准;开发了系列X90焊管与管件;形成了X90管道断裂控制技术,通过实物爆破试验验证了X90输气管道的安全性[41,62-68]。这些工作为X90管线钢与钢管的工程应用奠定了基础,具备了实施工程应用段建设的基本条件。

（4）持续深化油气管道失效控制、完整性理论与技术的研究及应用[69]。通过对典型管道失效案例的分析研究,归纳总结其失效模式、规律、原因,提出失效抗力指标、失效判据及具体的预防措施。油气管道失效控制可以从三个方面入手：①控制失效概率,提高安全可靠性,主要是通过安全可靠性设计与评价,减少偶发失效;②控制失效后果,减少失效影响,主要是通过风险分析评价与控制来实现;③控制失效发生的时间,延长使用寿命,主要是通过优化设计、合理选材、制造质量控制、严格检验验收,减少早期失效,通过过程检验、维修维护、科学管理,减少耗损失效,发展寿命设计评估与预测技术。安全评价、风险评价、可靠性评估、寿命预测等是油气管道服役安全与完整性技术的主要内容。完整性的本质与核心是全寿命周期的安全可靠性,包括完整性设计即安全可靠性及寿命设计(全寿命周期的安全可靠性设计)、完整性评价、完整性管理。应当强调指出,油气管道的完整性保障与失效控制,就是要保障油气管道全寿命周期的安全可靠性与经济性,而70%以上的管道失效是前期设计、选材、评价及不规范的焊接施工造成的。因此,加强油气管道优化设计、选材、试验评价,综合考虑管道设计寿命、失效概率、失效后果等因素,发展基于风险的油气管道可靠性设计新方法及配套技术,是控制管道失效、保障管道完整性的关键。近年来,中国石油天然气集团有限公司已初步研发形成基于可靠性的管道设计技术,但尚未纳入标准规范,拟进一步深化研究形成配套技术,并将研究成果纳入设计规范,强制推广应用。

4 结论及建议

（1）经过近20年的发展,中国X80管线钢与钢管的研发及应用取得显著成效,建成X80油气输送管道约17000km,单管输气能力达到$380×10^8 m^3/a$,X80管材生产与管道建设技术进入国际领跑者行列。

（2）X80管线钢与钢管组织分析鉴别与评定、强度试验与屈强比控制、断裂与变形控制等关键技术的突破,为X80管线钢与钢管的工程应用奠定了基础。

（3）中国已研发形成X80系列热轧板卷及大口径厚壁螺旋埋弧焊管制造技术、宽厚板及大口径厚壁直缝埋弧焊管制造技术、大应变管线钢及钢管制造技术、感应加热弯管及管件设计与制造技术,为西气东输二线、西气东输三线、中缅天然气管道、中亚C线、陕京四线、中俄东线等重大油气管道工程建设提供了有力的技术支撑。

（4）根据中国加快天然气开发利用与中长期油气管网规划新要求，以及大输量天然气管道长期安全高效运行面临的新挑战，提出进一步加强对制管用板卷或钢板的可焊性评价与控制、深化现场焊接技术与质量性能控制研究、持续有序推进高钢级管线钢与钢管在大输量管道建设中的应用、持续强化油气管道失效控制、完整性理论与技术研究及应用等建议。

参 考 文 献

[1] Chaudhari V, Ritzmann H P, Wellmitz G. German gas pipeline first to use new generation line pipe[J]. Oil & Gas Journal, 1995, 93(1)：40-47.
[2] Glover A G, Horsley D J, Dorling D V. Pipeline design and construction using higher strength steels[C]. Calgary：International Pipeline Conference, 1998：659-664.
[3] Glover A G. High-strength steel becomes standard on Alberta gas system[J]. Oil & Gas Journal, 1999, 97(1)：54-57.
[4] Gray J M. Recent X80 pipelines in the United States[C]. Xi'an：International Seminar on Fracture Control Technology for X80 high pressure gas transmission pipeline, 2007：1-17.
[5] 王晓香. 国内外超大输量天然气管道建设综述[J]. 焊管, 2019, 42(7)：1-9.
[6] 冯耀荣, 庄传晶. X80级管线钢管工程应用的几个问题[J]. 焊管, 2006, 29(1)：1-5.
[7] 冯耀荣, 陈浩, 张劲军, 等. 油气输送管道工程技术进展[M]. 北京：石油工业出版社, 2006：108-112.
[8] 李鹤林, 吉玲康, 田伟. 高钢级钢管和高压输送：我国油气输送管道的重大技术进步[J]. 中国工程科学, 2010, 12(5)：84-90.
[9] 王晓香. 关于管线钢管技术的若干热点问题[J]. 焊管, 2019, 42(1)：1-9, 16.
[10] 中国石油集团石油管工程技术研究院. 西气东输二线X80管材技术条件及关键技术指标研究：009-2009[R]. 西安：中国石油集团石油管工程技术研究院, 2009：13-29.
[11] 中国石油集团石油管工程技术研究院. 西气东输二线管道断裂与变形控制关键技术研究：016-2013[R]. 西安：中国石油集团石油管工程技术研究院, 2013：7-15.
[12] 姜昌亮. 中俄东线天然气管道工程管理与技术创新[J]. 油气储运, 2020, 39(2)：121-129.
[13] 蒲明, 李育天, 孙骥姝. 中俄东线天然气管道工程前期工作创新点及创新成果[J]. 油气储运, 2020, 39(4)：371-378.
[14] 程玉峰. 保障中俄东线天然气管道长期安全运行的若干技术思考[J]. 油气储运, 2020, 39(1)：1-8.
[15] 王晓香. 2016—2017年以来我国焊管产业的运行情况及技术进步[J]. 焊管, 2018, 41(6)：1-6.
[16] 张振永. 中俄东线X80钢级φ1422mm管道工程设计关键技术应用[J]. 焊管, 2019, 42(7)：64-71.
[17] 周亚薇, 张振永. 中俄东线天然气管道环焊缝断裂韧性设计[J]. 油气储运, 2018, 37(10)：1174-1179.
[18] 张宏, 吴锴, 刘啸奔, 等. 直径1422mm X80管道环焊接头应变能力数值模拟方法[J]. 油气储运, 2020, 39(2)：162-168.
[19] 赵新伟, 池强, 张伟卫, 等. 管径1422mm的X80焊管断裂韧性指标[J]. 油气储运, 2017, 36(1)：37-43.
[20] 张振永, 周亚薇, 张金源. 现行设计系数对中俄东线OD1422mm管道的适用性[J]. 油气储运, 2017, 36(3)：319-324.
[21] 张振永, 张文伟, 周亚薇, 等. 中俄东线OD1422mm埋地管道的断裂控制设计[J]. 油气储运, 2017, 36(9)：1059-1064.
[22] 毕宗岳. 新一代大输量油气管材制造关键技术研究进展[J]. 焊管, 2019, 42(7)：10-25.
[23] 刘迎来, 许彦, 王高峰, 等. 中俄东线-45℃低温环境油气管道工程用X80钢级φ1422mm×33.8mm感

应加热弯管研发[J].焊管,2019,42(7):48-54.
[24] 尤泽广,王成,傅伟庆,等.中俄东线站场用直径1422mm×1219mm三通设计[J].油气储运,2020,39(3):347-353.
[25] 黄昌武.中国石油2015年十大科技进展[N].中国石油报,2016-01-22[2020-04-10].
[26] 冯耀荣,霍春勇,吉玲康,等.我国高钢级管线钢和钢管应用基础研究进展及展望[J].石油科学通报,2016,1(1):143-153.
[27] 冯耀荣,高惠临,霍春勇,等.管线钢显微组织的分析与鉴别[M].西安:陕西科技出版社,2008:92-163.
[28] 吉玲康,霍春勇,冯耀荣,等.西气东输二线X80管材关键技术研究及产品质量状况//X80管线钢和钢管研究与应用文集[M].西安:陕西科技出版社,2011:39-52.
[29] 冯耀荣,李洋,吉玲康,等.屈强比对X80焊管力学性能和安全使用的影响//X80管线钢和钢管研究与应用文集[M].西安:陕西科技出版社,2011:60-68.
[30] 熊庆人,杨扬,许晓锋,等.切环法和盲孔法测试大口径厚壁X80钢级埋弧焊管的残余应力[J].机械工程材料,2018,42(12):27-30,67.
[31] Feng Y R, Huo C Y, Zhuang C J, et al. Fracture control of the 2nd West to East Gas Pipeline in China[J]. Procedia Structural Integrity, 2019, 22: 219-228.
[32] 杨坤,池强,李鹤,等.高钢级天然气输送管道止裂预测模型研究进展[J].石油管材与仪器,2019,5(4):9-14.
[33] 霍春勇,李鹤,张伟卫,等.X80钢级1422mm大口径管道断裂控制技术[J].天然气工业,2016,36(6):78-83.
[34] 蒋庆梅,张小强,钟桂香,等.中俄东线黑龙江穿越段管材关键性能指标对比与确定[J].油气储运,2020,39(1):92-98.
[35] Chen H Y, Ji L K, Gong S T, et al. Deformation behavior prediction of X80 steel line pipe and implication on high strain pipe specification[C]. Calgary: 7th International Pipeline Conference, 2008: IPC2008-64575.
[36] Chen H Y, Ji L K, Wang H T, et al. Test evaluation of high strain line pipe material[C]. Hawaii: International Offshore and Polar Engineering Conference, 2011: 581-585.
[37] 吉玲康,李鹤林,陈宏远,等.管线钢管局部屈曲应变分析与计算[J].应用力学学报,2012,29(6):758-762.
[38] SY/T 7042—2016 基于应变设计地区油气管道用直缝埋弧焊钢管[S].
[39] API SPEC 5L-2018 Specification for line pipe[S].
[40] SY/T 7318.3—2017 油气输送管特殊性能试验方法 第3部分:全尺寸弯曲试验[S].
[41] 中国石油新闻中心.中国石油国际石油2017年十大科技进展[E/OL].(2018-01-24)[2020-04-09].
[42] 赵新伟,罗金恒,张广利,等.0.8设计系数下天然气管道用焊管关键性能指标[J].油气储运,2013,32(4):355-359.
[43] 刘迎来,吴宏,井懿平,等.高强度油气输送管道三通验证试验研究[J].焊管,2014,37(3):28-33.
[44] 王国丽,赵乐晋,管伟,等.直径1422mm、压力12MPa、钢级X80管道输气方案可行性[J].油气储运,2014,33(8):799-806.
[45] 李丽锋,罗金恒,赵新伟,等.OD1422mm X80管道的风险水平[J].油气储运,2016,35(4):25-29.
[46] 黄维和,郑洪龙,李明菲.中国油气储运行业发展历程及展望[J].油气储运,2019,38(1):1-11.
[47] 孔君华,郑琳,刘小国,等.西气东输二线用X80管线钢热轧卷板的组织与韧性[J].焊管,2011,34(5):5-11.
[48] 崔天燊,张彦睿.太钢X80级管线钢热轧卷板技术开发与应用[J].焊管,2009,32(12):32-39.

[49] 杨忠文, 毕宗岳, 牛辉. 高钢级管线钢焊管研制[J]. 焊管, 2011, 34(4): 5-11.

[50] 李少坡, 姜中行, 李永东, 等. 超低碳贝氏体厚壁X80宽厚板的研发[J]. 钢铁, 2012, 47(4): 55-59.

[51] 陈小伟, 王旭, 李国鹏, 等. 我国高钢级、大直径油气输送直缝埋弧焊管研究进展[J]. 钢管, 2016, 45(5): 1-8.

[52] 谢仕强, 桂光正, 郑磊, 等. X80钢级大直径UOE直缝埋弧焊管的开发及应用[J]. 钢管, 2011, 40(4): 29-36.

[53] 冯耀荣, 吉玲康, 陈宏远, 等. 一种高钢级大应变管线钢和钢管的制造方法: 201010251848.3[P]. 2011-12-07.

[54] 张伟卫, 吉玲康, 陈宏远, 等. 一种X70或X80抗大变形钢管生产方法: 201010550557.4[P]. 2013-07-31.

[55] 刘文月, 任毅, 张禄林, 等. 一种X80抗大变形管线钢及制造方法: 201510336908.4[P]. 2018-08-31.

[56] 雷胜利, 杨忠文, 毕宗岳, 等. 管线钢用高强度埋弧焊丝: 201010110747.4[P]. 2012-02-01.

[57] 牛辉, 刘清友, 杨忠文, 等. 基于应变设计用厚规格X80管线钢组织与性能关系[J]. 钢铁, 2013, 48(8): 55-60.

[58] 冯耀荣, 刘迎来, 牛靖. 一种X80弯管和管件的制备方法: 201010199050.9[P]. 2012-05-30.

[59] 新浪网. 贵州一天然气管道发生爆燃[E/OL]. (2017-07-03)[2020-04-09].

[60] 搜狐网. 中石油中缅管道晴隆段"2018.6.10"燃爆事故调查报告[E/OL]. (2019-05-08)[2020-04-09].

[61] 隋永莉. 新一代大输量管道建设环焊缝自动焊工艺研究与技术进展[J]. 焊管, 2019, 42(7): 83-89.

[62] 史立强, 牛辉, 杨军, 等. 大口径JCOE工艺生产X90管线钢组织与性能的研究[J]. 热加工工艺, 2015, 44(3): 226-229.

[63] 刘刚伟, 毕宗岳, 牛辉, 等. X90高强度螺旋埋弧焊管组织性能研究[J]. 焊管, 2015, 38(10): 9-13.

[64] Kong L, Shuai J, Zhou X, et al. A universal method for acquiring the constitutive behaviors of API-5L X90welds[J]. Experimental Mechanics, 2015, 56(2): 165-176.

[65] 张继明, 吉玲康, 霍春勇, 等. X90/X100管线钢与钢管显微组织鉴定图谱[M]. 西安: 陕西科技出版社, 2017: 31-94.

[66] 王红伟, 吉玲康, 张晓勇, 等. 批量试制X90管线钢管及板材强度特性研究[J]. 石油管材与仪器, 2015, 1(6): 44-51.

[67] Kong L Z, Zhou X Y, Chen L Q, et al. CTOD-R curve tests of API 5L X90 by SENT specimen using a modified normalization method[J]. Fatigue & Fracture of Engineering Materials & Structures, 2017, 40(2): 288-299.

[68] Schmid S, Hahn M, Issler S, et al. Effect of frequency andbiofuel E85 on very high cycle fatigue behaviour of the high strength steel X90CrMoV18[J]. International Journal of Fatigue, 2014, 60: 90-100.

[69] 冯耀荣, 张冠军, 李鹤林. 石油管工程技术进展及展望[J]. 石油管材与仪器, 2017, 3(1): 1-8.

本论文原发表于《油气储运》2020年第39卷第6期。

大应变管线钢和钢管的关键技术进展及展望

冯耀荣[1]　吉玲康[1]　陈宏远[1]　姜金星[2]　王　旭[3]　任　毅[4]
张对红[5]　牛　辉[6]　柏明卓[7]　李少坡[8]

(1. 中国石油集团石油管工程技术研究院·石油管材及装备材料服役行为与结构安全国家重点实验室；2. 南京钢铁股份有限公司；3. 中国石油集团渤海石油装备制造有限公司；4. 鞍钢股份有限公司；5. 中石油管道有限责任公司；6. 宝鸡石油钢管有限责任公司；7. 宝山钢铁股份有限公司；8. 首钢集团有限公司)

摘　要：油气管道工程在途经地震带、滑坡带、矿山采空区、沉陷带等特殊地质环境时，使用的大应变管线钢和钢管的研制及应用配套技术是国际上研究的热点之一，也是我国重要油气管道工程必须破解的重大难题。为此，围绕大应变管线钢和钢管研发应用中的一系列关键技术难题，通过十余年的联合攻关，取得了多项理论技术创新。主要成果包括：(1)提出了应用多个不同的应力比、屈强比、均匀延伸率等参数联合表征和评价钢管变形行为的方法，建立了 X70HD/X80HD 大应变管线钢和钢管新产品技术指标体系和标准；(2)研发形成了 X70HD/X80HD 钢板制造成套技术，获得了兼备低屈强比、高均匀伸长率、高应力比、高强韧性的大应变钢板；(3)研发形成了 X70HD/X80HD JCOE 和 UOE 大应变直缝埋弧焊管制造技术，解决了钢管母材和焊缝性能合理匹配及成型、焊接、扩径、热涂覆过程性能劣化难题；(4)自主研制了钢管内压+弯曲大变形实物试验装置，研发形成了钢管实物模拟变形试验技术。X70HD/X80HD 大应变管线钢和钢管在西气东输、中缅管道等重大输气管道工程中实现了规模化应用，取得了良好的应用实效。根据复杂地质条件管道建设与长期安全运行的新要求，提出了进一步发展基于应变的管道设计新方法，研发或完善与环焊缝强匹配要求相适应的焊接方法、焊接材料、焊接工艺、环焊缝性能质量及缺陷控制要求等配套技术的建议。

关键词：大应变管线钢；大应变钢管；变形行为评价；X70；X80；环焊缝；中缅管道；西气东输管道工程

基金项目：国家重点研发计划项目"高应变海洋管线管研制"(编号：2018YFC0310300)，中国石油天然气股份有限公司重大科技专项"西气东输二线管道断裂与变形控制关键技术研究"(编号：2009E-0105)、"西气东输二线特殊焊管与管件研制"(编号：2009E-0104)，中国石油天然气集团有限公司科技基础条件平台建设项目"CNPC 石油管工程重点实验室科技基础条件平台建设"(编号：07H611)。

作者简介：冯耀荣，1960 年生，正高级工程师，博士，长期从事石油工程材料应用基础研究及重大工程技术支持工作，现任石油管材及装备材料服役行为与结构安全国家重点实验室主任。地址：陕西省西安市锦业二路 89 号(710077)，ORCID：0000-0002-3325-6067，E-mail：fengyr@cnpc.com.cn。

通信作者：吉玲康，1966 年生，正高级工程师，博士。地址：陕西省西安市锦业二路 89 号(710077)，E-mail：jilk@cnpc.com.cn。

为了满足我国国民经济发展对清洁能源日益增长的需求,推进"一带一路"倡议建设,迫切需要建设高压大输量油气长输管线,实现"西气(油)东输""北气南(油)运""陆气(油)出海""海气(油)登陆"。我国幅员辽阔,地质地貌复杂,大口径高压油气长输管道必然要经过大量地震断裂带、滑坡带、矿山采空区、沉陷带等复杂工况区。对于这种复杂工况管道,必须采用基于应变的极限状态设计方法代替传统的许用应力设计方法。这就对大应变管线钢管提出了巨大的需求。

大应变管线钢管是在保证钢管高强度和高韧性的同时具有低的屈强比($R_{t0.5}/R_m$)、高的均匀塑性变形延伸率(以下简称均匀延伸率,UEL)、高的形变硬化指数(n)和高的临界屈曲应变能力(ε_{ave})的新型钢管,适用于通过滑坡带、地震断裂带、沉陷带、矿山采空区等地质灾害地区的管道,以防止由于内压和轴向压缩产生的大应变而引起管道屈曲、失稳和延性断裂引发灾难性事故[1,2]。然而,随着油气输送管道向高压大口径和高强度方向发展,管线钢和钢管的主流钢级已从过去的 X52、X60 发展到 X70 和 X80,其屈强比已从过去的 0.80 增加至 0.90~0.93 或以上,过高的屈强比限制了管线钢和钢管的极限塑性变形能力,从而对管道的安全服役造成严重影响[3]。因此,发展具有较低屈强比和良好塑性变形能力的大应变管线钢和钢管已成为管道安全服役、特别是应变控制工况下安全服役的必然要求。大应变管线钢和钢管的关键技术难点是既要满足强度要求,又要满足韧性要求,同时还要具有在复杂工况下的良好塑性和大应变能力。而这些性能要求往往相互矛盾,如何兼顾是国际上面临的重大难题。这就需要建立大应变管线钢和钢管的技术指标体系和标准,系统研究揭示管线钢和钢管化学成分、组织结构、性能和服役性能、生产制造工艺之间的内在联系和影响规律,研发大应变管线钢和钢管生产制造及配套技术。

以我国油气战略通道中缅油气管线(以下简称中缅管线)和西气东输三线为例,设计压力介于 10~12MPa,全线采用 X70/X80 管线钢管,管径介于 813~1219mm,是国内地质地貌条件最为复杂的管线,为保证中缅管线、西气东输管线等重大输气管道工程顺利建成和安全运行,亟待开展 X70HD/X80HD 大应变管线钢管研发及应用关键技术攻关。大应变管线钢管的概念首先由日本 JFE 公司提出[4,5],并研制和生产了 X65 钢级产品在俄罗斯萨哈林管线上应用。随后日本新日铁和 JFE 公司又研发了 X70 和 X80 大应变钢管,在中缅管线和西气东输二线工程中应用。

国内 X70/X80 大应变管线钢和钢管基本上与日本同步研发,依托国家科技支撑计划课题"X80 管材规范关键技术研究"、中国石油天然气集团有限公司"特殊地区管道建设关键技术研究"等多个重大科技项目,经过十余年的研究攻关,形成 X70HD/X80HD 大应变管线钢管及应用关键技术[6-9],国产 X70HD/X80HD 大应变管线钢和钢管在中缅管线和西气东输管道工程中实现了规模化应用,从根本上解决了中缅管线、西气东输管线等重大工程地震断裂带、滑坡带、矿山采空区、沉陷带等复杂工况管道建设和安全运行关键技术难题。为此,及时总结我国大应变管线钢和钢管研发和应用的理论与技术成果及经验,对于进一步发展大应变管线钢和钢管及应用关键技术、提升管道设计和管道建设水平、保障管道长期安全运行等都具有重要的意义。

1 大应变管线钢管变形行为表征、评价、预测、控制技术

采用激光测量技术获得钢管精确形状,运用数字化逆向建模技术,引入钢管材料实际本构方程,对钢管内压+弯曲条件下的屈曲行为进行精确仿真(图 1),形成钢管屈曲应变容量

数值模拟技术[10-14]。

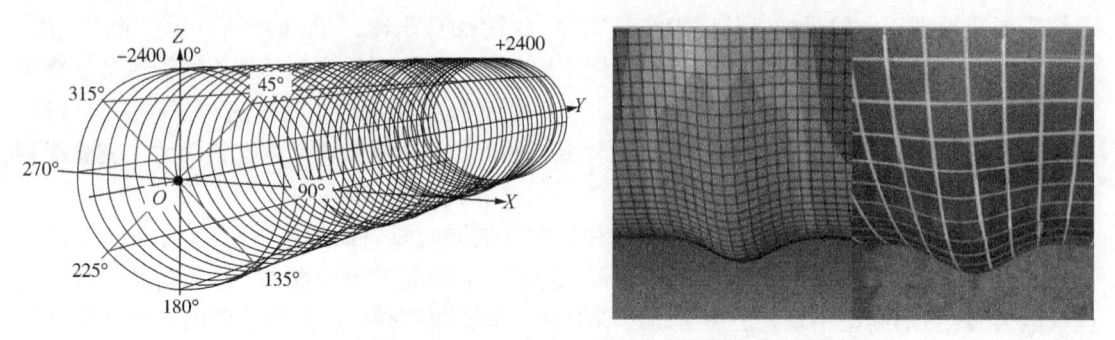

图 1 外形精确测量和数字化逆向建模、数值仿真计算图

提出了用材料应力比($R_{t5.0}/R_{t1.0}$、$R_{t2.0}/R_{t1.0}$、$R_{t1.5}/R_{t0.5}$)代替形变硬化指数(n)作为评价管线钢管应变行为的重要指标,揭示了材料应力比与钢管 2D 长度平均应变的相关性(图2),发明了用应力比确定钢管屈曲应变能力的方法[15-19]。

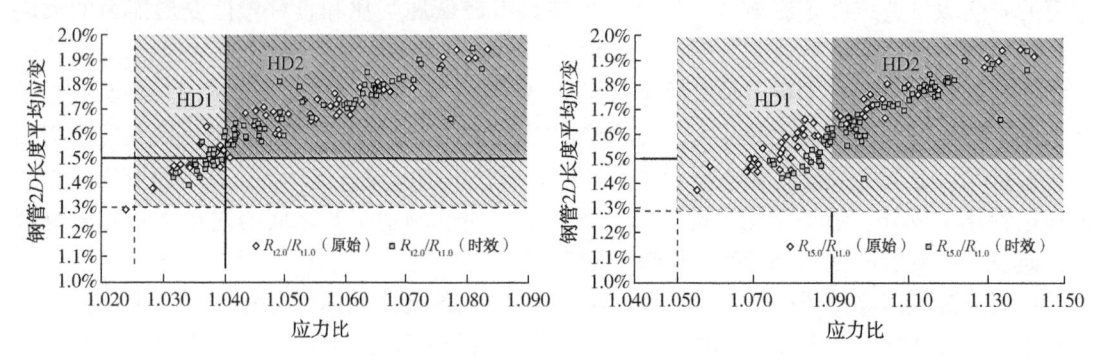

图 2 应力比与钢管 2D 长度平均应变的关系图

提出大应变管线钢管技术指标体系(表1),用多个不同的应力比、屈强比、UEL 等参数联合表征和评价钢管的变形行为(图3),可全面控制大应变管线钢管的质量和性能水平。制定了大应变管线钢管技术条件和标准,被美国石油学会(API)采纳作为标准附录发布[15-20]。

表 1 X70HD/X80HD 大应变钢管关键技术指标要求表

项目	$R_{t0.5}$(MPa)	R_m(MPa)	$R_{t0.5}/R_m$	$R_{t1.5}/R_{t0.5}$	$R_{t2.0}/R_{t1.0}$	$R_{t5.0}/R_{t1.0}$	A	CVN/J	$DWTT$	UEL	S—S 曲线
X70/X70HD 横向,时效前	485~635	570~760	≤0.930	—	—	—	≥22%	≥160	≥85%	—	无要求
X70HD 纵向,时效前	450~570	570~735	≤0.850	≥1.100	≥1.040	≥1.088	—	—	—	7.0%	拱顶型
X70HD 纵向,200℃时效后	450~590	570~735	≤0.860	≥1.070	≥1.040	≥1.088	—	—	—	6.0%	拱顶型
X80/X80HD 横向,时效前	555~690	625~825	≤0.930	—	—	—	API 5L 规定	≥220	≥85%	—	无要求

续表

项目	指标									
	$R_{t0.5}$ (MPa)	R_m (MPa)	$R_{t0.5}/R_m$	$R_{t1.5}/R_{t0.5}$	$R_{t2.0}/R_{t1.0}$	$R_{t5.0}/R_{t1.0}$	A	CVN/J	DWTT	UEL	S—S曲线
X80HD纵向,时效前	530~630	625~770	≤0.850	≥1.100	≥1.033	—	—	—	—	7.0%	拱顶型
X80HD纵向,200℃时效后	530~630	625~770	≤0.860	≥1.070	≥1.033	—	—	—	—	6.0%	拱顶型
X70/X80纵向	时效前后均无要求										

注：A 为延伸率；CVN 为冲击功，J；$DWTT$ 为落锤撕裂试验延性断裂面积百分比。

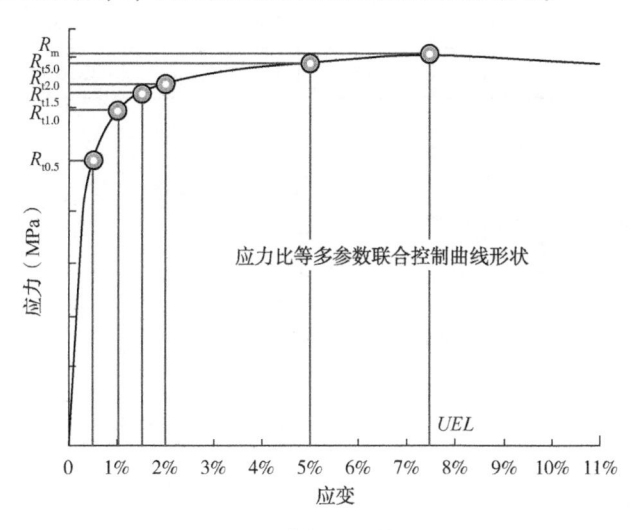

图3　应力比联合控制钢管变形行为图

2　X70HD/X80HD大应变钢板制造技术

为保证X70HD/X80HD大应变管线钢强度、塑性、韧性、可焊性、应变时效性能等要求，在成分设计方面，选用了低碳或超低碳含量，加入了Mn、Mo、Cr、Cu、Ni、Nb、Ti等元素，控制N含量，严格控制S、P、O_2、H_2等有害物的含量，采用Al、Si全脱氧，通过Ca处理控制夹杂物形状[21-24]。典型化学成分见表2。

表2　X70HD/X80HD大应变管线钢的成分设计表

项目	成分组成								
	C	Si	Mn	P	S	Cr+Mo+Ni+Cu	Nb	Ti	N
X70/X80	0.040%~0.060%	0.150%~0.300%	1.500%~1.800%	≤0.012%	≤0.003%	0.150%~0.800%	0.040%~0.100%	0.010%~0.020%	≤0.006%
X70HD/X80HD	0.040%~0.060%	0.150%~0.300%	1.500%~1.800%	≤0.012%	≤0.003%	0.150%~0.800%	0.030%~0.050%	0.010%~0.020%	≤0.004%

注：各组分含量以质量分数计。

应用了"多边形铁素体+贝氏体"双相组织设计,其中贝氏体(硬相)保障了管线钢必要的强度,铁素体(软相)保障了管线钢足够的塑性。系统研究揭示了块状铁素体体积分数对管线钢强度、屈强比、应力比、UEL等指标的影响规律。当块状铁素体为50%~75%时,可以实现X70HD/X80HD管线钢的强度、塑性和韧性的合理匹配。控制晶粒尺寸3~6μm,通过软硬相复合组织合理匹配,保证了钢板的屈强比不大于0.75、UEL不小于11%、高应力比和高强韧性(表3、图4至图6)[21,22,25-29]。

表3 不同铁素体含量管线钢板力学性能表

铁素体含量	S—S曲线	$R_{t0.5}$(MPa)	R_m(MPa)	$R_{t0.5}/R_m$	$R_{t1.5}/R_{t0.5}$	$R_{t2.0}/R_{t1.0}$	$R_{t5.0}/R_{t1.0}$	UEL	CVN/J	DWTT
34.2%	有拐点	540	688	0.800	1.080	1.050	1.156	7.7%	277	97%
50.8%	拱顶型	495	660	0.750	1.110	1.079	1.200	11.9%	294	98%
70.2%	拱顶型	445	645	0.690	1.134	1.107	1.263	13.9%	351	98%

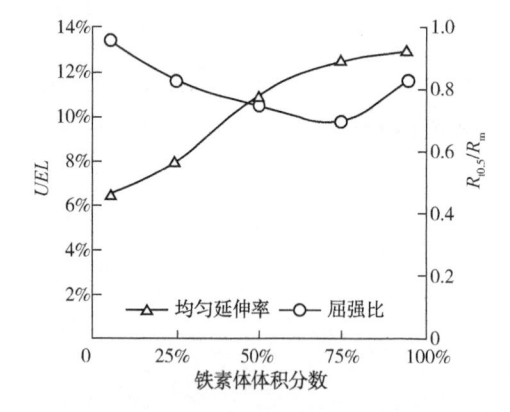

图4 铁素体含量对管线钢性能的影响图 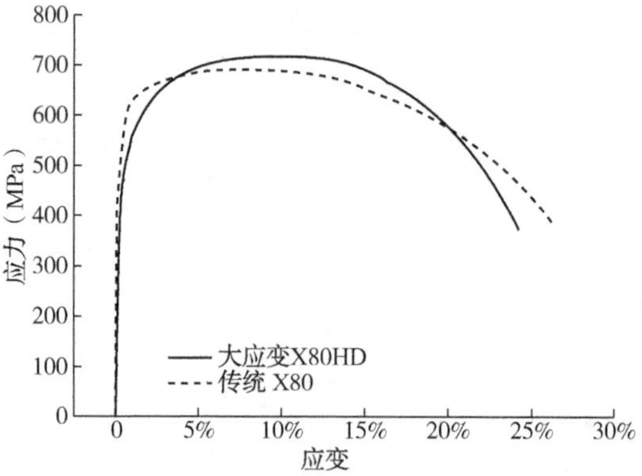 图5 管线钢X80HD与X80应力—应变曲线对比图

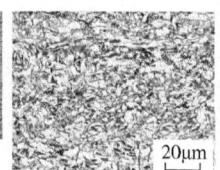

(a)普通X80的针状铁素体 (b)JFEX80HD的贝氏体+MA双相组织 (c)该项目X80HD的F+B双相组织 (d)该项目X80HD的铁素体 (e)该项目X80HD的贝氏体

图6 管线钢X80HD与X80显微组织比较图

发明了用热轧方法来制备X70HD/X80HD大应变管线钢的工艺技术(图7),突破了针状铁素体型X70/X80管线钢的制造工艺和JFE贝氏体+M/A大应变管线钢制造工艺,可以对"多边形铁素体+贝氏体"双相组织和性能进行精准调控[21,22,25-29]。

3 X70HD/X80HD大应变管线钢管制造技术

揭示了铁素体+贝氏体钢板在扩径及制管成型过程中各项关键力学性能的变化规律及其相关性,规范了钢板强度、应力比、韧性及塑性等指标要求(图8、图9)[21,22,30-33]。

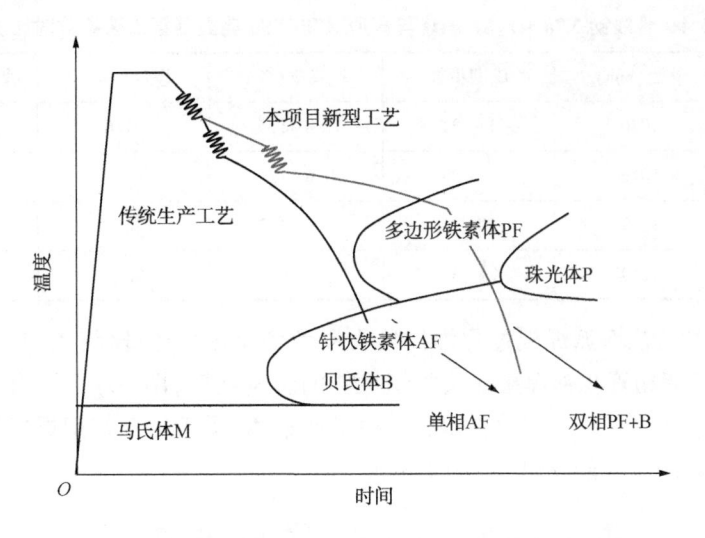

图 7　管线钢 X70HD/X80HD 与 X70/X80 工艺对比图

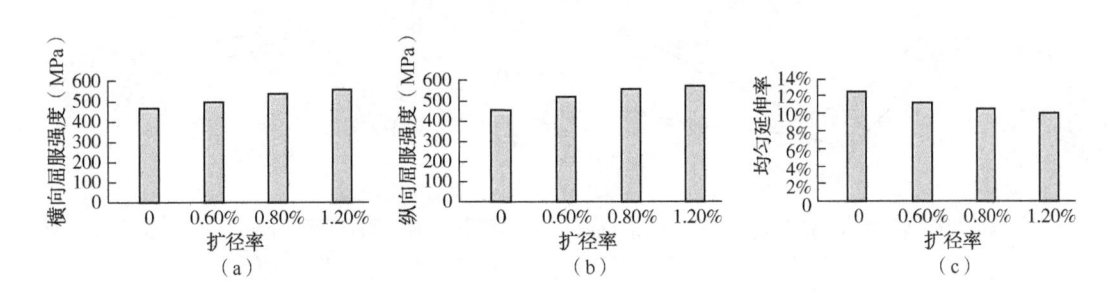

图 8　屈服强度、均匀延伸率和扩径率的关系图

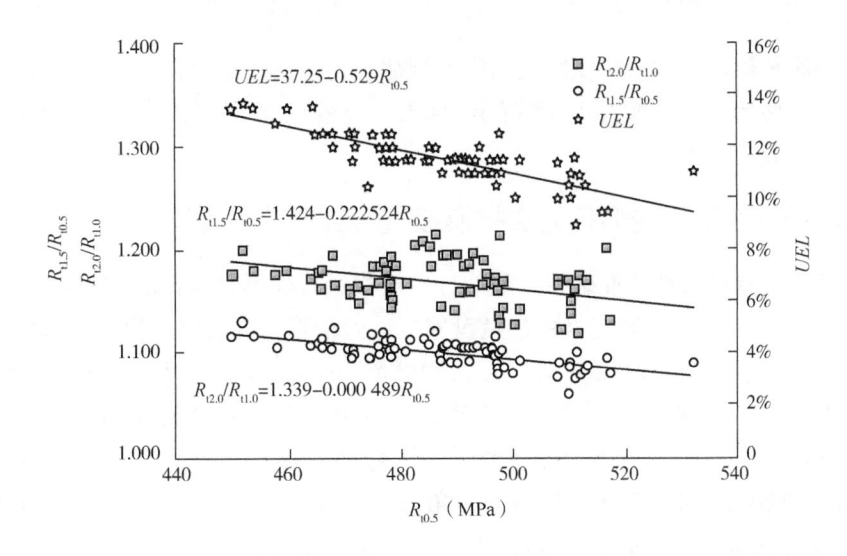

图 9　屈服强度 $R_{t0.5}$ 与 UEL 和应力比的关系图

研发了大应变钢管 JCOE 和 UOE 专有成型、扩径等关键工艺技术，优化了扩径率及成型步数(表 4)[21,22,30-33]。

表4 管线钢 X70HD/X80HD 与普通 X70/X80 典型制管工艺参数对比表

钢级	外径(mm)	成型步数	扩径率(%)	壁厚(mm)	焊接线能量(kJ/cm)
X70	1016	17~21	1.0~1.5	21.0	42
X70HD	1016	23~29	0.6~1.0	21.0	37
X80	1219	19~25	0.9~1.3	26.4	53
X80HD	1219	25~31	0.6~0.8	26.4	45

揭示了多丝共熔池埋弧焊接过程焊接接头温度场及热循环过程，掌握了焊接热输入对"铁素体+贝氏体"双相管线钢焊接接头组织性能的影响规律(图10)，发明了高强韧匹配焊接材料，自主研发了多丝埋弧焊接技术，解决了铁素体+贝氏体双相钢焊接热影响区易脆化的突出问题，热影响区性能及焊缝100%达标[33-35]。

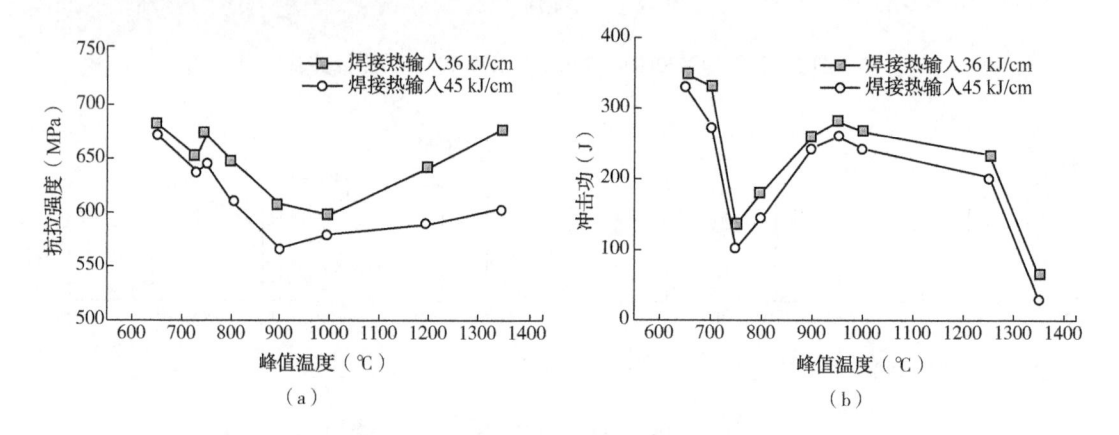

图10 焊接热输入、峰值温度对 X70 大应变钢管焊接接头抗拉强度和热影响区冲击功的影响图

明确了热涂覆温度对钢管应变时效的影响规律，优选出涂覆材料，形成了大应变钢管热涂覆工艺技术，使热涂覆温度由230℃降至200℃，有效控制了热涂覆过程钢管 UEL 的下降和屈服强度的升高[10,11,22,23,32-35]。

4 大口径管线钢管变形行为实物模拟试验装置及试验技术

自主研发了具有灵活性和准确性、可进行多规格钢管变形行为试验的实物钢管内压+弯曲试验系统[图11(a)][36]。可更换的法兰盘连接试验钢管，可以使实物内压+弯曲变形试验系统能进行外径介于 508~1219mm 的钢管实物试验[图11(b)]；可拆卸式承载框架模块，可以使实物内压+弯曲变形试验系统进行长度 6~12m 的钢管实物试验[图11(c)]；2自由度轴承式水平支撑结构，可以有效降低实物内压+弯曲变形试验系统阻力，提高试验精度[图11(d)]；试验管中部 2D 长度横截面转角测量装置，可以获得卧式实物试验中的试验管特定部位的横截面转角；钢管管端转角测量装置，可以有效测量试验钢管管端的弯曲变形试验结果。研究形成了管线钢管变形能力试验评价技术及标准[37]。实现了对管径介于 508~1219mm、壁厚不大于 26.4mm、长度介于 6~10m 的管材进行服役性能的实物试验评价。

(a) 实物图　　　　　(b) 可更换的法兰盘　　(c) 可拆卸式承载框架　(d) 2自由度轴承式水平支撑

图 11　全尺寸内压+弯曲变形试验装置图

5　大应变管线钢和钢管在我国重大管道工程中的应用

X70/X80 大应变管线钢和钢管及应用关键技术的突破，显著提高了我国高钢级管线钢管的生产技术和产品质量水平，为地震断裂带、滑坡带、矿山采空区、沉陷带等复杂工况基于应变设计地区管道失效控制与安全保障提供了系统配套的技术。2011 年，在中缅油气管道建设过程中，为克服通过 9 度区、5 条活动断裂带以及地质灾害易发地段等可能发生地面位移区段内管道的大应变难题，采用基于应变的设计方法，首次在上述地段应用了国产 X70HD 的 D1016mm×17.5mm（壁厚，下同）大应变钢管 115.0km，D1016mm×21mm 大应变钢管 5.0km，D813mm×14.7mm 大应变钢管 10.5km，D813mm×17.2mm 大应变钢管 7.5km，合计 59126t。由南京钢铁股份有限公司、宝山钢铁股份有限公司、中国首钢集团秦皇岛首秦金属材料有限公司、沙钢股份有限公司、湘潭钢铁集团有限公司等供应钢板，中国石油集团渤海石油装备有限公司、宝鸡石油钢管有限责任公司生产钢管，保障了中缅油气管道建设质量，并节省由于绕避上述地段引起的工程投资的增加。

2012 年，在西气东输三线东段工程设计中，首次在基于应变设计地区应用了国产 X80HD 的 D1219mm×26.4mm 大应变钢管 3.9km，由鞍钢股份有限公司等供应钢板、中国石油渤海石油装备有限公司生产钢管，实现了大应变钢管的国产化应用。2017 年，在陕京四线天然气管道的建设中，应用国产 X80HD 的 D1219mm×26.4mm 大应变管线钢管约 5000t，长度 3.5km，由鞍钢股份有限公司等供应钢板、中国石油集团渤海石油装备有限公司和宝鸡石油钢管有限责任公司生产钢管，保障了陕京四线输气管道工程质量。

6　总结与展望

我国大应变管线钢和钢管及应用关键技术取得了重大突破，建立了大应变管线钢管变形行为表征/评价/预测/控制技术、技术指标体系及标准，自主研发了 X70HD/X80HD 大应变钢板和钢管制造成套技术，建立了大口径管线钢管变形行为实物模拟试验装置及试验评价技术，为基于应变设计地区大应变管线钢管的进一步发展奠定了坚实基础。

近年来，复杂工况油气管道失效事故时有发生，原因涉及多个方面。应进一步应用并发展复杂地形地貌条件下基于应变的管道设计新方法，除采取必要的工程措施外，进一步优化大应变管线钢管成分设计和制造工艺、完善对组织性能的综合调控，持续完善大应变管线钢管技术与标准体系，加大与基于应变设计环焊缝强匹配要求相适应的焊接方法、焊接材料、焊接工艺、环焊缝性能质量及缺陷控制等配套技术攻关力度，开展管道环焊缝性能质量和服役性能评价，完善管道环焊缝性能质量指标体系，以实现对复杂工况油气管道失效的有效控制。

致谢：中国石油集团渤海石油装备制造有限公司陈小伟、中国石油天然气股份有限公司西南管道分公司刘雪光等参加了本文编写工作，特此致谢！

参 考 文 献

[1] 冯耀荣,高惠临,霍春勇,等.管线钢显微组织的分析与鉴别[M].西安：陕西科学技术出版社,2008.

[2] 李鹤林,李霄,吉玲康,等.油气管道基于应变的设计及抗大变形管线钢的开发与应用[J].焊管,2017,30(5)：5-11.

[3] 冯耀荣.X80管线钢和钢管研究与应用文集[M].西安：陕西科学技术出版社,2011.

[4] Okatsu M, Ishikana N, Shinmiya T, et al. Development of a high deformability linepipe with resistance to strain-agedhardening by heat treatment on-line process[J]. International Journal of Offshore and Polar engineering, 2008, 18(4)：308-313.

[5] Okatsu M, Shinmiya T, Ishikawa N, et al. Development of high strength linepipe with excellent deformability [C]. 24th International Conference on Offshore Mechanics and Arctic Engineering, 2005.

[6] 冯耀荣,张冠军,李鹤林.石油管工程技术进展及展望[J].石油管材与仪器,2017,3(1)：1-8.

[7] 冯耀荣,霍春勇,吉玲康,等.我国高钢级管线钢和钢管应用基础研究进展及展望[J].石油科学通报,2016,1(1)：143-153.

[8] 刘文月,任毅,高红,等.大变形管线钢研究进展[J].鞍钢技术,2016,401(5)：8-12.

[9] 中国石油信息网.2018年石油十大科技进展[EB/OL].(2019-01-24)[2020-07-23].

[10] 中国石油集团石油管工程技术研究院.西气东输二线X80管材技术条件及关键技术指标研究[R].西安：中国石油集团石油管工程技术研究院,2010.

[11] 中国石油集团石油管工程技术研究院."西气东输二线管道断裂与变形控制关键技术研究"技术报告[R].西安：中国石油集团石油管工程技术研究院,2013.

[12] Chen Hongyuan, Ji Lingkang, Gong Shaotao, et al. Deformation behavior prediction of X80 steel line pipe and implication on high strain pipe specification[C]. 7th International Pipeline Conference, 2008.

[13] Chen Hongyuan, Wang Haitao, Gong Shaotao, et al. Test evaluation of high strain line pipe material [C]. International Society of Offshore and Polar Engineers, 2011.

[14] 吉玲康,李鹤林,陈宏远,等.管线钢管局部屈曲应变分析与计算[J].应用力学学报,2012,29(6)：758-762.

[15] 吉玲康,李鹤林,赵文轸,等.X70抗大变形管线钢管的组织结构和形变硬化性能分析[J].西安交通大学学报,2012,46(9)：108-113.

[16] 陈宏远,吉玲康,黄呈帅,等.确定高钢级管线钢管均匀延伸率的方法及装置：中国,201310236663.9[P].2013-10-02.

[17] 吉玲康,陈宏远,李为卫,等.快速确定钢管屈曲应变能力的方法：中国,200910087054.5[P].2010-12-22.

[18] SY/T 7042—2016 基于应变设计地区油气管道用直缝埋弧焊钢管[S].

[19] API SPEC 5L—2018 American Petroleum Institute. Specification for line pipe[S].

[20] 冯耀荣,吉玲康,李为卫,等.中国X80管线钢和钢管研发应用进展及展望[J].油气储运,2020,39(6)：612-622.

[21] 冯耀荣,吉玲康,陈宏远,等.一种高钢级大应变管线钢和钢管的制造方法：中国,201010251848.3[P].2010-12-08.

[22] 张伟卫,吉玲康,陈宏远,等.一种X70或X80抗大变形钢管生产方法：中国,201010550557.4

[P].2012-05-23.

[23] 尹雨群，左秀荣，姜金星，等．一种热轧抗大变形管线钢及其制备方法：中国，201010266536.X[P].2010-12-15.

[24] 李少坡，查春和，李家鼎，等．韧性优良的 X70 级抗大变形管线钢板及其制备方法：中国，201210050196.6[P].2012-07-04.

[25] 孙磊磊，柏明卓，郑磊．两阶段冷却工艺对基于应变设计 X70 管线钢组织的影响[J]．钢铁，2014，49(9)：81-86.

[26] 李少坡，姜中行，李永东，等．超低碳贝氏体厚壁 X80 宽厚板的研发[J]．钢铁，2012，47(4)：55-59.

[27] 姜金星，于新攀，潘学福，等．碳、铬、钼元素对 X70 管线钢显微组织与性能的影响[J]．机械工程材料，2018，42(7)：37-40.

[28] 孙磊磊，柏明卓，郑磊，等．显微组织对 UOE 管线管时效敏感性的影响[J]．金属热处理，2018，43(9)：41-46.

[29] 刘文月，任毅，张禄林，等．一种 X80 抗大变形管线钢及制造方法：中国，201510336908.4[P].2017-01-11.

[30] 牛辉，刘清友，杨忠文，等．基于应变设计用厚规格 X80 管线钢组织与性能关系[J]．钢铁，2013，48(8)：55-60.

[31] 陈小伟，付彦宏，王旭，等．X70 抗大变形直缝埋弧焊管的开发[J]．焊管，2012，35(3)：71-75.

[32] 柏明卓，郑磊，张备，等．基于应变设计用 X70 大直径 UOE 管线管的研发[J]．焊管，2015，38(1)：21-27.

[33] 牛辉，张君，赵红波，等．X70 抗大变形钢管研发及批量生产[J]．焊管，2013，36(6)：26-31.

[34] 王旭，赵波，张红，等．一种多丝埋弧焊数值模拟热源模型参数确定方法：中国，201210132525.1[P].2012-08-15.

[35] 雷胜利，杨忠文，毕宗岳，等．管线钢用高强度埋弧焊丝：中国，201010110747.4[P].2010-06-16.

[36] 吉玲康，李彦锋，陈宏远，等．一种钢管弯曲变形试验系统：中国，201110446626.1[P].2012-07-04.

[37] SY/T 7318.3—2017 油气输送管特殊性能试验方法 第3部分：全尺寸弯曲试验[S].

本论文原发表于《天然气工业》2020年第40卷第9期。

316L衬里复合管道主要失效形式及其完整性检测技术研究

赵新伟 魏 斌 杨专钊 聂向晖 李发根 李为卫

(中国石油集团石油管工程技术研究院·石油管材及装备材料服役行为与结构安全国家重点实验室)

摘 要：316L衬里复合钢管在国内油气集输管线上已大批量推广应用，在高含CO_2/H_2S和Cl^-油气集输管道腐蚀控制方面取得了良好的效果，但也发生过多起泄漏、开裂失效的事故。简要介绍国内外316L衬里复合钢管应用现状，分析316L衬里复合管道在服役过程中发生的主要失效形式。讨论了316L衬里复合管道在完整性检测方面面临的技术挑战，试验验证并分析了漏磁检测技术和爬行机器人视频检测技术的适用性，研究提出了基于风险的316L衬里复合管道检测方法。

关键词：316L衬里复合管道；失效形式；内检测；基于风险的管道检测

当油气介质中CO_2和/或H_2S含量较高时，普通碳钢或低合金钢管难以满足内防腐要求。因此，选择技术可行、经济合理、安全可靠的油气集输管道材料是油气田地面工程领域的一项关键技术。双金属复合钢管兼顾了碳钢或低合金钢的高强度与不锈钢或耐蚀合金良好的耐蚀性，同时与纯不锈钢或耐蚀合金钢管相比更为经济，在高含CO_2和/或H_2S油气田的油气集输管道上已大量应用[1-5]。

双金属复合钢管的外层管(称为基管)，由碳钢或低合金钢焊管或无缝管构成，起到承压和支撑内层管的作用，用以保证管道的各项力学性能。内层材料为不锈钢、铁-镍基合金、镍基合金或其他耐蚀合金材料，若与基管之间为冶金结合，称为内覆层(Clad layer)；若与基管之间为机械结合，称为衬里层或内衬层(Lined layer)。内层材料的主要功能是提高管道耐腐蚀与抗冲刷性能，延长管道使用寿命。双金属复合钢管按基管与内层材料复合方式不同，分为机械复合和冶金复合，分别称为衬里耐蚀合金复合钢管和内覆耐蚀合金复合钢管[6]。20世纪90年代初，国际上开始应用双金属复合管，我国从21世纪初开始双金属复合管的开发及应用，目前已是双金属复合管的制造大国，在油气集输管线上的应用已超过1000km。在双金属复合管中，316L不锈钢衬里复合钢管的用量最大，占到80%以上，在高含CO_2和/或H_2S油气集输管道上取得良好的防腐效果。但是，316L不锈钢衬里复合管道也发生过多起泄漏、开裂失效事故，未能完全做到免维护，管道检测、修复等管道完整性技术面临一些技术难题和挑战。

作者简介：赵新伟(1969)，男，教授级高级工程师，主要从事油气输送管和管道完整性技术研究工作。地址：陕西省西安市锦业二路89号中国石油集团石油管工程技术研究院(710077)，E-mail：zhaoxinwei001@cnpc.com.cn。

本文综述316L不锈钢衬里复合管道的主要失效形式,分析316L不锈钢衬里复合管道在管道检测方面面临的技术挑战,并验证漏磁内检测技术、爬行机器人视频检测技术以及射线、超声常规无损检测技术在316L不锈钢衬里复合管道的适用性,研究提出基于风险的管道检测方法。

1 316L衬里复合管道主要失效形式

从20世纪90年代开始,国际上在油气输送领域应用双金属复合钢管,基管材料主要为X52和X65,耐腐蚀合金层主要为316L和镍基合金825,规格为$\phi114mm\sim\phi998mm$,主要用于含H_2S/CO_2和Cl^-苛刻腐蚀环境。在国内,2005年塔里木油田开始采用国产中小口径($\phi60mm\sim\phi168mm$)衬里复合钢管(20G/316L)用于高含CO_2和Cl^-油气集输,并有良好的防腐效果,这是复合钢管在国内油气田的首次使用。2011年,国产316L内衬复合钢管开始应用于海洋油气集输管道,先后在多个海上油气田应用超过100km,材料为X65/316L,规格$\phi168mm\sim\phi219mm$。2013年开始,大口径316L衬里复合钢管在塔里木油田集气干线得到试用与推广,规格主要为$\phi508mm$和$\phi355.6mm$,材料X65/316L。截至目前,双金属复合钢管在国内油气田的使用里程已近2000km[7],在中国石油、中国石化和中国海油等油气公司都有应用,而且以316L衬里复合钢管应用最多,占到双金属复合管应用量的80%。

尽管316L衬里复合钢管应用于油气集输管线有良好的防腐效果,但316L衬里复合管道也发生了多起失效事故。从316L衬里复合管道失效案例统计分析结果来看,主要的失效形式包括环焊缝腐蚀刺漏、环焊缝开裂、衬里层塌陷和衬里层腐蚀,尤其是前三种失效形式占比较高。

1.1 环焊缝腐蚀刺漏

环焊缝腐蚀刺漏主要发生于采用"封焊+多层焊"焊接工艺的管道,腐蚀主要发生于焊接热影响区以及熔合线部位,少量发生于焊缝中心部位,腐蚀形貌如图1所示。端部若采用封焊工艺设计(图2),封焊过程中不锈钢焊材不断往316L衬里层上熔焊,衬里层会反复受热,一旦封焊电流使用过大,则会造成管端衬里层热影响区晶粒粗大[8];现场环焊采用药芯焊丝,背面未充氩气保护,保护效果不佳,会导致衬里层局部严重贫Cr(例如失效样品焊缝及热影响区Cr含量分别仅为10.96%和5.77%)。上述因素使复合钢管焊缝及热影响区的耐腐蚀能力严重降低,现场检验手段又难以发现,最终导致焊缝刺漏。

(a)焊缝中心刺漏 　(b)热影响区刺漏

图1 环焊缝腐蚀刺漏宏观形貌　　图2 衬里复合钢管端部封焊示意

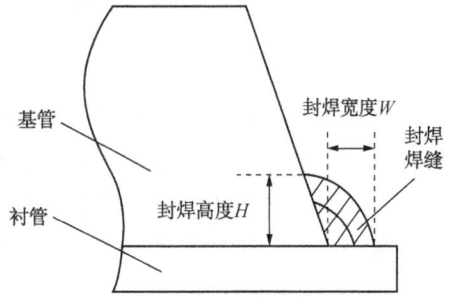

1.2 环焊缝开裂

环焊缝开裂失效的宏观形貌如图3所示。失效分析发现采用"封焊+多层焊"焊接工艺的

焊接接头，在基管、衬里层和封焊交界位置易出现孔洞和裂纹缺陷(图4)，构成了启裂源区，另外，焊缝存在高硬度(超过400HV10)的马氏体组织，韧性差，一旦有裂纹萌生就会很快扩展，最终导致环焊缝开裂失效。

图3 环焊缝开裂宏观形貌 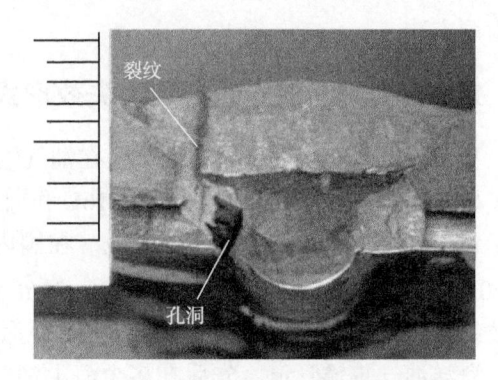　　　　图4 环焊缝孔洞缺陷及裂纹扩展路径

1.3 衬里层塌陷

国内油气田316L衬里复合钢管发生了较多的衬里层塌陷失稳现象，有两种情况：一种情况是发生在复合钢管3PE外防腐过程中，如图5(a)所示；另一种情况是塌陷发生在管道服役过程，如图5(b)所示。造成防腐施工中衬层塌陷的原因是，复合工艺管控不到位，基管和衬管间隙中存在气体、水分，外防腐过程中钢管温度升高造成间隙高压，从而导致衬里层失稳。

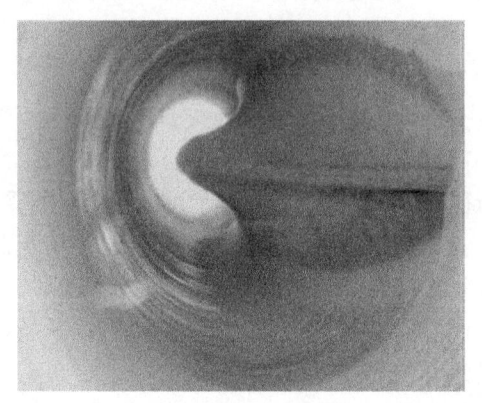

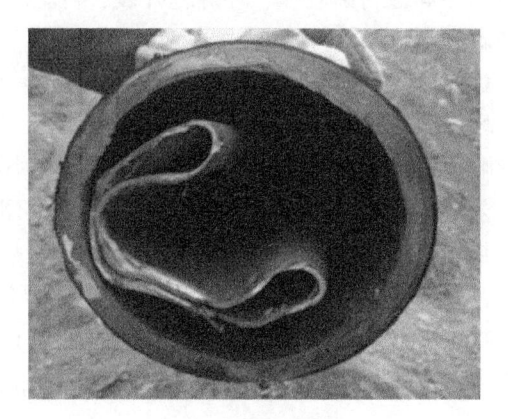

(a) 外防腐过程衬层塌陷　　　　(b) 运行中程衬层塌陷

图5 316L内衬层塌陷照片

运行中衬里层塌陷的机理和原因尚不明确，可能与以下因素有关：(1)衬里层径厚比设计与管材截面圆度等控制不到位，影响了衬里层结构稳定性和抗塌陷能力；(2)运行过程中，环焊缝刺漏导致高压输送介质进入基管与衬里层间隙，管线因停输检修快速泄压时，由于间隙中高压介质来不及释放，存在较高压差，从而导致衬里层塌陷失稳。

1.4 衬里层腐蚀

国内某油田使用的316L衬里复合钢管发生了衬里层腐蚀，典型的腐蚀形貌如图6所示。工况参数为：天然气中CO_2含量为0.59%~1.04%，CO_2分压最高为0.19MPa，Cl^-含量为125000mg/L，不含H_2S，运行过程中温度低于60℃。运行工况参数在国际上推荐的316L适用服役环境范围内[9]，理论上，316L 衬里层不应该出现点腐蚀。经排查管线服役历史，投产初期运行温度超过设计温度，最高达到 93℃，而且在酸化作业后有残酸进入管线，从而导致衬里层管体发生腐蚀。在实验室模拟投产初期工况进行 30 天的腐蚀试验，观察到 316L 材料表面有腐蚀坑存在，证实了上述结论。

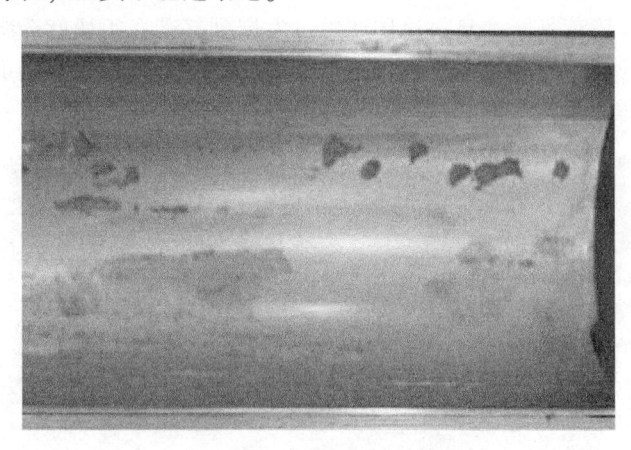

图 6　316L 衬里层腐蚀形貌

2　316L 衬里复合管道检测技术研究

因为 316L 衬里复合管道发生多起失效事件，必须重视其完整性管理。损伤缺陷检测技术是 316L 衬里复合管道完整性管理急需解决的关键技术。对于衬里层塌陷，可以通过清管通球或者通径检测发现。对于 316L 衬里复合管道环焊缝腐蚀、开裂及衬里层腐蚀的检测，面临技术挑战和困难，在现有内检测技术中，有些技术不适用于 316L 衬里复合管道，有些技术应用于 316L 衬里复合管道有局限性。如目前应用较广泛也较成熟的管道漏磁检测器（MFL），因为 316L 奥氏体不锈钢为非铁磁性材料，使该技术不能检测内衬管缺陷；超声内检测器因为耦合问题，不能应用于输气管线；电磁超声（EMAT）内检测器在国外已经开发成功且商业化[10,11]，但该技术在管道环向缺陷以及非轴向缺陷定量化检测方面存在局限性[11]，在国内，目前该技术尚处在开发阶段；爬行机器人视频内检测技术，只能检测内表面缺陷，而且必须停输。另外，由于 316L 衬里复合管道塌陷情况较普遍，存在内检测器无法通过的可能。以下试验验证漏磁检测技术和爬行机器人视频内检测技术在 316L 复合管道上应用的可行性，结合现有技术条件，研究提出基于风险的开挖检测方法，以满足 316L 衬里复合管道现场检测需要。

2.1　内检测技术的适用性试验验证

2.1.1　漏磁检测技术（MFL）

采用 φ508mm 三轴高清漏磁检测器在长度约 100m 的试验管段上进行牵拉试验。试验管段由普通碳钢管和 2 段 316L 衬里复合管道组成，中间用法兰连接。316L 衬里复合管道取自油田现场，规格为 φ508mm，基管材料为 L245，公称壁厚为 15mm，衬里层为 316L 不锈钢，公称壁厚为 2.5mm。2 个 316L 衬里复合管段均包含环焊缝，1#管段的环焊缝有一处腐蚀刺漏，在法兰连接处有衬里层塌陷，2#管段上无原始缺陷。在基管内外表面预制 8 个人工缺陷，缺陷直径为 30mm 和 50mm，深度为 1.7~7.0mm。在 316L 衬里层上预制 3 个人工缺陷，为环向刻槽，轴向宽度 1mm，环向长度 40mm，深度为 0.75~2.5mm。分别进行了 0.5m/s，

1m/s，2m/s，3m/s等4种速度下的牵拉试验，如图7所示速度为0.5m/s下牵拉试验给出的特征信号。图7中，1#~8#为基管上预制缺陷的特征信号，衬里层上预制的3个缺陷(编号为9#~11#)无特征信号显示。可见，MFL可检测到环焊缝位置、环焊缝腐蚀刺漏、衬里层塌陷、基管内外表面缺陷，但无法检测到316L不锈钢衬里层上的缺陷。实际工程中，由于油田集输管线服役环境苛刻，一旦衬里层腐蚀并发展到基管，很快就会腐蚀穿孔，因此，对于不锈钢衬里复合管道，及早检测发现衬里层缺陷比检测发现基管缺陷更有意义。另外，衬里层严重塌陷可能会导致MFL检测器无法通过。

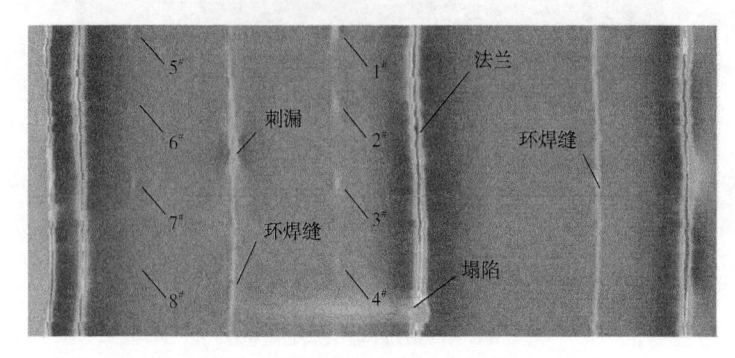

图7 MFL给出的缺陷特征信号

2.1.2 爬行机器人视频检测技术

采用爬行机器人视频检测技术对某油田集气干线的 ϕ508mm×(14.2+2.5)mm 316L衬里复合管道进行内检测，检测在停输条件下的2个管线开口处进行。检测发现，该集气干线复合管道的316L内衬层塌陷比较严重，而且多处环焊缝及其附近发生腐蚀。表1和表2列出爬行机器人视频检测的部分结果。图8为检测到的管道典型缺陷形貌。检测试验结果表明，爬行机器人视频检测技术可以有效地检测316L复合管道内部的塌陷及表面损伤情况，但管道内壁塌陷变形程度、内壁积液、机器人续航等因素对检测有一定的影响。由于该技术必须在管道停输条件下使用，限制了其应用。另外，衬里层塌陷严重时，检测器将无法通过。在管道建设期，可以采用该方法检查环缝表面质量和缺陷。最近，塔里木油田已将该技术应用于2205不锈钢管线施工质量检查中。

表1 爬行机器人视频检测发现的衬里层塌陷

缺陷类型	离开口处距离(m)	周向方位(钟点位置)	长度(m)
纵向塌陷	22.0	1：00	4.6
纵向塌陷	33.6	11：00	9.0
纵向塌陷	58.8	12：00	10.0

表2 爬行机器人视频检测发现的内表面腐蚀损伤

离开口处距离(m)	缺陷特征	离开口处距离(m)	缺陷特征
9.9	环焊缝表面腐蚀	60.8	衬里层腐蚀
22.0	环焊缝附近腐蚀	68.4	环焊缝附近疑似腐蚀穿孔
33.6	环焊缝附近腐蚀	81.0	环焊缝附近腐蚀
45.6	环焊缝附近衬里层损伤		

(a）衬里层塌陷　　　　　（b）衬里层腐蚀缺陷　　　　　（c）环焊缝腐蚀缺陷

图8　采用爬行机器人视频检测技术检测到的管道缺陷

2.2 基于风险的环焊缝开挖检测技术

316L衬里复合管道环焊缝腐蚀和环焊缝开裂，在不停输条件下，缺乏有效的内检测手段，根据管道检测技术发展现状，解决316L内衬复合管道检测在工程上可行的方法是基于风险的开挖检测，即在风险评估的基础上，对高风险区段的环焊缝进行开挖，然后采用射线（RT）、超声（UT）等常规无损检测手段进行环焊缝缺陷检测。

基于风险的管道环焊缝开挖检测的基础是管道风险评估。由于双金属复合管线应用时间还比较短，工程经验少，积累数据非常有限，不具备建立定量风险评估方法的基础，建立半定量的316L衬里复合管道风险评估方法（即评分法），在工程上更可行。在316L衬里复合管道失效调查的基础上，充分征询设计、制造、施工、运行管理等方面专家意见和经验，针对腐蚀刺漏和环焊缝开裂失效风险，从设计、制造、施工、运行和环境5个方面，分别识别出28个和17个风险因素，并给出了相应的打分权重和赋值方法，建立了半定量的风险评分法。该方法能够较全面地考虑管道设计、制造、施工和运行等全过程风险因素，打分权重和分值充分结合了管道设计、制造、施工及运行管理等方面的专家经验，工程适用性强，为开展基于风险的开挖检测奠定了基础。

以某油田KS2集气干线ϕ508mm的316L衬里复合管道的风险评估为例。该管线基管材料为L245，壁厚为14.2，20mm，316L衬里层厚度为2.5mm，管线长度为29.16km。在管道属性和服役环境调查基础上，划分了4个评价单元，在风险因素识别基础上，采用半定量风险打分方法对4个评价单元的环焊缝腐蚀刺漏和环焊缝开裂风险分别进行评估。作为算例，表3和表4列出评价单元1环焊缝腐蚀刺漏和环焊缝开裂风险因素识别及风险评分结果。表5列出KS2集气干线两种风险的综合评估结果。对环焊缝腐蚀刺漏和环焊缝开裂每一类风险，最优为10分，分值越低，风险越高。根据风险评估结果，管道风险由高到低的排序为单元1（11.61分）、单元4（12.35分）、单元2（12.95分）和单元3（13.55分），对应的历史环焊缝失效次数分别为8次，3次，1次和0次。风险评估结果和历史失效次数统计结果是一致的。

表3　KS2集气干线评价单元1环焊缝刺漏风险因素识别及评分结果

类别及权重	风险因素	二级风险因素	权重	调查结果	打分[①]	加权计分
设计因素（15%）	管端处理方式	—	60%	封焊	2	0.18
	衬层壁厚	—	40%	不符合设计标准要求	0	0

续表

类别及权重	风险因素	二级风险因素	权重	调查结果	打分[①]	加权计分
制造因素(25%)	封/堆焊工艺评定	—	35%	有工艺评定	10	0.88
	端部封/堆焊厚度	—	30%	有控制	10	0.75
	封/堆焊后PT检验	—	15%	未检验	0	0
	封/堆焊后RT检验	—	10%	检验合格	10	0.25
	是否内检测	—	10%	是	10	0.25
施工因素(35%)	组对情况	是否死口	7%	非死口	10	0.25
		工装情况	5%	有组对工装	10	0.25
	是否有管端修补	—	10%	无	10	0.25
	焊接工艺	—	40%	老工艺：R316LT1-5打底，ER309LMo过渡，E5015填充盖面	3	0.08
	是否有焊接工艺评定	—	9%	有焊评，结果合格	10	0.25
	施工环境条件	—	14%	冬季施工，有保护措施	5	0.13
	气保护措施	外表面气体保护	3%	有	10	0.25
		内表面气体保护	3%	有	10	0.25
	焊后检验情况	—	2%	有检验记录	10	0.25
	焊接返修情况	—	7%	焊接返修率在30%~50%	3	0.08
运行因素(20%)	输送介质(腐蚀性介质)	含水量	4%	无凝析水	10	0.08
		pH值	8%	在4~7之间	7	0.11
		CO_2含量	10%	CO_2分压<0.021MPa	10	0.20
		Cl^-含量	10%	小于10000mg/L	10	0.20
		H_2S含量	4%	H_2S含量≤20mg/m³	10	0.08
		介质流速	19%	介质流速≥3m/s	10	0.38
	压力波动和超压情况	—	5%	输送压力无显著波动、无超压现象	10	0.10
	历史类似案例30%	—	30%	曾经具有相似案例3起以上	0	0
	输送温度	—	5%	温度无波动或波动小于30℃	10	0.10
	线路类型	—	5%	干线	10	0.10
环境因素(5%)	管道走势(是否存在低洼)	—	100%	存在低洼1处	5	0.25
综合得分						5.95

① 最优为10分，最差0分。

表4　KS2集气干线评价单元1环焊缝刺漏风险因素识别及评分结果

类别及权重	风险因素	二级风险因素	权重	调查结果	打分[①]	加权计分
设计因素(15%)	管端处理方式	—	70%	封焊	2	0.21
	衬层壁厚	—	30%	不符合设计标准要求	0	0
制造因素(20%)	封/堆焊工艺评定	—	35%	有工艺评定	10	0.70
	端部封/堆焊厚度	—	30%	有控制	10	0.60
	封/堆焊后PT检验	—	15%	未检验	0	0
	封/堆焊后RT检验	—	20%	检验合格	10	0.40

· 40 ·

续表

类别及权重	风险因素	二级风险因素	权重	调查结果	打分[①]	加权计分
施工因素（50%）	组对情况	是否死口	6%	非死口	10	0.30
		工装情况	4%	有组对工装	10	0.20
	焊接工艺	—	30%	老工艺：R316LT1-5 打底，ER309LMo 过渡，E5015 填充、盖面	3	0.45
	施工环境条件	—	20%	冬季施工，有保护措施	5	0.50
	气保护措施	外表面气体保护	10%	有	10	0.50
		内表面气体保护	10%	有	10	0.50
	焊接检验情况	—	10%	有检验记录	10	0.50
	焊接返修情况	—	10%	焊接返修率在 30%~50%	3	0.15
运行因素（10%）	压力波动和超压情况	—	40%	输送压力无显著波动、无超压现象	10	0.40
	历史类似案例 30%	—	60%	曾经发生类似案例 3 起以上	0	0
环境因素（5%）	管道走势（是否存在低洼）	—	100%	存在低洼 1 处	5	0.25
综合得分						5.66

① 最优为 10 分，最差 0 分。

表 5　ϕ508mm316L 衬里复合管道风险评分结果

风险类型	评分			
	评价单元 1(8.36km)	评价单元 2(6.64km)	评价单元 3(7km)	评价单元 4(7.16km)
环焊缝腐蚀刺漏	5.95	6.40	6.70	6.10
环焊缝开裂	5.66	6.55	6.85	6.25
总分	11.61	12.95	13.55	12.35
历史失效次数（次）	8	1	0	3

基于风险评估结果，对该 KS2 集气干线的 ϕ508mm 316L 衬里复合管道中风险最高区段（评价单元 1）的 5 道环焊缝进行开挖检测，验证了 X 射线检测和超声检测两种方法对 316L 衬里复合管道环焊缝检测的适用性。如图 9 所示 ϕ508mm 316L 衬里复合管道环焊缝部分 RT 检测结果，图中的环焊缝腐蚀缺陷和开裂缺陷均为射线Ⅳ级片。另外，在同一位置，采用 UT 检测，发现缺陷回波信号位于Ⅲ区，超过判废线。现场 RT 和 UT 检测试验结果表明，采用 RT 和 UT 方法可以检测 316L 衬里复合管道环焊缝腐蚀和裂纹缺陷，在开挖条件下，RT 和 UT 检测是 316L 衬里复合管道环焊缝检测的有效手段。

（a）环焊缝腐蚀　　　　　　　　　（b）环焊缝开裂

图 9　采用 RT 方法检测到的 316L 衬里复合管道环焊缝缺陷

3 结语

(1) 316L衬里复合钢管在国内高含H_2S/CO_2和Cl^-油气集输管线上的应用取得了良好的防腐效果,但也发生了多次失效事件。主要失效形式包括环焊缝腐蚀刺漏、环焊缝开裂、衬里层塌陷和衬里层腐蚀等。

(2) 损伤缺陷检测技术是316L衬里复合管道完整性管理亟待解决的关键技术,目前缺乏有效的内检测手段。在管道上应用较成熟的漏磁检测(MFL)技术可以检测316L衬里复合管道环焊缝刺漏、内衬塌陷、基管损伤等缺陷,但无法检测316L不锈钢衬里层表面缺陷。管道爬行机器人视频检测技术可检测衬里层塌陷、表面损伤以及焊缝腐蚀缺陷,但必须在停输条件下进行。

(3) 基于现有技术条件,研究建立了基于风险的环焊缝开挖检测方法,可以满足316L衬里复合管道检测的需要。

参 考 文 献

[1] 魏斌,李鹤林,李发根. 海底油气输送用双金属复合管研发现状与展望[J]. 油气储运,2016,35(4):343-355.

[2] Wei B, Bai Z Q, Yin C X, et al. Research and application of clad pipe for gathering pipelines in yaha gas condensate field: the proceedings of NACE 2010, 2010[C]. The Proceedings of NACE 2010, 2010.

[3] 罗世勇,贾旭,徐阳,等. 机械复合钢管在海底管道中的应用[J]. 管道技术与设备,2012(1):32-34.

[4] Spence M A, Roscoe C V. Bimetal CRA lined pipe employed for north sea field development[J]. Oil & Gas Journal, 1999, 97(18): 80-88.

[5] 张捷,聂新宇,徐平,等. 模拟海洋复杂载荷条件复合管道性能测试装备[J]. 压力容器,2017,34(1):1-6.

[6] 国家市场监督管理总局,中国国家标准化管理委员会. 石油天然气工业用内覆或衬里耐腐蚀合金复合钢管:GB/T 37701—2019[S]. 北京:中国标准出版社,2019.

[7] 李发根. 高腐蚀性油气集输环境用双金属复合管[C]//压力容器先进技术——第九届全国压力容器学术会议论文集. 合肥:合肥工业大学出版社,2017:131-135.

[8] 李发根,孟繁印,郭霖,等. 双金属复合管焊接技术分析[J]. 焊管,2014,37(6):40-43.

[9] Craig B D, Smith L. Corrosion resistant alloys(CRAs) in the oil and gas industry-selection guidelines update, Nickel Institute Technical Series No. 10073(3rd Edition) [R]. Belgium: Nickel Institute, 2011.

[10] Roy van Elteren, Ian Diggory, Jochen Spalink, et al. Pipeline integrity framework: "Mind the gap!" [C]. 15th Pipeline Technology Conference, Hannover Germany: Euro Institute for Information and Technology Transfer in Environmental Protection Gmbh, 2020.

[11] Matt Romney, Dane Burden. Detection of non-axial stress corrosion cracking(SCC) using MFL technology [C]. The Proceedings of 15th Pipeline Technology Conference, Hannover Germany: Euro Institute for Information and Technology Transfer in Environmental Protection Gmbh, 2020.

本论文原发表于《压力容器》2020年第37卷第11期。

Constitutive Equation for Describing True Stress-Strain Curves over a Large Range of Strains

Cao Jun[1] Li Fuguo[2] Ma Weifeng[1] Li Dongfeng[1, 3] Wang Ke[1]
Ren Junjie[1] Nie Hailiang[1] Dang Wei[1]

(1. State Key Laboratory of Performance and Structural Safety for Petroleum Tubular Good Goods and Equipment Materials, Tubular Goods Research Institute of CNPC; 2. State Key Laboratory of Solidification Processing, School of Materials Science and Engineering, Northwestern Polytechnical University; 3. China University of Petroleum (East China))

Abstract: Full-range strain hardening behaviour, containing the post-necking stage, is difficult to account for in detail. Hence, a Swift-Voce(S-V) model was used to describe the full-range stress-strain behaviour of 7050-T7451 aluminium alloy under uniaxial-tension, tension-with-notch and pure-shear processes with a hybrid experimental-numerical framework. Since the S-V model is not able to describe accurately the behaviour of Ti-6Al-4V alloy, a combined Swift and 4th degree polynomial (S-P4) model is proposed. Results indicate that the S-V model can successfully describe large-deformation behaviour of the 7050-T7451 aluminium alloy, and the S-P4 model can successfully describe that of the Ti-6Al-4V alloy based on careful designing process of the constitutive equation.

Keywords: Stress-strain measurements; Numerical simulation; Tensile testing; Necking

1 Introduction

True stress-strain curves are indispensable in precision forming process because the accuracy of simulation depends mainly on the constitutive relation. In the large-deformation stage, the stress and strain distributions are no longer uniform, so it is difficult to estimate the true stress-strain relation within necking zone using an engineering stress-strain relation. Theoretical work in this regards has brought significant scientific thoughts into the essential understanding of the necking phenomena[1]. In this study, large deformation means local deformation of necking phenomenon.

Post-necking strain hardening behaviour was used to predict the limiting major strain for sheet metal formability[2]. Besides, it was also applied to predict plastic strain localisation of dual-phase steels using finite-element analyses (FEA)[3]. The post-necking strain hardening behaviour was

---

Corresponding author: Cao Jun, caojun1@ cnpc. com. cn, juncao1105@ 163. com.

especially required in the prediction of ductile damage[4].

Many efforts were made to obtain full-range true stress-strain curves using the finite-element method (FEM)[5-7] and the digital image correlation (DIC) method[8,9]. A testing method employed the DIC technique and iterative FEA method to obtain stress-strain curves including post-necking strain[10]. Although this method can obtain accurate stress-strain curves, the method is too complex to tackle complicated project problems. Tu et al.[11] summarised three groups of methods to describe post-necking strain hardening behaviour of metallic materials. The first group consists of analytic solutions derived with round bar specimens[12,13], of which the Bridgman method[13] is the most well known, although suitable only for smooth round-bar specimens. The second group consists of experimental-numerical iterative methods[9,14-17], which yield trustful stress-strain behaviour in the post-necking regime. A method of this second group is applied in the present study. The third group contains an inverse method[18-20], in which the correction formula applied in an inverse way is based on predefined hardening rules. The accuracy of this latter method is a controversial issue for an actual material[11].

In this paper, a Swift-Voce (S-V) model is used to describe the large-deformation behaviour of 7050-T7451 aluminium alloy under uniaxial-tension, tension-with-notch and pure-shear processes. Since the S-V model cannot describe the large-deformation behaviour of Ti-6Al-4V alloy, a combined Swift model and 4th degree polynomial (S-P4) model is proposed to better describe this.

2 Methods

2.1 Experiments

Uniaxial-tension, tension-with-notch and pure-shear specimens of 7050-T7451 aluminium alloy plate and a cylindrical uniaxial tension specimen of Ti-6Al-4V alloy were prepared to obtain force-displacement responses, as illustrated in Fig. 1. All specimens were processed along the rolling direction. Tests on the uniaxial-tension, tension-with-notch and pure-shear specimens were carried out on an Instron 3382 (Instron Inc., USA) machine. The tests were conducted at a constant crosshead velocity of 1 mm/min. The testing temperature was room temperature (18 ~ 25℃). Meanwhile, the specimens of 7050-T7451 aluminium alloy plates were tested with a DIC system to obtain displacement and local strain data. Since the specimen of Ti-6Al-4V alloy is cylindrical and so is hard to test with a DIC system, the displacement was obtained using an extensometer at small strains and a beam at large strains.

2.2 Simulations

Three-dimensional finite element (FE) models of uniaxial-tension, tension-with-notch and pure-shear specimens were performed using ABAQUS/Explicit software with a user material subroutine (VUMAT). Quarter geometric FE models were adopted for the specimens of uniaxial tension and tension with notch. All the specimens were meshed using reduced-integration eight-node solid elements (C3D8R).

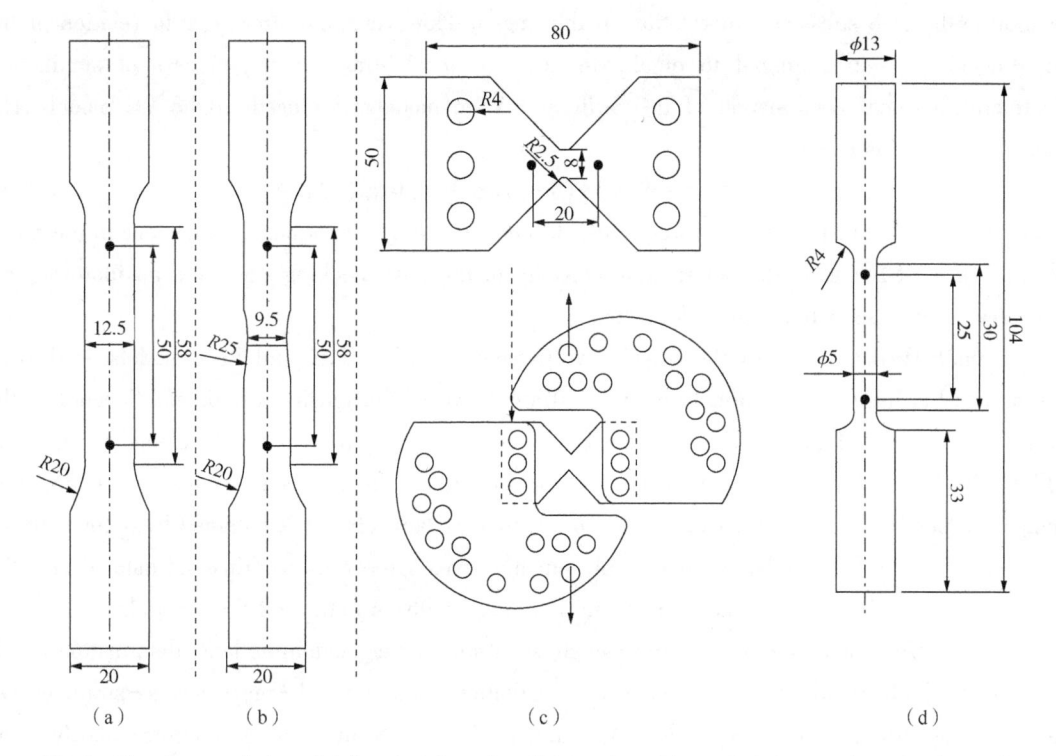

Fig. 1 The shape and size of (a) uniaxial tension (b) tension with notch (c) pure shear specimens for 7050-T7451 aluminium alloy plate and (d) uniaxial tension specimen for Ti-6Al-4Vcylindrical alloy specimen

2.3 Constitutive equations

Two common constitutive equations were applied to describe the stress-strain curves before the onset of necking. These equations can be written as

$$\sigma = K(\varepsilon_p + \varepsilon_0)^n \quad \text{(Swift equation)} \quad (1)$$

$$\sigma = \sigma_0 - \sigma_0 A \exp(-\beta \varepsilon_p) \quad \text{(Voce equation)} \quad (2)$$

where K is the strength coefficient, n is the strain hardening exponent and ε_0 is the pre-strain, σ_f and A are saturation stress and material coefficient, respectively. β is used to determine the rate at which the stress tends to reach a steady state value.

The Swift and Voce equations were used to fit the true stress-strain curves of the 7050-T7451 aluminium alloy plate and the Ti-6Al-4V alloy before the onset of necking. The constitutive parameters of K, n and ε_0 were obtained by fitting the true stress-strain curves of the two alloys with the Swift equation. σ_f, A and β were obtained by fitting the true stress-strain curves of the two alloys with the Voce equation. These constitutive parameters are listed in Table 1.

Table 1 The constitutive parameters of 7050-T7451 aluminium alloy and Ti-6Al-4V alloy

Material	K	n	ε_0	σ_f	A	β
7050-T7451	827	0.16316	0.0233	600	3.92	18.96
Ti-6Al-4V	1207	0.05638	0.0039	1082	0.1643	24.20

In the non-uniform deformation stage, the largest local deformation occurs in the post-necking region. The law of volume invariance cannot be used to transform from the engineering stress-strain relation to the true stress-strain relation in this region. However, the stress-strain relation of the post-necking stage is crucial to obtain an accurate local strain in the process of simulation. Therefore, a linear combination of the Swift and Voce models was used in the FE models. The special form of this model is

$$\sigma = q[K(\varepsilon_p+\varepsilon_0)^n] + (1-q)[\sigma_0 - \sigma_0 A\exp(-\beta\varepsilon_p)] \qquad (3)$$

where q is a weight factor, used to weight the difference of large-deformation behaviour in the post-necking stage. It relates to the deformation capacity in the post-necking stage and an increase of q indicates a decrease of the large deformation.

In the S-V model, firstly the constitutive parameters of the Swift and Voce models need to be determined by fitting the uniform true stress-strain curve of the tensile experiment. Secondly, the initial q value should be set. Thirdly, the S-V model is submitted to finite-element simulation (FES) by VUMAT, and the simulated full-range force-displacement curves are obtained to compare experimental data. Finally, the accurate stress-strain relation is obtained by optimising the value of q to make the simulated force-displacement curves agree with experimental data. Then, the accurate local strain can be obtained from the FE model with an optimised S-V model.

Can the S-V model describe full-range stress-strain curves containing large deformation for all materials? Which constitutive equation is more suitable when the full-range stress-strain curves cannot be described by the S-V model? A combination of the Swift model and a correction function is proposed to describe the full-range stress-strain curve. The correction function could bepolynomial since it is easily modified. The correction process needs to be based on a necking type of material. This is an iterative process obtained by adjusting the correction function and weight factor q to make a simulated force-displacement curve that accords with experiment. The constitutive equation is expressed as

$$\sigma = qK(\varepsilon_p+\varepsilon_0)^n + (1-q)f(\varepsilon_p) \qquad (4)$$

where $f(\varepsilon_p)$ is a correction function. The power-law type constitutive equation (Swift model) is suitable for body-centred cubic (BCC) metals[21], whereas the exponential-type equation (Voce model) is suitable for most face centred cubic (FCC) metals such as aluminium and copper[22]. Some materials (for example, titanium alloys) have a close-packed hexagonal (HCP) structure. For these materials, the power-law type and exponential-type constitutive equation are not suitable for describing the full-range stress-strain curve. To better describe the full-range stress-strain curve of HCP materials, the correction function $f(\varepsilon_p)$ is used to modify the shape of the stress-strain curve in the post-necking type; this is a modification for large-deformation behaviour induced by material type.

3 Results and discussion

Fig. 2(a) - Fig. 2(c) exhibits a comparison of force-displacement responses and local equivalent plastic strain between experiment and simulation under uniaxialtension, tension-with-notch and pure-shear processes, respectively. These simulated force-displacement responses of 7050-T7451 aluminium alloy were obtained from the FE model with an embedded S-V model, and the local equivalent plastic strain was obtained from the surface of the specimen. It is noted that the force-displacement responses of the FE model in the post-necking stage are sensitive to the constitutive equation.

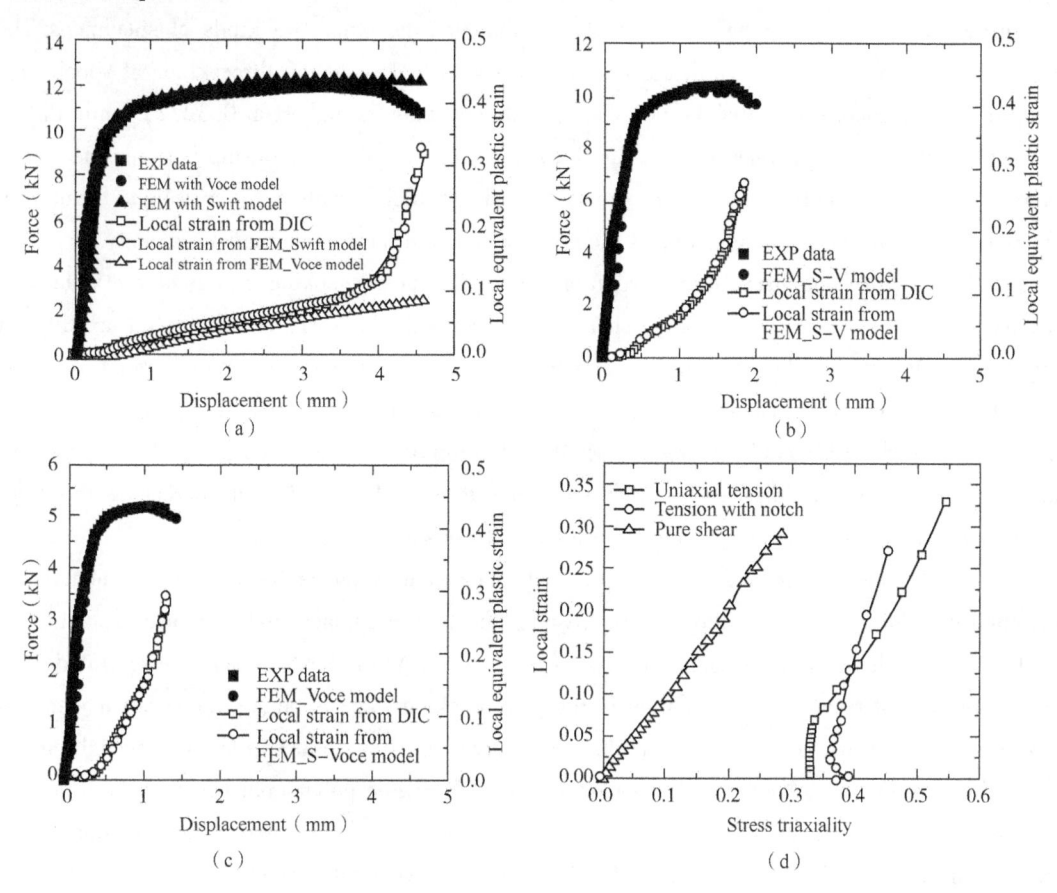

Fig. 2 The comparisons of force-displacement curve and local equivalent plastic strain between experiment and simulation with Swift model and Voce model for of 7050-T7451 aluminium alloy plate (a) uniaxial tension process, (b) tension with notch process and (c) pureshear; (d) the local strain evolution at necking region with the variation of stress triaxiality for the three kinds of specimens

In Fig. 2(a), the force-displacement responses and local strain obtained from the FE model with the Voce model are consistent with those obtained from experimental data with the DIC method. The weight factor q of the S-V model is 0. Nevertheless, the FE model with the Swift model does not predict the stress-strain curves and local strain very well. It indicates that the Voce model is suitable for describing the full-range stress-strain relation of 7050-T7451 aluminium alloy (FCC structure).

In Fig. 2(b), the force-displacement responses and local strain of FE model obtained with the S-V model are consistent with those of experiment data for tension-with-notch specimens, with a weight factor q is 0.7. It reflects the fact that the large-deformation capacity of the tension-with-notch process is less than that of the uniaxial-tension process. In Fig. 2(c), the force-displacement responses and local strain of simulation with the Voce model are consistent with those of experimental data for pure shear; the weight factor q of the S-V model is 0. This indicates that the Voce model is suitable for describing the large-deformation behaviour of the pure shear specimen.

The three post-necking types of uniaxial-tension, tension-with-notch and pure-shear processes are different, as illustrated in Fig. 2(a)-(c). The ratio of the post-necking stage to the uniform deformation stage for pure shear is larger than for the other two kinds of specimens. The uniform deformation stage of the uniaxial tension process is longest in the three kinds of specimens. In addition, the fracture strain of the uniaxial-tension specimen is highest at 0.3272, while that of the tension-with-notch specimen is lowest at 0.2726. Comparing the relationship between the weight factors q and three fracture strains, the fracture strain is higher with a lower weight factor q. It reflects the fact that the increasing weight factor q reduces the large-deformation capacity.

Fig. 2(d) shows the local strain evolution within the post-necking region of the uniaxial-tension, tension-with-notch and pure-shear specimens with the variation of stress triaxiality for 7050-T7451 aluminium alloy. As can be seen from Fig. 2(d), the large variation range of stress triaxiality leads to a high local strain. By recalling the q values for describing the constitutive behaviour of the three kinds of specimen, q is 0 for the uniaxial-tension and pure-shear specimens and q is 0.7 for the tension-with-notch specimen. It is to be noted that increasing the large deformation leads to an increasingly large variation of stress state.

Fig. 3(a) shows a comparison of the force-displacement responses between experimental and simulation data for Ti-6Al-4V alloy. In Fig. 3(a), the force-displacement response of simulation with the Swift model is not consistent with that of experiment. As the Swift model is unsaturated, the force-displacement response of simulation in the post-necking stage is largely inconsistent with that of experiment. How can one find a suitable constitutive equation to describe this special large-deformation behaviour? A combined Swift model and a 4th degree polynomial (S-P4) were used to describe the local deformation behaviour. Since the correction function needs two inflection points from yielding point to fracture point, a the 4th degree polynomial was chosen as the correction function. The S-P4 model can be expressed as:

$$\sigma = qK(\varepsilon_p+\varepsilon_0)^n + (1-q)(a\varepsilon_p^4 + b\varepsilon_p^3 + c\varepsilon_p^2 + d\varepsilon_p + e) \tag{5}$$

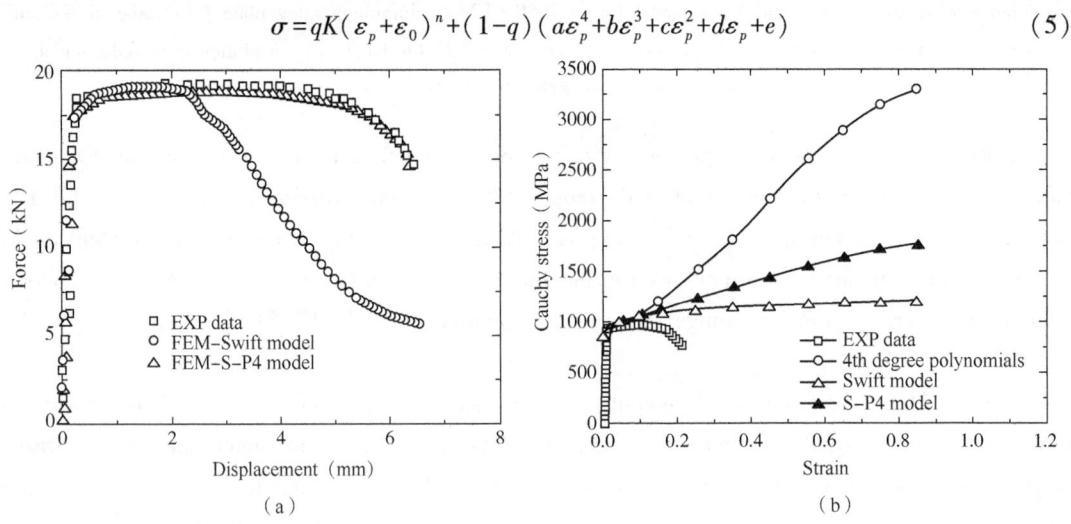

Fig. 3 (a) Comparisons of force-displacement responses between experimental data and simulation with different models for Ti-6Al-4V alloy; (b) the Cauchy stress-strain responses with different models of Ti-6Al-4V alloy

This 4th degree polynomial (P4) is necessary to better describe the local deformation of Ti-6Al-4V alloy. The procedure for determining S-P4 is as follows. Firstly, the constitutive parameters of the Swift model are determined by fitting the uniform true stress – strain curve of the tensile experiment, the Swift model being the dominant constitutive equation. Secondly, the Swift model is substituted into FES by VUMAT, and the simulated full-range force-displacement curve is obtained to compare with experimental data. Thirdly, a 4th degree polynomial needs to be designed based on the comparative result. The curve of P4 before necking needs to be consistent with the stress-strain curve of the tensile experiment. Then the concavity and convexity of the P4 curve from the beginning of necking to the end need to be changed from convexity to concavity, and an initial P4 equation should be established. Finally, the parameters of P4 and the weight factor q need to be adjusted iteratively until the experimental and numerical force-displacement curves are consistent.

As illustrated in Fig. 3(b), there are two stages of post-necking, namely a slow descent stage and a rapid descent stage. The type of necking of the Ti-6Al-4V alloy is different from that of the 7050-T7451 aluminium alloy. Therefore, the full-range of the constitutive equation containing the post-necking stage needs to be designed based on the correction P4 function. Fig. 3(b) shows the designed constitutive equation of S-P4 in the full-range of the deformation stage of the Ti-6Al-4V alloy. The slow-descent necking stage needs to be described by combining the Swift model and a convex P4 function. On the other hand, the rapid-descent necking stage needs to be described by combining the Swift model with a concave P4 function. Therefore, the S-P4 constitutive equation is designed as follows:

$$\sigma = 0.73 \times 1207(\varepsilon_p + 0.0039)^{0.05638} + 0.27 \times (-3282\varepsilon_p^4 + 383\varepsilon_p^3 + 3638\varepsilon_p^2 + 1452\varepsilon_p + 906) \quad (6)$$

The force-displacement response of simulation with the S-P4 constitutive equation is consistent with that of experiment, as shown in Fig. 3(a). The q of the S-P4 constitutive equation is 0.73, which reflects the fact that the Swift model is the dominant constitutive equation and the 4th degree polynomial an auxiliary correction function.

Fig. 4 shows a comparison of the local strain evolution within the necking region of the uniaxial tension specimens of the Ti-6Al-4V and the 7050-T7451 aluminium alloys as a function of stress triaxiality. As can be seen, the local strain with stress triaxiality has two stages for the Ti-6Al-4V alloy, while that of the 7050-T7451 aluminium alloy has only one stage in the local-deformation stage. The variation in the relationship also reflects the post-necking process. In addition, the local strain in the Ti-6Al-4V alloy is close to 1, because the large-deformation stage is a large proportion of the whole deformation stage.

4 Conclusions

An S-P4 model is proposed to describe the large-deformation behaviour of Ti-6Al-4V alloy because the S-V model cannot describe this successfully. A hybrid experimental-numerical method was performed to investigate the applicability of the S-V model and the S-P4 model in describing the full-range stress-strain behaviour of 7050-T7451 aluminium and Ti-6Al-4V alloys. The conclusions are as follows:

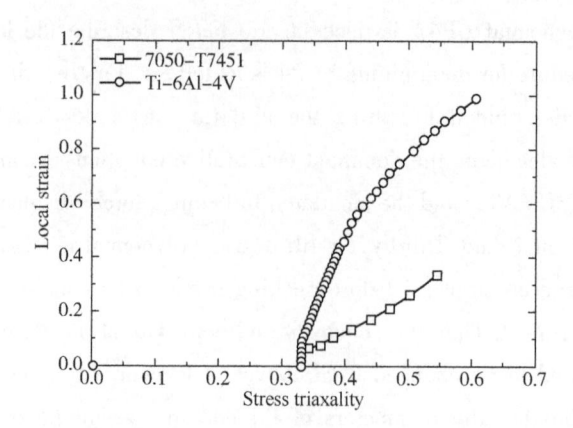

Fig. 4　The comparison of local strain evolution at necking region of uniaxial tension specimen between Ti-6Al-4V alloy and 7050-T7451 aluminium alloy with the variation of stress triaxiality

(1) The S-V model can successfully describe the full-range stress-strain behaviour of 7050-T7451 aluminium alloy containing a necking stage. Comparing the relationship between the weight factors q and three fracture strains of uniaxial-tension, tension-with-notch and pure-shear processes, the fracture strain is higher at lower weight factor q. Increasing the weight factor reduces the large-deformation capacity.

(2) The S-P4 model can successfully describe the full-range stress-strain behaviour of Ti-6Al-4V alloy under uniaxial tension whereas the S-V model does not.

(3) The S-P4 model needs to be carefully designed based on the post-necking type. The constitutive equation used to describe the full-range stress-strain curve of Ti-6Al-4V alloy in the post-necking stage needs to be designed using correction (P4) functions.

References

[1] M. K. Duszek and P. Perzyna, Int. J. Solids Struct. 27 (1991) pp. 1419-1443.
[2] L. Smith, R. Averill, J. Lucas, T. Stoughton and P. Matin, Int. J. Plast. 19 (2003) pp. 1567-1583.
[3] X. Sun, K. S. Choi, W. N. Liu and M. A. Khaleel, Int. J. Plast. 25 (2009) pp. 1888-1909.
[4] M. Luo, M. Dunand and D. Mohr, Int. J. Plast. 32 (2012) pp. 36-58.
[5] J. Isselin, A. Iost, J. Golek, D. Najjar and M. Bigerelle, J. Nucl. Mater. 352 (2006) pp. 97-106.
[6] H. D. Kweon, E. J. Heo, D. H. Lee and W. K. Jin, J. Mech. Sci. Technol. 32 (2018) pp. 3137-3143.
[7] M. Rossi, A. Lattanzi and F. Barlat, Strain 54 (2018) p. e12265.
[8] S. K. Paul, S. Roy, S. Sivaprasad and S. Tarafder, J. Mater. Eng. Perform. 27 (2018) pp. 4893-4899.
[9] L. Wang and W. Tong, Int. J. Solids Struct. 75-76 (2015) pp. 12-31.
[10] M. Kamaya and M. Kawakubo, J. Nucl. Mater. 451 (2014) pp. 264-275.
[11] S. Tu, X. Ren, J. He and Z. Zhang, Fatigue Fract. Eng. Mater. Struct. 43 (2020).
[12] W. W. Davidenkov, Proc. ASTM 46 (1946) pp. 1147-1158.
[13] P. W. Bridgman, Studies in Large Plastic Flow and Fracture, Vol. 177, McGraw-Hill, New York, 1952.
[14] S. Coppieters, S. Sumita, D. Yanaga, K. Denys, D. Debruyne and T. Kuwabara, Identification of post-necking strain hardening behavior of pure titanium sheet, in Residual Stress, Thermomechanics & Infrared Imaging, Hybrid Techniques and Inverse Problems Simon Quinn, Xavier Balandraud, ed., Springer, Orlando, 2016, p. 59-64.

[15] S. Coppieters, S. Cooreman, H. Sol, P. V. Houtte and D. Debruyne, J. Mater. Process. Tech. 211 (2011) pp. 545-552.

[16] S. Coppieters and T. Kuwabara, Exp. Mech. 54 (2014) pp. 1355-1371.

[17] J. -H. Kim, A. Serpantié, F. Barlat, F. Pierron and M. -G. Lee, Int. J. Solids Struct. 50(2013) pp. 3829-3842.

[18] W. J. Yuan, Z. L. Zhang, Y. J. Su, L. J. Qiao and W. Y. Chu, Mater. Sci. Eng. A 532 (2012) pp. 601-605.

[19] Z. Zhang, J. Ødegård and O. SØvik, Comput. Mater. Sci. 20 (2001) pp. 77-85.

[20] K. Zhao, L. Wang, Y. Chang and J. Yan, Mech. Mater. 92 (2016) pp. 107-118.

[21] H. S. Ji, H. K. Ji and R. H. Wagoner, Int. J. Plast. 26 (2010) pp. 1746-1771.

[22] B. K. Choudhary, E. I. Samuel, K. B. S. Rao and S. L. Mannan, Met. Sci. J. 17 (2001) pp. 223-231.

本论文原发表于《Philosophical Magazine letters》2020 年。

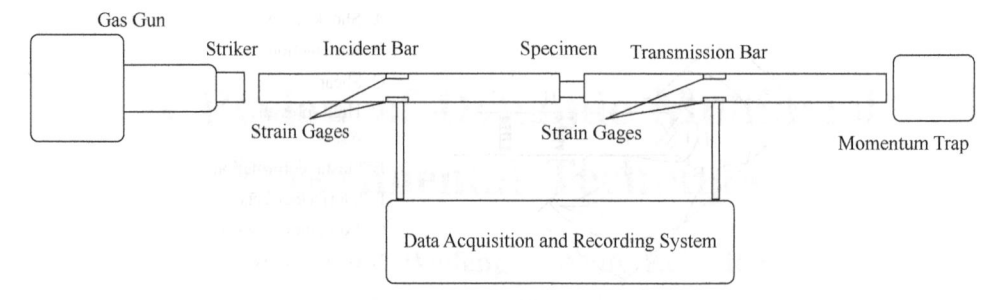

Fig. 2 Schematic diagram of split Hopkinson pressure bar device

techniques that are useful for dynamic rock tests with SHPB including the multiaxial loading techniques are introduced in detail, and various measurement techniques for rock tests in SHPB are fully discussed. In addition to the loading technology, data measurement technology is also a key point of dynamic testing. In 2017, Xing et al.[17] summarized the high-speed photography and digital optical measurement techniques for geomaterials in detail, which is worth learning and studying for readersIn order to study the dynamic tensile properties of the material, the Hopkinson tensile bar was developed based on the pressure version. Various Hopkinson tensile bars have been proposed by many researchers[18-20]. At present, the most widely used is the direct drawing Hopkinson tensile bar, which were improved by Qgawa based on J. Harding's device in the 1980s[21].

The torsional test has no three-dimensional problem, lateral inertia effect, and frictional effect on the ends of the specimen, thus the dynamic torsional experiment is paid more and more attentions. At present, the main loading methods of Hopkinson torsional bar equipment include prestored energy loading, explosive loading, direct impact loading, flywheel loading, and electromagnetic loading[14]. In a recent review of Yu[22], these five typical types of Hopkinson torsional bar were systematically reviewed, and interested readers can refer to it.

The successful development of the Hopkinson bar device indicates that the dynamic compressive, tensile and shear mechanical properties of the material at the strain rate of $10^2 \sim 10^4 s^{-1}$ can be measured separately. The loading process during the experiment is usually within tens to hundreds of microseconds. However, the single stress state is only an ideal condition, further study of the structure and material under the actual working conditions still requires multiaxial loading tests. The development of dynamic multiaxial testing equipment and experimental platform under high strain rate has always been the forefront of solid mechanics research and the dream of dynamists, especially in the high speed collision, penetration, protection engineering, fracture dynamics, plastic dynamics, material plastic molding and processing, explosion and impact and other fields of important basic issues. In recent years, with the efforts of some researchers, considerable progress has been made in the dynamic multiaxial testing, but there is still a lot of room for growth.

2 Dynamic multi-axial loading technique

2.1 Compression-shear dynamic biaxial loading devices

There are few researches about compression-shear dynamic biaxial loading devices. Most of the existing compression – shear dynamic biaxial loading techniques are based on falling weight experiment, direct impact experiment and high speed flying disc impact experiment.

Chung et al. [23,24] introduced biaxial loadings by limiting the lateral displacement of specimens in the traditional falling weight experiments. In the device, the specimen deformation is obtained by time integration of the acceleration signal of the falling weight. However, this data processing method may cause a large error.

Hong et al. [25] developed a set of direct impact dynamic compression-shear composite loading device on the basis of static oblique loading techniques. In this device, the air gun fires the plate to hit the loading conversion unit, and then hits the specimen. The specimen is fixed on the target plate, and the loading information is obtained through the multiaxial loading sensor. The defect of the direct impact device is that the impulse generated by impact loading is too short, and the deformation rate of specimen is not constant. In their device, Hong et al. designed a loading transfer unit, which applied the impact kinetic energy to the specimen through the inertia movement of the massive block, so as to correct and make up for the above defects.

The dynamic compression-shear test based on high speed flying disc appeared in the 1960s and 1970s, and the shear and high hydrostatic mechanical properties of materials can be tested at a very high strain rate ($10^5 s^{-1}$) [26]. A high speed air gun is used to shoot the elastic flying piece at a certain angle, and the specimen sheet is attached to it. The elastic flying piece carries the specimen to hit a thick elastic cutting plate, so as to produce the dynamic shear deformation. The flying strip shear experiment is applicable to the ultra-high strain rate, however due to the limitation of sample size, the experiment can only be conducted on materials less than the thickness of the specimen, which requires high requirements on the specimen, long preparation period and complex operation. At the beginning of the 21st century, many researchers are devoted to the research of compression-shear composite loading techniques. However, due to the limitation of loading equipment, the existing compression-shear composite loading is to use a uniaxial Hopkinson pressure bar to load a specially designed specimen, so that the specimen can be subjected to local compression–shear composite loading, or to modify the Hopkinson pressure bar to realize compression-shear composite loading.

Rittle et al. [27] proposed a compression-shear composite loading specimen, it has a groove in the middle of the specimen with a certain angle with the axial direction. In the experiment, when the split Hopkinson pressure bar is used to load along the axial direction of the specimen, the compression-shear composite load will be generated in the groove. However, the disadvantage of this method is that there is a large stress concentration in the groove and the specimen is easy to be destroyed along the root of the groove.

On the basis of the Hopkinson torsional Bar (TSHB, Torsional Split Hopkinson Bar), Huang et al. [28] established a compression-torsion composite loading device (Fig. 3). However, due to the propagation velocity difference between the compression wave and the torsional wave, the compression and torsional loadings cannot be applied to the specimen synchronously.

In order to realize synchronous loading of compression and shear, Zhao et al. [29] designed a Hopkinson compression-shear composite loading system with double transmission bars (Fig. 4). In the device, there are two symmetrical inclined planes at the end of the incident bar, the two

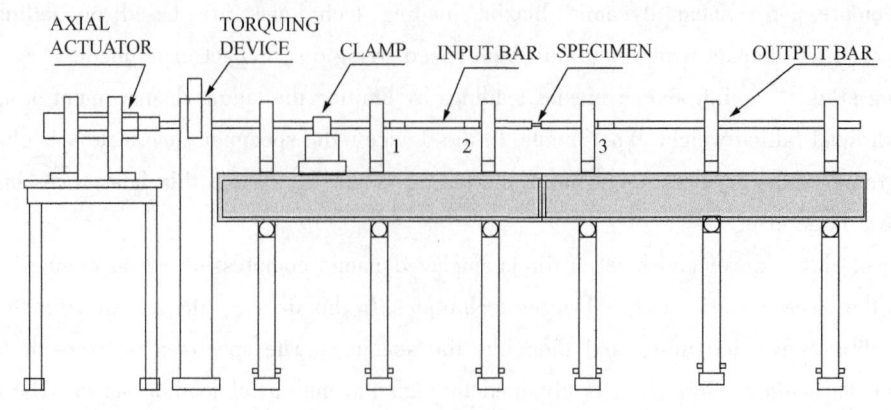

Fig. 3 The schematic diagram of dynamic compression-torsion composite bar by Huang et al.[28]
[Reproduced with permission from Int J Solids Struct. 41(11), 2821 (2004). Copyright 2004 Elsevies]

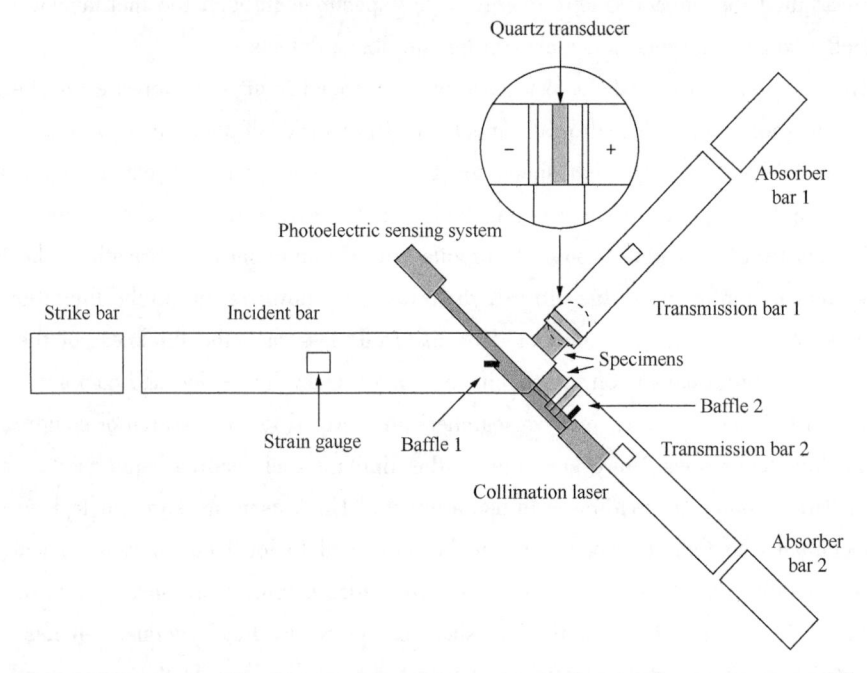

Fig. 4 Schematic diagram of dynamic compression-shear composite bar by Zhao et al.[29]
{Reproduced from [Zhao PD, Lu FY, Chen R, Lin YL, Li JL, Lu L, "A technique for combined dynamic compression-shear test," Rev Sci Instrum. 82(3): 35 (2011).], with the permission of AIP Publishing}

transmission bars form a certain included angle, and are opposite to the two inclined planes of the incident bar respectively. The two specimens are symmetrically placed between the two inclined planes of the incident bar and the end planes of the two transmission bars. Since the axial direction of the incident bar is at a certain angle to that of the transmission bar, when the stress wave is propagated into the transmission bar, the specimen will be subjected to compression-shear composite loading at the same time. In this experiment, the shear and compression stress wave signals on the incident and transmission bars were measured by piezoelectric sensors respectively. Similarly, Hou et al.[30,31] realized compression-shear composite loading by changing the end of the contact ends of the incident and transmission bars into inclined planes and installing the specimen between the two inclined planes (Fig. 5). However, in the experiment the shear force completely depends on the friction transfer between the specimen and the contact surface of the loading bars, while in order to ensure the uniform deformation of the specimen, the Hopkinson bar experiment requires the end friction to be as small as possible, so the friction between the specimen and the loading bars is limited, and the amplitude of the shear load will be limited.

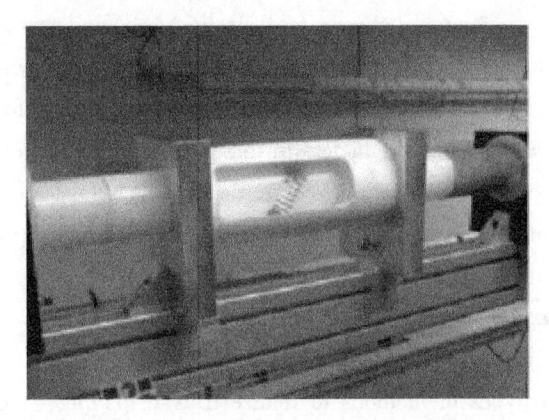

Fig. 5 Dynamic compression-shear composite loading device of Hou et al. [30]
[Reproduced with permission from Int J Solids Struct. 48(5), 687 (2011). Copyright 2011 Elsevier]

2.2 In-plane dynamic biaxial loading device

Compared with the dynamic compression-shear loading, the in-plane biaxial loading is more difficult, because the loads from both directions are hard to synchronously control within hundreds of microseconds.

Grolleau[32] developed a dynamic bulge test using a split Hopkinson bar to obtain an equal-biaxial tension. In this device, a movable "bulge cell" composed of a thick-walled steel cylinder and a die ring was designed to perform dynamic bulge tests in a SHPB system. During the test, a round sheet specimen is clamped between the cylinder and the die ring. The input bar is sealed insert the cylindrical cell, and a fluid is filled into the cell to transmit the pressure. An output bar with tubular cross-section is contact with the end of the die ring. When the incident stress pulse propagates to the input bar/fluid interface, the pressure wave is transmitted through the fluid and ultimately causes the bulging of the sheet specimen to form an equal-biaxial tension stress condition. This experimental procedure is more reliable than direct tension tests because it circumvents the inherent gripping and force measurement issues associated with direct tension tests. However, the sealing of the "bulge cell" is required to be very high, and the diameter of the incident bar and transmission bar is large, waveform dispersion effect is serious. Sealed fixtures also have safety hazards under high pressure loading.

Shimamoto et al. [33] developed and validated a biaxial testing rig to generate universal biaxial impact loading in the cruciform specimen. The testing device consisted of four actuators, which were orientated at 90° to each other. A programmable controller was used to control the circuits. Four hydraulic actuators operated independently generate dynamic loads on the four arms of the specimen and the center point was always maintained at the home position without movement. Tests under unequal biaxial stress (load ratio of 1 : 1 to 1 : 4) were also possible. However, due to the power limitation of the hydraulic actuators, the strain rate obtained in the experiment is much lower than that in the traditional split Hopkinson bar experiment.

Hummeltenberg et al.[34] invented a biaxial in-plane tensile split Hopkinson bar system. This system consists of two sets of separated Hopkinson bar systems which are perpendicular to each other. A cruciform sample is clamped between four elastic bars, and two strikers are driven by the electromagnetic driving force. A circuit is designed to control the synchronization of the two launchers. However, the two strikers need to move a certain distance before impacting the incident bar, and there are too many influencing factors in this process, as a result, the stress waves of the two systems cannot get completely synchronized in time, which leads to the bending load in the specimen arms. In addition, the cruciform specimen is loaded asymmetrically, which makes the center of the specimen unable to remain immobile.

2.3 Dynamic triaxial loading device

Triaxial loading experiments are mainly used to study the dynamic responses of underground rocks. There are two main types of methods to realize triaxial loading, i.e. displacement boundary conditions and pressure boundary conditions.

2.3.1 Displacement boundary conditions

Displacement boundary conditions are typically achieved through jacketing the cylindrical surface of the specimen using a shrink-fit metal sleeve or a passive thick vessel. This is a passive confining technique, the boundary conditions on the specimen lateral surface include both stress and displacement, thus the key to this technology is the material of the jackets. The boundary can be treated as nearly rigid if the jacket is too harder than the specimen, while the effect is closer to pressure boundary conditions if the jacket is plastically deformable during the experiment.

Chen and Ravichandran[35] firstly employed a metal jackets in the axial loading systerm to radially confine cylindrical brittle specimens, which was further performed by Rome et al.[36], Forquin et al.[37] and Nemat-Nasser et al.[38].

Gong and Malvern[39] provides another inexpensive passive confining jacket system to study experimentally the multiaxial compressive response of rock-like specimens. In their device, jackets made of steel and aluminum and a 76.2-mm-diameter split Hopkinson pressure bar system was used.

In order to explore the damage and failure processes of a transparent polycrystalline aluminum oxynitride (AlON), Paliwal et al.[40] developed an experimental technique, which enabled a controlled and homogeneous stress state with high lateral compressive stresses. In the device, a prismatic specimen (with a rectangular cross section) was statically precompress from two perpendicular directions and then subjected to axial dynamic compressive loading using a modified compression Hopkinson bar setup.

2.3.2 The pressure boundary conditions

Pressure boundary conditions are achieved through hydrostatic pressure in a triaxial test. In such a test, a specimen placed inside a pressure chamber is isotropically loaded by hydrostatic pressure. When an additional axial load is applied to the specimen which is in the constant hydrostatic pressure, the lateral deformation of the specimen results in the change of hydrostatic pressure and an additional shear stress will apply to the specimen. The specimen is isolated from the confining fluid through a soft seal membrane that is placed over the specimen. In such an experiment, the boundary condition on the lateral surface of the specimen is pressure only, making the stress state in the specimen clearly defined.

Christensen et al. [41] and Lindholm et al. [42] performed some of the most pioneering work in the dynamic tests of rocks under hydrostatic confinement in the early 1970s. The device is composed of an SHPB system with two hydraulic cylinders, and the sample is enclosed in the lateral confining cylinder. The transverse confining stresses and the axial confining stress will be generated by the action of the lateral confining cylinder and the confining cylinder applies.

Li [43] improved the devices of Christensen et al. and Lindholm et al. In this experimental design, the two pressure cylinders was connected with two tie-rods. The claimed that they can generate triaxial confinement, while the results they showed were only axial confinement in their work.

Frew et al. [44] used a very similar idea to improve the device of Christensen et al. and Lindholm et al. In their device, four tie-rods were connected with the two cylinders to apply hydrostatic confinement. The method to achieve such a confining state is to first expose the cylindrical rock sample to the confining fluid and then to maintain the same fluid pressure in both cylinders.

Cadoni and Albertini[45] designed a truetriaxial loading apparatus, this setup adopts hydraulic servo devices to apply 0~100MPa static load to the cube sample from three-directions independently, and then the split Hopkinson pressure bar is used to apply impact dynamic load on the specimen from one incident bar. However, the very short loading times do not enable one to carry out multiaxial dynamic loading in synchronicity.

3 Electromagnetic solution of synchronization problem in dynamic multiaxial loading

The traditional split Hopkinson pressure bar generates stress waves by means of striker impact. The striker is launched through the air valve switch. The striker needs to travel a distance before impact, so the time synchronization of multiple stress waves cannot be accurately controlled. In order to solve the synchronization problem, the traditional mechanical energy conversion method should be abandoned and the energy conversion method that can accurately control the time should be developed. For this purpose, some researchers are trying to find new mechanisms of stress waves. Silva et al. [46] proposed a new design for the compressive split Hopkinson bar that makes use of the intense pressure created in a transient magnetic field formed by the passage of a pulse of electric current through a series of coils. In this device, the striker is driven and accelerated by an instantaneous magnetic field, and finally impact the incident bar to produce a stress wave. The system has some advantages over the traditional Hopkinson bar, while the principle of the stress wave generation is still the impact method, the time synchronization problem is still not solved.

A new stress wave generation principle that is helpful to solve the stress wave synchronization problem was proposed by Nie et al.[47] in 2008. They developed a novel electromagnetic split Hopkinson pressure bar (ESHPB), which employs the electromagnetic energy conversion technique of LC circuit to generate directly the incident stress pulse (Fig. 6). This technique can generate easily compressive as well as tensile incident pulses. Compared with traditional pulse generation techniques by the impact of a projectile or by a sudden release of a pre-stressed section, the electromagnetic energy conversion technique can be accurately triggered within several microseconds. It is, therefore, a good candidate to supply the symmetrical and synchronous loads in bidirectional or biaxial split Hopkinson bar systems in the future.

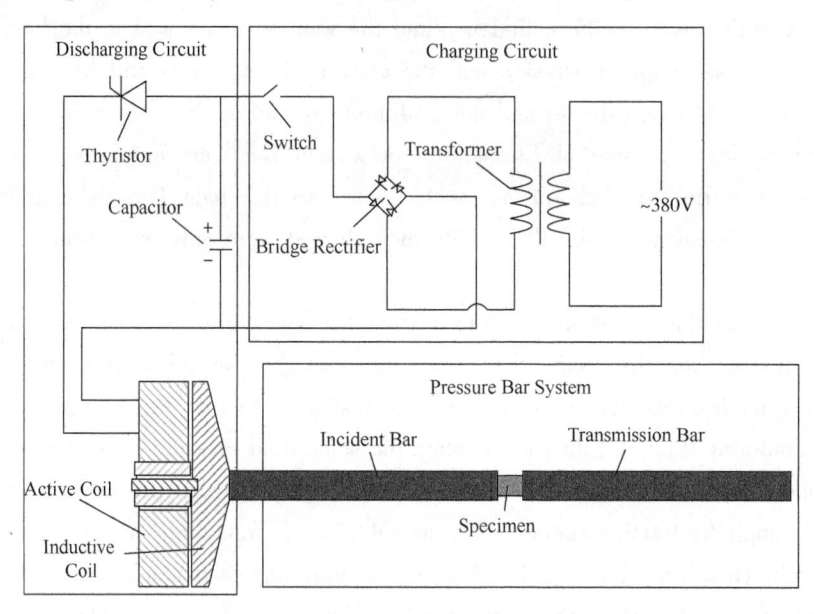

Fig. 6 Schematic diagram of the Electromagnetic SHPB setup[47] [Reproduced with permission from Int J of Impact Eng. 116, 94 (2018). Copyright 2018 Elsevier]

In fact, the electromagnetic split Hopkinson bar does show unique advantages in the problem of stress wave synchronization. At present, Nie et al. have successfully made use of electromagnetic split Hopkinson bar to develop a symmetrically loading Hopkinson bar[48] (Fig. 7). In order to obtain two synchronized stress pulses, two identical electromagnetic stress pulse generators connected to the same LC discharge circuit are used in this device. This symmetric impact loading configuration might be easily interchanged into a compressive as well as a tensile version because of the versatility of the electromagnetic stress pulse generators. They measured the stress waves in the two incident bars, and the results showed that the two incident waves basically reached the end faces of the specimen at the same time, with an error of less than 3 microseconds (Fig. 8).

The biggest disadvantage of the electromagnetic Hopkinson bar is the electromagnetic interference problem. Since the device will generate an instant electromagnetic pulse in the process of stress wave producing, which will interfere with the nearby data acquisition equipment, and the interference and stress wave signals will be both captured by the data collector almost at the same time, so it will affect the rising edge of the collected stress waves.

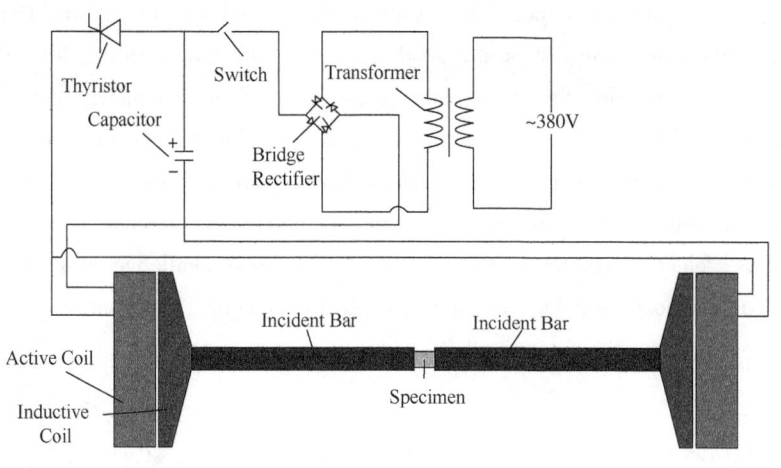

Fig. 7　Schematic diagram of the symmetric split Hopkinson compression bar[48] [Reproduced with permission from Int J of Impact Eng. 122, 73 (2018). Copyright 2018 Elsevier]

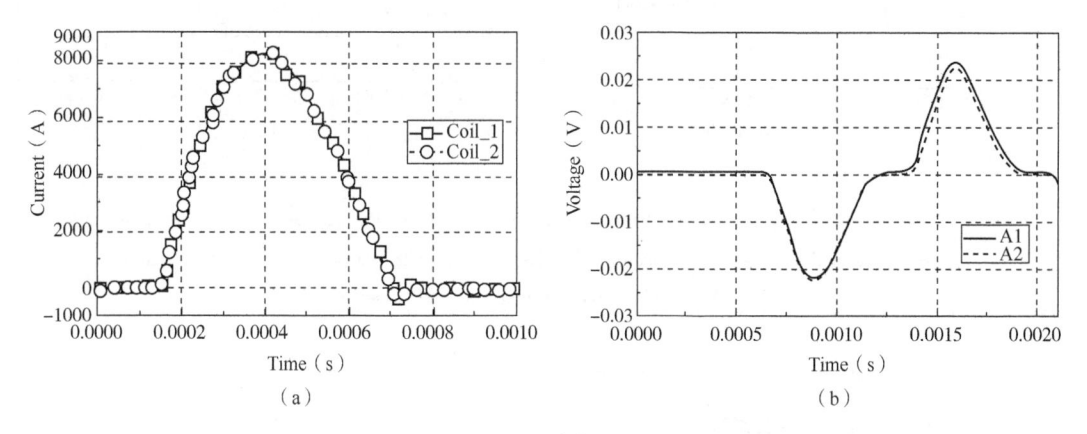

Fig. 8　Comparison of experimental results: (a) discharge currents of two active coils, (b) signals on the bars measured by the data collector[48] [Reproduced with permission from Int J of Impact Eng. 122, 73 (2018). Copyright 2018 Elsevier]

The distribution of pulse electromagnetic fields produced by unidirectional and symmetrical loading equipment are relatively simple and have a regular to follow, and the electromagnetic interference can be eliminated through some methods, such as the location optimization of data acquisition equipment, the special design of the wires and so on. In fact, we have eliminated electromagnetic interference in unidirectional and symmetric loading devices by the above method and obtained stress-strain curves of some common materials. For example, Fig. 9(a) shows a typical stress pulses obtained in symmetric compression experiment of copper. It is necessary to point out that the negative values in reflected waves are not due to an un-perfect contact between the specimen and the bars, they are the natural result of the superposition of waves. A detailed explanation can be found in reference[48]. The forces at the two sides [Fig. 9(b)] of the specimen are in a good equilibrium state. The results of copper symmetrical compression test are compared in Fig. 10(a) with that of traditional striking SHPB test and the single electromagnetic SHPB (ESHPB) test under the similar average strain rate of about 1200 1/s, since the same material was tested by using in the previous paper, the result is also added for comparison. The stress–strain curve obtained in the symmetric test is in a good agreement with that of the traditional SHPB test. It is noticed from Fig. 10(b) that the strain rate–strain curve of the symmetric test is not constant, which is a drawback of this device because the stress wave is almost half–sin shape unless the pulse shaper method is applied. However, if four devices discharge at the same time, the distribution of the electromagnetic field will be more complex, and the interference cannot be offset by the above method. Therefore, electromagnetic interference is still a great challenge to realize the dynamic biaxial or even triaxial loading device by using the electromagnetic Hopkinson bar.

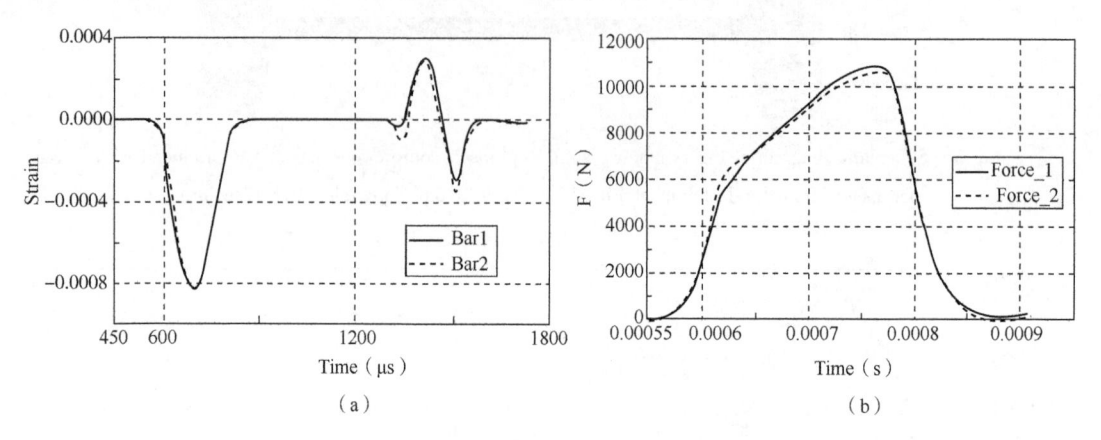

Fig. 9 Pulses recorded in a copper test using symmetric SHPB apparatus: (a) the strain signal in bars, (b) the forces at the sides of the specimen[48] [Reproduced with permission from Int J of Impact Eng. 122, 73 (2018). Copyright 2018 Elsevier]

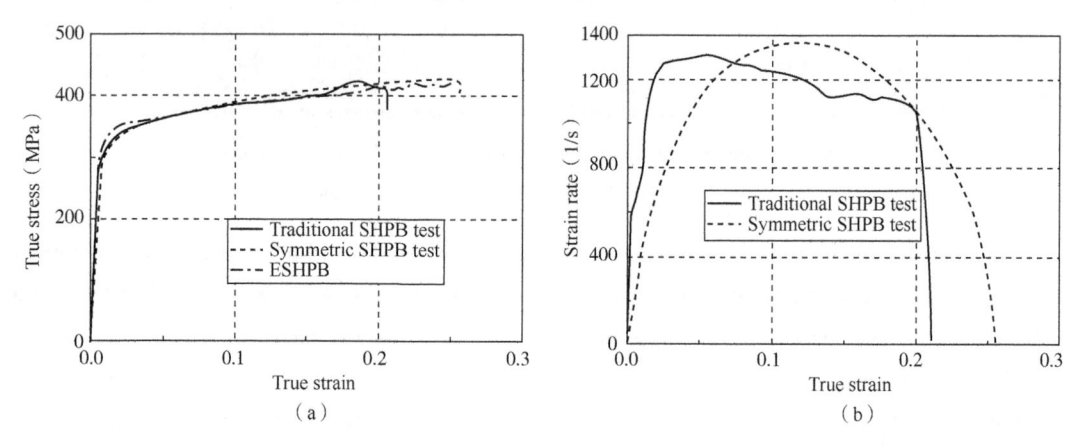

Fig. 10 Compression results of copper by using symmetric loadings, electromagnetic SHPB apparatus (ESHPB) and traditional SHPB apparatus: (a) stress–strain curves, (b) strain rate–strain curves[48] [Reproduced with permission from Int J of Impact Eng. 122, 73 (2018). Copyright 2018 Elsevier]

4 Summary and looking forward

After more than 60 years of efforts, the single–axial Hopkinson bar devices have played an important role in promoting the development of many disciplines. However, either by way of mechanical collision to produce compression and tensile pulse, or adopt the way of energy storage to release a loading impulse, its synchronicity is unable to effectively controlled within microseconds time scale, as a result, the development of tension – compression and tension – torsion coupled multiaxial Hopkinson bar devices have not made substantial progress.

Although many researchers want to develop a tension–torsion or compression–torsion multiaxial loading device by using the principle of stored mechanical energy release, satisfactory results have not been achieved due to the complexity of the release mechanism and the large difference in wave velocities between the compressive/tensile and shear waves.

Some researchers also used the uniaxial compression Hopkinson bar with inclined plane to conduct compression – shear coupling experiments. The friction between the specimen and the inclined plane of the Hopkinson bar can impose shear load on the specimen, but the compressive stress is not evenly distributed in the sample.

The development of dynamic multiaxial loading device under has not made a breakthrough, so far it is still in blank. The main reason is that the duration in dynamic tests is very short compared with that of quasi – static tests, which requires to precisely control the generating time of stress waves. However, the stress pulse generation principle of traditional Hopkinson bar is the conversion between one mechanical energy and another mechanical energy, that is, the use of impact method or energy storage method to generate stress wave, while the time control of mechanical method is unable to achieve the accuracy required by multiaxial loading. Therefore, in order to solve the time control problem of dynamic multiaxial loading technique, it is necessary to abandon the traditional stress pulse producing method in Hopkinson bar, so that the initial times of stress waves can be controlled precisely.

With the invention of the electromagnetic Hopkinson bar device, the synchronization problem of stress waves has been technically solved, and the dynamic symmetrical loading of the specimen has been successfully realized. It is believed that in the near future, it will be possible to achieve dynamic biaxial loading even more triaxial loading.

5 Data availability statement

The data that support the findings of this study are available from the corresponding author upon reasonable request.

Acknowledgement

This work was supported by the China Postdoctoral Science Fund (2019M653785).

References

[1] Wright TW, "A survey of penetration mechanics for long rods," Springer. (1983).
[2] Hoferlin E, Bael AV, Houtte P V, Steyaert G, Maré2 C, "The design of a biaxial tensile test and its use for the validation of crystallographic yield loci," Model Simul Mater Sc. 8(4), 423 (2000).
[3] Rotter, Julian B, "Generalized expectancies for internal versus external control of reinforcement," Psychol Monogr. 80(1), 1 (1966).

[4] Sutton R, Barto A, "Reinforcement Learning: An Introduction," MIT Press. (1998)

[5] Donovan PE, "A yield criterion for Pd40Ni40P20 metallic glass," Acta Metall, 37(2), 445 (1989).

[6] Shrivastava HP, Mr ÃZ, Dubey RN, "Yield Criterion and the Hardening Rule for a Plastic Solid," ZAMM-J Appl Math Mec. 53(9), 625 (1973).

[7] Jia P, Tang CA, "Numerical study on failure mechanism of tunnel in jointed rock mass," Tunn Undergr Sp Tech. 23(5), 500 (2008).

[8] Krebs F C, Norrman K, "Analysis of the failure mechanism for a stable organic photovoltaic during 10000 h of testing," Prostate. 15(8), 697 (2007).

[9] Apostolico F, Gammaitoni L, Marchesoni F, Santucci S, "Resonant trapping: A failure mechanism in switch transitions," Phis Rev E. 55(1), 36 (1997).

[10] Belov D, Yang MH, "Failure mechanism of Li-ion battery at overcharge conditions," J Solid State Electr. 12 (7-8), 885 (2008).

[11] Hopkinson J, "Further experiments on the rupture of iron wire," Original Papers-by the late John Hopkinson, Vol II, Scientific Papers. (1872).

[12] Hopkinson B, "A method of measuring the pressure produced in the detonation of high explosives or by the impact of bullet," Philos T R Soc A. 437, (1914).

[13] Kolsky H, "An investigation of the mechanical properties of materials at very high rates of loading," Proc Phy Soc B. 62(11), 676 (1949).

[14] Chen WW, Song B, "Split Hopkinson (Kolsky) bar: design, testing and applications," Springer Science & Business Media. (2010).

[15] Zhang Q, Zhao J, "A Review of Dynamic Experimental Techniques and Mechanical Behaviour of Rock Materials," Rock Mech Rock Eng. 47, 1411 (2014).

[16] Xia K, Yao W, "Dynamic rock tests using split Hopkinson (Kolsky) bar system-A Review," J Rock Mech Geotechnical Eng. 7(1), 27(2015).

[17] Xing H, Zhang Q, Braithwaite C, Pan B, Zhao J, "High-speed photography and igital optical measurement techniques for geomaterials: fundamentals and applications," Rock Mech Rock Eng. 50, 1611 (2017).

[18] Lindholm U, Yeakley L, "High strain-rate testing: tension and compression," Exp Mech. 8(1), 1(1968).

[19] Nicholas T, "Tensile testing of materials at high rates of strain," Exp Mech. 21(5), 177 (1981).

[20] Harding J, Wood E, Campbell J, "Tensile testing of materials at impact rates of strain," J Mech Eng. 2(2), 88 (1960).

[21] Ogawa K, "Impact-tension compression test by using a split-Hopkinson bar," Exp Mech. 24(2), 81 (1984).

[22] Yu X, Chen L, Fang Q, Jiang X, Zhou Y, "A Review of the Torsional Split Hopkinson Bar," Advances in Civil Engineering, 2018(PT9), 17(2018).

[23] Chung J, Waas AM, "Compressive response of circular cell polycarbonate honeycombs under inplane biaxial static and dynamic loading. Part I: experiments," Int J Impact Eng. 27(7), 729(2002).

[24] Chung J, Waas AM, "Compressive response of circular cell polycarbonate honeycombs under inplane biaxial static and dynamic loading—Part II: simulations," Int J Impact Eng. 27(10), 1015(2002).

[25] Hong ST, Pan J, Tyan T, Prasad P, "Dynamic crush behaviors of aluminum honeycomb specimens under compression dominant inclined loads," Int J plasticit. 24(1), 89 (2008).

[26] Espinosa H, Patanella A, Xu Y, "Dynamic compression-shear response of brittle materials with specimen recovery," Exp Mech. 40(3), 321 (2000).

[27] Rittel D, Lee S, Ravichandran G, "A shear-compression specimen for large strain testing," Exp Mech. 42 (1), 58 (2002).

[28] Huang H, Feng R, "A study of the dynamic tribological response of closed fracture surface pairs by Kolsky-bar compression-shear experiment," Int J Solids Struct. 41(11), 2821 (2004).

[29] Zhao PD, Lu FY, Chen R, Lin YL, Li JL, Lu L, "A technique for combined dynamic compression-shear test," Rev Sci Instrum. 82(3): 35 (2011).

[30] Hou B, Ono A, Abdennadher S, Pattofatto S, Li Y, Zhao H, "Impact behavior of honeycombs under combined shear-compression. Part I: Experiments," Int J Solids Struct. 48(5), 687 (2011).

[31] Hou B, Pattofatto S, Li Y, Zhao H, "Impact behavior of honeycombs under combined shear-compression. Part II: Analysis," Int J Solids Struct. 48(5), 698 (2011).

[32] Grolleau V, Gary G, Mohr D, "Biaxial testing of sheet materials at high strain rates using viscoelastic bars," Exp Mech. 48, 293(2008).

[33] Shimamoto A, Shimomura T, Nam, "The development of a servo dynamic loading device," Key Eng Mater. 243-244, 99 (2003).

[34] Hummeltenberg A, Curbach M, "Entwurf und Aufbau eines zweiaxialen Split-Hopkinson-Bars," Beton-Und Stahlbetonbau. 107, 394(2012).

[35] Chen W, Ravichandran G, "Dynamic compressive failure of a glass ceramic under lateral confinement," J Mech Phys Sol, 45(8), 1303(1997).

[36] Rome J, Isaacs J, Nemat-Nasser S, "Hopkinson techniques for dynamic triaxial compression tests," In: Gdoutos E editor. Redent Advances in Experimental Mechanics. Netherlands Springer, 2004.

[37] Forquin P, Gary G, Gatuingt F, "A testing technique for concrete under confinement at high rates of strain," Int J Impact Eng. 35(6), 425(2008).

[38] Nemat-Nasser S, "Introduction to high strain rate testing," In: Kuhn H, Medlin D editors. ASM Handbook: Volume 8: Mechanical Testing and Evaluation. Ohio, USA: ASM International. 427(2000).

[39] Gong JC, Malvern LE, "Passively confined tests of axial dynamic compressive strength of concrete," Exp Mech. 30(1), 55(1990).

[40] Paliwal B, Ramesh KT, McCauley JW, Chen M, "Dynamic compressive failure of AION under controlled planar confinement," J Am Ceram Soc. 91(11), 3619(2008).

[41] Christensen RJ, Swanson SR, Brown WS, "Split-Hopkinson-bar tests on rock under confining pressure," Exp Mech. 12(11), 508(1972).

[42] Lindholm US, Yeakley LM, Nagy A, "The dynamic strength and fracture properties of Dresser basalt," Int J Rock Mech Min. 11(5), 181(1974).

[43] Li XB, Zhou ZL, Lok TS, Hong L, Yin TB, "Innovative testing technique of rock subjected to coupled static and dynamic loads." Int J Rock Mech Min. 45(5), 739(2008).

[44] Frew DJ, Akers SA, Chen W, Green ML, "Development of a dynamic triaxial Kolsky bar," Measurement Science and Technology. 21(10), 105(2010).

[45] Cadoni E, Albertini C, "Modified Hopkinson bar technologies applied to the high strain rate rock tests," Advances in Rock Dynamics and Applications, New York. USA: CRC Press, 79(2011).

[46] Silva C, Rosa P, Martins P, "An innovative electromagnetic compressive split Hopkinson bar," Int J Mech Mater Des. 5(3), 281(2009).

[47] Nie H, Suo T, Wu B, Li Y, Zhao H, "A versatile split Hopkinson pressure bar using electromagnetic loading," Int J of Impact Eng. 116, 94 (2018).

[48] Nie H, Suo T, Shi X, Liu H, Li Y, Zhao H, "Symmetric split Hopkinson compression and tension tests using synchronized electromagnetic stress pulse generators," Int J of Impact Eng. 122, 73(2018).

本论文原发表于《Review of Scientific Instruments》2020 年。

Fatigue Failure Analysis of Dented Pipeline and Simulation Calculation

Luo Jinheng[1,2] Zhang Yani[3] Li Lifeng[1,2]
Zhu Lixia[1,2] Wu Gang[1,2]

(1. CNPC Tubular Goods Research Institute; 2. State Key Laboratory of Performance and Structural Safety for Petroleum Tubular Goods and Equipment Material; 3. Xi'an Shiyou University)

Abstract: Natural gas pipeline leaks may cause considerable damage to people, properties, and the surrounding environment. This study investigates a dent leak in a gas pipeline in China that has operated for six years. The failure pipe section meets American Petroleum Institute (API) specification 5L-2012 standard for carbon steel pipelines. The cause of leakage is evaluated not only through visual examination, micromorphology observation, structural deformation analysis, and residual stress measure of the dented region near the leak point of the pipe body, but also through finite element analysis on the deformation and stress of the pipe under different loads. The dent formation is related to rocks during pipeline laying. Under the combined effect of external load and internal pressure, the maximum equivalent stresses of the dented edge at the outer wall, bulge top of the inner wall, and dented edge of the inner wall are greater than the yield stress of the material. Moreover, micro-crack formation occurs due to the maximum deformation at the bottom of the dent and the maximum equivalent stress on top of the inner surface bulge. Under alternating stress, microcracks gradually expand until they penetrate the wall thickness, and thereby cause pipeline leakages.

Keywords: Pipeline; Leak; Dent; Finite element analysis; Fatigue failure

1 Introduction

Pipeline transportation is considered a safe and economical means to transport gas. However, the safe operation of pipelines is threatened by several factors, such as: the potential corrosive characteristics of the transport medium (natural gas may contain hydrogen sulfide, carbon dioxide, free water, dust, and other corrosive matter); geological, meteorological, and hydrological disasters; pipe and design defects; operational errors; and man-made damage. Xiao et al.[1] proposed that the failure pressure of pipelines is reversely proportional to the scratch length and

---

Corresponding author: Luo Jinheng, luojh@cnpc.com.cn

depth and that the dent depth exerts a strong influence on limit internal pressure. As a flammable and explosive substance, natural gas can cause considerable damage to people, properties, and the surrounding environment when pipeline leaks occur. Research on the safe transportation of natural gas mainly focuses on corrosion failures (including single- and multiple-, stress, and impact) and interaction models of corrosion and mechanics[2-12]. However, limited studies evaluate fatigue failure events of damaged pipelines[13-16].

The present study investigates a leak failure in a gas pipeline in China that has operated for six years. Excavation reveals the dent leak in the six o'clock pipe direction. The leak cause is evaluated on the basis of chemical composition analysis, mechanical performance test, microstructure observation, residual stress measure, and analysis of finite element simulation of the stress−strain distribution in different load states near the leak region. This study presents the pre−defect fatigue failure model to serve as reference for the safe operation and life evaluation of natural gas pipelines.

2 Pipeline profile

Thefailure pipeline was completed and operated in 2010. Its main characteristics are as follows: L485M carbon steel, ϕ813mm diameter, 8.8mm wall thickness, spiral − seam submerged − arc welding process, and three layers of polyethylene as anticorrosion coating. The design pressure and gas transmission capacity of the pipeline are 6.3MPa and $1200\times10^4 m^3/d$, respectively. The actual operating pressure of the pipeline is 2.5~3.8MPa (mostly 3.5MPa), and the gas supply capacity is $(300~600)\times10^4 m^3/d$ (mostly $500\times10^4 m^3/d$).

Figs. 1 and 2 show the scene of the failed pipeline after full excavation. The rock wall is located at the bottom left side of the pipeline in the gas flow direction. This hard stone is underneath the pipeline, and the leaking point is at the dented area of the pipe wall.

Fig. 1 Morphology of the failed pipeline site

Fig. 2 Rock and dent features at the bottom of the pipeline

The failed pipe segment is located at the top of a small hill in grades 1 and 2 areas, where the mountains are mainly composed of hard rocks. Mechanical slotting and stone backfilling are used during pipeline construction. The failed pipe is buried 3m deep with a large height difference from the surrounding houses. No third-party construction, geological disasters, and other phenomena are observed nearby.

3 Detection analysis

3.1 Visual inspection

Figs. 3 and 4 show themacromorphology of the failed pipe segment. The dent leak is located at the 6 o'clock position of the pipe body. The outer surface shows an approximately circular dent zone with axial length of 313mm and ring length of 244mm. The inner surface has a dent protrusion with axial length of 349mm and ring length of 280mm. Fig. 5 shows that the dent zone has a maximum depth of 40.48mm and the deepest region has penetrating cracks. Around the crack, the minimum wall thickness is 4.87mm. The maximum length and width of the crack are 54mm and 2mm from the outer surface measurements, respectively, and are 58mm and 3mm from the inner surface measurements, respectively. The widths of the internal and external surface cracks lead to an initial assessment that the crack starts from the inner surface.

Fig. 3 Outer surface dent morphology of the failed pipe segment

Fig. 4 Inner surface morphology of the failed pipe segment

Fig. 5 Macromorphology of surface cracks in the dent region

Fig. 6 shows the circumferential wall thickness and perimeter of the region near the dent zone at an interval of 100mm. Table 1 and Table 2 present the measurement results. The wall thickness and perimeter of the tube body meet the API 5L-2012 standard. The dent is found at the 6 o'clock position in sections B, C, D, and E.

Fig. 7 shows the intensive measurement of the wall thickness. A measurement grid is drawn at 50mm intervals in the axial B to E sections and at 53mm intervals in the circumferential direction from 5 to 7 o'clock positions. Table 3 presents the measurement results. The wall thickness clearly thins around the crack in the dent region but not in its surrounding areas.

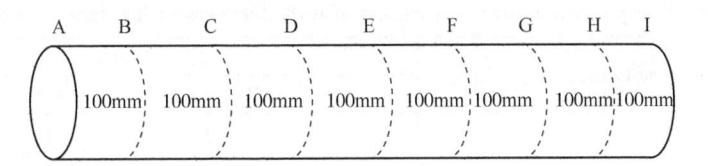

Fig. 6 Schematic of the wall thickness measurement of the failed pipe segment

Table 1 Measurement results of wall thickness (mm)

Measurement location	A	B	C	D	D	F	G	H	I
at 1 o'clock	8.68	8.72	8.69	8.67	8.78	8.73	8.79	8.72	8.73
at 2 o'clock	8.69	8.67	8.69	8.73	8.70	8.77	8.72	8.73	8.73
at 3 o'clock	8.69	8.74	8.78	8.60	8.56	8.72	8.69	8.70	8.70
at 4 o'clock	8.69	8.65	8.70	8.68	8.66	8.68	8.72	8.70	8.72
at 5 o'clock	8.68	8.65	8.80	8.67	8.66	8.71	8.71	8.65	8.74
at 6 o'clock	8.63	8.67	7.98	8.63	8.67	8.69	8.66	8.58	8.59
at 7 o'clock	8.75	8.71	8.68	8.73	8.68	8.66	8.70	8.71	8.74
at 8 o'clock	8.76	8.79	8.68	8.67	8.68	8.70	8.73	8.73	8.67
at 9 o'clock	8.67	8.76	8.72	8.73	8.78	8.70	8.69	8.74	8.76
at 10 o'clock	8.76	8.72	8.68	8.69	8.69	8.76	8.69	8.73	8.75
at 11 o'clock	8.69	8.65	8.75	8.73	8.71	8.71	8.69	8.73	8.76
at 12 o'clock	8.69	8.60	8.62	8.68	8.68	8.69	8.78	8.69	8.73

Table 2 Perimeter measurement result (mm)

Section location	A	B	C	D	E	F	G	H	I
Measurement result	2552	2554	2560	2560	2554	2553	2557	2551	2553

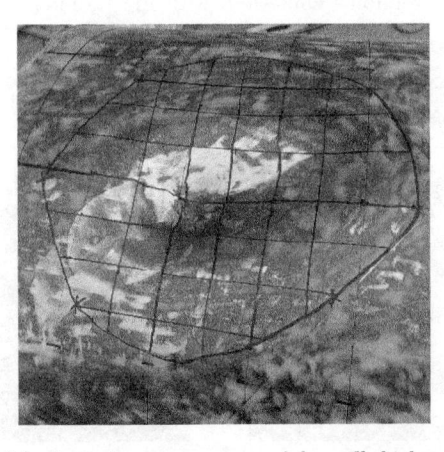

Fig. 7 Schematic of the intensive measurement of the wall thickness at the dent region

Table 3　Encrypted measurement results of wall thickness at the dent region (mm)

Measurement location	Dent area					
	B	B+50	C	C+50	D	D+50
5 o'clock to 6 o'clock equidistant at the ring direction	—	—	8.75	8.72	8.68	—
	—	8.70	8.75	8.68	8.66	8.73
	8.61	8.66	8.65	8.69	8.66	8.72
	8.66	8.66	8.53	8.67	8.62	8.74
at 6 o'clock	8.67	8.68	7.98	8.55	8.63	8.65
6 o'clock to7 o'clock equidistant at the ring direction	8.71	8.66	8.62	8.55	8.62	8.74
	8.66	8.69	8.71	8.69	8.67	8.69
	—	8.70	8.69	8.71	8.67	8.68
	—	—	8.68	8.81	8.71	—

3.2　Analysis of pipeline materials

In long-distance pipelines, the causes of deformation and cracking include chemical composition, microstructure, and material toughness[17-22]. In this study, the chemical composition analysis, microstructure observation, and mechanical properties test are processed in the dent segment of the failed pipe according to API standards. All of the indicators of the pipe body meet the requirements of API specification 5L-2012 standard for carbon steel pipelines.

3.3　Analysis of dentregion

3.3.1　Analysis of cracks in dentregion

Contour scanning is performed on the outer surface of the dent region with a Smartzoom5ultradepth digital microscope. Fig. 8 shows visible cracks at the bottom of the dent region, which is divided into six blocks for convenience in the microscopic examination. Fig. 9 shows visible cracks in blocks #1 and #2 while Figs. 10 and 11 show the crack micromorphology. The crack tip has gray matter but no branches are detected. The energy dispersive spectrometer (EDS) results show that the gray matter is mainly composed of Fe, O, C, and Mn elements (Fig. 12) and estimated to be iron oxide. Figs. 13 and 14 present the microscopic analysis, which shows that the body structures of the outer surface and internal surface around the dent region have deformation flow characteristics along the axis direction (consistent with the radian).

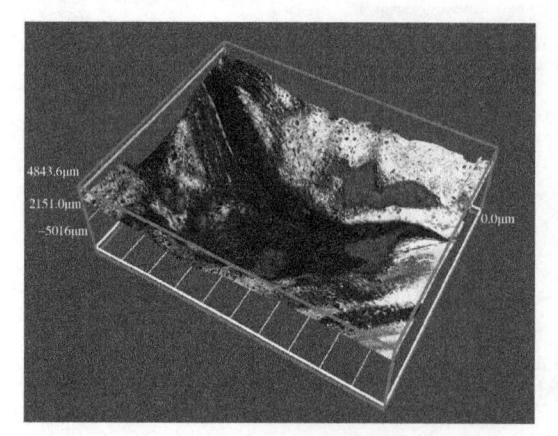

Fig. 8　Outer surface morphology of the dent region

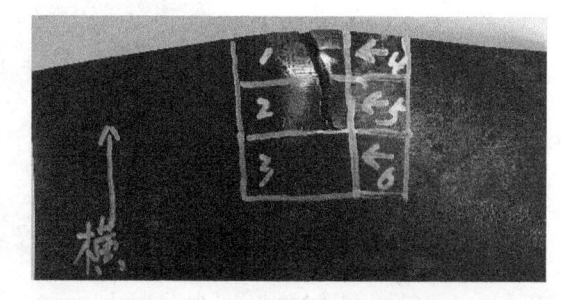

Fig. 9　Internal surface blocks of the dent region

Fig. 10 Crack morphology of fracture surface #2

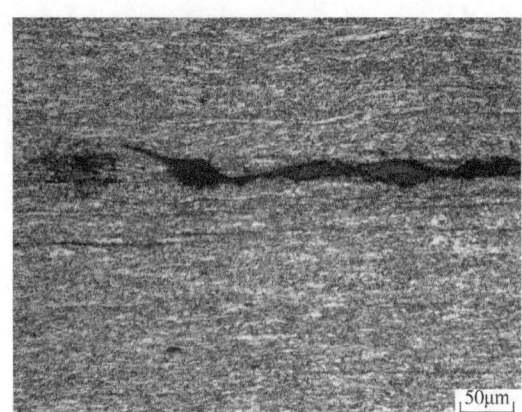

Fig. 11 Crack tip morphology of fracture surface #2

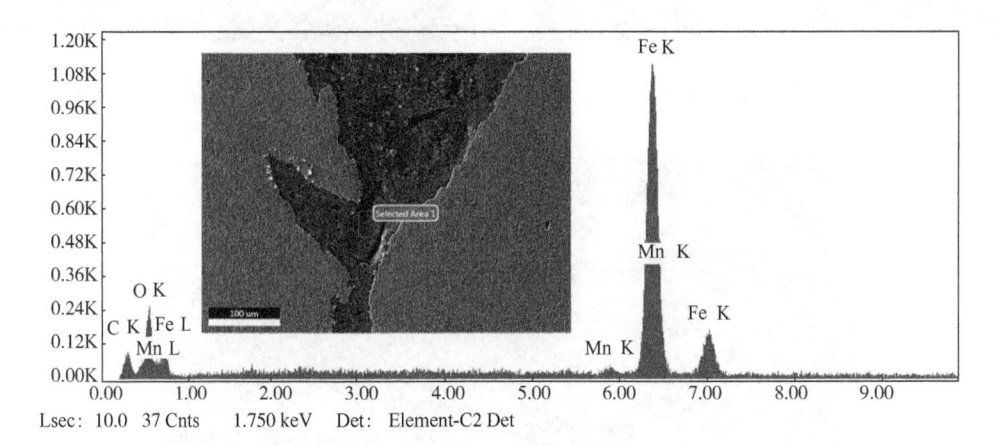

Fig. 12 Analysis results of the energy spectrum of gray matter in the crack

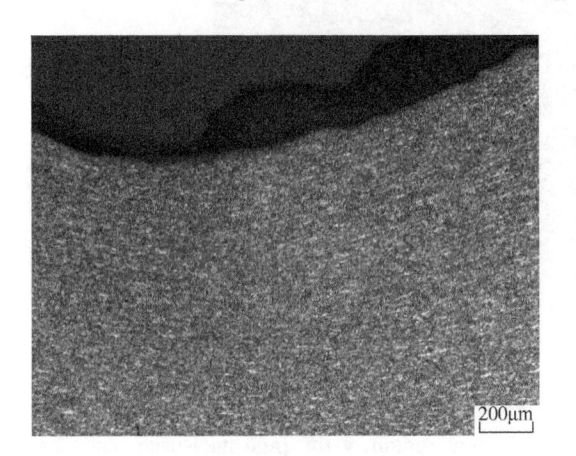

Fig. 13 Structural deformation around the outer surface of block #2

Fig. 14 Structural deformation around the internal surface of block #2

Visual inspection in Fig. 15 and 16 shows a fuzzy bay line on the inner surface of block #2 of the cracked section. The morphology is further investigated using a scanning electron microscopy

· 71 ·

(SEM). Figs. 17 and 18 show several steps or strips on the inner wall, bounded by the wall thickness center. High magnifications in Figs. 19 and 20 show a few fatigue striations that are parallel to each other but perpendicular to the direction of crack expansion.

Fig. 15　Macromorphology of fracture #1

Fig. 16　Macromorphology of fracture #2

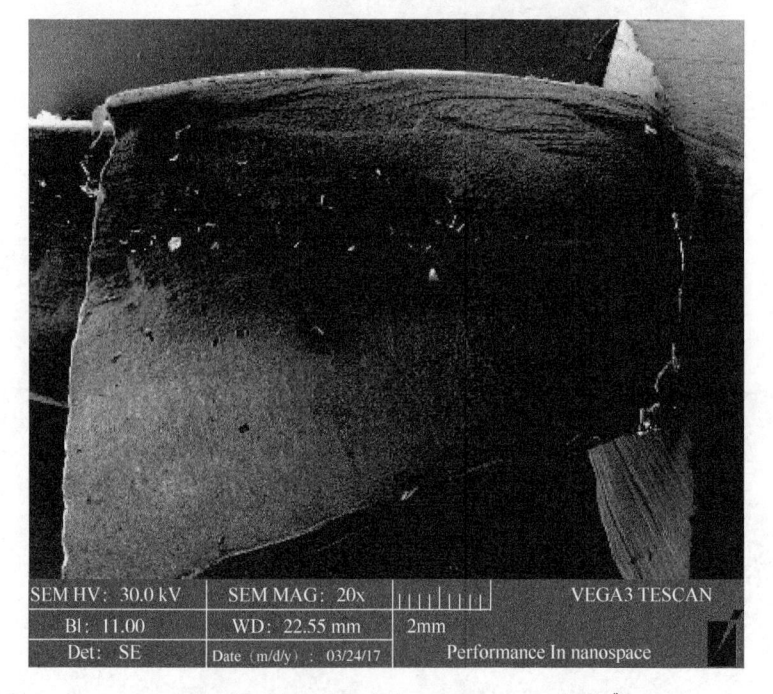

Fig. 17　SEM macromorphology of fracture surface #1

3.3.2　Residual stress measurement of dentregion

Table 4 lists the results of the residual stress at the dent region of the pipe measured using an MSF-3M X-ray stress analyzer. The dent bottom and edge are subjected to pressure and tensile stresses, respectively. Stress at the dent bottom is higher than that at the edge. In the dent bottom, the outer and inner walls of the pipe are subjected to pressure stress. The inner wall has higher stress than that of the outer wall, and has significantly higher ring stress than axial stress.

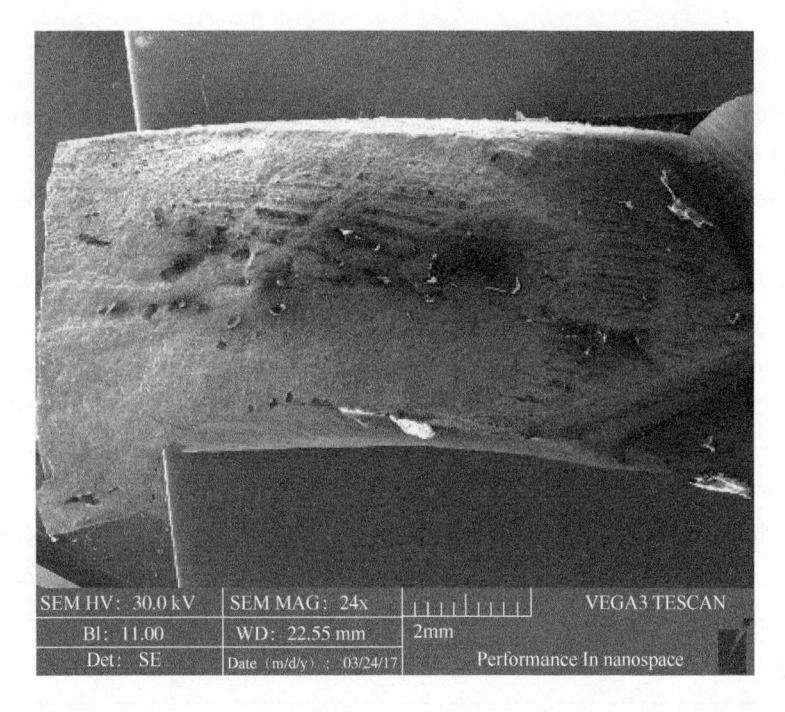

Fig. 18 SEM macromorphology of fracture surface #2

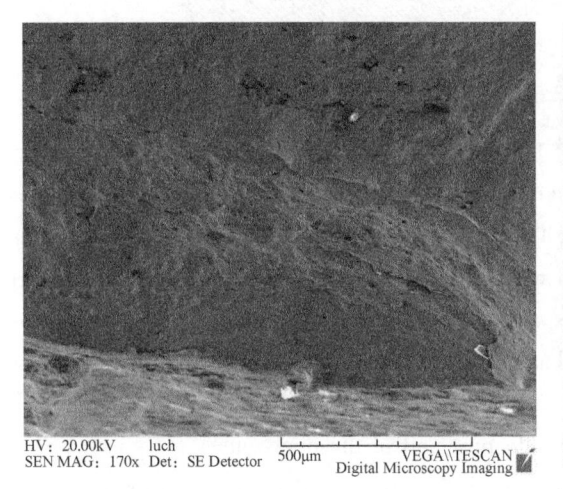

Fig. 19 Morphology of the arc-shaped steps in fracture surface #2

Fig. 20 Morphology of fatigue striations in fracture surface #2

Table 4 Residual stress test results (MPa)

Measurement location			Ring stress	Axial stress
Dent area	Dent bottom	Inner wall	−372.43	−207.47
	Dent edge	Outer wall	−249.27	−276.97
		Outer wall	244.68	79.73

· 73 ·

4 Analysis and discussion

The chemical composition, microstructure, and mechanical properties of the pipe body meet the requirements of the API specification 5L-2012 standard for carbon steel pipelines.

The dent leak is located at the 6 o'clock position of the pipe body and is approximately circular in shape. Outer surface measurements show that the dent zone has an axial length of 313mm and ring length of 244mm. Inner surface measurements show a protrusion zone with an axial length of 349mm and a ring length of 280mm. The maximum depth of the dent zone is 40.48mm. The failed pipe segment is located at the top of a hill that has numerous quarries, and the mountains are basically composed of hard rocks. Mechanical slotting and stone backfilling are used during pipeline construction. Excavation after the pipeline leak reveal a hard rock with approximate dimensions of 350mm×350mm (on-site estimation) at the 6 o'clock position of the dent region of the pipe body and no surrounding soil filling. The pipe wall dent is thus clearly caused by the pipeline laying, and the bottom of the pipeline (i.e., at 6 o'clock position) jammed by the rock is subjected to the dead weight and compression load from the backfill stone (Fig. 21). Thus, pipeline deformation is the cause of the dent.

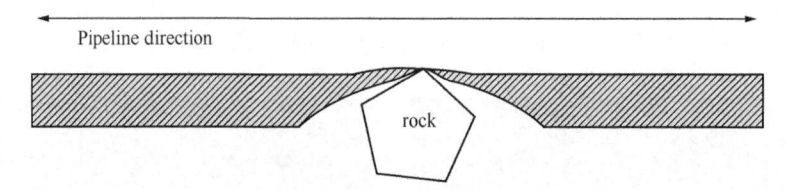

Fig. 21 Schematic of stuck pipeline caused by the rock

Fig. 5 shows that the minimum wall thickness around the crack is 4.87mm, which is approximately 55% of the nominal wall thickness. The maximum length and width of the crack 54mm and 2mm from the outer surface measurements, respectively, and are 58mm and 3mm from the inner surface measurements, respectively. The crack expands along the ring direction. The residual stress analysis shows that the dent zone of the pipeline has a wall ring stress that is higher than axial stress while the ring stress of the inner wall is higher than that of the outer wall. Therefore, the leak originated in the inner wall of the pipe body at the deepest part of the dent zone.

In the fracture morphology, visual examination of the inner surface shows cowrie pattern lines. Other typical fatigue characteristics such as steps, strips, and fatigue striations are also observed using SEM. However, branches and corrosive scales from the transport medium are absent at the crack tip. Hence, the pipeline dent leak is a type of fatigue failure and the crack starts at the inner wall of the pipeline.

In the fatigue failure of the failed pipeline segment, inner pressure fluctuation (role of fatigue load) is an important contributor to the formation of cracks in the dent region. Fitting of inner pressure fluctuation can be beneficial to the recovery of this region, without effect or only slight effects on the service life of pre-defects in the pipeline. Zheng[23] pointed out that the threshold of crack initiation increases with pre-deformation due to the strain hardening effect. In the present study, the deformation and force of steel pipe under different loads are simulated and calculated using finite element analysis to determine the effect of inner pressure on the pipe body recovery. The process includes (1) stress-strain distribution of dented pipelines; (2) stress-strain distribution of dented pipelines under operation pressure and removal of load and pressure; (3) internal and external surface stress states of dent zone, which demonstrate the propagation direction of micro-cracks.

The boundary conditions of finite element analysis models include an elastic modulus of L485M carbon steel of spirally submerged arc welding with 206GPa and Poisson's ratio of 0.3. Fig. 22 shows the true stress-strain curve of the pipe body. In the simulation, the hard rock holding the pipe at the 6 o'clock position is reduced to a (350×350)mm hemisphere, as shown in Fig. 23. A true stress-strain model is used for the simulation calculations.

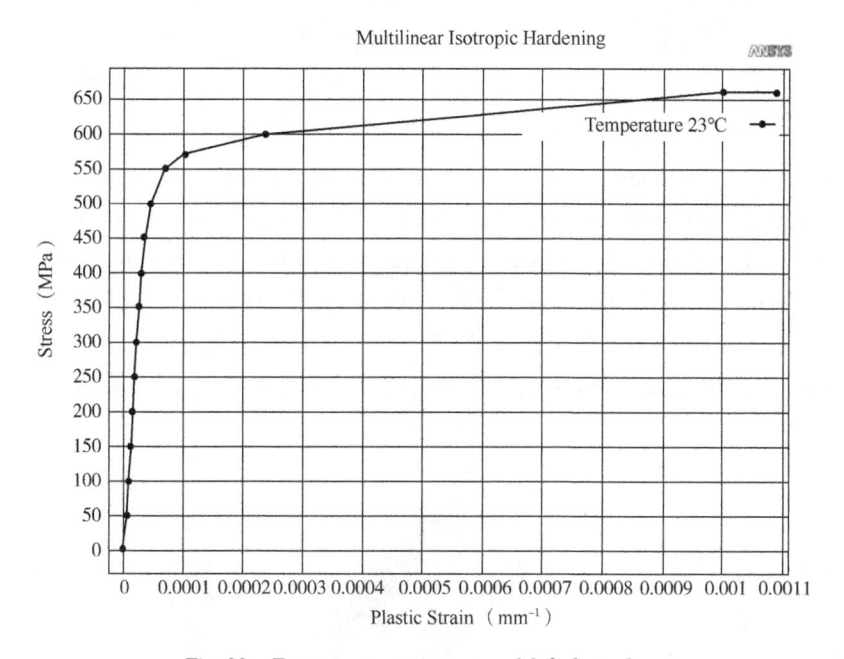

Fig. 22 True stress-strain curve of failed pipeline

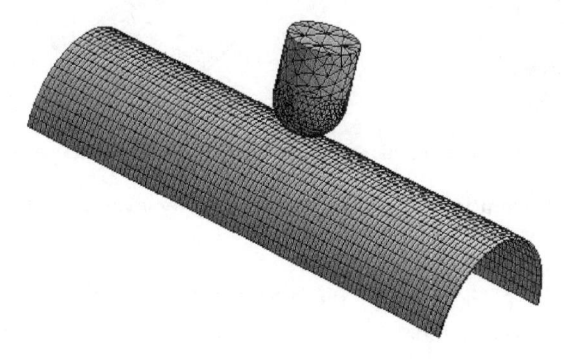

Fig. 23 Schematic of the geometric model of steel pipe and mesh

(1) Stress-strain distribution of dented pipeline.

Load is applied at the hemispherical indenters to produce a dent with a depth of 40mm and counterforce of 373 kN (equivalent to 37 tons of force on the pipe body surface to produce the dent). Fig. 24 presents the stress nephogram. The stress value of the loading region is significantly higher than that of the other parts of the pipe. The maximum stress occurs in the loading center of the inner surface of the steel pipe and the dent edge on its inner and outer surfaces. The maximum equivalent stress at the loading region is 657.48MPa, which exceeds the 531MPa yield stress value. Thus, the carbon steel pipe enters the plastic deformation stage.

Fig. 24 Stress-strain distribution of pipelines under 40mm depth dent

(2) Stress-strain distribution of dented pipelines under operation pressure.

Fig. 25 shows the operating pressure of 3.5MPa applied to the pipe with a 40mm dent to obtain the stress nephogram in the pipeline operation. Compared with Fig. 24, this operating pressure has a slightly smaller stress value at the dent region and a small effect on the dent recovery of the pipeline. The maximum stress still occurs in the loading center of the inner surface and in the dent edge in the inner and outer surfaces of the steel pipe. The maximum equivalent stress at the loading point is 657.22MPa.

Fig. 25 Stress-strain distribution of dented pipe under 3.5MPa operating pressure

(3) Stress – strain distribution of dented pipe after removal of external force and operating pressure.

Fig. 26 presents the stress nephogram of pipes with 40mm dent after removal of external force and operating pressure. The findings include irreversible plastic deformation occurs at the loading point, elastic deformation moving away from the loading point can be fully recovered, and the stress value is the simulation error. At the dent region, the maximum equivalent stress values of the inner and outer surfaces are 595.33MPa and 400MPa, respectively, which are located at the edge of the steel pipe dent. The equivalent stress value of the dent center ranges approximately 330.74~396.88MPa at the inner surface and approximately 198.44~264.59MPa at the outer surface of the steel pipe.

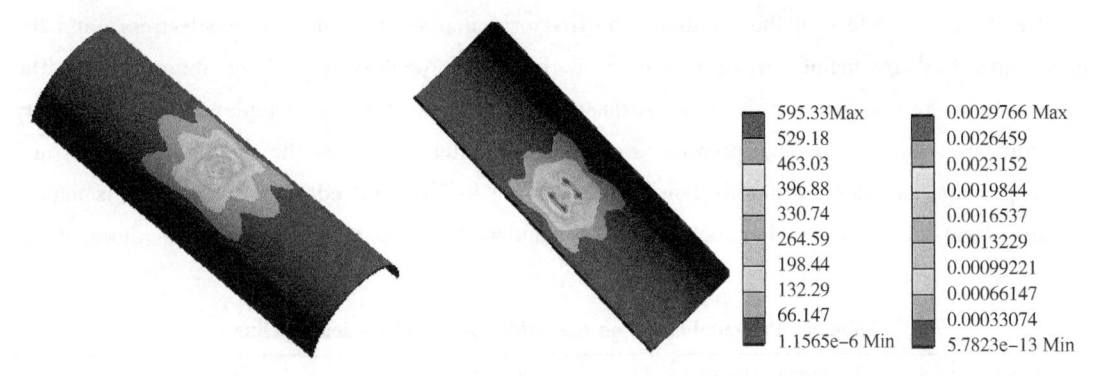

Fig. 26　Stress-strain distribution after removal of external force and operating pressure

(4) Stress simulation of the dented pipeline.

An operating pressure of 3.5MPa is applied to the pipe with a 40mm depression. Tables 5 and 6 show the stress nephograms of the first, second and third principal stresses at the inner and outer surfaces near the dent region, respectively. Figs. 27 and 28 show the stress direction schemes.

Table 5　Principal stress on the inner surface of dented pipeline

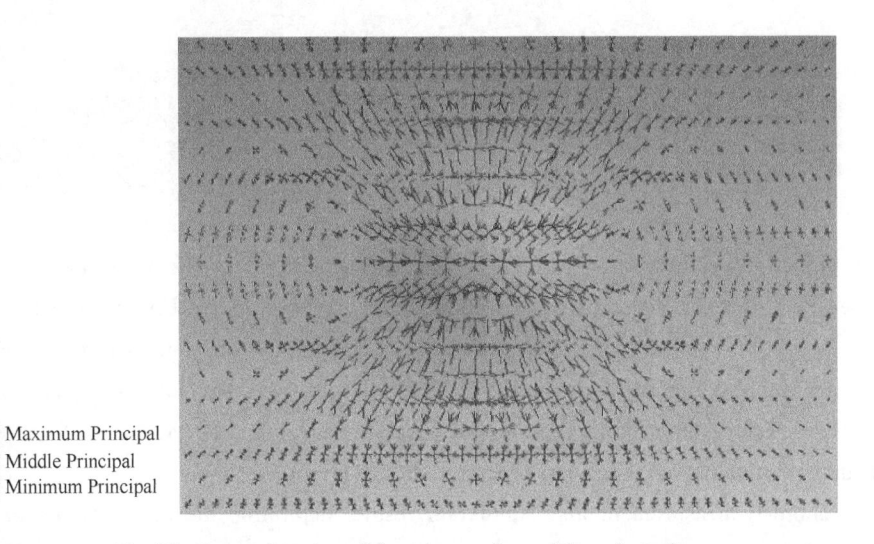

Location	First principal stress	Second principal stress	Third principal stress
Inner surface	223.59 Max 196.31 169.03 141.75 114.48 87.196 59.916 32.637 5.3576 −21.922Min	91.228Max 35.481 −20.265 −76.011 −131.76 −187.5 −243.25 −299 −354.74 −410.49Min	17.928Max −58.545 −135.02 −211.49 −287.96 −364.44 −440.91 −517.38 −593.86 −670.33Min

Fig. 27　Stress direction of the inner surface of dented pipeline

On the inner surface of the pipeline, the first principal stress is mainly tensilestress, and the second and third are mainly pressure stresses borne by the dent bottom at the inner surface. The stress state of the dent edge at the inner surface of the pipeline is highly complex, and the majority of regions are subjected to local pressure stress. On the outer surface of the pipeline, the first and second principal stresses are mainly tensile stress borne by the dent edge, and the third is mainly pressure stress borne by the dent bottom. These simulation results are consistent with those of the residual stress test.

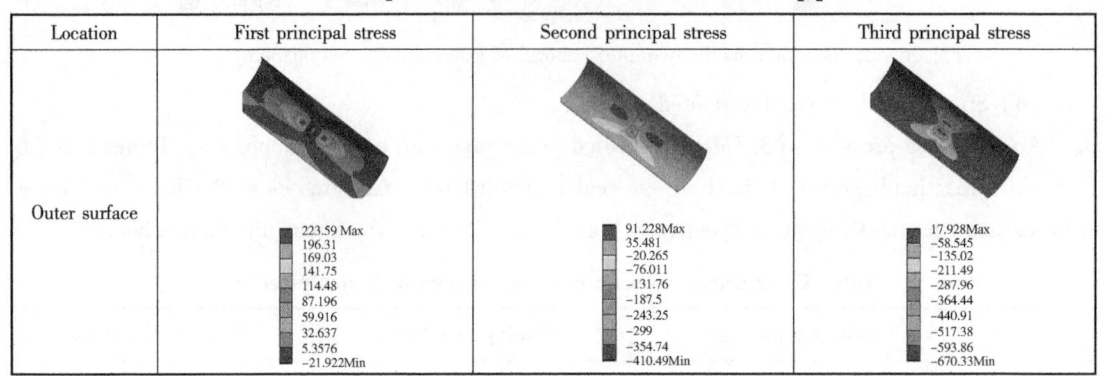

Table 6 Principal stress on the outer surface of dented pipeline

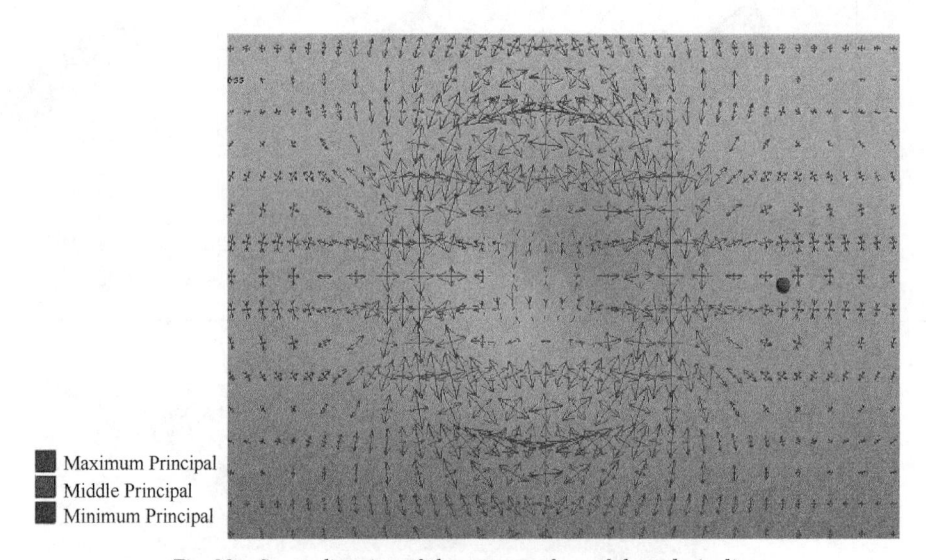

Location	First principal stress	Second principal stress	Third principal stress
Outer surface	223.59 Max 196.31 169.03 141.75 114.48 87.196 59.916 32.637 5.3576 −21.922 Min	91.228 Max 35.481 −20.265 −76.011 −131.76 −187.5 −243.25 −299 −354.74 −410.49 Min	17.928 Max −58.545 −135.02 −211.49 −287.96 −364.44 −440.91 −517.38 −593.86 −670.33 Min

Fig. 28 Stress direction of the outer surface of dented pipeline

The simulation results from steps (1) to (4) show that the dent effects on the stress distribution mainly focuses on the dent center and dent/bulge edge of the pipeline. Under the combined effect of external load and internal pressure, the maximum equivalent stress of dent edge of the outer wall and bulge top of the inner wall is greater than 650MPa, which exceeds the yield stress of the material. Micro-cracks then form because of the maximum deformation at the dent bottom/bulge top and maximum equivalent stress on top of the inner surface bulge.

The combined measurement and simulation results determine the cause of the dent leak. The dent formation is related to the rock during pipeline laying, during which the pipe body at the 6 o' clock position is jammed by the rock. Thus, this area is dented because of the pipe weight and compressed load of the backfill stone. The dent formation enhances the stress concentration on the dent bottom edge. However, the operating inner pressure has a small effect on the dent recovery of the pipeline. Micro-cracks form when the maximum equivalent stress at the dent bottom exceeds the yield stress of the material.

During practical service, the pipeline is subjected to alternating stress caused by the fluctuation of transport pressure (the actual operating pressure of the failed pipe segment ranges 2.5 ~ 3.8MPa), gas medium layering, external load vibration, and other factors. Micro-cracks appear on the pipe wall and gradually expand until they penetrate the wall thickness and result in pipeline fatigue failure.

5 Conclusions and suggests

On the basis of observation, testing, and analysis of failure, the conclusions are summarized as follows:

(1) The pipeline with dent leak meets the requirements of API specification 5L-2012 standard for carbon steel pipelines.

(2) The pipeline dent leak is a type of fatigue failure, and fatigue cracks started in the inner wall.

(3) The formation of pipe dent is related to the extruding deformation from a rock during pipeline laying. The dent region has clearly thinned wall thickness. The micro-cracks form at the deepest part of the dent region because of stress concentration. During practical service, the pipeline is subjected to alternating stress caused by the fluctuation of transport pressure, gas medium layering, external load vibration, and other factors. Micro - cracks appear on the pipeline and gradually expand until they penetrate the wall thickness and result in pipeline fatigue failure.

On the basis of these conclusions, the following measures are recommended to ensure the safe operation of pipelines:

(1) Before pipe laying, sundries and raffle should be removed, especially rocks with sharp corners, to ensure the evenness of the elementary layer.

(2) During the laying and backfilling processes, extra care should be used to prevent damage on the pipe surface. The pipeline should be slowly lowered via manual work and crane machine.

References

[1] X. Tian, H. Zhang, Failure pressure of medium and high strength pipelines with scratched dent defects, Engineering Failure Analysis 78(2017)29-40.

[2] M. Javidi, S. Bekhrad, Failure analysis of a wet gas pipeline due to localised CO_2 corrosion, Engineering Failure Analysis 89 (2018) 46-56.

[3] Q. Qiao, G. X. Cheng, W. Wu, et. al., Failure analysis of corrosion at an inhomogeneous welded joint in a natural gas gathering pipeline considering the combined action of multiple factors, Engineering Failure Analysis 64(2016) 126-143.

[4] X. Li, Y. Bai, C. L. Su, M. S. Li, Effect of interaction between corrosion defects on failure pressure of thin wall steel pipeline, International Journal of Pressure Vessels and Piping 138(2016)8-18.

[5] Y. F. Chen, H. Zhang, J. Zhang, et al., Failure assessment of X80 pipeline with interacting corrosion defects,

Engineering Failure Analysis 47 (2015) 67-76.

[6] Q. Qiao, G. X. Cheng, Y. Li, et. al., Corrosion failure analyses of an elbow and elbow-to-pipe weld in a natural gas gathering pipeline, Engineering Failure Analysis 82(2017)599-616.

[7] M. A. L. Hernández-Rodríguez, D. Martínez-Delgado, R. González, et al., Corrosive wear failure analysis in a natural gas pipeline, Wear 263 (2007) 567-571.

[8] E. Sadeghi Meresht, T. Shahrabi Farahani, J. Neshati, Failure analysis of stress corrosion cracking occurred in a gas transmission steel pipeline, Engineering Failure Analysis 18 (2011) 963-970.

[9] J. L. Sun, Y. F. Cheng, Modelling of mechano-electrochemical interaction at overlapped corrosion defects and the implication on pipeline failure prediction. Engineering Structures 212 (2020) 110466.

[10] B. Bedairi, D. Cronin, A. Hosseini, A. Plumtree, Failure prediction for Crack-in-Corrosion defects in natural gas transmission pipelines, International Journal of Pressure Vessels and Piping 96-97(2012) 90-99.

[11] Y. Y. Yang, M. j. Chai, Sealing failure and fretting fatigue behavior of fittings induced by pipeline vibration, International Journal of Fatigue 136 (2020) 105602.

[12] M. S. G. Chiodo, C. Ruggieri, Failure assessments of corroded pipelines with axial defects using stress-based criteria: Numerical studies and verification analyses, International Journal of Pressure Vessels and Piping 86 (2009) 164-176.

[13] B. C. Pinheiro, I. P. Pasqualin, Fatigue analysis of damaged steel pipelines under cyclic internal pressure, International Journal of Fatigue 31 (2009) 962-973.

[14] Y. Shuai, X. H. Wang, J. Shuai, et al., Mechanical behavior investigation on the formation of the plain dent of an API 5L L245 pipeline subjected to concentrated lateral load, Engineering Failure Analysis 108 (2020) 104189.

[15] N. I. I. Mansor, S. Abdullah, A. K. Ariffin, et al., A review of the fatigue failure mechanism of metallic materials under a corroded environment, Engineering Failure Analysis 42 (2014) 353-365.

[16] A. M. El-Batahgy, G. Fathyb, Fatigue failure of thermowells in feed gas supply downstream pipeline at a natural gas production plant, Case Stud. ies in Engineering Failure Analysis 1 (2013) 79-84.

[17] H. R. Hajibagheri, A. Heidari, R. Amini, An experimental investigation of the nature of longitudinal cracks in oil and gas transmission pipelines, Journal of Alloys and Compounds 741(2018)1121-1129.

[18] A. Fragiel, S. Serna, J. Malo-Tamayo, et al., Effect of microstructure and temperature on the stress corrosion cracking of two microalloyed pipeline steels in H2S environment for gas transport, Engineering Failure Analysis 105 (2019) 1055-1068.

[19] F. G. Liu, X. Lin, H. O. Yang, et al., Effect of microstructure on the fatigue crack growth behavior of laser solid formed 300M steel, Materials Science and Engineering: A 695(2017)258-264.

[20] H. S. Zhao, S. T. Lie, Y. Zhang, Elastic-plastic fracture analyses for misaligned clad pipeline containing a canoe shape surface crack subjected to large plastic deformation, Ocean Engineering 146 (2017) 87-100.

[21] O. I. Zvirko, S. F. Savula, V. M. Tsependa, et al., Stress corrosion cracking of gas pipeline steels of different strength, Procedia Structural Integrity 2(2016) 509-516.

[22] D. Q. Wu, H. Y. Jing, L. Y. Xu, et al., Theoretical and numerical analysis of the creep crack initiation time considering the constraint effects for pressurized pipelines with axial surface cracks, International Journal of Mechanical Sciences. 141(2018)262-275.

[23] M. Zheng, J. H. Luo, X. W. Zhao, et al., Effect of pre-deformation on the fatigue crack initiation life of X60 pipeline steel, International Journal of Pressure Vessels and Piping 82 (2005) 546-552.

本论文原发表于《Engineering Failure Analysis》2020 年第 113 卷。

Failure Analysis on Tee Pipe of Duplex Stainless in an Oilfield

Zhang Shuxin

(Tubular Goods Research Institute, China National Petroleum Corporation & State Key Laboratory for Performance and Structure Safety of Petroleum Tubular Goods and Equipment Materials; School of Materials Science and Engineering, Northwestern Polytechnical University)

Abstract: During the hydraulic test in the station of an oil field, a tee-pipe of S32205 duplex stainless burst into several fragments without reaching the target pressure. In order to clarify the failure cause, visual inspection, chemical composition, mechanical test, hardness test, Charpy impact test, heat treatment, and finite element analysis were carried out in this paper. The results show the chemical composition of tee meets the requirements of the standard for S32205 duplex stainless steel, and the tensile property, hardness, Charpy impact and ferrite content of harmful phase do not meet the requirements of reference standard. The presence of sigma phase in the tee which induced by the improper heat treatment after cold forming makes the material hard and brittle. In the case of micro defects, the pressure bearing capacity of tee joint will be reduced, resulting in brittle cracking. Therefore, the quality control of tee should be strengthened, the heat treatment should be implemented strictly and nondestructive inspection of tee can be carried out such as field micro-hardness test and metallographic examination. Considering the burst failure case of tee pipe was rarely reported so far, the findings of this paper would provide useful quality control advice in the future.

Keywords: Tee pipe; Duplex stainless steel; Cold forming; Burst failure

1 Introduction

With the continuous development and production of oil and gas fields, the water content in the oil and gas continues to rise, and the corrosion problem of transmission pipelines will become more and more serious[1]. In China's western oil field, due to its high contentof Cl^-, H_2S, CO_2 and high salinity in the hydrates of oil and gas, it brings great challenges to the equipment's integrity. At present, the corrosion mitigation methods commonly used in the station for the process pipeline are: adding corrosion inhibitor for carbon steel pipeline to alleviate the corrosion to some extent[2], adopting

---

Corresponding author: Zhang Shuxin, zhangshuxin003@cnpc.com.cn.

bi-metal clad pipe, using stainless steel namely 316L and S32205. While the carbon steel with corrosion inhibitor cost much, and mechanical clad pipe is prone to lining collapse[3] and circumferential weld failure, 316L austenitic stainless steel is prone to stress corrosion cracking with high content of Cl^{-}[4]. S32205, duplex stainless steel (ferrite+austenite, the recommended ratio is 1∶1)[5-7] have been a choice due to its excellent mechanical property and pitting resistance. The selection of S32205 new-type material equipment solves the corrosion problems but brings new problems.

Cesar Roberto de Farias Azevedo et al.[8] systematically summarized the duplex stainless steel failure cases. The S32205 material is mainly used in the field of petroleum pipelines and vessels. The main failure modes are pitting and stress corrosion cracking. The failures mainly occur at the weld joint, either in the heat affected zone or in the fusion zone. Almost 50% of failures are related to the precipitation of harmful phases, such as sigma-phase (σ), chi-phase (χ), carbide, and nitride[9-11]. There is merely no case of pipe fittings burst reported, especially tee-pipe. Tee-pipe is used to change the direction of the fluid. The amount of the tee is relatively small at the field. It has the problem of poor quality control and prone to failure. Tee burst, this kind of failure mode is different from pitting and stress corrosion cracking. The burst releases a large amount of energy and may even cause casualties. Therefore, it should be paid attention to.

The failure case in this study is that during the pipeline hydraulic pressure test in the station of an oil field, when the pressure rises to 22MPa, the tee exploded. The appearance of the failed tee is shown in Fig. 1. The tee is made of S32205 duplex stainless steel, DN300×10mm, the design pressure is class 900 (16MPa), the target pressure is 24MPa. The manufacturing process of the tee is cold forming from a seamless pipe and then solution treatment. In order to find out the root cause of the failed tee, the macro and micro morphology of the sample and the physical and chemical properties of the material were tested and

Fig. 1 The appearance of burst tee pipe in the oil field

analyzed, finite element analysis was conducted to find the load distribution, the validation heat treatment was conducted to evaluate its manufacturing process. Finally, the failure reasons of the sample were analyzed comprehensively.

2 Experimental procedure

In order to find out the burst causes of the tee pipe, a series of characterization method was adopted to evaluate the material property. The visual inspection was conducted to find the burst initiation point. The chemical composition were detected by ARL4460 photoelectric direct reading spectrometer, nitrogen content was measured by LECO TC600 oxygen/nitrogen determinator. The UTM-5305 and SHT 4106 Mechanical testing machine was used to test the tensile strength and yield strength of the tee pipe at room temperature. The PIT302D (300J) impact machine was used to test the Charpy impact energy of the main pipe of tee at different temperature, the Charpy impact energy of -40℃ was used to evaluate the detrimental intermetallic phase in duplex austenitic/ferrite stainless steels according to ASTM A 923-2014. The KB30BVZ-FA Vickers hardness and RB 2002 Rockwell hardness testing machine was used to test the hardness of the base material. Besides, the OLS-4100 laser confocal microscope was used to observe the microstructureof the base metal of the tee. The fracture surface and crack growth of the fracture were analyzed bymacroscopic morphology and TESCAN VEGA II scanning electron microscope (SEM). Finally, the finite element analysis was conducted to simulate the load distribution of the tee pipe and subsequent heat treatment was conducted to verify the manufacturing process.

3 Results

3.1 Visual inspection

The failed tee burst into 6 fragments, marked as 1#-1 ~ 1#-6, as shown in Fig. 2. Typical fragments, such as tee branch fragments, and main pipe abdominal fragments are shown in Fig. 3 and 4. From a macro perspective, the edge of the fragments has almost no plastic deformation characteristics. The fracture shows a chevron cracking pattern. It is a typical macroscopic feature of brittle fracture, and formed by the rapid propagation of the crack after the tee-pipe bursts. Crack propagation path is drawn along the direction of chevron pattern convergence, as shown by the red arrow in Fig. 2. Therefore, the initiation point of the crack can be identified as the neck of the branch and the abdomen of the main pipe.

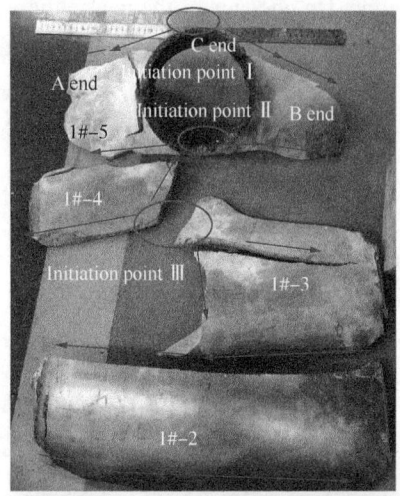

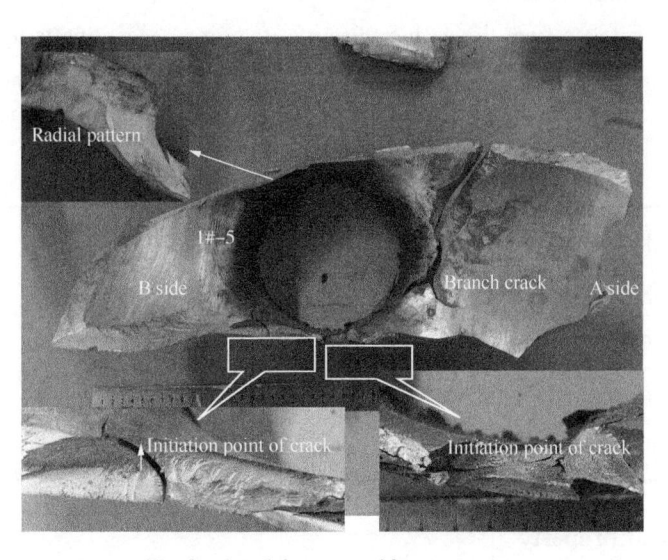

Fig. 2 Fragments of burst tee pipe for analysis

Fig. 3 1#-5 fragment of burst tee pipe

The macroscopic appearance of branch neck fragment is shown in Fig. 3. There are two longitudinal penetrating cracks on the broken fragment, and a horizontal branch crack on the longitudinal fracture on one side. Generally, the circumferential pressure of the pipeline is considered to be greater than the longitudinal pressure, so the tee pipe cracks along the longitudinal direction. On one side of the neck, there is a volume peeling on the outer wall of the neck with radial pattern converging to the neck. On the other side, there are 2 branch cracks besides the main crack, one has an angle with the thickness direction, and the other is almost parallel to the wall thickness direction. So it can be inferred that the internal coupling force of the wall thickness direction of the tee neck is low.

Fig. 4 shows themacroscopic appearance of $1^{#}-4$ bursting fragment. It can be seen that the fragment contains two crack initiation points, and the initiation point I and II have obvious chevron pattern, converging to the center of wall thickness. The direction of arrow indicates the direction of crack growth. Therefore, the crack of the tee pipe mainly originates from the abdomen of the main pipe and the neck of the branch pipe.

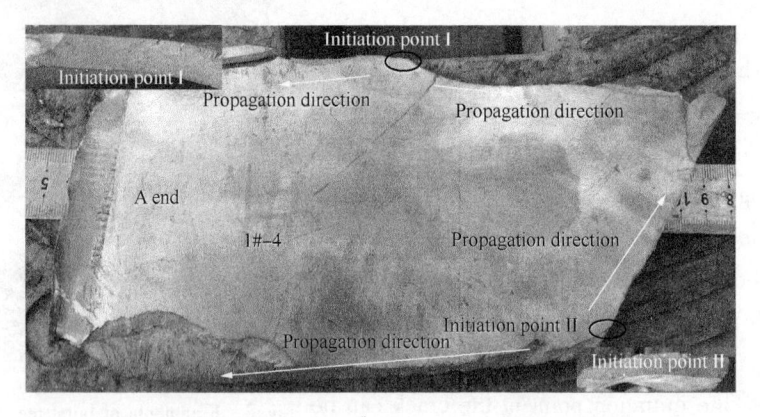

Fig. 4　Fracture view of $1^{#}-4$ fragment of burst tee pipe

3.2　Chemical composition

Three samples was taken from the each fragment of tee pipe, according to the standard ASTM A 751-14a, the chemical composition was tested (Table 1). Compared with the standardASTM A 790/A790M, the chemical composition were in accordance with the requirement of S32205 (Table 1).

Table 1　Chemical composition of the tee pipe (wt. %)

Sample area	C	Si	Mn	P	S	Cr	Mo	Ni	N
$1^{#}-1$	0.022	0.53	1.44	0.017	0.0012	23	3.1	5.6	0.18
$1^{#}-3$	0.024	0.53	1.44	0.017	0.0013	22.8	3.1	5.6	—
$1^{#}-5$	0.022	0.55	1.46	0.018	0.0012	22.6	3.2	5.7	0.19
S32205[12]	≤0.03	≤1.00	≤2.00	≤0.03	≤0.02	22.0~23.0	3.0~3.5	4.5~6.5	0.14~0.20

3.3　Mechanical property

In order to test the longitudinal and transverse tensile properties of the tee pipe, the tensile test was carried out according to the standard ASTM A370. Plate and rod tensile specimens were acquired from the failed tee pipe (Fig. 7). The test results are shown in Table 2. The tensile strength and yield strength of the failed parts met the requirements ofASTM A790/A790M. However, fracture elongation was much lower than the standard limit, indicating that the tensile properties of the failed tee did not meet the standard.

Table 2 Tensile test of the tee pipe

Sample	Diameter/Width×gauge (mm×mm)	Tensile strength R_m (MPa)	Yield strength $R_{t0.5}$ (MPa)	Yield strength tensile strength ratio	Elongation after break A (%)
1#-1 transverse sample	6.25×25	886	650	0.73	4
		879	641	0.73	5
		880	653	0.74	4
1#-1 longitudinal sample	25.4×50	Break at the clamping end			
		Break at the clamping end			
		862	677	0.79	2
S32205[12]		≥655	≥450	—	≥25

Fig. 5 shows the specimen of tensile test. It can be seen that the fracture of transverse and longitudinal specimen is straight and perpendicular to the direction of stress, without "necking" plastic deformation trace. One longitudinal tensile specimen is broken at the parallel section and two broken at the clamping end. Radial pattern can be observed from the fracture, converging to the sample surface.

(a)   (b)

Fig. 5 Transverse and longitudinal tensile specimen after break of 1#-1 tee pipe's fragment

3.4 Hardness test

According to ASTM E18, the specimen was taken from the 1#-1 fragment of the failed tee pipe. The hardness was tested along the longitudinal direction. The measured hardness results are shown in Table 3. The hardness of the tee piperanged from 25.7 HRC to 37.5 HRC and the average hardness was 32.9 HRC. The results show that the hardness of the main pipe is not uniform. Generally, the maximum allowable hardness of tee pipe should blow 290HBW (after conversion, 300HV10, 30HRC, approximately). It can be seen that the maximum hardness of the tee exceeds the requirements.

Table 3 Rockwell hardness test of the tee pipe (HRC)

Sample	1	2	3	4	5	6	7	8	9	10	11
1#-1	25.7	26.6	28.0	28.3	37.5	36.7	36.3	36.1	36.0	35.3	35.5
S32205[12]	≤290HBW (300HV10, 30HRC)										

3.5 Charpy impact test/detrimental phase analysis

A Charpy V-notch impact test was conducted to determine the impact toughness. It is generally believed that a more brittle material and a smaller impact energy correspond to a higher risk of fracture. In contrast, greater impact energy and better material toughness mean a lower risk of cracking.

Three V-notch transverse specimens were machined from the 1#-1 fragment of tee pipe. The transverse impact toughness test was performed according to the standard ASTM E23. The test results are shown in Table 4. The fracture of the processed charpy impact specimens are shown in Fig. 6.

Table 4 Charpy impact test result

Sample	Dimension(mm)	Temperature (℃)	Absorbed energy (J)			Average absorbed energy(J)	Minimum value of single specimen(J)
1#-1 transverse specimen	10×10×55	-40	1	1	1	1	1
		20	2	2	1	1.7	1
S32205[13]	10×10×55	-40	—			≥70	≥54

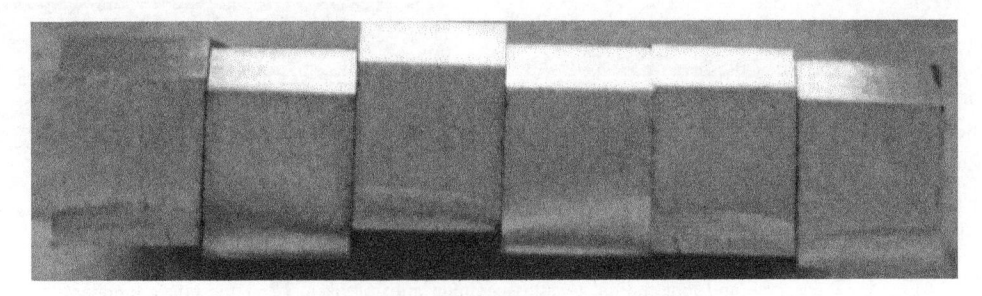

Fig. 6 1#-1 tee pipe Charpy impact fracture (room temperature)

The absorbed energy is really low at the room temperature and -40℃. The fracture surface is brittle and almost has no ductile zone.

According to method B in ASTMA923[13], the Charpy impact energy at -40℃ can be used to evaluate the detrimental phases in the S32205 duplex stainless steel. The minimum value required is 54J. This indicated that the tee pipe must have a certain amount of detrimental phases, and did not met the requirement for the detrimental phase.

3.6 Metallographic structure

The transverse metallographic specimen was taken from the 1#-1 main pipe. The specimen was etched by $K_4Fe(CN)_6$/KOH at 60℃ for 15min. Result is shown in Fig. 7. The metallographic structure of the main pipe is "ferrite (dark)+austenite (white)+sigma (brown)". At the boundary between the ferrite and austenite phases, a large number of brown σ phases are distributed, and the ferrite phase is greatly reduced[14]. The σ phase has a tetragonal phase structure and is precipitated from the ferrite phase, which is consistent with the reduction of the ferrite content. Furthermore, the tetragonal structure of σ phase is hard and brittle. A small amount of σ phase can greatly reduce the toughness and plasticity of the material, which is consistent with the poor impact toughness of the tee pipe.

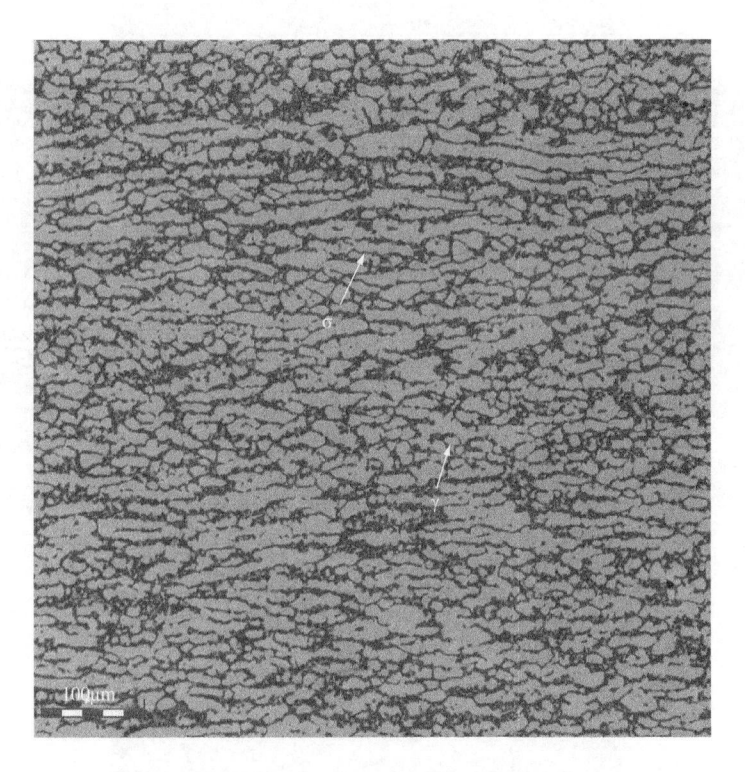

Fig. 7 Transverse metallography of 1#-1 tee pipe

3.7 Fracture analysis

Through the observation of the abdominal fracture of the main pipe of 1#-3 tee, it can be seen that the overall surface of the fracture is flat with radial crack growth pattern. On the inner side of the tee, the color of this area is blackened, showing typical oxide inclusion characteristic, as shown in the Fig. 8 (b). Observing the high magnitude image of the crack growth zone, "lamellar" and "crystalline" quasi cleavage characteristics can be seen, showing typical brittle fracture characteristics, as shown in the Fig. 8 (c). In addition, discontinuous micro-cracks with length 1mm, width 100μm can be observed along the wall thickness direction.

The results of energy spectrum analysis show that the near-surface zone of the fracture contains

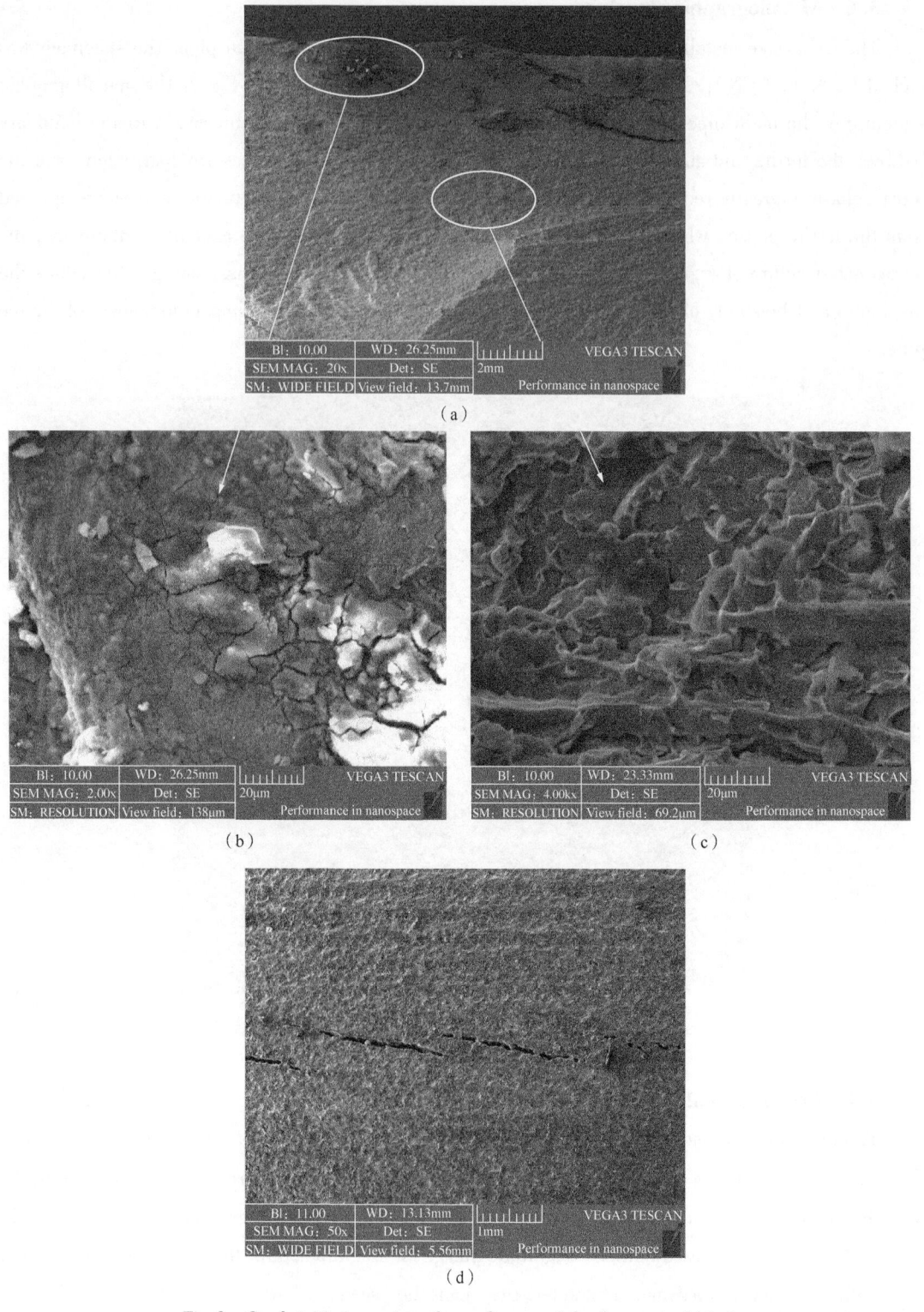

Fig. 8 Crack initiation pointand growth zone of the fragment of 1#-3

more O, Si, K and a small amount of Al elements, while the crack growthzone contains less O, Si, K and no Al element is found, as shown in Fig. 9 and Table 5.

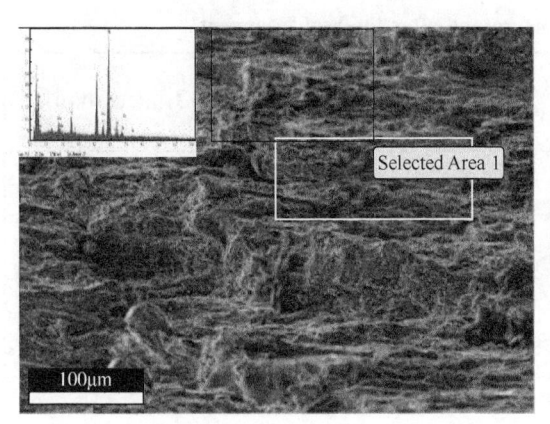

Fig. 9　Result of the EDS of fragment 1#-3

Table 5　Charpy impact test result

\multicolumn{3}{c	}{1#-3 crack growth zone}	\multicolumn{3}{c}{1#-3 initiation point}
Element	Weight(%)	Atomic(%)	Element	Weight(%)	Atomic(%)
O	8.65	24.44	O	41.94	68.87
Mo	2.64	1.24	Al	1.39	1.36
K	3.59	4.15	Si	2.28	2.13
Cr	23.87	20.75	Mo	3.83	1.05
Mn	0.93	0.77	K	12.92	8.68
Fe	56.26	45.53	Cr	5.89	2.98
Ni	3.85	2.96	Fe	31.75	14.94

3.8　Verification of heat treatment

Generally, S32205 has a typical ferrite+austenite dual-phase structure with good strength and toughness, but the structure, Charpy impact and hardness results of the tee-pipe indicate a brittle material property, so it is inferred that the heat treatment process is not suitable. Therefore, the transverse Charpy impact sample was used for solution treatment to verify the heat treatment process. The solution treatment process is : heating to 1030℃, holding for 30min, water cooling. Charpy impact test and hardness test are carried out on the sample after heat treatment. The results of the tests (Table 6) meet the requirements of the standard, indicating that the material performance was improved with the reasonable heat treatment process(Fig. 10).

Fig. 10　The sample and impact fracture after heat treatment

Table 6 The property difference before and after the heat treatment

Sample	Charpy impact energy(J) (test temperature: −40℃)	Shear area ratio(%)	Hardness (HRC or HV10)
Tee pipe before heat treatment	1.7	0	27.8~43.5 HRC
Tee pipe after heat treatment	258	100	260 HV10
Requirement of ASTM A923& ASTM A815/A815M for S32205	average⩾70	avergae⩾85	⩽290HV10(30HRC)

4 Discussion

The tee failed during the hydrostatic test. To analyze the root cause of the tee burst failure, thestress distribution, manufacturing process, and material properties of the tee should be analyzed.

4.1 Stress analysis

The pressure capacity of the tee can refer to the calculation formula of pipeline.

$$p = \frac{2t\sigma}{D}$$

Where: p refers to the bearing pressure, MPa. D refers to the diameter, 300mm. t refers to the thickness of the pipe, 10mm. σ refers to the yield strength, 650MPa.

The calculated bearing pressure of the tee pipe is about 43MPa, which is far exceed the bursting pressure 22MPa. Therefore, the hydraulic test pressure is not the root cause of the failure.

According to the structure of the failed tee, a 3D model of the tee is established by finite element analysissoftware. The tee is simplified as two ideal equal diameter hollow cylinders intersecting orthogonally and filleted. A uniform internal pressure load of 22MPa is applied on the inner surface of the model. Branch pipe and main pipe are fixed. The load distribution of the tee is shown in Fig. 11. The dangerous part of the tee bearing internal pressure is the inner surface of the neck of the branch tee, and the stress on the outer surface of the abdomen is high.

According to the results of macro fracture analysis, the crackinginitiation pointof burst tee is located in the neck of the branch and the abdomen of the tee pipe, which is consistent with the simulation results, indicating that burst tee cracks in the weak area.

4.2 Manufacturing process

The tee pipe was manufactured by cold forming method which is different from pipeline, as shown in Fig. 12. After cutting and placing raw pipe into a die, the pipe is pressed as hydraulic pressure pushes out the branch. And then, solution treatment was carried out. Generally, the solution

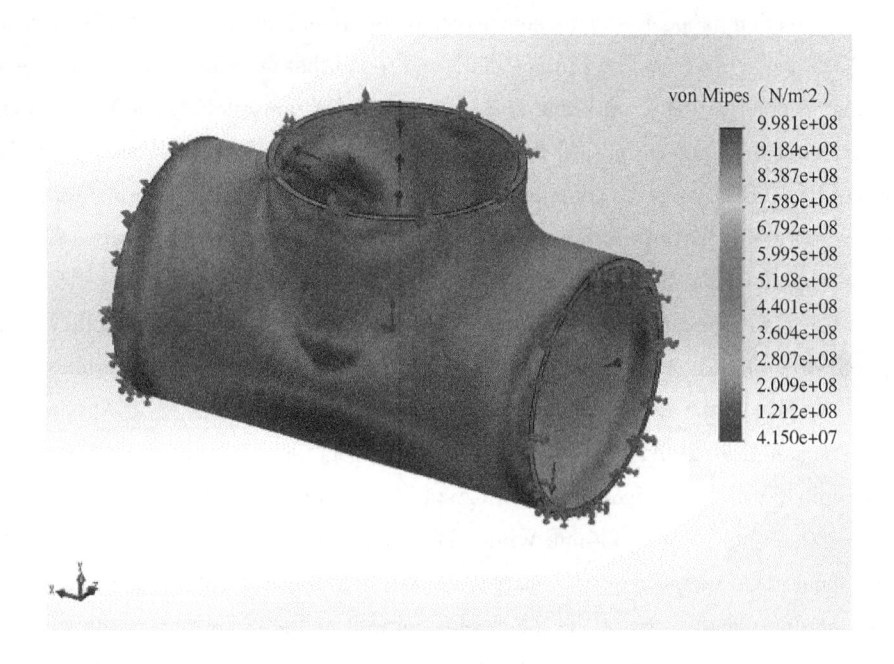

Fig. 11 VonMises load distribution of tee pipe

treatment process of S32205 duplex stainless steel is: heating to 1000~1100℃, holding for 30min, then cooling in water quickly. The "900℃ brittleness" and "475℃ brittleness" shall be avoided in the heat treatment process. When the duplex stainless steel is heated at 700~975℃, σ brittle phase is easy to precipitate and at 350~525℃, body centered cubic α' phase precipitates easily, resulting in material brittleness.

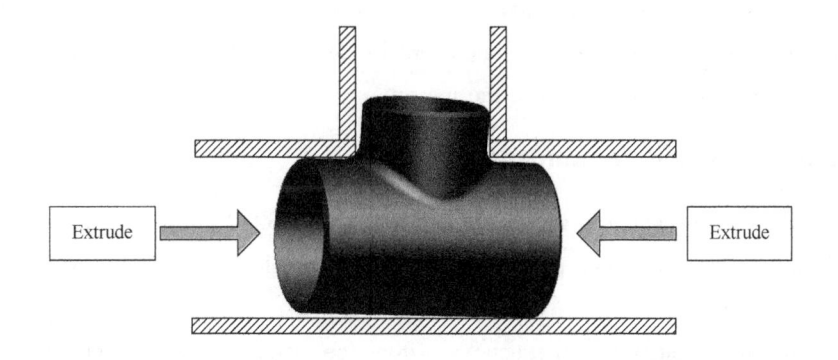

Fig. 12 The scheme of tee pipe manufacturing process

The cold forming of tee pipe was a process of extrusion. The tee pipes usually need to control the quality including shape and size, surface quality, structure. When the factors of process, mold, temperature, lubrication and other factors are not properly controlled, various defects will appear in the final tee. There is volume exfoliation in the neck at theinitiation point of the burst tee, and the cracks extend along the wall thickness. At the same time, the micro-cracks can be seen along the wall thickness direction under SEM observation. It can be inferred that the binding force through the wall thickness direction is poor and defects exist. Wang Jianze et al. studied the surface peeling defect of S32205 steel strip rolled by furnace coil, and found that the micro-cracks was generated by the uncoordinated deformation of austenite and ferrite at high strain rate[15]. Ma Yueyue etal. studied the failure of as-rolled S32205 cylindrical flange, multiple micro cracks were observed by SEM [16]. The defect may be the folding defect[17] generated during the tee pipe extrusion deformation process, or it may be that micro cracks have formed inside the material during the extrusion process.

From the experiments, it can be seen that due to the existence of the σ phase, the S32205 tee pipe material has high strength and poor plasticity, mainly manifested in high tensile strength, high hardness, and low impact energy. Hitchcock et al.[18] studied the micro hardness of ferrite, austenite, and σ phase, the results show that the σ phase has a hardness of 567kgf/mm^2, ferrite is 344kgf/mm^2, austenite is 290kgf/mm^2. σ phase is intermetallic compound based in the system of iron and chromium with a tetragonal crystallographic structure[19~21]. The lattice parameter of σ phase[22] is $a = 0.879$nm, $c = 0.434$nm. While the lattice parameter of ferrite is about 0.286nm, austenite is about 0.358nm. There is a large mismatch between σ phase and austenite, which increases the hardness and decrease the toughness, as well as the elongation of the steel, and even changes the fracture type from transgranular to intergranular. Børvik et al.[14] studied the impact toughness of S32205 at different temperatures and sigma levels. The result shows that the higher the sigma level is, the lower the impact toughness is. The burst tee pipe has an impact energy of 1J at room temperature, which indicate the amount of σ phase nearly reach the maximum. Combined with the heat treatment verification, it can be infer that the tee pipe was subjected the wrong heat treatment process. Further, the pipe was air-cooled, not the water-cooled.

From the above analysis, tee pipe generated micro cracks in the manufacturing process, and adopted improper heat treatment which led to the presence of σ phase, made the material hard and brittle. Under the hydraulic pressure, the micro cracks decreased the bearing capacity, and finally led to brittle cracking at the weak area of the tee pipe.

5 Conclusions and recommendations

In this study a failure analysis was performed ona burst S32205 duplex stainless steel tee pipe which failed during the hydraulic pressure test without reaching the target pressure. The main conclusions are as follows:

(1) The failed tee burst into 6 fragments, from the visual inspection from the fracture, the initiation point of the crack can be identified as the neck of the branch and the abdomen of the main pipe. Further, the finite element analysis verifies that these two areas' stress is high. The tee burst at the weak area.

(2) The chemical composition of tee meets the requirements of the standard for S32205 duplex stainless steel, while the results of tensile property, hardness, Charpy impact and ferrite content of harmful phase are not in accordance with the requirements of the standard and the results indicate the material is hard and brittle. The metallographic structure shows that there's sigma phase in the tee which can greatly reduce the toughness and plasticity of the material. Combined with the heat

treatment verification, it can be infer that the tee pipe was subjected the wrong heat treatment process. Further, the pipe was air-cooled, not the water-cooled.

(3) The visual inspection and SEM analysis demonstrate that there is micro-crack along the thickness of the tee, which result in a low internal coupling force. The defect may be the folding defect generated during the tee pipe extrusion deformation process.

(4) The root cause of the tee pipe burst failure isthat the improper heat treatment of tee joint makes the material hard and brittle, in the case of micro defects, the pressure bearing capacity of tee joint will be reduced, resulting in brittle cracking.

In order to avoid such failure from happening again, for the manufacturer, the heat treatment should be implemented strictly and ultrasonic testing could be adopted to detect the surface defect. For the user, the field micro-hardness test, metallographic examination, ferrite test by ferritescope can be carried out to guarantee the quality of the tee.

Acknowledgement

The authors are grateful to the fund support of National Key R&D Program of China (2017YFC0805804), andall members in Tubular Goods Research Institute who assisted in carrying out this failure analysis study. Our sincere thanks to Mr. Ding H. for his advice to share this work and Ms. Yan X. for her tremendous support to me.

References

[1] PengHu, Daoyi Chen, Mucong Zi, et al. Effects of carbon steel corrosion on the methane hydrate formation and dissociation, Fuel 230(2018)126-133.

[2] T. Hong, Y. H. Sun, W. P. Jepson. Study on corrosion inhibitor in large pipelines under multiphase flow using EIS, Corroso Scio, (2002) 44(1): 101-112.

[3] Lin Yuan, Stelios Kyriakides, Liner wrinkling and collapse of bi-material pipe under bending, International Journal of Solids & Structures, 2015, 51(3-4)(2014)599-611.

[4] You, Yiliang, Zhang, Zheng, Ma, Luoning. Cracking analysis of 316L stainless steel lining plates in alkaline environments, Engineering Failure Analysis, 2014, 39: 34-40.

[5] K. Devendranath Ramkumar, A. Bajpai, S. Raghuvanshi, et al., Investigations on structure - property relationships of activated flux TIG weldments of super-duplex/austenitic stainless steels, Mater. Sci. Eng., A 2015, 638: 60-68.

[6] J. C. de Lacerda, L. C. Cândido, L. B. Godefroid, Effect of volume fraction of phases and precipitates on the mechanical behavior of UNS S31803 duplex stainless steel, Int. J. Fatigue, 2015, 74: 81-87.

[7] H. B. Cui, G. M. Xie, Z. A. Luo, et al. Microstructural evolution and mechanical properties of the stir zone in friction stir processed AISI201 stainless steel, Mater. Des., 2016, 106: 463-475.

[8] C. R. F. Azevedo, H. B. Pereira, S. Wolynec, A. F. Padilha, An overview of the recurrent failures of duplex stainless steels, Eng. Fail. Anal., 2019, 97: 161-188.

[9] J. Nowacki, A. Łukojc, Microstructural transformations of heat affected zones in duplexsteel welded joints, Mater. Charact., 2006, 56: 436-441.

[10] R. Badji, M. Bouabdallah, B. Bacroix, et al. Effect of solution treatment temperature on the precipitation kinetic of σ-phase in 2205 duplex stainless steel welds, Mater. Sci. Eng., A 2008, 496: 447-454.

[11] S. M. Yang, Y. C. Chen, C. H. Chen, et al. Microstructural characterization of δ/γ/σ/γ2/χ phases in silver-doped 2205 duplex stainless steel under 800℃ aging, J. Alloys Compd. , 2015, 633: 48-53.

[12] ASTM A790/790M, Standard Specification for Seamless and Welded Ferritic/Austenitic Stainless Steel Pipe, 2018

[13] ASTM A923, Standard Test Methods for Detecting Detrimental Intermetallic Phase in Duplex Austenitic/Ferritic Stainless Steels, 2014

[14] T. Børvika, H. Lange, L. A. Marken, et al. Pipe fittings in duplex stainless steel with deviation in quality caused by sigma phase precipitation, Materials Science and Engineering A. , 2010, 527(26): 6945-6955.

[15] Wang J Z, Qian Z X, Ji X B. Study on surface peeling defect of 2205 duplex stainless steel. Hot Working Technology, 2016, 45 (17): 256-258.

[16] Ma, Y. -Y. , Yan, S. , Yang, Z. -G. , Qi, G. -S. & He, X. -Y. Failure analysis on circulating water pump of duplex stainless steel in 1000MW ultra-supercritical thermal power unit. Engineering Failure Analysis, 2015, 47: 162-177.

[17] Sudhakar Mahajanam, Hernan Rincon, Dale McIntyre. Metallurgical examination of defects in duplex stainless steel pipe fittings, Materials performance, 2010, 49(4): 56-60.

[18] Hitchcock, G. R. , Deans, W. F. , Thompson, D. S. & Coats, A. Pin-hole and crack formation in a duplex stainless steel downhole tool. Engineering Failure Analysis2001, 8: 213-226.

[19] Jordan, P. , Maharaj C. , Asset management strategy for HAZ cracking caused by sigma-phase and creep embrittlement in 304H stainless steel piping. Engineering Failure Analysis, 2020, 110: 104452.

[20] Bahrami, A. , Ashrafi, A. , Rafiaei, S. M. , & Mehr, M. Y. , Sigma phase-induced failure of AISI 310 stainless steel radiant tubes. Engineering Failure Analysis, 2017, 82: 56-63.

[21] Villanueva, D. M. E. , Junior, F. C. P. , Plaut, R. L. , & Padilha, A. F. , Comparative study on sigma phase precipitation of three types of stainless steels: austenitic, superferritic and duplex. Materials Science and Technology, 2006, 22(9), 1098-1104.

[22] Maj, P. , Adamczyk-Cieslak, B. , Nowicki, J. , Mizera, J. & Kulczyk, M. Precipitation and mechanical properties of UNS 2205 duplex steel subjected to hydrostatic extrusion after heat treatment. Materials Science and Engineering: A 734, (2018) 85-92 .

本论文原发表于《Engineering Failure Analysis》2020 年第 115 卷。

Cracking Analysis of a Newly Built Gas Transmission Steel Pipe

Ding Han[1,2]   Qi Dangtao[1]   Qi Guoquan[1,2]   Wei Bin[1]   Ding Nan[1]   Li Houbu[1]
Wang Fushan[3]   Yan Zifeng[3]   Feng Quan[3]   Zhang Zhihao[4]   Sun Yinjuan[4]

(1. Tubular Goods Research Institute, China National Petroleum Corporation&State Key Laboratory
for Performanceand and Structure Safety of Petroleum Tubular Goods and Equipment Materials;
2. Department of Applied Chemistry, School of Science, Northwestern Polytechnical University;
3. Tarim Oilfield Company, PetroChina Company Limited;
4. Xi'an Changqing Technology Engineering Co. Ltd.)

Abstract: A longitudinal crack was found in the welding seam during the X-ray inspection of a newly built L245NS gas transmission pipeline in China western oilfield. The crack causes were analyzed by the nondestructive tester (NDT), direct-reading spectrometer, tensile strength test machine, impact test machine, Vickers hardness tester, optical microscope(OM), macroscopic fracture morphology, scanning electron microscopy(SEM) and energy spectrum analysis (EDS) in this paper. The results show that the crack is caused by the original defect. First, the fracture surface of the crack is characterized by multi-source cracking, and the origin of the crack is containing foreign matters. Second, the surface of crack fracture is blue, which indicates that the crack fracture has experienced high - temperature burning, so the crack may be formed before heat treatment. The main reason for steel pipe crack is the low stress cracking at the pipe end by the "slag inclusion" formed in the pipe body when the pipe is processing.

Keywords: Steel pipe; Cracking; Weldingseam; Slag inclusion

1 Introduction

As a mature product, the steel pipeline is widely used in the oil and gas industry. Due to the harsh environment of oil and gas fields, the produced medium usually has strong corrosivity with high-temperature and high-pressure (HTHP)[1]. For steel pipes, it is faced with multiple cracking failure risks, such as stress corrosion cracking (SCC)[2-6], sulfide stress cracking (SSC)[7-9], hydrogen-induced cracking (HIC)[10], hydrogen assisted cracking (HAC)[11], and Corrosion fatigue crack propagation (CFCP)[12]. Etc.

The line pipe steel manufacturing operations, appropriate metallurgical processes leading to low

---

Corresponding author: dinghan@ cnpc. com. cn.

residual stress (6.2% YS), relatively low fraction of high angle boundaries (about 0.75) and predominant {110}<110> texture in the material (or in the near surface) are expected to greatly improve the stress corrosion cracking resistance of line pipe steels[2,3]. Wang[4] studied the influence of copper on the SCC of 304 stainless steel because of the passive film became weak in high temperature, and corrosion caused by cupric ions was the main reasons of pit corrosion. Majchrowicz[5] explored the susceptibility of P110 pipeline steel to SCC in CO_2-rich environments, which indicated that the susceptibility of P110 steel to SCC in CO_2-rich environments is associated with activity of anodic dissolution process and hydrogeninvolvement. Li[6] studied the effect of microstructural aspects in the heat-affected zone of high strength pipeline steels on the stress corrosion cracking mechanism in acidic soil environment. HIC and sulfide stress cracking SSC are two main types of hydrogen embrittlement (HE) phenomena. HIC crack can occur in the absence of external stress, while SSC failure results from the combined actions of corrosion and applied stress[8].

Also, many scholars[13-15] have studied the causes of cracking of other kind of pipe or equipment used in oil and gas fields, but there are few reports on the cracking of the newly builtsteel pipe. The failure mode and mechanism of thelong-term service pipeline closely related to the complex service conditions. However, the newly built pipeline failure mostly due to the internal defects. In this paper, the cracking reason fora newlybuilt gas pipeline in an oil and gas field whichdid not put into operation has been studied.

In this work, a longitudinal crack was found in the base metal around the weld joint of a gas transmission pipeline. The gas pipeline is a newly built project in the peripheral block of HLHT oilfield, it has not yet put into operation. The material of the pipe is L245NS PSL2 seamless steel, with the dimension of ϕ219mm×6.5mm. The reference standard of product quality certificate is GB/T 9711-2011[16]. The crack pipe with girth weld is marked as sample1#, the cracked end is side A, the non-cracked end is side B. For comparative analysis, we also took the same batch pipe sample with side Aof sample 1#, which is marked as sample2#, as shown in Fig. 1.

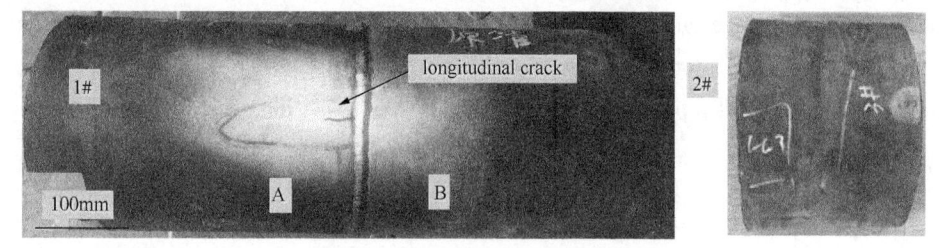

Fig. 1 Thesample visual appearance of steel pipe

2 Experimental

In order to find out the crack reasons of the newly built steel pipe, a series of experiments were carried outon the steel pipe, the crack reasons of the sample were analyzed comprehensively. The nondestructive test(NDT) of fluorescent magnetic particle detection was adopted near the crack. The chemical compositions were detected by ARL4460 photoelectric direct reading spectrometer. The UTM-5305 Mechanical testing machine was used to test the tensile strength and yield strength of the crack side of the samples at room temperature. The PIT752D-2(300J) impact machine was used to test the impact energy of the both weld side of sample 1# and sample 2# at room temperature and $-30^\circ\!C$ respectively. The KB30BVZ-FA Vickers hardness testing machine was used to test the HV_{10} hardness of the base material, weld, and the heat-affected zone (HAZ) of sample 1#crack side. Besides, the OLS-4100 laser confocal microscope was used to observe the microstructure of the base material, fracture, and crack. Finally, the crack and the fracture surface of the crack were analyzed by optical microscope (OM), scanning electron microscope (SEM), and energy spectrum analysis(EDS).

3 Results and discussion

3.1 Nondestructive testing

According to ASTM E709 - 2015[17], the results of fluorescent magnetic particle testing (FMPT) near the crack shown in Fig. 2. It can be seen from the test results that the axial linear indication with a length of 60mm is found near the base metal of the weld and that there were no other cracks or defects that were found in this area.

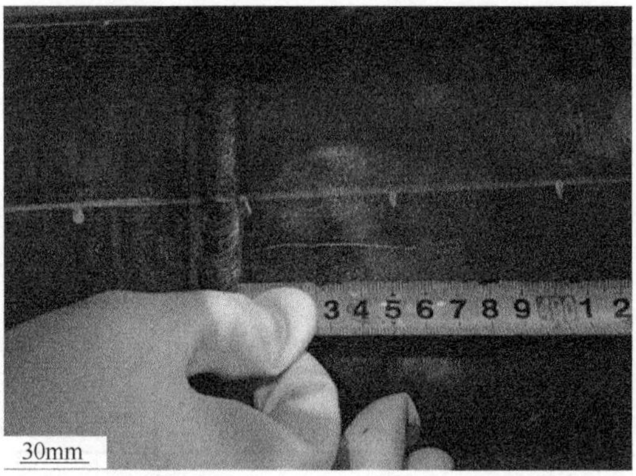

Fig. 2 Penetration detection for the sucker rod

3.2 Chemical composition

Spectral analysis for material composition at each side of sample 1#'s girth weld as A and B, and the sample 2# as shown in Fig. 1. The results show that the chemical composition of the side A and side B of sample 1#, and the sample 2# all meet the requirements of GB/T 9711—2011 for L245NS steel pipe(Table. 1).

Tab 1 Chemical composition of the steel pipe(wt. %)

Element	C	Si	Mn	P	S	Cr	Mo	Ni	Ti
GB/T9711-2011	≤0.24	≤0.40	≤1.40	≤0.025	≤0.015	—	—	—	≤0.04
Sample 1#(A)	0.097	0.31	1.29	0.013	0.0019	0.13	0.021	0.044	0.0019
Sample 1#(B)	0.10	0.31	1.30	0.014	0.0019	0.13	0.022	0.044	0.0019
Sample 2#	0.096	0.33	1.27	0.0064	0.0017	0.042	0.0042	0.032	0.0018

3.3 Mechanics property analysis

The tensile test results of the side A of sample 1# at room temperature are shown in Table 2. The results showed that the tensile strength, yield strength, and elongation are meet the requirements of the GB/T 9711—2011. The stress-strain curves of the tensile test shown in Fig. 3, the mechanical properties are consistent with each tensile test sample.

Table 2 Tensile test results at room temperature

Sample	Size(mm)	Tensile Strength(MPa)	Yield Strength(MPa)	Elongation(%)
Sample 1#(A)	25×50	543	397	37
		575	418	37
		535	394	37
GB/T 9711—2011		415~760	245~450	≥24

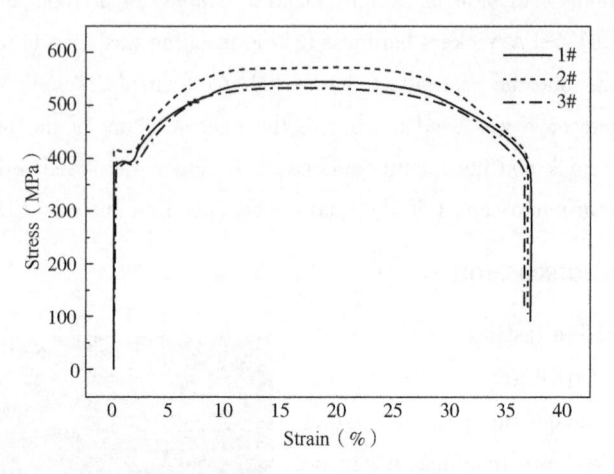

Fig. 3 Stress-strain curves of the tensile sample

The impact energy test results of the sample 1#(A), sample 1#(B), and sample 2# at 0℃ and -30℃ are shown in Table 3. According to the test results, the impact performance of the steel pipe meets the requirements of GB/T 9711—2011 standard for L245NS steel pipe.

Table 3 Impact test result (J)

Sample	Size(mm)	Notch	Temperature(℃)	Measured	GB/T 9711—2011
Sample 1#(A)	3.3×10×55	V	0	54	≥13.4J
				51	
				58	
Sample 1#(B)				56	
				57	
				59	
Sample2#				59	
				55	
				59	
Sample 1#(A)			-30	58	
				55	
				54	
Sample 1#(B)				55	
				55	
				54	
Sample 2#				54	
				57	
				57	

The HV$_{10}$ hardness test results of the base metal, heat-affected zone (HAZ), and weld are shown in Table 4. According to the Vickers hardness test results, we can clearly see the hardness of the base metal, HAZ and weld are gradually increased, but they are meet the requirement of the GB/T 9711—2011.

Table 4 Vickers hardness test results (HV$_{10}$)

Position	Measure				Average value
Base metal	170	190	160	183	176
HAZ	181	187	197	193	190
Weld	179	187	218	225	202
GB/T 9711—2011	≤250				

3.4 Macroscopic morphology analysis of the fracture

The sample was cut and prepared from the longitudinal crack area of the pipe after the sample was opened along with the crack, the fracture morphology as shown in Fig. 4. The fracture surface near the outer surface of the pipe is relatively flat, which is characterized by brittle cracking, and ductile cracking near the inner surface. Indicating that the crack propagates from the outside to the inside of the steel pipe's cross-section. According to fracture analysis theory, the center of arc is the source of fracture. At the same time, defects were found at the crack source, which also verified that the fracture originated from the outer surface of the steel pipe. It can be seen from Fig. 4, the crack sources are located on the outer surface of the steel pipe, that is characterized by multi-source cracking. According to the distance from the weld, we marked the two sources as origin 1# and origin 2#.

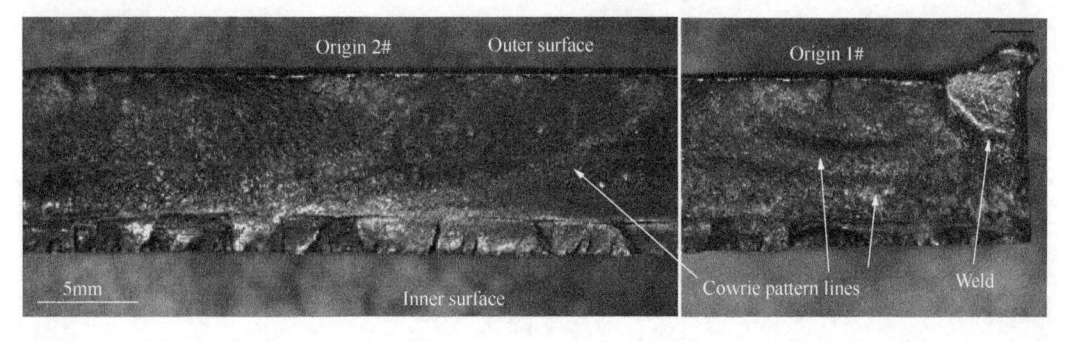

Fig. 4 Macroscopic fracture morphology

The morphology of the source area near the welding seam is amplified as shown in Fig. 5. Through further observation, we found obvious defects at the origin 1#. The surface of the fracture has the characteristics of "bluing" and attachment of oxidation products. The "bluing" character indicates that the fracture has undergone high-temperature oxidation, and it also indicates the fracture has been cracked for a long time. It can be inferred that the cracks may have appeared before the heat treatment of the steel pipe.

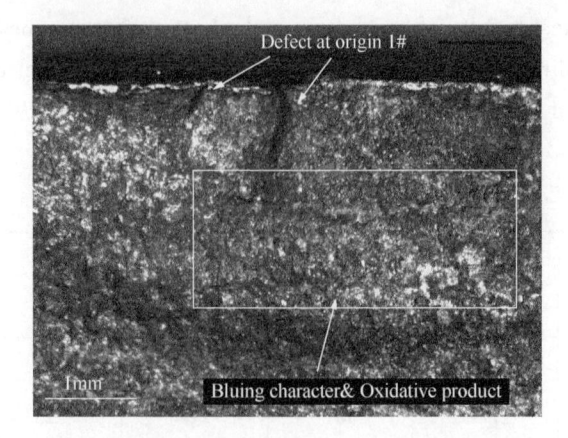

Fig. 5 Macroscopic morphology of the fracture source area and surface character

3.5 Metallographic analysis

The fractured sample was cut and prepared for the metallurgical examination. The metallographic structure at the side of the fracture is pearlite and ferrite, and the fracture metallographic structure is deformed locally. There has gray matter attached to the fracture surface, as shown in Fig. 6(a) and Fig. 6(b). Through further observation of the metallographic structure near the fracture, it was found

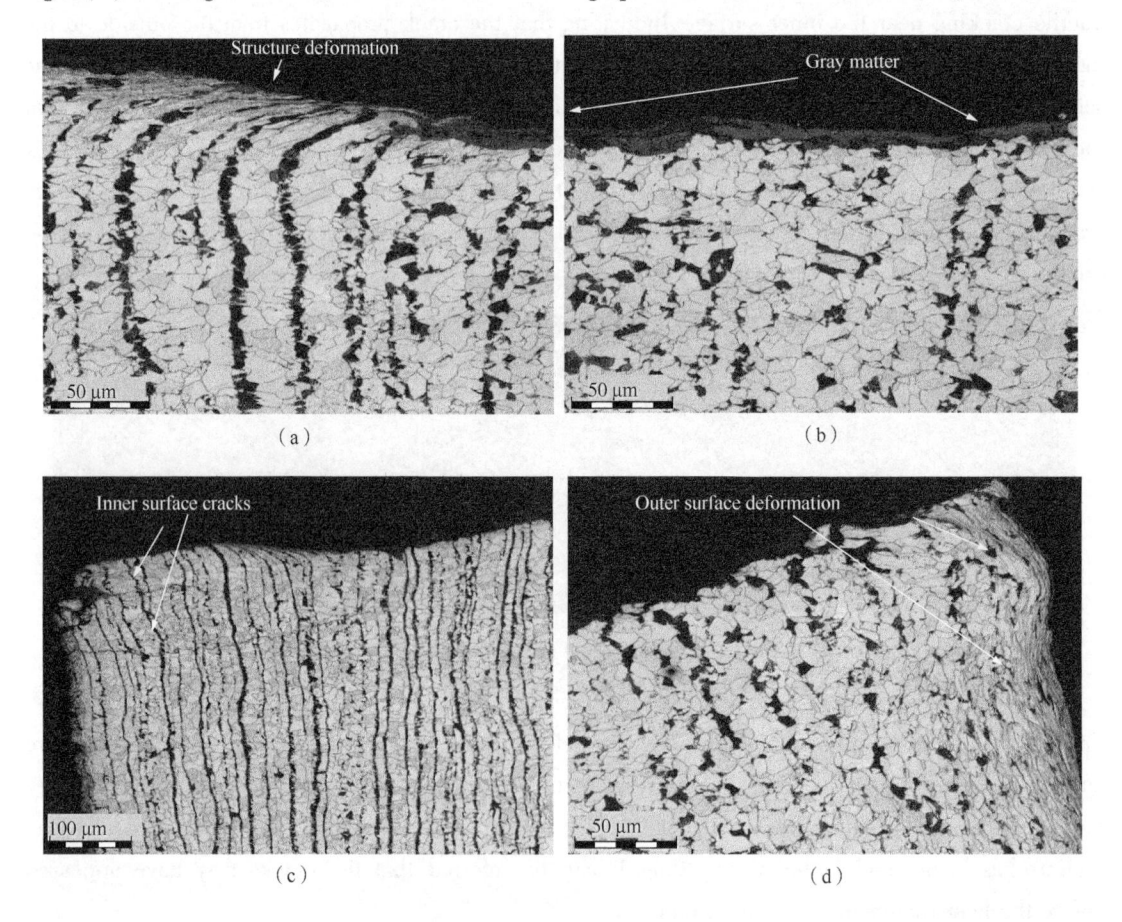

(a)

(b)

(c)

(d)

Fig. 6 Metallurgical structure of the fracture

that there were crack defects near the inner surface of the steel pipe and the metallographic structure deformation near the outer surface.

After enlarging the longitudinal crack tip of the steelpipe, it was found that there were also gray matter inclusions inside the crack, as shown in Fig. 7. Therefore, it can be inferred that the gray matter existed long before the steel pipe cracked.

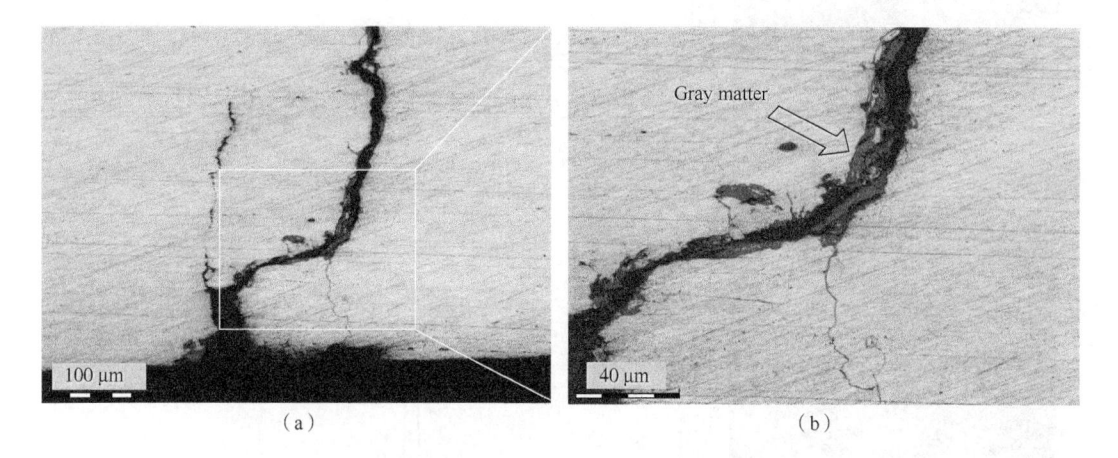

Fig. 7 Morphology of the longitudinal crack tip inside

3.6 Electron microscope analysis

The SEM photo of the fracture as shown in Fig. 8, a layer of dense oxide can be seen on the surface of the fracture, and a chapped appearance showing at high power. The fracture surface, the gray matter at the inner surface of the fracture, and the gray matter in the longitudinal crack tip were analyzed by EDS as shown in Fig. 9, Fig. 10, and Fig. 11, respectively. As a result, that the fracture surface is mainly composed of C, O, K, and Fe elements, the gray matter in the fracture is mainly composed of elements of C, K, Fe, Cl, and Si, the gray matter at the crack tip is mainly composed of elements of C, O, K, Fe, Na, Mn, and Si.

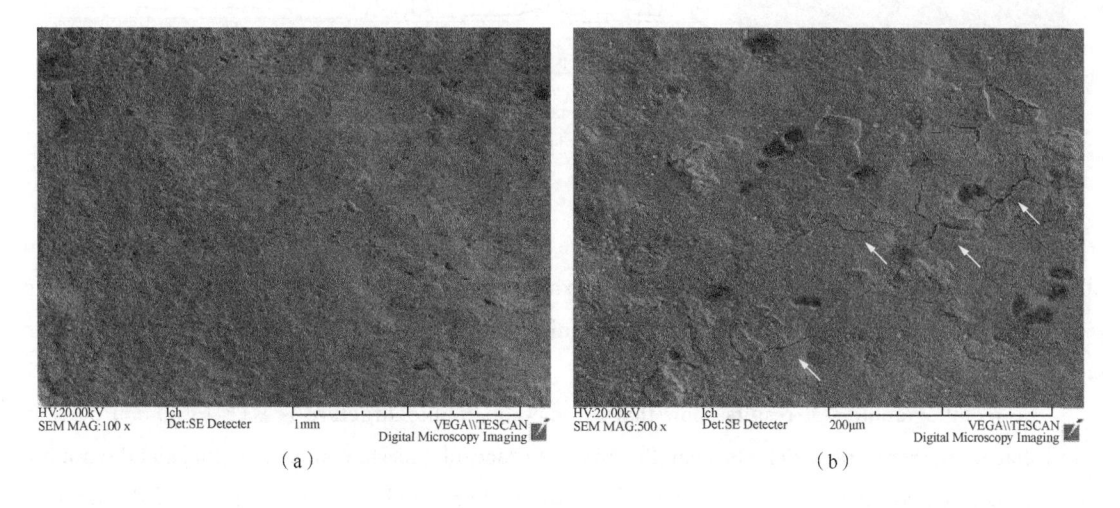

Fig. 8 SEM photo of the fracture surface

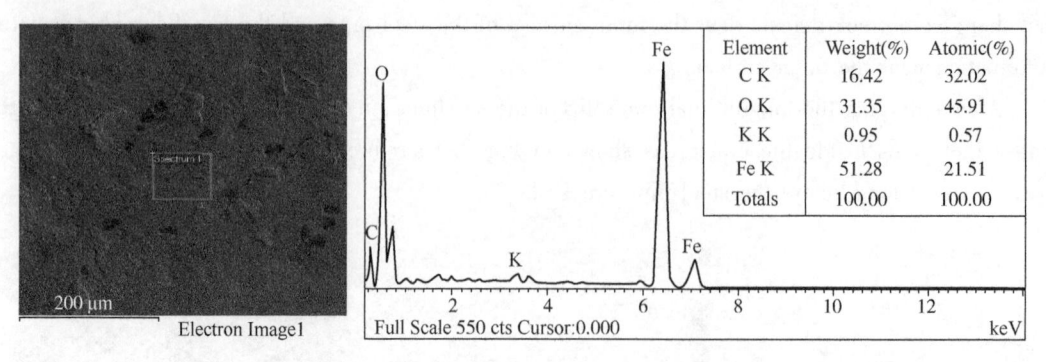

Fig. 9　EDS result of the fracture surface

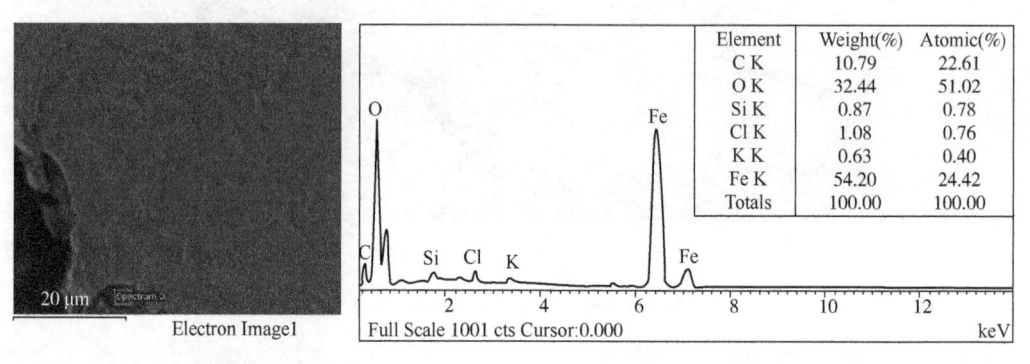

Fig. 10　EDS result of the gray matter on inner fracture surface

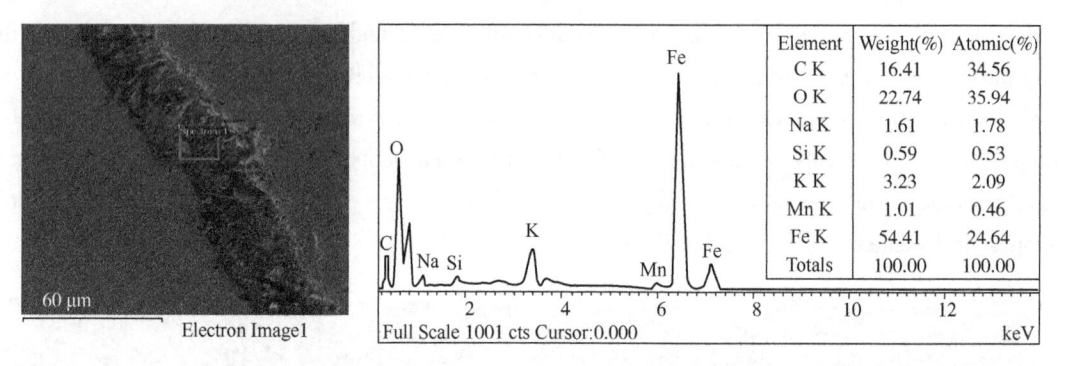

Fig. 11　EDS result of the gray matter in the longitudinal crack tip

4　Root cause analysis

Based on the test results and analysis above, the NDT test results showedthat there are no obvious defects on the outer surface of the pipe except for the cracking part. It can be seen from the test results that the chemical composition, mechanical properties of the pipe meet the requirements of GB/T 9711—2011 for L245NS pipeline steel.

The metallographic test results show that the metallographic structure is P+F (Fig. 6.), it was found that there were crack defects near the inner surface of the steel pipe and the metallographic structure deformation near the outer surface. The gray matter founded in the both side of the fracture surface and the longitudinal crack tip. It can be inferred that the gray matter existed long before the steel pipe cracked.

The gray matter at the inner surface of the fracture and the longitudinal crack tip are shown that there are a lot of C, O, K and Si elements besides Fe elements. By analyzing the source of each element, it can be inferred that the O element should come from the oxidation reaction with the foreign matter or oxygen in the air with iron at high temperature during the manufacturing process, and the K and Si element might come from the foreign matter (such as slag), indicating that there is "slag inclusion" in the cracked part of the failed pipe section. It is because of the existence of "slag inclusion" in the steel pipe, the structure of the steel pipe at this part is uneven, the intergranular bonding force of metal is poor, that iseasily to crack. The defect position becomes the crack initiation point, which may lead to the crack propagation in the subsequent transportation process.

It is also can be seen from the macro morphology of the fracture which just started at the slag inclusion part and showed the characteristics of multi-source cracking. The surface of the fracture has the characteristics of "bluing" and attachment of oxidation products. The "bluing" character indicates that the fracture has undergone high-temperature oxidation. The "bluing" character and the gray matter found in the longitudinal crack tip both indicated that the fracture has been cracked for a long time. It can be inferred that the cracks may have appeared before the heat treatment of the steel pipe.

To sum up, the chemical composition, mechanical properties, and metallographic structure of the pipe sample meet the requirements of GB/T 9711—2011 for L245NS pipeline steel. The cracking of the pipe section is caused by the slag inclusion in the steel pipe, which leads to the cracking.

5 Conclusions and recommendations

(1) The chemical composition and mechanical properties of the pipe samples submitted for inspection meet the requirements of GB/T 9711—2011.

(2) The original defects in the steel pipe are the main reasons for the cracking of the pipe.

(3) It is suggested to strengthen the supervision of product quality and check the same batch of pipe samples on-site.

Acknowledgement

The authors are deeply grateful to the financial support from Tubular Goods Research Institute, China National Petroleum Corporation & State Key Laboratory for Performance and Structure Safety of Petroleum Tubular Goods and Equipment Materials. Their permission to use the investigation results for publication is also acknowledged.

References

[1] A. Shadravan, M. Amani, What every engineer or geoscientist should know about high pressure high temperature wells, SPE Kuwait International PetroleumConference and Exhibition, 2012, Kuwait City, Kuwait, 2012.

[2] Olivier Lavigne, et al. Microstructural and mechanical factors influencing high pH stress corrosion cracking susceptibility of low carbon line pipe steel Engineering Failure Analysis, 2014, 45: 283–291.

[3] M. A. Mohtadi-Bonab. Effects of Different Parameters on Initiation and Propagation of Stress Corrosion Cracks in Pipeline Steels: A Review. Metals 9(2019) 590.

[4] Xuehan Wang, et al. The influence of copper on the stress corrosion cracking of 304 stainless steel. Applied Surface Science. 2019, 478: 492-498.

[5] Kamil Majchrowicz, et al. Exploring the susceptibility of P110 pipeline steel to stresscorrosion cracking in CO_2-rich environments. Engineering Failure Analysis. 2019, 104: 471-479.

[6] Xueda Li, et al. Effect of microstructural aspects in the heat-affected zone of high strength pipeline steels on the stress corrosion cracking mechanism: Part I. In acidic soil environment. Corrosion Science. 2019, 160: (11), 108167.

[7] Z. H. Zhang, et al. A systematical analysis with respect to multiple hydrogen traps influencing sulfide stress cracking behavior of API-5CT-C110 casing steel. Materials Science & Engineering A. 2018, 721: 81-88.

[8] M. Luo, et al. Effect of tempering temperature at high temperature zone on sulfide stress cracking behavior for casing steel. Engineering Failure Analysis. 2019, 105: 277-236.

[9] X. T. Wang, et al. Effects of chromium and tungsten on sulfide stress cracking in high strength low alloy 125 ksi grade casing steel. Corrosion Science 160 (2019) 108163.

[10] M. A. Mohtadi-Bonab, et al. Role of cold rolled followed by annealing on improvement of hydrogen induced cracking resistance in pipeline steel. Engineering Failure Analysis. 2018, 91: 172-181.

[11] E. Viyanit, et al. Hydrogen assisted cracking of an AISI 321 stainless steel seamless pipe exposed to hydrogen-containing hot gas at high pressure. Engineering Failure Analysis 100 (2019) 288-299.

[12] Ming Liu, et al. Corrosion fatigue crack propagation behavior of S135 high-strength drill pipe steel in H2S environment. Engineering Failure Analysis 97 (2019) 493-505.

[13] A Q Fu, et al. Failure analysis of girth weld cracking of mechanically lined pipe used in gasfield gathering system. Engineering Failure Analysis, 2016, 68: 64-75.

[14] Mohsen Dadfarnia, et al. Assessment of resistance to fatigue crack growth of natural gas line pipe steels carrying gas mixed with hydrogen. International Journal of Hydrogen Energy. 2019, 44: 10808-10822.

[15] Qi Guoquan, et al. Analysis of cracks in polyvinylidene fluoride lined reinforced thermoplastic pipe used in acidic gas fields. Engineering Failure Analysis. 2019, 99: 26-33.

[16] GB/T 9711—2011, Petroleum and natural gas industries-Steel pipe for pipeline transportation systems.

[17] ASTM E709—2015, Standard Guide for Magnetic Particle Testing.

本论文原发表于《Engineering Failure Analysis》2020 年第 118 卷。

Microstructure and Impact Toughness of X70-L245 Butt Girth Weld in Natural Gas Station

Ren Junjie[1,2] **Ma Weifeng**[1] **Hui Wenying**[3] **Tong Ke**[1]
Nie Hailiang[1] **Wang Ke**[1] **Dang Wei**[1] **Cao Jun**[1]

(1. State key Laboratory for Performance and Structure Safety of Petroleum Tubular Goods and Equipment Materials, CNPC Tubular Goods Research Institute;
2. School of Materials Science and Engineering, Northwestern Polytechnical University;
3. PetroChina West Pipeline Company)

Abstract: At present, the numerical simulation and safety evaluation of girth welds are based on the properties of tubular materials, which cannot truly reflect the properties of girth welds. Based on the investigation results of girth weld in the station, this paper took X70 (21mm) and L245 (12mm) joints with large material and wall thickness gap in the station as research objects, prepared girth weld according to the welding process in the project site, and studied the structure and impact toughness of the joints. The microstructure of weld zone was analyzed by optical microscope. Charpy impact toughness increased from the center of weld to both sides, and the center of weld was the lowest. On X70 side, the impact toughness of FGHAZ is the highest and the impact energy is about 200 J, which is significantly higher than the weld center. The impact toughness near the fusion line on side L245 is the highest, but significantly smaller than that on side X70. Grain refinement, interlacing structure and M/A structure are conducive to the improvement of impact toughness. In this kind of girth weld, the center of the welding seam is the weakest area, and the X70 side has a significantly higher toughness.

1 Introduction

After long service of oil and gas transmission pipeline, many cases of oil and gas transmission pipeline failure were caused by weld defects. In China, the failure of pipelines caused by the welds is particularly prominent. In recent 10 years, with the construction and production of high-steel large-caliber pipelines, more than 30 girth weld cracking and leakage accidents occurred in the pipeline pressure testing stage and the initial operation stage, more than 70% of which were caused by girth weld defects. In addition, girth weld failure accidents occur frequently in storage and transportation facilities of oil and gas transmission stations [1-3]. In 2011, girdle-weld butt joint

---

Corresponding author: Ren Junjie, renjunjie@cnpc.com.cn.

between flange and process pipelines of the compressor import and export in a natural gas pipeline compressor station burst during the pressure test. The whole workshop almost be destroyed, causing serious economic losses. It can be seen that the in-service safety and failure prevention and control technology of girth weld has become an important engineering problem facing the current pipeline safety in service. It can be seen that the in-service safety and failure control technology of girth weld in natural gas station yard has become an important link of current pipeline system safety.

Different from the girth welding joints of long-distance pipelines, the girth welds in the natural gas station have the characteristics of different steel grades, different wall thickness and different pipe fitting types due to a large number of joints between different pipes [4]. The investigation results show that the difference between the two butt joints of the reduced tee -- straight pipe butt girth weld is relatively larger, and the X70-L245 joint is one of the typical forms of the larger difference [5]. Compared with the same material and the same wall thickness butt joint, this butt joint form will cause different thermal fields on both sides under the same heat input condition, which will lead to different performance of heat affected zones (HAZ) on both sides. However, there are few researches on the strength and toughness of girth weld.

Impact toughness can better reflect the strength and toughness characteristics of girth weld. Based on the investigation results, this paper prepared the butt joint with different wall thickness X70-L245, investigated the distribution rule of impact toughness of girth welds, and analyzed it in combination with the organization structure. The results can provide a basis for failure analysis, numerical simulation and safety evaluation of girth weld in natural gas station.

2 Experimental procedure

2.1 Welds preparation

The girth weld was prepared by welding X70 pipe to a L245 pipe. Table 1 shows the details specifications of the materials. The butt joint was welded according to Q/SY GJX 0221 - 2012 "*Welding specification for station pipes of West - east gas transmission third line*" [6] with groove angle of 60 degree. The welding was done by manual semi-automatic welding using ER50-6 welding rod (ϕ2.5mm) on the root welding and the E6015-Ga welding rod (ϕ2.5mm) on the filling and cover welding.

Table 1 Material specifications for weld joints

Materials	X70	L245
Pipe diameter	ϕ1016mm	ϕ1016mm
Wall thickness	21mm	12mm
Manufacturing standards	API Spec 5L	GB/T 9711

2.2 Test

Because the weld seam area, especially the heat-affected area, has a small width in the pipeline axial direction, in order to better obtain the performance distribution in these areas, charpy impact test samples adopted a smaller thickness of 3.3mm, so as to minimize the occupation of grooves in different structure areas. The samples were cuted from three positions of near outer surface, middle of pipe wall and near inner surface. The metallographic analysis were carried out with a Leica MeF3A metallographic microscope and a Olympus confocal laser microscope.

3 Results and discussion

3.1 Metallographic analysis

Fig. 1 shows the macroscopic metallographic picture of the welded joint. It can be seen that there are obvious background welding, filling welding, cover welding layer and HAZ at the girth weld joint. The metallographic structure photos of welding seam backing weld, filling weld and covering weld layer are shown in Fig. 2.

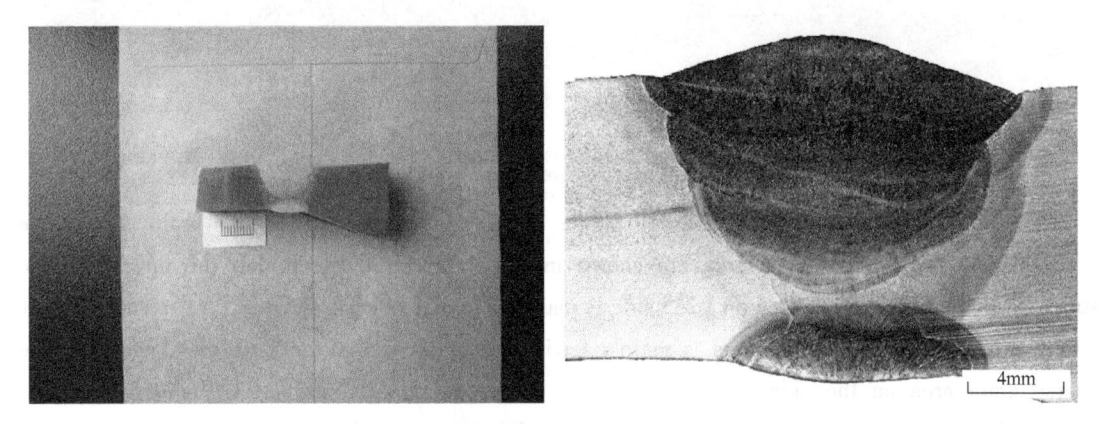

Fig. 1　Macrostructure of girth weld joint

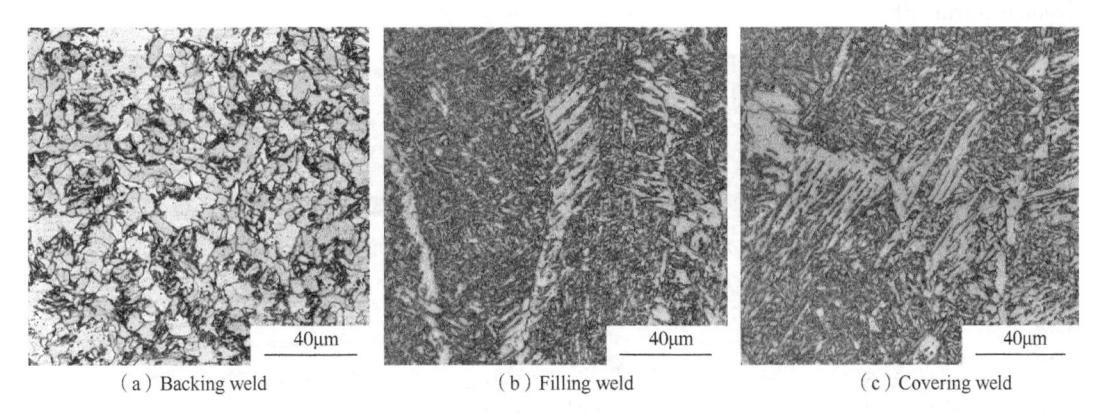

(a) Backing weld　　　(b) Filling weld　　　(c) Covering weld

Fig. 2　Macrostructure of girth weld joint

In order to study the properties of different areas in the heat-affected zone, the microstructure of the heat-affected zone of girth weld joints was observed. Figure 3 shows the metallographic structure of the HAZ on both sides of the weld. It can be seen that the HAZ can be divided into coarse crystal zone, fine crystal zone and two-phase zone according to the microstructure. The width of L245 side HAZ [Fig. 3(a)] is about 2mm, among which the coarse crystal zone is about 1.1mm, the fine crystal zone is about 0.6mm, and the two-phase zone is about 0.3mm. The coarse grain area is large, the fine grain area is small, and the two-phase area is the mixed structure of the fine grain and the original structure of the parent material. It can be seen from Fig. 3(b) that the width of heat affected zone on the side of X70 is about 1.5mm, which can also be divided into the of the coarse crystal zone can be used to infer the fusion line (L2/R2). By comparing Fig. 4(a) and Fig. 4(d), it can be seen that compared with X70 side, the coarse crystal zone on side L245 is mainly PF+P+GB, with interlaced structures and smaller grains, while the grain size on side X70 is significantly larger, so the impact toughness of L2 is higher than that of R2.

In L3 (R3) and L4 (R4) position, the impact toughness of X70 side of L245 side show different changes. Compared with R2, R3 and R4 continues to increase, but compared with L2, L3 and L4 decrease. This is mainly because of X70 side fine grain zone [Fig. 4(e)] containss clearly grain refinement, obvious structure crisscross martensite/austenite (M/A) islands structure, while grain refinement, interlacing structure and M/A structure are conducive to the improvement of impact toughness [8]. Relatively, grain size of L245 side fine crystal zone is larger [Fig. 4(b)], staggered characteristics is not obvious, the presence of P, whichare not conducive to the improvement of toughness. Therefore, the impact toughness of the L245 side fine crystal zone and the two-phase zone is lower.

As shown in Fig. 4(c), structure of L245 side base material is F+P. Grain size is relatively large, grain boundary is clear, and there is linear pearlite strucure, which are not conducive to the improvement of toughness. Structure of X70 side base material is GB+PF [Fig. 4(f)]. Grain size is relatively small, the two structure intersect each other, which are the mainly reason for its higher toughness.

On X70 side HAZ, impact toughness of the fine-grained zone is the highest, and the impact energy is about 200 J, which is significantly higher than the weld seam center. On L245 side HAZ, the impact toughness near the fusion line is the highest, but significantly smaller than that on side X70. The difference in the microstructure of the HAZ on both sides is the reason for the performance gap. Grain refinement, structure crisscross and martensite/austenite islands (M/A) structure are conducive to improvement of impact toughness, coarse grains, clear grain boundaries, Widmannstatten structure and pearlite are adverse factor. In the L245-X70 girth weld, the center of the welding seam is the weakest area, and the X70 side has a significantly higher toughness.

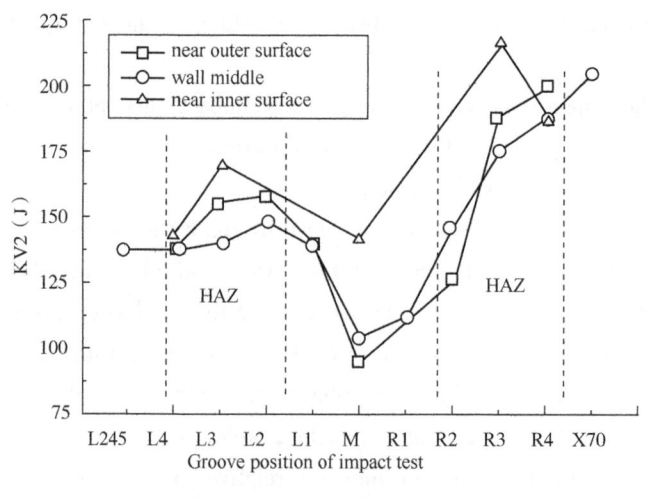

Fig. 6 Impact toughness of L245-X70 girth weld joint

4　Conclusion

Charpy impact toughness increased from the center of weld to both sides, and the weld seam center is the lowest. On the X70 side, Impact toughness of FGHAZ is the highest and the impact energy is about 200 J, which is significantly higher than the weld center. Impact toughness near the fusion line is the highest of in the L245 side HAZ, but significantly smaller than that on side X70. The main reason for the difference of the highert position in HAZ between L245 side and X70 side is that grain size of L245 side FGHAZ is larger, staggered characteristics is not obvious and the presence of P. Grain refinement, interlacing structure and M/A structure are conducive to the improvement of impact toughness. In this kind of girth weld, the center of the welding seam is the weakest area, and the X70 side has a significantly higher toughness. The value of impact toughness in weld zone would provide technical basis for the numerical simulation and safety evaluation of real girth weld in the station. The results of metallographic analysis agree with the performance distribution

Acknowledgment

This work was supported by the National Key R&D Program of China (2016YFC0802101).

References

[1] H. L. Li, X. W. Zhao, L. K. Ji, Failure analysis and integrity management of oil and gas pipeline[J], 2005Oil & Gas Storage and Transportation z1 1-7.

[2] X. W. Zhao, H. L. Li, J. H. Luo, C. Y. Huo, Y. R. Feng. Managerial technique for integrity of oil and gas pipeline and its progress[J], 2006 China Safety Science Journal. 01 129-135.

[3] A. L. Yao, Z. G. Zhao, Y. L. Li, D. Q. Li. The developing trend of oil and gas pipeline integrity management [J], 2009Natural Gas Industry 08 97-100.

[4] J. J. Ren, W. F. Ma, W. Y. Hui, J. H. Luo, K. Wang, Q. R. Ma, C. Y. Huo. Research status and prospect in performance of high grade oil &gas pipeline's butt girth welds[J], 2019Petroleum Engineering Construction. 45 1-5.

[5] A. Q. Chen, W. F. Ma, J. J. Ren, K. Wang, K. Cai, J. H. Luo, C. Y. Huo, Y. R. Feng. Study on rehabilitation of high steel pipeline girth weld defects[J], 2017Natural Gas and Oil 35 12-17.

[6] Q/SY GJX 0221-2012 Welding specification for station pipes of West - east gas transmission third line.

[7] ASTM A370 Standard Test Methods and Definitions for Mechanical Testing of Steel Products.

[8] K. Tong, C. J. Zhuang, Q. Liu, X. L. Han, L. X. Zhu, X. D. He. Microstructure characteristics of M/A islands in high grade pipeline steel and its effect on mechanical properties[J]. 2011 Materials for Mechanical Engineering 35 4-7.

本论文原发表于《Journal of Physics：Conference Series》2020 年第 1637 卷。

Dynamic Impact Damage of Oil and Gas Pipelines

Nie Hailiang[1,2]   Ma Weifeng[1]   Sha Shengyi[3,4]   Ren Junjie[1]
Wang Ke[1]   Cao Jun[1]   Dang Wei[1]

(1. Institute of Safety Assessment and Integrity, State Key Laboratory for Performance and Structure Safety of Petroleum Tubular Goods and Equipment Materials, Tubular Goods Research Center of CNPC; 2. Northwestern Polytechnical University; 3. China University of Petroleum(EastChina); 4. PetroChina Pipeline Company)

Abstract: Dynamic impact damage is one of the main causes of Oil and Gas pipeline failure, but there is no systematic research method and achievement on pipeline dynamic impact damage at present. As the weak link of pipeline, girth weld is the most important research object in practical engineering application, so it is urgent to study the damage behaviour of pipeline girth weld under dynamic impact load. This paper reviews the experimental technology of split Hopkinson Pressure Bar in the field of dynamic impact mechanics, the theoretical knowledge of stress wave propagation theory, dynamic mechanics and damage mechanics, as well as the finite element simulation analysis method, explains the application of theory and technology in practical engineering, introduces the research status of mechanical behaviour of pipeline girth weld under dynamic impact load, and looks forward to the theoretical and technical trend of dynamic impact mechanics of pipeline girth weld. It lays a foundation for the dynamic impact safety evaluation and the establishment of protection system of pipeline girth weld.

1 Introduction

Girth weld is an indispensable part of pipeline construction[1]. Long distance oil and gas pipeline is connected by girth weld, and the pipeline is connected by girth weld in the station. The toughness of girth weld material is poor, and it is easy to break and fail, which becomes the weak link of oil and gas pipeline. After a long period of service, there are countless cases of oil and gas pipeline failure caused by weld defects. For example, according to the statistics of the U.S. Pipeline Safety Office[2], between 1985 and 1996, the failure rates of dangerous liquid pipelines and natural gas pipelines due to weld problems accounted for 12% and 8%, respectively.

In many cases, the structure and material of oil and gas pipeline bear the action of explosion and impact load[3], such as the impact of third party construction, the impact of falling stone, the shock wave load caused by pipeline explosion, the dynamic load caused by the sudden change of

---

Corresponding author: Nie Hailiang, niehai liang@ mail. nwpu. edn. cn.

ground load, the hull impact load of offshore pipeline and the impact load caused by earthquake, and so on. Their common characteristics are that the pipeline material and structure bear the action of transient strong impact load and experience large elastic and even plastic deformation at high deformation rate[4]. Different from the static load, the dynamic impact load will produce stress wave in the pipeline material and structure and propagate along the pipeline. Therefore, the damage caused by thedynamic impact load is often the weak part of the loading point at a certain distance. Girth weld is the weak part of pipeline, so when the pipeline is subjected to impact load, it is often damaged at the girth weld. In order to design the impact protection of pipeline structure, it is necessary to fully understand and master the dynamic impact mechanical behavior of pipeline girth weld[5]. Therefore, the study of damage evolution mechanism and failure prediction of pipeline girth welds under dynamic impact load is a key problem to be solved in the field of safety evaluation and protection of oil and gas pipelines at present.

At present, the research on pipeline safety basically adopts quasi-static research theory and experimental technology[6], the research on high strain rate damage of pipeline structure is very few, and there is no systematic research method and results. At present, pipeline research only uses Charpy impact test (CVN), drop hammer tear test (DWTT), and fracture toughness test (KIC, CTOD, J integral) to qualitatively measure and characterize the ability of materials to resist fracture. However, the field of dynamic mechanics has developed mature theoretical basis (such as stress wave theory, dynamic fracture mechanics and damage mechanics)[7]. Experimental techniques (such as split Hopkinson Pressure Bar experimental technique for measuring stress – strain relationship and fracture strain at high strain rate) and finite element analysis (structural dynamic analysis module of simulation software such as ABAQUS, ANSYS) have not yet been introduced into the study of dynamic impact damage of pipeline[8], and the girth weld itself is an uneven material, which makes the quantitative study of dynamic impact damage of pipeline blank for a long time.

In this paper, the present situation of dynamic impact damage of girth welding joints in the world is summarized, the necessity of studying the dynamic damage mechanism of girth welding joints is analyzed, the experimental and theoretical results in the field of dynamic mechanics of materials and structures are summarized, and the history and application status of separated Hopkinson bar experimental technology are reviewed. Combined with the problem of dynamic impact damage faced by national pipeline safety and economy at present, the theory and experimental technology developed in the field of dynamic mechanics of materials and structures are introduced into the study of dynamic impact damage mechanism of pipeline girth welds. The stress wave propagation theory, dynamic fracture mechanics and damage mechanics theory are used, combined with the structural dynamic analysis module in finite element simulation software. It is of great engineering significance and application value to study the damage mechanism and evolution law of pipeline girth weld under dynamic impact load, which provides a certain reference for the study of dynamic mechanics of pipeline.

2 Development status of dynamic mechanics

At present, in the field of material science, the split Hopkinson Pressure Bar (SHPB) technology is the most widely used in measuring the dynamic mechanical properties of materials [13]. The typical split Hopkinson Pressure Bar device is shown in Fig. 1. The basic principle is that the short sample is placed between two compression bars, and the acceleration wave is generated by accelerating mass block, short bar impact or explosion, and the sample is loaded. At the same time, the wave signal is recorded by a strain gauge pasted on the compression bar and a certain distance from the end of the bar. If the compression bar remains elastic, the wave in the bar will propagate at the elastic wave velocity without dispersion. Based on the one-dimensional stress wave theory, the stress, strain and strain rate history in the specimen can be deduced by using the incident wave, the reflection wave and the transmission wave measured by the strain gaugepasted to the compression bars.

Hopkinson bar experimental technology was originally only used to measure the dynamic mechanical properties of ductile materials such as metals. Later, some scholars extended its testing technology to brittle materials [14], such as large diameter Hopkinson bars used to test rock mechanical properties, and some scholars developed nylon Hopkinson bars for testing soft materials [15]. Nowadays, Hopkinson bar has become one of the most widely used and mature technologies in the testing of dynamic mechanical properties of materials.

Hopkinson bar experimental technology has been more and more used in the dynamic mechanical design and protection of structures, and a lot of practical engineering applications have been made. Hopkinson pressure bar has made good progress in the study of spallation of concrete: Klepaezko [16] has used Hopikinson bar experimental technology to measure the spallation strength of a concrete, and Rubio [17] has also used similar technology to study the spallation strength of ceramic materials. Hu Shisheng [18] et al proposed that the spallation strength of concrete materials be studied by using large diameter Hop kinson compression bar and patch on the specimen, and better results have been obtained. The high G value accelerometer is mostly used to measure the impact load in the process of penetration and armour piercing. Because the usual calibration equipment can not produce a very high acceleration value, it is impossible to calibrate the acceleration sensor with high G value. In foreign countries, Hopkinson pressure bar technology has been applied to the calibration of high G value acceleration sensor [19,20], and there are also related application research in China [21]. At present, the traditional evaluation of the safety of gunwork products is to use Macheit hammering method, but the overload acceleration of Macheit hammering method can not reach the actual overload of pyrotechnics. Nanjing University of Science and Technology Zhang Xueshun et al. [22] have applied Hopkinson pressure bar technology to the dynamic target simulation test of gunwork products, which has solved the difficult problem of ultra-high G value environmental simulation of pyrotechnics. A similar method can be used to evaluate the safety and structural stability of fuze [23]. Due to the shortage of experimental methods, the research progress on shear initiation criteria of explosives is slow. Recently, an improved SHPB was designed by the United States Army Laboratory to carry out shear stamping experiments on explosive materials. The critical shear velocity and critical wave time of explosive initiation were determined by adjusting the shear speed and wave time by changing the bullet speed and length [24].

3 Development of dynamic mechanics of girth weld

In the field of pipeline research, the evaluation methods of residual strength are summarized as follows: (1) based on the semi-empirical formula obtained by hydraulic blasting test of a large number of defective pipe segments; (2) the analytical analysis method based on elastic-plastic mechanics and fracture mechanics theory; (3) the finite element numerical calculation method; (4) based on the failure criterion of defective pipeline, combined with probability and reliability theory, the probabilistic integrity evaluation method of defective pipeline is established.

The evaluation method of residual strength of defective pipeline has been studied since the early 1970s in the world, and many evaluation standards and norms have been formed at present. For the evaluation of volume defects, based on the semi-empirical fracture mechanics relationship established by Kiefner and Maxey et al.[25,26] in the 1970s, the evaluation criteria and norms represented by ASME B31G[27] and CAN/CSA Z144-M86[28] are formed. Since 1990s, in order to avoid the excessive conservatism of B31g method, a lot of research has been carried out on the evaluation method of local corrosion defects in the world, and new standards and methods have been promulgated, including chapter 5 of DNV RP F101[29] and API RP 579[30]. For the evaluation of planar defects, elastic-plastic fracture analysis methods are mainly used, including EPRI method[31], CEGB R6[32] method, ASEMXI appendix C "Austenite Pipeline defect acceptance Criterion and Evaluation Code"[33] and IWB-3650 Appendix C "Ferrite Pipeline defect acceptance Criterion and Evaluation Code"[34]. For the eighth chapter of geometric defect, API RP 579, the evaluation methods of pipe body non-circle, straight weld pout mouth and wrong edge are given. In chapter 6 of API RP 579, the evaluation method of point corrosion damage is given, but the evaluation process, especially the quantitative process of defects, is too complex to be operated and applied in engineering. In addition, only a rough guiding method is given for the evaluation of hydrogen bubbling. Although a lot of research has been carried out in the world since the end of 1980s, there is no systematic evaluation standard and standard for the evaluation method of mechanical damage defects.

In the aspect of dynamic impact of girth welds, the commonly used impact dynamics experimental methods are limited to the use of Charpy (CVN) impact test, drop hammer tear test (DWTT), and fracture toughness test (KIC, CTOD, J integral) to qualitatively measure and characterize the ability of materials to resist fracture. The results are also used to assist in the study of the calculation formula derived under quasi-static conditions, but there is no systematic research method. At present, most of the research on the dynamic impact behavior of pipeline girth weld is aimed at a specific accident, andthe dynamic impact process is roughly analyzed. Lu Guoyun et al.[35] in 2003, the impact failure of three-span continuous pressure pipeline was studied by using DH R-9401 impact loading tester, and the critical failure speed and corresponding failure mode under different working conditions were obtained. Emin Bayraktar et al.[36] used impact tensile test (impact tensile test, in 2004. ITT) and Charpy impact strength test (V-notch) were used to study the effect of strength mismatch on fracture position of laser welding girth weld of offshore pipeline. Wang Hong et al.[37] deduced the formula for calculating the impact force of falling stone on pipeline by means of theoretical mechanics in 2009, and the results are in good agreement with the actual situation. Tan Xuegang et al. [38] calculated the impact of blasting collapse on underground pipeline by using engineering mechanics theory in 2012. Liu Shuo [39] combined with chemical composition analysis technology, scanning electron microscope (scanning electron microscopy, SEM) and scattering energy spectrum measurement technique (energy dispersive spectrometry, in 2014. EDS) the impact strength of three kinds of self – shielded tubular cored girth welds is estimated. Hang Zhang [40] studied the impact effect of internal testing tool (Pipeline inspection gauges, PIG) on pipe girth welds by finite element simulation in 2015.

4 Conclusion and prospect

At present, the research on dynamic impact of pipeline girth weld at home and abroad is limited to obtaining the impact strength of weld by a certain experimental means, or using a certain theory to calculate the dynamic impact force, which has not yet formed a systematic research method from material to structure, from theory to experiment, and the mature theory and experimental technology in the field of dynamic impact mechanics have not been fully utilized. Because the pipeline girth weld is an uneven material, the propagation process of stress wave between girth weld and wood involves many complex reflection and transmission. It is difficult to accurately grasp the dynamic impact damage mechanism and evolution law of girth weld by single theoretical calculation or numerical simulation. Therefore, the applicant believes that the dynamic damage study of girth weld in the future needs dynamic mechanics of fusion material and dynamic mechanics of structure. The theoretical basis and experimental technology of impact dynamics and the structural dynamics simulation module of finite element simulation can be carried out in order to have a scientific and systematic understanding of the dynamic mechanical behaviour of girth weld and to provide a full and scientific theoretical basis for the engineering application of dynamic damage of girth weld.

Acknowledgment

This work was supported by the China Postdoctoral Science Fund (2019M653785) and the Basic Research and Strategic Reserve Technology Research Fund of China National Petroleum Corporation [2019D-5008 (2019Z-01)].

References

[1] Liu M, Wang Y, Horsley D. Significance of HAZ [heat affected zone] softening on strain concentration and crack driving force in pipeline girth welds. Offshore Mechanics and Arctic Engineering (OMAE 2005). Proceedings, 24th International Conference, Halkidiki, Greece, 12-17 June 2005.

[2] Bastola A, Wang J, Shitamoto H, et al. Investigation on the strain capacity of girth welds of X80 seamless pipes with defects. Eng Fract Mech, 2017: S0013794416306397.

[3] Hertelé S, O'Dowd N, Minnebruggen K V, et al. Fracture mechanics analysis of heterogeneous welds: Numerical case studies involving experimental heterogeneity patterns. Eng Fract Mech, 2015, 58: 336-350.

[4] Hoh H J, Pang J H L, Tsang K S. Stress intensity factors for fatigue analysis of weld toe cracks in a girth-welded pipe. Int J Fatigue, 2016: S0142112316000402.

[5] Paredes M, Ruggieri C. Engineering approach for circumferential flaws in girth weld pipes subjected to bending

load. Int J Pres Ves Pip, 2015, 125: 49-65.

[6] Fu AQ, Kuang X R, Han Y, et al. Failure analysis of girth weld cracking of mechanically lined pipe used in gas field gathering system. Eng Fail Anal, 2016, 68: 64-75.

[7] Lee CH, Chang K H. Failure pressure of a pressurized girth-welded super duplex stainless steel pipe in reverse osmosis desalination plants. Energy, 2013, 61(Complete): 565-574.

[8] Chang K H, Lee C H, Park K T, et al. Analysis of residual stress in stainless steel pipe weld subject to mechanical axial tension loading. Int J Steel Struct, 2010, 10(4): 411-418.

[9] Pan Jiahua. Fracture mechanics analysis of oil and gas pipeline. Petroleum Industry Press, 1989. (in Chinese)

[10] Starostin V. Pipeline disaster in theUSSR. Pipes Pipelines Int, 1990, 35(2): 7-8.

[11] Li H, Zhao X, Ji L. Failure analysis and integrity management of oil and gaspipelines. Oil and gas storage and transportation, 2005, 24 (S1): 1 - 7. (in Chinese)

[12] Dannemann K A, Chalivendra V B, Song B. Dynamic behavior of materials. Exp Mech, 2012, 52 (2): 117-118.

[13] Chen W W, Song, B. Split Hopkinson (Kolsky) bar: design, testing, and applications. US: Springer, 2011. C. Flaw distribut

[14] Greene, H, ions and the variation of glass strength with dimensions of the sample. J Am Ceram Soc, 2010, 39 (2): 66-72.

[15] Song B, Chen W W, Jiang X. Split Hopkinson pressure bar experiments on polymeric foams. Int J Vehicle Des, 2005, 37(2-3): 185-198.

[16] Klepaczko j, Brara a. An experimental method for dynamic tensile testing of concrete by spalling. Intl J Impact Eng, 2001, 25(4): 387-409.

[17] GálvezF, Rodrí Guez Pérez J, Sánchez V. The spalling of long bars as a reliable method of measuring the dynamic tensile strength of ceramics. Int J Impact Eng, 2002, 27(2): 161-177.

[18] Hu S, Zhang L, Wu N. Experimental study on spallation strength of concrete materials. Engineering Mechanics, 2004, 21 (4): 128 -132. (in Chinese)

[19] Robert D S. Testing techniques involved with the development of high shock acceleration sensors. In Range Commanders Council Twelfth Transducer Workshop, 1983.

[20] Togami T C, Baker W E, Forrestal M J. A split Hopkinson bar technique to evaluate the performance of accelerometers. J Appl Mech, 1995, 63(2): 353-356.

[21] Li Y, Guo W. Research on Calibration system of High g acceleration Sensor. Explosion and shock, 1997, 17 (1): 90-96. (in Chinese)

[22] Zhang X, Shen R. Research on Simulation Technology of dynamic targeting of initiating products. Gunwork, 2003, 4 (1-4). (in Chinese)

[23] Zhang M, Wu Z. Test technology of chamber explosion point. Journal of Detection and Control, 1994, 21 (2): 9-15. (in Chinese)

[24] Brian K, Oliver Be, Robert L, et al. Shear deformation and shear initiation of explosives and propellants. Symposium of 12th Detonation. 2003.

[25] Kiefner J F, Maxey W A, Eiber R J, et al. Failure stress levels of flaws in pressurized cylinders. Astm Special Technical Publication, 1973, 536, 461-481.

[26] Maxey W A, Kiefner J F, Eiber R J, et al. Philadelphia: ASTM, 1972.

[27] ASEM: Manual for Determining the Remaining Strength of Corroded Pipelines. 1991.

[28] CAN/CSA Z184-M86: Gas pipeline systems. 1986.

[29] DN V RP F101: Corroded Pipelines. 1999.

[30] API 579-2000: Recommended practice for fitness for service. 2000.

[31] Kumar V, German M D, Shih C F. Palo Alto: EPRI, 1981.

[32] Milne I, Ainsworth R A, Dowling A R, et al. Assessment of the integrity of structures containing defects. International Journal of Pressure Vessels and Piping, 1988, 32(1): 3-104.

[33] ASEM XIIWB - 3640 and Appendix H: Flaw evaluation procedures and acceptance criteria for austenic piping. 1986.

[34] ASEM XIIWB-3650 and Appendix C: Flaw evaluation procedures and acceptance criteria for ferritic piping [D]. 1986.

[35] Road N, Lei J, Wu Y, et al. Experimental study on lateral impact failure of multi-span thin-wall pressure pipeline. Explosion and shock, 2003, 23 (5): 454-459. (in Chinese)

[36] Bayraktar E, Hugele D, Jansen J P, et al. Evaluation of pipeline laser girth weld properties by Charpy (V) toughness and impact tensile tests. J Mater Process Tech, 2004, 147(2): 155-162.

[37] Wang H, Yu Z. Quantitative analysis of impact effect of falling stone on buried pipeline. Petroleum Engineering Construction, 2009, 35 (6): 5-8. (in Chinese)

[38] Tan X, May D, Tian Y, et al. Discussion on impact effect of blasting collapse on underground pipeline. Blasting, 2012, 29 (1): 23-26. (in Chinese)

[39] Liu S. Investigation of the impact toughness of self - shielded flux - cored wire girth welds for X80 pipelines. Baosteel Technical Research, 2014, 8(4): 53-60.

[40] Zhang H, Zhang S, Liu S, et al. Measurement and analysis of friction and dynamic characteristics of PIG's sealing disc passing through girth weld in oil and gas pipeline. Measurement, 2015, 64, 112-122.

本论文原发表于《Journal of Physics: Conference Series》2020 年第 1637 卷。

X70管道自保护药芯焊丝环焊接头力学性能及影响因素

何小东[1,2] 薛 如[3] 李为卫[1,2] 池 强[1,2] 高雄雄[1,2]

(1. 中国石油集团石油管工程技术研究院；2. 石油管材及装备材料服役行为与结构安全国家重点实验室；3. 中国石油西气东输管道公司)

摘 要：采用ER70S-G STT根焊与E551T8-K2自保护药芯焊丝进行热焊、填充和盖面焊接了X70管道环焊缝，测试并统计了环焊接头的力学性能，分析了影响力学性能的主要因素。结果表明，X70管道自保护药芯焊丝环焊接头具有较好的力学性能，影响其力学性能的主要因素是焊缝中的夹渣、气孔和未熔合等焊接缺陷；并指出焊接缺陷不仅与自保护药芯焊接的自身工艺特点有关，更重要的是与焊接操作水平和质量意识密切相关。

关键词：X70；管道；环焊缝；自保护药芯焊丝焊接；力学性能

焊接是当今油气长输管道连接的唯一方式。因此，焊接效率和焊接质量不仅对管道工程施工起着关键作用，也决定了管道环焊缝接头本质安全。目前，国内外长输管道常用的焊接方法主要有药皮焊条电弧焊、手工钨极氩弧焊、熔化极气体保护半自动焊、自保护药芯焊丝电弧焊、熔化极活性气体保护自动焊、自动埋弧焊和闪光对焊等[1]。药芯焊丝自保护焊(Self-shielded flux-cored arc welding, FCAW-S)和熔化极活性气体保护自动焊(Gas metal arc welding, GMAW)效率高、劳动强度低，在高强度、大口径、厚壁管道环焊缝的填充、盖面焊接中得到广泛应用。但这两种焊接工艺也有局限性。隋永莉等[2]从焊接原理、焊接材料、焊接设备、焊接坡口、管口组对、焊接施工及施工组织、焊接质量控制、经济效益等几个方面总结分析了半自动焊工艺和自动焊工艺的适用性及其优缺点。

为了满足高强度油气管道现场焊接需要，国内相关单位和机构开展了自保护药芯焊丝研制。有研究者[3]通过提高熔渣的碱度以及在药粉中加入多种氟化物提高脱氢能力；再增加Mn含量来提高强度，通过降低熔敷金属中的铝含量和硅含量，并加入微量合金元素和提高镍含量，获得了X80管道焊接用高强韧性自保护药芯焊丝。张敏等人[4]基于成分匹配与组织匹配的设计原则，设计并制成了X100管线钢匹配用自保护药芯焊丝，并选用合适的焊接工艺参数试焊，得到强度、韧性和组织与母材相匹配的焊接接头。但是，近年来X80管道环焊缝发生多起失效事故[5]，业界相关人士对高钢级管道现场焊接能否使用自保护药芯焊丝焊接持怀疑态度，甚至在相关技术文件中禁止使用此焊接工艺。

基金项目：国家重点研发计划项目(2018YFC0310300)。

作者简介：何小东，1970年出生，硕士，教授级高级工程师，主要从事管线钢焊接工艺、材料性能测试及表征研究，已发表论文40余篇。

针对某管道工程焊接工艺适用性评定，文中测试了焊接接头关键力学性能，并分析了其主要影响因素。研究结果有助于自保护药芯焊丝焊接工艺在高强度管道现场环焊缝焊接中的合理应用和焊接接头质量控制。

1 试验材料及方法

1.1 试样制备

环焊缝试样制备采用直径为813mm，钢级为X70的钢管，组对壁厚为14.3~14.3mm、17.5~17.5mm和17.5~19.1mm，焊接坡口形式为"V"形，坡口角度为25°。焊接工艺采用表面张力过渡(Surface tension transfer，STT)根焊，焊丝牌号为Bohler SG3-P(ER70S-G)，直径为1.2mm。FCAW-S热焊、填充焊和盖面焊的焊丝牌号为JC-30(E551T8-K2)，直径为2.0mm。环焊缝焊接分别由5组焊工完成，焊接工艺参数见表1。

表1 焊接工艺参数

焊道	焊接方法	焊接方向	极性	焊接电流 I(A)	电压 U(V)	送丝速度 v_F(m·min^{-1})	焊接速度 v_W(mm·min^{-1})
根焊	STT	下向	直流反接	峰值：380~440	14~18	3.30~4.06	160~240
热焊	FCAW-S	下向	直流正接	160~230	18~22	2.03~2.54	200~300
填充焊	FCAW-S	下向	直流正接	160~260	18~22	2.28~3.05	160~260
盖面焊	FCAW-S	下向	直流正接	140~240	18~22	2.03~2.79	160~220

1.2 试验方法

样品完成焊接后，采用HR-90A/3093-1634型X射线探伤机对环焊缝进行无损检测。对探伤合格的焊缝依据GB/T 31032—2014《钢质管道焊接及验收》进行评价。采用KB30BVZ-FA维氏硬度计测试了焊接接头内外表面附近的硬度分布。用SHT4106材料试验机测试了不同焊接位置的焊接接头在室温下的拉伸性能和塑性变形能力；截取55mm×10mm×10mm夏比冲击试样，缺口类型为"V"形(Charpy V-Notch，CVN)，深度为2mm，用PSW750冲击试验机测试了焊缝和热影响区在管道设计温度为-5℃下的夏比冲击吸收能量。在环焊缝接头的平焊、立焊和仰焊位置分别截取并制备金相试样，采用MEF4M金相显微镜及图像分析系统分析了焊接接头的宏观和组织微观组织。利用VEGA扫描电镜和NSS-300型能谱仪分析了断口形貌和成分组成。

2 试验结果与讨论

2.1 试验结果

图1是环焊缝接头的抗拉强度分布。统计结果表明，三种壁厚的环焊接头抗拉强度R_m分布范围为590~670MPa，均大于标准要求的最小抗拉强度570MPa，且服从正态分布，其平均值约为628MPa，标准差约为16MPa。拉伸试验中86.5%的试样均断于母材(BM)，只有13.5%的焊接接头拉伸试样断于热影响区(HAZ)，但断于热影响区的焊接接头其抗拉强

度也大于标准要求的最小抗拉强度570MPa。由于焊接接头拉伸试样绝大多数断于母材,其抗拉强度体现的是管材的纵向拉伸性能。

如图2所示是环焊缝接头夏比冲击吸收能量分布。图2表明,环焊缝中心的夏比冲击吸收能量单个值分布为80~260J,标准差约为35J;每组焊缝(WM)冲击吸收能量平均值范围为107~245J,标准差为29J。热影响区夏比冲击吸收能量单个值和组平均值的分布范围分别是77~325J、134~302J,其标准差分别约为49J和33J。因此,不管是焊缝还是热影响区的单个值和夏比冲击吸收能量值均满足标准规定的38J和50J要求。但是,从图2也可以看出,相比于焊缝,热影响区的冲击吸收能量离散程度略大。这主要是由于焊接热影响区的组织和结构不均匀性所致。一般地,油气管道环焊接头HAZ冲击试样取至壁厚中心,缺口位置为焊缝金属和热影响区母材金属各占50%。但由于焊接热影响区较窄,在冲击试样缺口加工过程中,难以完全保证焊缝金属和热影响区的母材所占比例完全一致。因此,热影响区的夏比冲击吸收能量值波动较大。从总体上看,在合理的焊接工艺下,虽然自保护药芯焊丝焊接的X70管道环焊缝接头夏比冲击吸收能量有一定的波动,但是焊缝和热影响区的冲击吸收能量单个值和平均值均满足标准要求。

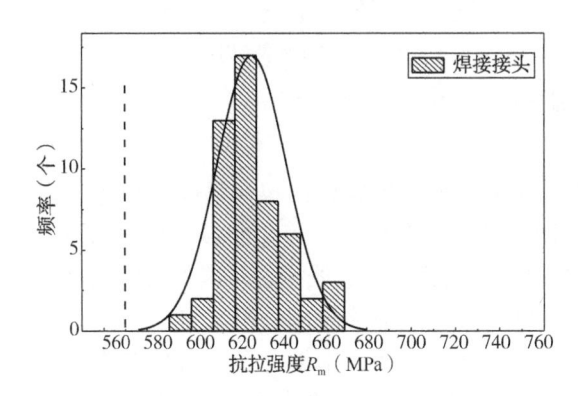

图1　环焊缝接头抗拉强度分布

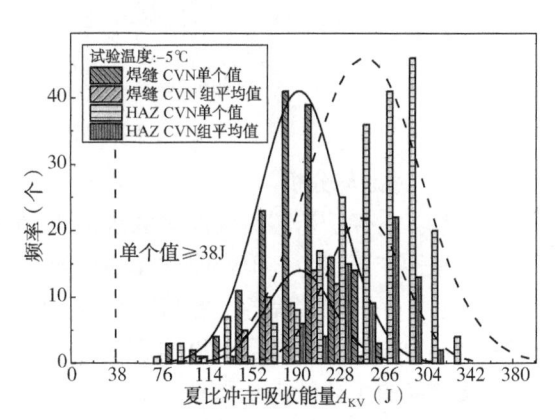

图2　环焊缝接头夏比冲击吸收能量分布

对环焊缝接头取样,按标准要求进行侧弯试验,弯轴直径为90mm,弯曲角度为180°。图3是侧弯后试样拉伸面形貌和缺陷统计。图3表明,约76%的弯曲试样拉伸面无缺陷,而有缺陷或断裂的试样占24%。按验收标准判定试样上缺陷性质或尺寸超出要求的仅有2%。因此,就整体环焊缝而言,自保护药芯焊丝焊接的X70管道环焊缝接头具有较好的塑性变形能力。

(a)侧弯后试样拉伸面形貌

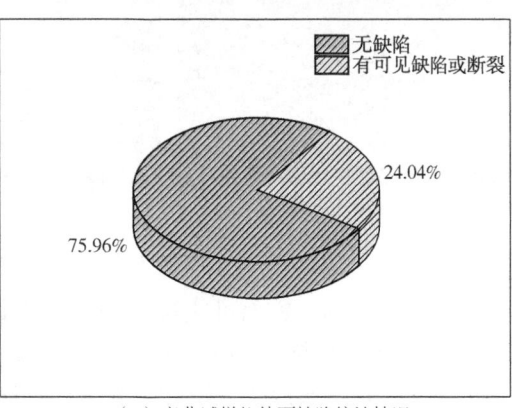

(b)弯曲试样拉伸面缺陷统计情况

图3　焊接接头弯曲后形貌及缺陷统计

图 4 是环焊缝接头的维氏硬度统计分布。图 4 表明，自保护药芯焊丝焊接的 X70 管道环焊缝接头维氏硬度呈正态分布，无论内表面还是外表面附近的维氏硬度均小于标准规定的 275HV10。但从图 4 也可以看出，管体和热影响区的硬度值分布稳定，变化相对较小，而外表面的盖面焊缝和内表面的根焊焊缝硬度值波动较大，分布范围也相对较宽；而且根焊焊缝的硬度低于热影响区和母材的硬度，而填充和盖面焊缝的硬度略高于热影响区和母材的硬度。

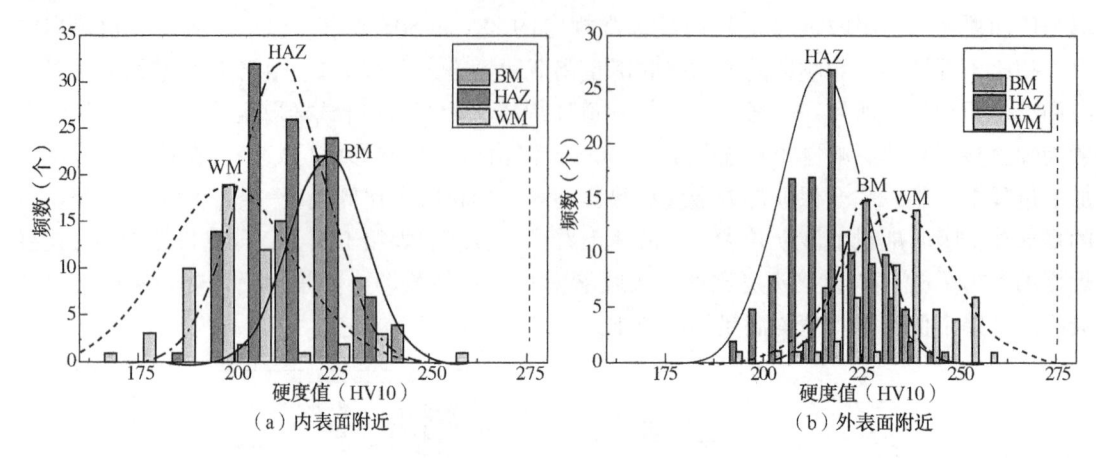

（a）内表面附近　　　　　（b）外表面附近

图 4　环焊接头维氏硬度分布

2.2　讨论与分析

虽然，自保护药芯焊丝焊接的 X70 管道环焊缝接头表现出较好的拉伸强度、夏比冲击韧性和塑性变形能力，但弯曲试样的拉伸面上仍出现较多的缺陷甚至发生断裂。究其原因主要是药芯自保护焊接焊缝上容易产生较多的焊接缺陷，而且通过射线探伤对这些缺陷难以判定或判定为合格。图 5 是 X70 自保护药芯焊丝环焊接头典型焊接缺陷。由于自保护药芯焊丝焊接自身的工艺特点所致[6]，这些焊接缺陷包括气孔、夹渣、未熔合等，而且夹渣的数量最多，其形貌及能谱分析如图 6 所示。自保护药芯焊丝焊接主要的是通过在药芯中加入造渣、造气、脱氧、脱氮的矿物粉来保证焊缝金属的组织、力学性能及成形性。从图 6 的能谱分析可以看出氧、铝、镁、钡等元素是自保护药芯焊丝熔渣主要成分[7]。因此，自保护药芯焊丝焊接缺陷与工艺自身特点密切相关。

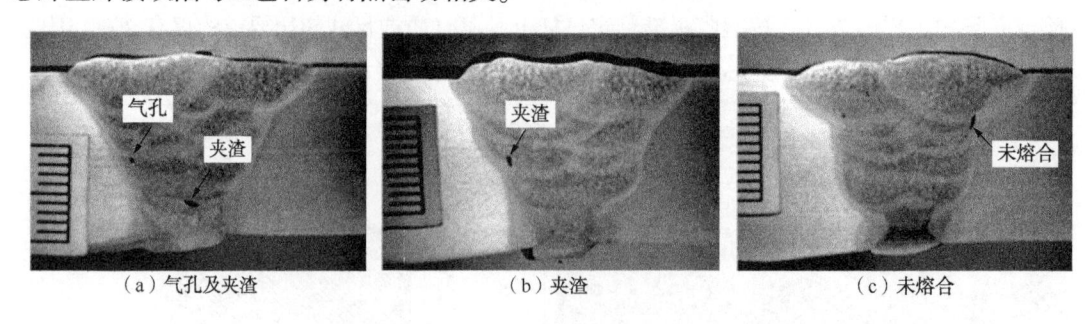

（a）气孔及夹渣　　　（b）夹渣　　　（c）未熔合

图 5　X70 药芯自保护焊接环焊缝接头典型焊接缺陷

这些焊接缺陷一方面可能位于冲击试样缺口根部启裂或裂纹扩展面上，对冲击韧性不利[8]，从而导致冲击韧性值离散，另一方面对焊接接头的抗载荷能力和弯曲变形能力也有影响，甚至会成为裂纹扩展源导致管道环焊缝接头失效。图 7 为金相横截面上宏观可见缺陷的数量与拉伸断于热影响区或弯曲拉伸面开裂数量的关系。从图 7 可以看出，二者呈线性相关，即自保护药芯焊丝焊接环焊接头中，缺陷数量越多，拉伸试样越有可能断于焊缝或热影响区，而且弯曲试样的拉伸面也容易出现较多的裂纹甚至断裂。实际上，自保护药芯焊丝环焊缝接头的焊接缺陷数量与焊工操作水平也密切相关，如图 8 所示。因此，为了保证管道安全，提升药芯自保护焊接环焊缝接头质量，焊接操作者的水平和质量意识起着至关重要的作用。

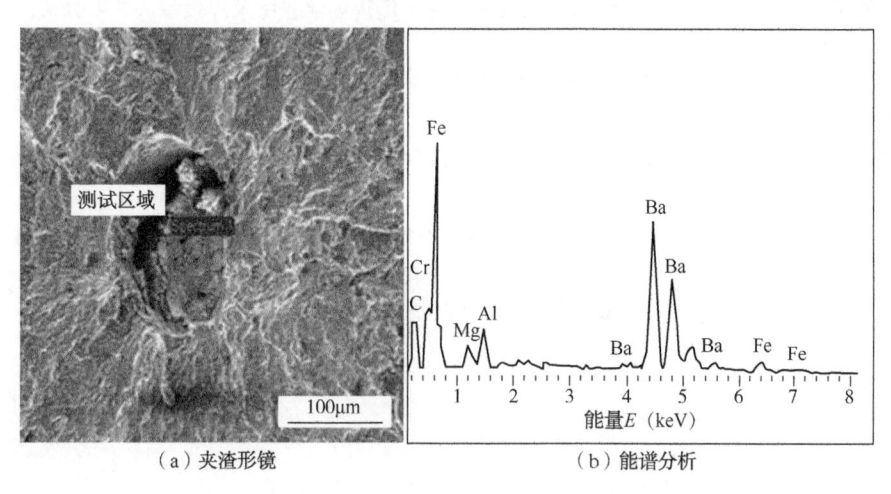

（a）夹渣形镜　　　　　　（b）能谱分析

图 6　药芯焊丝自保护焊接夹渣形貌及其能谱分析

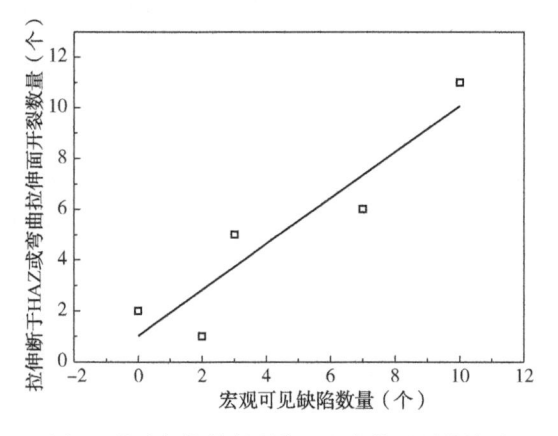

图 7　缺陷与拉伸断裂位置和弯曲开裂的关系

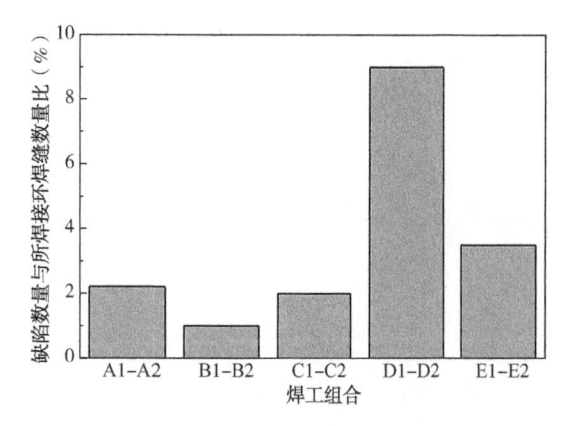

图 8　焊工组合与焊接缺陷出现概率的关系

3　结论

（1）采用 ER70S-G STT 根焊与 E551T8-K2 自保护药芯焊丝进行热焊、填充和盖面焊接 X70 高强度管道，环焊缝接头具有较好的力学性能；

（2）影响自保护药芯焊丝环焊接头力学性能的主要因素是焊缝中的夹渣、气孔和未熔合等焊接缺陷；

（3）焊接缺陷不仅与自保护药芯焊丝焊接的工艺特点有关，还与焊接操作水平和质量意识密切相关。

参 考 文 献

[1] 尹长华，隋永莉，冯大勇，等. 长输管道安装焊接方法的选择［J］. 焊接，2004（6）：31-34.
[2] 隋永莉，郭锐，张继成. 管道环焊缝半自动焊与自动焊技术对比分析［J］. 焊管，2009，36（9）：38-47.
[3] 侯杰昌. X80管道钢焊接用药芯自保护焊接焊丝研制［J］. 石油工程建设，2007，33（5）：51-52.
[4] 张敏，刘明志，陈阳阳，等. X100管线钢药芯自保护焊接焊丝研制及焊接接头性能分析［J］. 焊接学报，2014，35（10）：1-4.
[5] 胡美娟，刘迎来，朱丽霞，等. 天然气输送管道环焊缝泄漏失效分析［J］. 焊管，2014，37（2）：56-63.
[6] 何小东，仝珂，梁明华，等. 长输管道自保护药芯焊丝半自动焊典型缺陷分析［J］. 焊管，2014，37（5）：53-57.
[7] 李继红，程康康，舒绍燕，等. $CaCO_3$含量对自保护药芯焊丝脱渣性的影响［J］. 机械工程材料，2018，42（9）：6-10.
[8] 周志良. 多层自保护药芯焊丝电弧焊焊缝组织和性能［J］. 大连铁道学院学报，1997，18（3）：41-45.

本论文原发表于《焊接》2020年第3期。

埋地管道泄漏数值模拟分析

王俊[1]　封辉[1]　高琦[2]　王鹏[1]

(1. 中国石油集团石油管工程技术研究院·石油管材及装备材料服役行为与结构安全国家重点实验室；2. 中石油管道有限责任公司西部分公司)

摘　要：针对不同因素对管道泄漏工况的影响进行了模拟研究。管道的铺设方式一般为埋地铺设，长时间埋地管道会因为外力破坏或管道自身老化、腐蚀穿孔等因素造成管道泄漏。管道泄漏时会造成重大压力损失和管道流体的损失，管道大孔泄漏后容易在地面上被检测出来，小孔泄漏不容易被检测出来。因此采用数值模拟方法，通过模型简化，同时考虑计算精度和计算成本，建立了埋地管道小孔泄漏扩散模型。研究埋地输气管道泄漏扩散特性，对管道的日常维护以及应急救援具有重要的意义。分别研究泄漏压力、泄漏孔径、管道埋深、土壤性质、环境温度、泄漏孔形状和障碍物等因素对埋地管道泄漏扩散的影响。数值模拟结果表明，同一点甲烷浓度随时间的增加而增加，主要分为三个阶段：孕育阶段、快速增长阶段和缓慢增长阶段。最终得到泄漏量和浓度随着管道埋深，土壤温度，泄漏口形状，土壤孔隙度，土壤颗粒直径等因素变化的规律。

关键词：埋地管道；小孔泄漏；泄漏扩散；数值模拟

随着人类生活水平的提高和工业的快速发展，越来越多的长输管道应用在人类生活中的各个方面，管道泄漏后的维修、检测等安全问题也需要进一步的研究。这不仅仅是中国目前面临的一个严峻问题，也是全社会待解决的问题。埋地管道是输气运输的重要基础设施之一。管道口破裂造成的挥发性有机物泄漏一方面会引起灾难性的环境污染，另一方面会造成经济损失和人员伤亡，对国民经济构成严重威胁[1-4]。杜明俊等[5]通过FLUENT软件数值模拟了埋地输油管道泄漏口在不同位置时泄漏前后温度场的变化。史晓蒙等[6]建立了地面油品泄漏的计算流体力学仿真模型，得到了油品扩展速度关于泄漏流量的关系式。管道的铺设方式一般为埋地铺设，埋地管道泄漏后会对周围的建筑物、人类产生巨大的危害。因此，研究埋地输气管道小孔泄漏预测及扩散特性，对管道的日常维护以及应急救援和管道安全具有重要的意义。建立三维物理模型并进行网格划分，并建立埋地管道泄漏扩散模型。研究泄漏压力、泄漏孔径、土壤埋深、土壤性质、泄漏孔形状和障碍物等因素对埋地管道泄漏扩散的影响。研究结果可为埋地管道的理论研究及制定管道泄漏应急预案提供理论基础。

基金项目：国家重点研发计划（2016YFC0802101）。

作者简介：王俊（1989—），男，汉族，陕西西安人，工程师，博士。研究方向：输气管道焊接及结构完整性，E-mail：wangjun1003@cnpc.com.cn。

1 埋地管道泄漏几何模型

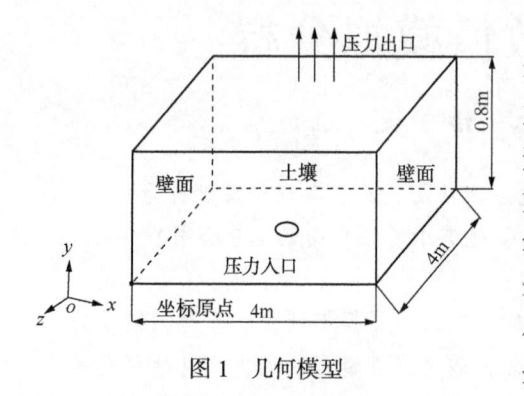

模拟的泄漏工况为圆孔泄漏，二维网格会将泄漏孔默认为狭缝泄漏，不能真实反映泄漏孔径对天气泄漏扩散的影响，而三维网格可将泄漏孔形状画为圆形。模拟的泄漏工况为圆孔泄漏，FLUENT中的二维网格会将泄漏孔默认为狭缝泄漏。试验现有的资源不能模拟成百上千米的管道泄漏，所以对埋地输气管道泄漏扩散模型进行简化[7]，模型所研究的区域大小为4m×4m×0.8m，泄漏孔位于模型底部的中心位置，几何模型如图1所示。

图1 几何模型

2 埋地管道泄漏控制方程

气体在土壤中泄漏扩散过程满足质量守恒、动量守恒和能量守恒三大守恒定律。三大守恒定律的数学描述分别是连续性方程、动量方程和能量方程。

2.1 连续性方程

气体在土壤中的泄漏扩散过程满足气体连续性方程：

$$\frac{\partial \rho}{\partial t} + \frac{\partial \rho u_x}{\partial x} + \frac{\partial \rho u_y}{\partial y} + \frac{\partial \rho u_z}{\partial z} = 0 \tag{1}$$

式中：ρ 为气体的密度，kg/m³；t 为时间，s；u_x、u_y、u_z 分别为 x、y、z 方向上的分速度，m/s。

2.2 动量方程

气体在土壤中的泄漏扩散过程满足动量方程：

$$\frac{\partial \rho u_x}{\partial t} + \nabla(\rho u_x \boldsymbol{u}) = -\frac{\partial p}{\partial x} + \frac{\partial \tau_{xx}}{\partial x} + \frac{\partial \tau_{yx}}{\partial y} + \frac{\partial \tau_{zx}}{\partial z} + \rho f_x \tag{2}$$

$$\frac{\partial \rho u_y}{\partial t} + \nabla(\rho u_y \boldsymbol{u}) = -\frac{\partial p}{\partial y} + \frac{\partial \tau_{xy}}{\partial x} + \frac{\partial \tau_{yy}}{\partial y} + \frac{\partial \tau_{zy}}{\partial z} + \rho f_y \tag{3}$$

$$\frac{\partial \rho u_z}{\partial t} + \nabla(\rho u_z \boldsymbol{u}) = -\frac{\partial p}{\partial z} + \frac{\partial \tau_{xz}}{\partial x} + \frac{\partial \tau_{yz}}{\partial y} + \frac{\partial \tau_{zz}}{\partial z} \rho f_z \tag{4}$$

式中：τ_{xx}、τ_{yy}、τ_{zz} 为黏性力的分量，Pa·s；f_x、f_y、f_z 为不同方向上的单位质量力，m/s；∇ 为哈密顿算子；$\boldsymbol{u}$ 为速度矢量；p 为泄漏压力。

2.3 能量方程

气体在土壤中的泄漏扩散过程满足能量方程：

$$\frac{\partial \rho E}{\partial t} + \nabla[\boldsymbol{u}(\rho E + p)] = \nabla\left[k_{\text{eff}} \nabla T - \sum_j h_j J_j + (\tau_{\text{eff}} \boldsymbol{u})\right] + S_h \tag{5}$$

式中：E 为控制体的总能量，J/kg；k_{eff} 为有效传热系数，W/(m·K)；h_j 为 j 组分的焓

值，J/kg；J_j为j组分的扩散通量，kg/(m·s)；S_h为化学反应热，J。

3 埋地管道泄漏数值模拟

3.1 网格划分和独立性检验

以管道埋深$h=0.8$m，泄漏孔径$d=4$mm为例，模型所研究的区域大小为4m×4m×0.8m，由于泄漏孔径相对较小，对泄漏口附近的网格进行加密处理，先对整体block划分O-Block网格，再对中心的block进行O-Block网格划分，泄漏孔向模型边界的网格线划分方式为Exponential 2，Spacing 2 = 0.0002，Ratio 2 = 1.135，网格数为376000，分别选择Determinant 2×2×2和Angle作为网格质量的判定标准，所有网格的Determinant 2×2×2大于0.7，其中大于0.85的占93.832%，所有网格的Angle大于40.5°，其中大于60°的占70.331%。生成的网格如图2所示。

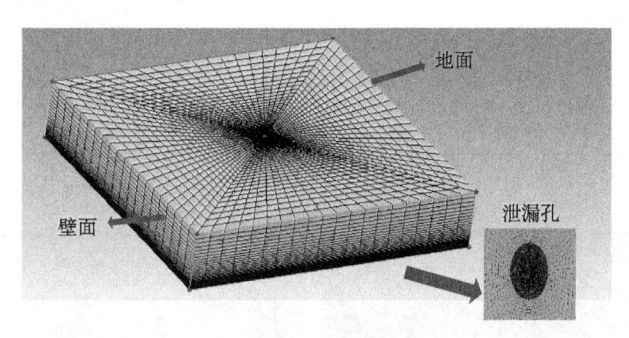

图2 模型网格划分

为了对网格进行独立性检验，通过设置最大网格尺寸、网格层数、第一层网格点与端点之间的距离等参数，得到不同网格划分方案，具体方案见表1。

表1 网格划分方案

方案	网格数
1	67800
2	376000
3	1029000

泄漏设置为压力入口边界条件，地面设置为压力出口边界条件，其余面设置为壁面边界条件，流体域设置为多孔介质，流体介质设置为空气，设置固体的密度为2650kg/m³，导热系数为1.512W/(m·K)。不同网格数对气体泄漏量影响不大，为了加快计算速度并考虑模型的准确性，选择网格数为$37.6×10^4$的模型来模拟气体在土壤中的泄漏扩散过程。

3.2 埋地管道泄漏后气体在土壤中扩散的影响因素分析

3.2.1 泄漏压力对气体在土壤中泄漏扩散的影响

城镇燃气设计压力(表压)分级[8]见表2，由于人口密集处一般为中低压燃气管道，高压燃气管道一般分布在城市周边或者长距离输气干线中，其危险性较小。不同泄漏孔径条件下，泄漏压力为0.005MPa、0.05MPa、0.1MPa、0.2MPa、0.3MPa、0.4MPa和0.5MPa，即次高压B和中低压管道在土壤中的泄漏扩散规律。

取土壤孔隙度为0.6，土壤颗粒直径为0.198mm，温度为300K，管道埋深为0.8m，泄漏方向朝上的圆孔泄漏。以泄漏孔径4mm，泄漏时间50s为例，不同压力下甲烷浓度分布云图如图3所示。从图3可以看出，同一时刻，泄漏孔径一定，泄漏压力越大，甲烷扩散的范围越大。

表2 城镇燃气设计压力(表压)分级

名称		压力(MPa)
高压燃气管道	A	$2.5 < p \leqslant 4.0$
	B	$1.6 < p \leqslant 2.5$
次高压燃气管道	A	$0.8 < p \leqslant 1.6$
	B	$0.4 < p \leqslant 0.8$
中压燃气管道	A	$0.2 < p \leqslant 0.4$
	B	$0.01 \leqslant p \leqslant 0.20$
低压燃气管道		$p < 0.01$

注：A、B为压力的两种级别。

（a）p=5kPa   （b）p=50kPa

（c）p=10kPa   （d）p=200kPa

（e）p=300kPa   （f）p=400kPa

（g）p=500kPa

甲烷的摩尔分数（%）

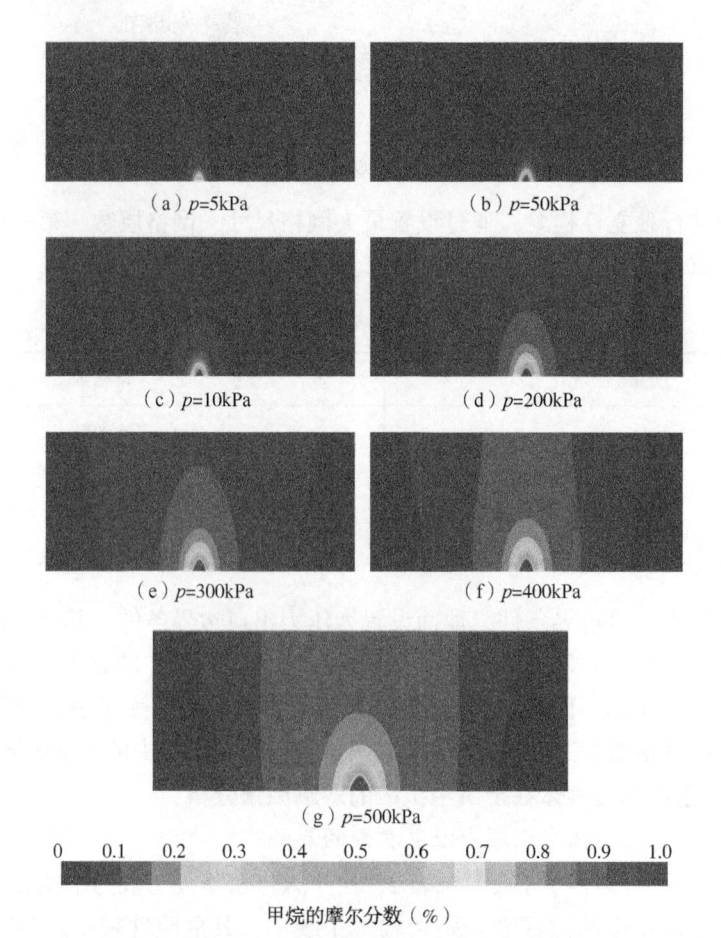

图3 d=4mm，t=50s不同压力下甲烷浓度分布云图

3.2.2 泄漏孔径对气体在土壤中泄漏扩散的影响

埋地输气管道泄漏，尤其是腐蚀穿孔引起的小孔泄漏，泄漏初期很难被发现与定位。当 $d/D \leq 0.2$ 时，可以视为小孔泄漏。欧洲输气管道事故数据组织（European Gas Pipe Line Incident Data Group，EGIG），将泄漏孔径 $d \leq 20mm$ 的情况定义为小孔泄漏。研究不同泄漏压力条件下，泄漏孔径为 2mm、4mm、6mm、8mm 和 10mm，天然气在土壤中的泄漏扩散规律。

取土壤孔隙度为 0.6，土壤颗粒直径为 0.198mm，温度为 300K，管道埋深为 0.8m，泄漏方向朝上的圆孔泄漏。以泄漏压力 100kPa，泄漏时间 50s 为例，不同泄漏孔径下甲烷浓度分布云图如图 4 所示。从图 4 可以看出，同一时刻，泄漏压力一定，泄漏孔径越大，甲烷扩散的范围越大。可以看出，相对于气体在大气和室内的自由紊流射流呈放射状[9,10]的特点，由于土壤有阻力作用，气体在土壤中的浓度分布近似呈圆形。

(a) d=2mm
(b) d=4mm
(c) d=6mm
(d) d=8mm
(e) d=10mm

甲烷的摩尔分数（%）

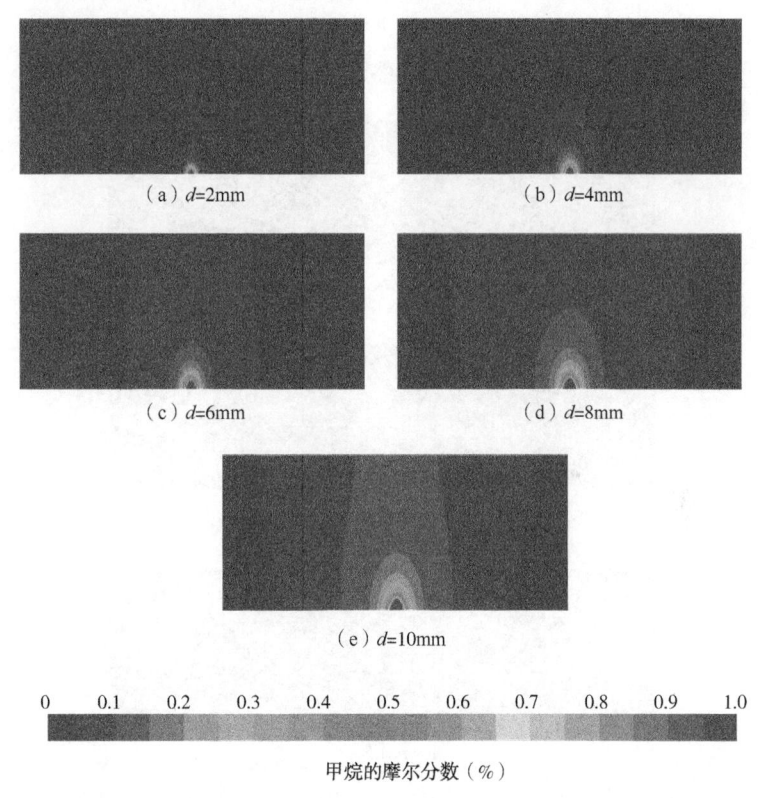

图 4　p=100kPa，t=50s 不同泄漏孔径下甲烷浓度分布云图

不同泄漏孔径条件下，泄漏量随泄漏压力的变化曲线如图 5 所示。由图 5 可知，泄漏孔径一定，天然气的泄漏量（Q）与泄漏压力（p）呈线性增长的关系。

3.2.3 埋深对气体在土壤中泄漏扩散的影响

管道埋深受地面负荷、最大冻土层深度和管道稳定性要求等影响，一般管道埋深为 0.8~1.2m，同时考虑地面荷载较大的特殊地段，取土壤孔隙度为 0.6，土壤颗粒直径为 0.198mm，温度为 300K，泄漏压力为 100kPa，泄漏方向朝上的圆孔泄漏。以泄漏孔径 4mm，泄漏时间 50s 为例，不同土壤埋深下气体浓度分布云图如图 6 所示。从图 6 可以看出，同一时刻，泄漏孔径一定，土壤埋深越大，气体扩散的范围越小。

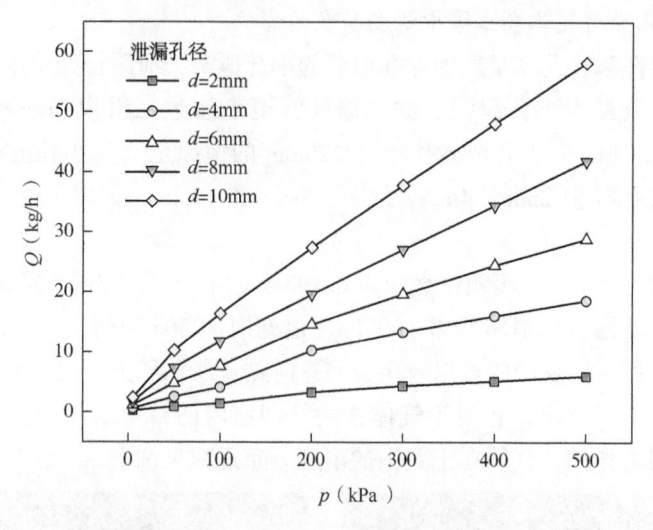

图5 不同泄漏孔径条件下，天然气泄漏量随泄漏压力的变化曲线

(a) h=0.8m

(b) h=1.2m

(b) h=1.5m

甲烷的摩尔分数（%）

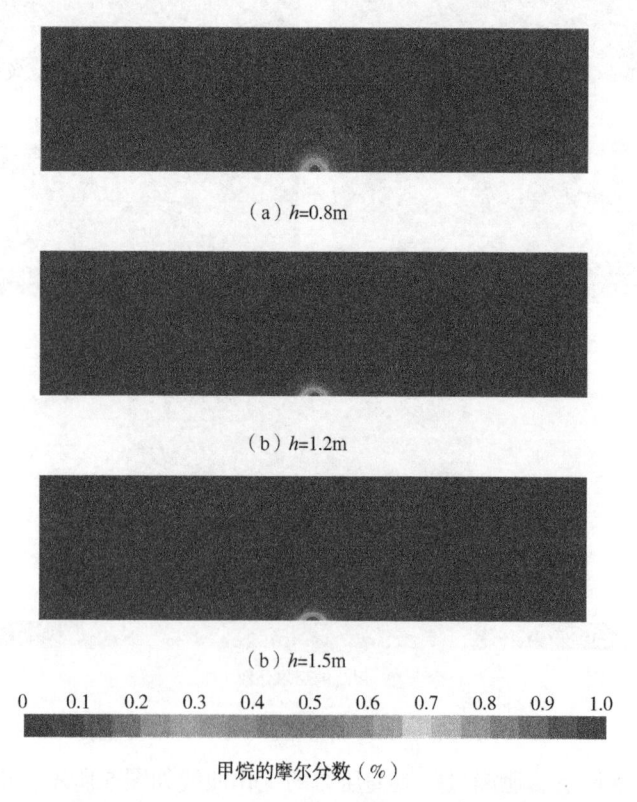

图6 d=4mm，t=50s 不同土壤埋深下甲烷浓度分布

以泄漏压力100kPa，不同泄漏孔径条件下气体泄漏量随管道埋深的变化曲线如图7所示，可以看出，管道的埋深几乎不影响天然气的泄漏量。

3.2.4 土壤性质对气体在土壤中泄漏扩散的影响

数值模拟过程中将土壤简化为粒径一致且各向同性的多孔介质模型多孔介质的黏性阻力和惯性阻力主要受土壤颗粒直径和土壤孔隙度影响。研究土壤颗粒直径对气体泄漏的影响。

取泄漏孔径为4mm，温度为300K，管道埋深为0.8m，泄漏方向朝上的圆孔泄漏。以泄

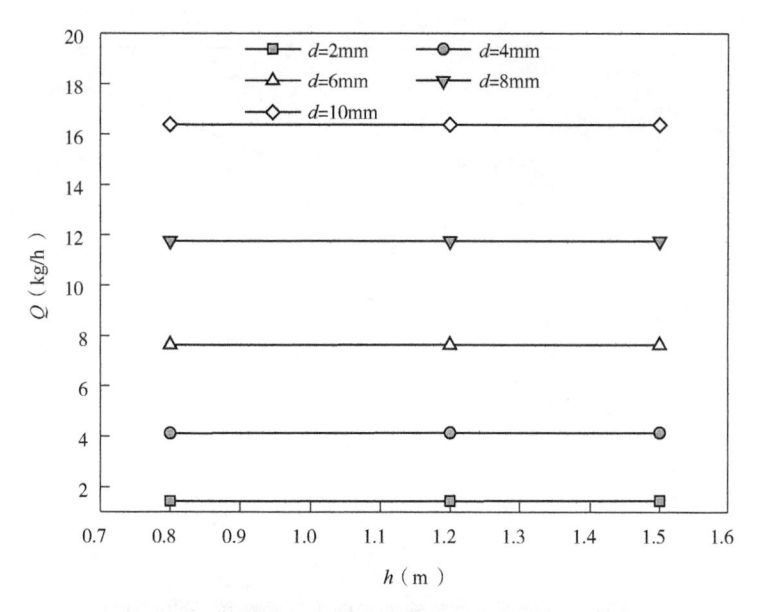

图7 不同泄漏孔径下甲烷泄漏量随管道埋深变化曲线

漏压力100kPa，泄漏时间50s为例，当土壤孔隙度 φ 为0.5，不同土壤颗粒粒径(d_s)条件下气体浓度分布云图如图8所示，由图8可以看出，土壤孔隙度一定，土壤颗粒直径越大，甲烷扩散的范围越大。

(a) d_s=0.04mm

(b) d_s=0.198mm

(c) d_s=0.5mm

甲烷的摩尔分数（%）

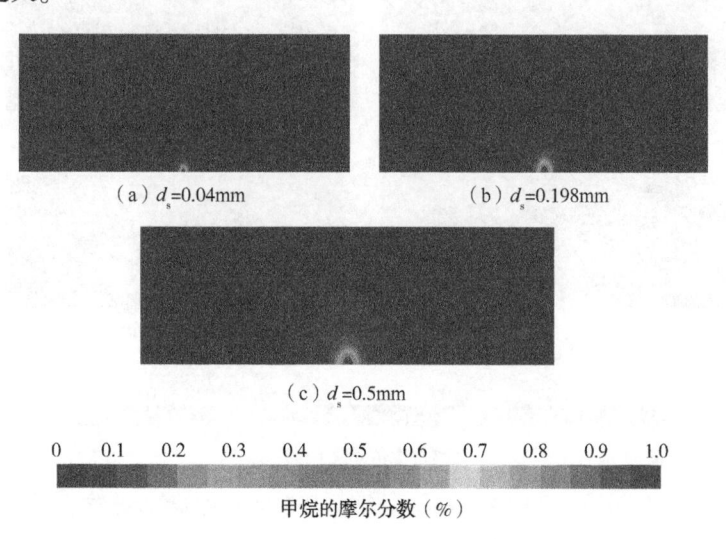

图8 φ=0.5，不同土壤颗粒粒径条件下甲烷浓度分布

以泄漏压力100kPa为例，不同孔隙度条件下，天然气泄漏量随土壤颗粒粒径的变化曲线如图9所示，从图9可以看出，土壤孔隙度一定，天然气泄漏量随土壤颗粒直径的增加而增加，且增加的幅度不断减小。

3.2.5 温度对气体在土壤中泄漏扩散的影响

埋地输气管道在不同季节发生泄漏主要表现在土壤温度的不同，而温度又对气体的性质有影响，从而对气体的泄漏扩散产生影响。

取土壤孔隙度为0.6，土壤颗粒直径为0.198mm，泄漏孔径4mm，管道埋深为0.8m，泄漏方向朝上的圆孔泄漏。以泄漏压力100kPa，泄漏时间50s为例，不同土壤温度(T)甲烷

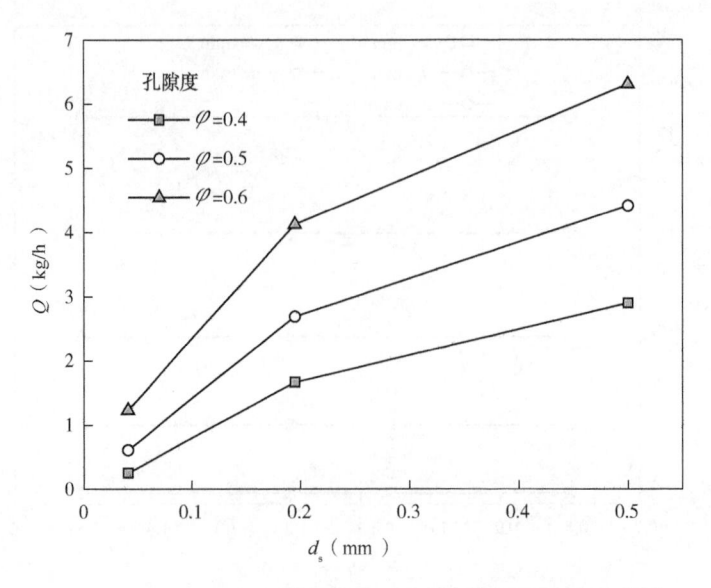

图9 不同孔隙度下天然气泄漏量随土壤颗粒粒径的变化曲线

浓度分布云图如图10所示，可以看出，同一时刻，泄漏压力一定，随着土壤温度的升高，气体扩散的范围略有增加，但影响不大。

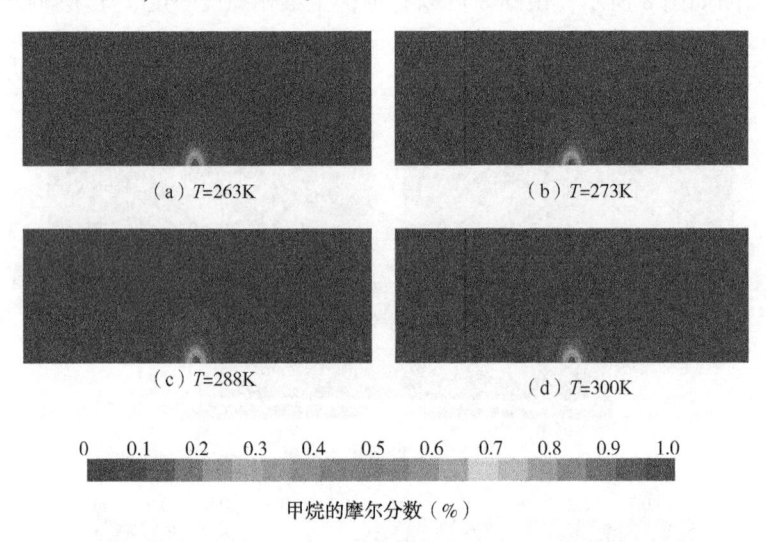

图10 $p=100kPa$，$t=50s$ 不同土壤温度下甲烷浓度分布

以泄漏压力100kPa为例，不同泄漏孔径条件下气体泄漏量随温度的变化曲线如图11所示，得出，土壤温度对甲烷泄漏量有一定的影响，温度越高，泄漏越小。

3.2.6 泄漏孔形状对气体在土壤中泄漏扩散的影响

上述模拟都是假设泄漏孔形状为圆孔泄漏，而实际埋地输气管道发生泄漏时泄漏孔的形状也不尽相同。取泄漏孔形状分别为圆形、正方形和三角形，研究泄漏孔形状对气体泄漏扩散的影响。设不同泄漏孔形状的面积相同。

取泄漏孔径为4mm，土壤孔隙度为0.6，土壤颗粒直径为0.198mm，温度为300K，管道埋深为0.8m，泄漏方向朝上。以泄漏压力500kPa，泄漏时间50s为例，令泄漏孔形状为圆孔时坐标值为1，当泄漏孔形状为正方形时坐标值为2，当泄漏孔形状为三角形时坐标值

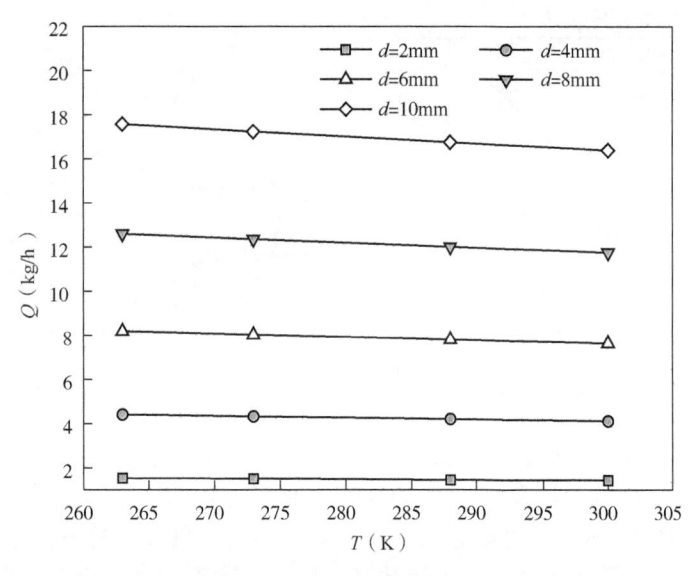

图 11　不同泄漏孔径下甲烷泄漏量随土壤温度变化曲线

为 3，不同监测点甲烷浓度随泄漏孔形状的变化曲线如图 12 所示。从图 12 可以看出，同一时刻，同一监测点甲烷浓度由大到小分别是圆孔泄漏、三角形孔泄漏和正方形孔泄漏。

（a）圆孔泄漏

（b）正方形孔泄漏

（c）三角形孔泄漏

甲烷的摩尔分数（%）

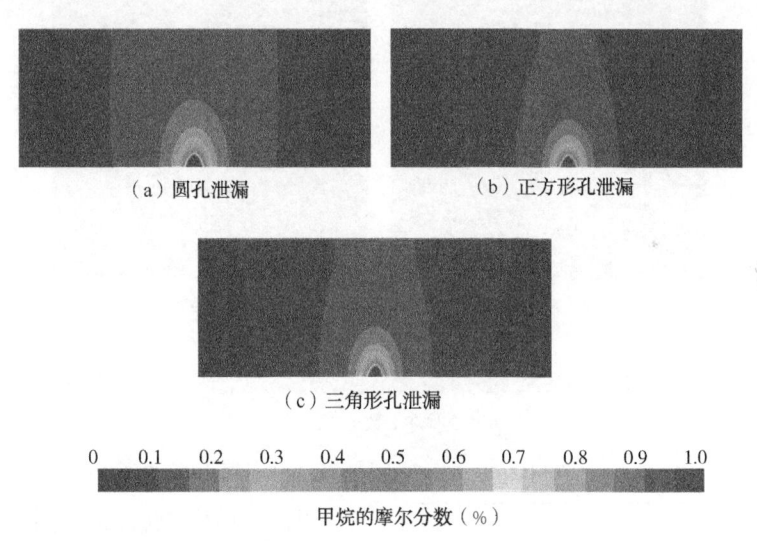

图 12　$p=500\text{kPa}$，$t=50\text{s}$ 时 不同泄漏孔形状甲烷浓度分布

3.2.7　障碍物对气体在土壤中泄漏扩散的影响

输气管道一般为埋地敷设，由于管道附近的地面可能会存在建筑物，而建筑物在地下会有地基，地基等障碍物会对管道泄漏扩散产生影响。障碍物的位置和尺寸都会对气体扩散浓度产生影响，只分析不同障碍物高度对天然气泄漏扩散的影响。假设障碍物形状为长方体，底面积为 0.2m×0.2m，障碍物高度 H 分别为 0.2m、0.4m、0.6m，与泄漏孔距离 0.2m，模型如图 13 所示，网格划分情况如图 14 所示，网格数为 694500，分别选择 Determinant 2×2×2 和 Angle 作为网格质量的判定标准，所有网格的 Determinant 2×2×2 大于 0.6，大于 0.85 的占 98.011%，所有网格的 Angle 大于 40.5°，大于 60°的占 97.065%。

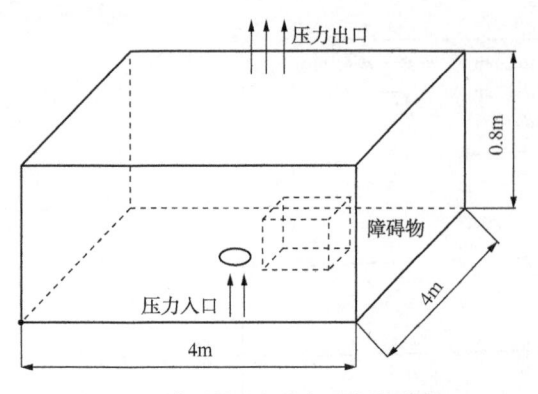

图 13　有障碍物的天然气泄漏扩散模型

图 14　有障碍物扩散模型的网格划分

取土壤孔隙度为 0.6，土壤颗粒直径为 0.198mm，泄漏孔径 4mm，管道埋深为 0.8m，泄漏方向朝上的圆孔泄漏。以泄漏压力 300kPa，泄漏时间 50s 为例，不同障碍物高度下气体浓度分布云图如图 15 所示，可以看出，存在障碍物时，气体会绕开障碍物，在障碍物两边形成浓度差，离泄漏孔较近的一侧甲烷浓度增加，而障碍物另一侧甲烷浓度减小。气体纵向扩散距离随着障碍物高度的增加而增加，即相同时间内扩散到地面的气体浓度增加。

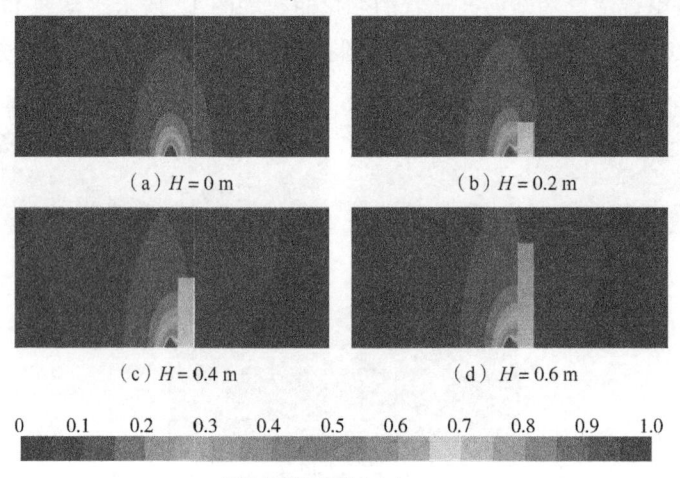

(a) $H = 0$ m

(b) $H = 0.2$ m

(c) $H = 0.4$ m

(d) $H = 0.6$ m

甲烷的摩尔分数（%）

图 15　$p = 300$kPa，$t = 50$s 时不同障碍物高度下甲烷浓度分布

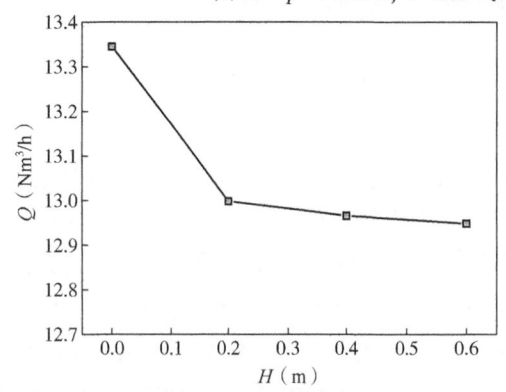

图 16　$p = 300$kPa，$t = 50$s 时天然气泄漏量随障碍物高度的变化曲线

以泄漏压力 300kPa，泄漏时间 50s 为例，天然气泄漏量随障碍物高度的变化曲线如图 16 所示。从图 16 可以看出，有障碍物存在时，天然气泄漏量会降低，随着障碍物高度的增加，气体泄漏量略有减小。

4　结论

通过模型简化，建立了埋地输气管道泄漏扩散模型，从而研究不同因素条件下气体在土壤中泄漏扩散规律，得出如下结论：

（1）同一点浓度随时间的增加而增加，

主要分为孕育阶段、快速增长阶段和缓慢增长阶段。

（2）管道埋深对天然气泄漏量影响不大，但随着埋深的增加，距离泄漏孔越远，浓度减小幅度越大。土壤温度对甲烷泄漏量有一定的影响，温度越高，气体泄漏量越小，而同一监测点气体浓度随土壤温度的增加略有增加。其他条件保持一致，气体相对分子质量越大，泄漏量越大。

（3）面积相同的不同泄漏口形状，当泄漏压力较低时，气体泄漏量从大到小分别为三角形泄漏孔、正方形泄漏孔和圆形泄漏孔，当泄漏压力较大时，气体泄漏量从大到小分别为圆形泄漏孔、三角形泄漏孔和正方形泄漏孔。

（4）土壤颗粒直径一定，天然气泄漏量随土壤孔隙度的增加而增加，且增加的幅度不断增大。土壤孔隙度一定，天然气泄漏量随土壤颗粒直径的增加而增加，且增加的幅度不断减小。

（5）存在障碍物时，气体会绕开障碍物，在障碍物两边形成浓度差，同一时刻，甲烷纵向扩散距离随着障碍物高度的增加而增加。

（6）考虑到计算成本和网格的复杂程度，对数值模拟模型进行了一定的简化，不能研究泄漏口方向和管道自身对天然气泄漏扩散的影响，需要进一步提高网格划分能力和计算机的处理能力来模拟天然气泄漏扩散工况。

参 考 文 献

[1] 朱庆杰，赵晨，陈艳华，等．埋地天然气管道泄漏的影响因素及保护措施[J]．环境工程学报，2018，12（2）：417-420．

[2] 慕园，余志峰，刘玉卿．埋地管道通过逆断层的有限元分析[J]．石油机械，2015，43（10）：112-115．

[3] 夏梦莹，张宏，王宝栋，等．基于壳单元的连续型采空区埋地管道应变分析[J]．油气储运，2018，37（3）：256-262．

[4] 刘洪飞，韩阳，冯新，等．埋地管道微小泄漏与保温层破坏分布式光纤监测试验[J]．油气储运，2018，37（10）：1114-1120．

[5] 杜明俊，马贵阳，高雪利，等 埋地输油管道泄漏影响区内大地温度场的数值模拟[J]．石油规划设计，2010，21（5）：24-26．

[6] 史晓蒙，吕宇玲，杨玉婷，等．地面输油管道泄漏流散数值模拟[J]．中国安全生产科学技术，2017，13（1）：90-96．

[7] 胡夏琦，含硫化氢高压天然气管道泄漏的数值模拟[D]．青岛：中国石油大学(华东)，2007．

[8] GB 50028—2006 城镇燃气设计规范 [S]．

[9] 臧子璇，黄小美，陈贝．燃气泄漏射流扩散模型及其应用[J]．煤气与热力，2011，31(6)：39-42．

[10] 黄小美，郭杨华，彭世尼，等．室内天然气泄漏扩散数值模拟及试验验证[J]．中国安全科学学报，2012，22(4)：27-31．

本论文原发表于《科学技术与工程》2020年第20卷第33期。

二、油井管与管柱失效预防

复杂工况油套管柱失效控制与完整性技术研究进展及展望

冯耀荣[1] 付安庆[1] 王建东[1] 王 鹏[1] 李东风[1]
尹成先[1] 刘洪涛[2]

(1. 中国石油集团石油管工程技术研究院·石油管材及装备材料服役行为与结构安全国家重点实验室;2. 中国石油塔里木油田公司)

摘　要：随着对深层碳酸盐岩、新区、页岩等油气勘探开发力度的加大,油气田的地层条件和介质环境变得更为苛刻复杂,油套管柱变形、泄漏、腐蚀、挤毁、破裂等失效事故时有发生;加上特殊结构井和特殊工艺井、特殊增产改造措施等对油套管柱提出的新要求,油套管柱面临着一系列新的挑战和难题需要破解。为此,围绕我国石油天然气工业增储上产的迫切需求和深井超深井、特殊结构和特殊工艺井、强酸/大排量高压力反复酸化压裂增产改造等复杂工况油套管柱失效频发的技术难题。历经十余年攻关。取得了一系列试验研究进展与重要技术成果,主要包括:(1)形成了基于气井全生命周期的油套管腐蚀选材评价、腐蚀控制和油套管柱完整性技术;(2)形成了油套管柱螺纹连接结构和密封可靠性设计评价及配套技术;(3)研发并形成了低渗透致密气井经济高效开发"API 长圆螺纹套管+CATTS101 高级螺纹密封脂"套管柱技术;(4)建立了高温高压等复杂工况油套管柱结构和密封完整性试验平台和评价技术。进而根据当前国家大边提升油气勘探开发力度的新要求和面临的新问题,提出了继续深入开展深层、酸性环境、页岩气等复杂工况油套管柱失效控制与完整性研究攻关的若干建议。结构认为,上述技术成果有力地支撑了我国重点油气田的经济有效开发。

关键词：复杂工况;套管;油管;失效控制;腐蚀控制;特殊螺纹;完整性评价;实物试验;井完整性

我国每年油套管消耗量介于 $(300\sim350) \times 10^4$ t,耗资 $(250\sim300)$ 亿元[1]。在油气勘探开发和生产过程中,油套管柱承受拉伸/压缩、内压/外压、弯曲等复杂载荷作用,同时会遭受油/气/

水、$H_2S/CO_2/Cl^-$等井下介质和温度作用。随着深井超深井、特殊结构和特殊工艺井、强酸/大排量高压力反复酸化压裂增产改造等工况条件日益复杂，油套管柱失效频发，严重制约油气田正常生产。近年来，我国油套管柱失效概率介于10%~20%，高温高压气井油套管柱泄漏一度超过40%。例如我国西部某油田2008—2012年油管柱发生腐蚀断裂失效123井(次)[2,3]，在完井过程中因油管柱失效造成的经济损失达7.24亿元，高产天然气井每口井修井费用高达(3000~5000)万元。每年因油套管柱失效造成的经济损失高达数十亿元[1]。然而现有的油套管柱失效控制技术只能解决常规油气井勘探开发过程中的失效问题，但是不能有效控制高温高压气井、非常规、特殊工艺和特殊结构井等复杂工况油套管柱的失效问题。复杂工况油套管柱的完整性与失效控制是国际上研究的热点和重大难题，也是我国石油天然气工业增储上产的瓶颈问题。这是一项十分复杂的系统工程问题，迫切需要通过系统研究加以解决。

针对我国复杂油气田开发中的油套管柱完整性技术需求，中国石油集团石油管工程技术研究院联合相关单位，依托"油井管柱完整性技术研究"等[4-6]多项重大科技项目，从油套管柱服役工况和失效分析入手，重点围绕西部高温高压气井油套管柱严重腐蚀泄漏和断裂、低渗透致密油气井油套管柱泄漏和经济高效开发等技术难题开展系统研究攻关，建立了复杂工况油套管柱结构和密封完整性试验平台，攻克了复杂工况油套管柱完整性技术与失效控制难题，为我国重点油气田高效勘探开发和安全生产提供了重要技术支撑。

但是随着深层碳酸盐岩、新区、页岩油气等勘探开发力度的加大，油气田的地层条件和介质环境变得更为苛刻复杂，油套管柱变形、泄漏、腐蚀、挤毁、破裂等失效事故时有发生，有的区块还十分严重；特殊结构井和特殊工艺井、特殊增产改造措施等对油套管柱提出了新的要求，油套管柱面临着一系列新的挑战和难题仍需要破解。

1 高温高压气井油套管柱腐蚀与防护技术

针对高温高压气井油套管柱均匀腐蚀、点蚀、缝隙腐蚀、应力腐蚀及化学—力学协同作用技术难题，形成了基于高温高压气井全生命周期的油套管腐蚀选材评价技术，研发了超级13Cr酸化缓蚀剂，自主研制了油套管实物应力腐蚀试验系统，形成高温高压气井油套管腐蚀与应力腐蚀控制技术、油套管柱完整性技术和管理规范[4-7]。

1.1 高温高压气井全生命周期的油套管腐蚀选材评价及控制技术

系统考虑油气井在作业生产过程中鲜酸酸化—残酸返排—凝析水—地层水四种典型的服役环境和具体的工况参数，以均匀腐蚀速率为参考、局部腐蚀速率为依据进行综合试验评价，形成油气井管柱全生命周期腐蚀评价新方法。首次揭示了13Cr油管、15Cr油管、V140低合金高强度套管等八种材料在鲜酸酸化—残酸返排—凝析水—地层水各个作业过程中的单环境和连续多环境中的腐蚀规律和协同作用机制。鲜酸—残酸—凝析水—地层水全过程的平均腐蚀速率及点蚀速率均大于各个独立过程之和，试验时间超过60天后实验数据趋于稳定[图1(a)]。揭示了温度、CO_2分压、Cl^-含量和流速等主要因素对油气井管材腐蚀的影响规律，构建了高温高压气井油管选材图[图1(b)]。基于多年的失效分析和试验研究，建立了高温高压气井油套管腐蚀数据库，包括井筒和工况数据采集、失效数据统计、失效数据分析，以及失效预警等四大功能。发明了油管腐蚀程度预测方法[8]，建立了含非均匀腐蚀缺陷油套管强度评价和寿命预测模型及软件，用于腐蚀油套管的完整性评价及安全预警。

1.2 超级13Cr油管系列酸化缓蚀剂

建立了适用于高温高压气井酸化增产工况的"空间多分子层多吸附中心"高温酸化缓蚀

模型(图2),该模型有效耦合多层大分子的屏蔽效应和小分子的填充效应,从而有效抑制酸化增产工艺中的酸腐蚀。通过百余次的喹啉季铵盐、曼尼希碱与多种金属离子(如 Cu^+、Ca^{2+}、Al^{3+} 等)复配试验,揭示了多分子多离子复配的酸化缓蚀剂的混合型缓蚀机理:季铵盐能有效抑制酸液中氢的作用,曼尼希碱能有效抑制酸液中氯的作用,金属离子则能提高缓蚀剂的耐温性和成膜性,因此复配后的酸化缓蚀剂呈现出同时抑制阴极和阳极的混合型缓蚀效果,如图3所示。应用上述模型和理论,研发了 TG201、TG201-Ⅱ、TG202 等超级 13Cr 油管系列酸化缓蚀剂产品[9],列入中国石油自主创新产品,对于超级 13Cr 油管酸化环境的缓蚀效率良好,从 2007 年起在塔里木库车山前高温高压气井应用 120 余井次,有效解决了塔里木油田高温高压气井酸化压裂过程中超级 13Cr 油管柱严重腐蚀问题。

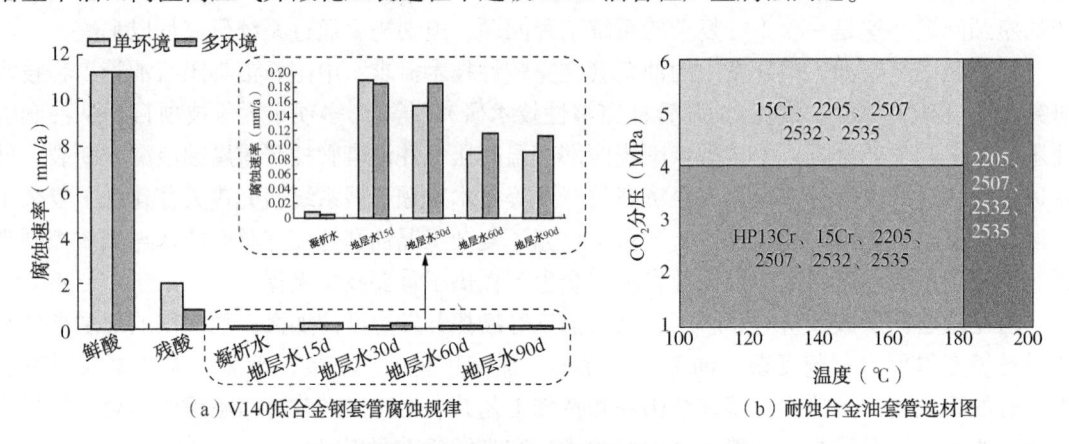

(a) V140低合金钢套管腐蚀规律　　(b) 耐蚀合金油套管选材图

图1　高温高压气井油套管腐蚀规律及选材图

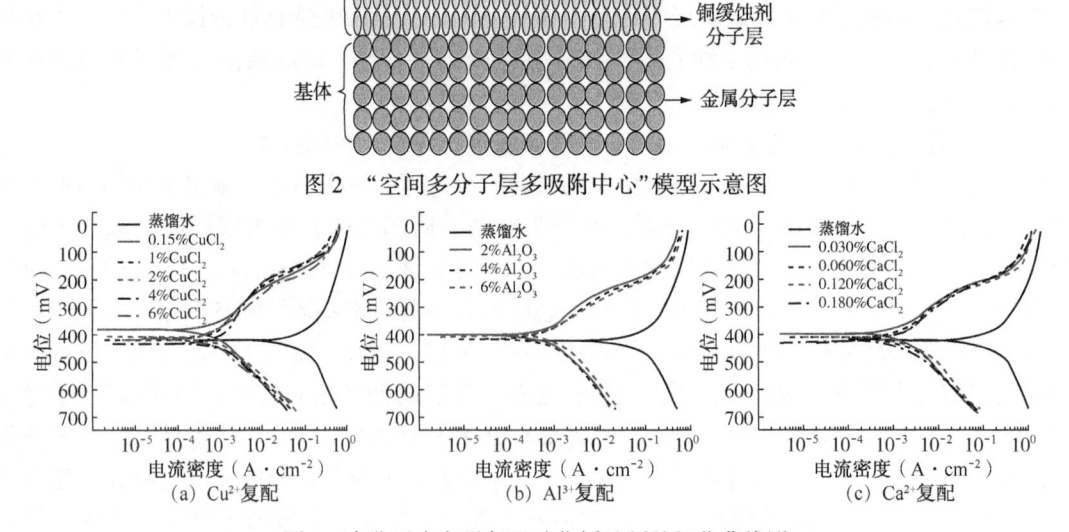

图2　"空间多分子层多吸附中心"模型示意图

(a) Cu^{2+} 复配　　(b) Al^{3+} 复配　　(c) Ca^{2+} 复配

图3　多分子多离子复配酸化缓蚀剂的极化曲线图

1.3　油套管全尺寸应力腐蚀试验系统

针对我国西部油气田超深、超高压、高含 CO_2 和 Cl^-、强酸大排量增产改造等严酷工况条件,自主设计研发了集环境介质、载荷、结构、材料于一体的全尺寸油套管应力腐蚀试验系统[10]，如图4所示，建立了油套管实物腐蚀试验流程和方法，实现了100MPa内压、1000kN拉伸载荷、200℃高温、油/气/水多相腐蚀介质等极端工况的多参量实验模拟，解决了油套管螺纹接头在应力作用下腐蚀和密封耦合作用试验评价难题及尺寸效应问题。系统开展了N80和V140低合金钢油套管、超级13Cr油管、W-Ni-P内涂层油管的实物应力腐蚀实验研究，首次发现了超级13Cr油管在模拟酸化压裂复杂工况下的点蚀—应力腐蚀失效过程和机制，如图5所示。研究揭示了V140低合金套管和超级13Cr油管的缝隙腐蚀特性，综合考虑腐蚀介质对密封面长度、表面洁净度、密封面接触压力的影响，提出采用折减系数来表征腐蚀介质对螺纹接头密封影响的方法，建立了腐蚀环境下特殊螺纹接头的密封评价准则。

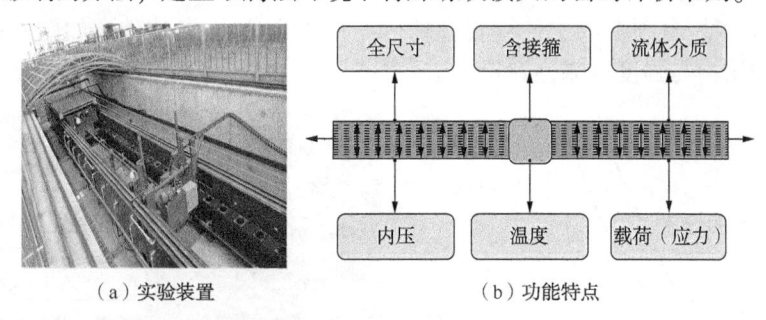

（a）实验装置　　　　　　　（b）功能特点

图4　油套管全尺寸应力腐蚀试验系统[11,12]图

1.4 超级13Cr油管应力腐蚀断裂控制技术

超级13Cr油管在甲酸盐环境中曾多次发生腐蚀穿孔或断裂，使用时间最短的只有11天。通过系统失效分析和实验研究，揭示了超级13Cr油管在磷酸盐体系中的失效规律、影响因素和断裂机理。其断裂机制表现为阳极溶解膜致损伤机理和裂纹沿马氏体多尺度结构界面扩展及有害第二相促进裂纹扩展[图5(b)]。研发了超级13Cr油管应力腐蚀断裂控制技术，以低开裂敏感的甲酸盐完井液体系替代磷酸盐完井液体系，2015年以来在塔里木油田应用36口井，至今未出现环空带压或油管柱腐蚀断裂失效。

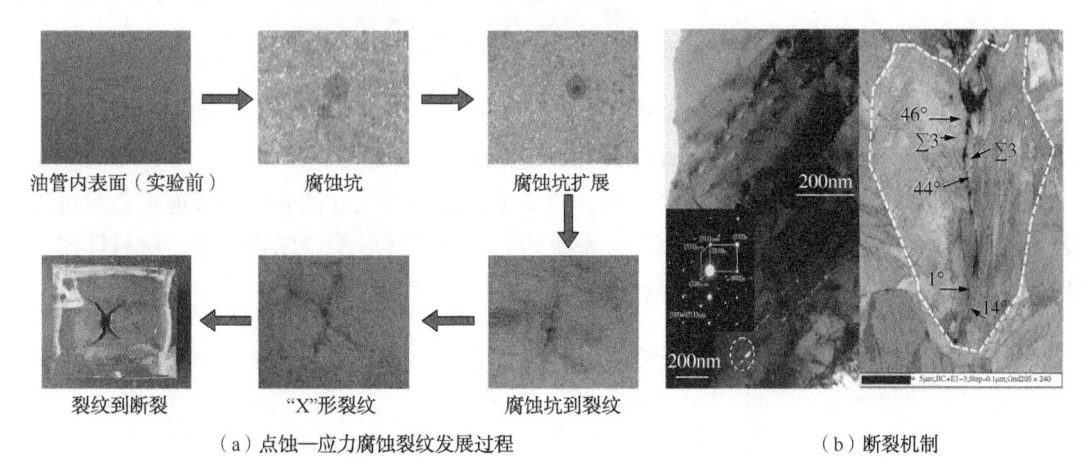

（a）点蚀—应力腐蚀裂纹发展过程　　　　　　　（b）断裂机制

图5　超级13Cr油管点蚀—应力腐蚀断裂过程[11,12]及机制图

1.5 油气井管柱完整性技术和管理规范

在系统总结高温高压及高含硫气井油套管柱研究成果和实践经验的基础上，制定了《油气井管柱完整性管理》行业标准[13]，编制了《高温高压及高含硫井完整性指南》[14]《高温高压及高含硫井完整性设计准则》[15]《高温高压及高含硫井完整性管理规范》[16]并推广应用，

塔里木油田井完整性从70%提高到79%。

2 油套管柱优化设计与可靠性技术

建立了套管柱结构和密封可靠性设计与评价方法，发明了水平井用新型特殊螺纹套管及制备技术，研发了低渗透致密气直井"API长圆螺纹套管+CATTS101高级螺纹密封脂"套管柱技术，在重点油气田得到应用[4-6]。

2.1 高温高压气井套管柱结构可靠性设计与评价方法

研究确定了高温高压气井套管柱的主要失效模式，构建了用故障树法计算套管柱失效概率的方法，建立了用分项系数法计算套管柱可靠性的方法[17]和软件，如图6所示，制定了《油气井套管柱结构与强度可靠性评价方法》[18]行业标准。

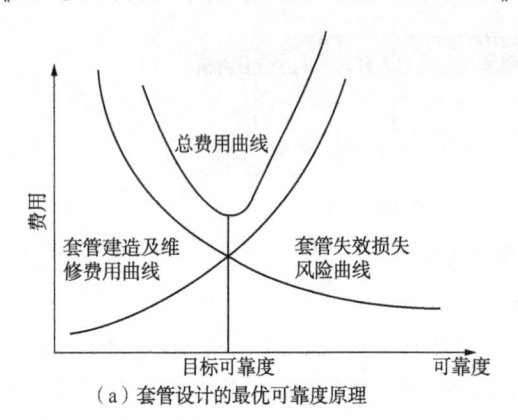

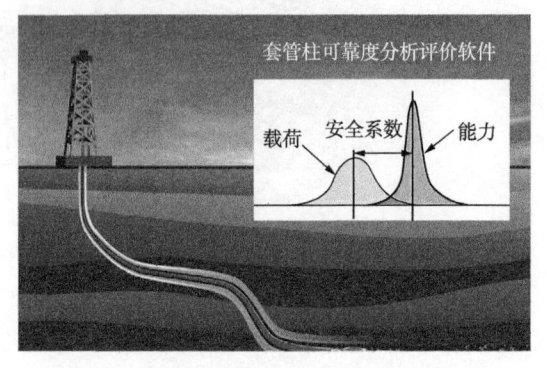

(a) 套管设计的最优可靠度原理   (b) 套管柱可靠性分析评价方法和软件

图6 套管柱可靠性设计分析与评价图

2.2 高温高压气井油套管柱密封可靠性设计与评价方法[4]

建立了特殊螺纹接头密封准则，即密封抗力(W_a)：

$$W_a = \int_0^l p_c^n(l)\,\mathrm{d}l \geq B\left(\frac{p_\text{gas}}{p_\text{atm}}\right)^m$$

式中：p_c为密封面接触压力，MPa；l为接触长度，m；p_gas为气密封内压力，MPa；p_atm为大气压力，0.1MPa；$n=1.4$；$B=0.01$；$m=0.838$。

采用径向基函数和蒙特卡洛模拟方法，建立了高温高压气井油套管螺纹连接密封可靠性设计的极限状态方程、计算程序及判据，系统研究揭示了套管螺纹结构尺寸、材料性能、工作应力等对螺纹密封抗力的影响规律(图7)。

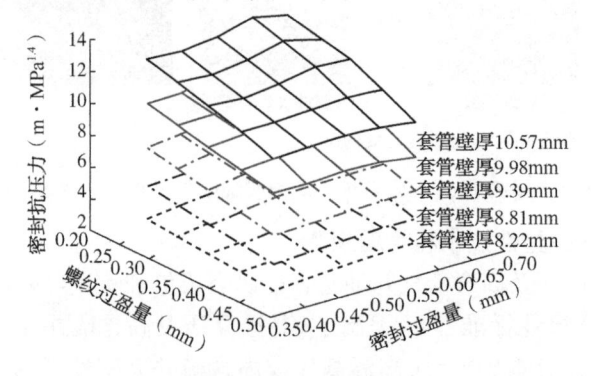

图7 套管壁厚、螺纹和密封过盈量对密封抗力的影响图

2.3 新型特殊螺纹套管设计研发及应用[4,19-22]

在深入研究油套管特殊螺纹密封机理、进行理论分析和试验研究的基础上，建立了考虑表面涂层和粗糙度影响的油套管特殊螺纹临界气密封压力计算方法（图8），用于特殊螺纹设计。

综合研究了拉伸/压缩、内/外压、弯曲等复合载荷作用下螺纹结构参数和公差、材料性能、密封面和螺纹过盈量等因素对其承载能力和密封可靠性的影响，设计研发了既安全又经济的水平致密气井用新型特殊螺纹套管及制备技术（图9），满足了4200m深、弯曲狗腿度20°/30m、液体压裂内压90MPa、气体生产压力50MPa、150℃水平井压裂改造和生产井工况下螺纹连接的强度和密封可靠性。

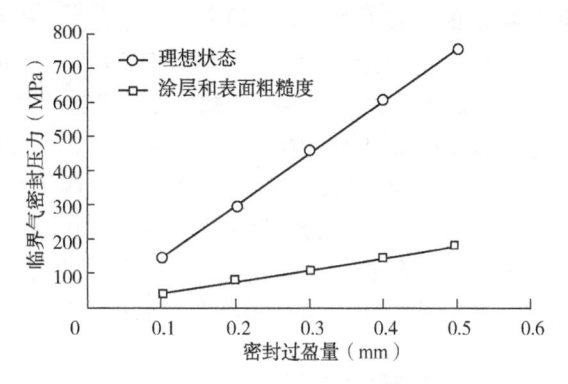

图8 临界气密封压力与密封过盈量的关系图

该套管螺纹密封面不易碰伤、螺纹易于加工和清洗，在宝鸡石油钢管有限责任公司、延安嘉盛石油机械有限责任公司等多个制造厂批量生产超过了$5×10^4$t，在长庆、延长、新疆等油气田推广应用，将长庆油田水平气井开发套管特殊螺纹由原来的7种统一为该螺纹，有效降低了管柱管理和使用成本。

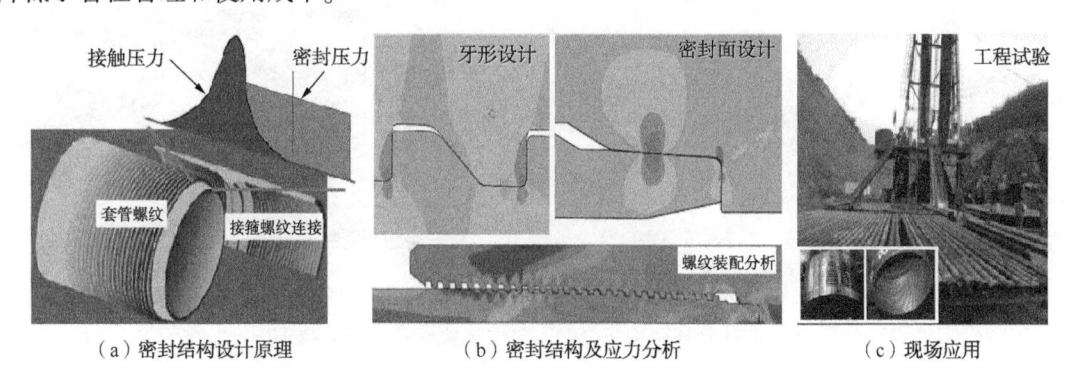

(a) 密封结构设计原理　　(b) 密封结构及应力分析　　(c) 现场应用

图9 新型特殊螺纹设计开发及现场应用图

2.4 "API长圆螺纹套管+CATTS101高级螺纹密封脂"套管柱技术[23,24]

系统开展了API长圆螺纹、偏梯形螺纹、特殊螺纹套管与API标准螺纹脂、CATTS101高级螺纹密封脂适用性试验研究，开发了"API长圆螺纹套管+CATTS101高级螺纹密封脂"套管柱技术，在中国石油长庆油田公司苏里格气田直井应用超过10000口井，在保证套管柱使用安全的前提下，套管成本降低了20%~30%。

3 油套管柱结构和密封完整性试验平台建设

建立了2500t全尺寸油套管复合载荷和环境试验系统，轴向+外压复合载荷挤毁试验系统，立式挤毁试验系统，热气循环试验系统，连接螺纹上卸扣试验系统等油套管柱结构与密封完整性试验平台，形成了高温高压气井、页岩气井、致密气井等油套管柱结构与密封完整性试验评价技术[7]。

该试验平台由9台试验系统构成（图10），是目前国际上最先进的平台之一，也是国内功能最为强大的试验平台，最大拉伸和压缩载荷为25000kN，最大内压为276MPa，最大弯矩为700kN·m，最大外压为210MPa，最高试验温度为500℃。可对外径介于88.9~406.4mm的所有钢级油套管在拉伸/压缩、内压/外压、弯曲复合载荷、温度循环条件下进行结构和密封完整性评价。其中，25000kN复合加载试验系统可进行高温外压试验，采用大口径单液缸加载方式，具有加载均匀、摩擦阻力小、寿命长等特点，将载荷轴直接作为力传感器，数据测量更为准确。发明了壁厚200mm抗外压至失效试验用挤毁缸的制造方法[25]，实现了挤毁试验机核心部件的国产化；发明了系列非API标准油套管实物评价试样制备方法[26-28]。该试验平台为塔里木、西南、长庆等十余个油气田和中国宝武钢铁集团有限公司、天津钢铁集团有限公司、湖南衡阳钢管（集团）有限公司等十余个制造厂提供试验评价技术服务超过15年，为油套管国产化、高性能油套管新产品研发及规模应用提供了技术支撑。

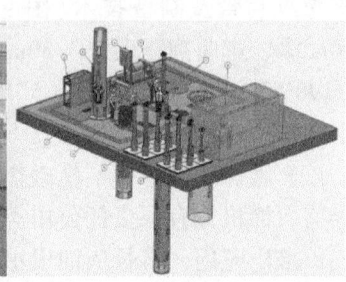

（a）25000kN复合加载试验系统　　（b）15000kN复合加载试验系统　　（c）立式快速挤毁试验系统

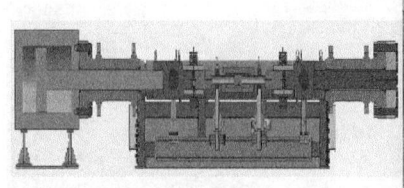

（d）25000kN复合加载试验系统结构图　　（e）复合挤毁试验系统　　（f）厚壁挤毁缸

图10　油井管柱结构和密封完整性试验平台及实验技术图

4　油套管柱面临的新挑战及建议

2018年以来，面对我国国民经济发展和人民生活对油气资源的巨大需求，党和国家领导人要求大力提升油气勘探开发力度、保障国家能源安全。国内4大石油天然气公司立即安排部署，实施了加大油气业务发展的计划，在塔里木盆地、准噶尔盆地、四川盆地、渤海等多个油气田取得了重大油气发现。随着古老碳酸盐岩、深层、新区、页岩油气等勘探开发力度的加大，油气田的地层条件和介质环境变得更为苛刻复杂，油套管柱变形、泄漏、腐蚀、挤毁、破裂等失效事故时有发生，在有的区块还十分严重，特殊结构井和特殊工艺井、特殊增产改造措施等对油套管柱提出了新的要求，油套管柱完整性面临着一系列新的挑战和难题仍需要破解。建议如下。

（1）持续深化复杂工况油套管柱失效控制与完整性理论和技术研究。持续发展油套管柱可靠性设计、应变设计方法及评价技术，油套管柱失效分析诊断预测预防技术，智能管柱技术（含专家系统），高强、高韧、高抗挤、耐腐蚀、特殊螺纹、铝（镁）合金、钛合金、复合材料等高性能油套管及应用技术，特种缓蚀剂和表面防护技术等研究攻关。

（2）持续开展复杂深层、高温高压、酸性气田等油套管柱失效控制与完整性技术研究。我国西部油气田是油气增储上产的主战场，7000m以上的超深井逐渐增多甚至超过8000m，井底温度大于180℃甚至超过200℃，压力大于100MPa甚至超过130MPa，且处于高含CO_2/Cl^-或H_2S/CO_2/Cl^-环境，面临复杂地质条件、强酸酸化、水平井大排量高压力反复压裂等复杂工况条件，对油套管柱的完整性和安全可靠性提出了更高要求，亟待深入开展高温高压、酸性气田等复杂工况油套管柱结构和密封完整性及适用性评价技术，复杂工况油套管柱腐蚀规律与机理、选材评价、腐蚀控制及预防技术等研究攻关。

（3）持续开展页岩气井套变机理与预防技术研究。页岩气开发是我国油气工业新的增长点，但深层页岩气开发井套管柱变形问题十分突出，套变率一度达到50%，导致桥塞等工具无法正常下入，局部区块丢弃长度达到1/3，严重影响了页岩气开发及产能建设。初步分析研究表明[6]，套管柱变形是套管柱使用性能、高压力大排量反复体积压裂工艺、地层裂缝等因素综合作用的结果。虽然近年来一直在持续研究，也采取了一些措施，但套损问题仍未得到很好的解决。从根本上来说，页岩气井套变取决于套管柱的承载能力与载荷的相对大小及分布。拟以套管柱为核心，优化套管柱设计，建立基于应变的套管柱设计新方法；提高套管柱的服役性能，选用抗剪切和抗外挤等使用性能优良的高钢级厚壁套管，发展模拟套管实际使用工况的使用性能评价技术；优化页岩气井井位选择、井距排布及井眼轨迹控制，减小天然裂缝、断层等对套变的影响；优化压裂工艺，减小断层滑移、天然裂缝扩展、水力裂缝扩展及相互干扰等对套变的影响。通过系统研究，建立相应的设计、选材、评价、工程应用和现场作业技术和标准体系。

（4）深井超深井复杂工况套管柱疲劳断裂失效及预防措施研究。2018年以来，新疆、长庆等油气田多次发生套管疲劳断裂事故，严重影响油气田开发。初步分析研究表明：套管断裂是由于水平井大排量高压力反复水力压裂及套管反复上提—下放引起的疲劳失效。需要进一步研究套管断裂的机理和原因、套管结构强度和寿命、作业工艺对断裂的影响、套管承受的动态载荷及控制措施、套管扣型选用评价方法、套管选用技术规范、现场作业规范等，系统解决套管柱的疲劳断裂失效问题。

（5）老油气田套损机理与防治技术研究。套损一直是制约油气田产量和效益的顽疾，多数油气田套损率超过15%。主要失效模式有变形、腐蚀、挤毁、泄漏、错断、破裂等，多是力学因素和/或化学因素对套管柱耦合作用的结果。老油气田复杂的地层和环境条件及其演变对套损有重要影响，稠油热采、火驱、CO_2驱、空气驱/空气泡沫驱等套管柱的变形和腐蚀问题仍然比较严重，这些问题需要针对性地持续研究加以解决。

（6）开展大数据和人工智能在油套管柱失效控制及预测预防中的应用研究。近年来，数据科学、人工智能、机器学习、材料信息学等发展迅速[29,30]，将这些学科的最新技术应用到石油管工程和材料服役安全领域，将会产生事半功倍的效果。重点是在本领域知识库的基础上，通过油套管柱材料性能和服役性能、力学状态分析、运行状态监检测等数据采集和失效数据库及案例库构建，数据解释，机器学习建模，模型评估，实验/计算设计，实验/计算，再反馈到前端的反复循环，辨识规律、机理及影响因素，提出控制和预防措施，确保油套管柱的运行安全。

应当强调指出，油套管柱的完整性保障与失效控制，就是要保障油套管柱全生命周期的安全可靠性和经济性，而70%以上的油套管柱失效是由于前期的设计/选材/评价和不恰当的工程作业及增产改造措施造成的。所以,加强油套管柱优化设计、选材、试验评价,综合考虑油套管柱设计寿命、失效概率、失效后果等因素,发展基于风险的油套管柱可靠性设计新方法及配套技术,是控制油套管柱失效、保障管柱完整性的关键。

致　　谢

中国石油集团石油管工程技术研究韩礼红、白真权、韩新利、刘文红、张娟涛、韩燕、龙岩等,中国石油大学(华东)闫相祯,加拿大C-FER公司谢觉人,中国石油塔里木油田公司谢俊峰等参加了相关试验研究工作,特此致谢!

参　考　文　献

[1] 冯耀荣,马秋荣,张冠军.石油管材及装备材料服役行为与结构安全研究进展及展望[J].石油管材与仪器,2016(1):1-5.

[2] 吕拴录.塔里木油田油套管失效分析及预防[C].塔里木油田井筒完整性会议,2013.

[3] 冯耀荣,韩礼红,张福祥,等.油气井管柱完整性技术研究进展与展望[J].天然气工业,2014,34(11):73-81.

[4] 中国石油天然气集团公司石油管工程重点实验室.油井管柱完整性技术研究[R].西安:中国石油集团石油管工程技术研究院,2014.

[5] 中国石油天然气集团公司石油管工程重点实验室.复杂工况气井油套管柱失效控制与完整性技术研究[R].西安:中国石油集团石油管工程技术研究院,2016.

[6] 中国石油天然气集团公司石油管工程重点实验室.苛刻服役条件油井管工程应用基础研究[R].西安:中国石油集团石油管工程技术研究院,2019.

[7] 冯耀荣,张冠军,李鹤林.石油管工程技术进展及展望[J].石油管材与仪器,2017(1):1-8.

[8] 王鹏,陈光达,宋生印,等.一种油管腐蚀程度预测方法及装置:中国,201210548324.X[P].2017-10-17.

[9] 尹成先,冯耀荣,白真权,等.一种用于含Cr油管的高温酸化缓蚀剂:中国,200710178677.4[P].2010-09-29.

[10] 白真权,韩燕,张娟涛,等.一种管材实物应力腐蚀试验机:中国,201210165009.9[P].2015-10-14.

[11] Lei XW, Feng YR, Fu AQ, et al. Investigation of stress corrosion cracking behavior of super 13Cr tubing by full-scale tubular goods corrosion test system[J]. Engineering failure analysis. 2015, 50: 62-70.

[12] 付安庆,史鸿鹏,胡垚,等.全尺寸石油管柱高温高压应力腐蚀/开裂研究及未来发展方向[J].石油管材与仪器,2017,3(1):40-46.

[13] SY/T 7026—2014油气井管柱完整性管理[S].

[14] 吴奇,郑新权,张绍礼,等.高温高压及高含硫井完整性指南[M].北京:石油工业出版社,2017.

[15] 吴奇,郑新权,张绍礼,等.高温高压及高含硫井完整性设计准则[M].北京:石油工业出版社,2017.

[16] 吴奇,郑新权,邱金平,等.高温高压及高含硫井完整性管理规范[M].北京:石油工业出版社,2017.

[17] 樊恒,闫相祯,冯耀荣,等.基于分项系数法的套管实用可靠度设计方法[J].石油学报,2016,37(6):807-814.

[18] SY/T 7456—2019油气井套管柱结构与强度可靠性评价方法[S].

[19] Wang Jiandong, Feng Yaorong. Economy and reliability selection of production casing thread connections for low pressure and low permeability gas field[J]. Procedia Environmental Sciences, 2011, 11: 989-995.

[20] 王建东,冯耀荣,林凯,等.特殊螺纹接头密封结构比对分析[J].中国石油大学学报(自然科学版),2010,34(5):126-130.

[21] 王建东,冯耀荣,林凯,等.高气密封油套管特殊螺纹接头:中国,200910092027.7[P].2012-11-14.

[22] 王建东, 杨力能, 冯耀荣, 等. 低压用气密封特殊螺纹接头: 中国, 201110229616.2[P]. 2015-12-2.

[23] 王建东, 林凯, 赵克枫, 等. 低压低渗苏里格气田套管柱经济可靠性优化[J]. 天然气工业, 2007, 27(12): 74-76.

[24] 王建东, 林凯, 赵克枫, 等. 低效气田套管经济可靠性选择[J]. 石油钻采工艺, 2007, 29(5): 98-101.

[25] 王蕊, 李东风, 张森, 等. 一种抗外压至失效试验用挤毁缸的制造方法: 中国, 201110251302.2[P]. 2014-09-03.

[26] 王蕊, 李东风, 韩军, 等. V150钢级油井管全尺寸实物试验试样制备方法: 中国, 201310109959.4[P]. 2016-07-13.

[27] 娄琦, 张森, 李东风, 等. 一种V140钢级油井管试验实物的制备方法: 中国, 2011100437217[P]. 2014-08-06.

[28] 李东风, 张森, 韩新利, 等. 一种G3合金油井管试验实物焊接制备方法: 中国, 200910092835.3[P]. 2011-4-20.

[29] 王鹏, 孙升, 张庆, 等. 力学信息学简介. 自然杂志, 2018, 40(5): 313-322.

[30] 李晓刚. 材料腐蚀信息学[M]. 北京: 化学工业出版社, 2014.

本论文原发表于《天然气工业》2020年第40卷第2期。

复杂压裂页岩气井套管变形机制及控制方法

韩礼红[1]　杨尚谕[1]　魏风奇[2]　叶新群[2]　王建军[1]　王　航[1]　潘志勇[1]　张华礼[3]
岳文翰[3]　谢　斌[4]　舒振辉[4]　张　平[5]　路彩虹[1]　尹　飞[6]

(1. 中国石油集团石油管工程技术研究院·石油管材及装备材料服役行为与结构安全国家重点实验室；2. 中国石油勘探与生产分公司；3. 中国石油西南油气田分公司；4. 中国石油新疆油田分公司；5. 中国石油川庆钻探工程有限公司；6. 成都理工大学能源学院)

摘　要： 针对西南长宁—威远页岩气井复杂压裂套管柱大量变形进行了套变机理分析，认为套管以局部径向变形为主，基本模式包括外挤和剪切，并以外挤为主。套变属于应变控制机制，主要影响因素是地质运移，包括储层注液膨胀、裂缝及页岩滑移等，导致套管柱承受显著的非均匀载荷及应变作用。对潜在套变控制措施进行了分析，提出通过井筒一体化协调变形缓解套管变形的新方法。通过添加高强度空心颗粒对水泥浆进行改性，实现了局部地层位移发生位置水泥环局部破碎容纳地层位移的预期效果。对不同颗粒配比条件下套管变形减缓进行了全井筒模拟试验分析，获得西南页岩气井工程可用的优化窗口。在优化窗口内，套管柱的变形显著减缓并允许桥塞等工具顺利通过，可保证工程压裂顺利进行。

关键词： 页岩气井；套管变形；分段压裂；控制方法

页岩油气已成为中国的油气开发重点，以川渝地区的页岩气和新疆吉木萨尔地区页岩油为代表，其核心技术是沿用北美页岩革命的成功做法，采用长距离水平井进行复杂水力压裂。在工程压裂作业中，射孔枪及桥塞等工具须下入套管柱进行作业，后者须保持有效的通径，保障工具顺利通过。自2015年页岩油气进入工业开发以来，水平段套管大量变形，例如长宁-威远页岩气国家示范区比例长期在30%左右。这些套管局部变形严重限制了工具下入，制约了压裂作业，造成丢段等，造成油气产能下降。大量的套管变形已构成中国页岩油气开发的重大瓶颈，急需有效的控制方法。

1　研究背景

传统的管柱设计采用强度方法，管柱是不允许屈服变形的[1]。随着井下工况的日益复杂，研究人员开始针对具体工况特征，研究管柱设计与选用方法，例如本文作者针对稠油热

基金项目：国家重点研发计划（项目编号：2019YFF0217500、2016ZX05022-005）；国家自然科学基金（项目编号：51574278、U1762211）；陕西省杰出青年科学基金（项目编号：2018JC-030）；中国石油集团科研项目（项目编号：2019A-3911）。

作者简介：韩礼红，男，1975年生，博士，教授级高级工程师，长期从事油气井管柱服役安全与控制技术研究。E-mail：hanlihong@cnpc.com.cn。

采井注蒸汽作业下的管柱屈服行为建立了系统的套管柱应变设计方法[2]。针对页岩气井管柱的后屈服行为，已经具有大量的基础理论研究，但在套管变形控制方面尚缺乏适用的做法及成熟的工业标准。

北美页岩油气开发中，也曾出现过套管变形问题，但针对性的研究成果较少。Carsero等人对全球页岩气开发中的套管变形进行了系统分析，认为储层所处构造特征是诱发套变的主要原因[3]。作者认为在构造特征明显地区，例如南美洲的 Nazca 平原、阿拉伯平原及印度平原等均毗邻山区地质构造环境下，套管变形明显，而中国四川地区构造特征尤其突出。这些地区地应力水平高且呈现明显的非均匀特征，页岩储层的层理面、断裂带及断层等易于激活，从而诱发套管变形。而北美的大多页岩气开发区所处构造特征并不明显，这与北美地区页岩油气井套管变形率不明显直接相关。与北美地区相比，中国西南地区页岩气普遍具有埋藏深、山区构造显著、地应力非均匀性强、地质裂缝、断层发育等特点，套管柱服役环境非常恶劣。

围绕西南地区页岩气套管变形，国内研究人员也已经进行了大量研究，包括压裂中复杂力学分析预测、地质裂缝预测、固井工艺优化、管柱力学校核等。练章华等认为预防套管变形不能仅仅依靠提高套管钢级与壁厚，并采用数值分析手段，对压裂工艺进行了优化，提出通过调整分级压裂间距等参数来减缓变形[4]。陈朝伟等认为，套管变形是压裂中地质裂缝滑移产生的，应重点进行地质裂缝的位置及活动行为预测，并通过提高水泥环强度、避免水泥环产生微环空等措施预防套管变形[5]。廖仕梦等针对变形段，提出多次射孔、分阶段压裂、变形区暂堵等工程作业措施，提供了工程继续作业措施[6]。李留伟等认为套管明显的剪切变形是由于高角度地质裂缝滑移造成的[7]。尹飞等通过数值仿真对裂缝滑移及其对套管变形影响进行了预测，结果与现场井套变测井数据基本一致[8]。麦洋、李奎等认为固井质量不好，产生的水泥环缺失会造成套管应力集中，诱发套管变形，并认为优化井眼轨迹有助于工具下入[9]。李勇、曾静、程小伟等认为可通过水泥环厚度、强度、韧性等调整来改善套管应力水平[10-12]。高德利等对页岩气套管柱力学行为进行了综合评述，认为页岩气套管变形与非均匀力学环境、固井质量差、地质裂缝等相关，需要综合考虑进行研究[13]。

综上所述，页岩油气井套管变形控制技术已成为研究热点与难点，迄今为止尚无系统性的解决方案。为缓解套管变形，保证顺利完成压力作业，工程上降低了压裂强度，包括压裂排水量、压力等，但套管变形率仍接近30%。本文以西南长宁—威远页岩气开发为例，对套变控制潜在技术进行了分析，并从套管柱的优化设计入手，将"管柱—水泥环—井筒"视为一体，通过水泥环的改性，实现井筒的协调变形，形成缓解甚至预防套管变形的新方法，获得全井筒模拟试验的有效验证。

2 套管变形模式与机理

2.1 套管变形模式

对大量页岩气套管变形井进行多臂井径检测，可以获得套管变形后的形貌特征，为预防变形提供研究依据。图1是典型的套变测井形貌，说明套管多段发生外挤变形，而在变形与非变形或外挤变形的过渡区则表现为剪切形貌。针对长宁—威远示范区的套变统计分析表明：外挤模式约占77%，而剪切约占23%，因此，在复杂压裂环境下，外挤变形是最主要的失效模式。

对套管变形位置的统计分析表明，变形在 A 点造斜段附近最集中，从 A 点到 B 点，变形比例逐渐减小，在 A 点到中点附近的套管变形比例超过 95%，如图 2 所示。这一分布规律和工程上压裂作业顺序的影响一致，工程压裂是从 B 点向 A 点逐步移动，每一次压裂造成的变形都会逐渐累积，越靠近 A 点，变形量越大。这一规律说明，套管变形的直接诱因就是压裂作业，而且套管变形具有明显的累积性。

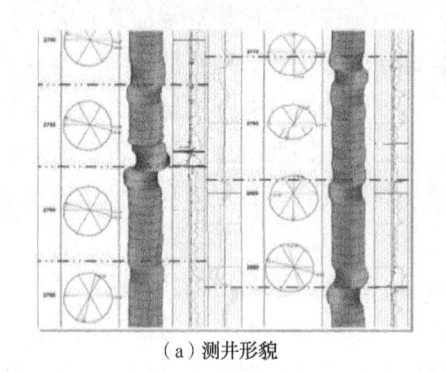

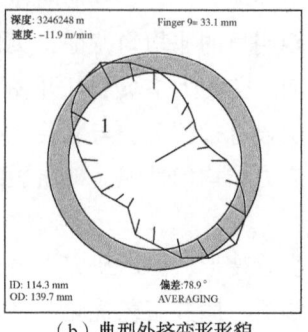

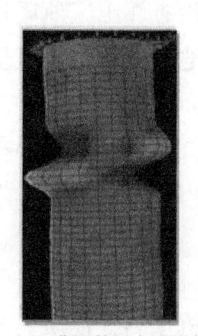

（a）测井形貌　　　　（b）典型外挤变形形貌　　　　（c）典型剪切变形形貌

图 1　页岩气水平井水平段套管变形典型形貌

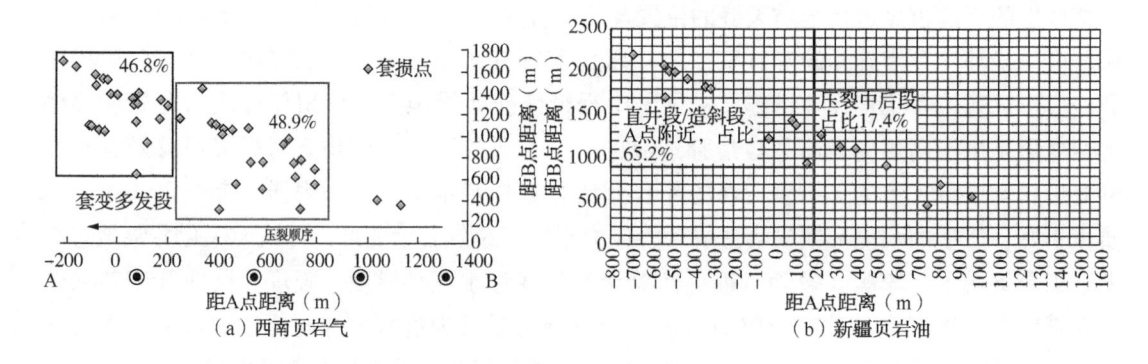

（a）西南页岩气　　　　（b）新疆页岩油

图 2　页岩油气井套变位置统计分布

2.2　套管的非均匀承载特性

长宁—威远示范区地质载荷非均匀性非常明显，对套管柱造成显著的非均布外挤载荷，与现有管柱设计标准采用的均匀外挤载荷分布差异显著，如图 3(a) 所示。在非均布外载条件下，超出均布载荷的区域极易发生挤毁，套管柱设计评价应依据套管非均匀承载特性规律进行。由于该地区地质裂缝发育［图 3(b)］，当裂缝滑移时将造成显著的定向集中载荷，造成剪切变形，这种情形可以视为极端的非均布载荷作用。对套管本身的承载特性研究表明，非均匀外载作用下，套管的等效应力分布具有显著的非均匀特征（图 4），该特征可以反映管柱真实服役能力，可用做管柱设计与评价的依据。

2.3　套管变形机制

套管为金属材料，具有典型的弹塑性行为特征，在低于屈服强度载荷作用下，管柱呈现出线弹性的应力—应变关系，这也是当前管柱强度设计方法的基本依据。当载荷超出屈服强度时，套管发生塑性变形，当变形过量时，套管将断裂。根据测井数据的套管变形反演结

(a）局部外挤造成的非均匀应力场

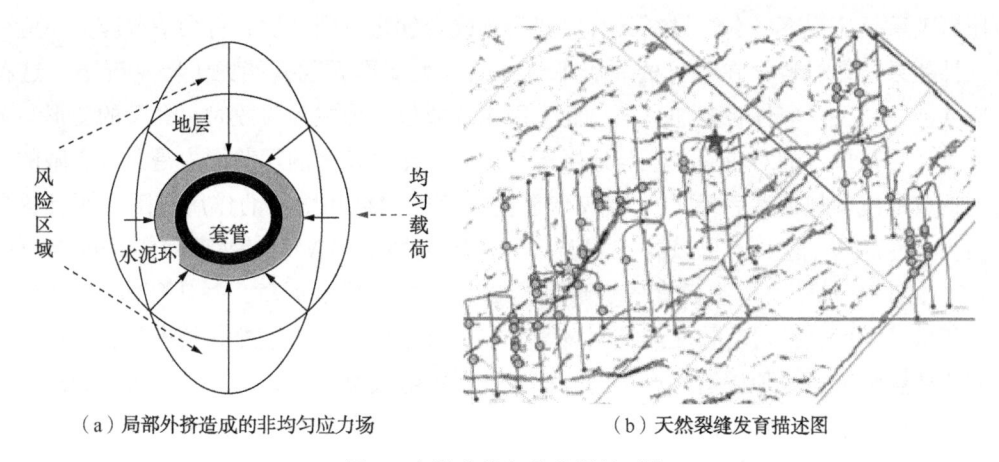

(b）天然裂缝发育描述图

图3 套管非均匀载荷服役环境

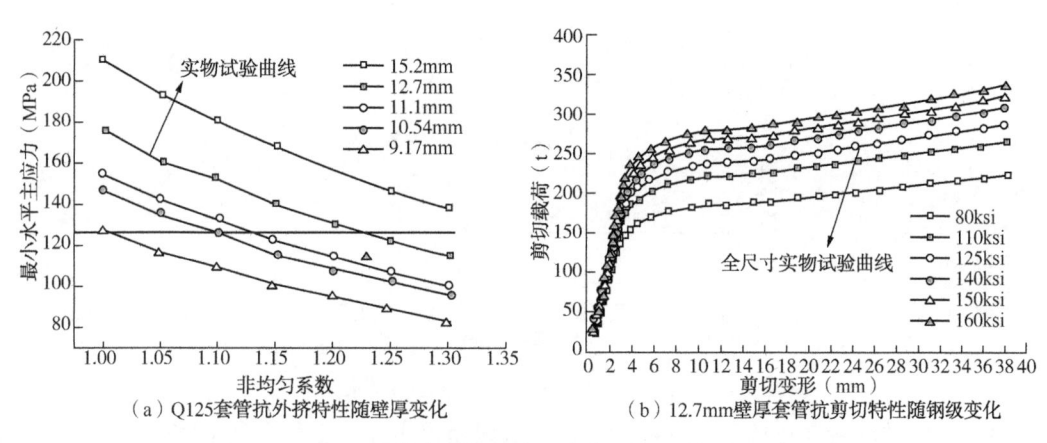

(a）Q125套管抗外挤特性随壁厚变化

(b）12.7mm壁厚套管抗剪切特性随钢级变化

图4 非均匀载荷环境下套管服役行为特征

果，结合工程用水泥环特性，通过大量试验，可以确定地层径向位移边界为40mm，套管径向变形量边界为25mm，可覆盖90%的套管变形点，如图5所示。

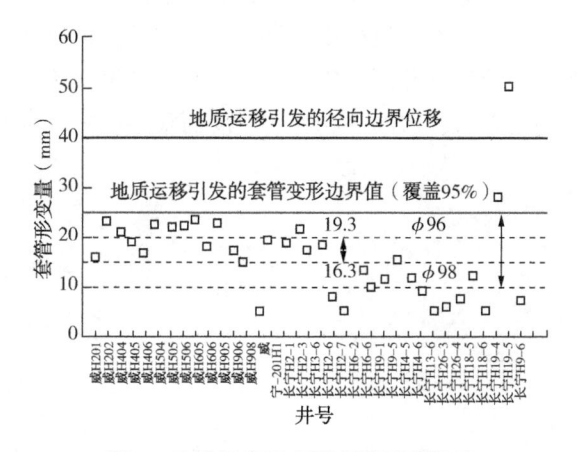

图5 地质运移及套管变形反演结果

依据工程井套管变形模式及变形量可以确定，压裂中套管发生显著的局部塑性变形。在这种模式下，套管的应力—应变曲线为非线性关系，需要采用材料的弹塑性本构来分析。在多级压裂及重复压裂下，套管材料具有应变强化或者软化效应，后者将会造成应力—应变的非单一性关系。在材料存在明显的屈服平台，或者大变形后承载能力下降情况下，这种应力—应变的非单一对应性将更加复杂。而采用应变或位移为主控参数时，套管的变形与地层的位移关系将是单一的对应关系。与地层运移效应相比，套管的承载能力是非常有限的，前者经常造成地震等规模性后果[14]，仅仅依靠提高套管钢级和壁厚的作用有限，而且容易造成井筒完整性问题，这一点已获得工程验证，油田现场采用 V140 钢级，15.2mm 壁厚的套管进行了 11 口井试验，压裂后套管变形依然显著，局部甚至更高，并伴随多口井漏发生，如图 6 所示。因此，针对套管变形控制技术研究，宜采用应变为主控参数进行管柱设计与评价，其主要影响因素是地层运移产生的位移，包括裂缝及页岩滑移，以及注入压裂液造成的储层体积膨胀，而在位移控制模式下，套管应在满足钻井及固井基础上降低其钢级与壁厚，以便更好地与地层协调变形。

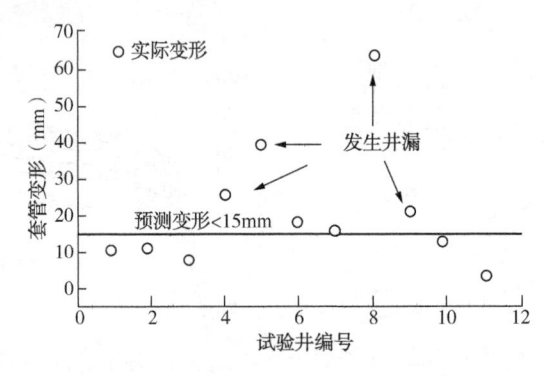

图 6 高钢级厚壁套管工程试验结果

页岩油气井套管变形属于位移控制，这一认识获得实际井工程大量案例验证，在大量套变井中，部分井在压裂作业前就发生了显著的套管变形，其共同特征就是附近的井经历了压裂作业，如图 7 所示。不同井之间存在显著的干涉效应，而他们之间唯一的物理纽带就是地层连通[15]。显然，压裂作业环境下，井筒所处的地层或储层发生了运移活动，对该区域内井下套管柱均造成位移作用。当然，载荷作用肯定是存在的，但是如前文所述其主控因素是应变或位移。工程套变井中的部分井固井质量较差，水泥环填充不足，甚至有局部管段几乎没有水泥环的支撑，但是压裂后，这部分套管变形反而很小，甚至没有，该现象恰恰验证了

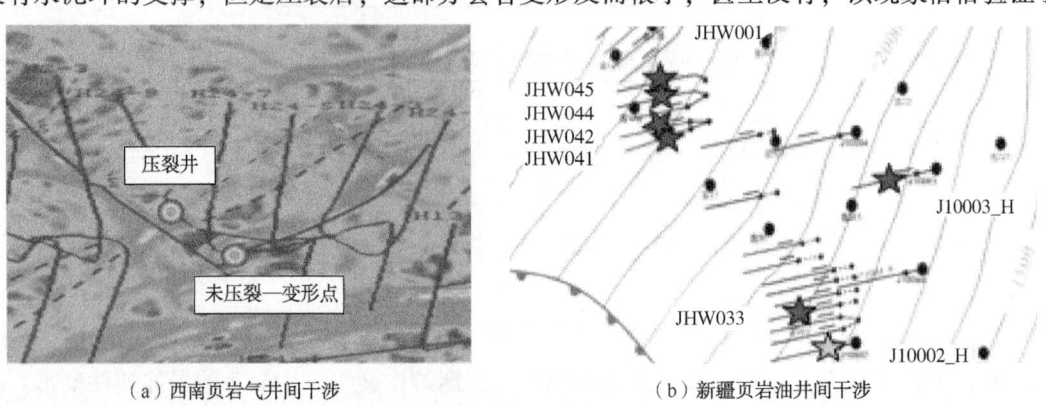

(a) 西南页岩气井间干涉　　(b) 新疆页岩油井间干涉

图 7 井间地质运移干涉案例

压裂造成地层运移活动，作用于套管，导致后者显著变形。

2.4 水泥环的影响

井筒包括套管柱和水泥环，前者通过后者与地层连接。地层与管柱的交互作用通过水泥环进行，但与套管相比，水泥环属于脆性材料，强度要低一个数量级，在套管变形前，水泥环已经破裂。如果水泥环破裂为分散状态，那么可以被地层作用压实，吸收部分地层位移而缓解套管变形，如果水泥环韧性较高，则破裂为大块状，会将地层位移直接传导至套管，并且因为应力集中加剧套管局部变形。对于油气井来说，核心是套管柱，其完整性与否直接影响工程作业，因此，从缓解套管变形角度来说，水泥环破碎呈分散状态是有利的，但水泥环填充质量较高时，其压实效果非常有限，因此，在满足工程作业条件下，大幅降低水泥环密度和韧性，同时适当提高水泥环厚度将有利于缓解甚至完全消除套管变形，其基本模型如图8所示。

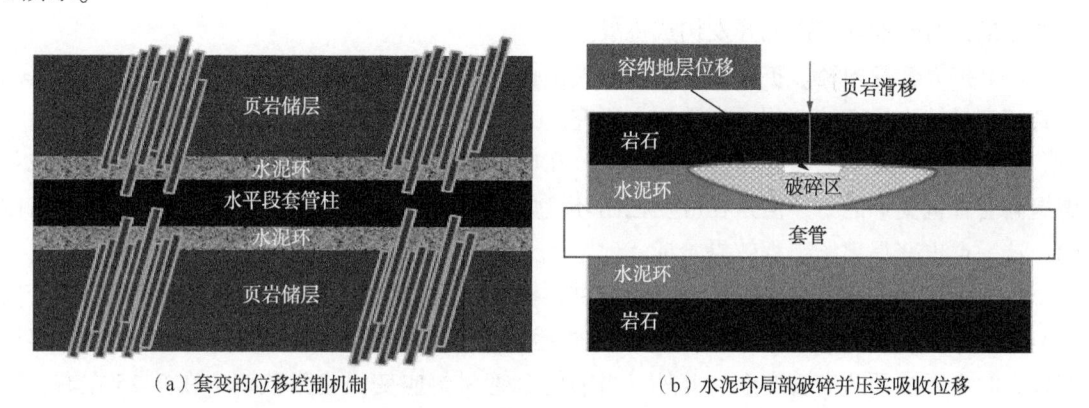

(a) 套变的位移控制机制　　(b) 水泥环局部破碎并压实吸收位移

图8　套管变形机理及控制思路

2.5 套管变形影响因素

页岩气井压裂作业涉及地质、工艺、环境等大量因素，与地质相关的参数存在极大的不确定性。然而，从套管分析入手，影响套管变形的直接参数为应力或应变。如上所述，两者相比，应变为主控参数。对于套管径向变形来说，井眼轨迹和作用到套管的径向位移是直接控制参数，因为两者直接导致套管的有效通径下降，而所有地质及工程因素对套管的影响均反映在这两方面。

3 潜在的套变控制方法

对页岩气井套管变形的控制，应从地质运移、作业工艺、井眼轨迹、水泥环及管材性能等方面进行变形来源及其量化的分析与评判。新的设计应确定管柱变形后有效通径，通过全井筒工况模拟试验验证，并满足固井工程作业需求。

3.1 潜在方向

(1) 量化地质运移，寻求适用套管。套管变形的主因是地质运移，在压裂作用下诱发。量化描述地质运移行为，选取适用套管是最直接的手段，工程上以及大量地质研究人员也在进行相关研究。然而，地质环境过于复杂，存在大量不确定性因素，诸如天然裂缝、地质分布及物性差异、地应力随机变化等，准确预测地质运移非常困难，数值模拟的准确度与可靠度都显著影响分析结果，而工程现场无法为此承担巨额的试验成本。

（2）井眼轨迹优化。通过井眼轨迹的优化，降低因井眼弯曲造成的套管有效通径降低，提高井下工具的可通过性。从现场作业来看，目前井眼狗腿度基本可以控制在 3°/30m，对于水平井来说，进一步降低的可行性及作用有限。

（3）井下工具的规格调整。对井下工具，如射孔枪、桥塞等规格调整主要为外径与长度，针对外径的调整最显著，可以直接提高工具可通过性。然而，相关工具的研发周期漫长，无法满足工程需求。

（4）压裂工艺的调整。复杂压裂的本质就是利用高压流体在储层内制造网格化的裂缝，最大程度获取资源。从理论上讲，降低压裂工艺可以降低地质运移强度，减缓甚至不会激活裂缝滑移等行为，从而使得套管服役环境重回传统的应力控制或者微量塑性变形范畴。此时，通过管柱强度力学设计，优化套管是可以缓解甚至消除套管变形的。然而，降低压裂工艺，如加砂量、注水量，以及调整分段压裂簇数及密度等，将直接影响压裂后裂缝网格化及产能效果，对于油气田产能开发构成挑战。

（5）扩大套管内径。扩大套管内径可直接提高其通径，有利于桥塞等通过，但是桥塞与套管的密封性、桥塞的适用性尚需深入研究。

（6）提高套管钢级与壁厚。理论上讲，在应力控制环境下，提高套管钢级与壁厚可以显著提高套管抗变形能力，但是在应变控制环境如地层显著滑移下，套管的抗变形能力是有限的，这一点也获得现场工程效果验证。

（7）采用抗大应变套管。从材料角度讲，提高套管材料的应变强化能力，可以实现套管"小应变大载荷"服役效果，这种措施在地质运移应变较小的情况下会有效，但地质运移应变较大时也无法预防套管变形。此外，当前的套管材料应变强化指数基本在 0.25 左右，追求高应变强化效果，尚需从材料成分及工艺设计入手，开展系统的攻关，需要较长的周期。

（8）双层套管设想。双层套管理论上为外层套管留出了变形空间，保护内层套管。除制造复杂外，其缺点是两层套管连接处将是瓶颈，易于变形，而井下管柱属于木桶效应，其完整性由瓶颈点主控。另外，在注水压裂过程中，压裂砂液将进入两层套管间隙，对内管造成外挤作用，用于分段堵塞的桥塞密封性风险显著。当内外管间隙填满砂后，将丧失原始设计的功能。

（9）裂缝暂堵。工程上采用地质裂缝暂堵的方式实施压裂，这种方法需要明确地层原始裂缝位置、取向及几何参数等。如果能实现压裂过程中暂堵的裂缝不开裂，不会激活裂缝滑移，套管变形将会降低甚至消除，相关的研究还需大量理论分析及工程试验进行验证。另外，地层裂缝的几何及二次激活载荷的随机性都对工程统一做法造成显著障碍。

（10）套管外包覆弹性体。在地质运移应变较小环境下，这种设想可以吸收部分变形，保护套管，对套变控制会有一定的效果，而当地层位移明显时，其外衣的弹性空间能否完全容纳地层位移是个问题。另外，其致命的弱点是由于储层滑移及膨胀等位置无法准确判定，水平段套管需要全部封装外衣，将对下套管作业带来额外的巨大摩阻，对水平段延伸极限，以及压裂后资源可获取程度都构成显著挑战。

（11）井筒协调变形。在井下密闭环境，套管—水泥环—地层连接为一体，地质运移应变通过水泥环直接传导至套管，产生变形。因此，地层运移应变由水泥环和套管共同承担，缓解甚至消除套管应变，可以从水泥环的改性吸收地层应变入手，通过井筒协调变形进行套变控制。这种思路需要满足井筒固井作业及水泥环的密封基本要求，而水泥环的位移吸收能力将可以通过增大厚度、提高空心度等手段调整，有望缓解甚至完全消除地质运移过程造成的套管变形。本文以下部分将按照该思路，通过试验对比方式，对其缓解套变效果进行阐述。

3.2 井筒协调变形控制套变基本方法

本文采用添加不同比例高强空心颗粒，对西南页岩气井工程固井用水泥浆进行混合改性。以工程上在用 $\phi139.7mm\times12.54mm$ 套管为主要对象，通过全井筒模拟试验装备施加径向外挤位移（最大位移采用前文测井反演的边界位移，即径向40mm），研究水泥环改性对套管变形影响，并与桥塞外径进行对比分析，以桥塞可通过变形套管为依据，提出工程上可用的井筒设计窗口。全井筒试样及模拟试验如图9所示。

（a）井筒全尺寸试样　　（b）试验后变形套管　　（c）全井筒试验全景图

图9　井筒全尺寸模拟试验

3.3 试验结果与分析

（1）水泥浆改性的作用。套管外水泥环的强度是远远低于套管的，因此，在井下发生明显的地质运移时，地层载荷首先作用于水泥环，使其破裂，之后载荷再传递给套管。而本文水泥浆改性的目的就是使其易于局部碎化，具有一定的可压缩空间，容纳地层位移，从而缓解套管变形。而在远离地层运移（如裂缝滑移）的位置，水泥环将保持完好。图10给出了模拟试验结果，初始完好的空心颗粒与水泥混合，固井后在外部位移作用下发生破碎，从而提高水泥环的可压缩性，而在远离外加位移的位置，水泥环保持完整。由于改性水泥环吸收了部分地层位移，套管变形得到明显缓解。

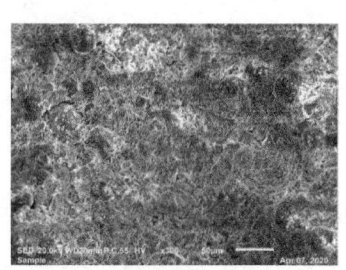

（a）空心颗粒　　（b）塑化剂颗粒破碎　　（c）水泥环局部破碎

图10　水泥环改性结果

（2）改性水泥浆的固井工艺性评价。水泥浆改性后须满足工程作业工艺条件，包括密度、流动性、稠化时间等技术要求，本文改性方案对作业工艺的适用性由西南页岩气开发主要作业单位——川庆钻探井下作业公司进行了系统的试验评价，各项参数均可满足工程作业。

(3)套管变形的缓解效果。对径向外挤位移作用下，不同配比水泥环与套管的变形特征分析表明：

① 在套管发生塑性变形前，水泥环局部破碎并被显著压实，可吸收部分地层位移（图11）；

② 当水泥环压实后，套管开始发生塑性变形，其变形量比水泥环改性前大幅降低（图11）；

③ 调整水泥环空心颗粒配比，可控制套管变形量，颗粒含量越高，套变越小，从而实现对套管有效通径的主动控制，获得优化设计窗口，满足工程压裂作业所需通径（图12）。按照该思路，适当加厚水泥环，可以完全预防套管变形。

（a）水泥环压实　　　　（b）吸收地层位移　　　　（c）套管变形减缓形貌

图11　水泥环改性后的作用

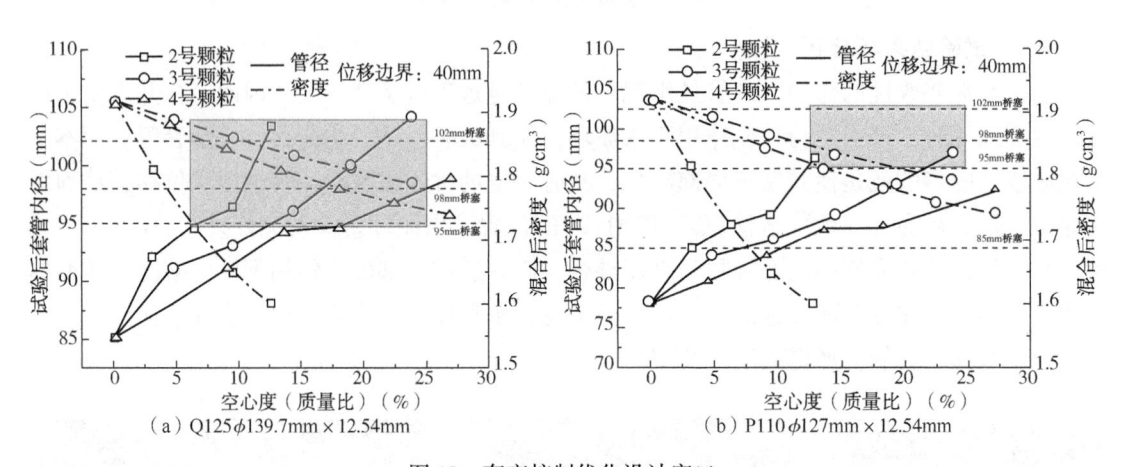

（a）Q125 φ139.7mm×12.54mm　　　（b）P110 φ127mm×12.54mm

图12　套变控制优化设计窗口

4　工业应用前景

页岩油气井压裂环境下，套管柱的变形属于局部径向变形。在地质运移环境下，套管难以完全抵挡地质的位移作用，产生变形不可避免。然而，对于工程作业来说，核心问题是保障射孔枪及桥塞等工具可以穿过套管，实现作业，因此，这种情况下的套管是允许变形的，但套管的变形量不能导致套管破裂、螺纹失效及井漏、井壁失稳坍塌等极端事件。因此，对套管局部变形量的控制是重点，工程上应重点围绕工具尺寸对套管有效通径的需求，开展套管变形缓解甚至消除方法的研究。类似的问题一直存在于油气井工程领域，例如注水井泥岩吸水膨胀造成套管挤毁及剪切、盐膏层蠕变造成的套管挤毁。

泥岩吸水膨胀效应与页岩层吸水膨胀本质是类似的，但在天然裂缝等地质环境方面差异显著，而盐膏层蠕变会造成长周期的定向外挤，其蠕变方向也可能随地质变化而变化，另外，盐膏层与上下岩层的物性差异显著。这些环境同样属于位移控制型特征，相关领域一直缺乏有效的预防方案，尤其是非均匀地质运移环境的模拟试验手段非常缺乏，大量理论及数值模拟分析无法获得验证。本文作者所属单位已针对页岩气井套变控制成功开发了非常规油气井筒全尺寸模拟试验装备，为位移控制型套变控制研究提供了物理手段。

上述三类典型的位移控制型套管变形，已构成中国油气井开发领域套管柱完整性控制技术的研究热点与重点。沿着本文提出的套变缓解思路，针对不同工况的需求进行井筒一体化优化设计，对套管变形量的控制具有通用性。

5 结论

本文对页岩气井压裂环境下套管变形机制进行了分析，提出井筒协调变形缓解套管变形的新方法，通过对水泥浆进行改性，实现了套管局部变形的主动控制，获得工程可用方案，获得全井筒模拟试验有效验证。

（1）页岩气井套管局部变形主要由地质运移诱发，以地层滑移与膨胀为主，控制套管变形应从优化井眼轨迹、改性水泥环、提高材料应变强化指数、井筒协调变形优化等方面入手。

（2）地质运移环境下套管变形属于位移控制，仅提高套管钢级与壁厚，对变形缓解及预防效果有限。地质运移环境下，套管可允许适量变形，满足工程作业通径需求即可。

（3）通过水泥环的空心化改性可以实现对地层集中位移位置的过量套变进行主动控制，改性的水泥环可吸收部分甚至全部地层位移，从而减缓或者完全预防套变。

参 考 文 献

[1] ISO. Petroleum and natural gas industries—Steel pipes for use as casing or tubing for wells：ISO 11960[S]. ISO copyright office，2014.

[2] Han L H, Wang H, Wang J J, et al. Strain-based casing design for cyclic-steam-stimulation wells[J]. SPE production & Operations，2018，33(2)：409-418.

[3] Casero A, Rylance M. The unconventional unconventionals：tectonically influenced regions，stress states and casing failures [J]. SPE 1997110-MS，2020.

[4] Lian Z H, Yu H, Lin T J, et al. A study on casing deformation failure during multi-stage hydraulic fracturing for the stimulated reservoir volume of horizontal shale wells [J]. Journal of Natural Gas Science and Engineering，2015，23：538-546.

[5] Chen Z W, Shi L, Xiang D G. Mechanism of casing deformation in the Changning-Weiyuan national shale gas demonstration area and countermeasures[J]. Natural Gas Industry B，2017，4：1-6.

[6] Liao S M, Sang Y, Song Y, et al. Research and field tests of staged fracturing technology for casing deformation sections in horizontal shale gas wells[J]. Natural Gas Industry B，2018，5：16-21.

[7] Li L W, Wang G C, Lian Z H, et al. Deformation mechanism of horizontal shale gas well production casing and its engineering solution：a case study on the Huangjinba block of the zhaotong national shale gas demonstration zone[J]. Natural Gas Industry B，2018，5：261-269.

[8] Yin F, Han L H, Yang S Y, et al. Casing deformation from fracture slip in hydraulic fracturing[J]. Journal of

Petroleum Science and Engineering, 2018, 166: 235-241.

[9] 麦洋, 莫丽, 傅栋, 等. 固井质量对页岩气井水平井段套管失效的影响[J]. 石油机械, 2019, 47(12): 123-130.

[10] 李勇, 刘硕琼, 王兆会. 水泥环厚度及力学参数对其应力的影响[J]. 石油钻采工艺, 2010, 32(4): 37-41.

[11] 曾静, 高德利, 王宴宾, 等. 体积压裂井筒水泥环拉伸失效机理研究[J]. 钻采工艺, 2019, 42(3): 1-5.

[12] 程小伟, 张高寅, 马志超, 等. 页岩气水平井油井水泥的原位增韧技术研究[J]. 西南石油大学学报(自然科学版), 2019, 41(6): 68-74.

[13] 高德利, 刘奎. 页岩气井井筒完整性若干研究进展[J]. 石油与天然气地质, 2019, 40(3): 602-615.

[14] Lei X L, Wang Z W. 页岩气开采水力压裂诱发四川盆地南部[J]. 世界地震译丛, 2020, 51(2): 144-160.

[15] Milena R, Coleen S, Iuciano M. Interference behavior analysis in vaca muerta shale oil development, Loma Campana Field, Argentina[J]. SPE 178620-MS, 2015.

本论文原发表于《石油管材与仪器》2020年第6卷第4期。

Investigation on Impact Absorbed Energy Index of Drill Pipe

Li Fangpo

(1. CNPC Tubular Goods Research Institute; 2. State Key Laboratory for Performance and Structure Safety of Petroleum Tubular Goods and Equipment Materials)

Abstract: Drill pipe is one of the most important tools for petroleum and natural gas drilling exploitation. The application of S135 drill pipe can significantly increase drilling depth, reduce drilling cost and improve drilling quality. Impact absorbed energy is the most important toughness index, which has a decisive influence on the service safety of drill pipe. The statistical results of 91 drill pipe failure cases show that the piercing is the main failure mode of S135 drill pipe, including oval – shaped piercing and slot piercing. The length of oval – shaped piercing hole is mainly distributed in 20 ~ 50mm, and the impact absorbed energy of failed drill pipe is dispersed in 42 ~ 156J. The length of slot piercing hole is mainly distributed in 60 ~ 90mm, and impact absorbed energy of failed drill pipe is concentrated in 76 ~ 150J. Analysis results show that there is certain correlation between the length of piercing hole and impact absorbed energy values of drill pipes. In this work, the impact absorbed energy index calculation formula of "leakage before fracture" failure mode for drill pipe is proposed. The results show that the impact absorbed energy value increases with the increasing of critical crack length, square of stress strength coefficient and strength level. The impact absorbed energy value of S135 drill pipe should be greater than 80J to ensure the service safety.

Keywords: Drill pipe; Failure analysis; Piecing failure; Impact absorbed energy; Service safety

1 Introduction

With the development of deep oil & gas resources and the application of advanced drilling technology, the drill pipe is subjected to more complicated loads and prone to fail[1,2]. Statistical data show that there are about 200 drill pipe failure cases in China every year, and the piercing is the major failure mode, which seriously affects the drilling speed and quality[3]. In order to prevent

---

Corresponding author: Li Fangpo, lifangpo@ 163. com.

drill pipe failure accidents and ensure the safety of the drilling process, it is necessary to control the proper properties of drill pipe. The design and production of S135 drill pipe is mainly based on the ISO 11961[4] or API Spec 5DP[5]. The impact absorbed energy is currently the only index which reflects the toughness of drill pipes in international standards and documents. The impact absorbed energy has a significant influence on the service performance and failure mode of drill pipe. Studies have shown that when the impact absorbed energy reaches a certain value, the fracture mode of drill pipe will change from brittle fracture to plastic fracture and become the piercing, known as "leakage before fracture". The basic requirement for impact absorbed energy of S135 drill pipe is 43 J (3/4 size) in ISO 11961 and API Spec 5DP. Reasonable impact absorbed energy value is essential to improve the service safety of S135 drill pipe and control its proper manufacturing cost.

2 Experiment

The chemical composition of S135 drill pipe was shown in Table 1. The tensile property test was performed at room temperature according to ISO 6892-1[6]. It was shown that yield strengths of all dill pipes were between 931MPa and 1138MPa. Impact absorbed energy test was conducted at a temperature of (21±3)℃ in accordance with ASTM A370[7]. Impact test was consisted of a set of three longitudinal specimens taken from a position one meter away from the failure of the drill pipe, as shown in Fig. 1. The average value of the three specimens was recorded as the impact absorbed energy of drill pipe. Impact absorbed energy of all drill pipes was shown in below. The crack in failure sample was opened mechanically and the surface was cleaned. The crack surface was observed by visual inspection and scanning electronic microscopy (SEM).

Table 1 Chemical composite of drill pipe

Element	C	Si	Mn	P	S	Cr	Mo	Ni
Content	0.24~0.28	0.21~0.30	0.85~1.10	≤0.02	≤0.015	0.95~1.03	0.40~0.45	≤0.1

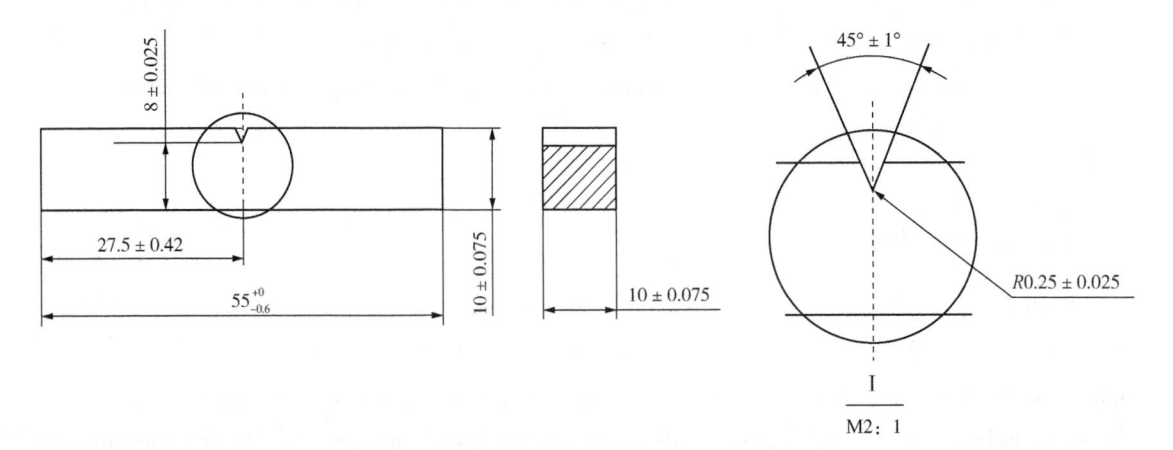

Fig. 1 Impact test specimen

3 Results and discussion

The statistical analysis results of a large number of failure cases show that the main failure modes of S135 drill pipe are piercing and fracture[3]. The piercing is the necessary and main stage of fracture failure, and fracture is the continuation of piercing failure. Whether the fracture will appear after formation of the piercing hole depends on the load conditions and performance of drill pipe. Therefore, the piercing damage and fracture damage of drill pipe body are the same to a certain content, but only at different stages of failure. According to the characteristics, the failure process of drill pipe can be described as: the formation of corrosion pit → the generation of cracks → the propagation of cracks through the wall thickness → the penetration of drilling fluid → the formation piercing hole or fracture. Cracks may be initiated on the outer surface or inner surface of drill pipe.

During the failure of piercing, the micro-crack is usually formed on the surface of drill pipe. Under the combined effect of stress load and drilling fluid scouring, the crack extends along the circumferential and radial direction, and finally penetrates the wall thickness. After the formation of piercing hole, a large amount of drilling fluid leaks into the annulus and result in the decrease of drilling fluid pressure consumption through the drill string, drilling bit and annulus[8]. As down-hole pressure consumption reduces, the ground pump pressure will decrease accordingly. The pump pressure is the total pressure consumption of the ground, drill string, annulus and drilling bit. If the performance of drill pipe is excellent, crack developing speed will be slow and the critical length of the crack will be large. As the piercing hole expands, the ground pump pressure begin to decrease gradually. When the piercing hole's length reaches certain value, ground pump pressure will decline significantly. In this condition, operators can determine drill pipe's piercing failure according to the pump pressure change, and take timely measures to avoid the down-hole fracture of drill pipe, and ensure that the failure mode is "leakage before fracture"[9]. On the other hand, if the performance of drill pipe is not good enough, the crack will propagate rapidly and its critical length will be small. The rapid propagation of cracks will cause the drill pipe to rupture before operators find the piercing failure, which is a very dangerous failure accident.

3.1 Analysis of the piercing failure characteristics

91 cases of S135 drill pipe piercing failure in recent years were analyzed. According to the characteristics of piercing hole, piercing failure of drill pipe can be divided into two types. One form is called "oval-shaped piercing" failure as shown in Fig. 2(b). The piercing hole is seriously washed by drilling fluid and the crack propagation tip is passivated. The "oval-shaped piercing" is usually formed under the condition that the crack propagation speed is quite slow. The other form is called "slot piercing" failure as shown in Fig. 2(c). The gap of piercing hole is 0.5~5.0mm wide and its crack propagation surface is relatively complete. In later propagation period, the crack begins to bifurcate and bend. With the further expansion of crack, when the load-carrying capacity of piercing section of drill pipe is insufficient, the drill pipe will break as shown in Fig. 2(d).

Although the macroscopic morphological characteristics of the two piercing failure forms are different, there is no absolute boundary, and they can transform into each other under certain conditions.

(a) failed drill pipe

(b) elliptical hole

(c) slot hole

(d) fracture surface

Fig. 2 Morphology of failed drill pipe

The piercing length of ninety-one failed S135 drill pipes was measured for statistical analysis. Fig. 3 shows that the length of fiftyfive "oval-shaped piercing" holes is mainly concentrated between 20mm and 50mm. Thirty-six "slot piercing" S135 drill pipes were measured, as shown in Fig. 4. The length of piercing holes is relatively dispersed, and distributed among 10~90mm.

It was concluded that crack propagation rate was relatively slow during the formation of oval-shaped piercing. The erosion of high pressure drilling fluid caused the crack tip to be passivated, thereby preventing the rapid propagation of the crack. When the piercing hole was very small, the pressure change of ground pump was not obvious and it was not easy to find the piercing hole. As the metal at the crack location was continuously washed by high-pressure drilling fluid, the size of the piercing hole increased gradually. When the piercing hole reached a certain size and the pump pressure dropped significantly, the ground detector could distinguish the piercing failure of drill pipe and replace the failed drill pipe from drill string. In this case, the "oval-shaped" piercing hole was relatively big. For the slot piercing hole, the length of short slot piercing holes was mainly concentrated between 10mm and 50mm. When the crack just penetrated through the wall thickness,

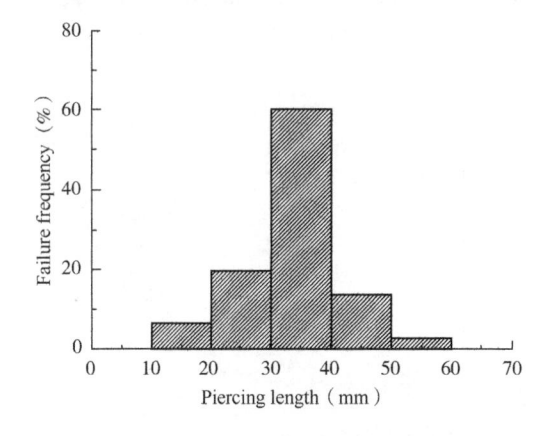

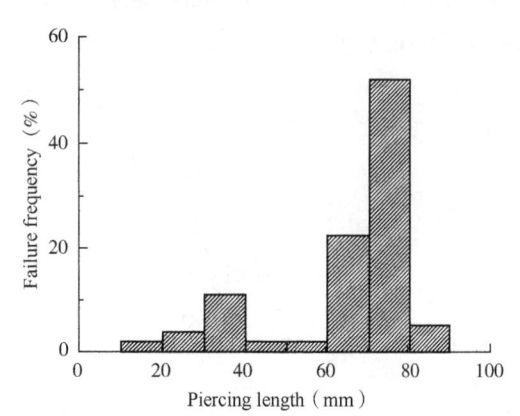

Fig. 3 Length of oval-shaped piercing

Fig. 4 Length of slot piercing hole

the piercing failure was found and its propagation section was not damaged seriously by drilling fluid. For the larger slot piercing hole, the length of large slot piercing holes was distributed between 60mm and 90mm. The crack propagation surface was flat in the stable expansion stage, where there were a large number of regular fatigue striation. Fatigue striation was mainly due to the alternating stress load that was subjected to the drill pipe during drilling process.

3.2 Impact absorbed energy of drill pipe

The impact absorbed energy is the most important toughness index of drill pipe in the international standards and documents, which plays an important role in ensuring the service safety of drill pipe. The impact absorbed energy of ninety-one piercing failed S135 drill pipes was tested, and the corresponding relationship between the impact absorbed energy and the piercing hole length of failed drill pipes was shown in Fig. 5 and 6. It can be seen that the impact absorbed energy of drill pipes with "oval-shaped piercing" holes is relatively dispersed and distributed between 42J and 156J. For the S135 drill pipes with "slot piercing" holes, when the piercing hole length is between 10mm and 50mm, the impact absorbed energy will be more dispersed between 56J and 147J. As piercing hole length is between 60mm and 90mm, the impact absorbed energy is relatively concentrated between 76J and 150J. It can be considered that when the impact absorbed energy increases, the critical length of piercing hole increases correspondingly, and the failure mode of drill pipe changes from fracture to piercing.

According to the theory of fracture mechanics, the mutual effect of material properties, service load and crack size determines the failure mode of drill pipe, and the critical crack length is an important parameter of drill pipe design. When crack extends through the wall thickness and forms a piercing hole, if the fracture condition is not reached, the drilling fluid will pierce along the crack channel. As the crack spread further and the drilling fluid leaks, the pump pressure will drop significantly so that ground operators can detect the drill pipe failure. The critical crack length can be used to design the drill pipe performance design to meet the "leakage before fracture" requirement. The piercing hole can be regarded as continuation of the crack. Based on the piercing hole length of ninety-one failed S135 drill pipes, it can be seen that the critical crack length of S135 drill pipes should not be less than 60mm in order to ensure the service safety of drill pipe.

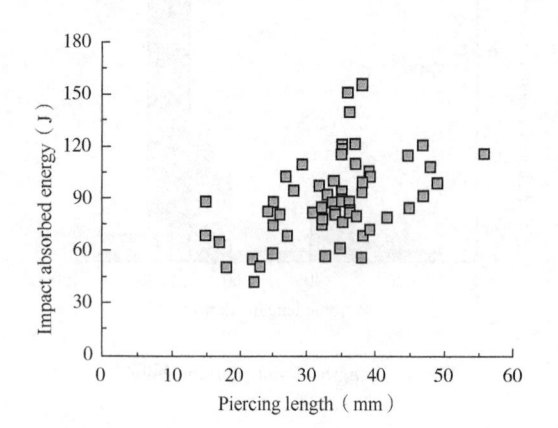

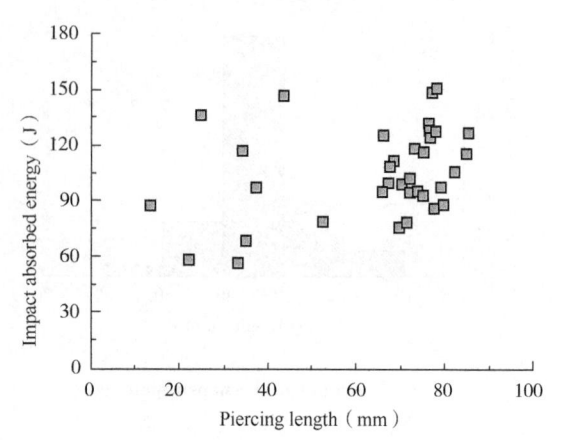

Fig. 5 Impact absorbed energy of elliptical piecing drill pipes

Fig. 6 Impact absorbed energy of slot piecing drill pipes

3.3 Stress intensity factor calculation

The morphological characteristics of the failure show that failure of drill pipe caused by the crack is open type crack, which is subjected to tensile stress acting perpendicular to the crack surface. From the linear elastic fracture mechanics, the stress intensity factor (K) of the crack tip is shown as formula (1).

$$K = \sigma\sqrt{\pi\alpha} \times F_t \tag{1}$$

Where F_t is a constant related to the component geometry, crack geometry and stress loading method. σ is tensile stress load, and α is half the length of the crack. Drill pipe can be considered as a generalized cylindrical pressure vessel, when there is circular crack in the drill pipe, F_t can be expressed as formula (2).

$$F_t(R/t,\ \theta/\pi) = 1 + A\left[5.3303\left(\frac{\theta}{\pi}\right)^{1.5} + 18.773\left(\frac{\theta}{\pi}\right)^{4.24}\right] \tag{2}$$

Where:

$$A = [0.125(R/t) - 0.25]^{0.25} \quad 5 \leqslant R/t \leqslant 10$$

$$A = [0.4(R/t) - 3.0]^{0.25} \quad 10 < R/t \leqslant 20$$

Where R is the radius of drill pipe, mm; θ is the angle of half-length crack, rad; t is the wall thickness of drill pipe, mm. According to the fracture mechanics theory, when the stress intensity factor of crack tip is up to the fracture toughness (K_{IC}) of the material, the crack will become unstable and frature immediately. Some scholars established an empirical relationship between the longitudinal impact absorbed energy and the plane strain fracture toughness (K_{IC}) of drill pipes based on a large number of test data. The empirical relationship is shown as formula (3).

$$K_{IC} = (0.5172 \times CVN \times \sigma - 0.0022\sigma^2)^{0.5} \quad (3)$$

Where K_{IC} is plane strain fracture toughness, MPa·m$^{1/2}$; CVN is Impact absorbed energy, J; σ is Yield strength, MPa.

When formula (2) and formula (3) are plugged into the formula (4), the impact absorbed energy of drill pipe can be calculated with formula (4).

$$CVN = \left[X \times \left(\frac{\sigma}{Y}\right)^2 + 0.00425\right] \times Y \quad (4)$$

Where X is a constant associated with the critical crack length and the wall thickness of drill pipe. β is defined as the stress intensity ratio coefficient, which is the ratio of tensile stress to yield strength. The drill string design is mainly based on the yield strength, and the current safety factor is not less than 1.4[10], and accordingly, the β value is not greater than 0.72. The required value of impact absorbed energy can be calculated with formula (5).

$$CVN = [X \times \beta^2 + 0.00425] \times Y \quad (5)$$

From the formula (5), it can be seen that the required value of the impact absorbed energy drill pipe increases with the increasing of drill pipe strength. The higher strength of drill pipe, the higher required value of matching impact absorption energy. The required value of impact absorbed energy increases with the increasing of tensile stress and square of stress intensity ratio coefficient. At present, four types of S135 drill pipe widely used with outer diameters are 127.0 mm and 139.7 mm are listed in Table 2, and the corresponding X values for different critical crack lengths are also shown in Table 2.

Table 2 X values of different drill pipes and critical crack length

Type of drill pipe	Critical crack length				
	20mm	40mm	60mm	80mm	100mm
127.0mm×9.19mm	0.067322	0.160815	0.298758	0.505086	0.816003
127.0mm×12.7mm	0.066538	0.155966	0.283768	0.469821	0.743875
139.7mm×9.17mm	0.06662	0.156436	0.284866	0.471056	0.742310
139.7mm×10.54mm	0.066334	0.154680	0.279491	0.458573	0.717208

For the 127.0mm×9.19mm S135 drill pipe with the maximum of X value, the variation of impact absorbed energy requirements with different stress intensity ratio coefficients and different critical crack lengths is shown in Fig. 7. As shown in Fig. 7, the required impact absorbed energy gradually increases with the increasing of the critical crack length. The value of 43J proposed in ISO/

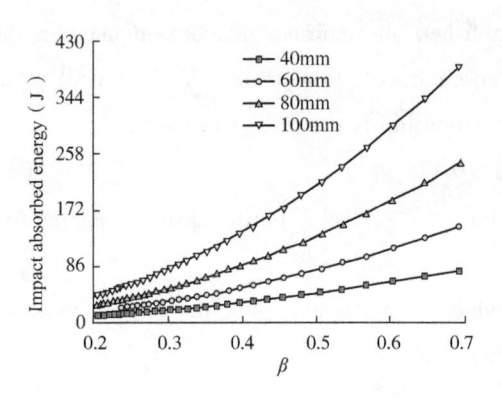

Fig. 7 Impact absorbed energy of S135 drill pipe

API standard can meet basic toughness requirements when the critical crack length is less than 40mm. Above statistical analysis shows that the critical crack length of the piercing failed S135 drill pipe should be between 60mm and 90mm currently. In order to ensure the "leakage before fracture" failure mode, the impact absorbed energy of S135 drill pipe should not be less than 80J according to the calculation results and impact absorbed energy test results shown in Fig. 5 and 6.

4 Summary

(1) The statistical analysis shows that the piercing failure modes of S135 drill pipe mainly include "oval-shaped piercing" and "slot piercing". The length of oval-shaped piercing hole is mainly distributed in 20~50mm, and its impact absorbed energy is dispersed in 42~156J. The length of slot piercing hole is mainly distributed in 60~90mm, and its impact absorbed energy is concentrated in 76~150J. There is a certain correlation between the length of piercing hole and the impact absorbed energy of drill pipe.

(2) The calculation formula for impact absorbed energy of drill pipe with the "leakage before fracture" failure mode is proposed. The impact absorbed energy of drill pipe increases with the increasing of the critical crack length, square of stress strength coefficient, and strength level. The impact absorbed energy value of S135 drill pipe should be greater than 80J in order to ensure the service safety of drill pipe.

Acknowledgement

Science and Technology Research Projects of CNPC (No. 2019B-4014, 2018E-2101, 2016A-3905) and Innovative Talents Promotion Plan in Shaanxi Province (No. 2017KJXX-01) sponsor this work.

References

[1] Sajad Mohammad Zamani, Sayed Ali Hassanzadeh-Tabrizi, Hassan Sharifi, Failure analysis of drill pipe: a review, Eng. Fail. Anal. 59 (2016) 605-623.

[2] Xinhu Wang, Fangpo Li, Yonggang Liu, et al., A comprehensive analysis on the longitudinal fracture in the tool joints of drill pipes, Eng. Fail. Anal. 79 (2017) 1-7.

[3] Fangpo Li, Yonggang Liu, Xinhu Wang, Analysis of manufacturing factors affecting drilling string failure, Drilling Prod. Technol. 6(36) (2013) 86-88 (in Chinese).

[4] ISO 11961[S], Drill pipe(ISO Standard) 2018.

[5] API Spec 5DP[S], Drill pipe(American Standard) 2020.

[6] ISO 6892-1[S], Metallic materials-tensile testing-part 1: method of test at room temperature (ISO Standard) 2016.

[7] ASTM A370[S], Standard Test Methods and Definitions for Mechanical Testing of Steel Products(American Standard) 2019.

[8] Xuncheng Song, Zhichuan Guan, Shaowei Chen, Mechanics model of critical annular velocity for cuttings transportation in deviated well, J. China Univ. Petrol. 1(33) (2009) 53-56 (in Chinese).

[9] BS7910[S], Guide to methods for assessing the acceptability of flaws in metallic structures (British Standard) 2013.

[10] Ting-gen Chen, Zhi-chuan Guan, Drilling engineering theory and technology, China, China University of Petroleum Press, 2006 (in Chinese).

本论文原发表于《Engineering Failure Analysis》2020 年第 118 卷。

Aging Treatment Effect on Microstructure and Mechanical Properties of Ti-5Al-3V-1.5Mo-2Zr Titanium Alloy Drill Pipe

Feng Chun[1]　Li Ruizhe[2,3]　Liu Yonggang[1]　Liu Le[2,3]
Song Wenwen[2,3]　Zhang Fangfang[2,3]

(1. State Key Laboratory for Performance and Structure Safety of Petroleum Tubular Goods and Equipment Materials, CNPC Tubular Goods Research Institute; 2. State Key Laboratory for Performance and Structure Safety of Petroleum Tubular Goods and Equipment Materials, CNPC Tubular Goods Research Institute; 3. School of Materials Science and Engineering, Xian Shiyou University)

Abstract: Titanium alloy drill pipe material used in the oil and gas industry has excellent characteristics such as highspecific strength, high-temperature resistance, and corrosion resistance. In this study, a systematic examination of the aging process of Ti-5Al-3V-1.5Mo-2Zr drill pipe material, and the microstructure evolution and properties was carried out. The results show that the optimum solution system is 930℃/1h/WQ (water quenching). After the aging treatment, the granular changes for α and β phases lead to an increase in the size of the initial lamellar α phase, the volume fraction of the secondary α phase, and the overall strength while the elongation decreases. As the aging temperature increases, the tensile strength and yield strength initially increase by 27 and 29.5%, respectively, and subsequently decrease. The elongation decreases to 9.5%, leading to an excellent performance. With the gradual decrease in the tensile strength and yield strength, the elongation increases gradually. When the heat treatment process is 930℃/1h/WQ+ 600℃/6h/AC, the elongation is 11.5%. TEM observations revealed that a large number of dislocation piles existed at the α/β phase interface after aging treatment, thereby, increasing alloy strength. The more the volume fraction of the secondary α phase precipitated, the smaller the size, leading to a more significant strengthening effect.

Keywords: Aging; Treatment; Mechanical; Properties; Microstructure; Titanium drill pipe

---

Corresponding author: Li Ruizhe, liruizhe 1212@ qq. com.

1 Introduction

Titanium alloys have high-specific strength, high-temperature resistance, corrosion resistance, and other excellent characteristics. They are not only widely used in aerospace applications, but also for complex conditions such as ultra-deep wells, extended wells, sulfur-containing wells, short radius wells, high temperature, and high-pressure wells in oil and gas well drilling[1-4]. The TA15 alloy is a type of near-α titanium alloy. The Russian equivalent grade is called the BT20. It is solid solution strengthened by the α-stabilizing element Al, and the neutral elements Zr and β-stabilizing elements Mo and V are added to improve the plasticity of the alloy. In order to address the harsh drilling environment (high temperature, high pressure, and high corrosion) and the other problems encountered in the development of the petroleum industry in recent years, Feng Chun and others have actively developed titanium alloy petroleum pipes. Based on the TA15 titanium alloy, a new type of Ti-5Al-3V-1.5Mo-2Zr titanium alloy petroleum drill pipe was developed[5,6].

At present, some well-known scholars have carried out research on the changes in microstructure and properties of mature grade TA15 titanium alloy during heat treatment. Most of them divided the heat treatment temperature interval into the (α+β) phase region and the β phase region. They focused on the evolution of the microstructure of the TA15 alloy after hot working, annealing, and solution aging, and its influence on strength and plasticity. Zhao et al.[7] studied the mechanical properties, microstructure, and fracture morphology of TA15 titanium alloy after annealing. The results show that annealing significantly improves the mechanical properties of the TA15 alloy. Increasing the annealing temperature increases the α/β phase interface and the secondary α phase, while reducing the volume fraction of the primary α phase results in increased strength and reduced toughness. After secondary annealing, more secondary α phases can be obtained with better overall performance. Sun et al.[8-10] studied the evolution of the primary lamellar α phase during the heat treatment of the TA15 alloy in the dual phase zone. The results show the transformation mechanism of the primary lamellar α phase during the heating process; the two-phase region below T_β becomes thinner, and then the decomposition begins above T_β. During the heat preservation process, the volume fraction of the primary lamellar α phase decreases, the average thickness increases, and the change trend of the average length is complicated. During the heat treatment of titanium alloys, the cooling rate has a significant influence on the volume fraction, size, and distribution of the α phase, and plays a decisive role in the performance of the material[11-15]. Zhu et al.[16] studied the influence of the cooling rate on the microstructure evolution of the primary α phase during the heat treatment of TA15 titanium alloy based on diffusion theory and mathemat-ical models. The above research mainly explores the application performance of titanium alloys in the field of aerospace, but it lacks relevant information on systemic properties for the evolution of the structure and performance of Ti-5Al-3V-1.5Mo-2Zr titanium alloy drill pipe materials for the petroleum industry, and the strength and toughness studies have only reported the corrosion resistance of titanium alloy petroleum pipes[17,18].

Generally, there are four typical microstructures for titanium alloys, equiaxed, bimodal, basket weave, and Widmanstatten. These microstructures are distinguished by the morphology and content of the α phase[19]. Wu[20] found that controlling the cooling rate from the β-phase region can produce lamellar α with different thickness. The mechanical properties can be effectively and favorably controlled by alloying and tailoring microstructure by heat treatment[21]. Lutjering[22] found that a fully lamellar microstructure in the IMI834 titanium alloy had a very high-fracture toughness but low ductility. Mantri[23] reported a very high-tensile strength (UTS 1200-1810MPa) in beta titanium alloy, but a low elongation of 4%~6.5% using solid solution and duplex aging treatments. Wu et al.[24] studied fine secondary α phase-induced strengthening in a Ti-5.5Al-2Zr-1Mo-2.5V alloy pipe with Widmanstatten microstructure, and found that the yield strength (YS) of the 920℃ solution and 450℃ aged sample was improved significantly to 1064MPa, and the elongation (EL) remained 10.5%.

In order to systematically understand the evolution of the microstructure and properties of Ti-5Al-3V-1.5Mo-2Zr titanium alloy petroleum drill pipe materials during the aging treatment, this study investigates the changes in the microstructure at different aging temperatures and aging times. The process controls the composition of the Ti-5Al-3V-1.5Mo-2Zr titanium alloy petroleum drill pipe material to achieve excellent service performance of petroleum pipes in high-temperature, highpressure, and high-corrosion environments. In light of the requirement of ASTM B861 grade 5 for the Ti-6Al-4V alloy drill pipe (YS of 828MPa, UTS of 895MPa, and EL of 10%)[25], this study shows potential for great improvement in its strength.

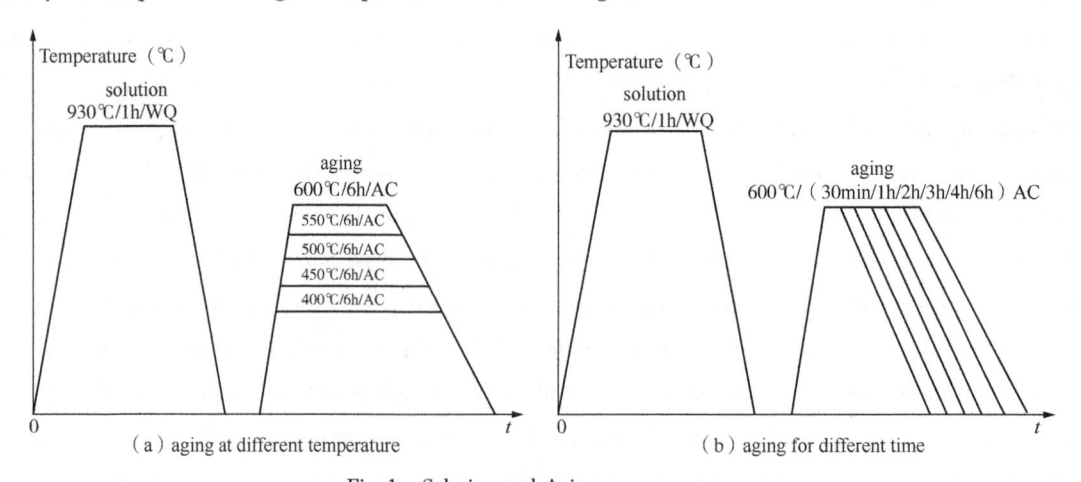

Fig. 1 Solution and Aging process routes

2 Experimental procedures

For the experiment, a Ti-5Al-3V-1.5Mo-2Zr titanium alloy pipe was used, with diameter and wall thickness of 88.9mm×9 9.35mm. A rectangular sample, 110mm×10mm×9.35mm, was taken from the pipe body. Its composition was (wt%): Al: 5%, Mo: 1.5%, V: 3%, Zr: 2%, and the balance was Ti. To prevent oxidation during the heat treatment, the surface of the sample was coated before heat treatment with a layer of Cr_2O_3 protective coating. The $T_β$ transformation point measured by the quenched metallographic method was approximately 970℃. The heat treatment process used in this study is shown in Fig. 1. The water cooling transfer time did not exceed 3s, and two samples were made under each group of processes. A box-type resistance furnace was used for the heat treatment, and the furnace temperature was calibrated using a UJ33D-2 digital potentiometer. Microstructural quantitative analysis was performed using Image Proplus software.

After the sample was subjected to a solution treatment, a tensile sample with diameter 5 mm was prepared, as shown in Fig. 2. The CMT5105 tensile tester was used to perform the tensile performance test according to GB/T228.1—2010 (China standard). Metallographic samples were polished to 1500$^#$ with finer sandpaper, polished twice to a mirror surface, etched with an etching solution of HF : HNO$_3$: H$_2$O = 1 : 3 : 7, and then examined for microstructure. Figure 3 shows the microstructure of the Ti-5Al-3V-1.5Mo-2Zr titanium alloy petroleum drill pipe body material after seamless forging and diagonal-rolling perforation formation. It can be seen that the primary lamellar α phase is relatively thin. There are colonies in the β grain with a colony size of approximately 40μm. The single phase α is distributed in a thin strip on the β phase matrix. The primary lamellar α phase volume fraction is approximately 68.3%, and the original β grains are coarse, the β grain size is about 160μm, β grain boundaries are clear, and the grain boundary α phase exists.

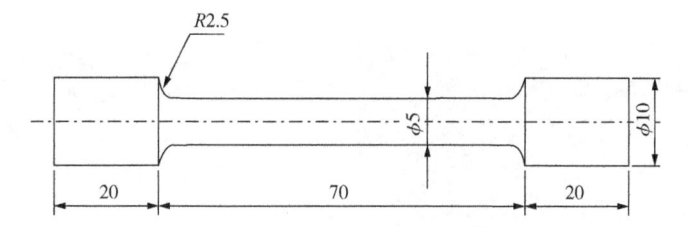

Fig. 2 Schematic of tensile specimen (unit, mm)

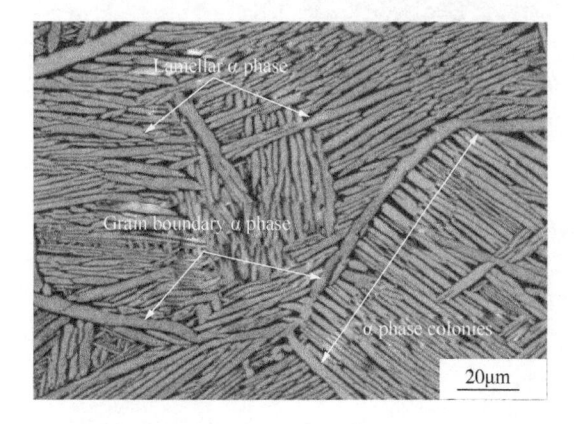

Fig. 3 Optical microstructure of the original specimen

3 Results and discussion

3.1 Effect of aging temperature on microstructure of Ti-5Al-3V-1.5Mo-2Zr titanium alloy drill pipe

Figure 4 shows the metallographic microstructures at different aging temperatures. It can be seen that compared with the original structure, the colony phenomenon of the primary lamellar α phase disappears after the aging treatment, and the grain boundary α phase disappears. The analysis shows that the acicular martensite α formed during solution treatment is a metastable phase, which is decomposed into secondary α and β phases during aging. There is a secondary α phase nucleation and growth during the aging process, and the β phase is nonuniformly nucleated at the α phase interface or substructure. In the later stage of aging, when the equilibrium β phase amount is small, the α parent phase occurs during crystallization. When the equilibrium β phase amount is large, the

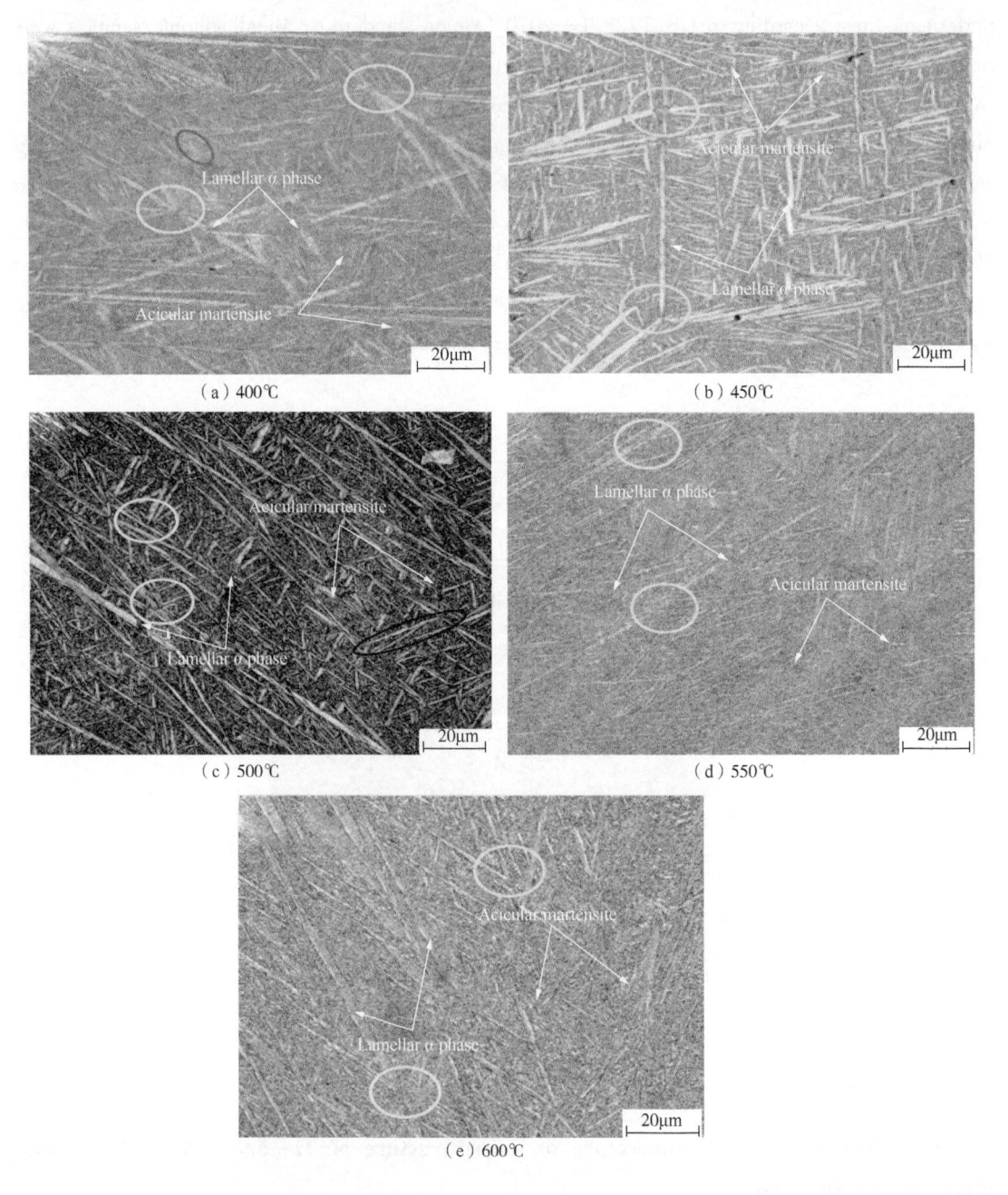

(a) 400℃

(b) 450℃

(c) 500℃

(d) 550℃

(e) 600℃

Fig. 4　Optical microstructure of samples aged at different temperatures β phase forms a continuous β layer on the α phase interface. As the aging temperature increases, the size of the primary lamellar α phase gradually increases, and the volume fraction of the secondary α phase gradually increases (as shown by the blue circle in Fig. 4). The size of the α phase gradually increases, and the α phase has no obvious colonies phenomenon. The β matrix is irregularly distributed and has no directionality. The wider α phase is the original lamellar structure, and the secondary α phase with a smaller size is formed by the metastable acicular martensite α′ transformation formed by solution treatment. The formation of the secondary α phase is the precipitation of a nucleating particle that relies on the boundary of the large-scale lamellar α phase (as shown by the yellow circle in Fig. 4). Sun et al.[26] also reached the same conclusion. When the aging temperature is 400℃ [Fig. 4(a)], the difference between the theoretical precipitation temperature and the actual aging temperature is large, the nucleation driving force is large, and the α phase is easy to nucleate, but it is not easy to diffuse and grow, so the structure is small, and the secondary α phase content is small. The α phase is irregularly distributed, the volume fraction of α is 46%, the size and width of the primary lamellar α phase are approximately 2μm, and the content of the fine secondary α phase of the acicular martensite α transformation formed by water cooling after solution treatment is small. When the aging temperature is 500℃ [Fig. 4(c)], the width of the primary lamellar α phase is 2.8μm, and the secondary α phase content is the largest. When the aging temperature is 600℃ [Fig. 4(e)], the size and width of the primary lamellar α phase is 4μm, and the secondary α phase begins to grow.

XRD analysis on samples aged at different temperatures.

It can be seen from the Fig. 5 XRD pattern that the α-phase volume fraction of the Ti-5Al-3V-1.5Mo-2Zr titanium alloy drill pipe material decreases after different solution and aging treatments, compared to the unheated sample. At aging temperatures of 500℃ and 600℃, the volume fraction of the α phase is lower than that of the original sample; however, at 500℃, the volume fraction of the secondary α phase is higher than that at 600℃. The reason is that over-aging occurs at 600℃.

3.2 Effect of aging time on microstructure and mechanical properties of Ti-5Al-3V-1.5Mo-2Zr titanium alloy drill pipe materials

During the aging treatment, titanium alloys rely on the metastable β phase or quenched martensite produced by the solid solution treatment to precipitate a dispersed α solid solution during aging, so that the alloy is strengthened. The decomposition method is closely related to the aging time. Aging time strongly affects the type, morphology, size, and distribution of the precipitated phases of the alloy, changes the microstructure, and eventually causes a large change in properties.

Figure 6 shows the metallographic microstructure photos for different aging times. As the aging time increases, the β grain size increases and the grain boundaries are clear. The α phase colonies phenomenon disappears, the grain boundary α phase disappears, the size of the α phase increases, and the volume fraction of the α phase gradually increases. The lamellar α phase mainly precipitates from the β grain boundary, and the precipitation ends at another β grain boundary. The nucleation

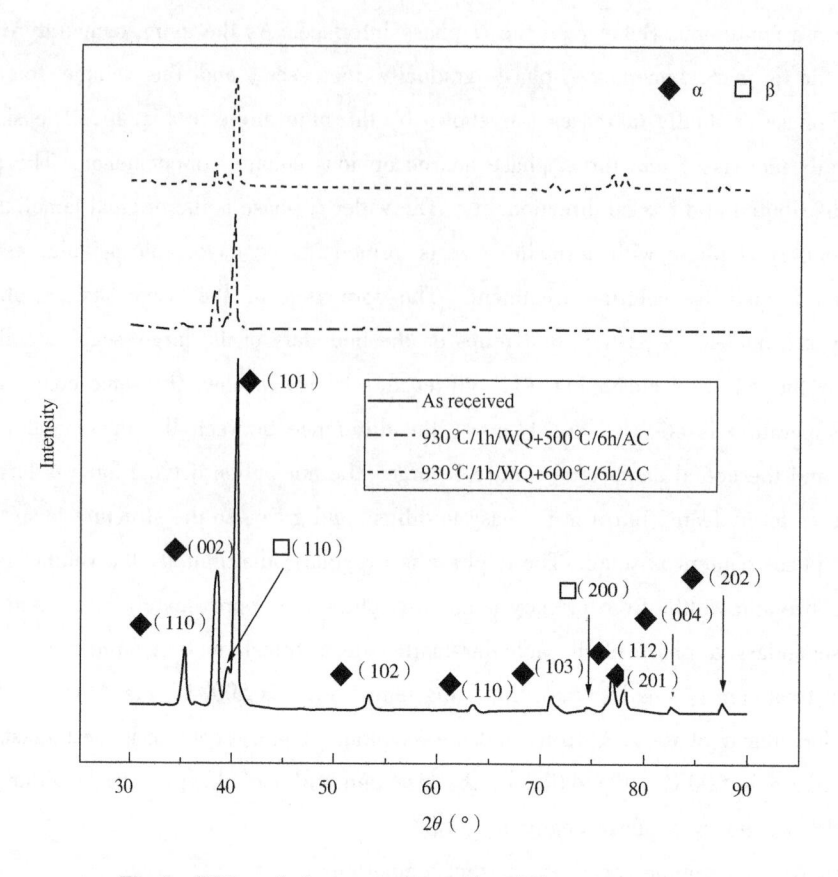

Fig. 5 XRD analysis on samples aged at different temperatures

and growth of the α lamellar structure must satisfy the Burgers orientation relationship, the α phase and the β matrix satisfy (011) β//(0001) α, and (111) β//(1120) α will be precipitated[31]. At this time, the interface between (011) β//(0001) α and (111) β//(1120) α has the lowest interface energy. The α phase grows faster in the [0001] and [1120] directions[32]. Because of the high energy of the original β grain boundary, large α lamellar nuclei preferentially originate from the original β grain boundary. The secondary α phase mainly precipitates from large lamellar α phase (shown in the yellow circle area in Fig. 6), and the α phase that formed first continues to grow until it meets other α lamellae. When the aging time is short [Fig. 6(a)], fine acicular secondary α phases are precipitated in the β grains, and the aging time increases. The length of the precipitated secondary α phase increases, the width also increases, and the secondary α phase content gradually decreases. This is because the aging time of the microstructure is completely consumed, and re-nucleation is relatively difficult. The volume fraction of the precipitated secondary α phase decreases, and the temperature range of the aging treatment is under the (α+β) phase range, the aging time is increased, and some unstable phases such as α and α″ begin to decompose.

XRD analysis on samples aged for different time.

It can be seen from the Fig. 7 XRD pattern that when the aging time is 30 min, the diffraction peak of the α phase is the lowest. The aging time is too short and the acicular martensite α phase

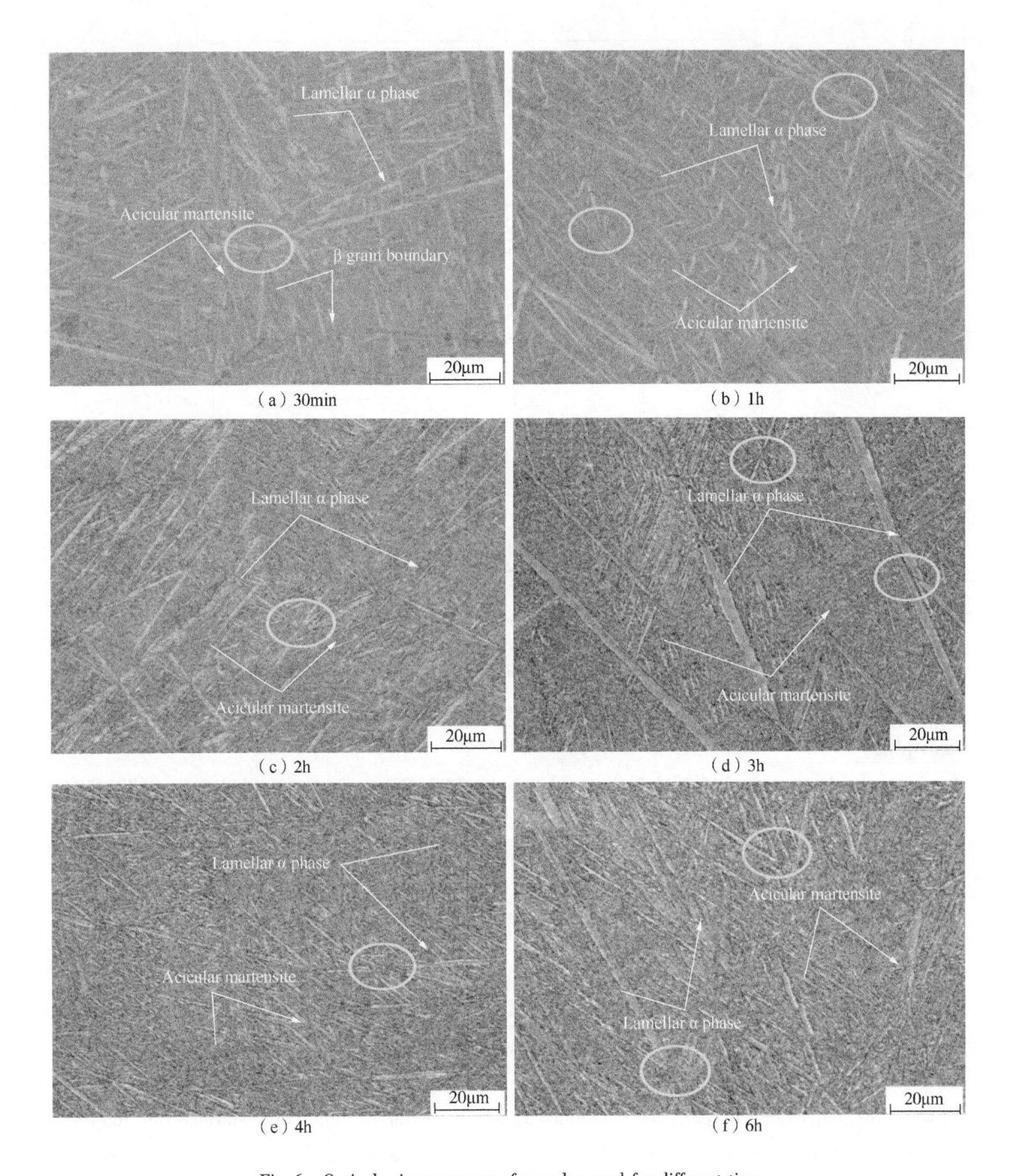

Fig. 6 Optical microstructure of samples aged for different time

has no time to decompose. When the aging time is 4h and 6h, the volume fraction of secondary α phase increases. The analysis shows that with the increase in the aging time, compared with the original sample, the diffraction peak intensity of α phase after heat treatment significantly decreases. The acicular martensite α′ phase produced by the rapid cooling from the β phase region, which transforms into the secondary α phase during the aging process. The size of the secondary α phase also increases with aging time.

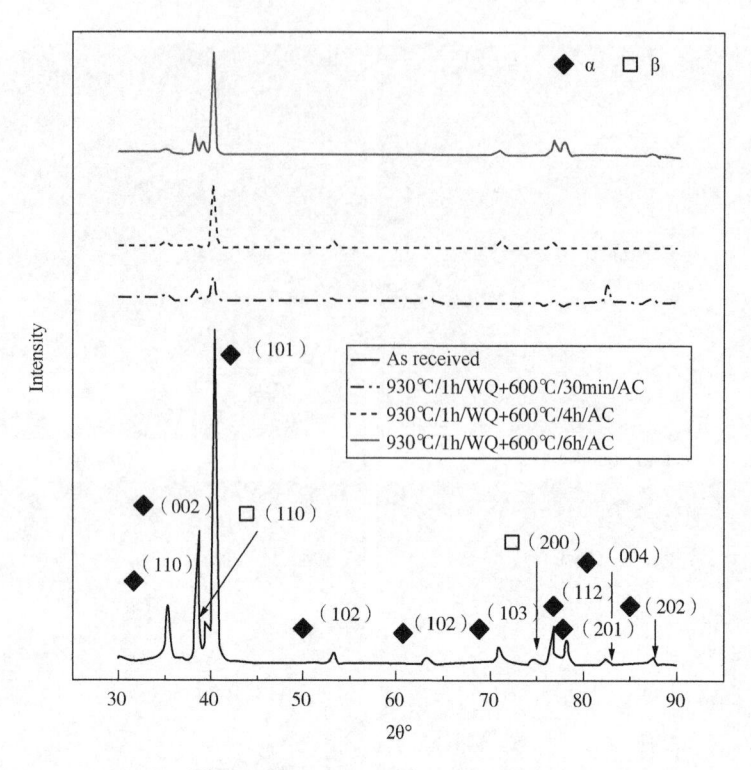

Fig. 7　XRD analysis on samples aged for different time

3.3　Effect of aging temperature on mechanical properties of Ti-5Al-3V-1.5Mo-2Zr drill pipe

The tensile strength (R_m) of the original sample is 972MPa, yield strength ($R_{0.2}$) is 885MPa, elongation (A) is 15%, hardness (HV) is 325, and dislocations in the α phase of the lamellae of the original sample are shown in Fig. 8(a). As shown in Fig. 9, after the aging treatment at different temperatures, the mechanical properties of the titanium alloy drill pipe material changed significantly while the hardness and strength increased, the elongation decreased. According to the analysis, aging hardening occurs during the aging process, and the products of precipitation, segregation, and ordering of the supersaturated solid solution during the aging process increases the resistance of dislocation movement. Under the interaction of dislocations and precipitated products, the hardening mechanism has dislocations shearing out the precipitated phase particles, the phase interfacial area between the matrix and the particles is increased, and the external force is transformed into interface energy. The dislocation movement produces an inverted domain boundary, so that the dislocation movement cannot be bent and bypassed to form a dislocation ring by precipitating the phase, and it can also harden[12]. It can be seen from the age hardening curve in Fig. 9(a), the hardness first increases and then decreases with the increase of aging temperature. When the aging temperature was 450℃, the hardness value increased the most, reaching a peak value of 374, an increase of 15%. When the temperature was 600℃, the hardness value increased the least, and the hardness was 340, an increase of 4.7%. It can be seen from the tensile stress-strain curve of Fig. 9(c) that the yield phenomenon of Ti-5Al-3V-1.5Mo-2Zr titanium alloy drill pipe material is not obvious,

(a) Original sample

(b) Aging at 500℃/6h/AC, secondary α phase

(c) α phase EDS observation

Elem	Weight(%)	Atomic(%)
Al K	05.40	09.30
Ti K	91.80	89.30
Zr K	02.90	01.50

(d) Dislocation structure

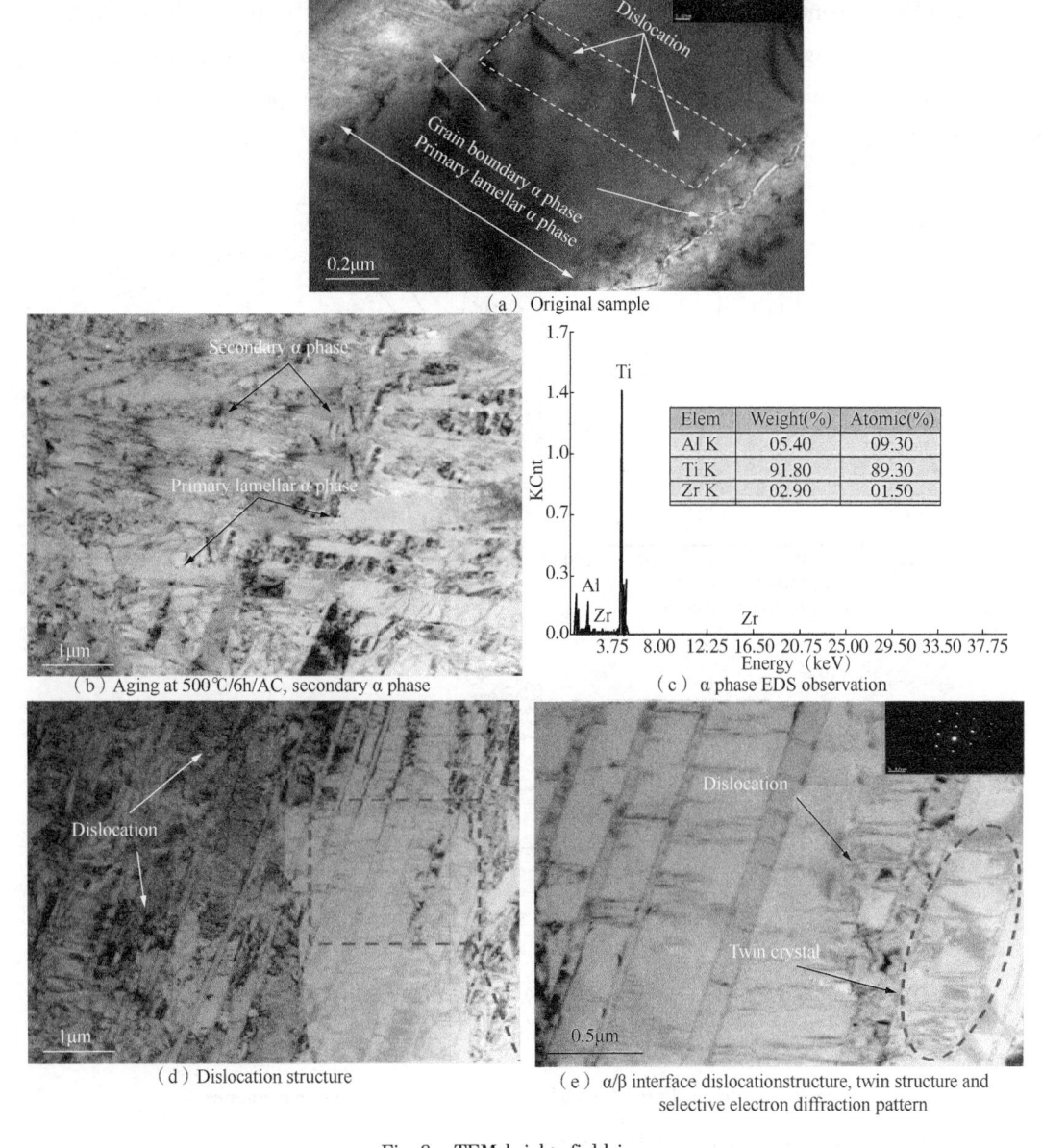

(e) α/β interface dislocationstructure, twin structure and selective electron diffraction pattern

Fig. 8　TEM bright-field images

and there is no obvious yield platform on the stress-strain curve. It can be observed from Fig. 9(b) that after aging treatment, compared with the original sample, the tensile strength and yield strength are significantly increased, the elongation is reduced, and the area reduction is reduced. As the aging temperature increases, the tensile strength and yield strength first increase and then decrease. This is because there is an α phase nucleation and growth process in the aging treatment. When the aging temperature is low, the supercooling (the difference between the theoretical precipitation temperature and the actual aging temperature) is large, the nucleation driving force is large, and the α phase nucleates easily. The aging treatment has an optimal temperature 500℃. At this optimal temperature, the fine secondary α phase with a relative high-volume fraction precipitates in dispersion, as shown in Fig. 8(b). Fig. 8(c) shows the EDS analysis of the α phase. It can be seen that alloy elements such as Al and Zr have been fully dissolved in the α phase. At 500℃, dislocation plugging occurs at the α/β interface, and α phase twins exist, [Fig. 8(d) and (e)], which has a good balance between strength and toughness. The aging temperature was then increased further to 600℃, the α phase grows up. The volume fraction of the secondary α phase t decreased.

(a) aging hardening curve

(b) mechanical property test curve

(c) stress-strain curve

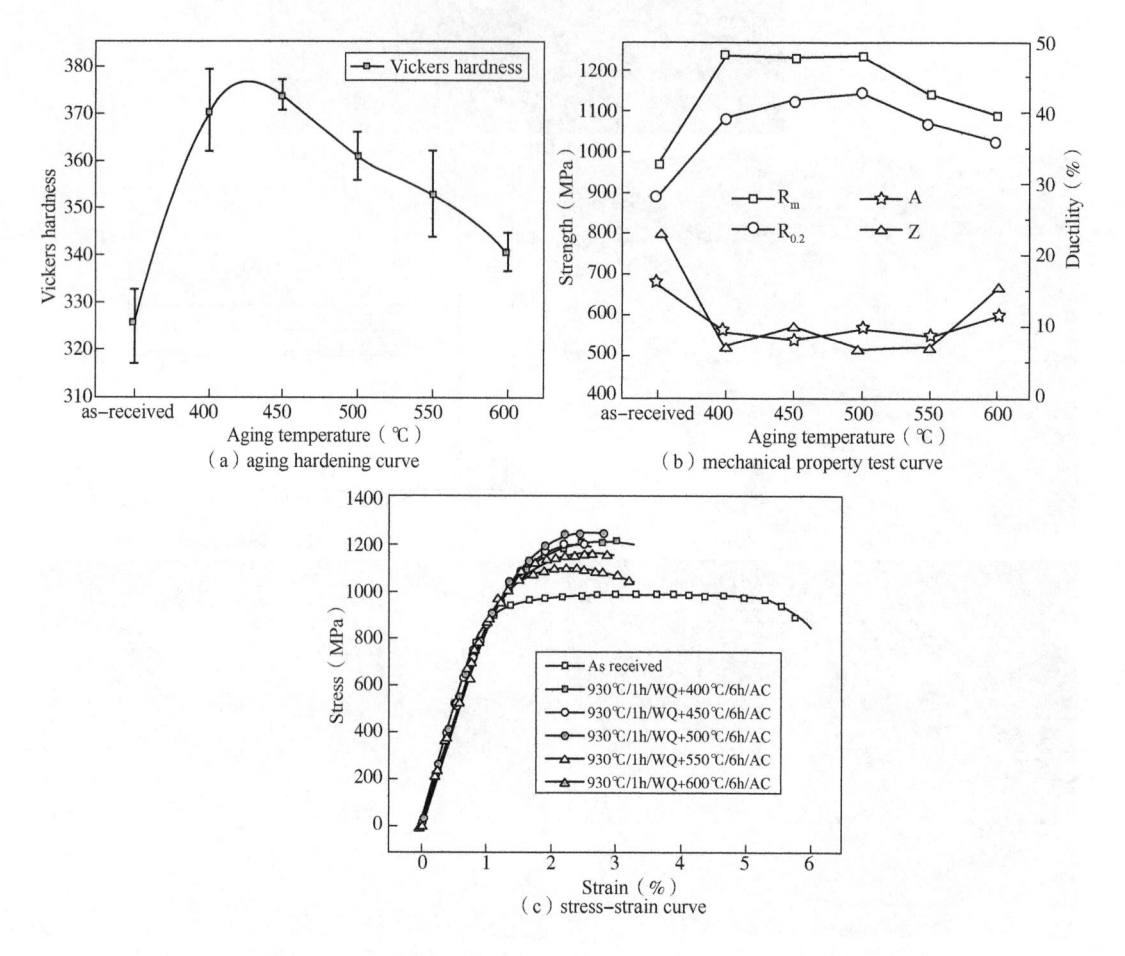

Fig. 9 Mechanical properties of samples aged at different temperatures

SEM fractography of samples aged at different temperature

As can be seen from Fig. 10, the fracture of the samples after different aging treatment shows brittle and ductile fracture with dimples and cleavage steps. However, as the aging temperature increases, the size and depth of the dimples change. When the aging temperature is lower than 500℃, the strength is higher, the elongation is low, and brittle fracture is dominant, as shown in Fig. 10(a) to Fig. 10(c). The dimple size is small, the number is small, the depth is shallow, and there is a cleavage plane. When the aging temperature is higher than 500℃, the strength decreases, the elongation is high, and ductile fracture is dominant, and as shown in Fig. 10(d) and Fig. 10(e), the dimple size is large and the number is large.

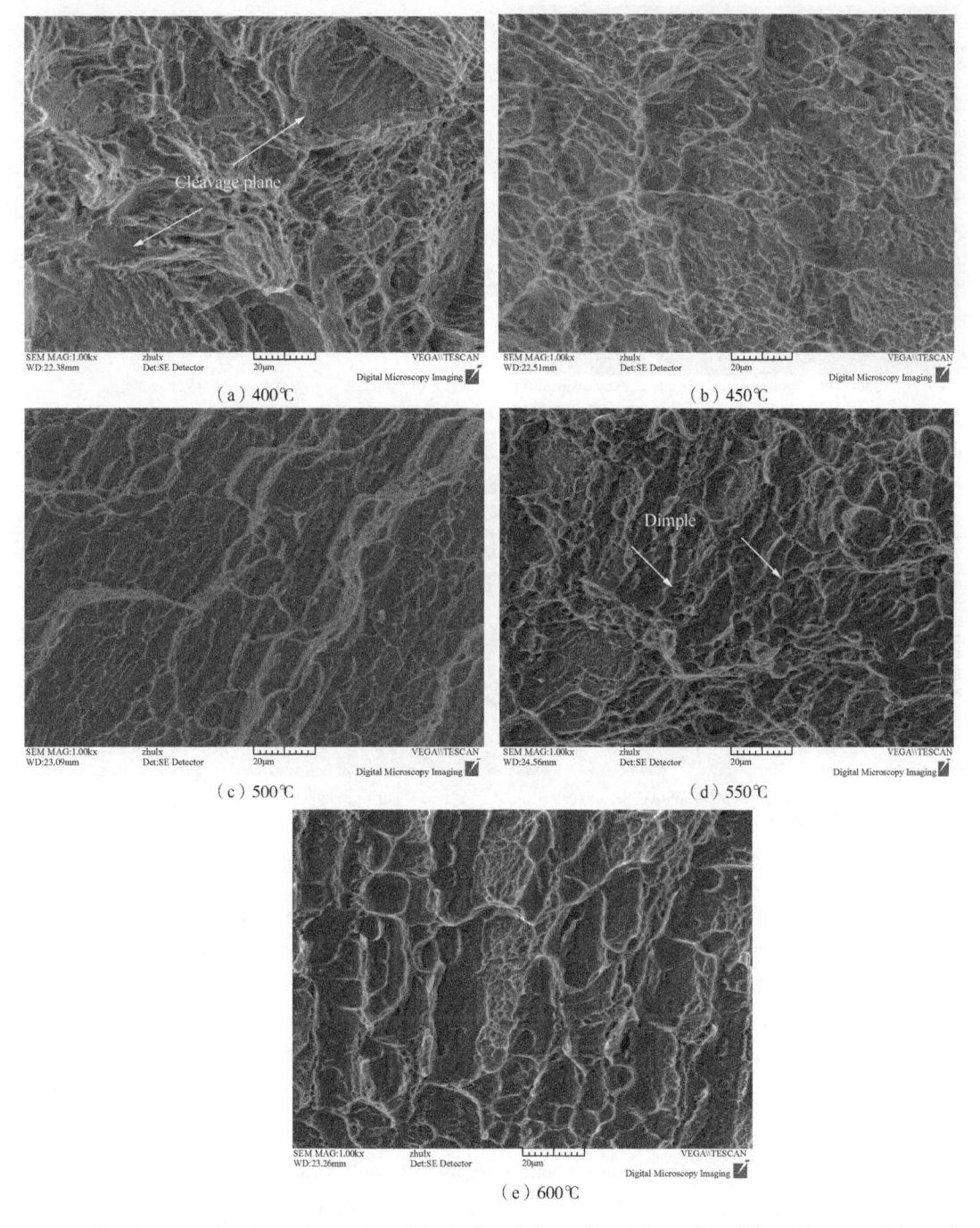

Fig. 10 SEM fractography of samples aged at (a) 400℃, (b) 450℃, (c) 500℃, (d) 550℃, and (e) 600℃

3.4 Effect of aging time on mechanical properties of Ti-5Al-3V-1.5Mo-2Zr titanium alloy drill pipe

The aging time is closely related to the amount, size, and distribution of the precipitated α phase. The changes in the volume fraction and size difference of the α phase have significant effects on the tensile strength and plasticity of the alloy. It can be seen from Fig. 12 that after aging treatment at different times, the mechanical properties of the titanium alloy drill pipe material also

(a) Aging at 600℃/30min amount of high-density dislocations

(b) β phase energy spectrum EDS analysis

Elem	Weight(%)	Atomic(%)
Ti K	89.00	89.00
V K	08.00	07.80
Zr K	02.50	01.40
Mo K	03.40	01.80

(c) Secondary α after aging at 600℃/6h/AC Selective electron diffraction pattern of phas α phase (SAD)

(d) Dislocation pile up of α/β phase interface

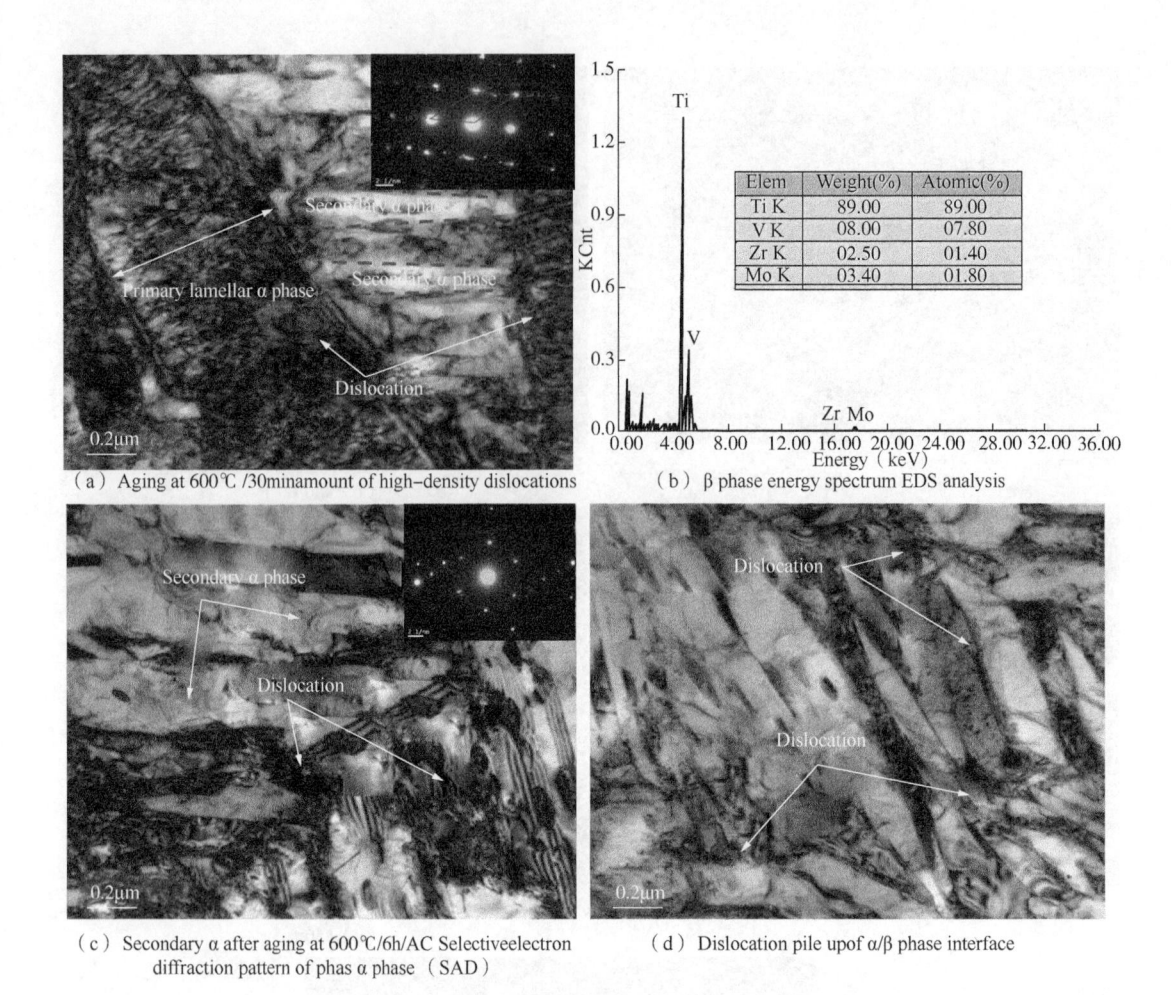

Fig. 11 TEM bright-field images

changed significantly, and the hardness and strength increased. This is primarily due to aging hardening. After the aging treatment, the structure is refined, the number of grain boundaries increases, and as the number of grain boundaries increases, and dislocation slip becomes difficult during the stretching process at room temperature, which can improve the tensile strength of the material. It can be seen from the hardness curve in Fig. 12(a) that with increase in the aging time, the hardness initially increased to reach the peak value (within 1h) of 363, an increase of 11.7%. Subsequently, as the aging time was prolonged (1~4h), the hardness value began to decrease. When the aging time was 4h, the hardness value dropped to the lowest value of 332, an increase of only 2.1%. The aging time continued to increase (more than 4h), and the hardness value increased again. When the aging time was 6h, the hardness value increased to 340, an increase of 4.6%. Fig. 12(b) and (c) show the tensile curve. After different aging treatments, the tensile strength and yield strength of titanium alloy drill pipe materials increased significantly and the elongation decreased. When the aging time was 30 min, R_m reached 1188MPa, which is an increase of 261MPa (28%) compared to the original, and $R_{0.2}$ reached 1105MPa, an increase of 220MPa compared to the original, an increase of nearly 25%. The rate A was 10%, which is 33% lower than the original rate. The reason for the increase in strength may be that the precipitation phase α

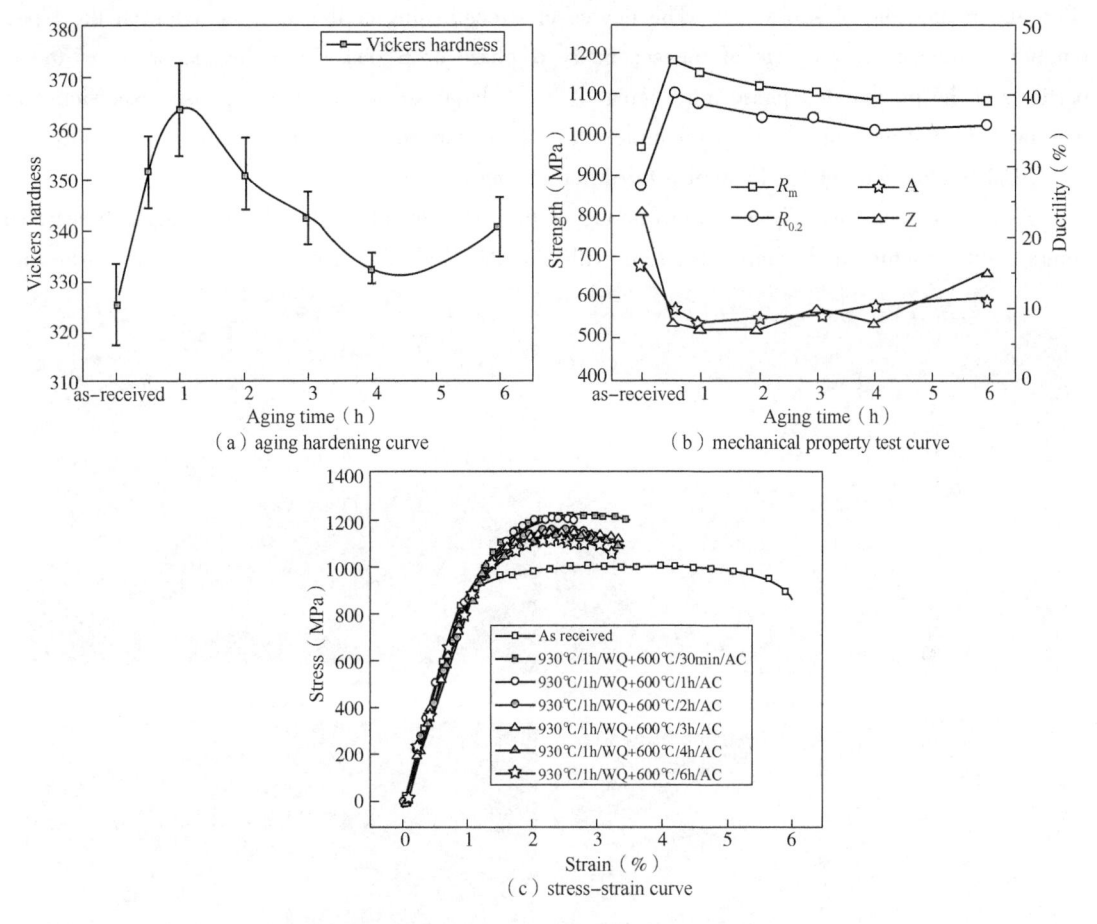

Fig. 12 Mechanical properties of samples aged for different time

produced by water cooling during the solution treatment stage has not completely decomposed during aging, and the internal microstructure of the α phase has a large number of dislocations, as shown in Fig. 11(a). Fig. 11(b) shows the EDS analysis of the β phase. It can be seen that V, Mo, Zr, and other elements were dissolved in the β phase, and the V content was the highest, and therefore, the strength was increased. As the aging time continues to increase, the α phase begins to dissolve, and the entire curve gradually decreases. When the aging time was 4h, R_m reduced to a minimum, which is the least increase compared to the original sample. Compared with the original sample, it increased by 11%. $R_{0.2}$ was 1011MPa, an increase of 14% as compared with the original sample. The elongation was 10.5%, reduced by 30% as compared with the original sample. As the aging time continued for 6 h, the strength began to increase slightly. At this time, R_m was 1091MPa, which was 12% higher than the original, $R_{0.2}$ was 1028MPa, which increased 16% compared to the original sample, and A was 11.5%, 23% lower than the original sample. Figure 11(c) and (d) shows TEM and selective electron diffraction (SAD) diagrams after 6h of aging. It can be seen that the size of the secondary α phase becomes larger after the aging time is prolonged, which is the reason for the increase in elongation, and dislocation pile up at the α/β phase interface increases the strength of the alloy. During the aging process, the tendency of the martensite phase to degrade the strength of the material interacts with the tendency of the precipitated dispersed phase to increase the strength of the alloy and reduce the plasticity [27]. The degree of strengthening of the alloy depends on the type, number, composition, and age of the secondary α phase or β phase and the dispersion of the α particles of the precipitated phase after aging[28, 29]. A large amount of fine α phase dispersions are precipitated, and the more the dispersion particles, the enhanced the aging strength and plasticity[30].

SEM fractography of aged samples for different time

As can be seen from Fig. 13, the fracture of the titanium alloy after different aging times also shows brittle fracture and ductile fracture with dimples and cleavage steps. However, with the

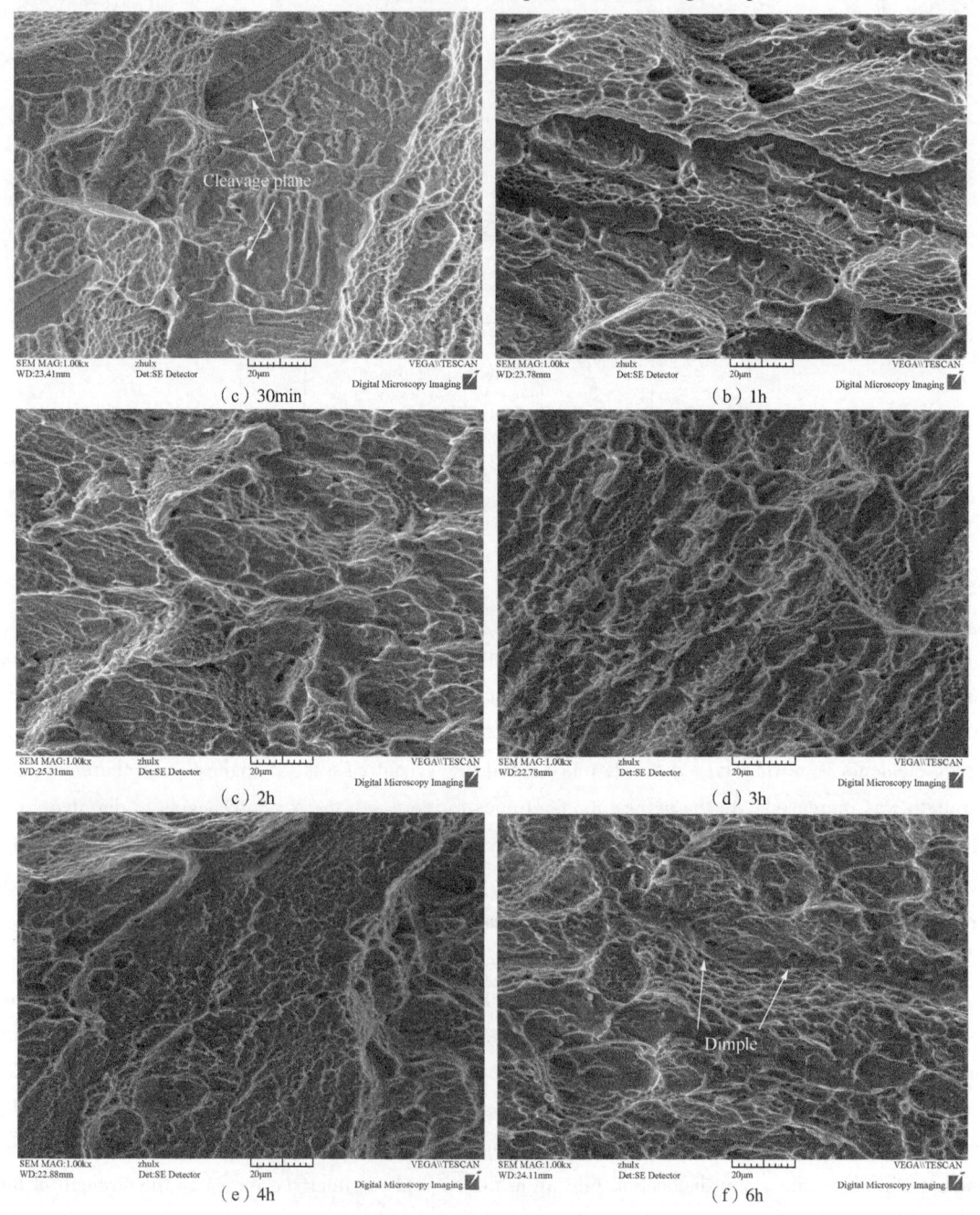

Fig. 13 SEM fractography of samples aged for different time

increase of the aging time, the size and depth of the dimples change. When the aging time was 30 min, the strength was the highest, the elongation was low, and brittle fracture was dominant. As shown in Fig. 13(a), the number of dimples is small, the depth is shallow, and there are many cleavage planes. When the aging time is increased, the strength is reduced, the elongation is high, and ductile fracture is dominant. The aging time is 6h, as shown in Fig. 13(f), the dimple size is large, and the number is large.

According to the analysis, the factors affecting the mechanical properties of the Ti-5Al-3V-1.5Mo-2Zr titanium alloy drill pipe are mainly the thickness, size, and β transformation content of the lamellar α phase. There are dislocations in the primary lamellar α phase; after the solution treatment, a martensite structure is generated. Then, the aging treatment is performed, and a large number of dislocation pile ups are found at the α/β phase interface, thereby, improving the strength of the alloy. When the aging time is short and the temperature is low, the martensitic structure transforms into a secondary α phase. As the aging time and temperature increase, the strength gradually decreases. This can be attributed to the growth of secondary α phase. The organizational transformation model is shown in Fig. 14. Therefore, the larger the number of precipitated secondary α phases, the smaller and the more significant the strengthening effect.

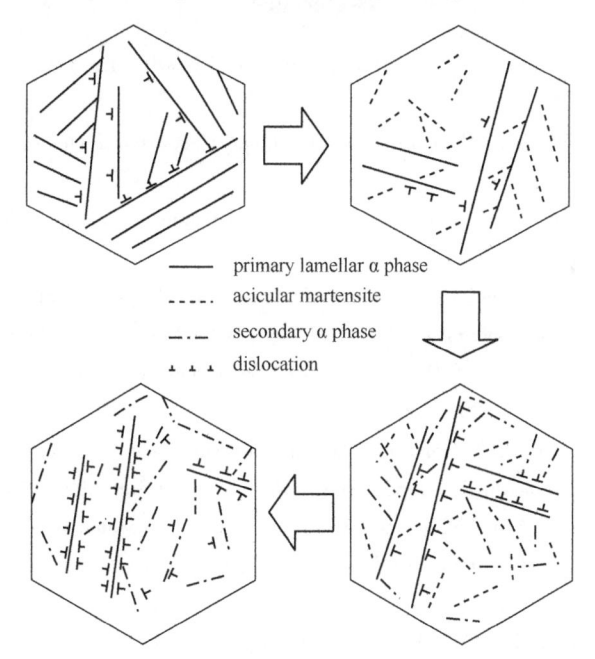

Fig. 14 Schematic microstructure evolution during solution and aging of Ti-5Al-3V-1.5Mo-2Zr titanium alloy

4 Conclusions

(1) After aging, the β grain size of Ti-5Al-3V-1.5Mo-2Zr titanium alloy increases, the grain boundaries are clear, the α phase colonies and the grain boundary α phase disappears. The size of primary lamellar α phase increases, and the volume fraction of secondary α phase gradually increases, resulting in an increase in strength and a decrease in elongation.

(2) As the aging temperature increases, the tensile and yield strength initially increase and then decrease, and the elongation decreases. After heat treatment of 930℃/ 1h/WQ + 500℃/6h/ AC, tensile strength is 1236MPa, an increase of 27%, yields strength is 1146MPa, an increase of 29.5%, and elongation is 9.5%, which shows excellent mechanical property balance. As the aging time increases, the tensile strength and yield strength gradually decrease, and the elongation gradually increases. After heat treatment of 930℃/1h/WQ + 600℃/6h/AC, the elongation reaches the higher value of 11.5%.

(3) After aging, dislocation piling at the α/β interface improved the strength of the alloy. The more and smaller secondary α phase precipitated, the more significant the strengthening effect.

Acknowledgement

The authors gratefully acknowledge the financial support provided by the National Science and Technology Major Project (No. 2016ZX05020 - 002) and Petro China Scientific Research and Technology Development Project (2018E 1808). National Key Research and Development Program of China (2019YFF0217504).

References

[1] J.C. Williams and E.A. Starke, Jr., Progress in Structural Materials for Aerospace Systems, Acta Mater., 2003, 51, 5775-5799.

[2] D. Banerjee and J.C. Williams, Perspectives on Titanium Science and Technology, Acta Mater., 2013, 61 (3), 844-879.

[3] R.W. Schutz and H.B. Watkins, Recent Developments in Titanium Alloy Application in the Energy Industry, Mater. Sci. Eng. A, 1998, 243(1-2), 305-315.

[4] J.E. Smith, R.B. Chandler, and P.L. Boster, Titanium Drill Pipe for Ultra-Deep and Deep Directional Drilling, SPE/IADC Drilling Conference, 27 Feb-1 Mar 2001 (Amsterdam), 236-249.

[5] R.Z. Li, C. Feng, L. Jiang, and Y.Q. Cao, Research Status and Development of Titanium Alloy Drill Pipes, Mater. Sci. Forum, 2019, 944, 903-909.

[6] C. Feng, W. Shou, H. Liu, D. Yi, and Y. Feng, Microstructure and Mechanical Properties of High Strength Al-Zn-Mg-Cu Alloys Used for Oil Drill Pipes, Trans. Nonferr. Met. Soc. China, 2015, 25(11), 3515-3522.

[7] H. Zhao, B. Wang, G. Liu, L. Yang, and W. Xiao, Effect of Vacuum Annealing on Microstructure and Mechanical Properties of TA15 Titanium Alloy Sheets, Trans. Nonferr. Met. Soc. China, 2015, 25(6), 1881-1888.

[8] Z. Sun, H. Wu, J. Sun, and J. Cao, Evolution of Lamellar α Phase During Two-Phase Field Heat Treatment in TA15 Alloy, Int. J. Hydrog. Energy, 2017, 42(32), 20849-20856.

[9] Z. Sun, L. Liu, and H. Yang, Microstructure Evolution of Different Loading Zones During TA15 Alloy Multi-cycle Isothermal Local Forging, Mater. Sci. Eng. A, 2011, 528(15), 5112-5121.

[10] Z.C. Sun, F.X. Han, H.L. Wu, and H. Yang, Tri-modal Microstructure Evolution of TA15 Ti-alloy Under Conventional Forging Combined with Given Subsequent Heat Treatment, J. Mater. Process. Technol., 2016, 229, 72-81.

[11] X.D. Zhang, P. Bonniwell, H.L. Fraser, W.A. Baeslack, III, D.J. Evans, T. Ginter, T. Bayha, and B. Cornell, Effect of Heat Treatment and Silicon Addition on the Microstructure Development of Ti-6Al-2Cr-2Mo-2Sn-2Zr, Mater. Sci. Eng. A, 2003, 343, 210-226.

[12] P. F. Gao, H. Yang, and X. G. Fan, Quantitative Analysis of the Microstructure of Transitional Region Under Multi-heat Isothermal Local Loading Forming of TA15 Titanium Alloy, Mater. Des., 2011, 32(4), 2012-2020.

[13] S. L. Semiatin, S. L. Knisley, P. N. Fagin, D. R. Barker, and F. Zhang, Microstructure Evolution During Alpha-Beta Heat Treatment of Ti-6Al-4V, Metall. Mater. Trans. A, 2003, 34, 2377-2386.

[14] Z. Sun, X. Mao, H. Wu, H. Yang, and J. Li, Tri-modal Microstructure and Performance of TA15 Ti-alloy Under Near-β Forging and Given Subsequent Solution and Aging Treatment, Mat. Sci. Eng. A, 2016, 654, 113-123.

[15] L. Shikai, H. Songxiao, Y. Wenjun, Y. Yang, and X. Baiqing, Effect of Cooling Rate on the Microstructure and Properties of TA15 ELI, Alloy, Rare Met. Mater. Eng., 2007, 36(5), 786-789.

[16] S. Zhu, H. Yang, L. G. Guo, and X. G. Fan, Effect of cooling rate on microstructure evolution during α/β heat treatment of TA15 titanium alloy, Mater. Charact., 2012, 70, 101-110.

[17] L. Xianghong, S. Ying, Z. Guoxian et al., Research and Application Progress of Ti Alloy Oil Country Tubular Goods, Rare Met. Mater. Eng., 2014, 43(6), 1518-1524.

[18] R. W. Schutz, Performance of Ruthenium-Enhanced α-β Titanium Alloys in Aggressive Sour Gas and Geothermal well Produced Fluid Brines, Corrosion 97, pp 9-14 Mar 1997 (New Orleans), NACE International, 1997.

[19] P. Guo, Y. Zhao, W. Zeng, and J. Liu, Effect of Microstructure on the Fatigue Crack Propagation Behavior of TC4-DT Titanium Alloy, J. Mater. Eng. Perform., 2015, 24, 1865-1870.

[20] H. Wu, Z. Sun, J. Cao, and Z. Yin, Microstructure and Mechanical Behavior of Heat-Treated and Thermomechanically Processed TA15 Ti Alloy Composites, J. Mater. Eng. Perform., 2019, 28, 788-799.

[21] R. Gaur, R. K. Gupta, V. AnilKumar, and S. S. Banwait, Effect of Cold Rolling and Heat Treatment on Microstructure and Mechanical Properties of Ti-4Al-1Mn Titanium Alloy, J. Mater. Eng. Perform., 2018, 27, 3217-3233.

[22] G. Lutjering, Influence of Processing on Microstructure and Mechanical Properties of (α+β) Titanium Alloys, Mater. Sci. Eng. A, 1998, 243(1-2), 32-45.

[23] S. A. Mantri, D. Choudhuri, T. Alam, G. B. Viswanathan, J. M. Sosa, H. L. Fraser, and R. Banerjee, Tuning the Scale of a Precipitates in b Titanium Alloys for Achieving High Strength, Scr. Mater., 2018, 154, 139-144.

[24] G. Wu, C. Feng, H. Liu, Y. Liu, and D. Yi, Fine Secondary α Phase-Induced Strengthening in a Ti-5.5Al-2Zr-1Mo-2.5V Alloy Pipe with a Widmanstatten Microstructure, J. Mater. Eng. Perform, 2020, 29, 1869-1881.

[25] "Standard Specification for Titanium and Titanium Alloy Seamless Pipe," B861-19, ASTM International, 2019.

[26] Z. Sun, S. Guo, and H. Yang, Nucleation and Growth Mechanism of α-Lamellae of Ti Alloy TA15 Cooling from an α + β Phase Field, Acta Mater., 2013, 61(6), 2057-2064.

[27] C. Li, J. Chen, W. Li, Y. J. Ren, J. J. He, and Z. X. Song, Effect of Heat Treatment Variations on the Microstructure Evolution and Mechanical Properties in α β Metastable Ti Alloy, J. Alloy Compound, 2016, 684, 466-473.

[28] Z. Du, S. Xiao, L. Xu, J. Tian, F. Kong, and Y. Chen, Effect of Heat Treatment on Microstructure and Mechanical Properties of α New β High Strength Titanium Alloy, Mater. Des., 2014, 55, 183-190.

[29] C. L. Li, L. N. Zou, Y. Y. Fu, W. J. Ye, and S. X. Hui, Effect of Heat Treatments on Microstructure and Property of a High Strength/ Toughness Ti-8V-1.5Mo-2Fe-3Al Alloy, Mater. Sci. Eng. A, 2014, 616, 207-213.

[30] J. Lu, Y. Zhao, P. Ge, Y. Zhang, H. Niu, and W. Zhang, Precipitation Behavior and Tensile Properties of New High Strength Beta Titanium Alloy Ti-1300, J. Alloy Compound, 2015, 637, 1-4.

[31] A.J. Williams, R.W. Cahn, and C.S. Barrett, The Crystallography of the β/α Transformation in Titanium, Acta Metall., 1954, 2(1), 117-128.

[32] D. Bhattacharyya, G.B. Viswanathan, R. Denkenberger, D. Furrer, and H. Fraser, The Role of Crystallographic and Geometrical Relationships Between α and β Phases in an α/β Titanium Alloy, Acta Mater., 2003, 51(16), 4679-4691.

本论文原发表于《Journal of Materials Engineering and Performance》2020 年第 13 卷第 847 期。

Tribological Properties of Ni/Cu/Ni Coating on the Ti-6Al-4V Alloy after Annealing at Various Temperatures

Luo Jinheng[1, 2]　Wang Nan[3]　Zhu Lixia[1, 2]　Wu Gang[1, 2]
Li Lifeng[1, 2]　Yang Miao[3]　Zhang Long[3]　Chen Yongnan[3]

(1. CNPC Tubular Goods Research Institute; 2. CNPC Key Laboratory for PetroChina Tubular Goods Engineering; 3. School of Materials Science and Engineering, Chang'an University)

Abstract: Diffusion reaction was a crucial route to enhance the wear resistance of Ti-6Al-4V alloys surface. In this work, the Ni/Cu/Ni composite layers were fabricated on the surface of Ti-6Al-4V alloy by electroplate craft, and then different annealing temperatures were applied to further optimize its tribological properties. The diffusion behaviors at various temperatures were systematically analyzed to reveal the physical mechanism of the enhanced tribological properties of the coatings. It was demonstrated that Cu_xTi_y and Ni_xTi_y intermetallic compounds with high hardness and strength were produced in the Ni/Cu/Ni coating, which acted as the reinforcing phases and improved the microhardness, reduced the friction coefficient, and lessened the wear rate. Specially, this effect reached the maximum when the annealing temperature was 800℃, showing excellent wear resistance. This work revealed the relationship between annealing temperatures andtribological properties of the Ni/Cu/Ni coating, and proposed wear mechanism, aiming to improve the surface performance of Ti-6Al-4V alloy by appropriately diffusion behavior.

Keywords: Ti-6Al-4V alloy; Ni/Cu/Ni coating; Phases transitions; Tribological properties; Wear mechanism

1 Introduction

Ti-6Al-4V alloys have the advantages of high specific strength, strong corrosion resistance and good biocompatibility, which have been widely used in aerospace, marine development and biomedical fields[1-3]. However, in some engineering applications, titanium alloys usually exhibit poor tribological properties, such as high friction coefficient, severe adhesive wear, low plastic shear resistance, weak work hardening ability and brittle oxide film on the surface[4,5], which greatly limit their application as friction components. Aydin et al.[6] suggested that the performance

---

Corresponding author: Chen Yongnan, chenyongnan@chd.edu.cn.

of matrix materials can be effectively reinforced by the thermal diffusion behaviour between different metals. Yao et al. [7] prepared the copper layer on Ti-6Al-4V alloy, exploiting diffusion between copper and titanium atoms, which can greatly improve the surface hardness and wear resistance. In addition, multilayer diffusion had been designed to form a variety of intermetallic compounds in coating by adding interlayer materials, optimizing the mechanical performance of the coating[8,9].

As well known, the atoms in surface coating will occur diffusion reaction and form the diffusion layer with a certain thickness during annealing[6,10,11]. Hu et al. [10] insisted that the Ni-Ti diffusion layer was mainly composed of the $NiTi_2$, Ni_3Ti and NiTiphases and hence it showed high hardness. Aydın et al. [6] provedthat the formed Cu_xTi_y intermetallic compounds during annealing process could apparently improve surface properties because of its high hardness. It can be concluded that the improvement of hardness in the diffusion layerwas strictly related to the intermetallic compounds, which can cause the wear mechanism transform to micro – adhesive wear from the adhesive wear, thereby improving the surface tribological properties. More importantly, the diffusion behavior of Ni and Ti atoms could be remarkably adjusted by the annealing temperatures [11]. Xia et al. [12] accurately analyzed the relationshipbetween annealing temperatures and the properties of Ti-6Al-4V alloy coatings, and the results confirmed the metallurgical bonding could be easily formed to strengthen its surface properties. Shen et al. [13] found that the depth of diffusion layer and the type of intermetallic compounds will be changed with the rise of annealing temperatures, which could observably improve the surface capability of the coatings. The above studies showed that the surface tribological properties of Ti-6Al-4V alloys can be improved by appropriate annealing of dissimilar metals.

However, the relationship between multi – layer diffusion behaviors, surface tribological properties and annealing temperatures were still indistinct. More importantly, a variety of intermetallic compounds may be dispersed in the diffusion layer without damaging the plasticity of the coating, which was essential for the study of wear mechanisms. Relevant research[2] had been done on Cu/Ni coating before, and a diffusion layer with certain properties has been obtained on the surface of Ti-6Al-4V alloy. But it was still not satisfactory in terms of the tribological properties. In thiswork, the diffusion behaviors and tribological properties of the Ni/Cu/Ni coating after annealing at various temperatures 600℃, 700℃ and 800℃ were studied. The formation mechanism of intermetallic compounds and the wear mechanism of diffusion layer were analyzed in detail. The purpose of this work was to study the diffusion layer evolution and wear resistance of the Ni/Cu/Ni coatings, obtaining the improved coating with a certain anti-wear effect by adjusting the annealing temperatures.

2 Experimental procedure

2.1 Materials synthesis

The commercial Ti – 6Al – 4V alloy was adopted for this study, and the chemical composition(wt.%) was 88.2Ti, 6Al, 4V, 0.6Fe, 0.5Mn, 0.4Si and 0.3Zn. The electroplated Ni/Cu/Ni composite coatings were synthesized from traditional electroplate method, the plating sequence of the Ni/Cu/Ni composite coating was first nickel plating for 10 minutes, then copper plating for 20 minutes, finally nickel plating for 10 minutes and the detail electrodeposition conditions and compositions of the bath were summarized in Table 1. The flow diagram of sample preparation process was shown in Fig. 1(a), and the macro picture of sample obtained at each step of the preparation process was shown in Fig. 1(b), showing the macro change process of sample. The schematic diagram of the cross section and surface of the prepared sample were shown in Fig. 1(c). It could be evidenced the structure of the Ni/Cu/Ni coating on the Ti-6Al-4Vsubstrate and the surface was uniform and dense. And then, annealing process was performed in a vacuum furnace (OTF-1200X), setting the vacuum level of 1×10^{-2} Pa. The plating samples were annealing at 600℃, 700℃ and 800℃ for 3 h, respectively, after that, cooled down to 25℃ in the furnace.

Table 1 The electrolyte and the operatingcondition

	Electrolyte composition	Operating condition
nickel plating	nickel sulfate hexahydrate($NiSO_4 \cdot 6H_2O$): 180g/L sodium sulfate(Na_2SO_4): 70g/L magnesium sulfate($MgSO_4$): 30g/L sodium chloride(NaCl): 30g/L boric acid(H_3BO_3): 30g/L	voltage: 3V temperature: 25℃ duration: 10min
copper plating	copper sulfate pentahydrate($CuSO_4 \cdot 5H_2O$): 210g/L sodium chloride(NaCl): 20mg/L sulfuric acid(H_2SO_4): 70g/L	voltage: 0.65V temperature: 25℃ duration: 20min

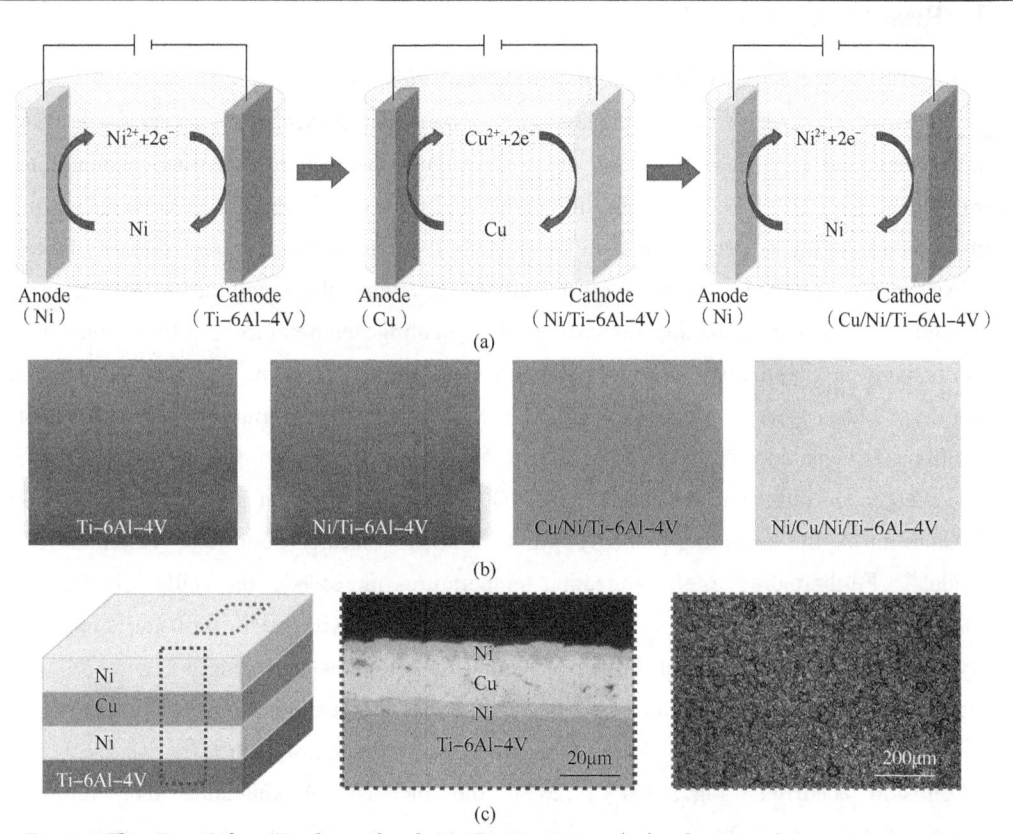

Fig. 1 The Ti-6Al-4V electroplated Ni/Cu/Ni coating (a) schematic diagram of preparation technology (b) macroscopic photographs of samples during preparation (c) schematic diagram and optical microscopy images and cross-sectional microstructure

2.2 Characterization after the thermaldiffusion

The microstructure analysis of the annealing sampleswas detected using a scanning electron microscopy(SEM, Hitachi-S4800). The thermal diffusion behaviors of Cu, Ni and Ti atoms were interpreted using an energy dispersive spectroscopy(EDS, Hitachi-S4800). The phases structures of the coatings were identified by X-ray diffraction (XRD, XRD D/M2500) with CuKa radiation(0.154nm wavelength) at tube voltage of 40kV, current of 40mA, an 8.0°/min scanning speed, a 20°~80° of 2θ range.

2.3 Tribological properties

The surface roughness and microhardness of the coating surface after annealing were measured by roughness tester (SJ-210) and microhardness tester (HV-1000A), respectively. For data validity, five tests were performed and then averaged. The wear resistance of the coatings was assessed with a ball-on-plate wear tester (MMQ-02G) using a 6mm GCr15 ball counterpart. A constant load of 3N was applied normally to the sample under non-lubricated condition atroom temperature. The abrasion resistance testswere performed on with a circular track of 3mm in diameter, a rotational speed of 100r/min and a total sliding distance of 37.70m. The worn morphologies of the coatings were explained using scanning electron microscopy(SEM, Hitachi-S4800) and Laser Confocal Microscope(LCM, Olympus OLS5000) to show the worn mechanism.

3 Results and discussion

3.1 The diffusion behaviorsand phases transitions

As shown in Fig.2(a), the morphology and composition of Ni/Cu/Ni layers on Ti-6Al-4V alloy after annealing were researched by cross-sectional observation with EDS examination. The element distribution of Cu, Ni and Ti in diffusion layers were clearly displayed. Obviously, the Cu, Ni and Ti atoms hardly occurred diffusion behavior without annealing, where the interfacesin the Ni/Cu/Ni layers were obvious with a clear outline. However, these interfaces gradually became blurred and disappeared with the increasing of annealing temperatures, which indicated that diffusion behaviorsbetween Cu, Ni and Ti atoms intensified step by step. It was noteworthy that the diffusion layer with a certain thickness was formed between Ni/Ti interfaces, which was distinctly different from Ni layer and Ti-6Al-4Vsubstrate. Moreover, the higher the annealing temperature was, the thicker the diffusion layer was[Fig.2(b)]. Actually, Ti atom had a higher diffusion rate than Ni atom during the annealing process, which resulted in a large number of Kirkendall voids on the Ti side[14]. Furthermore, as the annealing temperatures increased, the voids will connect with each other and then coalesce into large gaps, forming the Kirkendall diffusion channels[15,16]. Finally, they would become a fast diffusion channel for all atoms[Fig.2(c-d)].

Fig.3 displayedthe elements line scan diagrams across the coatings from its substrate to the top after annealing. It can be seen that without annealing, the interface of each layer was clear, and no atomic diffusion occurred [Fig.3(a)]. With the increase of annealing temperatures, the interdiffusion between Cu, Ni, and Ti atoms intensified, gradually forming a continuous one. In addition, the interdiffusion ability of the element in the Ni layer was better than that in the Cu layer. As shown in Fig.3(b)-(d), there were fewer Ni and Ti atoms in the Cu layer. In the Ni layer, Cu and Ti atoms were fully diffused and the diffusion reaction occurs, forming a series of intermetallic compounds and solid solutions.

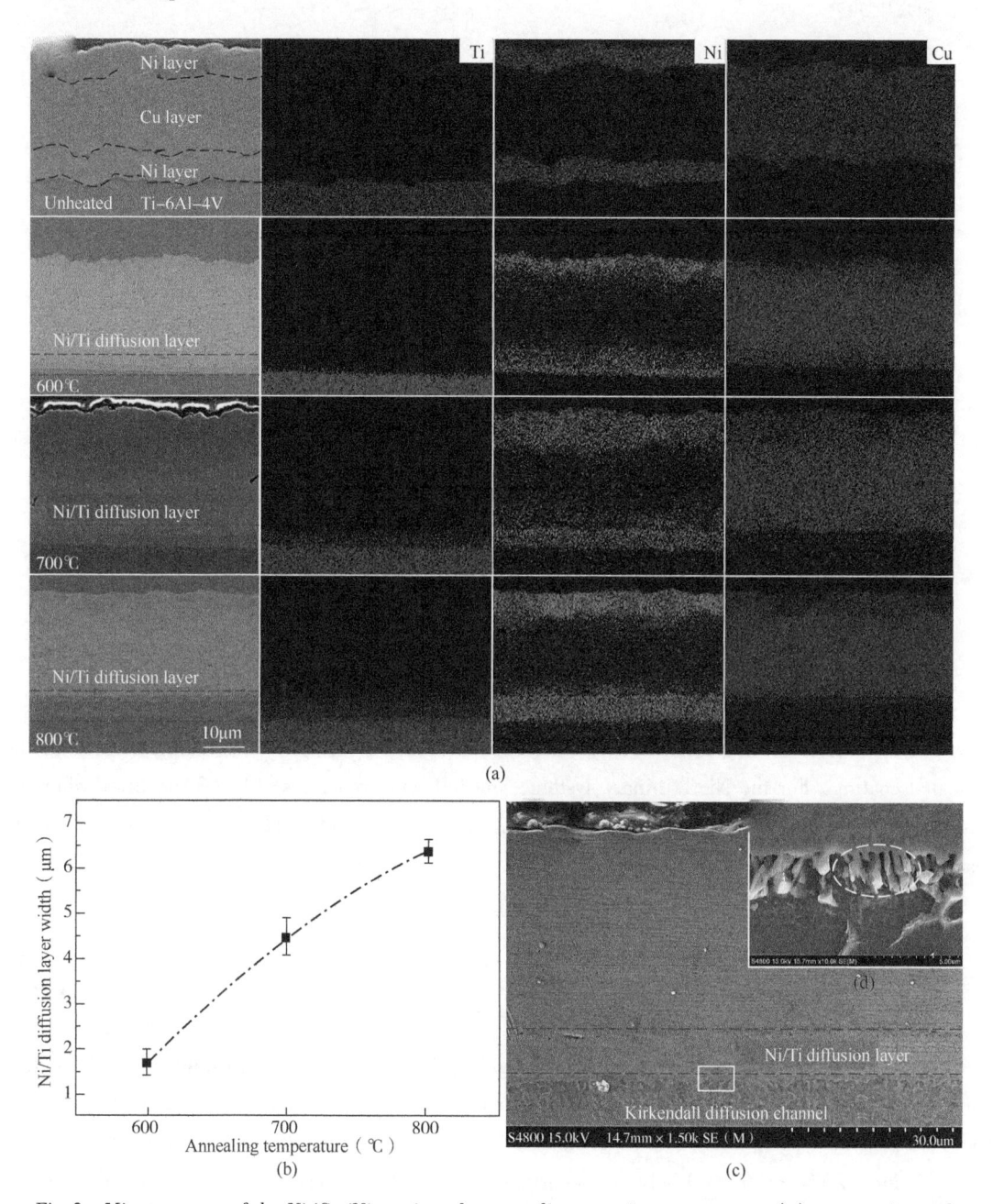

Fig. 2　Microstructure of the Ni/Cu/Ni coatingsafter annealing at various temperatures(a)cross section and corresponding element composition(b)Ni/Ti diffusion layer width(c)the diffusion layer structure of the Ni/Ti interface for 800℃. The obvious Ni/Ti diffusion layers was observed in the red line. (d) showed the Kirkendall diffusion channel, which was showed in the white area

Fig. 4(a) showed the Ni−Cu solid solution and the intermetallic compounds of Ni_xTi_y and Cu_xTi_y in the coatings when the annealing temperatures increased from 600℃ to 800℃. According to

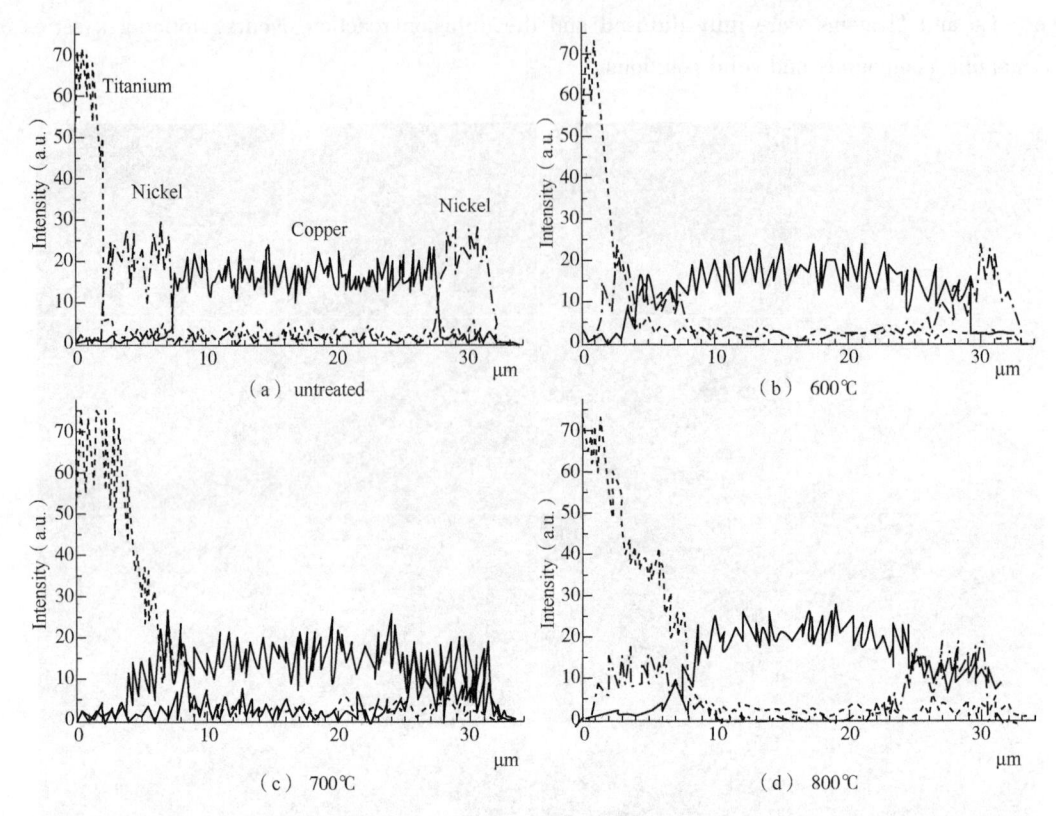

Fig. 3 The elements line scan diagrams across the coatings from its substrate to the top after annealing

the Ni-Cu equilibrium diagram, the α(Cu, Ni) solid solution (Cu$_{3.8}$Ni) was mainly formed in the annealed coatings. For the Ni-Ti binary system, the following reactions (1)-(3) occurred after the diffusion of Ni and Ti atoms, and the Ni$_3$Ti, NiTi$_2$ and NiTi phases were formed, respectively[17,18]:

$$3Ni + Ti = Ni_3Ti \qquad (1)$$

$$Ni + 2Ti = NiTi_2 \qquad (2)$$

$$Ni + Ti = NiTi \qquad (3)$$

In order to reveal the formation mechanism of Ni$_x$Ti$_y$ intermetallic compounds, the Gibbs free energies of the above reaction were summarized in Fig. 4(b). The Gibbs free energies of Ni$_3$Ti and NiTi$_2$ phases were substantially lower than that of NiTi phase, which illustrated that the reaction (1) and reaction (2) were more likely to occur than reaction (3). Bastin et al.[14] found that the nucleation and growth of NiTi phase caused a sharp decrease in NiTi$_2$ phase and Zhou et al.[19] also confirmed that the growth of NiTi phase considerably consume Ni$_3$Ti and NiTi$_2$ phases. According to previous analysis, the nucleation and growth of NiTi phase mainly depend on the Ni$_3$Ti and NiTi$_2$ phases. That is, the nucleation of the Ni$_3$Ti and NiTi$_2$ phases was earlier than that of NiTi phase, which was consistent with the order of the phase's formation calculated. The Ni$_3$Ti and NiTi$_2$ phases in this study were gradually formed in the Ni/Ti interfacial diffusion layer with the thermaldiffusion temperature increased, the absence of NiTi phase may be due to insufficient diffusion time or a small amount of formation[20,21].

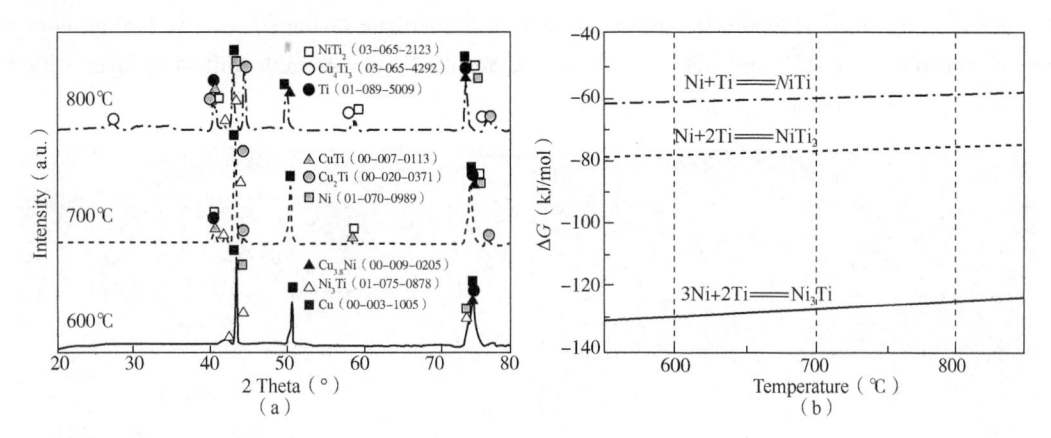

Fig. 4 Phases transitions of the Ni/Cu/Ni coatings after annealing at various temperatures (a) surface XRD images (b) Gibbs free energy of the Ni-Ti reaction

The atom diffusion direction in the Cu-Ti binary system could be theoretically analyzed by chemical potential. It is well known that the atom always spontaneously transfers from a high chemical potential to a low chemical potential, and the following theory is summarized by Wu et al.[22]:

$$\mu_i = \frac{\partial G}{\partial n_i} \quad (4)$$

Where μ_i is the chemical potential of component i, G is the Gibbs free energy and n_i is the atomic number of component i. It can be calculated that the chemical potential of Ti atoms was higher than that of Cu atoms. Under the driving force of the diffusion, the Ti atoms diffused toward the Cu side while the diffusion reaction occurred. Previous researches proved that the Ti atoms with greater diffusion capacity formed Cu_xTi_y intermetallic compounds on the Cu side [6,22]. Therefore, under the above synergistic effect, the Ti atoms rapidly diffused toward the Cu side, gradually forming the $CuTi$, Cu_2Ti and Cu_4Ti_3 intermetallic compounds, which can be expressed as follows [22,24]:

$$2Cu+Ti = Cu_2Ti \quad (5)$$

$$Cu+Ti = CuTi \quad (6)$$

$$4Cu+3Ti = Cu_4Ti_3 \quad (7)$$

Apparently, the intermetallic compounds of Ni_xTi_y and Cu_xTi_y were gradually formed in the coating with the annealing temperatures increased. These intermetallic compounds with strong atomic bonds and high hardness served as the reinforcing phases in the diffusion layer, which can effectively improve the tribological properties of the coating[25,26].

3.2 Tribological properties of the Ni/Cu/Ni coating on Ti-6Al-4V alloy

It is well known that surface topography is an important factor in describing tribological properties of the coating[27]. As depicted in Fig.5(a), the surface of untreated coating was characterized of continuous uniform grain size. As the annealing temperature increased, the coating surface gradually transformed to granulated and densified status[Fig. 5(b-d)]. It showed that a large number of fine particles like the Ni_xTi_y and Cu_xTi_y phases existed on the surface of the Ni/Cu/Ni coating after the thermal diffusion at 700℃ and 800℃, which was resulted from a stable diffusion layer with the diffusion reaction between Ni, Cu and Ti[26].

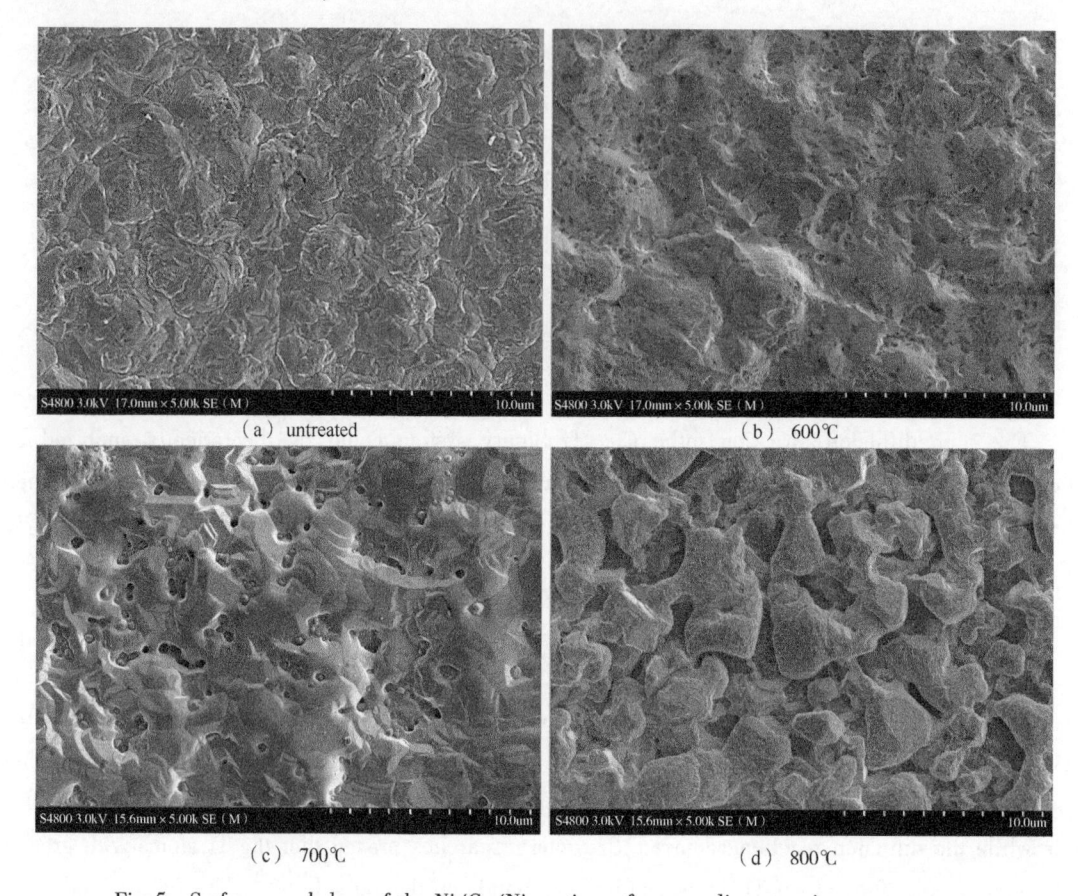

(a) untreated (b) 600℃

(c) 700℃ (d) 800℃

Fig. 5 Surface morphology of the Ni/Cu/Ni coatings after annealing at various temperatures

The microhardness of the annealed coatings was shown in Fig. 6(a). With the annealing temperatures up to 800℃, the microhardness progressively increased from 155HV to 357HV. The increased microhardness was ascribed to the formation of the intermetallic compounds and the strengthening effect of solid solution[25,28]. The Cu_xTi_y and Ni_xTi_y intermetallic compounds with high hardness can substantially enhance the surface microhardness of the Ni/Cu/Ni coating. At the same time, the annealed coatings had a lower amount of mass loss, and can even be reduced to half of the Ti-6Al-4V substrate. This was mainly because the hard phases dispersed on the coating surface can effectively enhance the coating's resistance to high contact stresses, thereby optimizing the coating surface wear behavior.

Fig. 6(b) displayedthe friction coefficient of the annealed Ni/Cu/Ni coatings. Clearly, the friction coefficient of the Ti-6Al-4V substrate was around 0.6, and it gradually decreased and stabilized at around 0.4 after annealing, which were all composed of the running-in stage and stabilization stage. It was noteworthy that some anomalous peaks appear during the stabilization stage after annealing at 600℃. This may be related to severe plastic deformation and adhesive wear caused by low microhardness. The process of adhesion, delamination, debonding and re-stabilization resulted in these anomalous peaks, which implied poor surface tribological properties of the coating. As for the Ni/Cu/Ni coating after annealing at 800℃, there was no obvious running-in stage and reached quickly to a relatively stabilization stage, which implied the increase of microhardness and the uniform dense diffusion layer can improve the surface tribological properties of the coating. Meanwhile, the plastic deformation ability of the Ni/Cu/Ni coating under stress concentration was also significantly reduced due to the increase of yield strength and microhardness[29]. Thus, the increase of thermal diffusion temperature was conductive to reducing the friction coefficient of the coating, which was put down to the strengthening effect of the hard particle during the diffusion process.

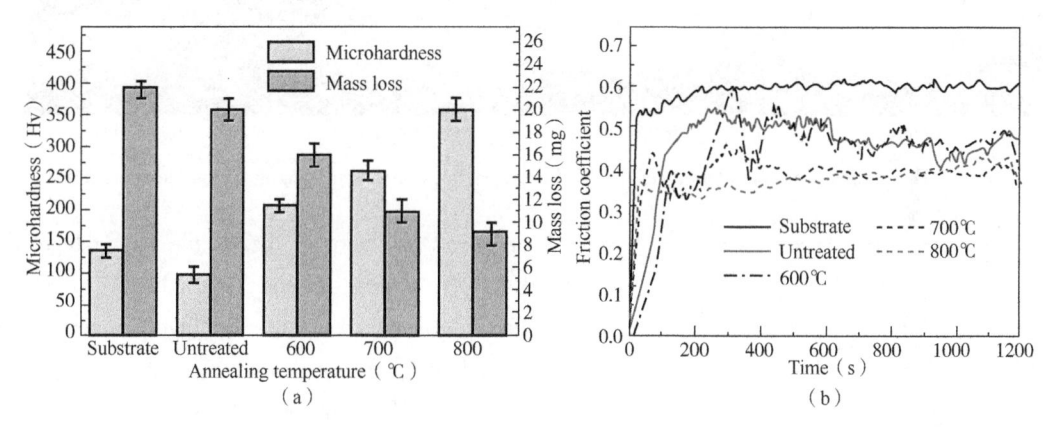

Fig. 6 The surface tribological properties of the Ni/Cu/Ni coatings after annealing at various temperatures (a) microhardness and mass loss (b) the change curves of friction coefficient with sliding time

3.3 Wear mechanism

The Ni/Cu/Ni coating without annealing was susceptible to plastic deformation and adhered to the contact surface to form adhesive wear, which caused continuous cutting penetrated on the coating and wore the Ti-6Al-4V alloy substrate. In this process, a large amount of grinding debris was formed on both sides of the grinding mark, as shown in Fig. 7(a), which indicated poor wear performance. After annealing, the coating was tightly bonded to the Ti-6Al-4V substrate and formed an effective diffusion layer and distributed a mass of intermetallic compounds, resulting in a prominent improvement in wear performance. The wear marks on the surface of the samples gradually became shallower and flatter [Fig. 7(b-d)], indicating that the adhesive wear gradually weakened. Especially, as the annealing temperatures raised up to 800℃, only slight wear was caused on the surface of the coating, showing the best tribological properties [Fig. 7(d)].

Obviously, the tribological properties of the coatings on Ti-6Al-4V alloy were improved after annealing, which could be stated using Archard's law[30,31]:

$$Q = KW/H \qquad (8)$$

where Q is the wear rate, K is the friction coefficient, W is the applied load, and H is the hardness. A low friction coefficient and a high hardness can result in a low wear rate under the same wear

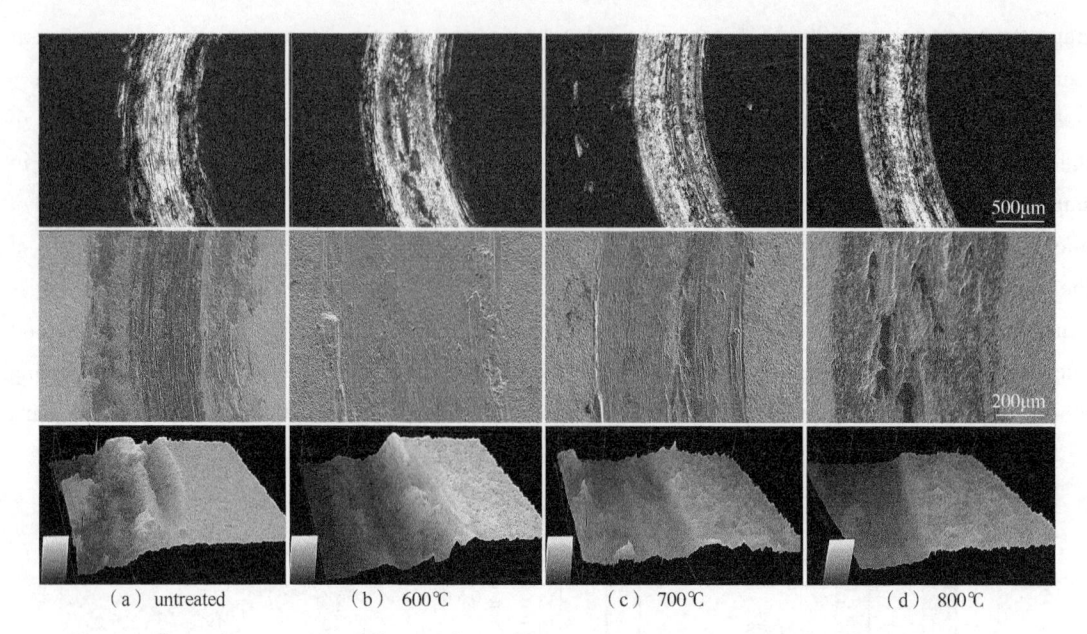

Fig. 7　The surface wear morphology of the Ni/Cu/Ni coatings after annealing at various temperatures

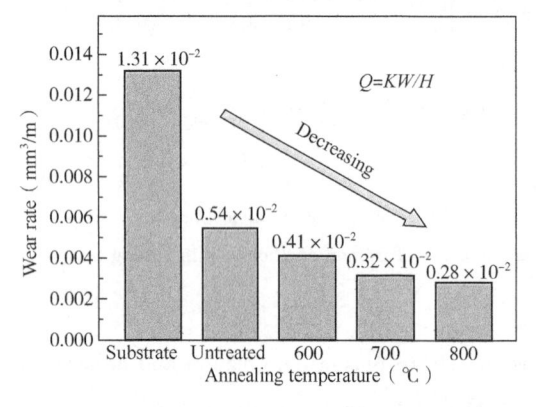

(a) untreated　　(b) 600℃　　(c) 700℃　　(d) 800℃

Fig. 8　The wear rare of the Ni/Cu/Ni coatings after annealing at various temperatures

conditions. As shown in Fig. 8, the wear rate of the Ni/Cu/Ni coating on the Ti – 6Al – 4V alloy was effectively reduced after annealing. Compared with Ti-6Al-4V substrate, the wear rate was reduced from 1.31×10^{-2} mm^3/m to 0.28×10^{-2} mm^3/m after annealing 800℃. It was mainly related to changes in the surface state of the coating. Wang et al. [11] studied that intermetallic compounds can increase the resistance of coatings to high contact stress due to their high microhardness. In addition, the Ni layer and the Cu layer had excellent ductility, improving the brittle fracture ability of the coating. During the adhesive wear process, the Ni layer and the Cu layer with lower hardness first underwent a certain degree of deformation, and the adhesive wear contact area was continuously increased. On the other hand, the Ni$_x$Ti$_y$ and Cu$_x$Ti$_y$ intermetallic compounds with high hardness could resist more wear loads and prevented the Ni layer and the Cu layer with low hardness from further deformation. Meanwhile, the good ductility of Ni layer and the Cu layer can alleviate cracks caused by the loading of the hard – intermetallic compounds. Therefore, the intermetallic compounds with the high hardness and the Ni layer and the Cu layer with good ductility can cooperate with each other to jointly improve the wear resistance of the Ni/Cu/Ni coating.

When theannealingtemperature was up to 800℃, the NiTi$_2$ and Cu$_4$Ti$_3$ phases could be formed in the coating, which had a higher microhardness than the Ni$_3$Ti, CuTi and Cu$_2$Ti phases[19,32]. Therefore, they would act as a supporting load to reduce the furrow effect of the counter ball to the coating during the wear test. The diffusion layer with a certain thickness reduced the area of adhesion in the friction surface, weakening the adhesion effect. Especially, the hard Ni-Ti diffusion layer of the Ni/Cu/Ni coating increased the yield strength of the contact surface and effectively reduces the tangential stresses and the interfacial stresses[33,34]. As a result, the adhesion effect and the furrow effect on the surface of the coating can be effectively suppressed by each otherafterannealing at 800℃.

4 Conclusions

The Ni/Cu/Ni coating were prepared on the surface of Ti-6Al-4V alloy by theelectrodeposition and subsequentlyannealing at 600℃, 700℃ and 800℃ for 3h, respectively. The microstructure and tribological properties of the diffusion layer were systematically investigated. Base on achieve results the following conclusions can be made:

The Cu, Ni, Ti atoms in coating displayed complex diffusion behaviors and formed the diffusion layers with a certain thickness. As the annealing temperature increased, the Kirkendall diffusion channel appeared, accelerating the diffusion behaviors of the atoms. In addition, the diffusion layers were mainly composed of the Ni_3Ti, $NiTi_2$, $CuTi$, Cu_2Ti and Cu_4Ti_3 intermetallic compounds and the $\alpha(Cu, Ni)$ solid solutionwith high hardness and strength, which acted as the reinforcing phases of wear resistant materials. The Ni_xTi_y and Cu_xTi_y intermetallic compounds were continuously and uniformly distributed in the Ni/Cu/Ni coating, which can tremendously strengthen the surface hardness, reduce the friction coefficient and lessen the wear rate. In wear process, the intermetallic compounds with the high hardness and the Cu layer and the Ni layer with good ductility can cooperate with each other to jointly improve the wear resistance of the coating. The synergistic effect was particularly remarkable at the annealing temperature of 800℃, showing excellent wear resistance.

References

[1] Wang, F.; Wang, D. Study on the Effects of Heat-treatment on Mechanical Properties of TC4 Titanium Alloy Sheets for Aviation Application. Titan. Ind. Progress. 2017, 45, 56-60.

[2] Chen, Y.; Liu, S.; Zhao, Y.; Liu, Q.; Zhu, L.; Song, X.; Zhang, Y.; Hao, J. Diffusion behavior and mechanical properties of Cu/Ni coating on TC4 alloy. Vacuum. 2017, 143, 150-157.

[3] Shen, Z. C.; Xie, F. Q.; Wu, X. Q.; Yao, X. F. Properties of Coating on TC4 Titanium Alloy by Copper Electroplating. China Surf. Eng. 2012, 25, 45-49.

[4] Zhang, Y.; Zhang, H. L.; Wu, J. H.; Wang, X. T. Enhanced thermal conductivity in copper matrix composites reinforced with titanium-coated diamond particles. Scripta. Mater. 2011, 65, 1097-1100.

[5] Jiang, P.; He, X. L.; Li, X. X.; Wang, H. M. Wear resistance of a laser surface alloyed Ti-6Al-4V alloy. Surf. Coat. Tech. 2000, 130, 24-28.

[6] Aydın, K.; Kaya, Y.; Kahraman, N. Experimental study of diffusion welding/bonding of titanium to copper. Mater. Design2012, 37, 356-368.

[7] Yao, X.; Xie, F.; Wang, Y.; Wu, X. Research on Tribological and Wear Properties of Cu Coating on TC4 Alloy. Rare. Metal. Mat. Eng. 2012, 41, 2135-2138.

[8] Petrovic, S.; Peruško, D.; Mitric, M.; Kovac, J.; Dražić, G.; Gaković, B.; Homewoodc, K. P.;

Milosavljević a, M. Formation of intermetallic phase in Ni/Ti multilayer structure by ion implantation and thermal annealing. Intermetallics2012, 25(3), 27-33.

[9] Wang, F. L.; Sheng, G. M.; Deng, Y. Q. Impulse pressuring diffusion bonding of titanium to 304 stainless steel using pure Ni interlayer, Rare Metals. 2016, 35(4), 1-6.

[10] Hu, L.; Xue, Y.; Shi, F. Intermetallic formation and mechanical properties of Ni - Ti diffusion couples. Mater. Design2017, 130, 175-182.

[11] Wang, Z.; He, Z.; Wang, Y.; Liu, X.; Tang, B. Microstructure and tribological behaviors of Ti6Al4V alloy treated by plasma Ni alloying. Appl. Surf. Sci. 2011, 257(23), 10267-10272.

[12] Xia, Y. X.; Li, D. Effects of Post Heat-treatment of Electroplating on Plating Adhesion for TC4 Titanium Alloys. Rare Metal Mat. Eng. 2001, 30(5), 390-391.

[13] Shen, Q.; Xiang, H.; Luo, G.; Wang, C.; Li, M.; Zhang, L. Microstructure and mechanical properties of TC4/oxygen - free copper joint with silver interlayer prepared by diffusion bonding. Mater. Sci. Eng. A. 2014, 596, 45-51.

[14] Bastin, G. F.; Rieck, G. D. Diffusion in the titanium-nickel system: I. occurrence and growth of the various intermetallic compounds. Metall. Trans. 1974, 5, 1817-1826.

[15] Puente, A. E. P. Y.; Dunand, D. C. Synthesis of NiTi microtubes via the Kirkendall effect during interdiffusion of Ti-coated Ni wires. Intermetallics2018, 92, 42-48.

[16] Fan, H. J.; Gosele, U.; Zacharias, M. Formation of nanotubes and hollow nanoparticles based on Kirkendall and diffusion processes: a review. Small2007, 3, 1660-1671.

[17] He, P.; Liu, D. Mechanism of forming interfacial intermetallic compounds at interface for solid state diffusion bonding of dissimilar materials. Mater. Sci. Eng. A. 2006, 437(2), 430-435.

[18] Vandal, M. J. H.; Pleumeekers, M. C. L. P.; Kodentsov, A. A.; Vanloo, F. J. J. Intrinsic diffusion and kirkendall effect in Ni-Pd and Fe-Pd solid solutions. Acta Mater. 2000, 48, 385-396.

[19] Zhou, Y.; Wang, Q.; Sun, D. L.; Han, X. L. Co - effect of heat and direct current on growth of intermetallic layers at the interface of Ti-Ni diffusion couples. J. Alloy. Compd. 2011, 509, 1201-1205.

[20] Lin, C. M.; Kai, W. Y.; Su, C. Y.; Tsai, C. N.; Chen, Y. C. Microstructure and mechanical properties of Ti - 6Al - 4V alloy diffused with molybdenum and nickel by double glow plasma surface alloying technique. J. Alloy. Compd. 2017, 717, 197-204.

[21] Simoes, S.; Viana, F.; Ramos, A. S.; Vieira, M. T.; Vieira, M. F. Reaction zone formed during diffusion bonding of TiNi to Ti6Al4V using Ni/Ti nanolayers. J. Mater. Sci. 2013, 48, 7718-7727.

[22] Akbarpour, M. R.; Moniri, J. S. Wear performance of novel nanostructured Ti-Cu intermetallic alloy as a potential material for biomedical applications. J. Alloy. Compd. 2017, 699, 882-886.

[23] Wu, M. F.; Yang, M.; Zhang, C.; Yang, P. Research on the liquid phase spreading and microstructure of Ti/Cu eutectic reaction. Trans. China Weld. Inst. 2005, 26(10), 68-71.

[24] Campo, K. N.; Lima, D. D. D.; Lopes, E. S. N.; Caram, R. Erratum to: On the selection of Ti-Cu alloys for thixoforming processes: phase diagram and microstructural evaluation. J. Mater. Sci. 2016, 51, 9912-9913.

[25] Semboshi, S.; Iwase, A.; Takasugi, T. Surface hardening of age-hardenable Cu-Ti alloy by plasma carburization. Surf. Coat. Tech. 2015, 283, 262-267.

[26] Zhang, X.; Ma, Y.; Lin, N.; Huang, X.; Hang, R.; Fan, A.; Tang, B. Microstructure, antibacterial properties and wear resistance of plasma Cu-Ni surface modified titanium. Surf. Coat. Tech. 2013, 232, 515-520.

[27] Zhou, X.; Shen, Y. Surface morphologies, tribological properties, and formation mechanism of the Ni-CeO$_2$ nanocrystalline coatings on the modified surface of TA2 substrate. Surf. Coat. Tech. 2014, 249, 6-18.

[28] Fan, D.; Liu, X.; Huang, J.; Fu, R.; Chen, S.; Zhao, X. An ultra-hard and thick composite coating metallurgically bonded to Ti-6Al-4V. Surf. Coat. Tech. 2015, 278, 157-162.

[29] Chen, H.; Zheng, L. J.; Zhang, F. X.; Zhang, H. X. Thermal stability and hardening behavior in super elastic Ni-rich Nitinol alloys with Al addition. Mater. Sci. Eng. A. 2017, 708, 514-522.

[30] Wang, L.; Gao, Y.; Xue, Q.; Liu, H.; Xu, T. Microstructure and tribological properties of electro deposited Ni-Co alloy deposits. Appl. Surf. Sci. 2005, 242, 326-332.

[31] Ranganatha, S.; Venkatesha, T. V.; Vathsala, K. Development of electroless Ni-Zn-P/nano-TiO$_2$, composite coatings and their properties. Appl. Surf. Sci. 2010, 256, 7377-7383.

[32] Cai, Q.; Liu, W.; Ma, Y.; Zhu, W.; Pang, X. Effect of joining temperature on the microstructure and strength of W-steel HIP joints with Ti/Cu composite interlayer. J. Nucl. Mater. 2018, 507, 198-207.

[33] Wang, N.; Chen, Y. N.; Zhang, L.; Li, Y.; Liu, S. S.; Zhan, H. F.; Zhu. L. X.; Zhu, S. D.; Zhao, Y. Q. Isothermal diffusion behavior and surface performance of Cu/Ni coating on TC4 alloy. Materials2019, 12, 3884.

[34] Dipak, T. W.; Chinmaya, K. P.; Ritesh, P. NiTi coating on Ti-6Al-4V alloy by TIG cladding process for improvement of wear resistance: Microstructure evolution and mechanical performances. J. Mater. Process. Tech. 2018, 262, 551-561.

本论文原发表于《Materials》2020年第13卷。

Application Analysis of Epoxy-Coated Tubing in an Oil Field

Zhu Lijuan　Feng Chun　Han Lihong　Lu Caihong
Yuan Juntao　Hang Wang　Wang Jianjun　Xu Xin

(State Key Laboratory for Performance and Structure Safety of Petroleum Tubular Goods and Equipment Materials, CNPC Tubular Goods Research Institute)

Abstract: The application of epoxy-coated tubing serviced in water injection wells for 19~80 months of J oil field was investigated. Testing indicated that one-third of the tubing did not met the technology agreement of users for continue use, the other tubing showed excellent corrosion resistance. The poor surface pretreatment of the tubing and poor high-temperature corrosion resistance of the epoxy coating were the main reasons for the failure of the epoxy-coated repaired tubing and the epoxy-coated tubing served in high working temperature respectively. Dissolve oxygen corrosion, carbon dioxide (CO_2) corrosion, and under-scale corrosion occurred on the tubing substrate with coating damage.

With the exploration of the oil field, the water injection tubing suffers increasing serious corrosion due to the deyradotion of the injected water quality[1]. In order to improve the water injection effect, protective epoxy coatings are widely used to improve the corrosion resistance of the water injection tubing due to its economic effectiveness[1,2]. However, tubing failure induced by unsuitable coating and improper construction of the coatings used in the oilfields occurred frequently in the recent years.

By the end of 2017, more than 36% of the API SPEC SCT³ P110 tubing was protected by epoxy coatings in the water injection wells of J oilfield in China, and the service life of tubing is significantly improved. However, a part of anticorrosion tubing's service life is less than two years. In the present work, application analysis of the epoxy-coated tubing serviced in water injection wells of J oilfield for 19~80 months was studied. The reason for failure of the epoxy coatings and the corrosion mechanism of the P110 tubing were both investigated.

1　Experimental work

The morphologies, metallographic structure, residual wall thickness, chemical composition and corrosion products of the epoxy-coated 73mm. tubing with a wall thickness of 5.51mm serviced in water injection wells for 19~80 months and the properties of the unused epoxy-coated tubing of

---

Corresponding author: Zhu Lijuan, zhulijuan1986@cnpc.com.cn; Feng Chun, fengchun003@cnpc.com.cn.

the same lot were observed and analyzed. The residual wall thickness was measured using an ultrasonic thickness measuring meter. The morphology and composition of the corrosion products were characterized by metallographic microscope, scanning electron microscopy, energy-dispersive spectroscopy(EDS) and X-ray power diffraction(XRD). The corrosion resistance of the unused tubing was conducted in an autoclave; the formation water with 6.5pH(composition: 9912.4mg.L^{-1} Cl^{-}, 104.2mg.L^{-1} Mg^{2+}, 110.8mg.L^{-1} Ca^{2+}, 1107.2mg.L^{-1} bicarbonate[HCO$_3^-$], 4.9mg.L^{-1} SO$_4^{2-}$, and balanced Na$^+$ and K$^+$)from J oilfield was used as the test solution.

2 Results and discussion

2.1 Visual inspection

Table 1 shows the data from nine joints of epoxy-coated tubing that were used in water injection wells for ~19 to 80 months and two joints of unused epoxy-coated tubing (No.10 and No.11) of the same lot sampled in J oil field. The epoxy coatings on the used tubing were damaged to varying degrees. One-third of the tubing's minimum residual wall thickness was <4.82mm, which did not meet the API SPEC 5CT standard for P110, nor the technology agreement of the users for continued use; however, the other six used tubing showed excellent corrosion resistance with only a little bit of coating damage. No corrosion products were observed on the tubing, which indicated the small amount of coating damage occurred in the sampling process. The No.2, No.3, and No.11 tubing in Table 1 were repaired tubing, which were recoated for downhole application after a period of service time. It means that the surface of the No.2, No.3, and No.11 tubing underneath the coating had been corroded.

Table 1 Information of epoxy-coated tubing used in the water injection wells of J oilfield

No.	Service time (month)	Well depth (m)	Working temperature (℃)	Minimum wall thickness (mm)	Coating damage area (percentage)
1	19	500	20	5.45	1%
2	22	500	20	3.7	100%
3	23	1000	36	4.7	50%
4	34	1000	36	5.5	0%
5	34	3000	82	5.5	0%
6	58	1000	36	5.5	1%
7	58	2000	60	5.5	1%
8	80	1000	36	5.51	0.5%
9	80	3900	102	4.2	100%
10	0	—	—	5.51	0
11	0	—	—	5.51	0

The macro-morphology of parts of the tubing samples described in Table 1 was shown in Fig.1. A heavy blistering and spalling of epoxy coating occurred on the inner surface of the No.2 repaired tubing; Reddish-brown and black corrosion products, which probably are iron oxides and iron carbonate (FeCO$_3$), respectively, were observed on the tubing where the coating had peeled; A large area of deposits was observed on the tubing, which would result in serious corrosion under the scale. No blisters and spallation appeared on the No. 5 tubing. However, large scale of coating blistering and spalling was observed on the No. 9 tubing.

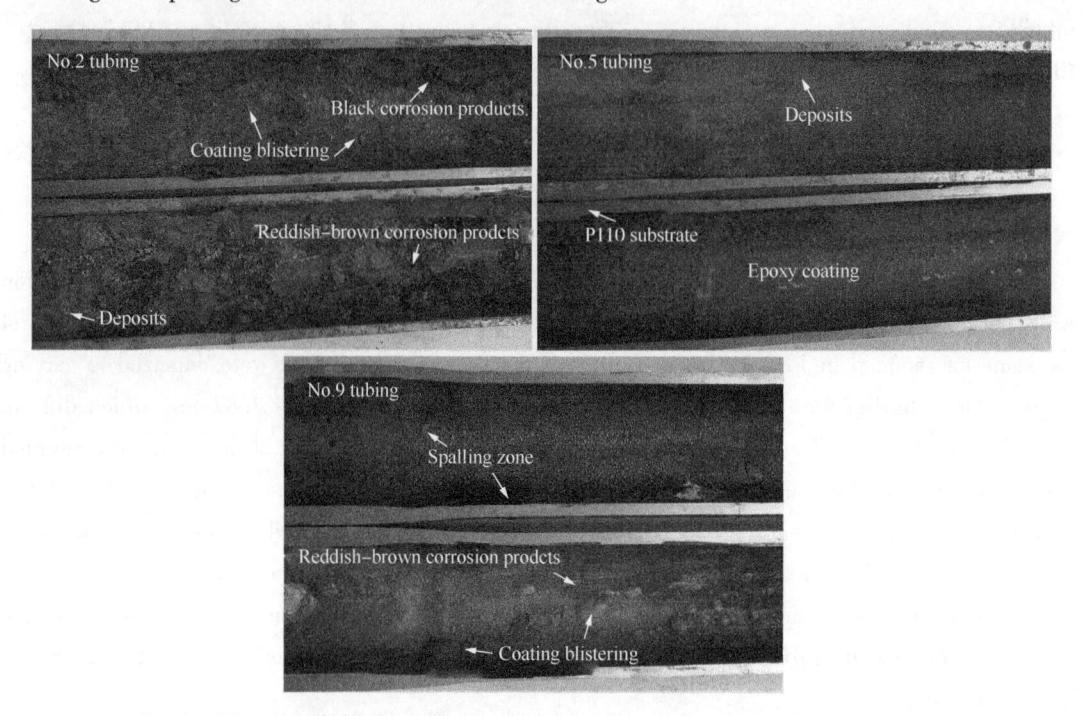

Fig. 1 Macromorphologies of the No. 2, No. 5 and No. 9 epoxy-coated tubing samples

2.2 Cross-sectional morphology and corrosion products

The cross-sectional morphology of parts of the tubing sampled from the non-blistering zone described in table 1 are shown in Fig. 2. The epoxy coatings on the No. 2 and No. 5 tubing samples were dense, and no crack was observed in the coatings. However, cracks were observed in the coating on the No. 9 tubing sample, and the structure of the epoxy coating became loose, which was quite different from those dense coatings on the No. 2 and No. 5 tubing samples. Therefore, the epoxy coating on No. 9 had its anticorrosive resistance. This is probably due to the high working temperature that reached 102℃ in a deep well of J oil field.

Corrosion products were detected at the coating/substrate interface of No. 2 repaired tubing and the No. 9 tubing sample. The results of EDS analysis showed that the corrosion products on No. 2 repaired tubing sample mainly contained Fe, O and C, no Ca was detected. It is worth noting that corrosion products can also be observed in the P110 carbon steel substrate beneath the coating/substrate interface of the No. 2 repaired tubing sample. As the epoxy coating on the No. 2 repaired tubing sample was still dense, it indicated that these corrosion products at the interface and in the P110 substrate had existed before the tubing was recoated, which would resulted in poor adhesion of the epoxy coating. Therefore, the surface pretreatment prior to the recoating of the No. 2 repaired tubing is poor. The results of EDS and XRD phase analysis showed that the corrosion products on the No. 9 tubing sample were mainly composed by FeO(OH), FeCO$_3$, FeO, Fe$_3$O$_4$, Fe$_2$O$_3$ and

CaCO$_3$ (Fig. 3), which indicated that dissolve oxygen corrosion, carbon dioxide corrosion and under-scale corrosion occurred on the P110 tubing of No. 9 tubing sample after the failure of the epoxy coatings.

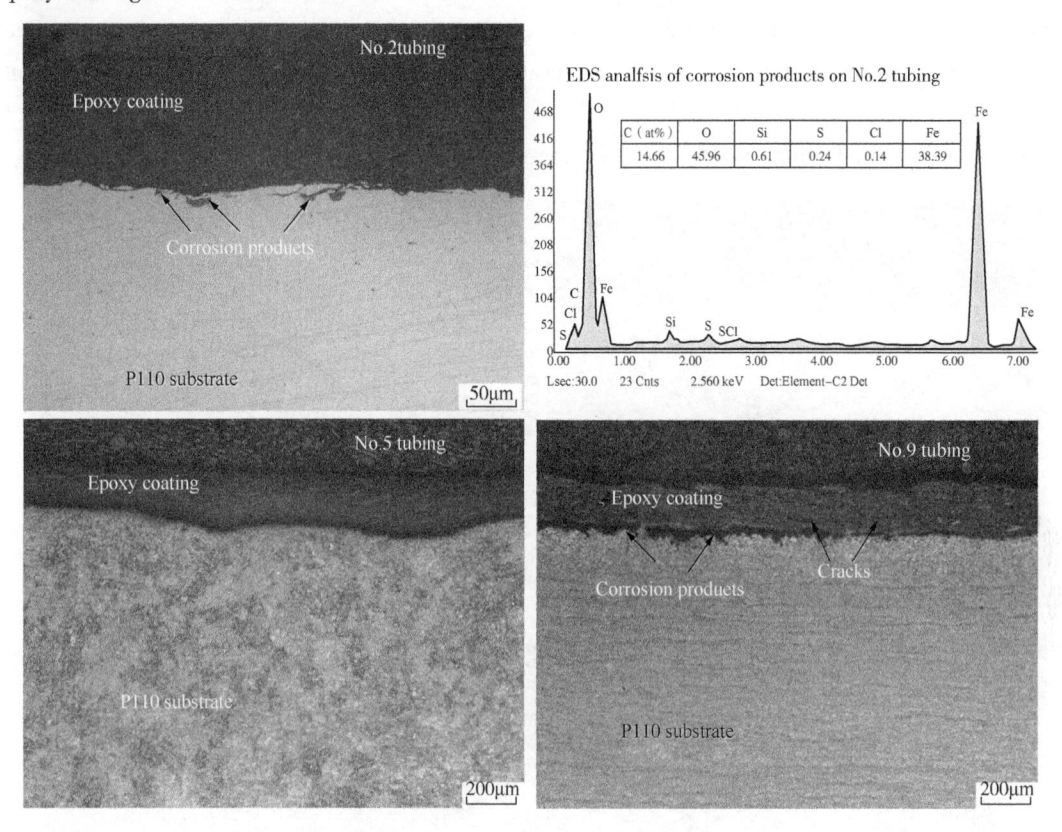

Fig. 2 Cross-sectional morphologies of No. 2, No. 5, No. 11 epoxy-coated tubing samples and EDS analysis of corrosion products in the P110 substrate beneath the coating/substrate interface of No. 2 tubing

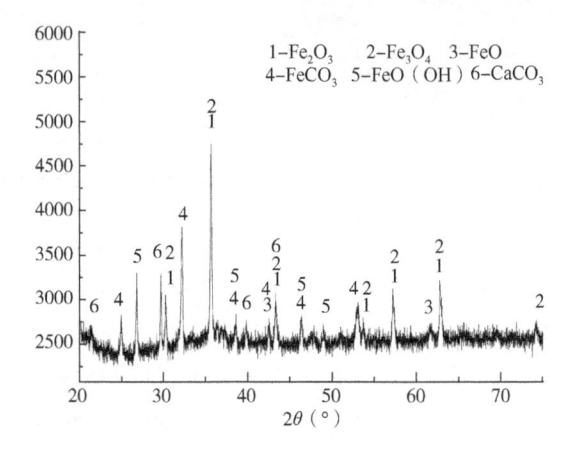

Fig. 3 XRD analysis of corrosion products on the No. 9 epoxy-coated tubing

2.3 High temperature-high pressure corrosion test

To verify the effect of the surface pretreatment on the adhesion and the working temperature on the anticorrosion resistance of the epoxy coating, 15-day high temperature-high pressure (HTHP) corrosion tests on epoxy-coated tubing samples at 90℃ and 120℃ under the pressure of 10MPa in the make-up water were carried out. Fig. 4 shows the surface morphology of the unused No. 10 tubing and No. 11 repaired tubing samples after the HTHP corrosion test at 90℃. No change was observed on the epoxy coating of the No. 10 tubing sample after testing at 90℃; however, large-scale coating blistering was observed on the epoxy coating of the No. 11 repaired tubing sample, due to the poor adhesion of the recoated epoxy coating. Therefore, the poor surface pretreatment processing prior to the recoating of tubing is the main reason for the failure of the No. 2 and No. 3 epoxy-coated repaired tubing in less than two years.

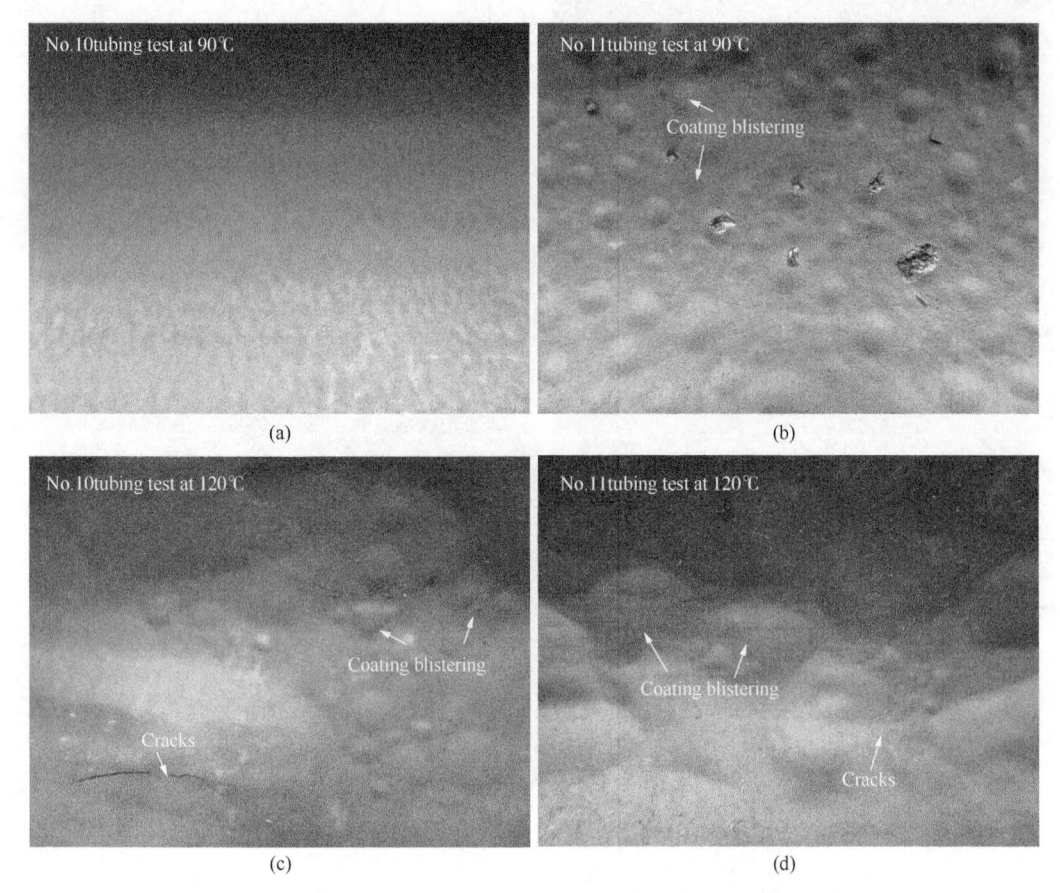

Fig. 4 The surface morphology of epoxy coatings on No. 10 and No. 11 repaired tubing samples after the HTHP corrosion test in the formation solution at 90℃ and 120℃

However, large-scale coating blistering occurred both on the No. 10 tubing and No. 11 repaired tubing samples after a test at 120℃ (Fig. 4), due to the poor high-temperature corrosion resistance of the epoxy coating. This is consistent with the failure of the No. 9 epoxy-coated tubing. Tempera-ture-dependent behavior of polymeric material systems subjected to thermal changes is well known.[4] Below glass transition temperature, spontaneous changes occur in polymer amorphous materials, which is ascribed to conformational changes of the polymer backbone and are related to physical aging. Physical aging and chemical degradation cause the loss of coating ability to relax stress. As the adhesive bond strength diminishes, it approaches a critical value of the local cohesive stress and adhesive damages occur.[5] As a result, the coating degradation created by conductive pathways in the coating causes cracks to fill with electrolytes as well as local delamination from the metal substrate.[6-8] Macroscopic failure occurs when a specific limit of accumulated damage is reached. Therefore, the poor hightemperature corrosion resistance of the epoxy coating is the main reason for the failure of the No. 9 epoxy-coated tubing.

After the failure of the epoxy coating, the corrosive solution directly contacted the P110 tubing substrate, resulting in the corrosion of the tubing. The corrosion mechanisms were DO corrosion, CO_2 corrosion, and under-scale corrosion.

3 Conclusions and recommendations

Based on the experimental results and discussion, the following conclusions and recommendations can be drawn.

(1) One-third of the used tubing's minimum residual wall thickness was less than 4.82mm, which did not met API SPEC 5CT standard and the technology agreement of users for continue use; however, the other six used tubing showed excellent corrosion resistance.

(2) The poor surface pretreatment processingprior to the recoating of tubing is the main reason for the failure of the epoxy-coated repaired tubing in less than two years, The poor high temperature corrosion resistance of the epoxy coating is the main reason for the failure of the epoxy-coated tubing served in the deep well with high working temperature in the J oil field.

(3) Dissolve oxygen corrosion, carbon dioxide corrosion, and under-scale corrosion occurred on the P110 tubing substrate after the failure of the epoxy coating.

(4) Suitable coating with careful application andproper surface pretreatment processing prior to the preparation of the coating are recommended.

References

[1] Huang Fei, Hu Jianxiu, Li Wentong, Xia Jing, "Corrosion of Produced Water in oilfield and Prevention," Corrosion & Protection in Petrochemical Industry, 27(4)(2010): pp 44-46. (In Chinese)

[2] L. Rongqiang, "Water Injection String Erosion Mechanism and Its Control," Petroleum Drilling Techniques, 36(4)(2008): pp 64-67.

[3] API SPEC 5CT, "Specification for Casing and Tubing"(Washington, DC: API).

[4] I. M. Hodge, "Enthalpy relaxation and Recovery in amorphous materials," J. Non-Cryst. Solids, 169(1994): pp 211-266.

[5] A. Miszczyk, K. Darowicki, "Effect of Environmental Temperature Variations on Protective Properties of Organic Coatings," Progress in Organic Coatings 46(2003): pp 49-54.

[6] L. Xiaowei, H. A. Hristov, A. F. Yee, D. W. Gidley, "Influence of Cyclic Fatigue on the Mechanical Properties of Amorphous polycarbonate" Polymer 36(1995): pp 759-765.

[7] M. E. Nichols, C. A. Darr, C. A. Smith, M. D. Thouless, E. R. Fischer, "Fracture energy of Automotive Clearcoats-I. Experimental Methods and Mechanics," Polym. Degrad. Stab. 60(1998): pp 291-299.

[8] H. Lee, S. Krishnaswamy, "Quasi-static propagation of sub interfacial cracks," ASME J Appl Mech, 67(2000): pp 444-452.

本论文原发表于《Materials Performance》第59卷第4期。

Failure Analysis of Casing Dropping in Shale Oil Well during Large Scale Volume Fracturing

Wang Hang[1,2] Zhao Wenlong[3] Shu Zhenhui[3]
Zhao Qiang[3] Han Lihong[1,2]

(1. Tubular Goods Research Institute of CNPC; 2. State Key Laboratory of Performance and Structural Safety for Petroleum Tubular Goods and Equipment Materials; 3. Xinjiang Oil Field Company of CNPC, Kelamayi 834000)

Abstract: Shale oil and gas are regarded as typical unconventional reservoir, their recovery often impedes by casing damage. Failure analysis is performed for casing dropping in shale oil well via optical microscope(OM), scanning electron microscope(SEM), full size test and finite element modeling. Morphology observation reveals that thread remains intact under as-received condition after make up and break out. Meanwhile, galling and improper make-up position occur for slipping thread. Full size test exhibits that strength of connection declines from 783.4Kips to 697.1Kips after 9th stage fracturing, with loss of 11.0%. Finite element modeling demonstrates that contact stress reaches to high values at two ends and low values at middle part in the pin and box threads under complex stress. Casing dropping can be attributed to insufficient strength of connection due to improper make-up position, which accelerates by fracturing operation. Finally, dynamic damage effect is suggested to introduce in design of casing string in shale oil well.

Keywords: Casing damage; Galling; Connection strength; Fracturing operation; Shale oil

1 Introduction:

Shale oil and gas belong to strategic energy, revolution of shale gas has succeeded in North America[1,2] using horizontal well and multi-stage fracturing. Currently, output gets into rapid growth in China, such as shale gas in South-West area, shale oil in Xinjiang area[3-5]. Because of rigorous work condition, such as high pressure, large displacement and multi-stage fracturing, geo-stress field alters notably around borehole. As a result, casing damage occurs frequently[6-9].

Corresponding author: Wang Hang, wanghang008@cnpc.com.cn; Shu Zhenhui, shuzh@petrochina.com.cn; Han Lihong, hanlihong@cnpc.com.cn.

Statistical data exhibits that there are 11 horizontal wells in total 24 ones and 23 stages in total 596 ones for casing deformation in Xinjiang area of China[5]. The same problem exists in North America, 32 wells in total 62 ones in shale gas in Marcellus oil field of USA. In addition, 28 wells in Utica shale reservoir in Quebec [10], 11 wells in total 14 ones, with 19 deformation points in Duvernay shale reservoir of Canada [11].

Although many works have been done on casing damage by far[12-17], little focuses on thread slipping during multi-stage fracturing operation[18-20]. The objective of this paper is to discover the failure mechanism of casing dropping in shale oil well using large scale volume fracturing.

2 Engineering background

An abnormal fluctuation occurs in operation curve during 10th stage fracturing. Operation parameters include displacement of $10m^3/min$ and pump pressure of 70.0MPa. Casing damage is identified after tripping at 885~890m, as shown in Fig. 1. The structure of shale oil well is vertical-like borehole, three-open configuration, well depth of 3793.9m and cementing up to 3255.0m. The steel grade of casing is P110, specification of (139.7× 10.54)mm and API LC thread. Field fracturing operates in the form of fine layered and large scale volume injection.

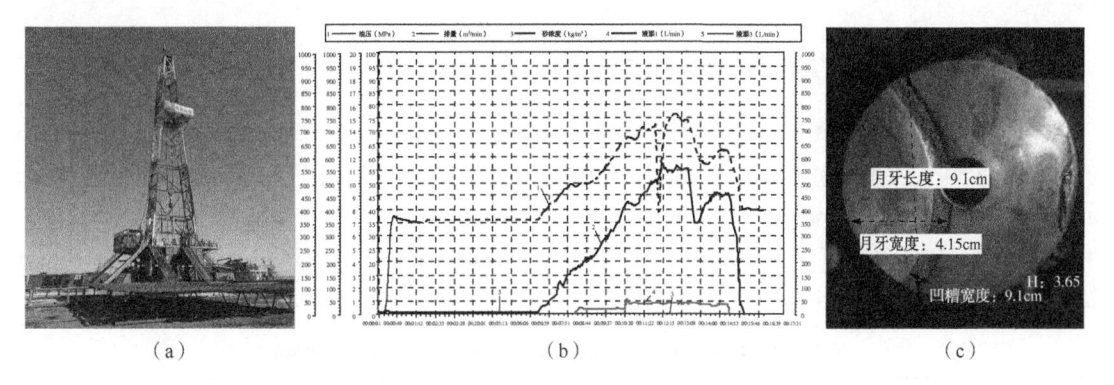

Fig. 1 Failure of casing dropping: (a)field site, (b)10th stage fracturing curve and(c)casing stamping

3 Property of materials

Chemical composition and mechanical property meet the requirement of standard of API Specification 5CT for dropping casing, including chemical content tensile strength and impact toughness, as shown in Table 1-3.

Table 1 Chemical composition of dropping casing with steel grade of 110ksi

Element	C	Si	Mn	P	S	Cr	Mo	Ni	Nb	V	Ti	Cu
Body	0.27	0.26	1.32	0.014	0.012	0.13	0.0054	0.01	0.001	0.0038	0.0017	0.011
Coupling	0.26	0.26	0.96	0.017	0.0055	0.96	0.14	0.013	0.0016	0.007	0.003	0.009
API 5CT	—	—	—	≤0.03	≤0.03	—	—	—	—	—	—	—

Table 2 Tensile property of dropping casing with steel gradeof 110ksi

Specimen	Size(mm)	Steel grade (ksi)	Tensile strength (MPa/MPa)	Yield strength (MPa)	Elongation (%)
Body	φ6.25×65.0	110	934.0	857.6	21.7
Coupling	φ6.25×65.0	110	1030.7	959.0	19.3
API 5CT		110	≥862.0	758.0~965.0	≥15.0

Table 3 Impact toughness of dropping casing with steel grade of 110ksi

Specimen	Size(mm)	Notch shape	Temper(℃)	Steel grade(ksi)	Impact energy (J)	Shear fracture ratio(%)
Body	10×10×55	V	0	110	96.7	100
API 5CT				110	≥41.0	—
Coupling	10×10×55	V	0	110	103.3	100
API 5CT				110	≥26.0	—

4 Thread detection and Non-destructive examination(NDE)

The parameters of pin and box threads are listed in Table 4 and Table 5. These results also meet the requirement of standard, based on API Specification 5B. NDE results reveal that there are a lot of deformation traces in the middle part of slipping thread. Meanwhile, a number of pits appear in the local region of thread without slipping, which is salvaged from downhole, as shown in Fig. 2.

Table 4 Parameters of pin thread as-received condition

Type	Close distance (mm)	Taper (mm/m)	Screw pitch deviation(mm/in)	Thread height deviation(mm)	L_4(mm)	D (mm)
Test	1.085	64.0	−0.01	+0.01	88.80	141.00
API 5B	−3.09~+3.27	59.9~67.7	−0.08~+0.08	−0.10~+0.05	82.55~88.91	139.00~141.10

Table 5 Parameters of coupling thread as-received condition

Type	Close distance (mm)	Taper (mm/m)	Screw pitch deviation(mm/in)	Thread height deviation(mm)	Q(mm)	q(mm)
Test	10.2	65.0	0	0	142.56	12.88
API 5B	6.31~12.67	59.9~67.7	−0.08~+0.08	−0.10~+0.05	142.0~143.66	12.70~13.49

Fig. 2　Non-destructive examination using magnetic powder: (a)slipping thread, (b)no slipping thread

5　Morphology observation

The typical morphology is characterized by galling for slipping thread, as shown by dotted box in Fig. 3(a). Based on macroscopic feature, these regions can be divided into three parts, i. e., initiation segment from 1st to 8th thread, intermediate segment from 9th to 18th thread and tail segment from 19th to 26th thread. On a close examination, indentation from clamps is observed in outside surface of coupling neighbor to field-end, as shown by arrow in Fig. 3(b).

Fig. 3　The macroscopic morphology of dropping casing: (a)pin thread, (b)corresponding coupling

Specimens are cut along axial direction and examined using optical microscopy. Metallographic observations reveal that deformation traces occur in the top of thread in initiation segment, as shown by arrow in Fig. 4(a). Meanwhile, some of these threads collapse evidently in the intermediate segment, as shown in Fig. 4(b). However, deformation trace does not even appears in the tail segment, as shown in Fig. 4(c). These evidences indicate that make-up position is not in place for slipping thread, its make-up position only reaches to the intermediate segment. SEM observations reveal that there are obvious friction scratches in the initiation segment, as shown in Fig. 5. It is seemed that these scratches are produced by interference contact between threads during make-up operation. However, this type of scratch does not presents in the tail segment, as shown in Fig. 6. These SEM results verify that make-up position is actually not in place for slipping thread. Chemical composition is analyzed for surface residue using energy disperse spectrum(EDS), which consists of K, Ca, Na, Mn, S, Cl, O, and so on, as shown in Fig. 7.

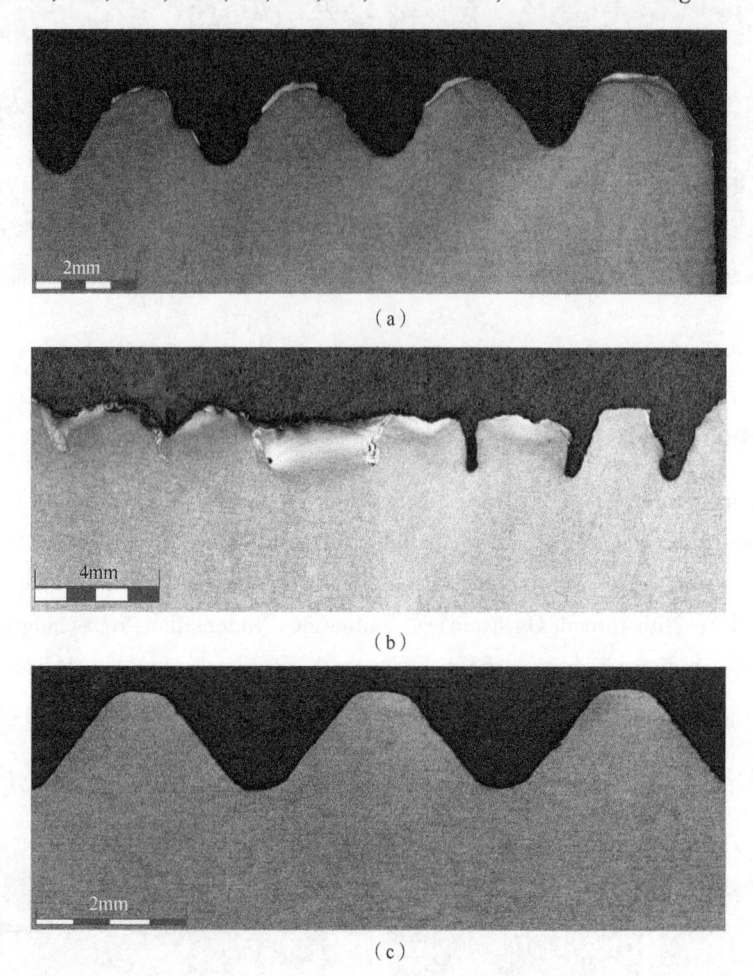

Fig. 4 The morphology of slipping thread indifferent regions:
(a) initiation (b) intermediate and (c) tail part

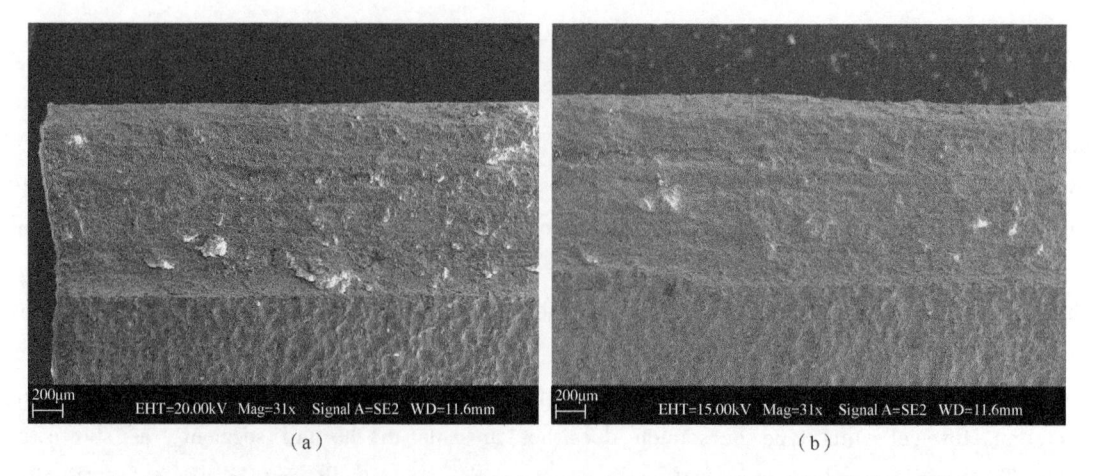

Fig. 5 The morphology feature of guide surface in the 3rd thread: (a) left-side and (b) middle region

(a)　　　　　　　　　　　　　　(b)

Fig. 6　The morphology feature of guide surface in the 21th thread:
(a) middle region and (b) left-side one

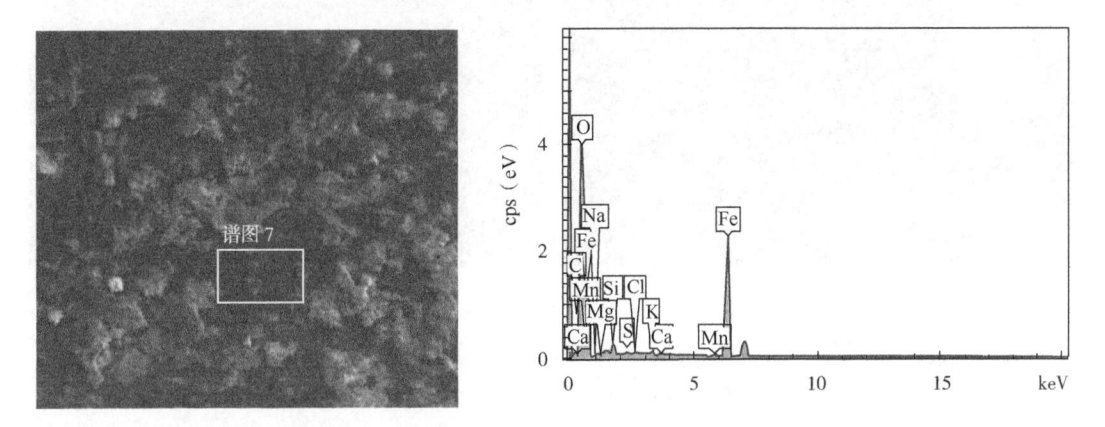

Fig. 7　EDS analysis of the residue in guide surface of the 3rd thread

6　Full scale test

According to standard of API Specification 5B, make up and break out tests are performed to evaluate galling resistance property of thread under as-received condition. The macroscopic morphology characteristics indicate that thread remains intact after three times make-up and two times break-out, as shown in Fig. 8. The morphology of thread is examined after 9th stage fracturing in downhole, as shown in Fig. 9. The box thread remains intact basically, except a little of mechanical knocking in local region. While damage appears in the form of pit and deformation for the pin thread. Tensile to failure tests are employed to evaluate strength of connection of casing thread with different conditions. The experimental results show that strength of connection is 783.4kips for thread under as-received condition, i.e. 3484.5kN, as shown in Fig. 10. After 9th stage fracturing, this strength is 697.1kips, i.e. 3100.7kN, as shown in Fig. 11. These results demonstrate that operation of multi-stage fracturing reduces strength of connection remarkably for thread with proper make-up position.

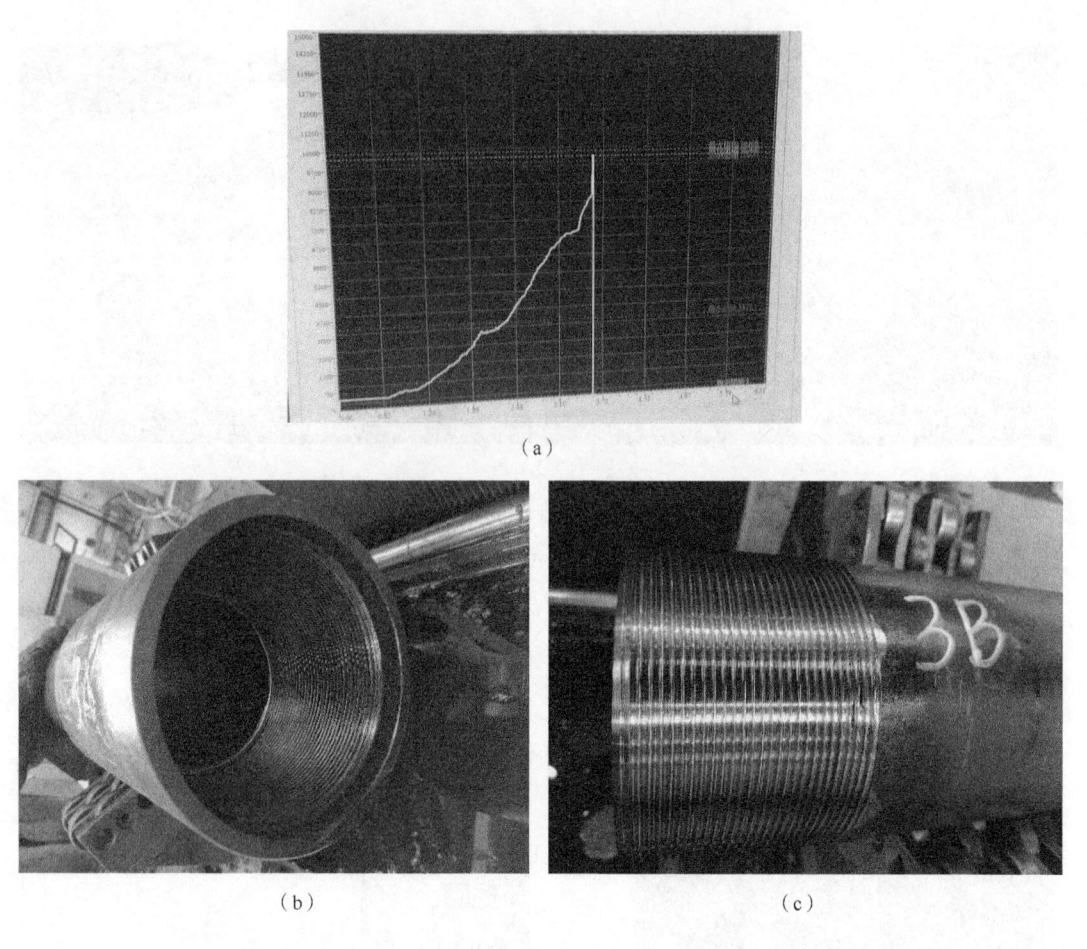

Fig. 8　The torque vs time curve(a)and morphology of box(b) and(c)pin threads after galling resistance test

Fig. 9　The morphology of thread without slipping after 9th stage fracturing: (a)box and(b)pin threads

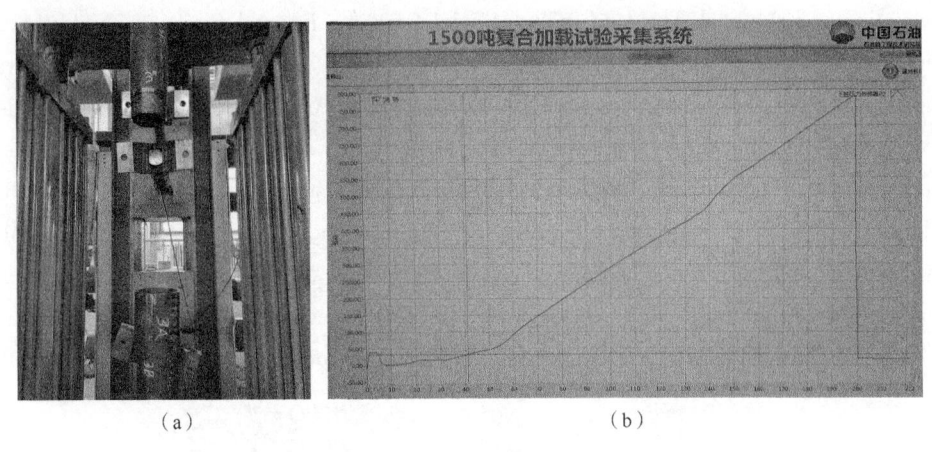

Fig. 10 Tensile to failure test(a) and corresponding loading curve(b) for casing thread as-received condition

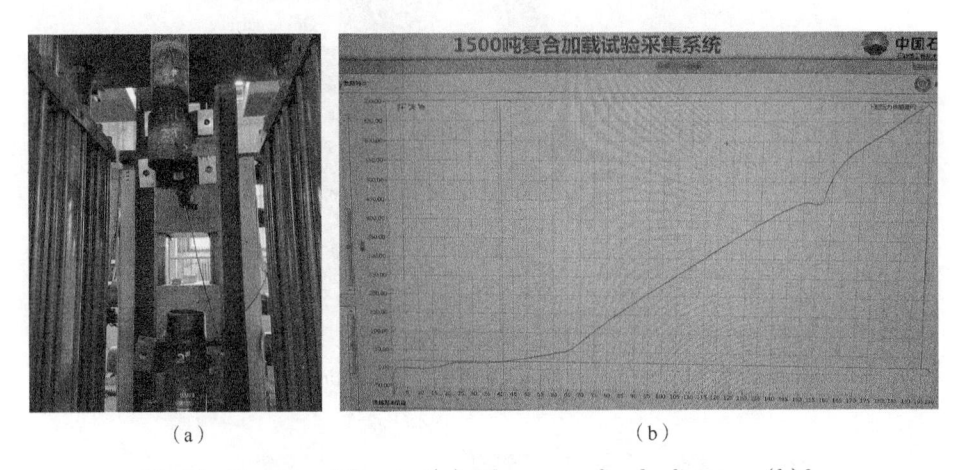

Fig. 11 Tensile to failure test(a) and corresponding loading curve(b) for casing after 9th stage fracturing

7 Finite element analysis

Stress distribution is analyzed for thread type with APILC using finite element modeling. The parameters of material include steel grade of P110, specification of $\phi(139.7\times10.54)$ mm, elastic module of 210GPa, poisson ratio of 0.3, density of 7850kg/m^3, and yield strength of 857MPa. Simplification and assumption are as follows[21]: (1) using structure with axial symmetry and ignoring the factor of lead angle, (2) coefficient of friction is supposed to be 0.02, (3) two end of model is elongated with size of 1/3 thread length to eliminate the boundary effect.

Finite element model is showed in Fig. 12, as well assubdivision mesh and local magnification, of which unit number is 9434 with node number of 20935 for casing after subdivision[22]. Meanwhile, this unit number is 9775 with node number of 21614 for coupling. As for make-up simulation, connection torque is applied by means of altering the magnitude of interference between the pin and box threads. As for API LC with 8 thread, its magnitude of interference is 0.099mm along axial direction after machine tight with 1 cycle. Displacement constraint is applied for the top end of coupling in FE model. The load includes axial tension, internal pressure, and other stress.

Finite element analysis reveals that contact stress has high values at two ends and low values at middle part in the pin and box threads under make-up torque. As a whole, this stress level is higher in the pin thread than that in the box one, as shown in Fig. 13. Moreover, this stress distribution is analyzed under make-up torque, internal pressure and tensile stress, as shown in Fig. 14. Based on FE modeling, stress level increases obviously for coupling under internal pressure of 50MPa and three cycle machine tight, the maximum of contact stress is 580MPa in small-end interference thread of coupling. In addition, the maximum of Von Mises stress is 624MPa under 100MPa tension stress[20]. It is believed that contact compression stress is proportional with loading area of thread.

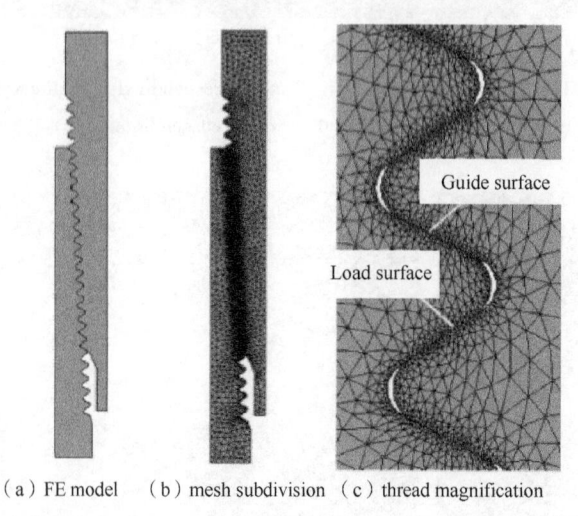

(a) FE model   (b) mesh subdivision   (c) thread magnification

Fig. 12 Finite element model of thread connection in casing[20]

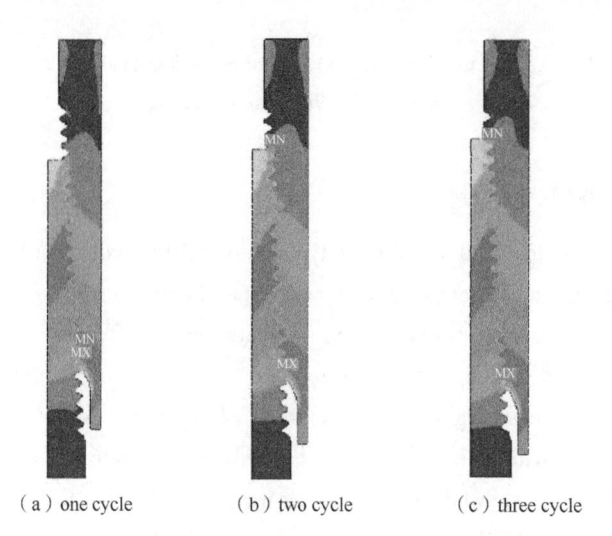

(a) one cycle   (b) two cycle   (c) three cycle

Fig. 13 Von-Mises equivalence stress in thread connection during make up[20]

Meanwhile, strength of connection is a function of loading area of thread. It means that contact stress has an important effect on strength of connection of casing thread. Therefore, strength of connection is much higher for thread with proper make-up position because of much larger loading area in the region of two ends.

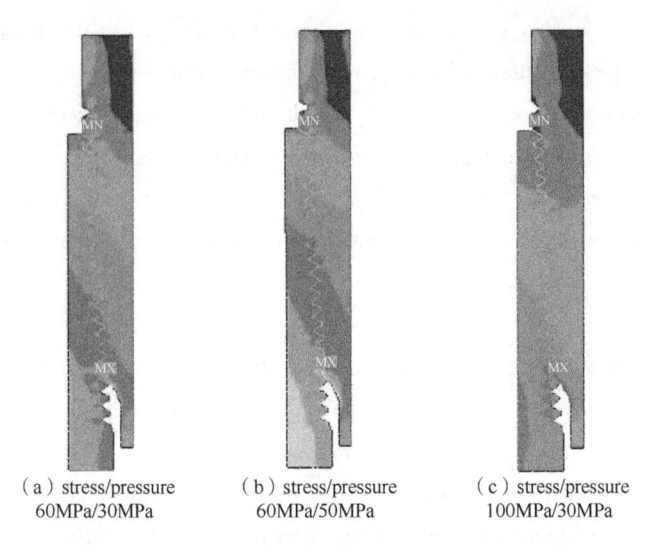

Fig. 14 Von-Mises stress of thread connection after 3 cycle machine tight under complex loading[20]

(a) stress/pressure 60MPa/30MPa
(b) stress/pressure 60MPa/50MPa
(c) stress/pressure 100MPa/30MPa

8 Discussion of results

Chemical content and mechanical property meet the requirement of standard for dropping casing, based on API specification 5CT. Therefore, material factor is not responsible for thread slipping in shale oil well during large scale volume fracturing.

Strength of connection is one of key parameters in design of casing string and maintaining structural integrity of casing string, which is associated with three aspects: galling resistance property, make-up position and fracturing operation. Make-up and break-out tests confirm that this thread type has a fine galling resistance property. Metallographic and SEM observation reveal that make-up position is not in place for slipping thread. On the basis of FEA, contact stress is high values at two ends and low values at middle part in the pin and box threads under complex load [20]. These FE modeling results indicate that strength of connection is higher for thread with proper make-up position due to larger loading area of thread. In addition, full size test exhibits that strength of connection decreases remarkably after 9th stage fracturing, with loss of 11.0%. Consequently, cause of failure is insufficient strength of connection for thread slipping.

Traditional design method of casing string involves in static loading for convention oil and gas wells, which includes hanging load and hole-wall friction. Multi-stage fracturing is used in shale oil well during recovery, casing undergoes tension, internal pressure and collapse in the region of horizontal well without cementing. In addition, there are much higher stress in casing thread due to stress concentration. In this case, plastic deformation would occur when stress reaches to yield strength. Therefore, service behavior can be characterized by low cycle fatigue during multi-stage fracturing, which leads to reduction in strength of connection continuously. This process could be also represent with dynamic damage effect. Based on analysis, it is speculated that degradation of property of material is dominant mechanism for dynamic damage effect.

The similar feature is also documented in thermal well[23]. Plastic deformation happens for thermal casing when injection steam with high temperature, which induces low cycle fatigue during multi-cycle operation of injection and recovery. As a result, plastic strain cumulates continuously and displays dynamic damage effect during this process. This dynamic damage effect finally induces casing damage when cumulative strain exceeds strain capacity in thermal well. Therefore, it can be seen that dynamic damage effect could be explained in terms of degradation of property of material.

9 Conclusions

(1) Chemical composition and mechanical property meet the requirement of standard of API Specification 5CT for dropping casing, including tensile strength and impact toughness, then material factor is not responsible for thread slipping.

(2) Casing dropping can be attributed to insufficient strength of connection due to improper make-up position, which induces loss of loading area of thread. In addition, reduction of strength of connection is accelerated by multi-stage fracturing in field operation.

(3) Dynamic damage effect has an important effect on reduction of strength of connection, this effect is induced by degradation of property of material. As a result, it is suggested to introduce dynamic damage effect in design of casing string in shale oil and gas well.

Acknowledgment

The authors gratefully acknowledge the financial support of CNPC science and technology development project(No. 2019B-4013).

References

[1] E. K. George, Thirty years of shale gas fracturing: What have we leaning? [C]//SPE Annual Technical Conference and Exhibition. Florence: Society of Petroleum Engineering, 2010.

[2] L. Wang, J. L. Ma, F. R. Su, et al. Shale gas factory fracturing technology in North America[J]. Drilling & Production Technology, 35(2012)48-50.

[3] C. J. Xue. Technical advance and development proposals of shale gas fracturing[J]. Petroleum Drilling Techniques, 39(2011)24-29.

[4] C. B. Yin, D. S. Ye, G. B. Duan, et al. Research about and application of autonomous staged fracturing technique series for horizontal well stimulation of shale gas reservoirs in the Sichuan Basin[J]. Natural Gas Industry, 34(2014)67-71.

[5] H. Liu, L. C. Kuang, G. X. Li, F. Wang, X. Jin, J. P. Tao, S. W. Meng. Considerations and suggestions on optimizing completion methods of continental shale oil in China[J]. Acta Petrolei Sinica, 41(2020)489-496.

[6] H. L. Zhang, Z. W. Chen, L. Shi, et al. Mechanism of how fluid passage formed and application in Sichuan shale gas casing deformation analysis[J]. Drilling & Production Technology, 41(2018)8-11.

[7] Q. Dai. Analysis of production casing damage during testing and completion of shale gas well[J]. Drilling & Production Technology, 38(2015)22-25.

[8] Y. Xi, J. Li, G. H. Liu, et al. Overview of casing deformation in multistage fracturing of shale gas horizontal wells[J]. Special Oil and Gas Reservoir, 26(2019)1-6.

[9] Z. L Tian, L. Shi, L. Qiao. Research of and countermeasure for wellbore integrity of shale gas horizontal well[J]. Natural Gas Industry, 35(2015)70-76.

[10] X. L. Guo, J. Li, G. H. Liu, et al. Research on casing deformation for shale gas wells based on focal mechanism[J]. Fault-Block Oil and Gas Field, 25(2018)665-669.

[11] Z. H. Lian, H. Yu, T. J. Lin, J. H. Guo. A study on casing deformation failure during multi-stage hydraulic fracturing for the stimulated reservoir volume of horizontal shale well[J]. Journal of Natural Gas Science and Engineering, 23(2015)538-546.

[12] Pablo. Cirimello, Jose. Luis Otegui, Alberto. Aguirre, Guillermo. Carfi. Undetected non-conformities in material processing led to a failure in a casing hanger during pre-fracture operation[J]. Engineering Failure Analysis, 104(2019)203-215.

[13] Y. Li, W. Liu, W. Yan, J. G. Deng, H. T. Li. Mechanism of casing failure during hydraulic fracturing: lessons learned from a tight-oil reservoir in china[J]. Engineering Failure Analysis, 98(2019)58-71.

[14] K. H. Deng, W. Y. Liu, T. G. Xia, D. Z. Zeng, M. Li, Y. H. Lin. Experimental study the collapse failure mechanism of cemented casing non-uniform load[J]. Engineering Failure Analysis, 73(2017)1-10.

[15] Pablo. G. Cirimello, Jose L. Otegui, Guillermo Carfi, Walter. Morris. Failure and integrity analysis of casing used for oil well drilling[J]. Engineering Failure Analysis, 75(2017)1-14.

[16] K. H. Deng, Y. H. Lin, W. Y. Liu, H. Li, D. Z. Zeng, Y. X. Sun. Experimental investigation of the failure mechanism of P110SS casing under opposed line load[J]. Engineering Failure Analysis, 65(2016)65-73.

[17] C. A. Cheatham, C. F. Acosta, D. P. Hess. Tests and analysis of secondary locking features in threaded inserts[J]. Engineering Failure Analysis, 16(2009)39-57.

[18] Satoshi Izumi, Takashi Yokoyama, Atsushi Iwasaki, Shinsuke Sakai. Three-dimensional finite element analysis of tightening and loosening mechanism of threaded fastener[J]. Engineering Failure Analysis, 12(2005)604-615.

[19] M. Y. Zhang, D. F. Zeng, L. T. Lu, Y. B. Zhang, J. Wang, J. M. Xu. Finite element modeling and experimental validation of bolt loosening due to thread wear under transverse cyclic loading[J]. Engineering Failure Analysis, 104(2019)341-353.

[20] G. J. Yu, X. Q. Chen, A. Q. Duan. Finite element analysis of casing thread connection under impact load[J]. China Petroleum Machinery, 45(2017)14-20.

[21] Z. G. Wang, Y. Zhang. Finite element analysis for stress in tubing connection with API LC under machine tight and tension[J]. Steel Pipe, 30(2001)20-25.

[22] J. T. Xi, G. Nie, X. S. Mei. Finite element analysis of contact of connection property for casing thread connection[J]. Journal of Xi'an Jiao Tong University, 33(1999)63-66.

[23] L. H. Han, H. Wang, J. J. Wang, B. Xie, Z. H. Tian, X. R. Wu. Strain-based casing string for cyclic steam stimulation well[J]. SPE PRODUCTION & OPERATIONS, 33(2018)409-418.

本论文原发表于《Engineering Failure Analysis》2020年第118卷。

Numerical Analysis of Casing Deformation under Cluster Well Spatial Fracturing

Wang Jianjun[1]   Jia Feipeng[1,2]   Yang Shangyu[1]
He Haijun [3]   Zhao Nan[4]   Zhang Le[1,2,]

(1. CNPC Tubular Goods Research Institute, State Key Laboratory of Performance and Structural Safety for Petroleum Tubular Goods and Equipment Materials;
2. College of Mechanical Engineering, Xi'an Shiyou University;
3. Daqing Oilfield Production Engineering & Research Institute;
4. Engineering Technology Research Institute of Xinjiang Oilfield Company)

Abstract: The large-displacement hydraulic fracturing technology widely used in the process of shale gas exploitation, which causes formation slip and serious casing damage under non-uniform load. It seriously restricts the exploitation and utilization of shale gas in China. In this paper, the writer employs the casing pipe with outer diameter of 139.7mm and grade N80 steel as an example and establishes the finite element model of stratum-cement ring-casing to analyze the influence rule of non-uniformity of stratum load, the difference of internal and external pressure and casing wall thickness to casing stress under non-uniform load. The analysis shows that: with the increase of the non-uniformity of formation load, the stress on the casing is gradually increased. Even the external load is far less than the casing's extrusion strength, but the non-uniformity of external load increases, the casing stress will reach its yield limit and lead to casing failure. Increasing the wall thickness is beneficial to reducing the stress value of casing under load. When the non-uniform coefficient of load decreases to 0.4, the increase of wall thickness has little effect on the casing's stress. When the formation non-uniform load is constant, the stress of casings decreases with the increase of the inside pressure of casings. Under the non-uniform load, reducing the pressure difference between the inside and outside of the casing is conducive to reducing the casing damage. The location where the casing failure occurs first is on the casing inner wall in the direction of the minimum horizontal stress. That is, the place where in the direction of the maximum horizontal in-situ stress on 90 degrees is the casing stress risk area.

Keywords: Non-uniform load; Casing; Numerical analysis; Damage

---

Corresponding author: Wang Jianjun, wangjianjun005@cnpc.com.cn; Yang Shangyu, yangshangyu@cnpc.com.cn.

1 Introduction

China's shale gas is characterized by deep burial and great difficulty in exploitation. Used in shale gas cluster horizontal well multi – stage spatial fracturing with large displacement, high pressure, effect time is long, etc, leads to formation fracturing construction of slip. As a result, an unusually high non-uniform load is generated on the outer wall of the casing, which causes serious casing damage and makes the well tools blocked. It severely restricts the exploitation and utilization of the shale gas in China.

At present, the speed of casing loss increases year by year, causing huge loss of manpower, material and financial resources. Casing damage has become one of the hot issues in petroleum engineering. It is an extraordinarily important theoretical and engineering significance to study casing stress under non-uniform load.

Dezhi Zeng has made use of the knowledge of elastic mechanics to solve and analyze the stress of thick wall casing under non-uniform load. His analysis suggests that the danger zone existed along the direction of minimum stress. By means of finite element analysis, Jun Fang finds that the ellipticity of casing weaken its strength and stiffness under non-uniform load.

Based on the existing research, this paper analyzes the influence of non-uniform coefficient on the stress of casing, and the influence law of casing wall thickness and casing pressure difference on casing load under the action of non-uniform load.

2 Analysis of casing force under non-uniform load

According to statistics, 74% of casing damage of oil and water Wells inDaqing oilfield occurs at the bottom of the second floor, where is the interface between mudstone and other strata. One of the main causes of casing damage in Daqing oilfield is interlayer slip caused by mud and rock water absorption creep. In oilfields developed by water injection, when the pressure of water injection exceeds the fracture pressure of formation, micro-cracks will occur near the interface of mudstone layer. The injected water will penetrate into the mudstone layer along the micro-cracks, and the mudstone will absorb water, expand and soften, resulting in ground slip. At the same time, interfacial micro-cracks gather to form macroscopic interfacial cracks and begin to expand under the push of formation pressure difference. In the process of mudstone creep and crack propagation at stratum interface, the stress near the interface is highly concentrated. The deformation energy of the crack tip field is much higher than that of other parts of the formation. The outer load of oil and water well casing is concentrated near the formation interface, which leads to the occurrence of casing damage near the formation interface frequently. With the increase of water absorption of mudstone layer, the stiffness value and interface friction coefficient of mudstone layer gradually decrease. The non-uniform load at the interface of mudstone layer increases gradually and finally causes the formation sliding and the casing damage of shale gas well.

Under the action of non – uniform ground stress, according to the knowledge of elastic mechanics, the phenomenon of the maximum horizontal ground stress and the minimum horizontal ground stress acting on the outer wall of casing in the same horizontal direction shows an elliptic change, as shown in Fig. 1. The non-uniform load distribution model acting on the outer wall of casing can be expressed as follows:

$$p(\theta) = p_0 + p_1 \cos(2\theta) \quad (1)$$

Where $p(\theta)$ is the radial load of acting on the casing, MPa; p_0 and p_1 is the equivalent external load of ground Stress, MPa; θ is the Angle between the radial load acting on the casing and the X-axis.

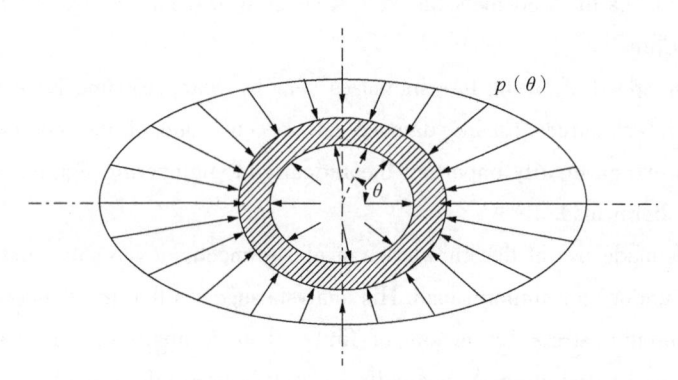

Fig. 1 Non-uniform load distribution on the casing

The inhomogeneity of load on casing is described by the ratio of maximum horizontal stress to minimum horizontal stress. The non-uniformity coefficient is calculated by the following equation:

$$n = p_{min}/p_{max} \quad (2)$$

Where p_{max} is the maximum horizontal stress, MPa; p_{min} is the minimum horizontal stress, MPa.

Under the plane stress condition, the effective stress of the casing under heterogeneous load under the Mises stress criterion is:

$$\sigma_e = \sqrt{(1-\mu+\mu^2)(\sigma_1^2+\sigma_2^2) - \sigma_1\sigma_2(1-2\mu-2\mu^2)} \quad (3)$$

Where σ_1, σ_2 is the main stress of formation acting on casing, MPa; μ is the poisson's ratio of stratigraphic rocks.

Under the action of non-uniform external load, Taking $\sigma_1 = k\sigma_2$, the maximum Mises stress on the casing is:

$$\sigma_2 = \frac{(1-2\mu)\bar{\sigma}}{\sqrt{(n^2+1)(1-\mu+\mu^2) - n(1+2\mu-2\mu^2)}} \quad (4)$$

Where $\bar{\sigma}$ is the formation's uniform geostress under uniform loading, MPa.

3 Finite element modeling under non-uniform load

In order to eliminate the influence of the edge effect on the simulation results, the edge length of the formation modelis set as 3m. The stratum-cement ring-casing model as shown in Fig. 2 is established by using ABAQUS software. All the models use tetrahedral elements, add ZSYMM constraints to the model, and adopt sweeping mesh generation technology to generate CPS4R elements, and make the following assumptions in the model.

(1) Casing, cement ring and formation cementation are intact, there is no cement ring missing, casing eccentric phenomenon.

(2) The axial length of casing is much larger than its radial dimension. The longitudinal deformation of casing under the action of in-situ stress is ignored, and the problem is simplified to plane strain problem.

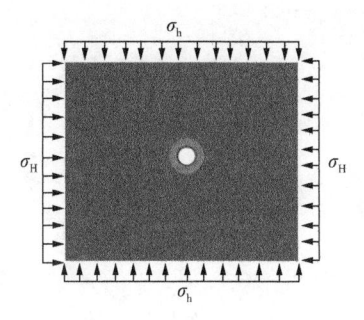

Fig. 2 Casing force model under non-uniform load

(3) Ignoring the initial defects such as casing ellipticity and uneven wall thickness, the casing end face is an ideal round end face.

4 Numerical calculation and analysis

According to the established finite element model, and the influence law of casing wall thickness and internal and external pressure difference on casing stress is simulated and calculated by changing the non-uniform load on the model and the parameters of the finite element model structure are changed (Table 1).

Table 1 Calculation parameters of the casing model

Object	External diameter(mm)	Elasticity modulus(MPa)	Poisson's ratio	Wall thickness(mm)
Stratum		5×10^3	0.22	
Cement ring	249.7	1.0×10^4	0.18	55
Casing	139.7	2.06×10^5	0.3	9.17、7.72、10.64

4.1 The influence of different non-uniformity coefficients on the stress of casing

Under uniform load ($n=1$), the maximum stress of casing is 445.2 MPa. With the increase of the degree of non-uniformity of formation non-uniform load, the maximum stress of casing gradually increases. When the coefficient of non-uniformity decreases to $n=0.4$, the maximum stress of casing is 633.1 MPa, which increases by 42.2% compared with that of casing under uniform load. According to the API5CT standard, the yield strength of N80 casing is 552 MPa. That is, under the external load far less than the compressive strength of casing, when the non-uniform degree of external load increases, the casing will also reach its yield limit, leading to casing failure. The Mises stress cloud diagram of casing under different heterogeneity coefficients is shown in Fig. 3.

4.2 The effect of casing wall thickness on casing stress under different non-uniform coefficients

According to the analysis in Fig. 4, when the casing wall thickness is smaller, the stress value of casing is larger. When the non-uniformity coefficient $n=0.8$, the wall thickness is 7.72mm, 9.17mm and 10.64mm, and the stress on the casing is respectively 551.8MPa, 507.4MPa and

n=1   n=0.8   n=0.5

Fig. 3　Casing strain diagram with different non-uniform coefficients

474.0MPa. Compared with the 7.72mm casing, the stress value of the casing with the wall thickness of 9.17mm is 7.9 % lower and the casing with the wall thickness of 10.64mm is 16.3% lower. That is, improving the wall thickness is helpful to reduce the stress of casing load values under non-uniform loading. With the increase of load non-uniform degree, the influence of wall thickness of casing stress value is reduced. When the coefficient of non-uniformity decreases to 0.4, the increase of wall thickness has little effect on casing stress under non-uniform load.

4.3　The influence of non-uniformity coefficient on casing stress under different pressure difference

According to the analysis in Fig. 5, with the increase of casing pressure under certain non-uniform load in formation, the stress of casing is reduced. That is, under the action of non-uniform load, reducing the pressure difference inside and outside casing is conducive to reducing the stress of casing.

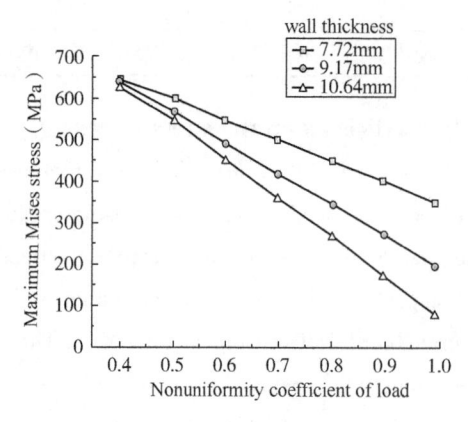

Fig. 4　Influence of non-uniform coefficient on casing stress under different wall thickness

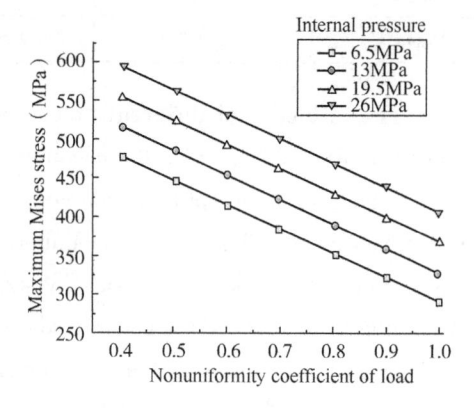

Fig. 5　Effect of casing non-uniformity coefficient on casing stress under different pressure difference

4.4　The stress distribution of casing under non-uniform load

Under non-uniform load, the annular stress of inner wall of casing(10.64mm), distance from outer wall of casing(7.08mm), distance from outer wall of casing(3.54mm) and outer wall of casing(0mm) were calculated respectively.

According to the analysis, under the action of non-uniform load, the stress of outer wall of casing is larger than that of inner wall of casing in the direction of larger horizontal stress, and that of outer wall of casing in the direction of smaller horizontal stress is larger than that of outer wall of casing. On the circumference of the casing, the first place where the casing failure occurs should be on the inner wall of the casing in the direction of the minimum horizontal in-situ stress, that is, where the maximum horizontal in-situ stress shows 90 is the danger area of casing stress(Fig. 6).

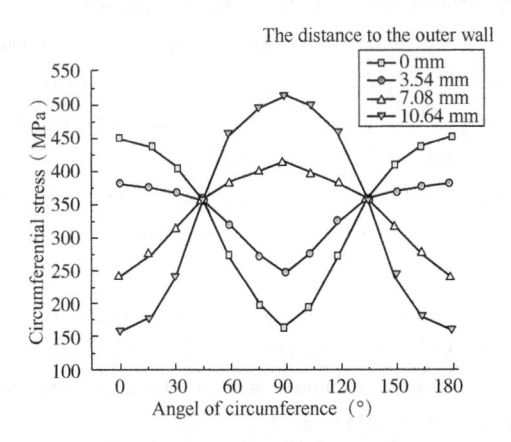

Fig. 6 Stress distribution map at different places away from casing outer wall

5 Conclusion

Through the above analysis, it can be concluded that the formation load non-uniformity coefficient has a significant impact on the casing stress.

(1) With the increase of non-uniformity of formation load, the local stress of casing is higher, that is, casing is easy to be damaged.

(2) When the non-uniform coefficient of load decreases to 0.4, the increase of wall thickness has little effect on casing stress under non-uniform load.

(3) On the inner wall of casing with the direction of the minimum horizontal ground stress. That is, the position of 90 degrees with the maximum horizontal ground stress is the first place where the casing failure occurs.

In actual production, if conditions permit, the steel grade of casing material and its wall thickness can be increased to improve the extrusion resistance of casing and reduce the risk of casing damage.

Acknowledgment

The project is supported by National key Technologies R&D Program of China (2017ZX05009003-004), the Natural Science Foundation of China(U1762211), CNPC Basic Research Project(2019A-3911) and Shaanxi Outstanding Youth Fund(2018JC-030).

References

[1] LIAN Z H, YU H, LIN T J, et al. A study on casing deformation failure during multi-stage hydraulic

fracturing for the stimulated reservoir volume of horizontal shale wells [J]. Journal of Natural Gas Science and Engineering, 2015, 23: 538-546.
[2] Zhiming Chen, Xinwei Liao, Xiaoliang Zhao, Xiangji Dou, Langtao Zhu. Performance of horizontal wells with fracture networks in shale gas formation[J]. Journal of Petroleum Science and Engineering, 2015, 133.
[3] Oyarhossien, M., & Dusseault, M. B. (2017). Risk Associated with Hydraulic Fracture Height Growth. In 70TH CANADIAN GEOTECHNICAL CONFERENCE. Ottawa, Ontario.
[4] Xi yan, Li jun, Liu gonghui, Zeng yijin, Li jianping. Research review on casing deformation during multi-stage fracturing of shale gas horizontal Wells [J]. Special oil and gas reservoirs, 2019, 26(01): 1-6.
[5] Liu, K., Gao, D., Wang, Y., 2017, Effect of local loads on shale gas well integrity during hydraulicfracturing process, J. Nat. Gas Sci. Eng., 37, 291-302.
[6] Wang jianjun, Yang shangyu, Ji haitao, Han lihong, Tian zhihua, Wang hang. Optimization of heavy oil thermal production well casing based on strain numerical analysis [J]. Petroleum pipes and instruments, 2019, 5(01): 42-45.
[7] Zeng dezhi, Lin yuanhua, Li shuanggui, Du rende, Shi taihe. Analysis of compressive strength of thick wall casing under non-uniform load [J]. Natural gas industry, 2007(02): 60-62+153.
[8] Fang jun, Gu yuhong, Mi fengzhen. Numerical analysis of casing extrusion failure under non–uniform load [J]. Petroleum machinery, 1999(07): 34-37+59-60.
[9] LAST N, MUJICA S, PATTILLO P, et al. Evaluation, impact, and management of casing deformation caused by tectonic forces in the Andean Foothills, Colombia[R]. SPE 74560, 2002.
[10] Yu hao, Lian zhanghua, Xu xiaoling, Yu wenming, Wang fuhui, Zuo jinju. Numerical simulation of casing failure during SRV fracturing in shale gas vertical Wells [J]. Petroleum machinery, 2015, 43(03): 73-77.
[11] Zhou xiong, Lin guoqing, Zhong yingying, Fang liping, Liang jinlu. Analysis on the compressive resistance of oil casing under non-uniform load and internal pressure [J]. Prospecting engineering(geotechnical drilling engineering), 2017, 44(08): 76-80.

本论文原发表于 International Field Exploration and Development Conference 2020。

Fracture Failure Analysis of C110 Oil Tube in a Western Oil Field

Zhu Lixia[1,2] Kuang Xianren[1,2] Xiong Maoxian[3]
Xing Xing[3] Luo Jinheng[1,2] Xie Junfeng[3]

(1. CNPC Tubular Goods Research Institute; 2. State Key Laboratory for Performance and Petroleum Tubular Goods and Equipment Materials; 3. Tarim Oil Field)

Abstract: There was an oil pipeline fracture found in the unlock process of Z well in western oilfield. The reason to make the oil tube fracture was systematically studied using macroscopic analysis, physical and chemical property test, scanning electron microscope with an energy dispersive X-ray spectrometer and X-ray diffraction analysis in the present study. The results suggested that the requirement of relevant standard for C110 steel grade was satisfied with the chemical component and mechanical property of the oil tube. The perforation of oil tube made H_2S in the fluid medium enter the oil set of connected rings. In addition, the high-pressure gas used in gas lift contained certain H_2S and O_2 so that the tube was fractured and invalid under the combined action of sulfide stress corrosion and oxygen corrosion.

Keywords: Oil tube; Fracture; Stress corrosion; Oxygen corrosion

1 Introduction

Oil tube is one of the most widely used tubing materials in petroleum industry and plays an important role in production. In recent years, with the development of oil and gas fields in harsh geological and environmental conditions, corrosion, bending, surface damage and cracking[1-4] of tube in work are increasing day by day, which causes serious economic losses. In February 2018, during the unlock process of Z well in western oilfield, it was found that the second tubing was fractured at the distance of 1.15m from the master buckle collar, and the specifications of the failed tube is ϕ88.9mm × 7.34mm, and the steel grade was C110. This oil tube began service in 2012. During the oil test, the packer test-retrofit and completion-integrated tubular column was put into the well, and the production zone was activated by gas lift. Then the high pressure gas was obtained. The high pressure gas used in gas lift contained H_2S gas. The high pressure gas used in gas lift contains H_2S gas. During the blowout production of this well, the H_2S concentration at the sampling port was 2100mg/L, and the H_2S concentration in the mouth of the tank was 7~21mg/L.

Corresponding author: Zhu Lixia, zhulx@cnpc.com.cn.

In this paper, the failure process and reasons of the tube were analyzed in combination with the field conditions, in order to provide technical references for the selection and application of the oil tube in oil and gas field.

2 Analysis method

2.1 Macro analysis

The macroscopic morphology of failed tube was shown in Fig. 1. Serious corrosion could be observed on the inner and outer surfaces of the tube. The fracture surface was flat and no obvious plastic deformation was observed after preliminary cleaning of the floating rust on fracture surface, it could be seen that the surface of the fracture was dark gray and could be divided into two areas: shear lip zone (final fracture zone) and flat zone (propagation source area and crack area). Slight radial pattern could be seen in the flat zone and converges to the outer surface. The fracture is stepped with multi-source characteristics.

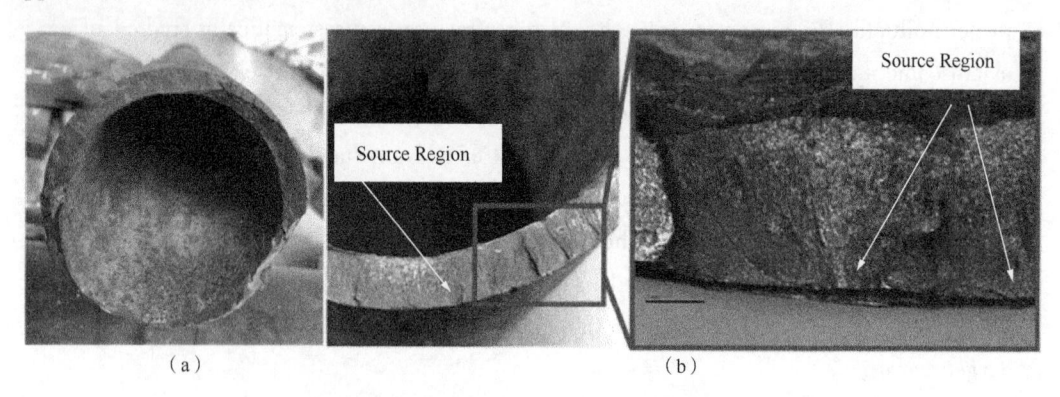

Fig. 1 (a) Macroscopic morphology of fracture.
(b) Multi-source characteristics of fracture source area

2.2 Chemical composition analysis

The samples were taken from the tube body, and chemical composition was analyzed by ARL 4460 direct reading spectrometer. The results were shown in Table 1, and the chemical composition analysis results conformed to the technical requirements.

Table 1 Results of chemical composition analysis (wt×10^{-2})

The sample location	C	Si	Mn	P	S	Cr	Mo	Ni	Nb	V	Ti	Cu	B	Al
The tube body	0.28	0.21	0.44	0.0077	0.0010	0.49	0.85	0.030	0.033	0.074	0.0034	0.047	0.0002	0.024
Technical requirements	—	—	—	≤0.015	≤0.005	—	—	—	—	—	—	—	—	—

2.3 Mechanical properties analysis

In the sample of tube body, tensile test, Charpy impact test and hardness test were carried out by UTM5305 material testing machine, PIT302D impact testing machine and RB2002 Rockwell hardness tester, and the results were shown in Table 2. It could be seen from the analysis results that the mechanical properties of the tube conformed to the requirements of the technical requirements.

Table 2 Results of mechanical properties test of the tube

Item	Tensile properties					Impact properties				Hardness
	Diameter ×distance (mm)	Yield strength (MPa)	Tensile strength (MPa)	Elongation (%)	Test temperature (℃)	Impact energy (J)		Shear rate (%)		HRC
						Single	Average	Single	Average	
Results	19.1×50	777	826	23	−10	80 80 78	79	100 100 100	100	24.0 24.7 25.1 25.4
Technical requirements	Tube body	758~828	≥793	≥16	−10	≥44		—		≤30

2.4 Metallographic analysis

Microscopic analysis was carried out on the sample of tube body and fracture. The microstructures of the specimens are fine tempered sorbite with grain size of 11.0 grade. Fig. 2 showed that there were secondary cracks in the fracture, and there were many corrosion pits on both inner and outer surfaces of the sample. The corrosion pits were filled with gray materials, and no abnormal structure was found. The bottom of the corrosion pit on both inner and outer surfaces is sharp, especially on the inner surface. The corrosion pit bottom extended like cracks. From the perspective of corrosion depth, the maximum depth of inner surface corrosion pit and bottom crack was 0.13mm, and the maximum depth of outer surface corrosion pit was 0.24mm. The morphology of outer surface corrosion pit bottom was wider, indicating that the corrosion on the outer surface of tube was more serious than that on the inner surface. The fracture crack analysis of the sample showed that the cracks propagate from the outer surface to the inner surface, and the crack was bifurcated and no abnormal structure was observed around the cracks.

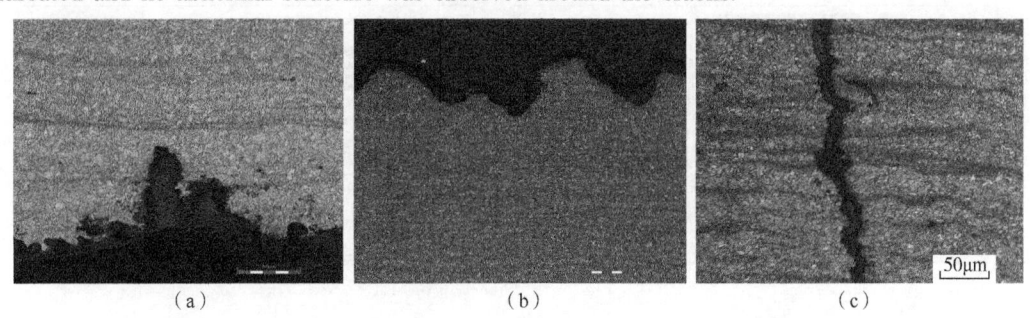

Fig. 2 (a) Microstructure around corrosion pit on inner surface.
(b) Microstructure around corrosion pit on inner surface.
(c) Fracture crack morphology

2.5 SEM and EDS analysis

The fracture source of failed tube was sampled and analyzed by SEM and EDS. The low magnification morphology of the source region was shown in Fig.3(a). It could be seen that the crack originated from the outer surface, which was consistent with the results of the fracture macroscopic and metallographic analysis. Fracture source zone, crack area and final fracture zone were observed at the low power of fracture. High magnification observation showed that the surface of the fracture was muddy and covered with corrosion products, as shown in Fig.3(b). After the surface corrosion products of the fracture were removed by mechanical and chemical methods, the morphology of the crack area was observed as shown in Fig.3(c), and the fracture was characterized by intergranular cracking.

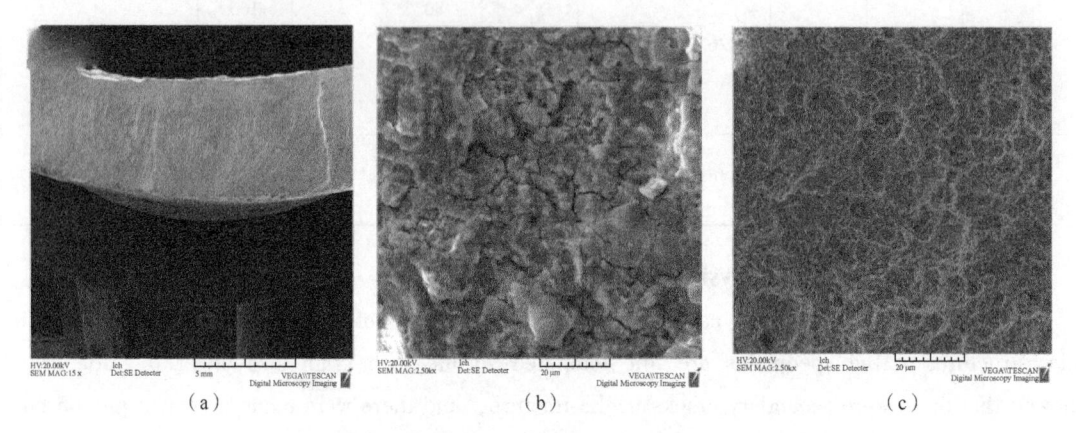

Fig.3 (a) Low magnification morphology of fracture source zone.
(b) Highmagnification morphology of fracture source zone.
(c) Fracture morphology of crack area

The results of EDS analysis of corrosion products on the outer surface, fracture and bottom of the corrosion pit are shown in Fig.4. All of corrosion products contained Fe, O, S and other elements. According to the elemental composition of energy spectrum analysis, the corrosion on fracture surface is was mainly composed of oxides and sulfides. After chemical cleaning, the fracture still contained certain S content.

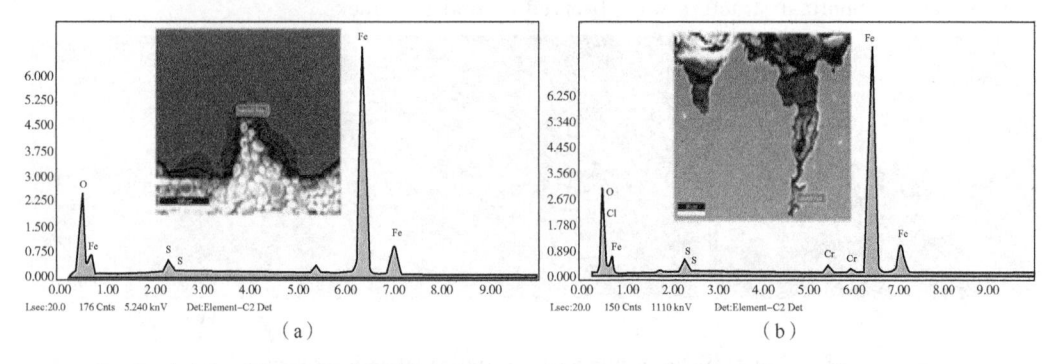

Fig.4 (a) Energy spectrum analysis of corrosion products on outer surface and bottom of pit.
(b) Energy spectrum analysis of corrosion products in inner surface cracks of corrosion pit Bottom

The corrosion products from the inner and outer wall of the failed tube were stripped off and fully ground into powder samples. XRD analysis method was used to analyze the phase of corrosion products, and the results were shown in Fig. 5. The results showed that the corrosion products of the inner and outer walls were mainly calcium carbonate (CaCO$_3$), ferrous sulfide (FeS) and iron oxide (Fe$_2$O$_3$). According to the peak phase of corrosion products, the content of calcium carbonate in the inner wall was the highest, followed by ferrous sulfide and iron oxide. The content of ferrous sulfide in the outer wall was the highest, followed by iron oxide and calcium carbonate. The analysis showed that calcium carbonate originates from the scaling of well fluid rather than corrosion products.

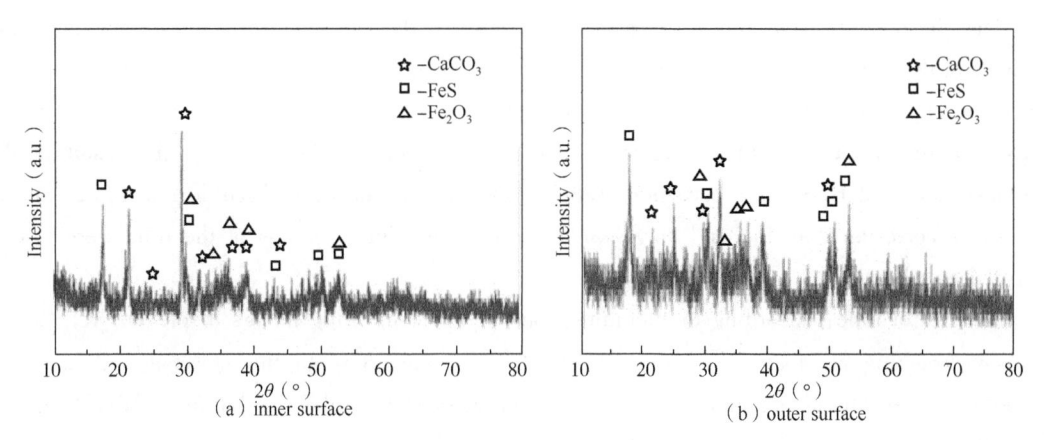

Fig. 5　XRD analysis of corrosion products

3　Results and discussion

The chemical composition analysis and mechanical properties test results of tube could be concluded that the chemical composition, tensile properties, impact toughness and hardness test results of ϕ88.9mm×7.34mm C110 tube used in the well accorded with the requirement of technical requirements; Metallographic analysis showed that the structure of the tube was tempered Sorbite with fine grain size of 11.0 grade. Charpy impact test results showed that the tube has good toughness at low temperature. According to the above test results, the types and causes of tube fracture were analyzed.

3.1　Macroscopic analysis results

Firstly, the macroscopic analysis results of the failed tube showed that the whole tube fracture showed no obvious plastic deformation and showed brittle fracture characteristics. The fracture surface was gray black with obvious radial pattern on the surface, indicating that the fracture originated from the outer surface of the tube. At the same time, the fracture was characterized by step-like and multi-source crack initiation. According to macroscopic characteristics of tube fracture, it was preliminarily judged that tube fracture belonged to H$_2$S stress corrosion cracking[5,6].

3.2　High magnification morphology of fracture

Secondly, the morphology could be seen from the high magnification of fracture. The fracture surface was covered by corrosion products, showing a muddy pattern. After the surface corrosion the products were removed by mechanical and chemical methods, obvious intergranular fracture characteristics were observed in the local areas. The results of energy spectrum analysis on the fracture surface of tube showed that Fe · O and S were the main elements which had characteristic of oxygen corrosion. The fracture microstructure was characterized by H_2S stress corrosion cracking. In addition, metallographic analysis of fracture showed that there were corrosion pits on both inner and outer surfaces of the tube. In terms of corrosion depth, the corrosion pit depth on the outer surface was greater than that on the inner surface, which indicating that the corrosion on the outer surface of the tube was more serious than that on the inner surface.

3.3 Phase analysis of the inner and outer walls

The phase analysis of the inner and outer walls of the oil tube showed that the corrosion products of the inner and outer walls contained FeS and Fe_2O_3. Therefore, the results of macroscopic and microscopic characteristics analysis, cross section energy spectrum analysis and physical phase analysis of the fracture could be determined that the fracture of the tube was the result of the combined action of H_2S stress corrosion cracking and oxygen corrosion. According to the factors of H_2S stress corrosion cracking and corrosion medium, the failure reasons of the tube were briefly analyzed as follows.

Material of tube: according to literature, one of the controlling factors to prevent H_2S stress corrosion cracking of steel is hardness ≤ HRC22[7,8]. According to the physical and chemical properties test results of the tube, the Rockwell hardness of the tube was between HRC 23.6 and HRC 25.4, indicating that the material of tube was sensitive to stress corrosion cracking.

Stress: according to the string structure of Z well, the failed tube was the second tube attached to the well tubing, with a total length of 6092.42m. The self-weight of the lower part of the tube produced a large axial tensile load. So, the tube had the stress factor of stress corrosion cracking sensitivity. Combined with the results of metallographic analysis, corrosion occurred on both inner and outer surfaces of the tube, with the maximum corrosion depth of 0.24mm on the outer surface and 0.13mm on the inner surface, equivalent to a maximum wall thickness reduction of 5%, which reduced the stress load that the tube could bear. At the same time, combined with phase analysis results, the corrosion products of the inner and outer surface of the tube are basically the same. But due to the more serious corrosion on the outer surface, the cracking source is located on the outer wall of the tube.

Corrosion medium: the oil and gas composition of the well shows that the Z well is a typical oil and gas well containing H_2S, and with H_2S concentration of 2100mg/L. In the well, tube packer was added at the lower end of the tube at 5920m, and annulus protection fluid was added to the oil set of annulus. The failed tube was the second tube attached to the well tubing, which was located above the packer. Normally, the outer surface of the short tube could not contact the H_2S medium, which meant there should be no risk of stress corrosion cracking. However, the well's operating history showed that the well was perforated through the tube at 3900~3901m and converted to gas lift production. Visibly, although the failed tube is above packer, the tube perforating made the H_2S in the tube internal fluid into the casing and tube annulus, and to the data showed that high pressure gas contained H_2S gas used in the gas lift. It could be seen that the oil casing annulus of this well had content of higher H₂S corrosive medium, indicating that the tube corrosion factors of the stress corrosion cracking sensitivity. In addition, according to the results of phase analysis, the corrosion products of the inner and outer walls of the oil pipe contained not only FeS but also Fe_2O_3. According to the construction conditions on site, the high-pressure gas used in the gas lift contained certain O_2, which caused oxygen corrosion of the oil pipe.

4 Conclusion

The failed tube had a high Rockwell hardness and was in the second tube attached to the well tubing, which itself bear a large tensile stress. In addition, the presence of H₂S corrosion medium and certain O_2 in the annulus lead to H₂S stress corrosion and oxygen corrosion, resulting in fracture.

(1) The chemical composition, tensile properties, impact toughness and hardness test results of the fractured tube conform to the technical requirements. The microstructure of the tube is tempered Sorbite with the grain size of 11.0 grade.

(2) The perforation of the tube makes H₂S in the fluid medium enter the oil casing annulus. In addition, the high-pressure gas used in the gas lift contains certain H₂S and O_2, making the tube fracture failure under the combined action of sulfide stress corrosion and oxygen corrosion, and the source of the fracture is located on the outer wall of the tube.

(3) It is recommended to treat high-pressure gas used in gas lift and to reduce H₂S and O_2 in injected gas as much as possible.

References

[1] Yang Long, Li Helin. Failure analysis prediction & prevention and integrity management[J]. Heat Treatment of Metals, 2011, 36: 15-16.

[2] Long Yan, Li Yan, Ma Lei, et al. The Fracture Reason of Repaired Tubing in a Western Oilfield[J]. Corrosion & Protection, 2018, 39(05): 359-364.

[3] Zhong Bin, Chen Yiqing, Meng Fanlei, et al. Perforation failure analysis of N80 pipe[J]. Corrosion & Protection, 2018, 39(08): 647-650.

[4] Lin Anbang, Peng Bo, Du Jinnan, et al. Corrosion failure analysis of a oil well tubing[J]. Total corrosion control, 2017, 31(10): 42-48.

[5] Li Helin, Li Pingquan, Feng Yaorong. Failure analysis and prevention of oil drill string[M]. Beijing: Petroleum Industry Press, 1999.

[6] Sun Zhi, Jiang Li, Ying Zhanpeng. Failure Analysis-Foundation and Application[M]. Beijing: China machine press, 2009.

[7] Yi Tao, Zhang Qing, Wan Qiang, et al. Stress corrosion of hydrogen sulfide and control of hardness[J]. Journal of Xinjiang petroleum institute, 2002, 14(2): 77-80.

[8] NACE International. NACE MR0175-2000. Sulfide stress cracking resistant metallic materials for oilfield equipment[S].

本论文原发表于《Materials Science Forum》2020年第993卷。

石油管用 Ti-6Al-4V-0.1Ru 钛合金高温流变行为及预测模型研究

刘 强[1,2]　白 强[2]　田 峰[2]　于 洋[1]　惠松晓[1]　宋生印[2]

(1. 有研科技集团有限公司·有色金属材料制备加工国家重点实验室；
2. 中国石油集团石油管工程技术研究院·石油管材及装备材料
服役行为与结构安全国家重点实验室)

摘　要：通过使用 Gleeble-3500 热模拟试验机进行等温单轴压缩试验，研究了 Ti-6Al-4V-0.1Ru 钛合金在温度 800℃ 到 1100℃，应变速率 $0.01s^{-1}$ 到 $10s^{-1}$ 条件下的高温流变行为。结果表明，Ti-6Al-4V-0.1Ru 钛合金的峰值应力随着变形温度的降低以及变形速率的增大而增大，软化机制在 950℃ 以下为动态回复，在 950℃ 以上为动态再结晶。通过使用线性回归的方法建立了 Ti-6Al-4V-0.1Ru 钛合金的 Arrhenius 本构模型，计算得到该合金的热激活能为 720.477kJ/mol，应变速率敏感指数为 4.809。通过引入应变对材料常数 α、n、A 和 Q 的影响，建立了考虑应变的流变应力预测模型，通过对试验值和预测值的比对，相关系数达到 96.9%，说明该模型具有较好的预测精度。

关键词：石油管；Ti-6Al-4V-0.11Ru 钛合金；流变应力；本构方程；显微组织

随着油气开发不断向深井超深井、高温高压、高腐蚀环境和海洋油气等非常规油气资源拓展[1,2]，对石油管材的要求不断提高。油井管作为开发油气的主要管柱材料，在井下不仅要经受高温高压，还要受到硫化氢、二氧化碳、高浓度氯离子甚至单质硫的综合腐蚀[3,4]，对管材的耐蚀性能提出更高的要求。钛合金材料由于具有高的比强度，低弹性模量，高韧性，优异的抗疲劳性能和耐腐蚀性，已经成为高端石油管材料研究的热点方向[5-7]。Ti-6Al-4V-0.1Ru 钛合金材料是 20 世纪末由美国 RMI 公司开发出来的新一代高耐蚀双相钛合金，除了具有较高的强度外，该材料在油气开发环境中耐缝隙腐蚀和应力腐蚀开裂的最高温度可高达 330℃[8-10]，已经成为新一代钛合金石油管的首选材料。

但是 Ti-6Al-4V-0.1Ru 钛合金的加工制备窗口较窄，在制备钛合金石油管时较为困难，特别是在端部镦粗等热加工过程成材率较低。钛合金的热变形行为，特别是钛合金在高温热变形过程中的流变应力、应变变化规律及热变形本构模型，一直是国内外钛合金材料制备加工研究的重点，Seshacharyulu 等研究了 Ti-6Al-4V 钛合金在不同相区的流变应力和本构方程，并绘制了热加工图[11]；Qu YD 等研究了不同热加工参数对 Ti-4Al-2.5V-1.5Fe 的流变

基金项目：国家科技重大专项(2016ZX05020-002)；陕西省创新能力支撑计划(2018KJXX-046)。
作者简介：刘强，男，1983 年生，博士，高级工程师。电话：029-81887814, E-mail: liuqiang030@cnpc.com.cn。

应力和组织变化的影响[12];徐勇等研究了TC4合金的流动软化行为和Arrhenius本构方程,并计算了激活能[13];朱晓弦等对2种TC4-xFe合金的热变形行为进行研究,建立了以变形温度、应变速率和真应力为参数的本构方程[14]。但对于Ti-6Al-4V-0.1Ru钛合金的热加工变形行为研究较少,Ti-6Al-4V-0.1Ru钛合金在热加工过程中的温度、加工速率及应变对合金流变应力和软化机制的影响尚不清楚,这些对于Ti-6Al-4V-0.1Ru钛合金油井管的制备加工和组织控制有着重要的影响。

本工作使用等温单轴压缩试验方法,对退火态的Ti-6Al-4V-0.1Ru钛合金在800~1100℃温度下以及0.01~10s^{-1}应变速率条件下进行高温流变行为研究,建立本构方程,并通过引入应变对材料常数α、n、A和Q的影响,优化得到了考虑应变的流变应力预测模型,为该合金的制备加工提供指导。

1 实验

试验用的Ti-6Al-4V-0.1Ru钛合金材料使用3次VAR熔炼、锻造和热轧的工艺制成管材,经过730℃退火处理后,测试化学成分见表1,原始组织如图1所示,从图1上可以看出材料的微观组织可见粗大的β相晶界,以及在原始晶界内分布着大量不同取向的针状$\alpha+\beta$双相魏氏组织。

表1 试验用Ti-6Al-4V-0.1Ru钛合金的化学成分 单位:%(质量分数)

Al	V	Ru	O	C	H	Ti
5.99	4.01	0.10	0.08	0.0089	0.0029	89.8082

沿钛合金管纵向取ϕ10mm×15mm的热模拟试样,使用Gleeble-3500热模拟试验机分别在温度800℃到1100℃以50℃为温度间隔,在0.01s^{-1}、0.1s^{-1}、1s^{-1}、10s^{-1}应变速率下进行单轴压缩试验,总变形量约为60%,压缩完成后迅速进行水冷以保留变形组织。将压缩后的试样沿轴向中心剖开,使用砂纸打磨和抛光后,使用KROLL市集进行腐蚀,然后使用MEF3A和MEF4M金相分析系统进行观察分析。

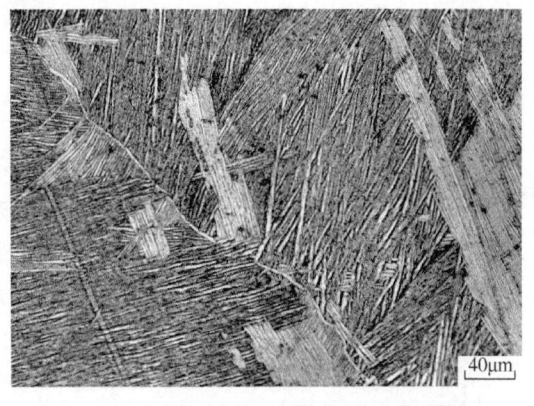

图1 Ti-6Al-4V-0.1Ru钛合金初始显微组织

2 结果与分析

2.1 流变应力特性

钛合金在高温下的流变应力受变形速率和变形温度影响较大,图2为Ti-6Al-4V-0.1Ru钛合金在不同变形温度和应变速率下的真应力—应变曲线,从图2中可以看出,在这些变形条件下,变形初期流变应力随着应变的增加迅速增大,当达到一个峰值后逐步减弱或先降低后趋于稳定。峰值应力随着变形温度的升高而降低,并且随着变形速率的增大而增大,这些流变应力特性与其他的双相钛合金基本一致[9,15,16]。从图2(b)至图2(d)中还可以发现,随着温度的升高,流变应力的特性明显不同。当变形温度小于950℃时,流变应力随着应变的增大迅速下降,而当温度高于950℃时,流变应力基本保持在一个稳定的状态,这种流变应力的差别与合金的动态软化机制有关。

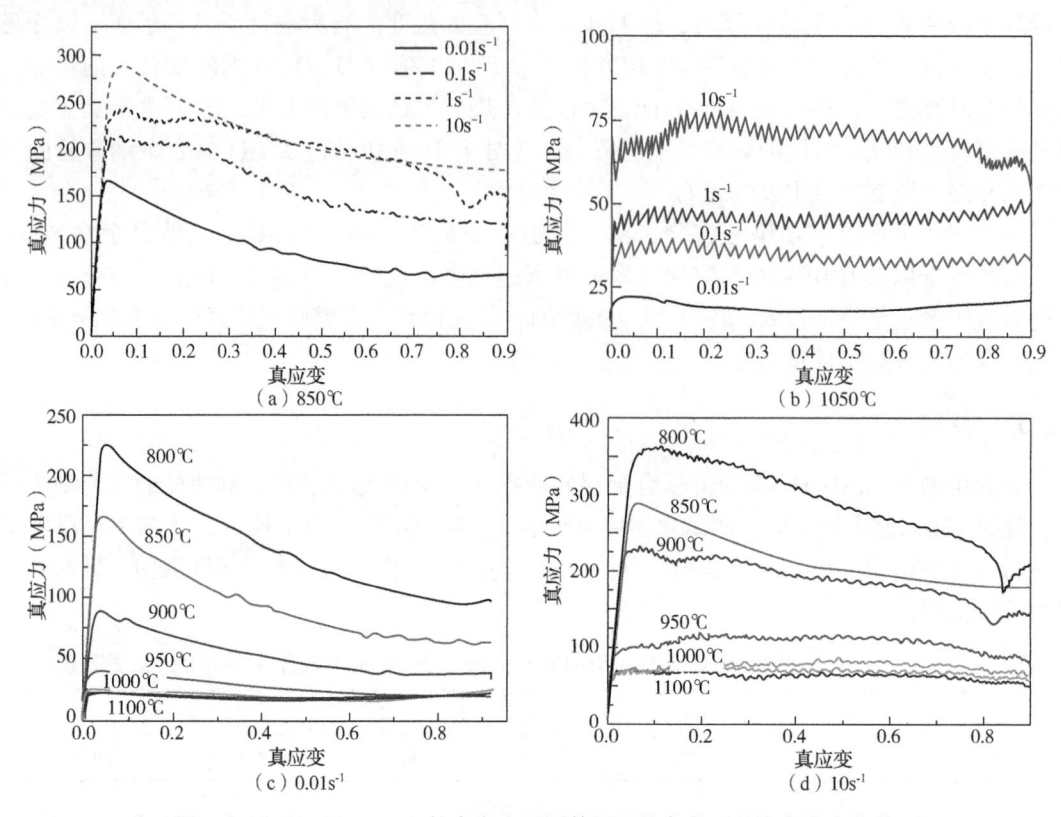

图 2 Ti-6Al-4V-0.1Ru 钛合金在不同等温压缩条件下的流变应力曲线

2.2 高温软化行为及组织分析

图 3 为 Ti-6Al-4V-0.1Ru 钛合金在热变形中峰值应力随着变形温度和变形速率的变化图，图 3 中可见在同一变形速率下，峰值应力先随着温度的升高不断降低，而下降的速度却不断在减慢，当变形温度高于 950℃ 时，峰值应力基本保持在一定范围内，无明显降低。这是由于当变形温度低于 950℃ 时，Ti-6Al-4V-0.1Ru 钛合金的显微组织为原始的粗大魏氏组织被拉长变形，在原始 β 相晶界处分布的部分针状 α 相发生了动态回复，如图 4 所示，从图中可以看出，当变形温度较低时，在原始基体的 α 和 β 相的小角度亚晶界位置出现了少量的多边形化现象，

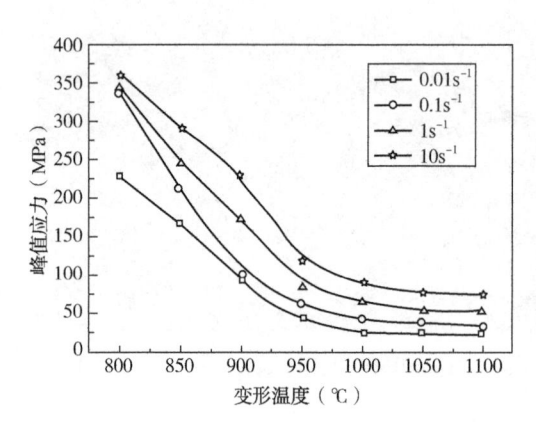

图 3 Ti-6Al-4V-0.1Ru 钛合金在不同热变形速率中峰值应力随着温度的变化

如图 4(a)中箭头所示，这是由于亚晶界附件的同号位错相互抵消或重新排列，形成了明显的胞状组织，但是数量较少，随着变形温度增大到 850℃ 时，同样变形速率下，发生回复的组织数量增多，亚晶粒长大程度更为明显，如图 4(b)所示。相同温度下不同变形速率对回复程度的影响，如图 4(b)和图 4(c)所示，对比可知随着变形速率的增大，发生动态回复亚晶粒的数量降低，亚晶粒尺寸增大明显，当应变速率为 0.01s^{-1} 时，亚晶粒尺寸为 5~15μm，

当变形速率增大10倍时，亚晶粒尺寸长大到50~80μm，如图4(d)所示。由于晶粒变形和动态回复的软化速度大于由于应变的增加而产生的加工硬化效应，流变应力在达到峰值后迅速下降，并且随着温度的不断升高，原子的动能不断增强，使得能开动晶界滑移面不断增多，部分第二相的融解使得对位错运动的钉扎作用不断减弱，在不同应变速率下所产生的峰值应力不断降低。

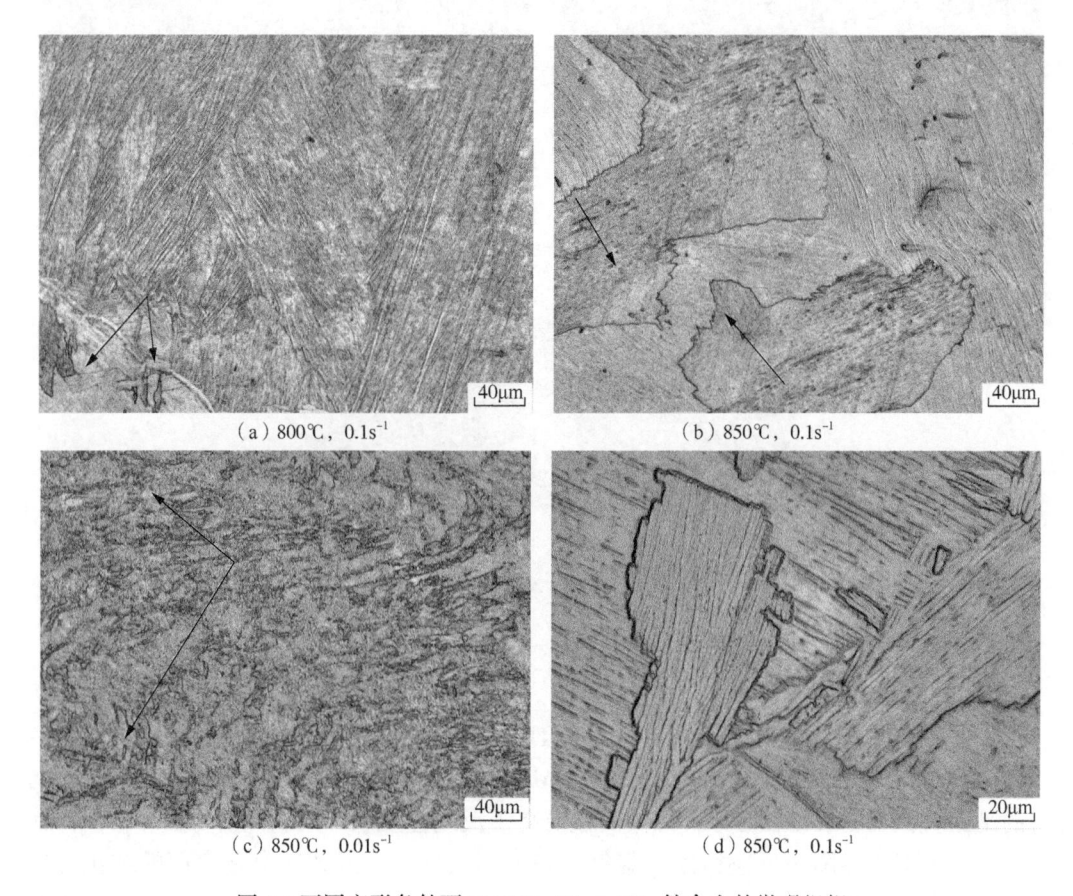

(a) 800℃，0.1s^{-1}

(b) 850℃，0.1s^{-1}

(c) 850℃，0.01s^{-1}

(d) 850℃，0.1s^{-1}

图4　不同变形条件下 Ti-6Al-4V-0.1Ru 钛合金的微观组织

当变形温度等于或高于950℃时，此时合金的显微组织发生较大变化，在原始晶界和晶内有大量的再结晶 β 相形核，当变形温度为950℃时，由于 α 和 β 相的晶界两侧存在较大的应变能差，从而减低了新晶粒形核的自由能壁垒，从而在晶界处形成较多数量的细小动态再结晶，如图5(a)所示，再结晶新晶粒约占原始组织的20%，新结晶的晶粒大小为 3~10μm，并伴随有少量的动态回复现象。当温度升高到1100℃时，有近60%的密排六方结构的 α 相转变为体心立方结构的 β 相，如图5(b)所示，并且再结晶晶粒沿原始针状 α 相方向发生长大，长大的晶粒尺寸为10~30μm。可以看出高于950℃时的软化机制主要以动态再结晶为主。由于温度较高同时体心立方结构的 β 相具有更多的滑移系，位错更容易产生和开动，因此高温塑性变形的流变抗力相对较低，并且随着应变速率的降低而进一步降低，此时的加工硬化效应被动态再结晶等产生的软化机制不断平衡，因此会产生图2中的流变应力稳态现象。再结晶的程度和再结晶晶粒大小与变形速率也有关系，在较高变形速率下再结晶的程度要较低，如图5(c)所示。

(a) 950℃, 0.01s^{-1}

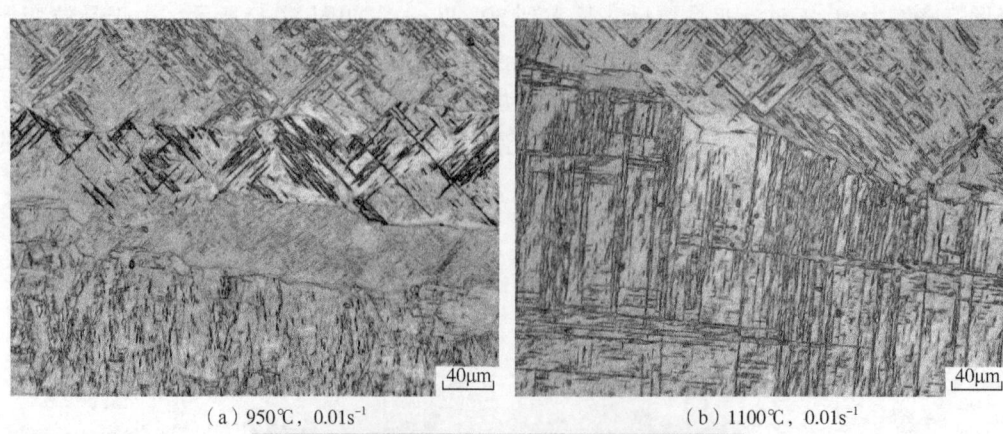

(b) 1100℃, 0.01s^{-1}

(c) 1100℃, 0.11s^{-1}

图5 不同变形条件下Ti-6Al-4V-0.1Ru钛合金的微观组织

跟Ti-6Al-4V合金相比，由于Ti-6Al-4V-0.1Ru钛合金加入了微量的钯组元素——钌，显微组织和热力学特性与不加钌的Ti-6Al-4V合金有所不同。该合金加入钌元素的主要目的是提高钛合金表面在还原性酸中阴极氢离子反应速度（H$^+$→H$_2$或者HER），使得钛合金阴极反应大于临界阳极反应密度，避免了临界阳极回路，保证钛合金表面的钝化层的稳定，从而提高钛合金材料在油气开发环境中耐缝隙腐蚀和应力腐蚀开裂的能力[17,18]。在显微组织和性能方面，钌元素与钯组其他钯、铂和铱元素不同，钌在钛合金中是β相稳定元素并且可以有限固溶于β相，同时钌元素在β相中的富集程度是α相中的4~5倍，因此钌的加入可以细化晶粒，从而提高钛合金的强度和塑性[19]，Lingen和Sandenbergh研究发现在含钌钛合金中，钌和Fe元素将形成独特的Ti-12Ru-5Fe相，经过标定发现这种相属于bcc结构，大小在0.3~1μm，在晶粒内部和晶界都有分布，起细化晶粒和钉扎位错的作用[20]。将Ti-6Al-4V-0.1Ru钛合金的热变形行为与Ti-6Al-4V合金相比[13]，可以发现相同热变形条件下，Ti-6Al-4V-0.1Ru钛合金要具有更高的峰值流变应力并且晶粒组织更细小。

2.3 本构方程的建立

为了描述金属及合金在热变形过程中的应变速率、温度和应力之间的关系，Sellars和Tegart等建立的Arrhenius方程被广泛用来解释材料的热激活稳态变形行为[21]，在高温变形下，修正Arrhenius的关系可由变形激活能Q和温度T的双曲正弦形式来描述：

$$\dot{\varepsilon} = Af(\sigma)\exp(-Q/RT) \tag{1}$$

式中：$\dot{\varepsilon}$ 为应变速率，s^{-1}；A 为与材料油管相关的常数；R 为气体常数，8.314kJ/mol·K；T 为热力学温度，K；Q 为材料的激活能，J/mol。

$f(\sigma)$是材料应力的函数，可用以下几种方程来表示[22,23]：

$$f(\sigma) = \sigma^{n_1} \quad (\alpha\sigma < 0.8) \tag{2}$$

$$f(\sigma) = \exp(\beta\sigma) \quad (\alpha\sigma > 1.2) \tag{3}$$

$$f(\sigma) = [\sinh(\alpha\sigma)]^n \quad (\text{所有 } \alpha\sigma) \tag{4}$$

式中：n_1，β，α 和 n 是与材料有关的常数，在本研究中 σ 取各变形条件中的峰值应力。双曲正弦形式的式(4)在高温变形中更适用于所有应力状态，因此结合式(1)和式(4)，可以得到：

$$\dot{\varepsilon} = A[\sinh(\alpha\sigma)]^n \exp(-Q/RT) \tag{5}$$

式中：α 应力乘数是一个非常重要的参数，可决定其他几个重要参数，α 可以通过求不同温度下的 $\ln\dot{\varepsilon}$—σ 和 $\ln\dot{\varepsilon}$—$\ln\sigma$ 曲线的平均斜率来求得[24-26]，如图6所示，计算得出 α 为 0.011MPa^{-1}。

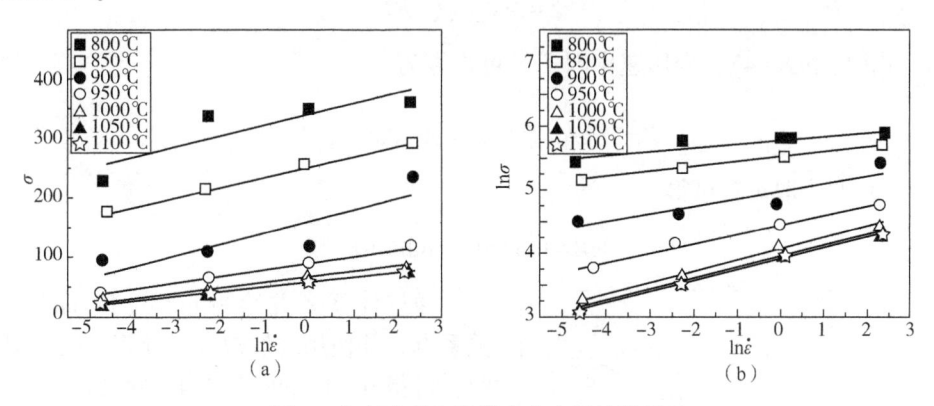

图6 应变速率和峰值应力之间的关系

对式(5)取自然对数，可以得到：

$$Q/RT = \ln A - \ln\dot{\varepsilon} + n\ln[\sinh(\alpha\sigma)] \tag{6}$$

当温度和应变速率一定时，对式(6)偏微分可得到热变形过程中的重要参数热激活能 Q：

$$Q = R\left\{\frac{\partial \ln\dot{\varepsilon}}{\partial \ln[\sinh(\alpha\sigma)]}\right\}_T \left\{\frac{\partial \ln[\sinh(\alpha\sigma)]}{\partial(1/T)}\right\}_{\dot{\varepsilon}} = RnS \tag{7}$$

图7为不同温度下 $\ln\dot{\varepsilon}$-$\ln[\sinh(\alpha\sigma)]$ 的关系曲线，计算 $\ln\dot{\varepsilon}$-$\ln[\sinh(\alpha\sigma)]$ 的平均斜率即可得到 $1/n$，计算可得 $n = 4.8089$，和其他文献中报导的结果基本吻合[15,27]。

S 可以通过应变速率一定下 $1000/T$ 和 $\ln[\sinh(\alpha\sigma)]$ 的关系得出，图8为不同应变速率下 $1000/T$-$\ln[\sinh(\alpha\sigma)]$ 的关系曲线，计算平均斜率可得 $S = 18.020$，由式(7)可以计算出 $Q = 720.477\text{kJ/mol}$。

图7 不同温度下 $\ln\dot{\varepsilon}$ 和 $\ln[\sinh(\alpha\sigma)]$ 的关系曲线

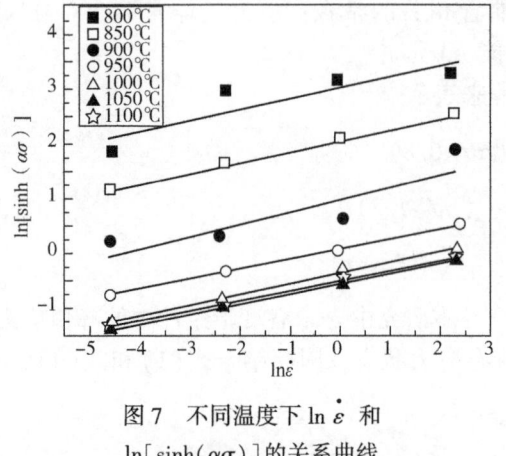

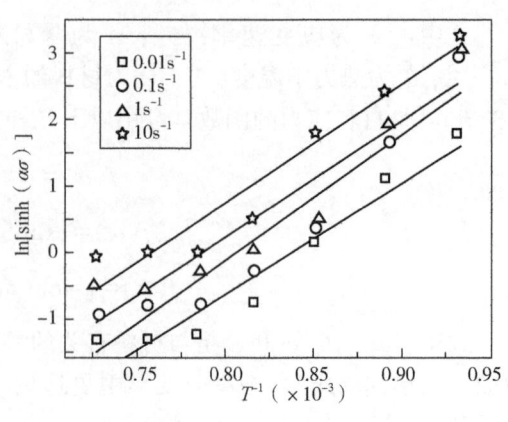

图8 不同应变速率下 $1000/T$ 和 $\ln[\sinh(\alpha\sigma)]$ 的关系曲线

Zener 和 Hollomon 研究发现用温度补偿的变形速率因子参数 Z 是表征变形温度和应变速率对材料变形行为影响的重要参数,变形速率因子 Z 可用下式来表示:

$$Z = \dot{\varepsilon}\exp(Q/RT) \tag{8}$$

结合式(1)和式(4),变形速率因子 Z 可推导为

$$Z = \dot{\varepsilon}\exp(Q/RT) = A[\sinh(\alpha\sigma)]^n \tag{9}$$

对式(9)两边取对数可得

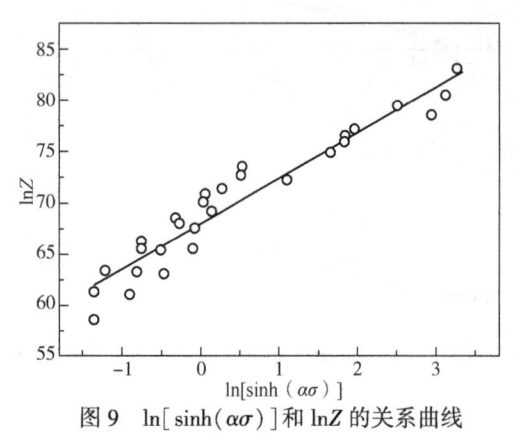

$$\ln Z = \ln A + n\ln[\sinh(\alpha\sigma)] \tag{10}$$

通过计算 Z 并绘制 $\ln Z$ 与 $\ln[\sinh(\alpha\sigma)]$ 的关系点,并使用线性拟合后计算直线的斜率和截距,可得 $n = 4.4596$ 和 $\ln A = 68.1129$,拟合相关系数(R^2)为 0.915,具有较好的一致性,如图9所示。

因此,基于以上的计算结果,Ti-6Al-4V-0.1Ru 钛合金峰值应力的本构方程可以表示为

$$\dot{\varepsilon} = 3.81\times10^{29}[\sinh(0.011\sigma)]^{4.809}\exp(-720.477/RT) \tag{11}$$

图9 $\ln[\sinh(\alpha\sigma)]$ 和 $\ln Z$ 的关系曲线

2.4 本构模型的修正及检验

由于 Arrhenius 模型只考虑了变形温度和变形速率对流变应力的影响,而忽视了另外一个重要的参数应变,而在实际变形中,由于材料的差异,不同应变区域内应变对应力的影响非常明显[13,15],因此需要在考虑应变的情况下对模型进行修正优化。

依据上文的算法分别计算从 0.1~0.916 不同应变下的材料常数(n,Q,α 和 A),并分别绘制材料常数 n,Q,α 和 A 与应变之间的关系,如图10所示,可以看出应变对材料常数具有非常大的影响。对图9中应变与材料常用 Q,α,A 的关系进行4次多项式拟合,拟合

相关系数（R^2）均达到 0.99 以上，具有非常好的相关性，而对材料常数 n 与应变采用了更高阶的多项式拟合方法，拟合相关度为 0.761，相关性一般。因此，将拟合修正后的材料常数（n，Q，α 和 A）代入 Arrhenius 模型，根据式(9)可得到基于试验数据并考虑应变的修正流变应力预测模型：

$$\sigma = \frac{1}{\alpha_x}\ln\left<\left[\frac{\dot{\varepsilon}\exp\left(\frac{Q_x}{RT}\right)}{A_x}\right]^{1/n_x} + \left\{\left[\frac{\dot{\varepsilon}\exp\left(\frac{Q_x}{RT}\right)}{A_x}\right]^{2/n_x} + 1\right\}^{1/2}\right> \quad (12)$$

式中：

$\alpha_x = 0.02832\dot{\varepsilon}^4 - 0.04096\dot{\varepsilon}^3 + 0.016\dot{\varepsilon}^2 + 0.00505\dot{\varepsilon} + 0.00989$

$n_x = -93.877\dot{\varepsilon}^5 + 300.492\dot{\varepsilon}^4 - 324.240\dot{\varepsilon}^3 + 147.080\dot{\varepsilon}^2 - 28.451\dot{\varepsilon} + 6.485$

$Q_x = 2610.020\dot{\varepsilon}^4 - 4466.171\dot{\varepsilon}^3 + 3429.950\dot{\varepsilon}^2 - 1868.214\dot{\varepsilon} + 934.088$

$\ln A_x = 239.811\dot{\varepsilon}^4 - 406.927\dot{\varepsilon}^3 + 315.464\dot{\varepsilon}^2 - 175.957\dot{\varepsilon} + 89.039$

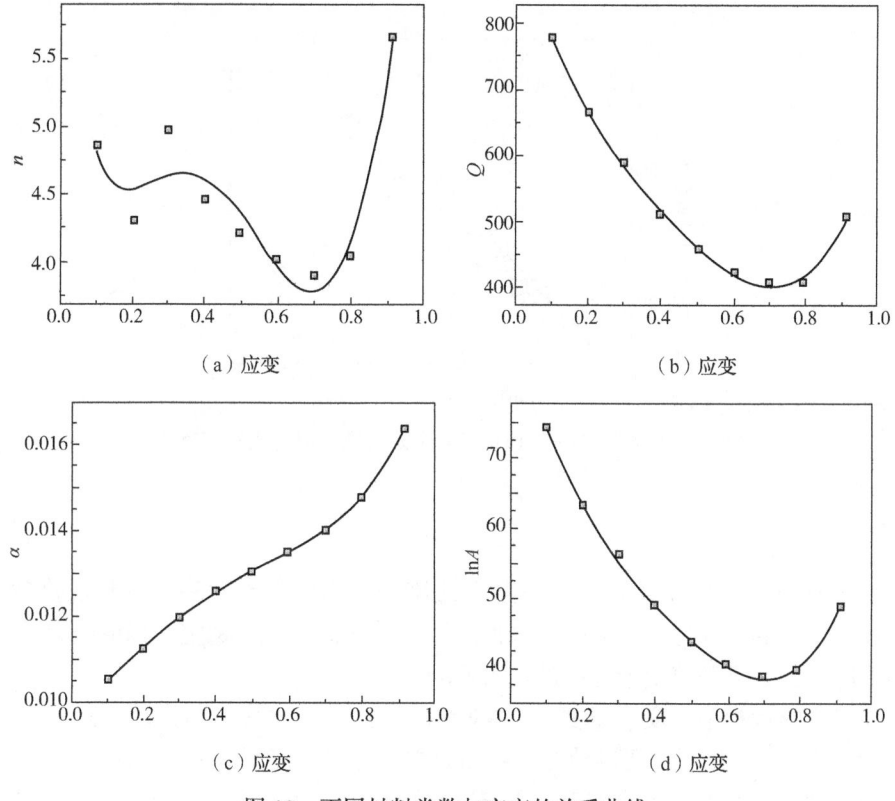

图 10 不同材料常数与应变的关系曲线

根据式(12)计算 900℃和 1000℃不同应变速率下的流变应力，并和实验测的真实流变应力进行对比，结果如图 11 所示。从图上可以看出，该修正的流变应力预测模型可以较好地预测 800~1100℃之间、应变速率从 $0.01s^{-1}$ 到 $10s^{-1}$ 之间的流变应力，当温度低于 1000℃时，全应变范围都能很好地与试验值相吻合，当温度高于 1000℃时，低应变下该模型预测值与实际值有一定差距，但高应变（高于 0.5）下预测精度较高。

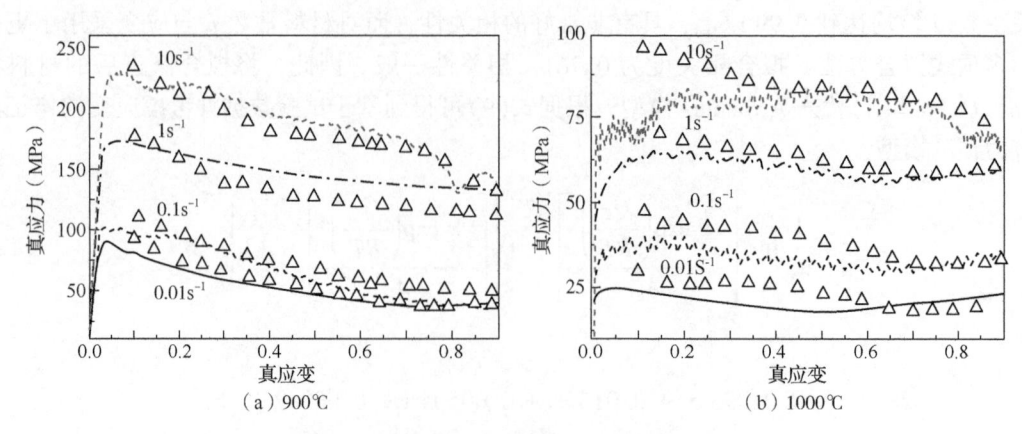

图11 流动应力的计算值与实验结果对比

图12为Ti-6Al-4V-0.1Ru钛合金根据式(12)预测的流变应力和试验值的综合对比分析，结果可见相关系数达到96.9%，说明本预测模型可以精确地预测Ti-6Al-4V-0.1Ru钛合金在热变形中的流变应力。

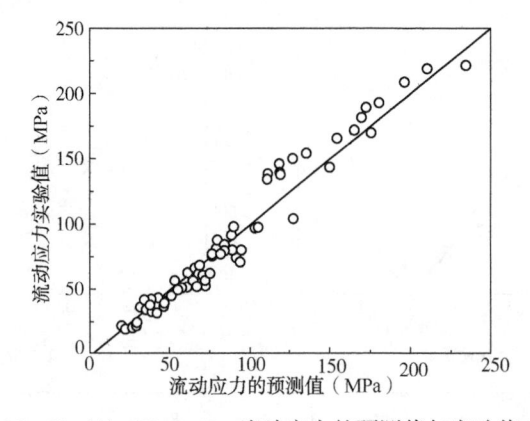

图12 Ti-6Al-4V-0.1Ru流动应力的预测值与实验值对比

3 结论

（1）Ti-6Al-4V-0.1Ru钛合金在高温下的流变应力对变形速率和变形温度较为敏感，流变应力随着应变的增加先升高到一个峰值后逐步减弱而后趋于稳定。峰值应力随着变形温度的升高而降低，并且随着变形速率的增大而增大。

（2）在热变形过程中，变形温度低于950℃时，主要的软化机制为动态回复，当变形温度高于950℃时，软化机制主要以动态再结晶为主，并产生流变应力稳态现象。

（3）通过使用线性回归的方法建立了Ti-6Al-4V-0.1Ru钛合金的Arrhenius本构模型，计算得到该合金的热激活能为720.477kJ/mol，应变速率敏感指数为4.809。

（4）在Ti-6Al-4V-0.1Ru钛合金热压缩试验数据的基础上建立了引入应变参数的流变应力预测模型，可较为准确的预测变形温度在800~1100℃之间，应变速率0.01s^{-1}到10s^{-1}之间Ti-6Al-4V-0.1Ru钛合金的流变应力，准确率达到96.9%，对于Ti-6Al-4V-0.1Ru钛合金加工制备石油管材具有重要的指导意义。

参 考 文 献

[1] 谷坛, 霍绍全, 李峰, 等. 石油与天然气化工[J]. 2008, 37(S): 63.
[2] 叶登胜, 任勇, 管彬, 等. 天然气工业[J]. 2009, 29(3): 77.
[3] 杜伟, 李鹤林. 石油管材与仪器[J]. 2015, 1(5): 1.
[4] 叶登胜, 任勇, 管彬, 等. 天然气工业[J]. 2009, 29(3): 77.
[5] Schutz R W, Watkins H B. Mater Sci Eng A[J], 1998, 243: 305.
[6] 刘强, 宋生印. 石油矿场机械[J]. 2014, 43(12): 89.
[7] 高文平, 吕祥鸿, 谢俊峰, 等. 稀有金属材料与工程[J]. 2018, 47(1): 151.
[8] Schutz R W. NACE Corrosion/97[C]. Houston, TX: NACE International, 1997: 32.
[9] Schutz R W. Platinum Metals Review[J], 1996, 40(2): 54.
[10] Schutz R W, Horrigan J M, Bednarowicz T A. NACE Corrosion/98[C]. Houston, TX: NACE International, 1998: 261.
[11] Seshacharyulu T, Medeiros S C, Frazier W G. Mater Sci Eng A[J]. 2000, 284: 184.
[12] Qu Y D, Wang M M, Lei L M et al. Mater Sci Eng A[J], 2012, 555: 99.
[13] 徐勇, 杨湘杰, 何毅, 等. 稀有金属材料与工程[J], 2017, 46(5): 1321.
[14] 朱晓弦, 常辉, 谢英杰, 等. 稀有金属材料与工程[J], 2017, 46(6): 205.
[15] Meng Q G, Bai C G, Xu D S. J Mater Sci Technol[J], 2018, 34: 679.
[16] Luo J, Li M Q, Yu W X et al. Mater Sci Eng A[J], 2009, 504: 90.
[17] Tomashov N D, Altovsky R M, Chernova G P. J Electrochem Soc[J], 1961, 108(2): 113.
[18] Uhlig H H. The Corrosion Handbook[M]. New York, NY: J Wiley& Sons, 1948: 1144.
[19] Murray J L. Phase Diagrams of Brine Titanium Alloys, Monograph Series on Alloy Phase Diagrams (Materials Park)[M]. OH: ASM International, 1987: 240.
[20] Lingen E van der, Sandenbergh R F. Corrosion Sci [J], 2001, 43: 577.
[21] Sellars C M, Tegart W J. Acta Metall[J], 1966(14): 1136.
[22] Liu X Y, Pan Q L, He YB et al. Mater Sci Eng A[J], 2009, 500: 150.
[23] Chen Z Y, Xu S Q, Dong X H. Acta Metallurgica Sinica[J], 2008, 21(6): 451.
[24] Dong Y, Zhang C S, Guan G Q. Mater Des[J], 2016 (92): 983.
[25] Gangolu S, Rao A G, Sabirov I. Mater Sci Eng A[J], 2016, 655: 256.
[26] Shi Z X, Yan X F, Duan C H. J Alloy Compd[J], 2015, 652: 30.
[27] Zhao J W, Ding H, Zhao W J et al. J Alloy Compd [J], 2013, 574: 407.

本论文原发表于《稀有金属材料与工程》2020年第49卷第1期。

油气开采用钛合金石油管材料耐腐蚀性能研究

刘 强[1,2]　惠松骁[1]　汪鹏勃[2]　叶文君[1]　于 洋[1]　宋生印[2]

(1. 有研科技集团有限公司·有色金属材料制备加工国家重点实验室；
2. 中国石油集团石油管工程技术研究院·石油管材及装备材料服役行为与结构安全国家重点实验室)

摘　要：钛合金因为具有高强低密、低弹性模量、优异的韧性、疲劳性能和耐蚀性，已经成为严酷工况环境下油井管和海洋开发工具的热门可选材料。由于我国油气开发工况更为恶劣，部分钛合金材料制备的油井管在服役时会发生一定的腐蚀问题，特别是缝隙腐蚀问题最为明显，严重者会导致钛合金油井管在服役过程中的接头密封失效而带来巨大的经济损失。本文选取5种油气开常用钛合金材料(Ti-6Al-4V、Ti-6Al-4V-0.1Ru、Ti-6Al-2Sn-4Zr-6Mo、Ti-3Al-8V-6Cr-4Zr-4Mo和Ti-5.5Al-4.5V-2Zr-1Mo)为研究对象，使用高温高压釜模拟国内典型严酷服役工况环境，研究了不同钛合金材料耐均匀腐蚀、局部腐蚀、点蚀、应力腐蚀开裂(SCC)及缝隙腐蚀的性能，通过使用扫描电镜和能谱分析等手段对腐蚀形貌和腐蚀产物进行了分析，并使用电化学方法对不同合金的耐腐蚀机理进行了研究。结果显示，在所测试工况条件下，所有钛合金材料腐蚀反应均为阳极控制过程，均匀腐蚀速率均低于0.001mm/a，并且对应力腐蚀开裂均有良好的抗力。Ti-6Al-4V和Ti-5.5Al-4.5V-2Zr-1Mo合金出现明显的点蚀和缝隙腐蚀问题。对腐蚀机理研究表明，在工况条件温度下，随着pH值的降低，所有钛合金均发生的自腐蚀电位降低，极化电阻减小，腐蚀电流增大，耐腐蚀性能下降，其中Ti-6Al-4V耐腐蚀性能下降的最为明显，研究结果为油气开发工况下钛合金石油管的选材和缝隙腐蚀问题防治提供理论基础。

关键词：钛合金；石油管；腐蚀性能；服役工况；腐蚀机理；缝隙腐蚀

随着世界范围内对油气资源消耗量的不断增长，促使油气勘探开发逐步向深井超深井、高温高压、高腐蚀环境和海洋油气等非常规油气资源拓展[1-3]。由于我国油气资源圈盖构造的特殊性，高产油气资源的开发具有高温、高压、高腐蚀的"三高"特点。例如塔里木区块的气井服役工况已经达到超深(>6000m)，超高温(>160℃)，超高压(井口压力达到110~130MPa)，高腐蚀(高含Cl^-达到$1.5×10^5$mg/L，CO_2超过1MPa)[4]；我国含硫气田也占相当大的比例，特别是2006年发现的川渝地区大型整装海相气田，埋藏深度为4500~5700m，气藏压力为55~57MPa，井底温度为130~150℃，H_2S的平均含量为15.37%，CO_2的平均

基金项目：国家科技重大专项(2016ZX05020-002)；陕西省创新能力支撑计划(2018KJXX-046)。
作者简介：刘强，男，1983年生，博士，高级工程师，E-mail：liuqiang030@cnpc.com.cn。

含量为 8.26%，Cl⁻含量约为 100000mg/L，天然气的 pH 值约等于 3[5]。罗家寨气田的 H_2S 含量为 7.13%~13.74%，CO_2 含量为 5.13%~10.41%，井底温度为 85~105℃，是目前我国也是世界上腐蚀环境最恶劣的气田之一[6]。

严酷的开发环境对石油管材的高耐腐蚀性和高性能提出了更高的要求，钛合金材料由于本身的高强度、低密度、耐蚀性优异、抗海水冲刷性好、低弹性模量和高耐疲劳性能，成为高端石油管材料研究的热点方向，国际上在 20 世纪 80 年代起就开始对钛合金在油气开发应用进行了研究。Schutz 等学者对钛合金在油气工况下的性能进行详细测试，并证明钛合金在油气开发有巨大的应用潜力[7,8]，Kane 等人对高温高压、天然气环境下钛合金的各项性能进行了模拟工况评价，指出并不是所有的钛合金在这些环境下都可以使用，要进行细致的评价和筛选[9,10]，RMI 公司等通过在部分钛合金中加入贵金属铂族元素等对钛合金的耐蚀性能进行提升，成功地研制出强度超过 1000MPa 的钛合金油套管、连续管、海洋钻井隔水管等产品，1999 年起在美国多个区块的油气井、墨西哥湾的 Oryx 海王星钻井项目和 Mobile Bay Field(莫比尔湾油田)热酸性油气井成功应用[11-14]。国内钛合金石油管研究起步较晚，中国石油集团石油管工程技术研究院最早对钛合金用于油井管的可行性和腐蚀性能进行了评价，证明钛合金在油气勘探开发中有着良好的应用前景[15]；西安石油大学研究了 TC4 钛合金的耐腐蚀性能[16]，东方钽业、天津钢管集团、攀钢集团成都钢钒有限公司等制造厂家对钛合金管材进行了试制研究，初步开发出了不同的钛合金油井管产品[17-19]。

由于我国的油气开发环境与国际上相比更为恶劣，国际上钛合金的评价试验结果并不完全能用于我国环境，不同的钛合金在我国典型服役工况下的适用性还是未知。特别是钛合金在酸性环境中易发生应力腐蚀开裂和缝隙腐蚀等问题[9]，这些问题对于石油管安全可靠性非常重要，特别是由于井下油井管的连接和气密封主要依靠螺纹和密封结构，这些都会产生大量细小的缝隙，缝隙腐蚀性能的好坏直接决定了油井管的密封完整性和结构完整性，选材不当甚至会带来整口井的失效报废，而国内外在此方面的研究极少。本研究通过研究几种油气开发常用的钛合金在国内典型腐蚀工况下的腐蚀性能，重点研究不同合金成分、不同缝隙条件对钛合金缝隙腐蚀性能的影响，探讨在服役工况下的腐蚀性能机理，为今后钛合金石油管的选用提供数据和理论支撑。

1 实验

试验选用的几种钛合金名义化学成分和拉伸性能见表 1，均使用 3 次 VAR 熔炼、锻造和热轧的工艺制成 φ88.9mm×6.45mm 规格的管材。

表 1 试验用钛合金化学成分和性能

样品	成分	YS(MPa)	UTS(MPa)
1#	Ti-6Al-4V	758	827
2#	Ti-6Al-4V-0.1Ru	758	827
3#	Ti-6Al-2Sn-4Zr-6Mo	965	1034
4#	Ti-3Al-8V-6Cr-4Zr-4Mo	896	965
5#	Ti-5.5Al-4.5V-2Zr-1Mo	850	952

模拟工况均匀腐蚀实验在管材纵向上取样尺寸为 40mm×10mm×3mm 的腐蚀挂片试样，经过 480#，600#，1200# 砂纸打磨后抛光，表面粗糙度小于 1.6 μm。模拟工况应力腐蚀开裂试验用试样在管体纵向截取 115mm×15mm×5mm 四点弯曲试样，经过 1200# 砂纸沿试样长边进行打磨抛光后，划痕均平行于长边方向，依照 NACE TM0177—2005 标准，采用四点弯曲方式进行应力加载，加载应力为 100% YS_{min}（758MPa），如图 1 所示。缝隙腐蚀试样在管体上截取 38mm×38mm×3mm 方片，用 1200# 砂纸打磨抛光后在片中心钻 φ10mm 圆孔，使用 0.3mm 的 15mm×15mm 聚四氟乙烯垫片，垫在方片之间以形成一定的缝隙，最后用和试样一样材质的钛合金螺栓将每种成分的钛合金缝隙腐蚀试片串起来，使用扭矩扳手统一施加 86cm·kgf 的扭矩将试片串上紧，组成缝隙腐蚀试样串，如图 2 所示。

图 1　钛合金四点弯曲应力腐蚀开裂试样形貌　　　　图 2　钛合金缝隙腐蚀试样

选取我国西部某油田典型腐蚀工况条件作为钛合金均匀腐蚀、应力腐蚀开裂（SCC）和缝隙腐蚀试验条件，模拟井深为 6000 米，井底温度为 160℃，试验时间为 720h，详见表 2。将按照表 2 的工况配成的模拟工况液体装入 Cortest 公司制造的 34.4MPa 高温高压釜内，试验前先通入高纯氮 10h 以上除氧，再装上试样并将高压釜密封，继续通入高纯氮除氧，然后再通入介质气体压力、升温至所需温度，开始计时试验。腐蚀试验后试样表面使用 TESCAN-VEGAⅡ 扫描电镜和 OXFORD-INCA350 型能谱仪进行观察和分析，腐蚀产物使用 D8 Advance 型 X 射线衍射仪进行定性分析。

表 2　钛合金抗腐蚀性能评价试验条件

条件	SCC	裂缝腐蚀
	外加应力 758 MPa	裂缝尺寸 0，0.3mm
总压力(MPa)	12(进口 N_2)	
温度(℃)	160	
H_2S 分压压力(MPa)	≤1.2	
CO_2 分压压力(MPa)	6	
试验时间(h)	720	
组分(mg/L)	CO_3^{2-}/0, HCO_3^-/189, OH^-/0, Cl^-/128000, SO_4^{2-}/430, Ca^{2+}/8310, Mg^{2+}/561, K^+/6620, Na^+/76500	
pH 值	2.5	

电化学试样在管体上截取 φ15mm×3mm 圆片并用环氧树脂封装在特制圆环中，工作面积为 1.76cm²。试验前，工作电极要用水砂纸逐级打磨至 1000#，然后用蒸馏水清洗、丙酮除油，再用酒精清洗制成试样。电化学测量采用美国 AMETEK 公司的 273A 恒电位仪、K0047 电解池辅助，恒温加热仪器为国产 CS501-3C 型恒温水浴加热器，电极选用大面积石墨惰性电极，参比电极为饱和甘汞参比电极（SCE），极化曲线测试时动电位扫描范围为开路电位-250~1600mV，动电位扫描速率为 0.5mV/s。电化学阻抗试验在开路电位测试后进行，扫描频率的范围为：0.01~10⁵Hz，信号幅值为±10mV，波形为正弦波。

2 结果与分析

2.1 均匀腐蚀及点蚀

依照 GB/T 18590 标准对不同钛合金试样在模拟工况条件下对均匀腐蚀性能进行计算，结果见表3。从表3中可以看出，在试验工况条件下5种钛合金均具有良好的耐蚀性能，年平均腐蚀速率均不超过 0.001mm/a，其中 2# 试样的腐蚀速率在实验室条件下基本为 0，依据 NACE 标准 RP-0775-91[20]对腐蚀程度的规定，在本试验条件之下所有钛合金均属于轻微腐蚀范畴。

表3 试样腐蚀速率计算结果

样品	测试前平均质量(g)	测试后质量(g)	质量损失(g)	样品表面积(mm²)	腐蚀速率(mm/a)
1#	4.8751	4.8748	0.0003	1075.38	0.0008
2#	4.8448	4.8448	0.0000	1068.034	0
3#	5.2134	5.2130	0.0003	1082.73	0.0007
4#	5.4921	5.4917	0.0004	1063.70	0.0009
5#	4.9849	4.9846	0.0003	1070.72	0.0007

对试样表面进行放大观察，发现 1# 试样的点蚀坑数量稍多且分布较为分散，点蚀坑面积较大且深度较深，如图3(a)和图3(b)所示；5# 试样的点蚀坑数量较多，点蚀坑面积小且深度浅，如图3(c)所示。2#、3# 和 4# 钛合金试样表面未发现明显点蚀形貌。从以上结果可以看出，试验的几种钛合金试样虽然具有良好的抗均匀腐蚀性能，但是在局部腐蚀方面 1# 和 5# 试样均发生不同程度的点蚀问题，其中 1# 钛合金试样的点蚀最为严重。

2.2 应力腐蚀开裂性能

对试验后的应力腐蚀开裂试样表面形貌如图4所示，从图中可以看出，在试验工况下5种钛合金试样经过 720h、758MPa 应力的加载下均未发生开裂，对试样表面进行观察也均未发现明显的裂纹，由此可以说明这几种钛合金试样在模拟工况条件下均具有良好的抗应力腐蚀开裂性能。

2.3 抗缝隙腐蚀性能

对试验后的不同试样缝隙腐蚀试样表面进行观察，结果如图5所示，试样片中的正方形区域为聚四氟乙烯垫片覆盖区域，此处缝隙大小为 0mm，从图5中可以看出 1# 试样无论是垫片下还是缝隙处均发生较为严重的腐蚀，腐蚀产物厚而并且致密，其中在 0.3mm 缝隙中腐蚀更为严重，试片周围有部分材料已经断裂缺失。2#~4# 试样在垫片和 0.3mm 缝隙处均未发现明显的腐蚀形貌，5# 试样在垫片下部分区域有腐蚀痕迹，在 0.3mm 缝隙处有多处疑似腐蚀痕迹。

(a) 1#

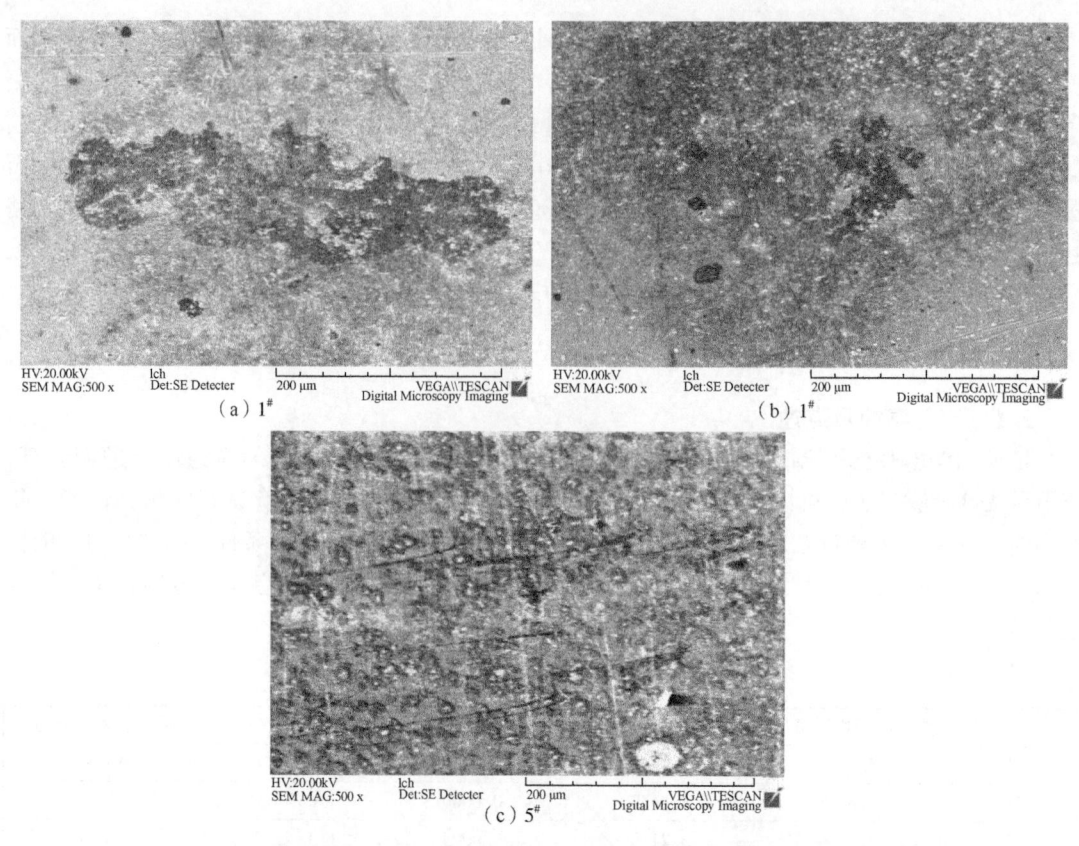

(b) 1#

(c) 5#

图 3 试验后几种钛合金试样表面点蚀形貌

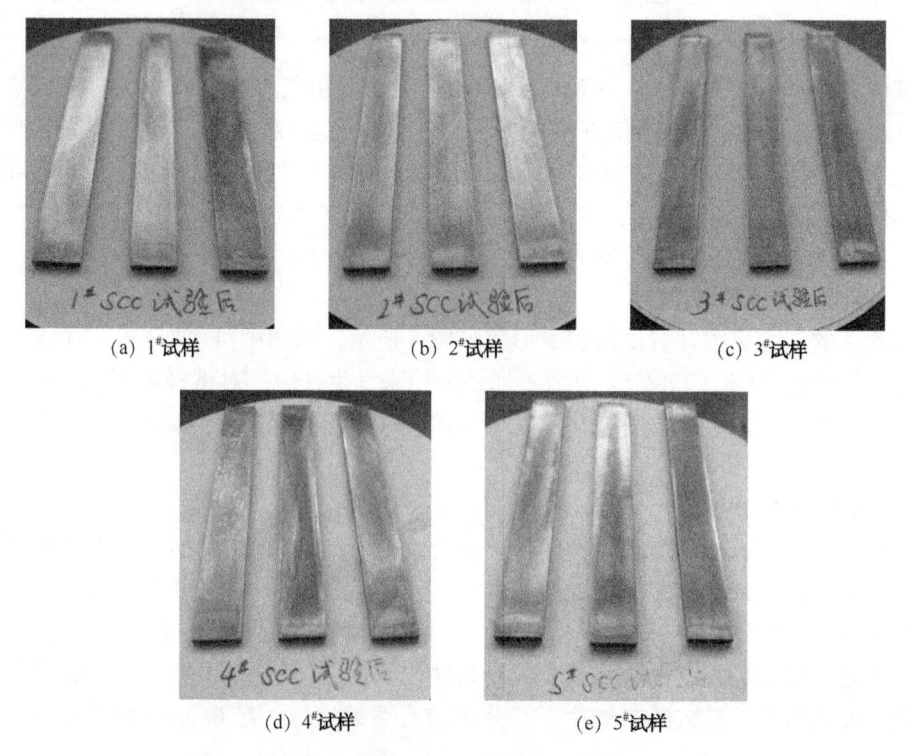

(a) 1#试样

(b) 2#试样

(c) 3#试样

(d) 4#试样

(e) 5#试样

图 4 应力腐蚀开裂试验后不同钛合金表面宏观形貌

(a) 1#试样　　(b) 2#试样　　(c) 3#试样

(d) 4#试样　　(e) 5#试样

图 5　缝隙腐蚀试验后不同钛合金 0.3mm 缝隙试样表面形貌

对发生腐蚀问题的 1#试样和 5#试样在不同缝隙大小下的腐蚀形貌进行分析,结果如图 6 所示,从图 6(a)和图 6(b)中可以看出,1#试样随着缝隙尺寸的增大,腐蚀越来越严重,腐蚀产物也迅速增多,在聚四氟乙烯垫片的边缘出现有较深的腐蚀坑和微裂纹,在垫片下部(0mm 缝隙)腐蚀情况有所减轻,对 1#试样表面的腐蚀产物进行能谱分析,可知主要的腐蚀产物为氧化铝和氧化钛,如图 7 所示和见表 4;而 5#试样正好相反,在聚四氟乙烯垫片的边缘产生出轻微的缝隙腐蚀形貌,在沿晶部位腐蚀较为明显,已经形成了明显的腐蚀坑,如图 6(d)所示,但是在垫片下部(0mm 缝隙处)5#试样的缝隙腐蚀较为严重,出现了较深的蚀坑和大量的微裂纹,如图 6(c)所示,对蚀坑内的腐蚀产物进行能谱分析表明较浅色相的主要成分为钛的氧化物,如图 8 所示和见表 5。

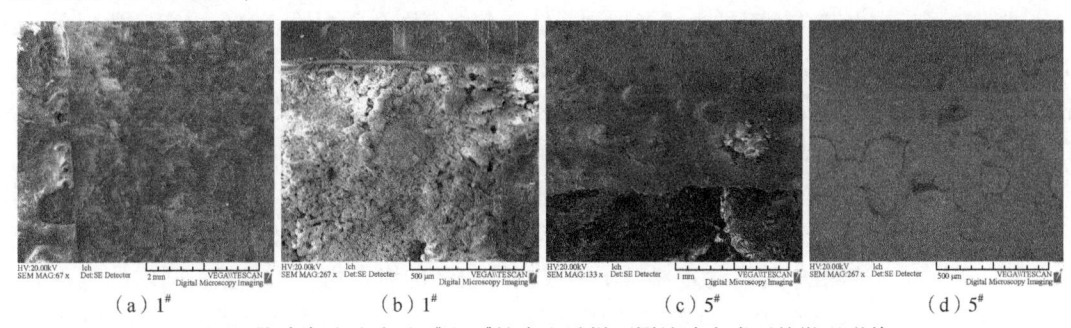

(a) 1#　　(b) 1#　　(c) 5#　　(d) 5#

图 6　缝隙腐蚀试验后 1#和 5#钛合金试样不同缝隙大小下的微观形貌

表 4　缝隙腐蚀试验后 1#钛合金试样腐蚀产物能谱试验结果

元素	质量分数(%)	原子百分数(%)	元素	质量分数(%)	原子百分数(%)
O	53.16	73.80	V	3.96	1.73
Al	12.76	10.50	合计	100.00	
Ti	30.11	13.96			

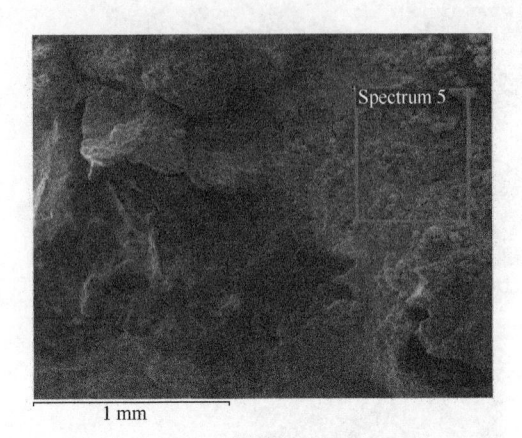

图7　缝隙腐蚀试验后 1# 钛合金试样腐蚀产物形貌

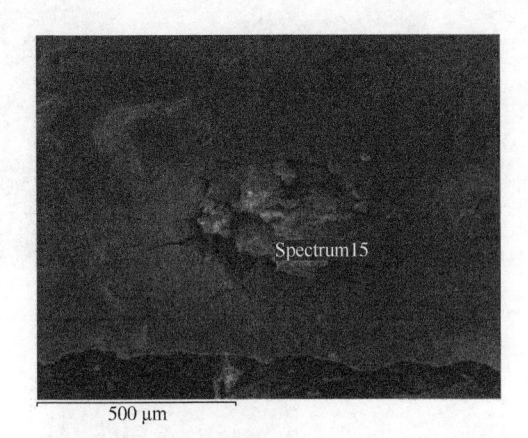

图8　缝隙腐蚀试验后 5# 钛合金试样腐蚀产物形貌

表5　缝隙腐蚀试验后 5# 钛合金试样腐蚀产物能谱试验结果

元素	质量分数(%)	原子百分数(%)	元素	质量分数(%)	原子百分数(%)
O	24.98	59.32	V	3.29	2.45
Al	1.12	1.58	Mo	0.89	0.35
Cl	0.37	0.40	合计	75.90	
Ti	45.26	35.90			

3　分析与讨论

钛合金耐蚀的本质是由于钛是一种热力学不稳定的元素，标准电极电位只有 -1.63V（标准氢电极 HSE），因此使得钛及钛合金在空气甚至水中极易形成一种连续、致密同时又非常薄的表面氧化膜（TiO_2），氧化膜覆盖在钛合金的表面使得基体作为电极进行活性溶解的面积大为减少，或阻碍了反应电荷传输而减少或者抑制了钛合金在腐蚀介质中的溶解，出现钝化现象。钛的钝化膜又具有非常好的自愈性，当其钝化膜遭到破坏时，能够迅速修复，弥合形成新的保护膜。这个保护性的氧化层/钝化层在很大范围的电位和 pH 值内都保持稳定，从而保证钛合金在盐水、卤水、海水、氧化性酸、温和还原性酸和碱中具有很强的耐腐蚀性。

但是钛合金石油管在使用过程中，钛合金管之间依靠螺纹进行连接，当内外螺纹上扣后，会存在大量细小的连接缝隙，考虑到在缝隙的小空间内，随着钛逐渐被溶解氧化为钝态氧化物，进一步与溶液中的高含量氯离子形成能量较高的络合物，当缝隙空间内的氧逐渐耗尽，钝化膜的溶解速度增大，会造成钛离子数量的增多，当缝隙内的腐蚀产物（如钛离子等）的数量到达一定的浓度时，会发生如下的水解反应：

$$Ti^{3+} + 3H_2O \longrightarrow Ti(OH)_3 + 3H^+ + 3e^-$$

或

$$Ti^{4+} + 4H_2O \longrightarrow Ti(OH)_4 + 4H^+ + 4e^-$$

由此造成缝隙空间内 pH 值的迅速降低[21]，因此需要对模拟工况介质环境中不同 pH 值条件的钛合金腐蚀行为和腐蚀机理进行研究，定量的分析钛合金耐腐蚀性能的影响因素和特征。

3.1 开路电位行为

开路电位可以初步反映出电极/溶液界面发生的电化学行为。图9为5种钛合金试样在模拟实际工况介质环境不同条件下开路电位随着浸泡时间的变化曲线。从图9中可以看出,在室温(23℃)和pH值为3条件下,所有钛合金试样的开路电位在浸泡开始的时候均迅速增大,并朝正方向移动,随着浸泡时间的增加所有钛合金试样的开路电位均逐步稳定,表明在这些试验钛合金的表面钝化膜均迅速形成并稳定。其中4#钛合金的最终开路电位最高,稳定在大约-226mV,说明4#钛合金的钝化膜最稳定,2#及5#钛合金最终开路电位稍低,分别大约稳定在-278mV和-286mV,1#及3#钛合金最终开路电位最低,分别为-303mV和-301mV。

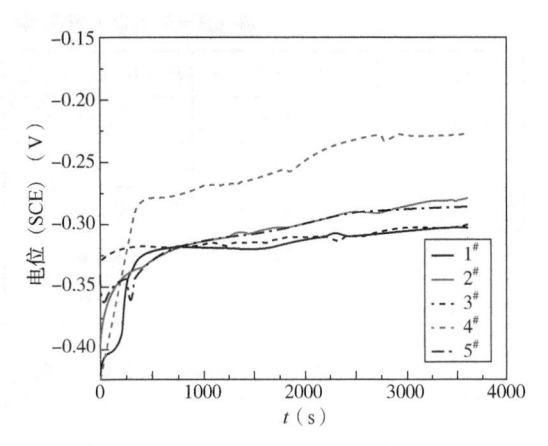

图9 模拟工况介质环境23℃,pH=3条件下5种钛合金试样开路电位随时间的变化

3.2 电位极化曲线

图10为不同钛合金试样在模拟工况介质环境中不同pH值条件时的极化曲线测试结果。由图10可知,5种钛合金试样在阳极极化曲线中当电位超过0V时,腐蚀电流几乎不发生明显变化,存在明显的钝化区。表6为pH=3试验工况条件下不同钛合金试样拟合的电化学参数,从表6中可以得到,当pH=3试验的5种钛合金试样中,2#钛合金试样的自腐蚀电位最高为-262mV,3#钛合金试样的自腐蚀电位最低且自腐蚀电流最大,但5种钛合金试样的自腐蚀电流均相差不大,并且极化电流曲线均从阳极极化直接进入钝化区,说明这5种钛合金试样在pH=3时的耐腐蚀性能相当,拟合的年腐蚀速率均不大于0.003mm/a。但是如果将环境pH值降低到1.5时,不同钛合金拟合的电化学参数见表7,从表7中可以看出,虽然五种钛合金均可以保持钝化区,但是所有合金的自腐蚀电位均发生降低,且自腐蚀电流均有明显的升高,说明随着pH值的降低,钛合金的耐腐蚀性能均发生不同程度的下降。横向对比可以发现1#和5#钛合金的自腐蚀电流增大最为显著,比其他3个合金要高出1个数量级,说明1#和5#钛合金的耐蚀性下降最快,当这两种钛合金处于缝隙空间内低pH值环境中时,钝化膜的溶解速度较高,特别是1# Ti-6Al-4V钛合金,耐蚀性能最低,这与前面腐蚀实验结果保持一致。

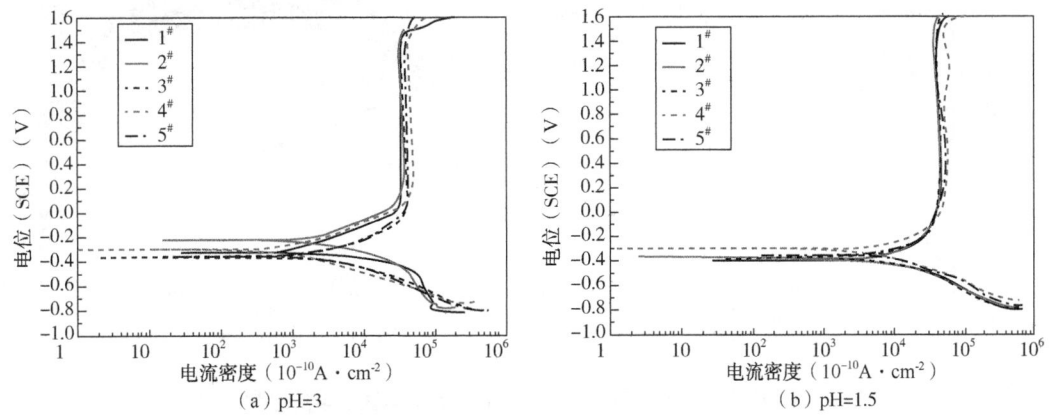

(a) pH=3

(b) pH=1.5

图10 模拟工况介质环境中在不同条件下5种钛合金试样的极化曲线

表6 在pH=3试验工况条件下不同钛合金极化曲线参数拟合结果

样品	E_{corr}(mV)	I_{corr}(10^{-7}A·cm^{-2})	B_a(mV·dec^{-1})	B_c(mV·dec^{-1})	腐蚀速率(mm/a)
1#	−359	1.058	212.29	−61.056	0.00124
2#	−262	1.878	222.43	−137.67	0.00221
3#	−406	2.209	228.55	−128.59	0.00259
4#	−348	8.267	221.53	−153.26	0.0009
5#	−399	1.838	227.57	−130.7	0.0021

表7 在pH=1.5试验工况条件下不同钛合金极化曲线参数拟合结果

样品	E_{corr}(mV)	I_{corr}(10^{-7}A·cm^{-2})	B_a(mV·dec^{-1})	B_c(mV·dec^{-1})	腐蚀速率(mm/a)
1#	−356	2.219	1408.8	−324.06	0.026101
2#	−335	8.454	414.27	−190.15	0.01723
3#	−341	9.013	422.7	−190.15	0.01060
4#	−465	7.002	315.56	−218.5	0.00823
5#	−312	1.205	517.36	−218.05	0.01418

3.3 交流阻抗特性

为了进一步研究5种钛合金在模拟工况介质环境中的电化学特性,对同一工况介质环境中不同pH值条件下的电化学阻抗(EIS)进行了测试,测试及拟合结果如图11和图12所示。从图11(a)中的Nyquist图可以看出,5种钛合金材料具有相似的容抗弧特征,图11(b)中的Bode图显示相位角在从低频到高频只有1个较宽的明显峰,说明该阻抗具有一个时间常数,并且5种材料的最大相位角均超过了80°,因此具有典型的电容特征,符合致密钝化膜的等效电路[22,23],如图13所示,在等效电路中,R_s为溶液电阻,表示实验体系所使用的溶液的电阻阻值,相位角常数CPE_p表示溶液和合金表面形成的双电层的电容,R_p表示合金表面钝化膜的电阻。表8为试验工况pH=3条件下不同钛合金阻抗拟合结果,可以看出,2#钛合金的钝化膜电阻最高,为1011.3kΩ·cm^2,其次为1#钛合金,说明这两种合金在试验工况pH=3条件下均具有较稳定的钝化膜,其余合金的钝化膜电阻相差不大。

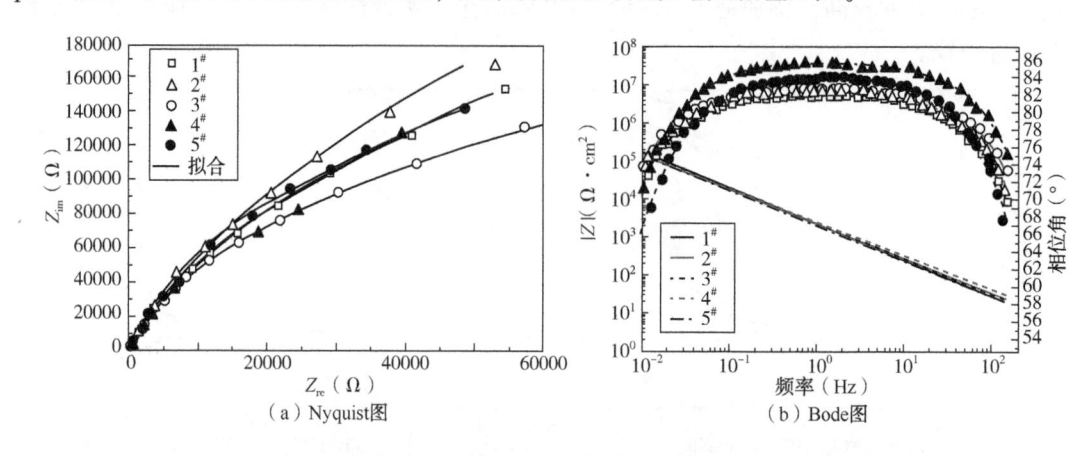

(a) Nyquist图　　(b) Bode图

图11 模拟工况介质环境中在pH=3条件下不同钛合金的电化学阻抗

(a) Nyquist图　　　　　　　　　　　　　　(b) Bode图

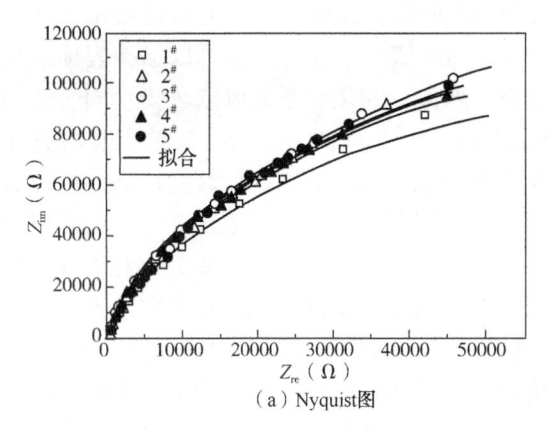

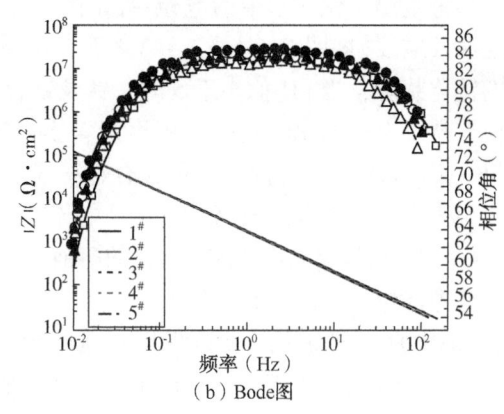

图 12　模拟工况介质环境中在 pH=1.5 条件下不同钛合金的电化学阻抗

但是当这些钛合金处于缝隙条件下的较低 pH 值时，从图 12 可以看出，容抗弧特征和相位角等阻抗特征变化不大，因此等效电路也基本相同，但是通过对模拟工况介质中 pH=1.5 条件下不同钛合金阻抗结果进行拟合发现（表 9），钛合金的钝化膜电阻 R_p 均随着酸性的增强而显著下降，其中 1# 钛合金的钝化膜电阻 R_p 从 pH=3 时的 816.04kΩ·cm² 下降到 pH=1.5 时的 255.56kΩ·cm²，在 5 种中最低，说明钝化膜的溶解程度最高，5# 钛合金的钝化膜电阻 R_p 值也从 587.85kΩ·cm² 下降到 309.75kΩ·cm²，仅次于 1# 钛合金，结合前面的腐蚀试验结果，可以说明这两种钛合金的耐蚀性受 pH 值影响最大，而 2# 钛合金虽然钝化膜电阻 R_p 降幅较多，但依然高达 623.58kΩcm²，表明钝化膜具有较高的稳定性，耐蚀性能最强（图 13）。

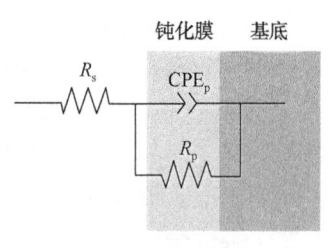

图 13　钛合金阻抗分析的等效电路

表 8　试验工况 pH=3 条件下不同钛合金阻抗拟合结果

样品	$R_s(\Omega \cdot cm^2)$	$R_p(k\Omega \cdot cm^2)$	$Q_p(10^{-5}F \cdot cm^{-2})$	n
1#	6.106±0.24	816.04	7.4396	0.91268
2#	4.534±0.25	1011.3	7.1233	0.91899
3#	6.234±0.10	453.97	8.342	0.93742
4#	7.777±0.10	484.22	1.1496	0.93383
5#	3.236±0.10	587.85	8.8634	0.95304

表 9　试验工况 pH=1.5 条件下不同钛合金材料阻抗拟合结果

样品	$R_s(\Omega \cdot cm^2)$	$R_p(k\Omega \cdot cm^2)$	$Q_p(10^{-4}F \cdot cm^{-2})$	n
1#	2.798	255.56	1.1287	0.92969
2#	2.887	623.58	1.1982	0.93543
3#	5.13	354.61	1.0132	0.9297
4#	3.714	393.75	1.1493	0.94422
5#	3.3	309.75	1.13	0.94551

参考室温下钛在水中的电位—pH图[24]，如图14所示，在图14(a)上可以看出，在水的还原反应线(即图14中的B线)之下，钛在23℃、pH值大于5时，都可以保持表面氧化膜的钝化状态，当pH值小于5时，钛及钛合金是否能保持钝化膜不溶解取决于不同合金的电位，从以上的电化学性能分析可以看出，当pH=3时，5种钛合金的自腐蚀电位均处于-406~-262mV之间，此时5种钛合金材料均处于图14(b)中的TiO_2钝化区，从表8可以看出合金表面钝化膜电阻较高，自腐蚀电流较低，均具有较好的耐蚀性。赵永新等人的研究表明[21]，TA2纯钛材料在NaCl+HCl的溶液环境中，无论原溶液的pH值是多少，发生缝隙腐蚀后缝隙部位的溶液pH值均降到0.9~1.0之间，只是降到此值的时间长短不同。因此当环境(缝隙中)的pH值进一步降低时，钛合金表面氧化膜TiO_2由钝化状态逐渐向Ti^{3+}和Ti^{2+}活化溶解区域转变，在23℃，pH=1时，钝化膜活化溶解的临界电位已经升到-350mV附近，已经非常接近1#钛合金的稳态开路电位，并且从图14(b)上可以看出，随着温度的升高，钝化膜活化溶解区域逐渐扩大，不同钛合金钝化膜的溶解程度虽然不同，但均发生不同程度的溶解，造成自腐蚀电流的升高和钝化膜电阻的下降，在本试验所测试的5种钛合金中，1# Ti-6Al-4V钛合金耐蚀性能下降最快，引发了局部腐蚀以及严重的缝隙腐蚀，5# Ti-5.5Al-4.5V-2Zr-1Mo合金次之，发生了轻微的点蚀和缝隙腐蚀，但随着时间的延长，腐蚀会不断加剧，其余3种合金在试验工况下未发生较明显的腐蚀问题，因此在不同油气开发工况中使用钛合金时，需谨慎选择钛合金，并对缝隙腐蚀问题进行特别的关注。

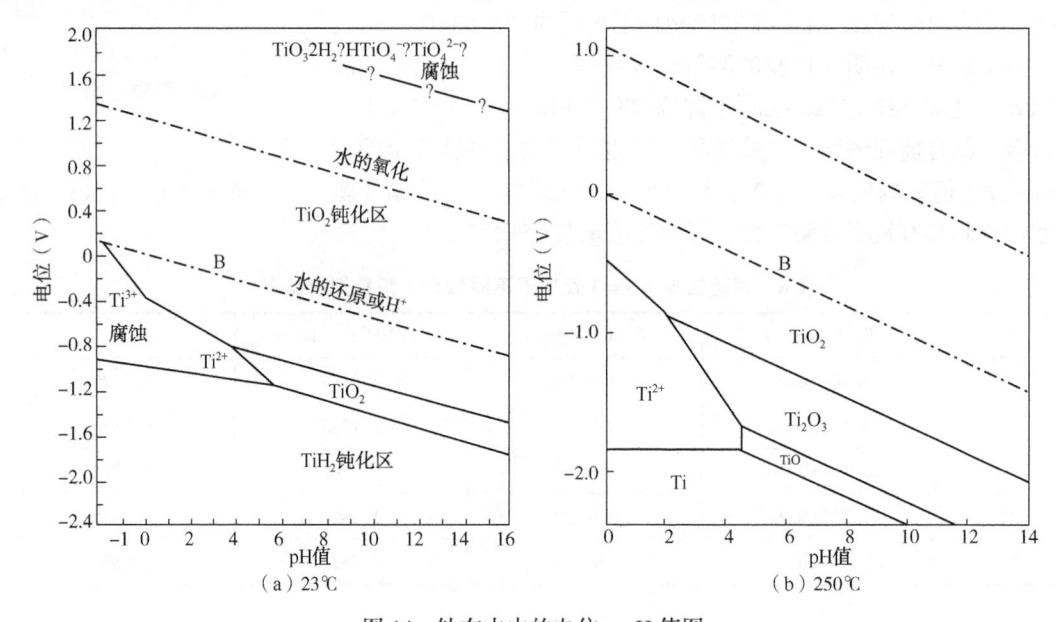

图14 钛在水中的电位—pH值图

4 结论

(1)在本试验所模拟的油气开发工况条件下，Ti-6Al-4V、Ti-6Al-4V-0.1Ru、Ti-6Al-2Sn-4Zr-6Mo、Ti-3Al-8V-6Cr-4Zr-4Mo和Ti-5.5Al-4.5V-2Zr-1Mo 5种钛合金均具有良好的抗均匀腐蚀性能和良好的抗应力腐蚀开裂性能，年平均腐蚀速率均不超过0.001mm/a。

(2)在试验条件下，Ti-6Al-4V-0.1Ru、Ti-6Al-2Sn-4Zr-6Mo和Ti-3Al-8V-6Cr-

4Zr-4Mo 钛合金具有良好的抗点蚀和缝隙腐蚀性能，但是 Ti-6Al-4V 和 Ti-5.5Al-4.5V-2Zr-1Mo 钛合金均发生不同程度的点蚀和缝隙腐蚀问题，其中 Ti-6Al-4V 钛合金的点蚀和缝隙腐蚀最为严重，出现了较深的蚀坑和大量的微裂纹，腐蚀产物主要成分为钛的氧化物。

（3）在工况条件温度下，试验用 5 种钛合金腐蚀反应均为阳极控制过程，随着 pH 值的降低，钛合金均发生自腐蚀电位降低，极化电阻减小，腐蚀电流增大，耐腐蚀性能下降，其中 Ti-6Al-4V 钛合金的钝化膜溶解最为严重，缝隙内较低的 pH 值和较高的温度引发了局部腐蚀的发生以及严重的缝隙腐蚀。

（4）在不同油气开发工况中使用钛合金时，需谨慎选择钛合金，并对缝隙腐蚀问题进行特别的防护。

参 考 文 献

[1] 谷坛, 霍绍全, 李峰, 等. 石油与天然气化工[J], 2008, 37(S1)：63.
[2] Shadravan A, Amani M. Energy Science and Technology[J], 2012, 4(2)：36.
[3] 叶登胜, 任勇, 管彬, 等. 天然气工业[J], 2009, 29(3)：77.
[4] 张水昌, 张宝民, 李本亮, 等. 石油勘探与开发[J], 2011, 38(1)：1.
[5] 马永生, 蔡勋育, 李国雄. 地质学报[J], 2005, 79(6)：858.
[6] 张桂林. 石油钻探技术[J], 2009, 37(6)：6.
[7] Schutz R W, Watkins H B. Materials Science and Engineering A[J], 1998, 243(1-2)：305.
[8] Schutz R W. 24th Offshore Technology Conference[C]. Houston：Omnipress, 1992：319.
[9] Kane R D, Craig B, Venkatesh A. NACE Corrosion 2009 Conference & EXPO[C]. Houston：Omnipress, 2009：09078.
[10] Kane R D, Srinivasan S, Craig B et al. NACE Corrosion 2015 Conference & EXPO[C]. Houston, TX：NACE International, 2015：5512.
[11] Schutz R W. Platinum Metals Review[J], 1996, 40(2)：54.
[12] Schutz R W, Poter R L, Horrigan J M. Corrosion[J], 2000, 56：1170.
[13] Schutz R W, Watkins H B. Materials Science and Engineering A[J], 1998, 243：305.
[14] Schutz R W, Jena B C. NACE Corrosion 2015 Conference & EXPO[C]. Houston, TX：NACE International, 2015：5794.
[15] 刘强, 宋生印, 李德君, 等. 石油矿场机械[J], 2014, 43(12)：89.
[16] 高文平, 吕祥鸿, 谢俊峰, 等. 稀有金属材料与工程[J], 2018, 47(1)：151.
[17] 李永林, 朱宝辉, 王培军, 等. 钛工业进展[J], 2013, 30(2)：28.
[18] 吴欣袁, 张恒, 徐学军, 等. 石油化工应用[J], 2016, 35(11)：105.
[19] 史雪枝, 周小虎, 乔智国. 石油机械[J]. 2016, 44(8)：11.
[20] NACE Standard RP0775-99[S], 1999.
[21] 赵永新, 姚禄安, 甘复兴, 中国腐蚀与防护学报[J], 1993, 10(3)：252.
[22] Wang Z B, Hu H X, Liu C B et al. Electrochimica Acta[J], 2014, 135：526.
[23] Li Yu, Xu Jian. Electrochimica Acta[J], 2017, 233：151.
[24] Schutz R W, Xiao M, Bednarowicz T A. 47th NACE Annual Conference[C]. Houston：Omnipress, 1992：51.

本论文原发表于《稀有金属材料与工程》2020 年第 49 卷第 4 期。

钛合金油套管抗挤毁性能计算与实验

刘 强[1] 申照熙[1] 李东凤[1] 张春霞[2] 祝国川[1] 宋生印[1]

(1. 中国石油集团石油管工程技术研究院·石油管材及装备材料服役行为与结构安全国家重点实验室；2. 宝钢股份钢管条钢事业部)

摘　要：钛合金材料由于具有高的比强度、低弹性模量，优异的韧性、疲劳性能和耐蚀性，已经成为严酷工况环境下油井管和海洋开发工具的热门候选材料，但采用API、ISO的标准体系中的油套管挤毁计算方法得出的钛合金油套管抗挤毁强度缺乏验证。为此，在考虑制造缺陷的条件下，通过对比ISO 10400标准中计算方法、外压强度挤毁准则计算方法和有限元模拟3种不同方法下计算出的钛合金油套管的抗挤毁强度，并与同等条件下钢制油套管抗挤毁强度对比，得出了钛合金油套管的抗挤毁强度变化规律，最后选取了4种规格的钛合金油套管进行了实物外挤毁试验验证。研究结果表明：(1)钛合金油套管抗挤毁强度均随着径厚比(D/t)的升高而下降，在D/t值较低时，钛合金油套管和钢制油套管的抗挤毁强度相差不大，当D/t值较高时，钛合金油套管抗挤毁强度显著低于钢制油套管；(2)强度挤毁准则方法计算出来的钛合金油井管抗挤毁强度和实际试验值较为接近，同时在$D/t>15$时，计算结果乘以0.9的系数，具有较好的安全性。结论认为，研究结果为钛合金油套管的设计、使用和管理提供重要的技术参考。

关键词：高温高压；天然气开发；钛合金油套管；径厚比；抗挤毁；力学计算；有限元分析；实物试验

随着我国油气开发的不断深入，深井超深井高温、高压、高腐蚀的"三高"井和大位移水平井等非常规油气资源地开发对钻完井石油管材的要求不断提高[1,2]。油管和套管作为油气资源的主要输送通道及防护屏蔽，在井下不仅要经受高温高压带来的复合载荷作用，还要受到服役工况中的H_2S、CO_2、高浓度氯离子甚至单质硫的腐蚀作用[3,4]，一旦发生失效事故，不仅会造成深层油气开发的事故，带来巨大的经济损失，严重的还会造成生命危险和环境破坏，因此，油井管的选材、安全可靠性和使用寿命对石油工业影响极其重大。钛合金由于具有高强度、低密度，低的弹性模量，优异的抗疲劳性能、耐腐蚀性和耐高温性能，已经成为高端石油管材料研究的热点方向[5,6]。

国际上在20世纪末就开始对钛合金用于油气开发的可行性和适用性开展了研究，对钛

基金项目：国家科技重大专项"深井超深井高效快速钻井技术及装备"（编号：2016ZX05020）；陕西省创新能力支撑计划"油气开发用低成本高耐蚀钛合金油井管材料开发"（编号：2018KJXX-046）；中国石油天然气集团公司基础研究和战略储备技术研究基金项目（编号：2017Z-05）。

作者简介：刘强，1983年生，高级工程师，博士。地址：陕西省西安市锦业二路89号管研院503室(710077)，E-mail：liuqiang030@cnpc.com.cn。

合金在油气工况下的性能进行详细的评价,得出钛合金材料在油气开发环境中具有巨大的技术优势和应用前景[5,7]。美国、日本等国家开发出了多种钛合金材料的油井管,例如钛合金油管、套管、钛合金钻杆、钛连续管和海洋钻井隔水管等系列化产品,在勘探开发领域开始工业化应用[7-10],美国RMI公司成功地研制了热旋转—压力穿孔管材轧制工艺,开发出钛合金油管、套管和海洋立管等产品,于1999年起在美国多个区块的油气井、墨西哥湾的Oryx海王星钻井项目和Mobile Bay Field(莫比尔湾油田)热酸性油气井成功应用[11,12]。Chevron公司研发的钛合金油套管材料,已在墨西哥湾一些高压高温、超高压高温井进行了应用[13]。中国在近十年才开始进行钛合金用于油气开发的研究,中国石油石油管工程技术研究院(以下简称管研院)对钛合金材料用于油气开发的可行性进行了综合分析,证明钛合金管材在油气开发领域有着良好的应用前景[6],同时对钛合金石油管的选材和适用性进行了深入的研究[14],攻克了一些制约钛合金油套管应用的技术瓶颈[15],并制定了相关的标准[16]。东方钽业等制造厂家优化了钛合金管材的加工制备技术[17];天津钢管集团、攀钢集团成都钢钒有限公司等企业均试制出了钛合金油管产品[18,19]。

我国油气资源的开发具有"三高"特点。其中塔里木盆地的气井服役工况已经达到超深(>6000m,局部达到8000m)、超高温(>160℃,局部190℃)、超高压(井口压力达到110~130MPa)、高腐蚀(高含Cl^-达到$15×10^4$mg/L,CO_2分压超过1MPa),特别是最新在塔里木盆地克拉苏构造带西部发现的超深、高压、高产、优质整装达$1000×10^8m^3$的博孜9凝析气藏[20],完钻深度达到了近8000m,井底压力达到了近140MPa,在此高温高压的服役工况下使用钛合金油套管完井是一种新的解决方案,但是对油套管的抗外挤毁提出了较高的要求。然而在目前石油工业中,对石油管的评价体系和指标都是基于钢的弹性模量计算的,而钛合金材料无论是弹性模量还是泊松比都与钢铁材料有较大差别,以油井管最为关键的抗外挤毁强度为例,对于钛合金管材没有任何的方法和指标可参考,钛合金油套管的真实抗挤毁能力与API、ISO的标准体系中的油套管挤毁计算方法和指标的差距也不得而知。美国Bob Hargrave等[13]在个别外径的Ti-3Al-8V-6Cr-4Zr-4Mo(Beta-C)钛合金管上进行的实物挤毁试验,数据表明Ti-3Al-8V-6Cr-4Zr-4Mo钛合金管的抗挤毁强度只能达到API 5C3标准[21]理论计算值的90%,但该研究数据较少且计算方法有待商榷,难以在实际设计和应用中借鉴。因此,笔者旨在考虑制造缺陷的条件下,通过对比国际标准中外挤毁强度计算方法、外压强度挤毁准则计算方法和使用有限元模拟计算3种不同方法下得出的钛合金油套管抗挤毁强度,并和同等条件下钢制油套管抗挤毁强度进行了对比,研究钛合金油套管的抗挤毁强度变化规律,并采用实物外挤毁试验验证的方法,以期得出钛合金油套管外挤毁性能变化规律和最精确的预测方法,为钛合金油套管的设计和应用提供参考。

1 外挤毁强度计算方法

1.1 国际标准计算方法

目前国际标准中针对油套管外压挤毁强度的计算方法主要由API BULL 5C3中和ISO/TR 10400中给出。其中API BULL 5C3中的抗挤强度公式的技术基础都是20世纪60年代早期发展起来的,多年来已证明具有较多的局限性,包括:对小D/t比(外径/壁厚比)计算精度不高,不能适用于非API钢级和材料(如特殊耐酸性环境钢级管子)等,不能体现不同生产制造工艺和热处理对管材抗挤毁强度的影响等,而ISO/TR 10400中给出了适用于非API产品抗挤毁强度的计算方法,因此在笔者的计算中采用标准ISO/TR 10400的附录H的方法[22]，即 Klever-Tamano 极限状态公式：

$$p_{ult} = \frac{(p_{e\,ult}+p_{y\,ult}) - [(p_{e\,ult}-p_{y\,ult})^2 + 4p_{e\,ult}p_{y\,ult}Ht_{ult}]^{1/2}}{2(1-Ht_{ult})} \quad (1)$$

其中

$$p_{e\,ult} = \frac{k_{e\,uls}2E}{(1-\nu^2)(D_{ave}/t_{c\,ave})(D_{ave}/t_{c\,ave}-1)^2}$$

$$p_{y\,ult} = k_{y\,uls}2f_y \frac{t_{c\,ave}}{D_{ave}}\left(1 + \frac{t_{c\,ave}}{2D_{ave}}\right)$$

$$Ht_{ult} = 0.127ov + 0.0039ec - 0.440\left(\frac{rs}{f_y}\right) + h_n 且 Ht_{ult} \geq 0$$

$$ec = \frac{100(t_{c\,max}-t_{c\,min})}{t_{c\,ave}}$$

$$ov = \frac{100(D_{max}-D_{min})}{D_{ave}}$$

式中：因子 $k_{e\,uls}$ 和 $k_{y\,uls}$ 分别为不同状态油套管产品挤毁试验的经验参数，目的是保证输入每个变量，实际挤毁压力与计算挤毁压力有良好的拟合性；E 为弹性模量；ν 为泊松比；Ht_{ult} 为折减因子；rs/f_y 为残余应力相关参数，当管材为具有很高压缩残余应力（$rs/f_y < -0.5$）的壁厚很小的管子[$\lg(p_y/p_e)>0.4$]时，式（1）不适用；h_n 为由挤毁试验数据得到的经验系数。对于计算用的钛合金管材应力—应变曲线（SSCs）屈服明显，因此不必进行修正（$h_n = 0$）；ec 为壁厚不均度；ov 为椭圆度。

1.2 外压强度挤毁准则计算方法

中国石油集团石油管工程技术研究院对多年来积累的油套管挤毁实物实验数据进行分析比对表明，油套管材料在纯外压条件下挤毁时，均是在外压到达极限值时发生突然的失效，这是由于实际工程中的油套管外形并不是理想的纯圆形，导致非圆的油套管在圆周方向上的环向应力并不是不均匀分布，即有附加弯矩效应，当外部压力不断地增加时油套管在最大压缩环向应力处达到屈服；当屈服达到极限值时，油套管由于强度承载力不足而失效，导致发生强度或者失稳挤毁，因此建立了在外压作用下的强度挤毁准则[23]。该准则计算方法不区分弹性挤毁或塑性挤毁，即：

$$P = \frac{N}{R(1+e)} = \frac{-3(e-e_0)(C+\rho D)}{(1+e)(Re+B)(R-t/2-3eR)} \quad (2)$$

其中

$$e = \frac{-K + \sqrt{K^2 - 4H(D-t\mu\sigma_\alpha+tE)}}{6R^2(D-t\mu\sigma_\alpha+tE)}$$

$$A = F\left(\frac{t\mu}{2R}-1\right)(\rho-t/2)\left(t+\rho\ln\frac{2\rho-t}{2\rho+t}\right)$$

$$B = \frac{\mu t^2}{12R}-\left(\frac{a^2 b\ln\frac{b}{a}}{b^2-a^2}-\frac{a}{2}\right)$$

$$C = t(E+\sigma_{\theta 2}-\mu\sigma_\alpha)\left(\rho-\frac{t}{2}\right)$$

$$D = F(\rho - t/2)\ln\frac{2\rho - t}{2\rho + t}$$

$$F = E + \sigma_{\theta 2} - \mu\sigma_{\alpha}$$

$$H = CB - 3RDBe_0 + t(\mu\sigma_\alpha - E)(R - t/2)B$$

$$K = 3AR + CR - 3RDe_0R + 3RDB + t(\mu\sigma_\alpha - E)(R - t/2)R - 3RBt(\mu\sigma_\alpha - E)$$

具体参数定义和公式推导过程中见参考文献[22]。

1.3 有限元模拟计算方法

利用 ANSYS Workbench 有限元分析软件，建立有限元三维实体油套管模型，为了避免端部效应对仿真分析结果的影响，所建立的模型长度/外径比均大于 10，将真实应力—应变带入材料模型，为了和实物评价试验条件相一致，在所建立油套管模型的两端施加固定约束边界条件，在计算模型外表面施加外压力，如图 1 所示。基于非线性屈曲分析理论[24]，通过采用有限元逐步增加载荷来得到结构模型发生失稳时的临界载荷分析模型，分析非均匀载荷下钛合金油套管的抗外挤能力，如图 2 所示。

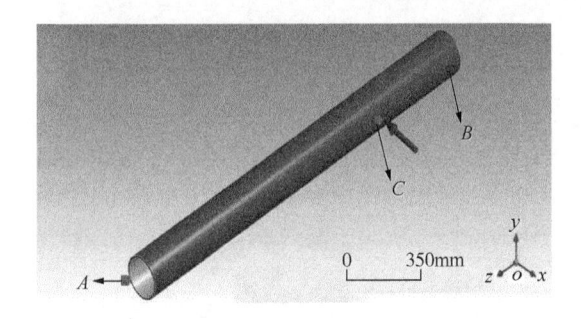

图 1 加载了边界条件和载荷的计算模型图

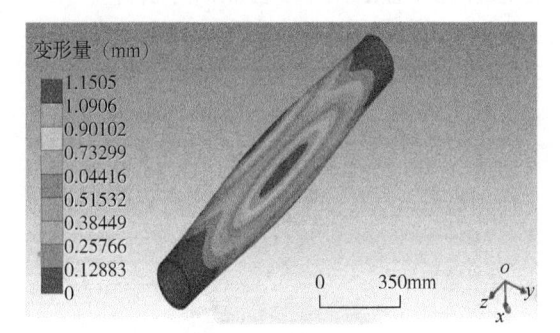

图 2 均匀载荷下钛合金油套管抗外挤毁非线性屈曲变形图

2 钛合金外挤毁强度对比分析

本文在所有抗挤毁计算中均假设管子均不受轴向载荷的影响，外径及壁厚均采用名义值，外径椭圆度设为 0.1%，无残余应力，壁厚均匀，屈服应力为 110ksi 强度级别的最小名义屈服强度即为 758MPa，钢制油套管材料的弹性模量取 207GPa，泊松比取 0.33，钛合金的弹性模量取 110GPa，泊松比取 0.3。从 API Spec 5CT 中选取典型外径从 73.02~339.72mm 的 28 个规格油套管进行计算和对比分析，选取的规格见表 1。

表 1 从 API Spec 5CT 中所选取的典型计算规格表

外径(mm)	壁厚(mm)	D/t	外径(mm)	壁厚(mm)	D/t
73.02	5.51	13.25	139.7	9.17	15.23
73.02	7.01	10.42	139.7	10.54	13.25
73.02	7.82	9.34	177.8	9.19	19.35
88.9	5.49	16.19	177.8	10.36	17.16
88.9	6.45	13.78	177.8	11.51	15.45
88.9	7.34	12.11	193.67	9.53	20.32

续表

外径(mm)	壁厚(mm)	D/t	外径(mm)	壁厚(mm)	D/t
88.9	9.52	9.34	193.67	10.92	17.74
101.6	5.74	17.70	244.47	11.05	22.12
101.6	6.65	15.28	244.47	11.99	20.39
101.6	8.38	12.12	273.05	11.43	23.89
114.3	6.35	18.00	273.05	12.57	21.72
114.3	7.37	15.51	273.05	13.84	19.73
114.3	8.56	13.35	339.72	12.19	27.87
139.7	7.72	18.10	339.72	13.06	26.01

依据标准 ISO/TR 10400 中 Klever-Tamano 极限状态公式方法计算、外压强度挤毁准则方法计算和有限元非线性屈曲分析计算结果分别如图 3(a) 至图 3(c) 所示。

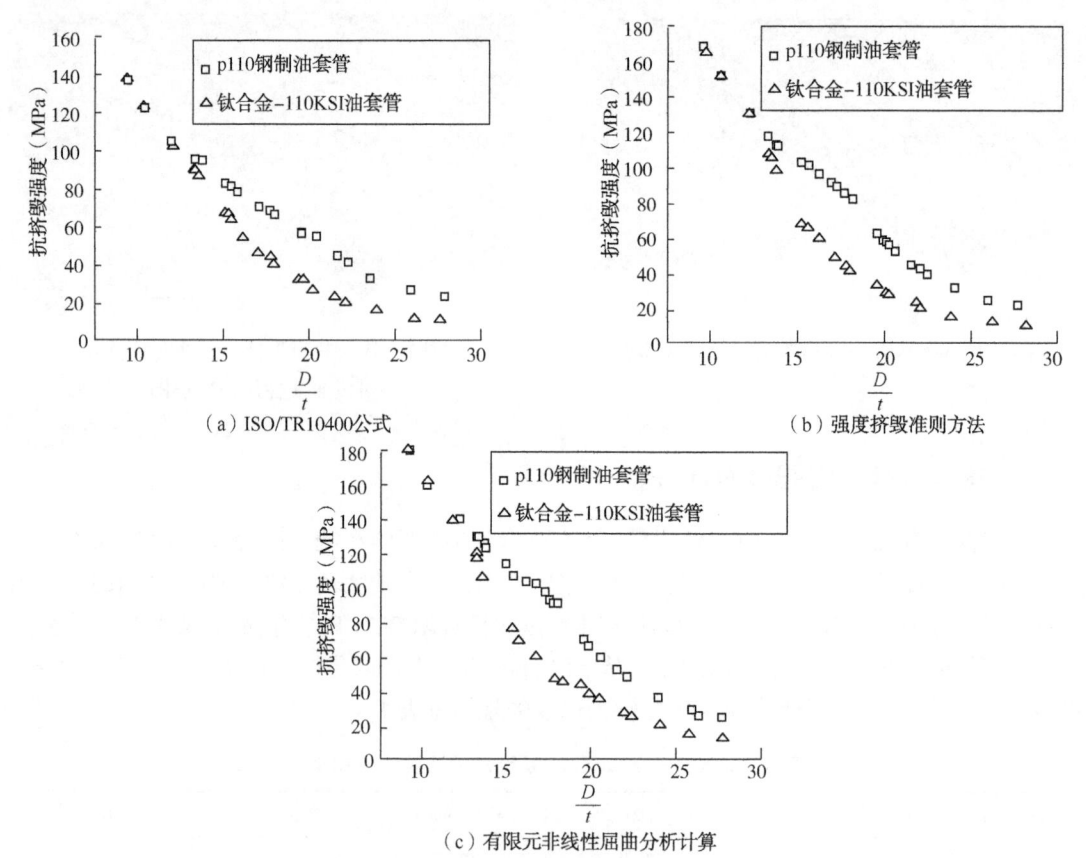

(a) ISO/TR10400公式

(b) 强度挤毁准则方法

(c) 有限元非线性屈曲分析计算

图 3 钛合金油套管和钢制油套管抗挤毁强度对比图

从图 3 中可以看出，3 种方法计算出的钛合金油套管抗挤毁强度均随着 D/t 的升高而下降，和钢制油套管对比，钛合金油套管的挤毁强度在 D/t 比(外径/壁厚比)较低时和钢制油套管的挤毁强度相差不大，这是由于当 D/t 比较低时，此时的油套管具有小外径大壁厚的特征，当受到外压作用时，主要发生屈服强度挤毁或者塑性挤毁，此时材料的弹性模量和泊松比差异对外挤毁强度影响不大，主要影响因素为材料的屈服强度，因此在这个低 D/t 比区域，钛合金油套管和钢制油套管的抗挤毁强度基本相同。当 D/t 比逐步增大时，钛合金油套管的抗挤毁强度明显小于钢制油套管的值，并且抗挤毁强度差值随着径厚比不断增大而变化。

从图 3 中深入分析可以看出，不同计算方法计算得出的钛合金抗挤毁强度和钢制油套管抗挤毁强度差值具有不同的特征，如图 4 所示，使用标准 ISO/TR 10400 中 K-T 极限状态公式方法计算得出的抗挤毁强度差值，当 D/t 小于 14 时，钛合金抗挤毁强度相比于钢制油套管下降量小于 0.3MPa，挤毁强度下降百分比小于 5%，根据以上的分析，此时主要发生屈服强度挤毁或者塑性挤毁，因此 $D/t=14$ 可以初步作为塑性挤毁和其他挤毁形式的分界值。随后挤毁强度下降百分比随着 D/t 比的增大急剧上升，在 $D/t=19.35$ 时，钛合金抗挤毁强度下降量达到最大为 32.54MPa；当 D/t 比进一步上升时，钛合金抗挤毁强度差值呈指数趋势下降，而当在 $D/t=20$ 以上时，挤毁强度下降百分比趋于稳定，介于 46%~47%，这是由于在大径厚比条件下，油套管主要发生纯弹性挤毁，此时钛合金和钢制油套管材料的弹性模量比约为 53.2%，下降量约为 46.8%，因此按标准 ISO/TR 10400 中 K-T 公式方法计算中，$D/t=20$ 可以初步作为纯弹性挤毁开始的分界值。

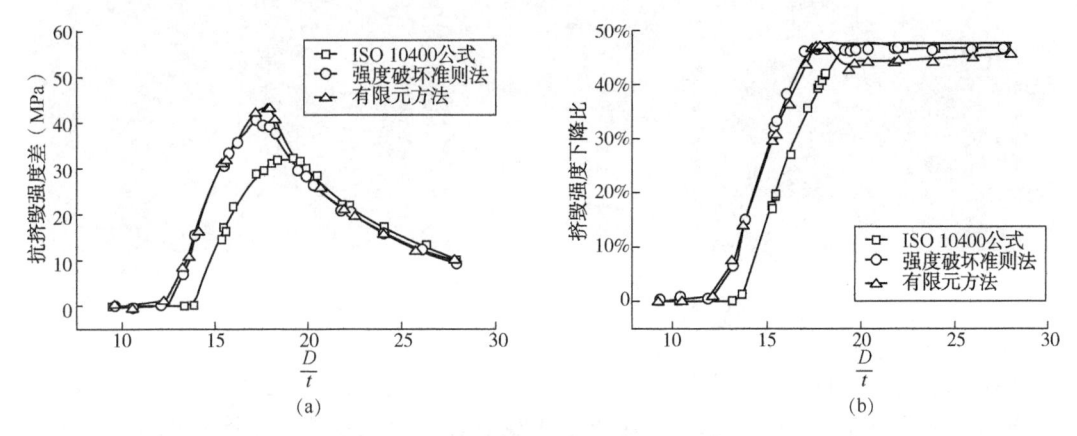

图 4　不同计算方式下钛合金和钢制油套管挤毁强度差值随 D/t 比变化对比图

将强度挤毁准则方法和有限元非线性屈曲计算的结果与标准 ISO/TR 10400 中 Klever-Tamano 公式计算结果相比较可以看出，这两种挤毁计算方法计算出的挤毁强度下降百分比和下降量的变化趋势与 ISO/TR 10400 中 Klever-Tamano 公式计算结果基本相同，但纯塑性挤毁的临界值均在 D/t 等于 12 左右，纯弹性挤毁开始的临界值均在大约 D/t 大于 17.5 左右，当 D/t 大于 17.5，强度挤毁准则方法计算出的挤毁强度下降百分比也稳定在 47%~48%，但有限元非线性屈曲方法计算出的下降百分比在 45% 左右，这可能是由于有限元计算的误差造成的。在 D/t 比介于 12~17.5，强度挤毁准则方法和有限元非线性屈曲计算的抗挤毁强度下降量均大于 Klever-Tamano 公式计算结果，差值介于 12.5~15.5，下降百分比较为稳定，介于 12%~13%。

3　实物外挤毁试验验证

为了验证 3 种挤毁强度计算结果与真实钛合金油套管挤毁强度的准确度，需要进行钛合金油套管的实物外挤毁试验。本文从国内不同厂家收集到不同规格的钛合金油套管产品，并对其进行外形尺寸测量和拉伸强度测试，所用的试验用钛合金试样信息及性能见表 2。

表2 实物外挤毁试验用钛合金油套管规格及性能表

钛合金管材	规格(mm)	实测规格(mm)	屈服强度 $R_{p0.2}$ (MPa)	抗拉强度 R_m (MPa)	延伸率
1号	φ88.9×6.45	φ89.4×6.60	951	1 061	15%
2号	φ88.9×6.45	φ89.0×6.46	860	935	13%
3号	φ177.8×9.19	φ177.9×9.19	850	952	14%
4号	φ244.5×11.99	φ245.66×12.64	835	895	13%

实物抗挤毁试验参照标准 API 5C5 及 ISO13679《Procedures for Testing Casing and Tubing Connections》(套管和油管螺纹连接试验程序),使用管研院全尺寸挤毁试验装置对所试验钛合金油套管两端进行固定,在外压实验舱内均匀施加液压至管体失稳/失效,由于实验设备条件的限制,无法模拟井眼轨迹对油套管的影响,在室温下采用管研院的挤毁试验装置进行试验,挤毁前后的样品形貌如图5所示。

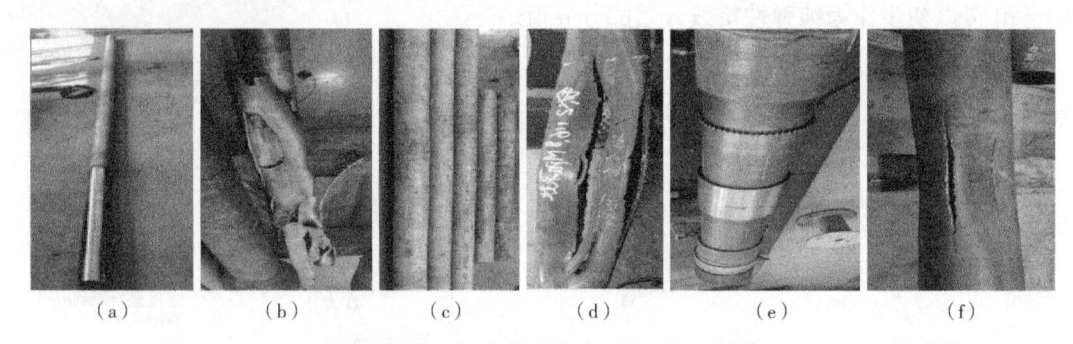

图5 钛合金油套管试样挤毁试验前后形貌

将钛合金油套管的实物外挤毁试验结果与3种计算方法得出的数值进行对比,见表3,可以看出强度挤毁准则方法和标准 ISO/TR 10400 中 K-T 公式在试验样条件下的计算结果基本相差不大,而有限元非线性屈曲方法计算出的抗挤毁强度均高于其他两种算法,同时有限元方法计算出的挤毁强度结果相比于实验值波动较大,其中3号试样的有限元计算结果已经高于实际值近9MPa,这可能是由于有限元方法在计算过程中网格划分的精度和计算误差造成,因此有限元计算方法在使用时会造成较高的预测风险;而强度挤毁准则方法和标准 ISO/TR 10400 中 K-T 公式计算出的挤毁强度值均接近并小于实物实验结果,在 D/t 比较高时具有较好的预测精度。

表3 不同方法下钛合金油套管挤毁结果对比图　　　　　　单位:MPa

管材编号	管径比	强度挤毁准则法	ISO/TR 10400法	有限元法	挤毁试验法	钢质油套管挤毁强度[22]
1号	13.55	98.15	98.71	124.35	135.00	93.20
2号	13.78	92.12	91.90	112.60	117.30	93.20
3号	19.36	32.10	33.86	40.86	32.50	42.90
4号	19.43	32.68	33.53	39.47	41.90	36.50

将钛合金油套管的实物挤毁试验结果放到 3 种挤毁强度计算结果的分布趋势图(图 6)中可以看出,当 $D/t>15$ 时,3 种计算方法所得出的钛合金油套管抗挤毁强度相差不大,其中有限元非线性屈曲方法计算值偏高甚至高于了实物试验结果。由于本次实物挤毁试验收到试验样本数量的限制,数据不够全面反映真实油套管挤毁强度的波动,因此强度挤毁准则方法和标准 ISO/TR 10400 中 K-T 公式计算方法所获得的挤毁强度与实物试验结果之间缺乏一定的安全空间;当 $D/t<15$ 时,ISO/TR 10400 中 K-T 公式计算结果相比于强度挤毁准则方法和有限元非线性屈曲计算的结果预测强度偏低,且明显低于实物试验结果,说明 ISO/TR 10400 中 K-T 公式在低 D/t 比范围内预测精度不如强度挤毁准则方法;而强度挤毁准则方法不但更接近于实物实验结果且具有较好的安全性。

因此,在综合以上分析,强度挤毁准则方法计算出来的钛合金油井管抗挤毁强度和实际值较为接近,同时在 $D/t>15$ 时,计算结果乘以 0.9 的系数,具有更好的安全性,如图 7 所示。

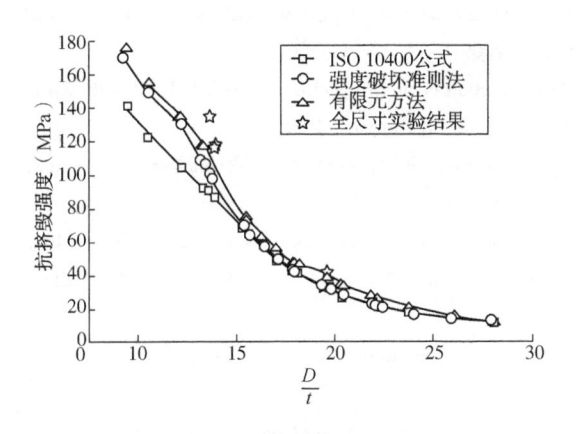

图 6 钛合金油套管实物挤毁试验结果分布情况图

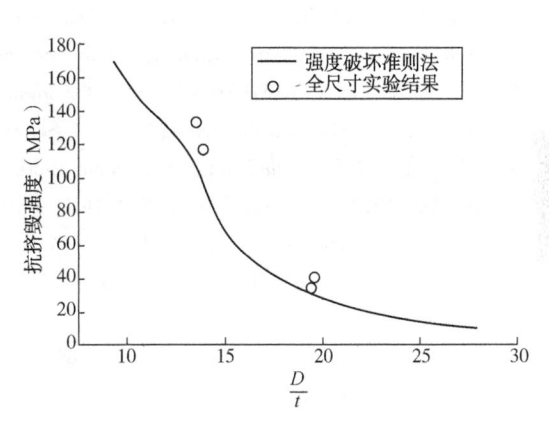

图 7 优化后的钛合金油井管强度挤毁准则法与实验法对比图

4 结论

通过研究 ISO/TR 10400 中 K-Tamano 公式、外压强度挤毁准则方法和有限元分析方法对钛合金油套管抗挤毁性能,并与钢制油套管计算结果对比,得出以下结论:

(1) 钛合金油套管抗挤毁强度均随着 D/t 比的升高而下降,在 D/t 值较低时,钛合金油套管和钢制油套管的抗挤毁强度相差不大,当 D/t 比较高时,钛合金油套管抗挤毁强度显著低于钢制油套管。

(2) 使用 ISO/TR 10400 中 K-T 极限状态公式方法计算时,$D/t=14$ 可以初步作为纯塑性挤毁的临界值,$D/t=20$ 可以初步作为纯弹性挤毁开始的临界值;当使用强度挤毁准则方法和有限元非线性屈曲计算时,纯塑性挤毁和纯弹性挤毁的临界值分别变为 $D/t=12$ 和 $D/t=17.5$。

(3) 实物实验表明,强度挤毁准则方法计算出来的钛合金油井管抗挤毁强度和实际值较为接近,同时在 $D/t>15$ 时,计算结果乘以 0.9 的系数,具有较好的安全性。

参 考 文 献

[1] 谷坛, 霍绍全, 李峰. 酸性气田防腐蚀技术研究及应用[J]. 石油与天然气化工, 2008, 37(增刊1): 63-72.

[2] 叶登胜, 任勇, 管彬, 等. 塔里木盆地异常高温高压井储层改造难点及对策[J]. 天然气工业, 2009, 29(3): 77-79.

[3] 杜伟, 李鹤林. 海洋石油装备材料的应用现状及发展建议(上)[J]. 石油管材与仪器, 2015(5): 1-7.

[4] 彭建云, 周理志, 阮洋, 等. 克拉2气田高压气井风险评估[J]. 天然气工业, 2008, 28(10): 110-112.

[5] Schutz R W, Watkins H B. Recent developments in titanium alloy application in the energy industry[J]. Materials Science and EngineeringA, 1998, 243(1/2): 305-315.

[6] 刘强, 宋生印, 李德君, 等. 钛合金油井管的耐腐蚀性能及应用研究进展[J]. 石油矿场机械, 2014, 43(12): 88-94.

[7] Kane R D, Craig B, Venkatesh A. Titaniumalloys for oil and gas service: A review[C]//NACE-International Corrosion Conference Series. Atlanta: NACE International, 2009.

[8] Love W W. The use of Ti-38644 titanium for downhole production casing in geothermal wells[C]//Sixth World Conference on Titanium Proceedings. France: SocieteFrancaise de Meallurgie, 1988.

[9] Schutz R W. Effective utilization of titanium alloys in offshore systems[C]//Offshore Technology Conference, 4-7 May 1992, Houston, Texas, USA. DOI: 10.4043/6909-MS.

[10] Kane R D, Srinivasan S, Craig B, et al. A comprehensive study of titanium alloys for high pressure high temperature (hpht) wells[C]//Corrosion Conference and Expo 2015, 15-19 March 2015, Dallas, Texas, USA.

[11] Schutz R W, Lingen e V. Characterization of the Ti-6Al-4V-Ru for application in the energy industry[C]//Proceedings of Eurocon'97 Congress. Tapir: 1997.

[12] Smith J E, Schutz R W, Bailey E I. Development of titanium drill pipe for short radius drilling[C]//IADC/SPE Drilling Conference, 23-25 February 2000, New Orleans, Louisiana, USA. DOI: 10.2118/59140-ms.

[13] Hargrave B, Gonzalez M, Maskos K, et al. Titanium alloy tubing for HPHT OCTG applications[C]//CORROSION-National Association of Corrosion Engineers Conference and Exposition, 14-18 March 2010, San Antonio, Texas, USA.

[14] 刘强, 惠松骁, 宋生印, 等. 油气开发用钛合金油井管选材及工况适用性研究进展[J]. 材料导报, 2019, 33(5): 841-853.

[15] 刘强, 范晓东, 宋生印, 等. 钛合金油管表面抗粘扣处理工艺研究[J]. 石油管材与仪器, 2017, 3(4): 26-31.

[16] 国家能源局. 石油天然气工业特种管材技术规范第3部分: 钛合金油管: SY/T 6869.3—2016[S]. 北京: 石油工业出版社, 2016.

[17] 李永林, 朱宝辉, 王培军, 等. 石油行业用TA18钛合金厚壁管材的研制[J]. 钛工业进展, 2013, 30(2): 28-31.

[18] 孟祥林, 何佳持. 攀钢钛合金油管有望填补国内市场空白[N]. 世界金属报导, 2014-12-09(1).

[19] 杨冬梅. 抗腐蚀钛合金油管井首次在超深高含硫气井应用[J]. 钢铁钒钛, 2015, 36(3): 15.

[20] 田军, 杨海军, 吴超, 等. 博孜9井发现与塔里木盆地超深层天然气勘探潜力[J]. 天然气工业, 2020, 40(1): 11-19.

[21] 5C3 A. Bulletin on formulas and calculations for casing[Z], 2008.

[22] General Administration of Quality Supervision, Inspection and Quarantine of the People's Republic of China, Standardization Administration. Petroleum and natural gas industries - equations and calculations for the

properties of casing, tubing, drill pipe and line pipe used as casing or tubing: ISO/TR 10400: 2007[S]. Beijin: Standards Press of China, 2012.

[23] 申昭熙. 套管外压挤毁强度分析与计算[J]. 应用力学学报, 2011, 28(5): 547-550.

[24] 李富平, 张冠林, 李振, 等. 非均匀椭圆载荷下套管抗外挤能力仿真分析[J]. 石油矿场机械, 2016, 45(7): 33-36.

本论文原发表于《天然气工业》2020年第40卷第8期。

三、腐蚀防护与非金属材料

Facile Fabrication of SnO$_2$ Modified TiO$_2$ Nanorods Film for Efficient Photocathodic Protection of 304 Stainless Steel under Simulated Solar Light

Zhang Juantao[1,3] Yang Hualong[2] Wang Yuan[1] Cui Xiaohu[2]
Wen Zhang[2] Liu Yunpeng[1,3] Fan Lei[1] Feng Jiangtao[3]

(1. NPC Tubular Goods Research Institute, State Key Laboratory for Performance and Structure Safety of Petroleum Tubular Goods and Equipment Materials;
2. Petro China Traim Oilfield Company; 3. Department of Environmental Science & Engineering, Xi'an Jiaotong University)

Abstract: SnO$_2$/TiO$_2$ nanorods (TNRs) composite film was successfully fabricated. The results showed that the SnO$_2$ nanoparticles were deposited onto the surface of the TNRs film, and this composite film exhibited excellent light absorption ability. Meanwhile, the SnO$_2$/TNRs composite film exhibited the superior photoelectrochemical performance. With white light illumination, this composite film made the potential of the coupled 304 stainless steel (304SS) in a 0.5mol/L NaCl solution drop by 240mV, exhibiting more efficient photocathodic protection (PCP) effect. What's more, it could provide delay protection for the steel in darkness attributing to the "electron pool" effect of SnO$_2$.

Keywords: TiO$_2$ Nanorods array; SnO$_2$; 304 Stainless steel; Photocathodic protection

1 Introduction

304SS is widely used as engineering steel in industrial production due to its outstanding anti-corrosion property. Nevertheless, pitting and crevice corrosion on the metals are barely to be avoided in the presence of Cl$^-$[1-3]. So far, many anti-corrosion technologies have been developed, such as anti-corrosion coatings, corrosion inhibitors, electrochemical protection, etc[4,5]. These traditional protection technologies are limited because of the consumption of material and energy, and the environmental pollution. On behalf of green steel-protection methods, the PCP is commonly

---

Corresponding author: Liu Yunpeng, liuyunpeng1994@stu.xjtu.edu.cn; Fan Lei, damifan198802@163.com; Feng Jiangtao, fjtes@xjtu.edu.cn.

considered as the one of the most effective technology[6]. When the photoanode is irradiated, the photoexcited electrons are immigrated to the connected steel, and the surface potential of the protected steel shifted negatively below its corrosion potential, resulting in the cathodic protection effect for metals[7].

TiO_2 has attracted many attentions in field of photocatalysis, owing to its low cost, low toxicity and large specific surface area properties. Compared with traditional TiO_2 particles and TiO_2 nanotubes (TNTs) array, TNRs array has been extensively exploited in pollution degradation, hydrogen production and solar cells due to one-dimensional(1D) electron transmission shortcut and excellent electronic mobility[8]. However, TNRs still suffers the inherent flaws of TiO_2, namely, the wide band gap (3.1eV) of rutile TiO_2 limits its PCP application only in the UV region and decrease the utilization ratio of solar light[9]. Meanwhile, the rapid recombination of photogenerated electrone/hole(e^-/h^+) pairs will happen after cutting off the light, resulting in the lose effectiveness of the cathodic protection in darkness[10]. Thus, in order to enhance the PCP performance, it is necessary to sensitize TNRs as a visible light absorber. At present, the combination of TNRs and narrow band-gap semiconductors (such as Bi_2S_3, WO_3, CdSe, etc.) have been proved to be effective methods to hinder with these problems[11].

SnO_2, a kind of stable and nontoxic n-type (3.5eV) semiconductor, has be used to combine with TiO_2 to form photoanode in some PCP researches[12]. The stagger band structure between SnO_2 and TiO_2, benefits the separation of the e^-/h^+ pairs to enhance photoelectric performance. In addition, since the conduction band (CB) of TiO_2 is more positive than that of SnO_2, SnO_2 can serve as an "electron pool" to storage e- in the binary photoanode material, to ultimately provide delay protection for the coupled steel in darkness[13]. Hu et al. have synthesized SnO_2 nanoparticles modified TiO_2 nanotube films, and provided an effective cathodic protection for 403 stainless steel[9]. Zhang et al. have fabricated nanoflower like SnO_2-TiO_2 nanotubes photoanode, and the delay protection for 304SS was achieved in darkness[13]. However, as the excellent succedaneous materials of TNTs film, there have been no reports on the PCP efficiency effects of the SnO_2/TNRs composite for 304SS.

Therefore, in this work, the main objective is to synthesize the SnO_2/TNRs film via hydrothermal treatment and electrodeposition method on the conductive fluorine-doped tin oxide (FTO) substrates, to achieve excellent PCP efficiency for 304SS. Meanwhile, their photoelectrochemical properties and PCP mechanism for 304SS were also proposed.

2 Materials and methods

2.1 Chemicals

Hydrochloric acid (HCl), stannic chloride ($SnCl_2$), sodium hydroxide (NaOH), sodium chloride (NaCl), sodium sulphide (Na_2S), sodium nitrate ($NaNO_3$) and titanium butoxide ($C_{16}H_{36}O_4Ti$) were all purchased from Sinopharm Chemical Regent Co. Ltd, (Shanghai, China). All the chemicals were analytical reagent without any purification and distilled water (DIW) was used in the preparation of all solutions.

FTO glass ($7\Omega \cdot cm^{-2}$, 2.2mm–thick) was bought from Guluo Glass Co. Ltd, (Luoyang, China). Before using of the testes, the FTO glasses were cleaned ultrasonically by DW, followed by acetone and alcohol for 30 min each. In the end, those structures were air–dried in ambient condition.

2.2 Synthesis of SnO$_2$/TNRs composite films

2.2.1 Preparation of TNRs film

The TNRs array film was fabricated on FTO substrate by the traditional hydrothermal reaction following these steps. In the first place, 60mL HCl aqueous solution (6mol/L) injected with 1.0 mL $C_{16}H_{36}O_4Ti$, was poured into a 100mL Teflon–lined stainless–steel autoclave. Then a sheet of FTO structure was put at an angle against the wall of the above autoclave with the conducting side facing down, and being heated at 150℃ in an electric oven (for 10 h). Finally, the film was washed with DW, and placed in a muffle furnace at 500℃ in air (for 2h).

2.2.2 Preparation of SnO$_2$/TNRs composite film

SnO$_2$/TNRs composite films were prepared in a traditional threeelectrode system. The work electrode (WE), the reference electrode(RE) and the counter electrode (CE) are corresponded to the TNRs film, Ag/AgCl electrode and Pt foil, respectively. The electrodeposition solution comprised 25mmol/L SnCl$_2$ and 50 mmol/L HNO$_3$. The electrodeposition method was carried out potentiostatically (CHI660D, Shanghai Chinstruments co., Ltd, China) at 0.5V under room temperaturecondition. After thoroughly rinsed with deionized water and dried in a vacuum drying oven at 50℃ (for 12h), these as–deposited composite films were annealed at 450℃ under air atmosphere (for 2h). Fig. 1 display the schematic illustration for preparing these composites. To compare the PCP performances of the SnO$_2$/TNRs composite films with different deposition time, the SnO$_2$/TNRs composite films with different electrodeposition time (0.5h, 1h and 1.5h) was synthesized, denoting as SnO$_2$/TNRs – 0.5h SnO$_2$/TNRs – 1h and SnO$_2$/TNRs – 1.5h, respectively.

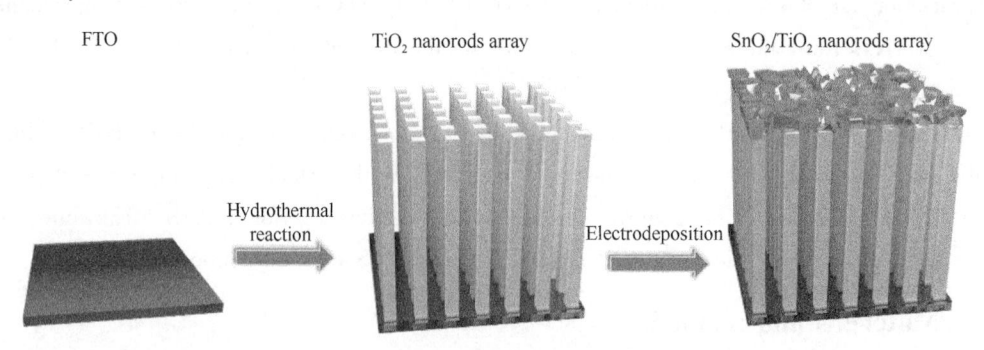

Fig. 1 Schematic illustration of the construction process of SnO$_2$/TiO$_2$ nanorods array composite

2.3 Characterizations

All the as–prepared samples were characterized by scanning electron microscope (SEM Zeiss Gemini SEM 500, Germany), X–ray diffraction instrument (Bruker D8 ADVANCE, Germany), Raman spectra (LabRAM HR Evolution with an excitation of 325 nm laser light, France), X–ray photoelectron spectroscopy (XPS, Thermo Scientific EscaLab 250Xi., USA), UV–vis diffuse reflectance absorption spectra (DRS, PE Lambda950, USA). The photoluminescence (PL) spectra were measured using an Edinburgh FLS9 luminescence spectrofluophotometer with a Xe lamp presenting an excitation light of 340 nm. The time resolved photoluminescence (TRPL) was also studied on a FLS9 series of fluorescence spectrometers.

2.4 Photoelectrochemical tests

In the photoelectrochemical experiments (Fig. 2), the test rig consisted of a corrosion cell (3.5 wt% NaCl solution) and a photoelectrode cell (0.1mol/L Na_2S + 0.2mol/L NaOH mixed solution), and the two cells were connected by a salt bridge (saturated KCl in agar gel). In the corrosion cell, the 304SS electrode, Ag/AgCl electrode and Pt foil were connected to WE, RE and CE poles of the CHI660D electrochemical workstation, respectively, and the as-prepared photoelectrode as the photoanode was immersed in photoelectrochemical cell, which was coupled with the steel electrode via a copper wire, as described previously[3]. The photoanodes with an effective irradiation area of 1 cm² were radiated by a 500 W Xenon arc lamp (CHF-XM-500W, Beijing Changtuo technology Co., Ltd. Beijing, China) coupled with an AM 1.5 G filter calibrated to 100mW cm^{-2}. The scanning rates of open circuit potentials (OCP) and Tafel curves were 1mV/s. Meanwhile, no bias voltage was applied in the test of photocurrent density of the photoelectrode. The Tafel result was fitted to calculate the corrosion potential (E_{corr}) and the corrosion current density (I_{corr}). The efficient area of 304SS was 1 cm², which was ground with 2000-mesh abrasive paper and rinsed with acetone.

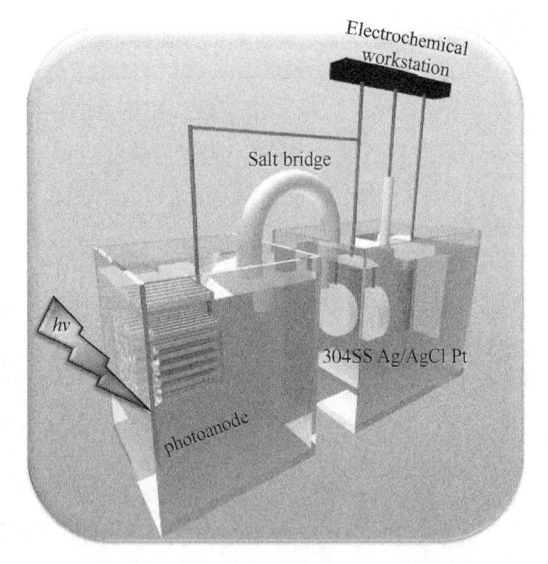

Fig. 2 Schematic of photoelectrochemical test device

3 Results and discussion

3.1 Structural characteristics

The structural information was provided by their Raman spectra and XRD patterns. As displayed in Fig. 3(a), three strong characteristic Raman peaks (the curve of TNRs) were found to located at 230, 446 and 610cm^{-1}, ascribed to the B_{1g}, E_g, A_{1g} modes of rutile TiO_2, respectively[14]. Moreover, any the characteristic peaks of anatase TiO_2, cannot be observed, indicating pure rutile phase was obtained after hydrothermal treatment. In the curve of STNRs-1.0, these characteristic peaks of SnO_2 were overlapped by the rutile TiO_2 signal for this composite film, so there is no difference between TNRs and STNRs-1.0. Furthermore, to confirm the composition of the STNRs-1.0 film, the XRD results of the TNRs and STNRs-1.0 were displayed in Fig. 3(b). The diffraction peaks of the TNRs at 27.5°, 36.1°, 39.3°, 41.3°, 54.5° 62.9° and 69.8° correspond well to the (1 1 0), (1 0 1), (2 0 0), (1 1 1), (2 1 1), (0 0 2) and (1 1 2) planes of rutile TiO_2 (JCPDS card no: 65-0912), respectively, which is consistent with the above Raman result[15]. After the electrodeposition of SnO_2 on the surface of TNRs film, the XRD curve of the STNRs-1.0 exhibits new diffraction peaks at 33.7°, 61.6° and 65.5°, which matches well with the characteristic XRD peaks [(1 1 0), (3 1 0) and (3 0 1)] of cassiterite SnO_2(JCPDS card no: 41-1445)[16]. The XRD result demonstrates the successful deposition of SnO_2 on the surface of TNRs film.

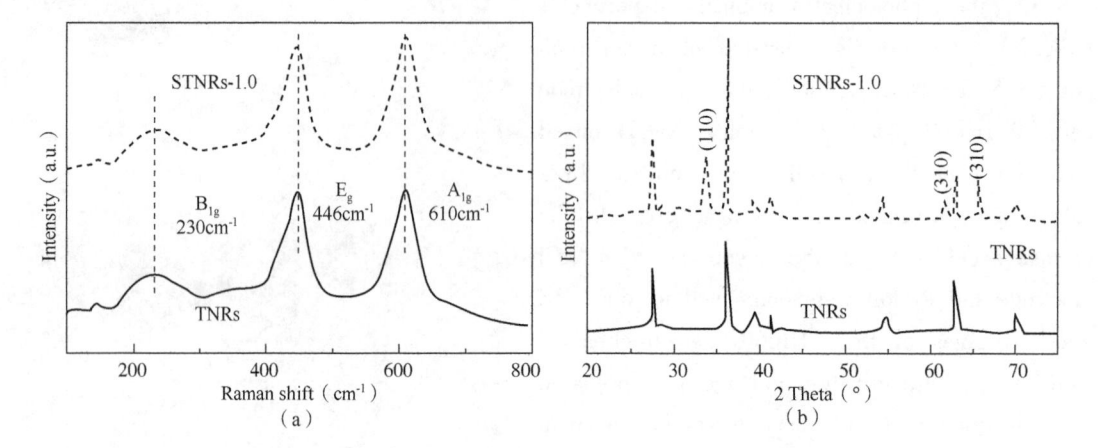

Fig. 3 The Raman spectra (a) and the XRD patterns (b) for the TNRs and STNRs-1.0

Fig. 4 displays the SEM images with two different magnifications of TNRs film before and after SnO_2 loading. From the images of TNRs [Fig. 4(a) and (b)], the TNRs grew uniformly and densely on the surface of FTO glass, and the mean diameter of nanorods is about 150 nm. After electrodeposition treatment, there are amount of SnO_2 nanoparticles on the surface of TNRs film as shown in Fig. 4(c). Moreover, the rod-like structure of TNRs maintains good property during the electrodeposition process [Fig. 4(d)]. This result also indicates that the SnO_2 nanoparticles are successfully loaded on the TNRs film.

To further confirm the chemical compositions and elemental chemical statuses of the composite film, the as-synthesized STNRs-1.0 film was characterized by XPS, and the result is displayed in Fig. 5(a-d). From the survey spectra shown in Fig. 5(a), the Ti, Sn and O were present in the sample, and the C 1s peak in the figure is due to the surface pollution of C. High resolution XPS spectra for Sn 3d, O 1s and Ti 2p are displayed in Fig. 5(b), c and d, respectively. The two peaks at 495.6 and 487.2eV are ascribed to Sn $3d_{3/2}$ and Sn $3d_{5/2}$, respectively [Fig. 5(b)], implying that the main valence state of Sn in this sample is +4[17]. As shown in Fig. 5(c), binding energies at 530.4 eV and 531.8eV, are corresponding to the contribution of lattice oxygen in TiO_2 and SnO_2 crystal lattice, and surficial hydroxyl groups (O—H), respectively[18,19]. As shown in the XPS comparison curves of Ti 2p of TNRs and STNRs-1.0 [Fig. 5(d)], the binding energies at 463.9 eV and 458.2 eV of TNRs are ascribed to Ti $2p_{1/2}$ and Ti $2p_{3/2}$, respectively[20]. Compared with the previous reports[21], both of the Ti 2p peaks of the TNRs slightly shifted to the lower banding energy, which is ascribed to the destroy of Ti—O—Ti linkages of the TiO_6 octahedra in strong acid solutions, generating a small amount of oxygen vacancies (OVs) in the synthesis process of nanorods[22]. However, the content of OVs is not sufficient to impact the pure rutile phase of TNRs,

Fig. 4　SEM images of the TNRs (a and b) and STNRs-1.0 (c and d)

according to the results of Raman and XRD spectra. After electrodeposition of SnO_2, all the Ti 2p peaks of the STNRs-1.0 films shifted negatively, indicating the interaction between TiO_2 and SnO_2. With the formation of heterojunction, the redistribution of valence electrons between TiO_2 and SnO_2 occurs[23], and the positive shift of Ti 2p indicates that the transfer tendency from SnO_2 to TiO_2.

UV-vis DRS was employed to study the optical properties of TNRs and SnO_2/TNRs films. As displayed in Fig. 6, the TNRs shows a typical absorption band-edge at about 390 nm, which agrees well with the band gap of rutile TiO_2 (E_g = 3.10eV)[3,24]. After deposition of SnO_2, the as-prepared STNRs-1.0 composite shows much stronger absorption ability to the UV light, and slight red-shift of the absorption edge to the longer wavenumber region compared with the pure TNRs. The measured band-gap value of STNRs-1.0 composite is narrowed to be about 3.06eV. This result demonstrates that the combination of TNRs and SnO_2 not only improve the solar-energy utilization efficiency of this photoanode, but also efficiently suppress the bulk recombination and surface recombination of the e^-/h^+ pairs, and finally promote its PCP properties for 304SS. Meanwhile, SnO_2 nanoparticles cannot absorb visible light due to its relatively wide band gap[25], resulting in the very slight optical absorption red-shift of STNRs-1.0 composites.

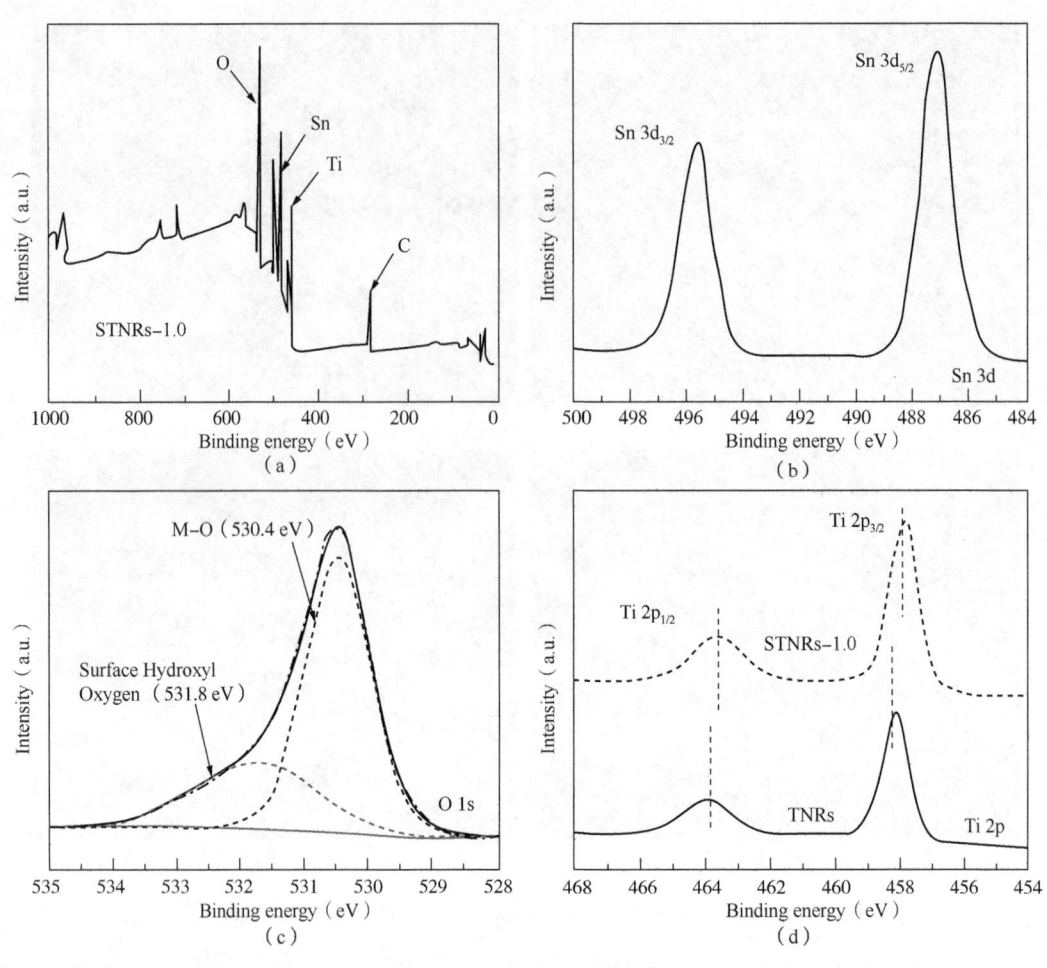

Fig. 5 Full scan survey XPS spectrum (a) and high-resolution spectra of (b) Sn 3d and (c) O 1s of STNRs-1.0 composite film, and the XPS comparison spectrum of Ti2p (d) of TNRs and STNRs-1.0 films

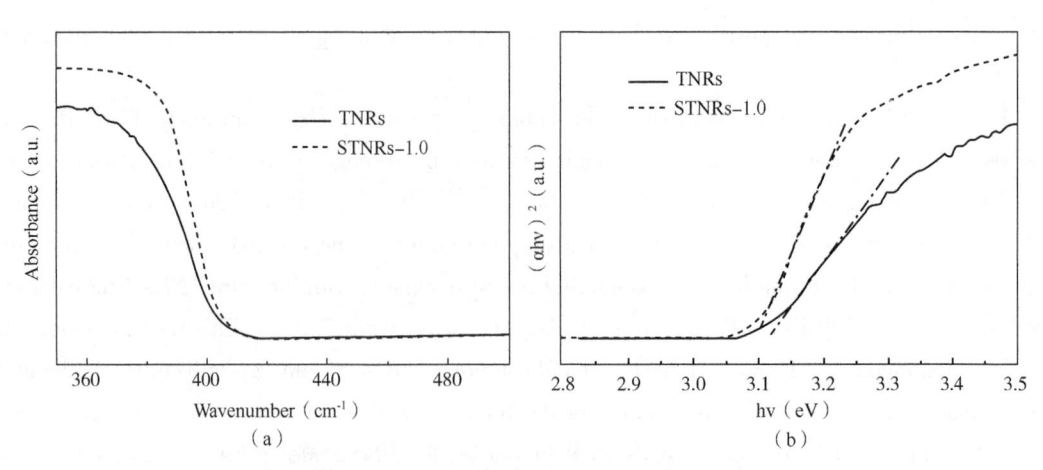

Fig. 6 UV-vis absorption spectra (a) and the plots of $(\alpha h\nu)^2$ versus $h\nu$ (b) for the TNRs and STNRs-1.0 films

In order to study the influence of the SnO_2 nanoparticles, PL emission spectrum and TRPL spectrum measurements were applied to reveal the migration, transfer and recombination processes of the e^-/h^+ pairs in the interface between SnO_2 and TNRs. As shown in Fig. 7(a), the main emission peak is located at 435 nm for the pure TNRs, which can be contributed to the band-to-band recombination process of the e^-/h^+ pairs[26]. The higher PL intensity, the faster recombination rate of photocarriers. The PL peak intensity of the STNRs-1.0 was much lower for comparison with that of the pure TNRs, implying that the separation efficiency of the e^-/h^+ pairs in the composite film has been enhanced because of the deposition of SnO_2. For more insights into the photogenerated charges, TRPL analysis was displayed in Fig. 7(b), and the lifetime of STNRs-1.0 (8.30ns) is higher than that of pure TNRs (5.34ns), indicating rapid charge immigration between SnO_2 and TNRs. These results demonstrated that the formation of built-in electric field in the interface can efficiently improve the transfer of photo-induced electrons and simultaneously hindered the recombination of the e^-/h^+ pairs.

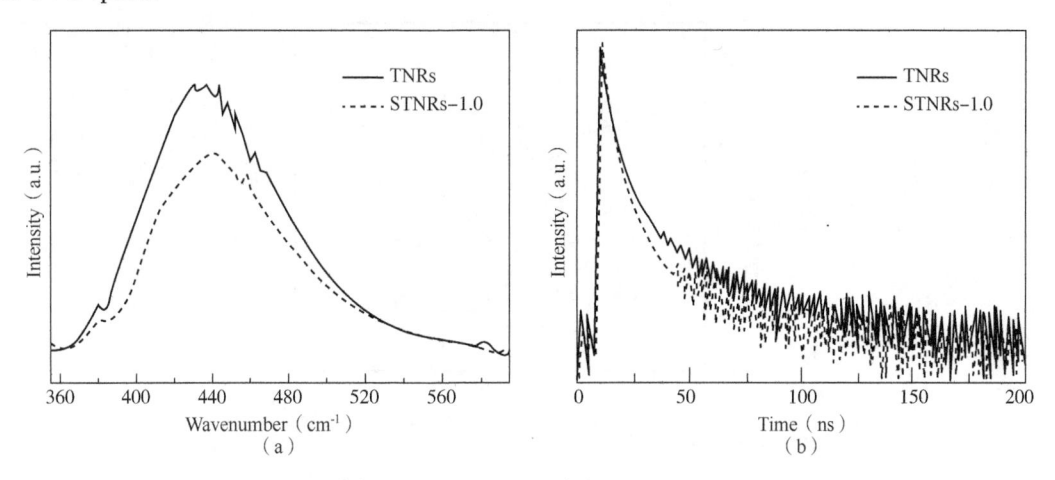

Fig. 7 PL emission spectra (a) and TRPL spectra (b) of TNRs and STNRs-1.0 composite films

3.2 Photoelectrochemical properties of TNRs and SnO_2/TNRs film

The transfer of photoinduced electrons play an important role in PCP process, so it's essential for assessing the PCP efficiency of a photoanode to investigate the photocurrent density[27]. Fig. 8(a) shows the photocurrent density of the TNRs, and STNRs-0.5, STNRs-1.0 and STNRs-1.5 films connected with metal with intermittent white light irradiation. Before illumination, the photocurrent values of the four samples were almost zero. After the light source was turned on, the photoelectric response of SnO_2/TNRs composite ($31-84\mu A \cdot cm^{-2}$) were higher compared with the unmodified TNRs film ($27\mu A \cdot cm^{-2}$), indicating that photon-to-current conversion efficiency of SnO_2/TNRs film was outstandingly improved. Namely, more electrons can be photoexcited in this composite film under white light illumination, suggesting that the SnO_2/TNRs film might provide more effective cathodic protection for 304SS than the TNRs film. From Fig. 8(a), the value order of photocurrent density is as follows: STNR-1.0>STNR-1.5>STNR-0.5, suggesting that the STNR-1.0 might exhibit superior PCP performance on 304SS. In addition, the SnO_2/TNRs photoanode has a stable and reproducible photoelectrochemical property during three on-off cycles of the illumination. Therefore, the SnO$_2$/TNRs film has an advantage for an excellent photocatalytic performance.

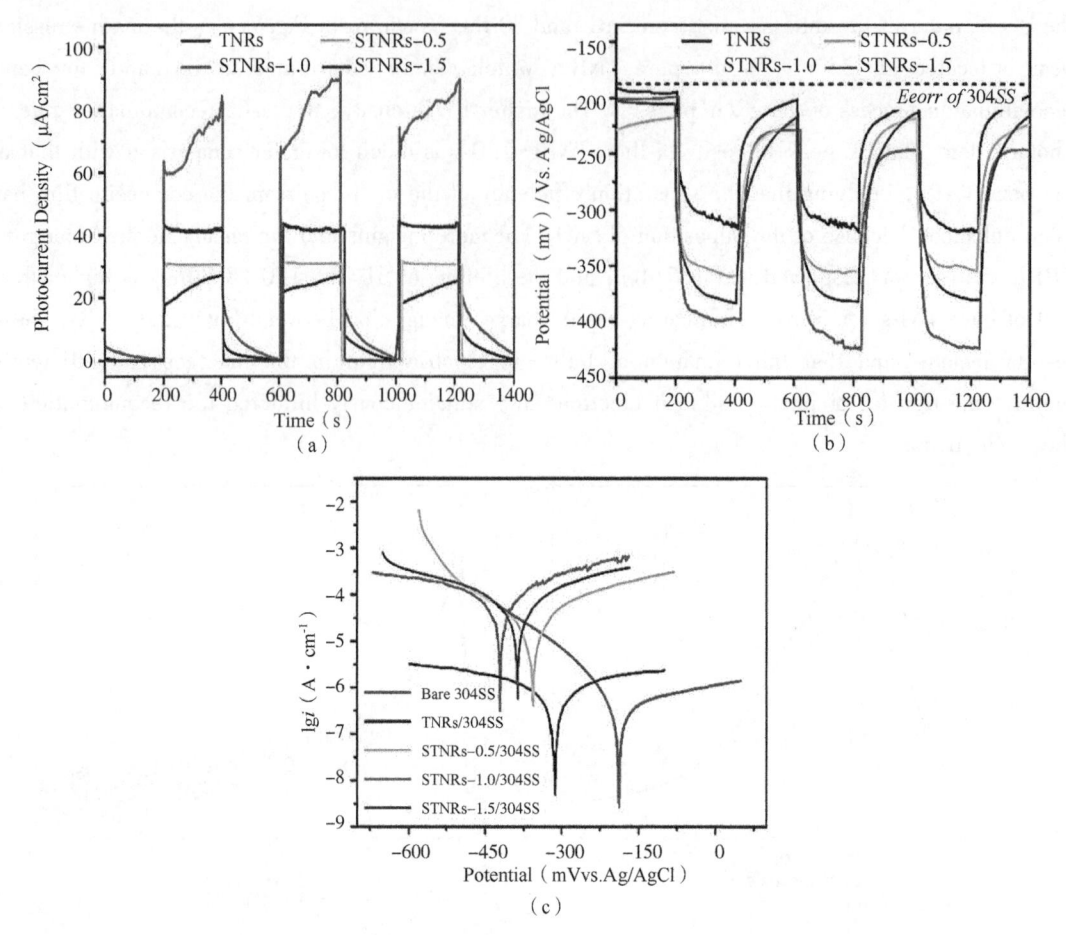

Fig. 8 (a) The photoinduced current density between the as-prepared samples and 304SS under intermittent light illumination. (b) The potential change of the 304SS electrodes coupled with the as-prepared films. Polarization curves (c) of bare 304SS, and 304SS coupled with TNRs, and SnO$_2$/TNRs films electrodes under illumination

To assessing the PCP performances of as-prepared photoelectrodes on the steel, the materials (TNRs, STNRs-0.5, STNRs-1.0 and STNRs-1.5 films) serve as photoanode in the photo-induced OCP tests via monitoring the change of steel surface potential under white light illumination [Fig. 8(b)]. The result shows the corrosion potential of bare steel is about -180mV vs. Ag/AgCl, meanwhile, the OCP value of 304SS coupled to TNRs film shifts slightly to negative values (-320mV vs. Ag/AgCl). Nevertheless, for SnO$_2$/TNRs films, the much more negative OCP value (-400mV vs. Ag/AgCl) of the steel can be observed when this photoanode is irradiated by white light firstly. After then, the surface potential of the steel connected to SnO$_2$/TNRs films go on dropping slightly, and sustain finally a negative value (-350 to -420mV vs. Ag/AgCl). As explained above, the formation of n-n heterojunction between SnO$_2$ and TiO$_2$ can improve the separation efficiency of e$^-$/h$^+$ pairs, and then more photoelectrons are immigrated to the connected steel to decrease its surface potential, resulting in superior PCP efficiency. Additionally, after cutting off the light, the OCP value of 304SS connected to TNRs film returns to the initial level before illumination (about 190mV vs. Ag/AgCl), indicating that it can't provide valid delay protection for 304SS. Nevertheless, it is obvious that all the surface potential of the steel coupled with SnO_2/TNRs composite films are more negative than the corrosion potential of 304SS in darkness (-210 to -230mV vs. Ag/AgCl), which can be ascribed to the effect of "electron pool". Meanwhile, the potential of 304SS coupled with STNRs-1.0 is significantly lower than that when it is coupled with pure STNRs-0.5 and STNRs-1.5 with light illumination, implying that STNRs-1.0 is optimal for anti-corrosion of 304SS. This result could be attributed to the appropriate deposition content of SnO_2 nanoparticles coordinated with the sustainability of the 1D electron transmission shortcut of the TNRs, therefore enhancing the photon-toelectron conversion efficiency and insuring abundant generation and rapid immigration of photoelectrons. For STNRs-0.5, because of the low deposition content of SnO_2 nanoparticles, it is unsatisfactory to generate sufficient photoelectrons to the maximum extent. For STNRs-1.5, owing to the excessive coverage of SnO_2 nanoparticles onto the surface of TNRs film, the light-absorption ability is reduced, and sectional SnO_2 nanoparticles may be the recombination center of the e^-/h^+ pairs. More interestingly, the OCP value of 304SS coupled to the STNRs-1.0 composite keeps going down to a fixed state (approximately -250mV vs. Ag/AgCl) after cutting off the light source for the second and third time. When the STNRs-1.0 film is irradiated for the first time, extra electrons was temporarily stored in SnO_2 and then released to achieve dark-state protection. These unconsumed electrons in darkness will be continued to transferred to the coupled steel when the light is provided again, resulting in more negative surface potential of 304SS. After the second on-off cycle, the dark-state OCP shift more negatively to a fixed value due to no need of more electrons to be stored. Therefore, the STNRs-1.0 film can serve as a potential photoelectrode in the PCP applications for 304SS.

The Tafel polarization curves of 304SS in the 0.5mol/L NaCl solution were also employed to evaluate the PCP efficiency of the aspreared samples under white light illumination (Fig. 8c). It is obvious that the E_{corr} value of 304SS connected to the as-prepared sample is more negative than that of bare 304SS (-187mV vs. Ag/AgCl), and the E_{corr} of 304SS coupled with STNRs-1.0 film exhibits the most negative value (-421mV vs. Ag/AgCl), which is in agreement with the OCP results. This result indicate that STNRs-1.0 film possess the higher photoelectric conversion efficiency than pure TNRs film, and more electrons are immigrated to the coupled steel under white light irradiation. It is well known that the higher I_{corr} value, the more effective anti-corrosion property will be. The I_{corr} of STNRs-1.0 is markedly higher than those of the other samples/304SS and bare 304SS, since superior solar light utilization capacity and separation efficiency of the e^-/h^+ pairs. According to the above analysis, the STNRs-1.0 composite could be supposed to display remarkable PCP efficiency for 304SS.

3.3 The stability evaluation of STNRs-1.0 film in PCP system of 304SS

As shown in Fig. 9(a), the relationship between the illumination time of white light and the dark-state delay PCP performance of 304SS coupled with STNRs-1.0 is investigated. After coupled with the STNRs-1.0 composite, the 304SS potential negatively shifts to around -200mV vs. Ag/

AgCl before illumination because of the galvanic effect. When provided with white light irradiation, all the potential values rapidly dropped to −440mV vs. Ag/AgCl, which are consistent with the above results. When the STNRs−1.0 film was only irradiated with white light for 5min, the OCP value of the 304SS recovered to its the initial potential (−200mV vs. Ag/AgCl) after 20min (about 0.34h). After extending the illumination time to 30 min, the duration of the delay protection could be effectively increased to 4.71h. However, when the illumination time was extended to 60 min further, the duration of the delay protection (about 4.92h) is almost identical to that of the above result (illumination for 30min). These results demonstrate that the increase of illumination time is in favor of the dark-state protection of 304SS, nevertheless, the enhancement extent of delay protection is limited and the STNRs−1.0 film can provide certain degree of delay protection even when the light source is turned off during this period.

In order to estimate the photoresponse stability of SiO_2/TNRs film in the PCP system, the OCP value changes of the 304SS coupled with STNRs−1.0 film to intermittently white light irradiation were measured [Fig.9(b)]. This result displays that the STNRs−1.0 film can prevent continuously and effectively 304SS from corrosion, and the OCP value (around −220mV vs. Ag/AgCl) of 304SS is still lower for comparison with its initial value (around −200mV vs. Ag/AgCl) and corrosion potential under 8h' intermittent light irradiation, displaying distinguished stability of this composite films. The SEM image of STNRs−1.0 film measured in the above stability evolution experiment, are exhibited in Fig.9(c). It can be obviously seen that a mass of SnO_2 are still steadily anchored on the surface of TNRs film, and there isn't much mass SnO_2 to strip from the substrate. Therefore, such composite film has good potential in PCP application.

In the PCP system, the S^{2-} of the electrolyte of the photoelectrochemical cell plays an extremely key role to eliminate h^+ [Eq.(2)], aiming to promote the separation of photocarriers and further enhance the PCP efficiency[28]. However, the accompanying elemental sulfur(S) may deposit onto the surface of the composite film to hinder the immigration of free carrier. The XPS survey spectrum of the STNRs−1.0film after PCP experiment is shown in Fig.9(d), and there is no peaks for elemental S, which suggests that the reaction-generated elemental S forms dissolvable polysulfide ions in the solution [Eq.(3) and (4)][29]. Therefore, the elemental S originated from the oxidation reaction of S^{2-} does not exist onto the surface of the electrode, and have little influence on the PCP performance.

$$S^{2-} + 2h^+ \longrightarrow S \qquad (1)$$

$$(x-1)S + S^{2-} \longrightarrow S_x^{2-} \ (x=2-5) \qquad (2)$$

$$S + S_x^{2-} \longrightarrow S_{x+1}^{2-} \ (x=2-4) \qquad (3)$$

3.4 Mechanism

After evaluating the PCP performances of the TNRs and SnO_2/TNRs films, the e^-/h^+ separation and immigration mechanism for this SnO_2/TNRs composite electrode with and without white light irradiation in the PCP application for 304SS were described in Fig.10. The band gap of SnO_2 is wider than that of rutile TiO_2, and the CB and valance band (VB) of SnO_2 can match well with those of TiO_2[14]. When the white light illumined this composite film, the photogenerated electrons in the VB of TiO_2 and SnO_2 were photoexcited and immigrated to their CB, respectively. Meanwhile, a majority of photoelectrons were rapidly transferred to the 304SS through FTO substrate because of a lower potential (304SS)[30], resulting in that the protected steel was in the state of thermodynamic stability. Compared with the pure TNRs photoanode, both of the photoexcited electrons of two kinds materials could immigrated to the coupled steel, leading to a better PCP performance.

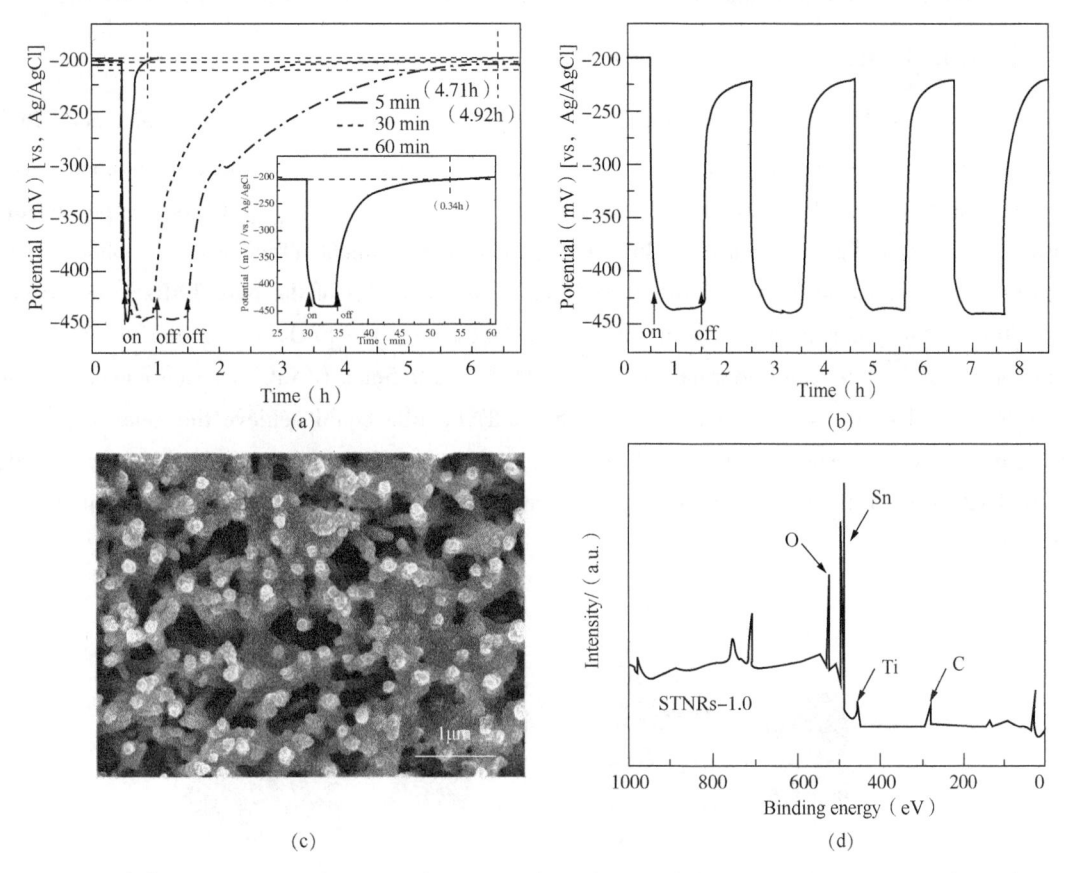

Fig. 9 (a) The potential change of the 304SS electrodes coupled with the STNRs-1.0 film under different-time intermittent visible light illumination. (b) The potential change of the 304SS electrodes coupled with the STNRs-1.0 film under 4 light on/off cycles. The SEM images (c) and the full scan survey XPS spectrum (d) of the STNRs film, which was used in the above stability evaluation experiments for 8 h under intermittent visible light illumination

At the same time, sectional electrons in TiO_2 CB could be immigrated to SnO_2 CB under white light illumination, because the CB of the latter is more negative than that of the former. And the photogenerated holes assembled in TiO_2 VB in the opposite direction are vanished by S^{2-} at the photoanode/electrolyte interface. The ordered band-gap structure is favour of the separation of e^-/h^+ pairs, resulting in that more electrons of SnO_2 and TiO_2 are immigrated to the steel surface. Meanwhile, sectional electrons could be served in SnO_2 CB temporarily, due to its inherent property via a reduction reaction as follows[13,31-33]:

$$SnO_2 + xe^- + xM^+ \rightleftharpoons M_xSnO_2 \ (M = Na\ or\ H,\ x = 1\ or\ 2) \tag{4}$$

with white light illumination, the electrons will be saved in SnO_2 temporarily. After cutting off the light, above reaction reverses to discharge the photoelectrons. Meanwhile, there are a small amount of OVs in the TNRs, and the energy levels of OVs have been reported to be about 1.18 eV below the CB of the rutile TiO_2[34]. Therefore, the electron transitions from the CB of SnO_2 to the OVs levels of TiO_2 contributed to the electron transfer from SnO_2 to TiO_2 along the 1D electron shortcut. And this is the principal factor to achieving the delay protection for 304SS in the dark.

4 Conclusion

In this research, SnO_2/TNRs composite film was successfully synthesized with enhanced PCP performances and electrons storage ability via hydrothermal treatment combined with electrodeposition method. Compared with the pure TNRs film, the SnO_2/TNRs composite film possesses superior light response ability and photoelectrochemical efficiency. The photocurrent density of this composite film is almost three times higher than that of the pure TNRs based on the deposition of SnO_2 nanoparticles. With white light irradiation, the SnO_2/TNRs film could exhibit superior cathodic protection performance for the 304SS in a 0.5mol/L NaCl solution compared with the pristine TNRs film. More importantly, the SnO_2/TNRs film could achieve the delay protection owing to the electrons storage property of SnO_2 material, even after cutting off the light. In a word, SnO_2/TNRs composite can be regarded as a potential photoelectrode in the PCP applications for 304SS.

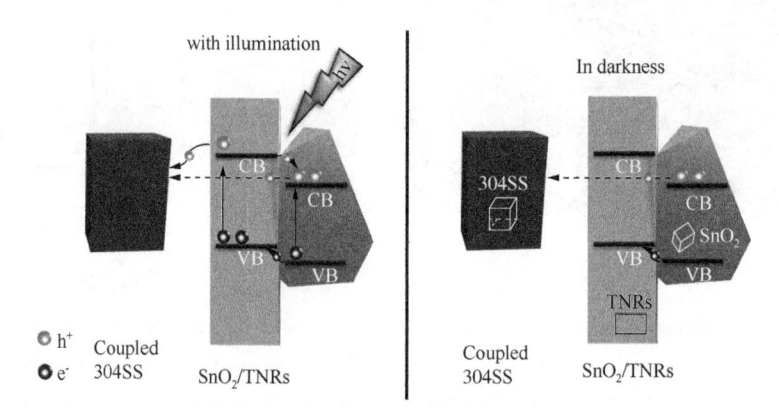

Fig. 10 The sketch of the proposed charge photogeneration and transfer mechanism in the SnO_2/TNRs film for the PCP application with and without solar light illumination

Acknowledgement

The authors gratefully acknowledge the Basic Research and Strategic Reserve Technology Research Fund of China National Petroleum Corporation Project (Grant No. 2018Z-01), Shaanxi Key research and development projects, China (Grant No. 2017SF-386) and the Fundamental Research Funds for the Central Universities of China.

References

[1] J.Zhang, R.G.Du, Z.Q.Lin, Y.F.Zhu, Y.Guo, H.Q.Qi, L.Xu, C.J.Lin, Highly efficient CdSe/CdS co-sensitized TiO_2 nanotube films for photocathodic protection of stainless steel, Electrochim. Acta 83 (2012) 59-64.

[2] H.Li, X.Wang, L.Zhang, B.Hou, Preparation and photocathodic protection performance of CdSe/reduced graphene oxide/TiO_2 composite, Corros.Sci.94 (2015) 342-349.

[3] Y.Liu, C.Zhao, X.Wang, H.Xu, H.Wang, X.Zhao, J.Feng, W.Yan, Z.Ren, Preparation of PPy/TiO_2 core-shell nanorods film and its photocathodic protection for 304 stainless steel under visible light, Mater.Res.Bull. 124 (2020) 110751-110756.

[4] H.Xu, W.Liu, L.Cao, G.Su, R.Duan, Preparation of porous TiO_2/ZnO composite film and its photocathodic protection properties for 304 stainless steel, Appl.Surf.Sci.301 (2014) 508-514.

[5] L.Xu, Y.F.Zhu, J.Hu, J.Zhang, Y.Z.Shao, R.G.Du, C.J.Lin, TiO_2 nanotube films prepared by anodization in glycerol solutions for photocathodic protection of stainless steel, J.Electrochem.Soc.161 (2014) C231-C235.

[6] Y.-F.Zhu, L.Xu, J.Hu, J.Zhang, R.G.Du, C.J.Lin, Fabrication of heterostructured $SrTiO_3$/TiO_2 nanotube array films and their use in photocathodic protection of stainless steel, Electrochim.Acta 121 (2014) 361-368.

[7] S.Cui, X.Yin, Q.Yu, Y.Liu, D.Wang, F.Zhou, Polypyrrole nanowire/TiO_2 nanotube nanocomposites as photoanodes for photocathodic protection of Ti substrate and 304 stainless steel under visible light, Corros.Sci. 98 (2015) 471-477.

[8] S.S.Ge, Q.X.Zhang, X.T.Wang, H.Li, L.Zhang, Q.Y.Wei, Photocathodic protection of 304 stainless steel by MnS/TiO_2 nanotube films under simulated solar light, Surf.Coat.Technol.283 (2015) 172-176.

[9] J.Hu, Q.Liu, H.Zhang, C.D.Chen, Y.Liang, R.G.Du, C.J.Lin, Facile ultrasonic deposition of SnO_2 nanoparticles on TiO_2 nanotube films for enhanced photoelectrochemical performances, J. Mater. Chem. A 3 (2015) 22605-22613.

[10] J.Hu, Y.F.Zhu, Q.Liu, Y.B.Gao, R.G.Du, C.J.Lin, SnO_2 nanoparticle films prepared by pulse current deposition for photocathodic protection of stainless steel, J.Electrochem.Soc.162 (2015) C161-C166.

[11] H.Li, X.Wang, L.Zhang, B.Hou, CdTe and graphene co-sensitized TiO_2 nanotube array photoanodes for protection of 304SS under visible light, Nanotechnology 26(2015) 155704-155714.

[12] W.Liu, K.Yin, F.He, Q.Ru, S.Zuo, C.Yao, A highly efficient reduced graphene oxide/SnO_2/TiO_2 composite as photoanode for photocathodic protection of 304 stainless steel, Mater.Res.Bull.113 (2019) 6-13.

[13] J.Zhang, Z.Ur Rahman, Y.Zheng, C.Zhu, M.Tian, D.Wang, Nanoflower like SnO_2-TiO_2 nanotubes composite photoelectrode for efficient photocathodic protection of 304 stainless steel, Appl.Surf.Sci.457 (2018) 516-521.

[14] T.H.Huy, D.P.Bui, F.Kang, Y.F.Wang, S.H.Liu, C.M.Thi, S.J.You, G.M.Chang, V.V.Pham, SnO_2/TiO_2 nanotube heterojunction: the first investigation of NO degradation by visible light-driven photocatalysis, Chemosphere 215 (2019) 323-332.

[15] Y.Zhang, Q.Lin, N.Tong, Z.Zhang, H.Zhuang, X.Zhang, W.Ying, H.Zhang, X.Wang, Simple fabrication of SnO_2 quantum-dot-modified TiO_2 nanorod arrays with high photoelectrocatalytic activity for overall water splitting, Chemphyschem 19 (2018) 2717-2723.

[16] B.Sun, Y.Chen, L.Tao, H.Zhao, G.Zhou, Y.Xia, H.Wang, Y.Zhao, Nanorods array of SnO_2 quantum dots interspersed multiphase TiO_2 heterojunctions with highly photocatalytic water splitting and self-rechargeable battery-like applications, ACS Appl.Mater.Interfaces 11 (2018) 2071-2081.

[17] Y.Li, Q.Zhang, L.Niu, J.Liu, X.Zhou, TiO_2 nanorod arrays modified with SnO_2-Sb_2O_3 nanoparticles and application in perovskite solar cell, Thin Solid Films 621(2017) 6-11.

[18] S. V. Mohite, V. V. Ganbavle, K. Y. Rajpure, Photoelectrochemical performance and photoelectrocatalytic

degradation of organic compounds using Ga：WO_3 thin films, J.Photochem.Photobiol.A 344（2017）56-63.

［19］ Y.Liu, W.Zhang, C.Zhao, H.Wang, J.Chen, L.Yang, J.Feng, W.Yan, Study on the synthesis of poly(pyrrole methane)s with the hydroxyl in different substituent position and their selective adsorption for Pb^{2+}, Chem.Eng.J.361（2019）528-537.

［20］ P.Zhang, S.Zhu, Z.He, K.Wang, H.Fan, Y.Zhong, L.Chang, H.Shao, J.Wang, J.Zhang, C.N.Cao, Photochemical synthesis of SnO_2/TiO_2 composite nanotube arrays with enhanced lithium storage performance, J.Alloys.Compd.674(2016) 1-8.

［21］ H.Zhu, J.Tao, X.Dong, Preparation and photoelectrochemical activity of Cr-Doped TiO_2 nanorods with nanocavities, J.Phys.Chem.C 114（2010）2873-2879.

［22］ P.K.Giri, B.Santara, K.Imakita, M.Fujii, Microscopic origin of lattice contraction and expansion in undoped rutile TiO_2 nanostructures, J.Phys.D-Appl.Phys.47(2014) 215302-215315.

［23］ W.Cui, J.He, H.Wang, J.Hu, L.Liu, Y.Liang, Polyaniline hybridization promotes photo-electro-catalytic removal of organic contaminants over 3D network structure of rGH-PANI/TiO2 hydrogel, Appl. Catal. B 232（2018）232-245.

［24］ Q.Shi, Z.Li, L.Chen, X.Zhang, W.Han, M.Xie, J.Yang, L.Jing, Synthesis of SPR $Au/BiVO_4$ quantum dot/rutile-TiO_2 nanorod array composites as efficient visiblelight photocatalysts to convert CO_2 and mechanism insight, Appl.Catal.B 244(2019) 641-649.

［25］ M.K.Singh, M.C.Mathpal, A.Agarwal, Optical properties of SnO_2 quantum dots synthesized by laser ablation in liquid, Chem.Phys.Lett.536（2012）87-91.

［26］ L.Zhang, Y.Li, Q.Zhang, H.Wang, Well-dispersed Pt nanocrystals on the heterostructured TiO_2/SnO_2 nanofibers and the enhanced photocatalytic properties, Appl.Surf.Sci.319（2014）21-28.

［27］ Y.Yang, Y.F.Cheng, Factors affecting the performance and applicability of $SrTiO_3$ photoelectrodes for photoinduced cathodic protection, J.Electrochem.Soc.164(2017) C1067-C1075.

［28］ J.Ren, B.Qian, J.Li, Z.Song, L.Hao, J.Shi, Highly efficient polypyrrole sensitized TiO_2 nanotube films for photocathodic protection of Q235 carbon steel, Corros.Sci.111（2016）596-601.

［29］ A.A.Anani, Electrochemical production of hydrogen and sulfur by low-temperature decomposition of hydrogen sulfide in an aqueous alkaline solution, J.Electrochem.Soc.137（1990）2703-2709.

［30］ Q.Wei, X.Wang, X.Ning, X.Li, J.Shao, H.Li, W.Wang, Y.Huang, B.Hou, Characteristics and anticorrosion performance of WSe_2/TiO_2 nanocomposite materials for 304 stainless steel, Surf.Coat.Technol.352（2018）26-32.

［31］ R.Subasri, T.Shinohara, K.Mori, TiO_2-based photoanodes for cathodic protection of copper, J.Electrochem.Soc.152（2005）B105-B110.

［32］ Raghavan Subasri, Tadashi Shinohara, Kazuhiko Mori, Modified TiO_2 coatings for cathodic protection applications, Sci.Technol.Adv.Mater.6（2005）507-507.

［33］ H.Li, X.Wang, Y.Liu, B.Hou, Ag and SnO_2 co-sensitized TiO_2 photoanodes for protection of 304SS under visible light, Corros.Sci.82（2014）145-153.

［34］ N.Wei, Y.Liu, M.Feng, Z.Li, S.Chen, Y.Zheng, D.Wang, Controllable TiO_2 coreshell phase heterojunction for efficient photoelectrochemical water splitting under solar light, Appl.Catal.B 244（2019）519-528.

本论文原发表于《Corrosion Science》2020 年第 176 卷。

Effect of CO_2/H_2S and Applied Stress on Corrosion Behavior of 15Cr Tubing in Oil Field Environment

Zhao Xuehui[1,2] Huang Wei[3] Li Guoping[4] Feng Yaorong[2] Zhang Jianxun[1]

(1. School of Materials Science and Engineering, State Key Laboratory for Mechanical Behavior of Materials, Xi'an Jiaotong University; 2. State Laboratory for Performance and Structure Safety of Petroleum Tubular Goods and Equipment Materials, CNPC Tubular Goods Research Institute; 3. Oil and Gas Engineering Research Institute of PetroChina Changqing Oilfield Company; 4. Research Institute of drilling and Production Technology of Qinghai Oilfield)

Abstract: The corrosion behavior of a 15Cr-6Ni-2Mo tubing steel (15Cr tubing) in a CO_2/H_2S environment was investigated by conducting high-temperature/high-pressure immersion tests combined with scanning electron microscopy and metallographic microscopy. The presence of H_2S decreased the corrosion resistance of the 15Cr tubing steel. The critical H_2S partial pressure (P_{H_2S}) for stress corrosion cracking of 15Cr tubing steel in the simulated oil field environment with a CO_2 partial pressure of 4 MPa and an applied stress of 80% σ_s was identified. The 15Cr tubing steel mainly suffered uniform corrosion with no pitting and cracking when the P_{H_2S} was below 0.5 MPa. When the P_{H_2S} increased to 1 MPa and the test temperature was 150℃, the pitting and cracking sensitivity increased. The stress corrosion cracking at higher P_{H_2S} is attributed to the sulfide-induced brittle fracture.

Keywords: Tubing steel; Stress corrosion cracking; High temperature High pressure; Pitting corrosion

1 Introduction

Corrosion is a difficult problem and a pervasive issue encountered during the development and exploitation of oil and gas fields. In particular, with the deep exploitation and the aging of oil fields, the environment for oil and gas exploitation are becoming more complex and harsher for the metal pipes, leading to their severe deterioration[1]. Therefore, the corrosion damage of oil casing tubes has increased significantly, causing frequent production halts or oil spill accidents, which leads to substantial economic losses[2-5]. Therefore, effective anti-corrosion technology and measures are

---

Corresponding author: Zhao Xuehui, zhaoxuehui@cnpc.com.cn.

urgently needed. So far, various anti-corrosion measures have been developed, which play an important role in the prevention and mitigation of oil field corrosion. At present, the main anti-corrosion measures include the use of corrosion resistant alloys, coatings and corrosion inhibitors, and their combinations[6-13]. A comparative analysis of the economic situation[14] indicates that in terms of long-term cost investments, the application of corrosion resistant materials showed distinct superiority. The use of corrosion resistant materials during the oil and gas exploitation not only prevents initial investments but also reduces maintenance, repair, and other associated expenses[15-17]. The main objective of developing a new type of oil pipe material is to further reduce the cost. The material 15Cr is a relatively new type of martensitic stainless steel material, and compared with 13Cr stainless steel[18-20], it has an increased content of Cr and Ni, which possibly increases the corrosion resistance of the material. To date, the basic properties of 15Cr stainless steel and its corrosion resistance in the oil field acidification environment have been reported[21-23]. The influence of pH and Cl$^-$ content on the corrosion resistance of 15Cr has also been investigated and the threshold P_{H_2S} of 15Cr SS is 1.5 psi; above this value 15Cr might suffer sulfide stress cracking[24,25]. Yasuhide Ishiguro et al.[26] studied the stress corrosion cracking of 15Cr-125ksi materials with a 100% SMYS applied stress in a H$_2$S environment, and showed no SSC at P_{H_2S} = 0.1MPa. However, there are few reports on the corrosion behavior of 15Cr tubing with applied stress under the co-existence of CO$_2$ and H$_2$S in simulated oilfields, and the critical value of H$_2$S affecting 15Cr-110ksi tubing under certain partial pressure of CO$_2$ still needs to be further determined.

In this study, the corrosion behavior of 15Cr martensitic stainless steel tubing under applied stress was studied under the co-existence of H$_2$S and CO$_2$, and the environmentally assisted cracking susceptibility of the materials under stress and stress-free conditions was compared. Moreover, the threshold values of the H$_2$S concentration for the 15Cr tubing steel in the H$_2$S and CO$_2$ environment were obtained.

2 Experimental

2.1 Materials and methods

The samples were cut from 15Cr-6Ni-2Mo (UNS S42625) martensite stainless steel commercial tubing, with the chemical compositions 0.024 wt.% of C, 0.002 wt.% of S, 0.013 wt.% of P, 0.28 wt.% of Si, 15.22 wt.% of Cr, 0.18 wt.% of Mn, 6.31 wt.% of Ni, 2.11 wt.% of Mo, 0.41 wt.% of Cu, 0.02 wt.% of N, and the remaining (up to 100) wt.% of Fe, and an actual yield strength (σ_s) of 825 MPa was used. The samples were divided into two groups: with and without applied stress. The samples without applied stress were cut to the size of 40mm×10mm×3mm (length × width × thickness), and the size of the samples for the stress corrosion test was 57mm × 10mm×3mm. Prior to each experiment, all the samples were carried out in accordance with the treatment procedures specified in the standard "GB/T 15970.2—2000 (ISO 7539-2)", so as to avoid and reduce the existence of residual stress. For the samples with applied stress, the longitudinal edge was polished to prevent stress concentration caused by the edges and corners. After polishing, the samples were cleaned with distilled water and ethanol, dried under cool air, and stored in a dry N$_2$ atmosphere.

The tests were carried out in an autoclave at 90 and 150℃ at the CO$_2$ partial pressure of

4MPa. The H₂S partial pressures (P_{H_2S}) were 0.1, 0.5, and 1 MPa, respectively. The test solution with the Cl⁻ concentration of 150g · L⁻¹, simulating the formation water in oil field, was prepared from analytical grade sodium chloride and deionized water.

2.2 Immersion tests

The immersion tests were conducted in an autoclave. Prior to the weight loss tests, the samples were cleaned with distilled water and acetone, dried, and then weighed using a balance with precision of 0.1 mg. The weight values were recorded as the original weight (W_{0i}, i = 1, 2, ⋯). In order to ensure the reproducibility of the results, four parallel samples were taken from each group. The solution was deoxygenated by bubbling pure nitrogen to ensure that the low oxygen concentration was below 10 ppb (according to the requirement of oil field water quality). Then, H₂S, CO₂, and heat were introduced. The total pressure was brought up to 10 MPa by purging nitrogen through the system.

After the completion of the tests, the corroded samples were divided into two groups. One sample was used to observe the corrosion morphology, and the other three were used to calculate the average corrosion rate to ensure the reproducibility of the results. The corrosion products were removed via the chemical method according to GB/T 16545—1996 (idt ISO 8407: 1991), then rinsed and dried, and then finally reweighed to obtain the final weight (W_{1i}). The corrosion rate (V_i) was calculated in mm/y (average corrosion thinning depth in years) from the weight loss by using Equation (1)[27] as follows:

$$V_i = 8.76 \times 10^6 \times (W_{0i} - W_{1i}) / (t \times \rho \times S) \quad (i = 1, 2, \cdots) \quad (1)$$

Where W_{0i} and W_{1i} are, respectively, the original and final weight of the samples in g; S is the exposed surface area of the samples in mm²; t represents the immersion time in h; and ρ is the steel density equaling 7.8 × 10⁻³ g · mm⁻³. An average corrosion rate of the three different samples for each test condition was reported as an overall corrosion rate for each set of conditions.

2.3 SCC testing

Stress corrosion cracking testing was carried out by the immersion method under high temperature and high pressure conditions in an autoclave. To ensure the reproducibility of the results, there were three parallel specimens in each stress test, and the test was only effective when the specimen broke or cracked in the working area. In these tests, all sample surfaces require precision machining and a surface roughness Ra ⩽ 0.2 μm. These stress corrosion cracking test were adopted a four - point bending loading method[28]; this method was operated according to the standard GB/T 15970.2—2000 (ISO 7539-2: Corrosion of metals and alloys - Stress corrosion testing—Part 2: Preparation and use of bent-beam specimens). The schematic diagram of the four - point bending loading device is shown in Fig. 1.

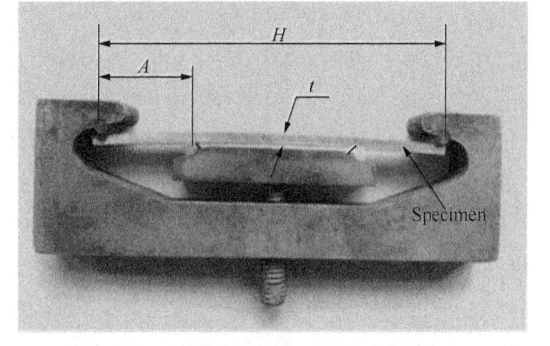

Fig. 1 Schematic presentation of the four-point bending loading method

In this study, according to the engineering requirements, the actual yield strength (825 MPa) value of the material was used to calculate the stress value loaded on the sample rather than the specified minimum yield strength

(SMYS) (758 MPa), so an appropriate stress ratio coefficient of 80% was used. Stress loading can be achieved by Euqation (2). as follows:[29]

$$\sigma = 12Ety/(3H^2-4A^2) \quad (2)$$

where H represents the distance between the two outermost fulcrums, E is the elastic modulus, A represents the distance between the inner and outer fulcrums, t is the sample thickness, and σ represents the stress value of the loading. Using these known parameters to calculate the maximum deflection y between the outer fulcrums, the stress is loaded by detecting the deflection change of the specimen.

2.4 Characterization

After the corrosion tests, the samples used for the surface analysis were removed and rinsed with deionized water, followed by an alcohol rinse, and then dried.

The corrosion products and cracks were studied using scanning electron microscopy (SEM) and optical microscopy (OM). EDS was used to analyze the compositions of the corrosion products.

3 Results and discussion

3.1 Average corrosion rate

The average corrosion rates under different H_2S partial pressure are listed in Table 1. When P_{CO_2} and P_{H_2S} equal 4MPa and 1MPa, respectively, the specimen breaks and the fracture surface becomes corroded, and thus the corrosion rate cannot be calculated accurately.

Table 1 The average corrosion rates of samples under high temperature and high pressure

Condition Number	Test Conditions	V_{corr}(μm/y), w—Range			
		90℃		150℃	
		0%	80%σ_s	0%	80%σ_s
1	P_{H_2S} = 0.1 MPa	0.73(w = 0.20)	1.30(w = 0.35)	1.24 (w = 0.36)	1.91(w = 0.06)
2	P_{H_2S} = 0.5 MPa	1.0 (w = 0.36)	1.60 (w = 0.40)	1.7 (w = 0.20)	3.2 (w = 0.42)
3	P_{H_2S} = 1 MPa	11.03 (w = 0.20)	15.20 (w = 0.43)	29.03 (w = 0.4)	Fracture

The corrosion rates of the samples with 80% σ_s were higher than those of the samples without applied stress. The average corrosion rate of the samples increased gradually with the increase in the H_2S partial pressure and test temperature. Under the high H_2S partial pressure, the sample under the applied stress cracked, indicating that the sulfide-caused stress corrosion cracking gradually became more dominant with the increase in the H_2S concentration[24]. H_2S dissolves in water and reacts chemically to form metal sulfides and hydrogen atoms. The hydrogen atom diffuses into the material at the crack tip with the highest tensile stress. The diffusion and accumulation of hydrogen on the lattice, the surface of the lattice, and the grain boundary reduce the plastic deformation ability of the material, and cause hydrogen embrittlement, makeing it easier the crack to expand.

On the other hand, H_2S is a hydrogen recombination poison, which significantly affects brittle fracture of materials[30]. The hydrogen ions formed on the cathode cannot effectively form hydrogen molecules due to the poisoning effect of the HS^- and S^{2-} ions[31,32], and thus the hydrogen concentration on the surface of the steel increases, resulting in the diffusion of hydrogen into the steel. Steel enrichment by hydrogen eventually makes the steel brittle. Moreover, under the applied stress, the surface activity of the samples gets enhanced, and fractures are more likely to occur[33]. The H_2S partial pressure also directly affects the pH of the solution[34]. When the partial pressure increases, the corrosion products on the metal surface become loose and they gradually increase the corrosion rate[35,36]. At the same time, when the loose product film falls off, there is a potential difference between the exposed substrate and the product film with good adhesion, and thus many tiny corrosion couples are formed on the surface of the sample which promote the accelerated corrosion in the anode region with a low potential.

Under the co-existence of CO_2 and H_2S, the co-rrosive medium has a competitive and synergistic effect on the corrosion behavior of the materials. With the increase in the test temperature, the solubility of the CO_2/H_2S corrosion gas decreases, yet the reaction rate in the solution system accelerates promoting corrosion. Therefore, the overall outcome depends on the synergistic effect among the temperature, stress, and corrosive gases.

3.2 Observation of corrosion morphology

After the high-temperature and high-pressure tests, the surface corrosion morphology was inspected. Fig. 2 and Fig. 3 show the microscopic morphologies of the corroded samples without applied stress. The corrosion scales were even. Polishing marks on the surface indicate that the corrosion film was thin (Fig. 2). Fig. 3 shows that with the increase in the test temperature, the surface corrosion morphology was relatively rough, and when at P_{H_2S} = 1 MPa, similar pitting morphology was found on the surface of the samples at 150℃ [Fig. 3(c)]. The results show that the pitting corrosion sensitivity of 15Cr stainless steel increases due to the synergistic action between a high temperature and high H_2S concentration.

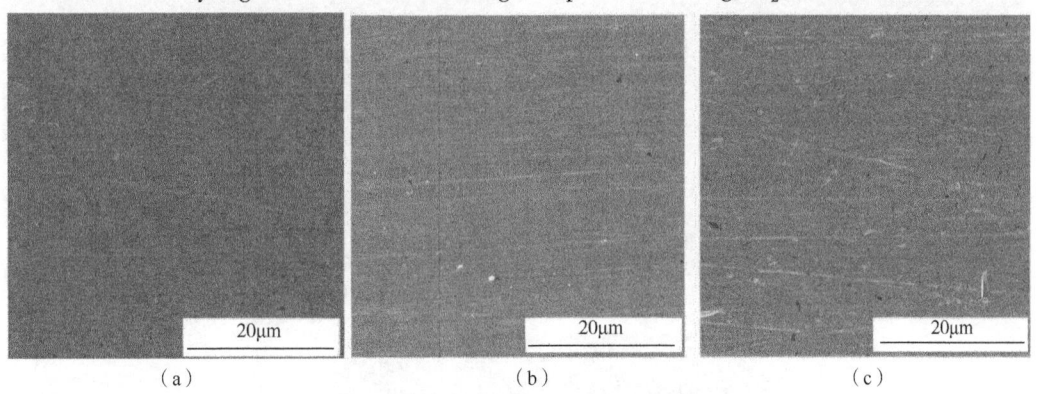

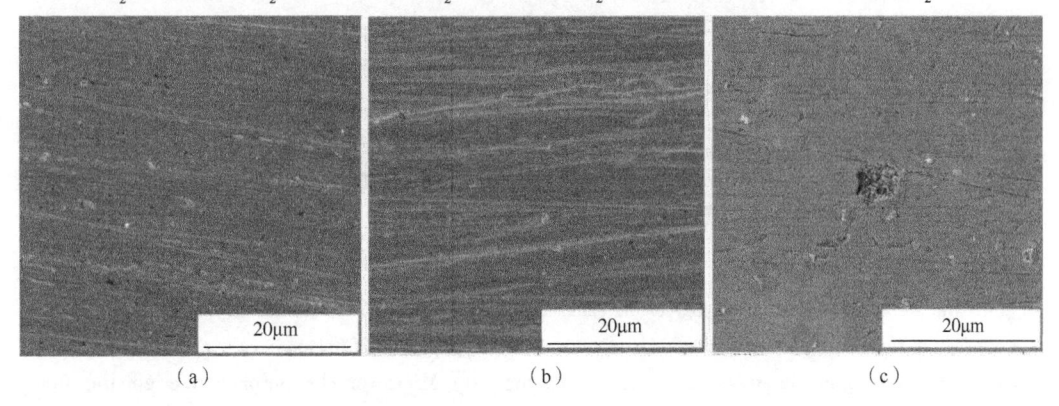

Fig. 2　Micro-corrosion morphology of the samples without stress at 90℃, with different experimental conditions
　　(a) P_{H_2S} = 0.1 and P_{CO_2} = 4MPa, (b) P_{H_2S} = 0.5 and P_{CO_2} = 4MPa and (c) P_{H2S} = 1 and P_{CO_2} = 4MPa.

Fig. 3　Micro-corrosion morphology of the samples without stress at 150℃, with different experimental conditions
　　(a) P_{H_2S} = 0.1 and P_{CO_2} = 4MPa, (b) P_{H_2S} = 0.5 and P_{CO_2} = 4MPa and (c) P_{H_2S} = 1 and P_{CO_2} = 4MPa

Fig. 4 shows the surface corrosion morphology (as shown in the red box area) of the samples under the condition of applied stress. At $P_{H_2S} \leq 0.5$ MPa and with an increase in the temperature, the corrosion film on the samples surface is similar to that on the surface of the samples without stress, and no cracking or fractures are observed [Fig. 4(a)], indicating that the applied stress of 80% σ_s under the experimental conditions has no obvious effect on the surface activity of the samples. However, at P_{H_2S} of 1MPa and a test temperature of 150℃, the samples exhibit crack and fracture at their center region. Nonetheless, there are more micro-cracks around the main crack [Fig. 4(b)].

(a) Macroscopic corrosion morphology (P_{H_2S}=0.5MPa P_{CO_2}=4MPa 150℃)

(b) Macroscopic mporphology of fracture (P_{H_2S}=1MPa P_{CO_2}=4MPa 150℃)

(c) Cracks and pitting (P_{H_2S}=1MPa P_{CO_2}=4MPa 150℃)

(d) Micro-cracks, 20X magnification (P_{H_2S}=1MPa P_{CO_2}=4MPa 150℃)

(e) Micro-cracks, 50X magnification (P_{H_2S}=1MPa P_{CO_2}=4MPa 150℃)

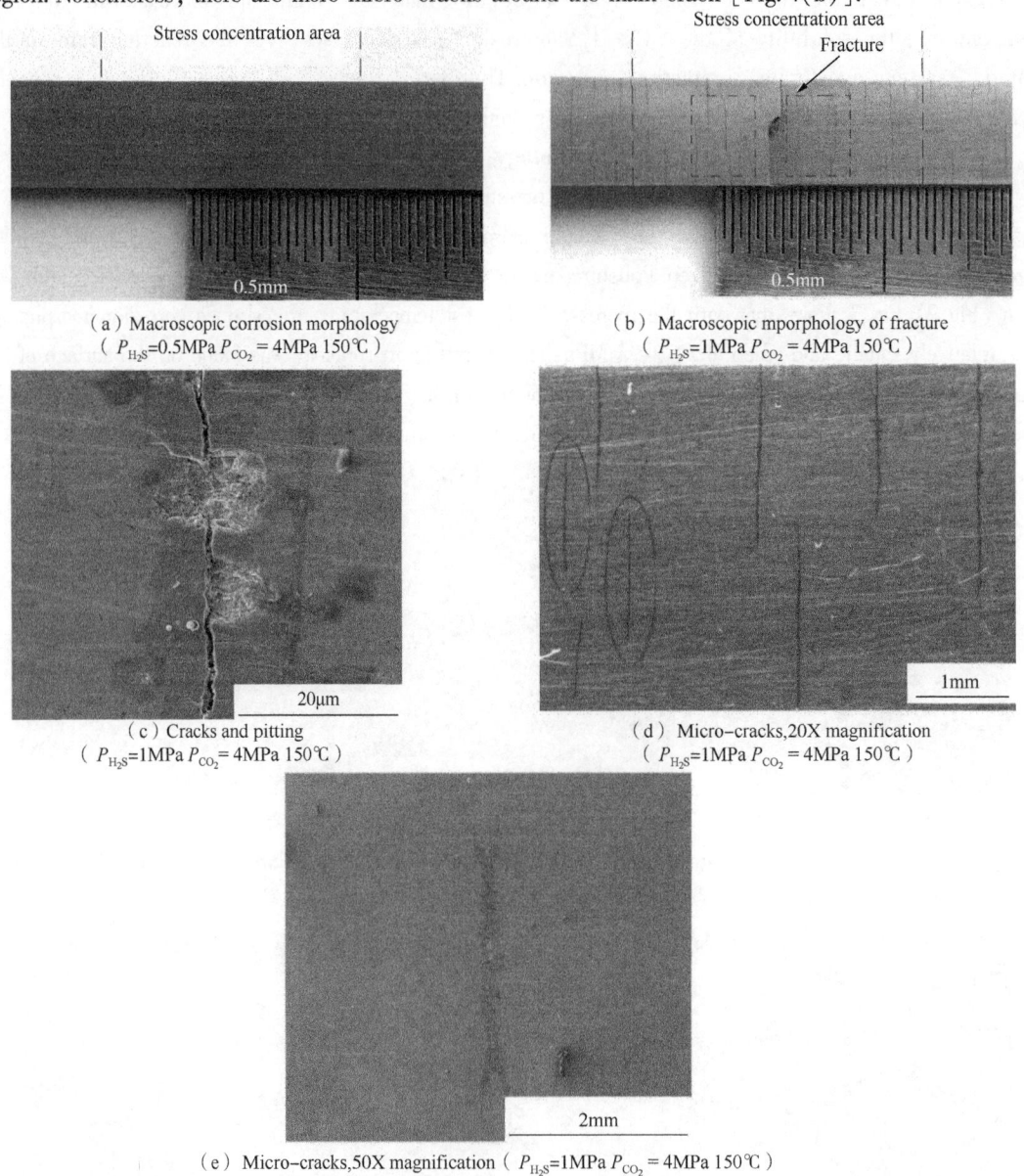

Fig. 4 The surface corrosion morphology of the samples under the applied stress at 150℃: (a) Macro-corrosion morphology of stress concentration surface, (b) Macro-corrosion morphology of fracture area, (c) Micro-morphology of crack and pitting in the stress concentration region, (d) Micro-cracks morphology originating from the center of the sample in the stress concentration region (20×magnification) and (e) Micro-crack morphology under 50×magnification

Fig. 4(c) exhibits that the pitting corrosion occurred around the crack, indicating that when the P_{H_2S} equals 1MPa, the pitting susceptibility of the material under the stress state reaches its critical value. Pitting corrosion plays an important role in stress corrosion cracking. Surface cracks were also observed on the fractured samples. Clearly, some micro-cracks originated from the center of the sample along its width direction and extended to both ends of the sample, and the two red rings in Fig. 4(d) indicate the origin of the cracks in the middle of the samples. Fig. 4(e) is a high magnification of the cracks marked with red rings, and the typical characteristics of sulfide stress corrosion cracking at both ends of the crack can be observed, which indicates that the stress corrosion cracking sensitivity of 15Cr stainless steel is higher and that cracking failure occurs under these simulated test conditions.

The cross-sections of the fracture side morphologies are shown in Fig. 5. The main crack propagated almost through the thickness of the sample. Many smaller sub- and micro-cracks were located around the main crack. Distinct dendritic cracks were found at the crack tips, showing typical sulfide stress corrosion features [marked by the red circles in Fig. 5(a, b)]. Thus, 15Cr stainless steel tubing is not suitable for use in the simulated oil field environment as H_2S-related corrosion occurs easily.

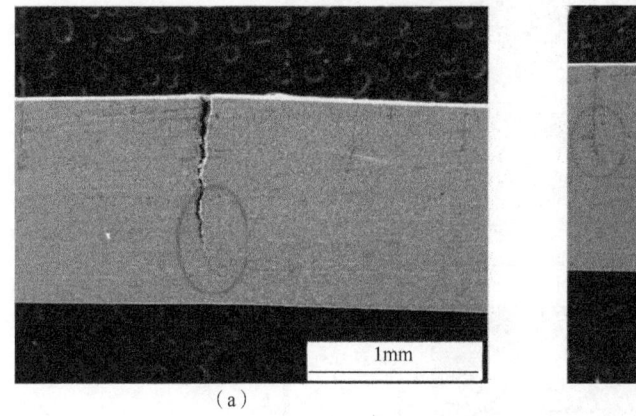

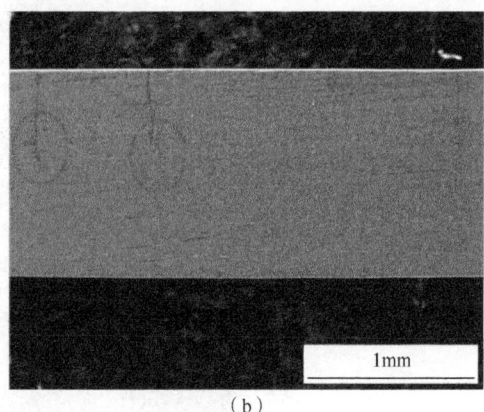

Fig. 5 The side morphology of the fractured sample:
(a) Main crack tip propagation morphology and (b) Growth morphology of sub-cracks tip

The color of the corrosion products near the cracks [marked as the "A" region in Fig. 6(a)] was different from that of the other no-cracking areas (marked as the "B" region): the "A" region appears dark gray and the "B" region has a lighter color.

The EDS analysis [shown in Fig. 6(b)] revealed that the "A" and "B" regions were mainly composed of S, C, O, and Fe elements. the However, sulfur content in the "A" region was higher than in "B" region. This indicates that the adsorption of H_2S on the active surface and the anodic dissolution of the matrix promote the occurrence of stress cracking. However, the higher content of sulfide at the fracture region further demonstrates that the H_2S can elevate the cracking susceptibility of the samples under applied stress[37,38]. Thus, the surface activity of the stress-concentrated region is much higher, and the chemical reaction between the alloy and H_2S is favored over the reaction of CO_2 and the material. Moreover, at high temperature, H_2S dissolves in the solution and the H atoms get adsorbed and penetrate the sample surface, accelerating The chemical reaction between S^{2-} and the material[39]. Owing to this effect, increased amounts of sulfides were generated in the stress-concentrated regions of the samples. Therefore, the synergistic effect of the applied stress and H_2S corrosion medium contributed to much more severe localized corrosion in the stress-concentrated region, prompting the formation and extension of the crack[38], and finally inducing the sample fracture. The results are consistent with the stress corrosion cracking characteristics of the preceding broken sample, indicating that the failure of the specimens mainly belongs to the sulfide stress cracking mode. The mechanism for the reaction between H_2S and the materials in the test solution is shown by Equations (3)–(7)[40]:

$$Fe + H_2S \rightarrow Fe + H_2S_{adsorbed} \quad (3)$$

$$Fe + H_2S_{adsorbed} \rightarrow Fe + HS^-_{adsorbed} + H^+_{adsorbed} \quad (4)$$

$$Fe + HS^-_{adsorbed} + H^+_{adsorbed} \rightarrow FeHS^-_{adsorbed} + H^+_{adsorbed} \quad (5)$$

$$FeHS^-_{adsorbed} + H^+_{adsorbed} \rightarrow FeHS^+_{adsorbed} + H_{adsorbed} + e^- \quad (6)$$

$$FeHS^+_{adsorbed} + H_{adsorbed} + e^- \rightarrow FeS_{adsorbed} + 2H_{adsorbed} \quad (7)$$

(a) Micromorphology of corrosion product film

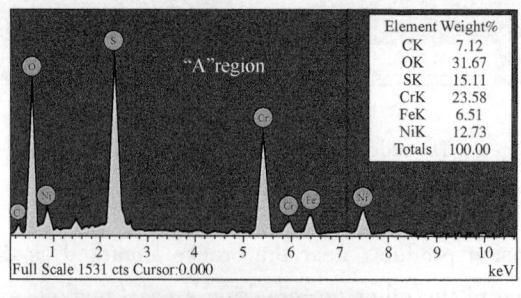

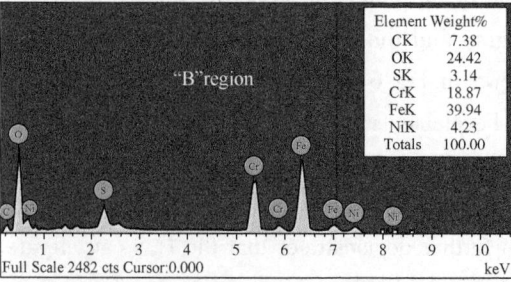

"A" region

Element	Weight%
CK	7.12
OK	31.67
SK	15.11
CrK	23.58
FeK	6.51
NiK	12.73
Totals	100.00

"B" region

Element	Weight%
CK	7.38
OK	24.42
SK	3.14
CrK	18.87
FeK	39.94
NiK	4.23
Totals	100.00

(b) EDS of corrosion product film (A and B region)

Fig. 6 Surface morphologies (a) corrosion products of two different regions near the crack and (b) (Energy-dispersive spectroscopy) EDS analyses

The fractured samples are separated along the fracture direction. Subsequently, the corrosion morphologies of the fracture surface are shown in Fig. 7. The Low magnification of the fracture surface does not reveal metallic luster [Fig. 7(a)]. Higher magnification microscopy images show athick corrosion film covering the fracture surface [Fig. 7(b)], and as a result, the exact fracture morphology of the crack is hard to detect. This corrosion film is mainly composed of S, Fe, C, and O elements [Fig. 7(c)]. This can infer that the sample fractured during the test, and the fresh fracture surface was further corroded when exposed to corrosive medium.

(a) Fracture morphology

(b) Corrosion product of fracture surface

Element	Weight%
CK	4.79
OK	13.01
SK	12.66
CrK	14.69
FeK	13.17
NiK	10.47
Totals	100.00

(c) EDS of corrosion product film

Fig. 7 (a) Low and (b) high magnification of the fracture morphology, and its (c) EDS analysis

Fig. 8 shows the micro-morphology of the fracture surface after cleaning the corrosion products, and intergranular corrosion characteristics can be clearly observed (Fig. 8). Secondary cracks along the grain boundaries can also be seen [Fig. 8 (b)]. This further illustrates that the samples loaded with 80% σ_s sustained brittle sulfide stress-corrosion cracking at a 1MPa H_2S partial pressure. Thus, this material has a high sensitivity to stress corrosion cracking at the given experimental condition, and therefore is not suitable for oil field environment with similar conditions.

4 Conclusions

In this study, the corrosion and stress corrosion cracking behavior of a 15Cr tubing steelwith and without applied stress were studied by the weight loss and four-point bending methods in a high-temperature and high-pressure environment. The main results and conclusions are as follows:

(a) Micromorphology of intergranular corrosion on the fracture surface

(b) Secondary crack morphology on fracture surface

Fig. 8　Intergranular corrosion characteristics of fracture surfaces

(1) When the P_{CO_2} was 4MPa and the P_{H_2S} was less than or equal to 0.5MPa, the average corrosion rate of the 15Cr tubing increased in the simulated oil field environment with the increase of H$_2$S pressure. The 15Cr tubing steel mainly showed uniform corrosion. No clear signs of pitting were found, and no crack or fracture occurred when the sam ples were loaded with 80%σ_s.

(2) At 150℃, and at P_{CO_2} = 4MPa and P_{H_2S} = 1MPa, the pitting sensitivity of the 15Cr tubing steel was higher, and cracking occurred when the samples were loaded at 80%σ_s.

Acknowledgements: The authors greatly acknowledge the financial support from the National Key R&D Program of China and the CNPC Science and Technology Project, and also acknowledge the state key laboratory for its support of high temperature and high pressure experiments.

References

[1] Zhao, X. H.; Han, Y.; Bai, Z. Q.; Wei, B. The experiment research of corrosion behavior about Ni-based alloys in simulant solution containing H$_2$S/CO$_2$. Electrochim. Acta. 2011, 56, 7725.

[2] Wang, H. T.; Han, E. H. Simulation of metastable corrosion pit development under mechanical stress. Electrochim. Acta 2013, 90, 128.

[3] Zanotto, F.; Grassi, V.; Balbo, A. Stress corrosion cracking of LDX2101 duplex stainless steel in chloride solutions in the presence of thiosulphate. Corros. Sci. 2014, 80, 205.

[4] Bueno, A. H. S.; Moreir, E. D.; Gomes, J. A. C. P. Evaluation of stress corrosion cracking and hydrogen embrittlement in an API grade steel. Eng. Fail. Anal. 2014, 36, 423.

[5] Lu, L. S.; Song, W. W.; Yang, X. T. Corrosion Causes of Premium Connection S13Cr Tubing in a well. Corros. Prot. 2015, 36, 76.

[6] Fajardo, S.; Bastidas, D. M.; Criado, M. Electrochemical study on the corrosion behavior of a new low-nickel stainless steel in carbonated alkaline solution in the presence of chlorides. Electrochim. Acta. 2014, 129, 60.

[7] Machuca, L. L.; Bailey, S. I.; Gubner, R. Systematic study of the corrosion properties of selected high-resistance alloys in natural seawater. Corros. Sci. 2012, 64, 8.

[8] Zhao, X. H.; Bai, Z. Q.; Feng, Y. R.; Wei, B.; Yin, C. X.; Wang, J. Z. Effects of heat treatment and precipitated phase on corrosion resistance of Ni-based alloy. Trans. Mater. Heat Treat. 2011, 33, 39.

[9] Sun, Y. L.; Bai, Z. Q.; Zhang, G. C.; Wei, B.; Zhu, S. D. Research Statue on Anticorrosion Properties of Bimetallic Composite Tubes in Oil and Gas Field. Total Corros. Control 2011, 25, 10.

[10] Zhao, S.; Lan, W. Present status and research progress of anticorrosion technology in pipeline. Surf. Technol. 2015, 44, 112.

[11] Palimi, M. J.; Peymannia, M.; Ramezanzadeh, B. An evaluation of the anticorrosion properties of the spinel nanopigment filled epoxy composite coatings applied on the steel surface. Prog. Org. Coat. 2015, 80, 164−175.

[12] Prabhu, R. A.; Venkatesha, T. V.; Shanbhag, A. V. Inhibition effects of some Schiff's bases on the corrosion of mild steel in hydrochloric acid solution. Corros. Sci. 2008, 50, 3356.

[13] Sumithra, K.; Kavita, Y.; Manivannan, R.; Noyel, V. S. Electrochemical investigation of the corrosion inhibition mechanism of Tectona grandis leaf extract for SS304 stainless steel in hydrochloric acid. Corros. Rev. 2017, 35, 111−121.

[14] Xiang, R.; Luo, D. D.; Wei, D. Economic evaluation of corrosion protection measures based on corrosion status of typical gathering pipelines. Saf. Environ. 2013, 13, 207.

[15] Ningshen, S.; Sakairi, M.; Suzuki, K.; Ukai, S.; The corrosion resistance and passive film compositions of 12% Cr and 15% Cr oxide dispersion strengthened steels in nitric acid media. Corros. Sci. 2014, 78, 322.

[16] Jiménez, H.; Olaya, J. J.; Alfonso, J. E.; Pineda-Vargas, C. A. Corrosion resistance of Ni-based coatings deposited by spray and fuse technique varying oxygen flow. Surf. Coat Technol. 2017, 321, 341.

[17] Mesquita, T. J.; Chauveau, E.; Mantel, M. Corrosion and metallurgical investigation of two supermartensitic stainless steels for oil and gas environments. Corros. Sci. 2014, 81, 152.

[18] Zhang, Z.; Zheng, Y. S.; Li, J.; Liu, W. Y.; Liu, M. Q.; Gao, W. X.; Shi, T. H. Stress corrosion crack evaluation of super 13Cr tubing in high-temperature and high-pressure gas wells. Eng. Fail. Anal. 2019, 95, 263.

[19] Nacéra, S. M.; Brian, C.; Russell, K.; Tanmay, A.; James, S. Sour Service Limits of Martensitic Stainless Steels: A Review of Current Knowledge, Test Methods and Development Work. In CORROSION 2013; NACE International: Houston, TX, USA, Paper No. 2639.

[20] Qiu, Z. C.; Liu, X.; Zhang, N. Corrosion Behavior of 13Cr Steel in Different Temperature. Adv. Mater. Res. 2014, 37, 168.

[21] Lv, X. H.; Zhang, F. X.; Yang, X. T. Corrosion Performance of High Strength 15Cr Martensitic Stainless Steel in Severe Environments. J. Iron. Steel Res. Int. 2014, 21, 774.

[22] Wan, J. Q.; Ran, Q. X.; Li, J.; Xu, Y. L. A new resource-saving, low chromium and low nickel duplex stainless steel 15Cr-xAl-2Ni-yMn, Mater. Des. 2014, 53, 43.

[23] Meng, J.; Chambers, B.; Kane, R.; Skogsberg, J.; Kimura, M.; Shimamoto, K. Environmentally Assisted Cracking Testing of High Strength 15 Cr Steel in Sour Well Environments. CORROSION 2011, Houston, NACE International 2011, Paper no. 100.

[24] Brian, C.; James, S.; John, M.; Mitsuo, K. Evaluation of Environmentally Assisted Cracking Resistance of High Strength 15Cr Steel in Sour Well Environments. In CORROSION 2012, Houston, USA, No. C0001353.

[25] ISO 15156-3-2015, Petroleum and natural gas industries− Materials for use in H2S-containing environments in oil and gas production − Part 3: Cracking-resistant CRAS (corrosion- resisitant alloys) and other alloys ISO: Genera Switzerland 2005.

[26] http://eurocorr.efcweb.org/2014/abstracts/10/7368.pdf

[27] GB/T 7901-2008: Metal materials− Uniform corrosion−Methods of laboratory immersion testing.

[28] GB/T 15970.2-2000: Corrosion of metals and alloys − Stress corrosion testing − Part 2: Preparation and use of bent-beam specimens.

[29] G; Hinds; L; Wickstrom; K; Mingard; Turnbull, A. Impact of surface condition on sulphide stress corrosion cracking of 316L stainless steel. Corros. Sci. 2013, 71, 43−52.

[30] Jiang, W. J. Piping Material Selection for Wet H2S Environment of Hydrotreating Unit. Shandong Chem. Ind. 2019,

48, 118.

[31] Kittel, J.; Ropital, F.; Grosjean, F.; Sutter, E. M. M.; Tribollet, B. Corrosion mechanisms in aqueous solutions containing dissolved H_2S. Part 1: Characterisation of H_2S reduction on a 316L rotating disc electrode. Corros. Sci. 2013, 66, 324.

[32] Stephen S. N. Current Understanding of Corrosion Mechanisms Due to H_2S in Oil and Gas Production Environments. In CORROSION 2015. , NACE, Dallas, Texas, paper 5485.

[33] Bao, M. Y.; Ren, C. Q.; Hu, J. S.; Liu, B.; Li, J.; Wang, F.; Guo, X. Stress Induced Corrosion Electrochemical Behavior of Steels for Oil and Gas Pipes. Chin. Soc. Corros. Prot. 2017, 37, 504.

[34] Plennevaux, C.; Kittel, J.; Frégonèse, M.; Normand, B.; Ropital, F.; Grosjean, F.; Cassagne, T. Contribution of CO_2 on hydrogen evolution and hydrogen permeation in low alloy steels exposed to H^2S environment, Electrochem. Commun. 2013, 26, 17.

[35] Sun, W.; Nesic, S.; Papavinasam, S. K. Kinetics of Iron Sulfide and Mixed Iron Sulfide/Carbonate Scale Precipitation in CO_2/H_2S Corrosion. In Corrosion 2006, San Diego, California. 12−16 March 2006.

[36] Choi, Y.S., Nesic, S., Ling, S. Effect of H_2S on the CO_2 corrosion of carbon steel in acidic solutions. Electrochim. Acta. 2011, 56, 1752.

[37] Zhang, N. Y.; Zhang, Z.; Zhao, W. T.; Liu, L.; Shi, T. H. Corrosion Evaluation of Tubing Steels and Material Selection in the CO_2/H_2S Coexistent Environment. In CORROSION 2018 , Phoenix, Arizona, USA, 15−19 April 2018.

[38] Lu, Y.; Zhang, Y.; Liu, Z. G.; Zhang, Y.; Wang, C. M.; Guo, H. J., Corrosion control in CO_2/H_2S-produced water of offshore oil fields, Anti−Corros. Method. M. 2014, 61, 166.

[39] Jingen, D.; Wei, Y.; Xiaorong, L.; Xiaoqin, D. Influence of H2S Content on CO2 Corrosion Behaviors of N80 Tubing Steel. Pet. Sci. Technol. 2011, 29, 1387.

[40] G. A. Zhang, Y. Zeng, X. P. Guo, F. Jiang, D. Y. Shi, Z. Y. Chen. Electrochemical corrosion behavior of carbon steel under dynamic high pressure H_2S/CO_2 environment. Corros. Sci. 2012, 65, 37.

本论文原发表于《Metals》2020 年第 10 卷第 409 期。

Corrosion Behavior of Cr-Bearing Steels in CO_2-O_2-H_2O Multi-Thermal-Fluid Environment

Yuan Juntao[1] Zhu Kaifeng[2] Jiang Jingjing[3] Zhang Huihui[4] Fu Anqing[1]
Li Wensheng[1] Yin Chengxian[1] Li Fagen[1]

(1. State Key Laboratory for Performance and Structure Safety of Petroleum Tubular Goods and Equipment Materials, CNPC Tubular Goods Research Institute; 2. PetroChinaDagang Oil and Gas Field Company; 3. Research Institute of Natural Gas Technology, PetroChina Southwest Oil and Gas Field Company; 4. School of Materials Science and Engineering)

Abstract: Multi-thermal fluid is used to enhance heavy oil recovery, but it also causes severe corrosion failure of metallic tubing and its mechanism is still open to debate. In the present work, the corrosion behavior of commercial steels with different Cr content (3 wt.%, 9 wt.%, and 25 wt.%) in simulated CO_2-O_2-H_2O multi-thermal fluid environments were studied by immersion corrosion test and microstructural characterization. The results exhibit that the Cr content of steel determines the type of corrosion products. The less protective corrosion scale is mainly composed of $FeCO_3$, $(Fe, Cr)_3O_4$, and Fe_2O_3, which is formed on the surface of 3Cr and 9Cr steels. The protective corrosion scale is mainly Cr_2O_3, which is formed on the surface of 25Cr steel. Only when the Cr content in steel is sufficient (i.e. 25wt.%) to repair the serious depletion of Cr in the metal near the scale/metal interface in time, can protection of the corrosion scale be sustained, to improve its ability to resist the corrosion of multi-thermal fluids.

Keywords: Stainless steel; Cr content; Corrosion; Exfoliation; Multi-thermal fluid; Corrosion scale

1 Introduction

Direct injection of thermal fluids into the oil reservoir has been used as an effective way to extract heavy oil worldwide[1,2]. Usually, thermal fluids are generated by the burning of fuel and oxidizer. For such wells, metallic tubing would be corroded by corrosive species such as CO_2, O_2, and H_2O. In China, several corrosion failure accidents (Fig. 1) have been taken place, which caused serious consequences. In these cases, the corrosion mechanism is much more complexowing to the competition and/or synergy between CO_2 and oxidizing gas like O_2 and water vapor.

Corresponding author: Yuan Juntao, yuanjuntaolly@163.com; Fu Anqing, fuanqing@cnpc.com.cn.

Fig. 1   Corrosion failure of metallic tubing used in multi-thermal fluid injected well

In the oil and gas industries, CO_2 corrosion has been widely studied[3-7]. The corrosion mechanism has been clear, and some prediction models have been established. Due to the complex composition of the multi thermal fluid, the corrosion mechanism of metallic materials in such an environment is still unclear. In addition to the complex multi – component characteristics, the special temperature range is beyond the scope of published literature. For example, the research temperature of CO_2 corrosion in the oil and gas industry is mostly below 150℃, while that in high-temperature combustion environment is more than 500℃[8,9]. There is a rare study in the intermediate temperature region (150 ~ 500℃). Electrochemical corrosion occurs on metallic materials in the CO_2-containing oil and gas production environment (usually lower than 150℃) and forms ferrous carbonate. Once oxygen is introduced into the CO_2-containing environment, the corrosion will be accelerated and oxides will be formed. In the high-temperature combustion environment (usually beyond 500℃), oxidation takes place on the metallic materials, where CO_2 reacts directly with metals to form oxides. In contrast, corrosion caused by the multi thermal fluid at a temperature range of 150~500℃ Cis still open to debate.

Table 1   Chemical compositions of Cr-bearing steels in the present work (wt. %)

Steel	C	Si	Mn	P	S	Cr	Mo	Ni	Nb	Fe
3Cr	0.056	0.18	0.38	0.0089	<0.002	2.96	<0.0009	0.0050	0.026	Bal.
9Cr	0.11	0.35	0.38	0.014	0.0051	8.82	0.96	0.061	—	Bal.
25Cr	0.06	0.40	1.20	—	—	25.0	—	20.0	0.45	Bal.

Based on the above considerations, the present work was performed to investigate the corrosion behavior of Cr-bearingsteels which is potentially applied as tubing in multi-thermal fluid injected well. On the basis of the characterization of microstructure, chemical compositions, and phases, the effect of Cr content on the corrosion were discussed.

2 Materials and Methods

Commercial steels with different Cr content (3wt. %, 9 wt. %, and 25wt. %) were used in the present work, and their chemical compositions are shown in Table 1. The metallographic analysis showed that the microstructures of the three steels are pearlite+ferrite, martensite and austenite, as shown in Fig. 2.

Rectangular specimens with dimensions of 50mm×10mm×3mm were cut from above Cr-bearing steel tubes, and a suspension hole with a diameter of 5mm drilled near an edge. All specimens were ground to a 1200 grit finish, and then cleaned ultrasonically in deionized water and anhydrous ethanol. After that, they were dried out with could air, and weighed with an electronic balance with an accuracy of 0.1mg.

(a) 3Cr steel (b) 9Cr steel (c) 25Cr steel

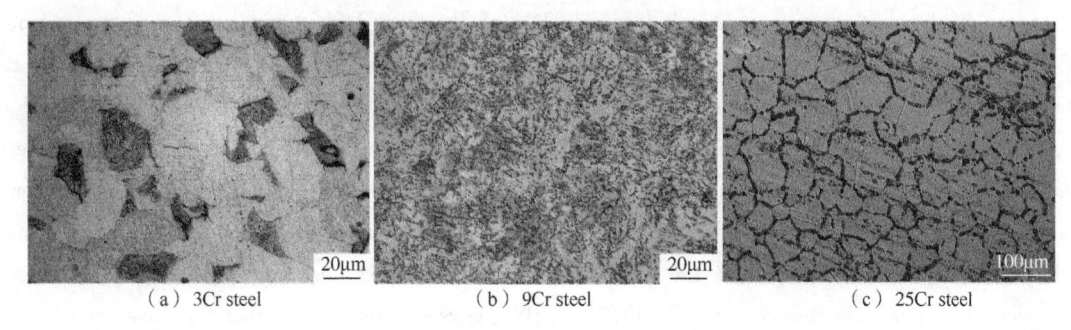

Fig. 2 Metallographic microstructures

An autoclave with a capacity of 3L were used to carry out corrosion tests. Prior to testing, 1 L simulated formation water was disposed according to Table 2, and then high purity nitrogen was introduced to remove oxygen from the solution. After that, the autoclave was heated up to 240℃, and then $CO_2-O_2-N_2$ mixture gas was introduced until the total pressure reached 10MPa. In this case, the p_{CO_2} (partial pressure of carbon dioxide) was 2MPa, the p_{H_2O} (partial pressure of water vapor) was 3.3MPa, and the p_{O_2} (partial pressure of oxygen) was 0.5MPa. During the whole test, the temperature was kept at 240℃, and the total pressure was kept at 10MPa.

Table 2 Ion compositions of simulated formation water (mg/L)

Ions	Ca^{2+}	Mg^{2+}	Na^+	K^+	HCO_3^-	Cl^-	SO_4^{2-}
Content	77.6	35.3	2000	13.4	1186.3	2637.8	22.2

In order to calculate the corrosion rate, three identically tested specimens for each steel were rinsed successively withdeionized water, acid solution, sodium hydroxide solution, deionized water and anhydrous ethanol, as described in ASTM G1 standard. Average corrosion rate (CR) was obtained by Eq. (1),

$$CR(mm/y) = \frac{8.76 \times 10^4 \times W}{A \times T \times D} \quad (1)$$

Where, W is the mass loss (g), T is the exposure time (h), A is the surface area (cm^2), and D is the density of the metal (g/cm^3).

Scanning Electron Microscope (SEM) was applied to examine surface morphology and cross-section morphology of tested specimens, where SecondaryElectron (SE) mode was used for the surface morphology investigation while Backscattered Electron (BSE) mode was used toobserve the cross-section morphology. The typical acceleration voltage of SEM observation was 20kV. Energy Dispersive Spectrometer (EDS) coupled with SEM was applied to analyze the chemical composition of corrosion scales. In addition, the phases of corrosion products wereidentified by X-ray Diffraction (XRD) operating with Cu-Karadiation with a wavelength of 1.5405Å at 40kV and 40mA in the 2θ range of 10°~90° and a scan rate of 0.1 degree per second.

3 Results

3.1 Corrosion of 3Cr steel

The uniform corrosion rate of 3Cr steel is 2.5365mm/y, which is very large compared with most cases. Fig. 3 shows the surface morphology of 3Cr steel after 240h corrosion. There are several distinguishing features. First, exfoliation of surface corrosion products happened in the local region, as shown in Fig. 3(a). Second, at high magnification, cracks can be seen near the edge of the exfoliation region, as shown in Fig. 3(b). Third, coarse granular corrosion products with sharp edges and corners can be seen in the exfoliation region, as shown in Fig. 3(c). In this region, the EDS analysis resultshows that the corrosion scales are composed of oxygen (32.23 wt.%), chromium (20.46 wt.%), and iron (the balance). Fourth, except for the local cracks, the outermost corrosion products show a more compact appearance, as shown in Fig. 3(d). The EDS analysis result reveals that this layer consists of oxygen (33.75 wt.%), chromium (12.10 wt.%), and iron (the balance). The chromium content is significantly lower than that in the exfoliation region.

(a) exfoliation at lowmagnification

(b) near the exfoliation regionat high magnification

(c) exfoliation region at high magnification

(d) the outermost corrosion products at high magnification

Fig. 3 Surface morphology of 3Cr steel after 240h exposure in multi-thermal fluid,

Fig. 4 shows the cross-section morphology [as shown in Fig. 4(a)] and the EDS element profile [as shown in Fig. 4(b)] of 3Cr steel after 240h corrosion. There are several distinguishing features for the corrosion scale. First, the outer corrosion products appear dark gray, the medium layer appears bright gray, and the inner layer appears dark brown. Generally, under the BSE mode, the contrast is related to the average atomic mass of corrosion products, and the brighter contrast indicates the corrosion products with a relatively larger average atomic mass. Second, beneath the outermost layer with dark gray contrast, the corrosion products show coarse grains similar to that in Fig. 3(c). In some regions, the coarse grains show bright gray contrast while some of them near to the outermost layer show dark gray contrast which is similar to the contrast of the outermost layer. Third, close to the interface of the corrosion scale and substrate, a very thin corrosion layer with dark brown contrast can be seen, as shown in Fig. 4(c) at higher magnification. And the EDS point analysis as shown in Fig. 4(d) indicates that the innermost corrosion products would be Cr-rich hydro/oxide such as $Cr(OH)_3$.

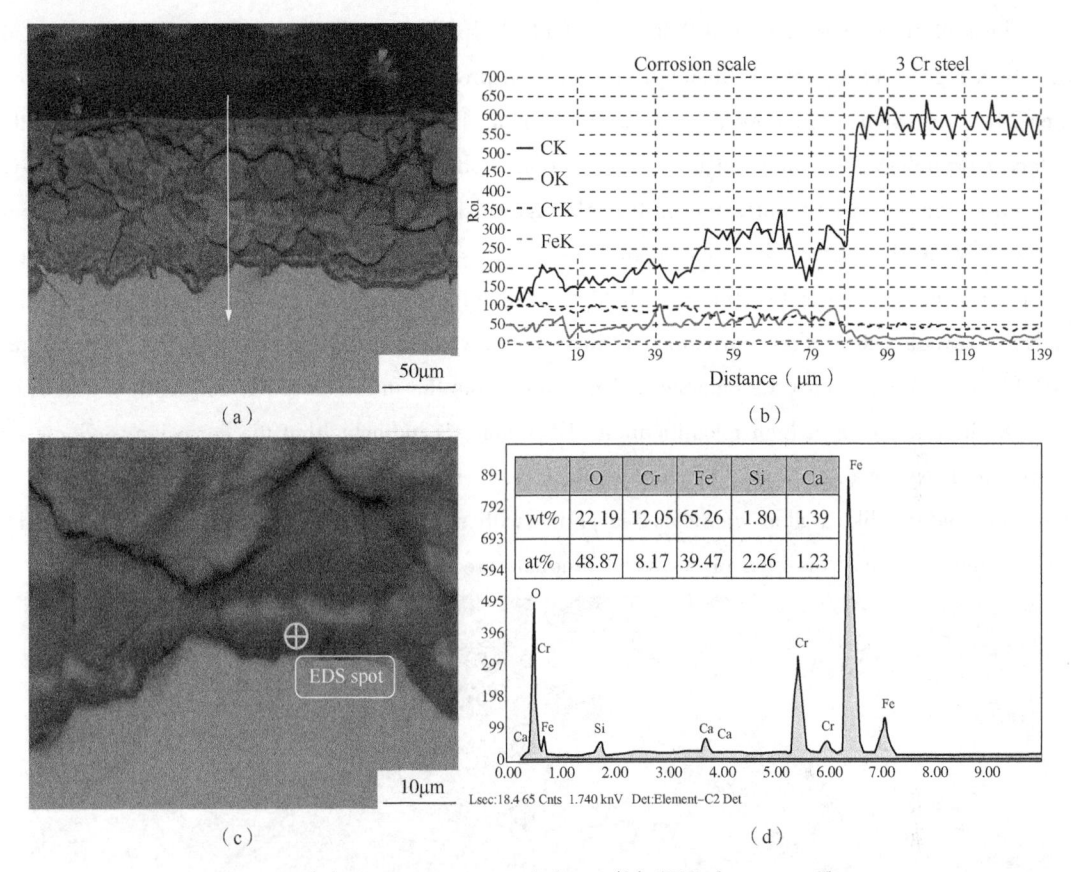

Fig. 4 (a) Cross-section morphology, (b) EDS element profile, (c) the innermost layer at high magnification, and (d) EDS spot analysis result

From the XRD pattern of corroded 3Cr steel, as shown in Fig. 5, it can be seen that the corrosion scales mainly consist of magnetite and iron carbonate. Combined with the above analysis, the outermost layer with dark grey contrast would be iron carbonate, while the layer with bright grey contrast would be magnetite.

It can be speculated that 3Cr steel suffered from serious corrosion in the multi-thermal fluid at 240°C, and formed thick multi-layer corrosion products containing magnetite, iron carbonate, and Cr-rich hydro/oxides. Clear boundaries (or cracks) can be seen through the whole scale,

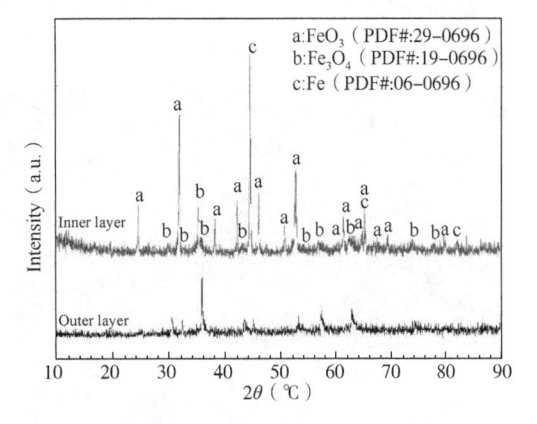

Fig. 5 XRD patterns of 3Cr steel specimens after 240h exposure in multi-thermal fluid

indicating that the corrosion scale is not a good barrier to the corrosive medium. It is too low for the Cr content in 3Cr steel to form a dense and protective Cr-rich layer in the studied multi-thermal fluid environment, thereby leading to poor corrosion resistance to multi-thermal fluid.

3.2 Corrosion of 9Cr steel

The uniform corrosion rate of 9Cr steel is 11.0140mm/y, which is almost 4.5 times that of 3Cr steel. Fig. 6(a) shows the blister-like corrosion products appeared on the surface of 9Cr steel after 240h corrosion, and they are prone to exfoliate. Employing the SE mode to investigate the surface morphology, the image at low magnification is shown as Fig. 6(b). It can be seen that 9Cr suffered extremely serious corrosion. Extremely significant exfoliation and long enough cracks can be observed on the surface. Besides, nodule-like corrosion products covered the larger part of the surface. Fig. 6(c) indicates the morphology of exfoliation region at high magnification. EDS analysis reveals that the corrosion scales contain oxygen (26.71 wt.%), chromium (19.93 wt.%) and iron (the balance). Fig. 6(d) shows the interface of flat and nodule. Fig. 6(e) presents the flat region at high magnification, EDS analysis indicates that the corrosion products are composed of oxygen (32.12 wt.%), chromium (26.73 wt.%), and iron (the balance). Fig. 6(f) shows the nodule-like region where coarse grains with sharp edges and corners can be seen. EDS analysis indicates that this kind of corrosion products are composed of oxygen and iron.

(a) Blister-like corrosion products investigated by optical microscope

(b) image at low magnification

(c) image in the exfoliation region,

(d) interface of flat and nodule

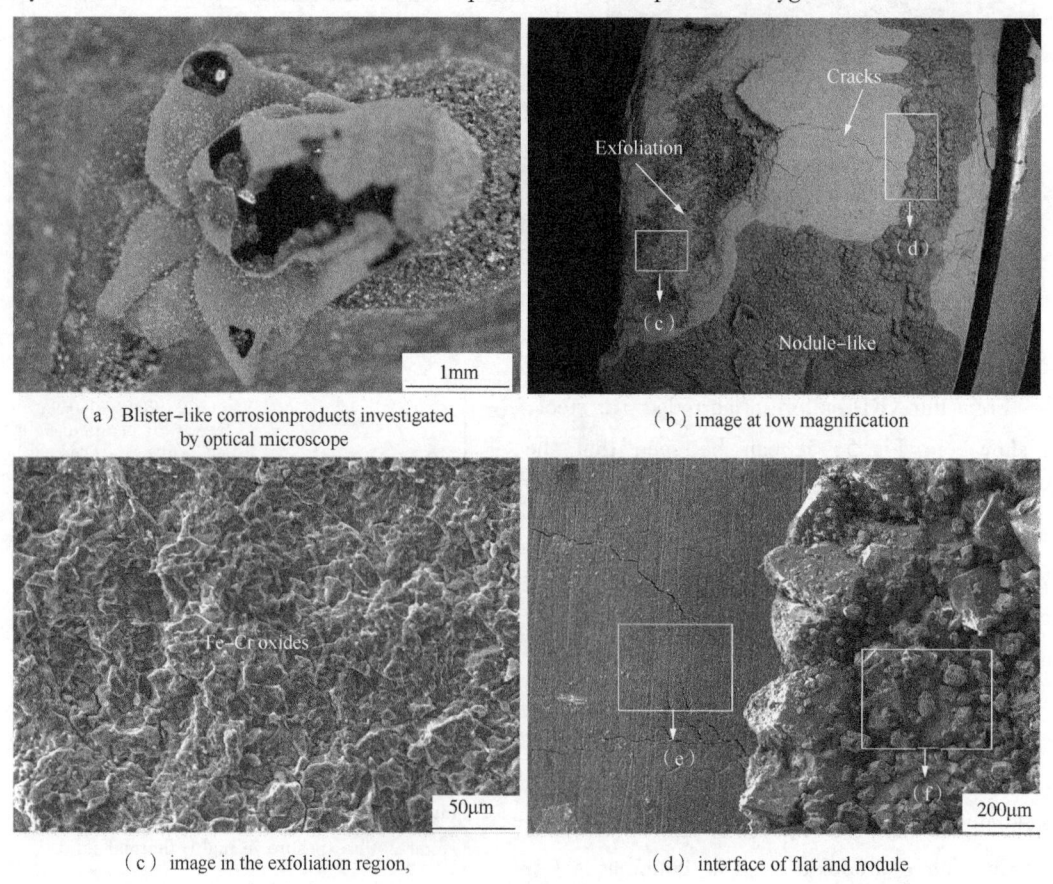

Fig. 6 Surface morphology of 9Cr steel after 240h exposure in multi-thermal fluid

(e) flat region with cracks,   (f) nodule at high magnification

Fig. 6　Surface morphology of 9Cr steel after 240h exposure in multi-thermal fluid(continued)

Fig. 7(a) shows the full view of the cross-section of 9Cr steel after 240h corrosion. There are several features. First, the thickness of the corrosion scale is nonuniform, and the largest thickness is about 600μm. Second, there are two different contrasts in the corrosion scale, and the outer layer is much thicker than the inner layer. Third, a large number of vertical and penetrating cracks (along the thickness direction) can be observed. Fourth, the corrosion scale separates from the substrate (a clear "gap"). As the corrosion scale is very thick, three regions [as denoted as b, c, and d in Fig. 7(a)] were selected for investigation at higher magnification. The morphology is shown in Fig. 7[(b)-(d)]. Fig. 7(e) shows the EDS element profile across the corrosion scale, where the scale is divided into three regions (R-I, R-II, and R-III) according to the relative content of Fe and Cr. Combined with the surface morphology discussed above, it is not difficult to conclude that the outmost corrosion products rich in Fe are not completely displayed on the cross-section morphology due to falling off. Therefore, the corrosion scale observed from the cross-section is mainly the Fe-Cr oxides.

Fig. 8 shows the XRD pattern for the corrosion scale formed on 9Cr steel. The main phases are $(Fe, Cr)_3O_4$, $FeCO_3$, and Fe_2O_3.

It needs to point out that the corrosion scale formed on 9Cr steel as shown in Fig. 7 is somewhat different from the observation in Fig. 6. The main difference is the absence of the outermost iron oxides in Fig. 7. This may be ascribed to the exfoliation of the outermost nodule-like iron oxides. As indicated in Fig. 6(b), the exfoliation region is composed of oxygen, chromium, and iron. From this perspective, the outer layer as shown in Fig. 7 is the corrosion scale in the exfoliation region as shown in Fig. 6(b).

For 9Cr steel, it suffered from much more serious corrosion in the multi-thermal fluid than 3Cr steel, although the chromium content in 9Cr steel is greater than that in 3Cr steel. It is generally considered that the increase of chromium content would improve the corrosion resistance of steels. In the present work, when the chromium content increased from 3 wt.% to 9 wt.%, corrosion became much more severe. There may be two possible reasons. First, 9 wt.% chromium in 9Cr steel cannot form a protective Cr-rich scale. Second, vertical cracks through the corrosion

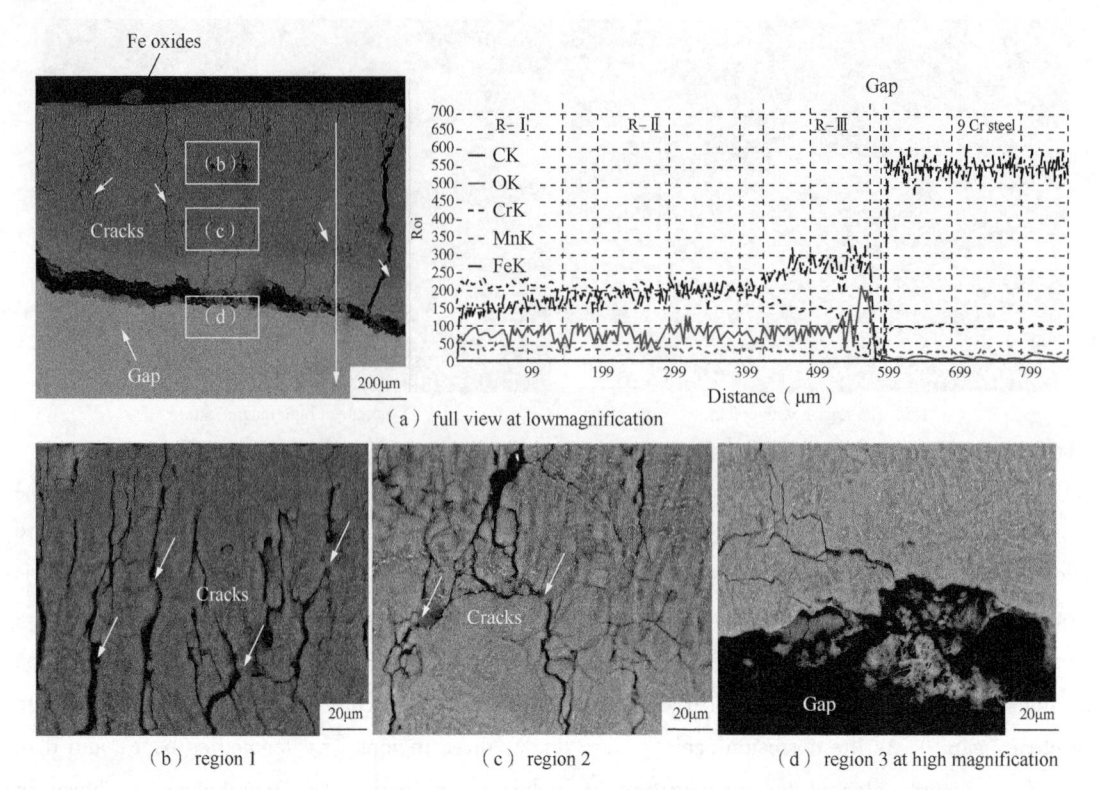

Fig. 7　Cross-section morphology of 9Cr steel after 240 h exposure in multi-thermal fluid

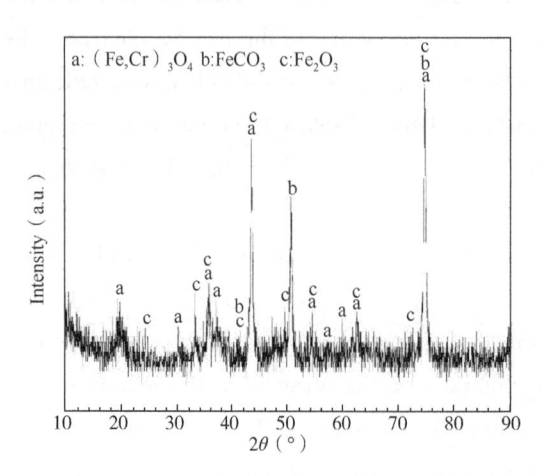

(a) full view at lowmagnification
(b) region 1
(c) region 2
(d) region 3 at high magnification

Fig. 8　The XRD pattern of 9Cr steel after 240h exposure in multi-thermal fluid

a: (Fe,Cr)$_3$O$_4$　b: FeCO$_3$　c: Fe$_2$O$_3$

scale would provide effective ways for the inward penetration of the corrosive medium to contact with the substrate, so that the corrosion continues at a fast rate.

3.3　Corrosion of 25Cr steel

The uniform corrosion rate is 0.0953mm/y, which indicates that the corrosion is very slight. Fig. 9(a) shows the surface morphology of 25Cr steel after 240h corrosion. The surface is uniform, where no evident cracks, exfoliation, and nodules can be seen. Besides, the grinding scratches are clear, indicating the corrosion scale is extremely thin. Fig. 9 (b) shows the cross-section morphology of 25Cr steel after 240h corrosion. A thin film with a thickness of 1 ~ 2μm can be seen. EDS spot analysis indicated that the film mainly consists of chromium and oxygen.

(a) top view   (b) cross-section view.

Fig. 9 Morphology of 25Cr steel after 240 h exposure in multi-thermal fluid

4 Discussion

Based on the above results, it can be summarized that the corrosion of Cr-bearing steels in multi-thermal fluids is different from that in CO_2-containing solutions below 150℃. Also, it is different from that in wet CO_2 at high temperatures.

4.1 Formation of corrosion products

In CO_2-containing aqueous solutions, corrosion products on carbon steels are commonly composed of iron carbonate[3-6] which is usually via the combination of Fe^{2+} and CO_3^{2-}. As temperature increases, the water vapor would become important corrosive species, which would result in the formation of iron oxides and/or iron hydroxides[10]. For the low Cr steels, the formation of amorphous $Cr(OH)_3$ (also containing $FeCO_3$) enhances the protective and self-repairing properties of the corrosion scale, which significantly reduces the corrosion rate and local corrosion tendency of low Cr steels in CO_2-containing solutions[11,12]. When the Cr content in steels is sufficient, the formation of $FeCO_3$, $Fe_{3-x}Cr_xO_4$, and Cr_2O_3 is dominated[13,14].

While in high-temperature carbon dioxide environment at temperatures above 500℃, iron oxides, iron-chromium spinel oxides, chromia, and undergoing carburization would form on Fe-Cr steels[15]. And the addition of water vapor would enhance the corrosion of Fe-Cr steels and cause breakaway corrosion. It means that the water vapor would break down the protective Cr-rich oxides, and then prompt the formation of less-protective iron oxides.

In the present work, the temperature is greater than that in conventional CO_2-containing solution (<150℃) while lower than that in conventional high temperature dry/wet CO_2 environments (>500℃). From the characterization of corrosion scales, it can be speculated that the corrosion scales formed in the multi-thermal fluid are closer to those formed in high temperature wet CO_2 environments. However, differences can be found that the vertical cracks and coarse iron oxides as observed on 9Cr steel.

Different from the conventional CO_2-containing solutions, iron carbonate was absent on surface corrosion products of 3Cr and 9Cr steels. In our previous work, the stability of iron carbonate, hematite, and magnetite was discussed from the thermodynamical perspective, and it was concluded that the iron carbonate is unstable above 200℃ due to the decomposition into iron oxides[16]. Thermodynamical calculation of the standard free energy changes ΔG^0 for the corrosion of iron and chromium with each single corrosive species (including CO_2, H_2O, and O_2) at the studied temperature also indicates that the reaction with oxygen and heated steam is easier to take place. Thus, the corrosion caused by oxygen and heated steam may dominate the whole process. And for the corrosion products of Cr, $Fe_{3-x}Cr_xO_4$ seems more possible than Cr_2O_3 and $Cr(OH)_3$ from thermodynamical and kinetic aspects[17]. As a result, Fe_2O_3/Fe_3O_4 and $Fe_{3-x}Cr_xO_4$ are the main corrosion products for steels with insufficient chromium (e.g. 3Cr and 9Cr steels in the present work), while Cr_2O_3 is dominant for 25Cr steel.

4.2 Effect of chromium content

Generally, an increase of chromium content insteels would improve their corrosion resistance due to the formation of the Cr-containing corrosion scale. In CO_2-containing solutions, when the chromium content in low Cr steels increases from 1 wt.% to 6.5 wt.%, the corrosion rate would be reduced to a great extent due to the protectiveness of $Cr(OH)_3$ which could block the path of iron's dissolution, keep apart the corrosion species and decrease the number of active sites of the iron dissolution[18]. However, it seems that 9 wt.% Cr in steels cannot form and sustain a protective Cr_2O_3 film on the surface which is the most ideal film for stainless steel with high corrosion resistance. The influence of chromium concentration on high-temperature corrosion of chromium-bearing steels follows a similar law, where the increase of chromium content would prompt the formation of Cr_2O_3 and/or $Fe_{3-x}Cr_xO_4$[19-22].

In the present work, a significant difference can be seen on 9Cr steel, which formed the thickest corrosion scale with vertical cracks and a large gap at the scale/substrate interface. These cracks are similar to the findings in Ref. [23] where the mud cracking morphology was ascribed to dehydration. Generally, scale cracks are resulted from intrinsic oxide growth stresses[24] and/or thermal stress, where the growth stress is developed by volume change while thermal stress is developed by temperature change. In addition, the segregation of elements such as Ni and Mo may lead to the formation of depleted area where suffer internal stress and generate micro-crack initiation[25]. Since the $Fe_{3-x}Cr_xO_4$ appears harder and more brittle than Fe_3O_4, a large number of cracks would form within the $Fe_{3-x}Cr_xO_4$ scale, and then accelerate the corrosion of 9Cr steel.

4.3 Possible corrosion mechanism

Fig. 10 shows the schematic illustration of the corrosion scale growth behavior of Cr-bearingsteels in $CO_2-H_2O-O_2$ multi-thermal fluid. For 3Cr steel, the chromium content is too low to form a protective chromium-rich film initially. Therefore, less-protective iron-chromium spinel film would form after immersion. Due to the consumption of chromium, the Cr-depleted zone would form near to the metal surface, leading to the continuous growth of the iron-chromium scale and the formation of the outermost iron-oxide scale. For 9Cr steel, the chromium content is also insufficient to form a protective Cr_2O_3 film initially. Similar to 3Cr steel, iron-chromium spinel film would form after immersion. However, the chromium content in the scale is greater than that for 3Cr steel, due to the relatively higher chromium concentration in 9Cr steel. Under the internal and/or external stress, cracks can nucleate from defects (e.g. pores in the outer scale) and propagate in the corrosion scale by connecting the adjacent cracks[26]. The formation of through-cracks and interface-cracks would damage the protectiveness of the corrosion scale (e.g. Fe - Cr spinel). Corrosion reaction occurs when the corrosive medium reaches the exposed Cr - depleted substrate surface through cracks, which further reduces the protection of the iron-chromium spinel scale. After that, iron in the substrate would reach the outer surface of the iron-chromium spinel scale by solid-phase diffusion, react with corrosive medium, and then form the outer iron oxide scale. For 25Cr steel, the chromium concentration is sufficient to form and sustain a protective Cr_2O_3 scale, so that the corrosion rate is extremely low

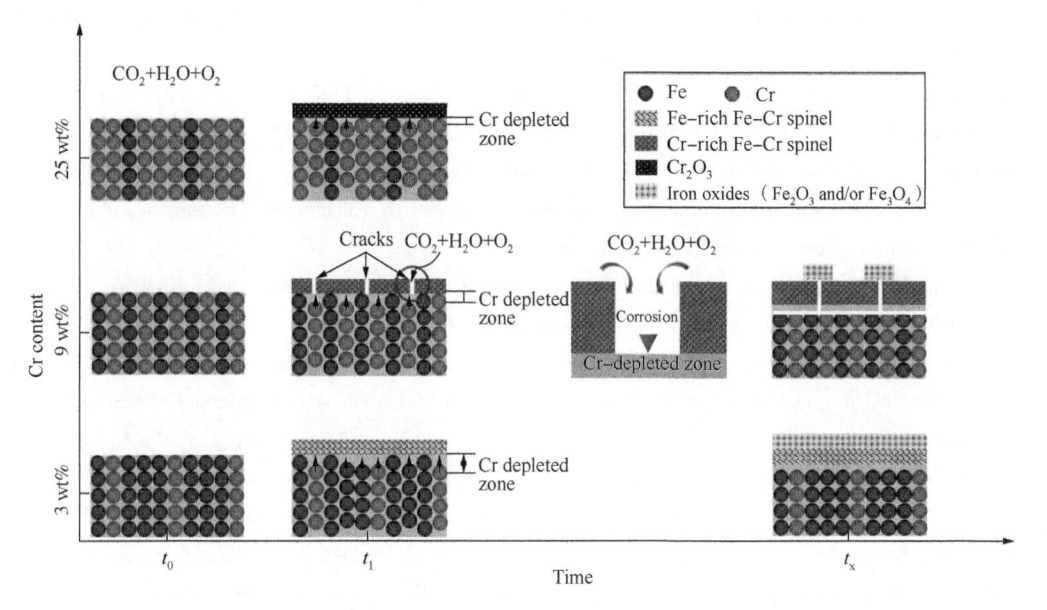

Fig. 10 Schematic illustration of the corrosion scale growth behavior of Cr-bearing steels in CO_2-H_2O-O_2 multi-thermal fluid

5 Conclusions

Corrosion behavior of commercial steels with different Cr content (3 wt.%, 9 wt.%, and 25 wt.%) in simulated CO_2-O_2-H_2O multi-thermal fluid environment was studied by immersion corrosion test and microstructural characterization. Based on the results, the following conclusions can be made.

(1) 3Cr and 9Cr steels were corroded to form less-protective compounds (i.e. $FeCO_3$, Fe_3O_4, $Fe_{3-x}Cr_xO_4$) instead of protective Cr-rich products [i.e. $Cr(OH)_3$ and Cr_2O_3], so it showed poor corrosion resistance.

(2) Presence of through cracks in the corrosion scale formed on 9Cr steel is the main reason for the deterioration of its corrosion resistance to multi-thermal fluid.

(3) For 25Cr steel, the Cr content is sufficient to repair the serious depletion of Cr in the metal near the scale/metal interface in time, so that the protection of the corrosion scale can be sustained. As a result, it exhibits excellent resistance to corrosion of multi-thermal fluids.

Acknowledgement

The authors are grateful for the financial supports from the National Key Research and Development Project (No. 2016YFC0802101), Innovation Capability Support Program of Shaanxi (No. 2019TD-038 and 2020KJXX-063), and Key Research and Development Program of Shaanxi Province (No. 2018ZDXM-GY-171).

References

[1] Xiaohu Dong, Huiqing Liu, Zhaoxiang Zhang, Changjiu Wang. The flow and heat transfer characteristics of multi-thermal fluid in horizontalwellbore coupled with flow in heavy oil reservoirs. J Petrol SciEng2014, 122, 56-68.

[2] Xiaohu Dong, Huiqing Liu, Zhanxi Pang, Changjiu Wang, Chuan Lu. Flow and heat transfer characteristics of multi-thermal fluid in a dual-string horizontal well. Numer Heat Tr A-Appl2014, 66, 185-204.

[3] Mobbassar Hassan Sk, Aboubakr M. Abdullah, Monika Ko, Bridget Ingham, Nick Laycock, Rakesh Arul, David E. Williams. Local supersaturation and the growth of protective scales during CO_2 corrosion of steel: Effect of pH and solution flow. CorrosSci2017, 126, 26-36.

[4] Lining Xu, Hui Xiao, Weijing Shang, Bei Wang, Jinyang Zhu. Passivation of X65 (UNS K03014) carbon steel in $NaHCO_3$ solution in a CO_2 environment. CorrosSci2016, 109, 246-256.

[5] Tatiana das Chagas Almeida, Merlin Cristina Elaine BaNDEIRA, Rogaciano Maia Moreira, Oscar Rosa Mattos. New insights on the role of CO_2 in the mechanism of carbon steel corrosion. CorrosSci2017, 120, 239-250.

[6] MdMayeedul Islam, ThunyalukPojtanabuntoeng, Rolf Gubner. Condensation corrosion of carbon steel at low to moderate surface temperature and iron carbonate precipitation kinetics. CorrosSci2016, 111, 139-150.

[7] D. Burkle, R. De Motte, W. Taleb, A. Kleppe, T. Comyn, S. M. Vargas, A. Neville, R. Barker. In situ SR-XRD study of $FeCO_3$ precipitation kinetics onto carbon steel in CO_2-containing environments: The influence of brine pH. ElectrochimActa2017, 255, 127-144.

[8] ThuanDinh Nguyen, Alexandre La Fontaine, Limei Yang, Julie M. Cairney, Jianqiang Zhang, David J. Young. Atom probe study of impurity segregation at grain boundaries in chromia scales grown in CO_2 gas. CorrosSci2018, 132, 125-135.

[9] D. Huenert, A. Kranzmann. Impact of oxyfuel atmospheres $H_2O/CO_2/O_2$ and H_2O/CO_2 on the oxidation of ferritic-martensitic and austenitic steels. CorrosSci2011, 53, 2306-2317.

[10] T. Tanupabrungsun, B. Brown, S. Nesic. Effect of pH on CO_2 corrosion of mild steel at elevated temperatures. NACE - International Corrosion Conference Series, Corrosion Conference and Expo 2013, Orlando, Florida, United states, March 17-21, 2013, Paper No. 2348.

[11] M. Du, S. Zhu, X. Zhang, J. Li, S. Song. Research progress in formation and formation mechanism of CO_2 corrosion scale on Cr containing low alloy steel. CorrosSciProtTechnol2019, 31, 335-342.

[12] S. Guo, L. Xu, L. Zhang, W. Chang, M. Lu. Characterization of corrosion scale formedon 3Cr steel in CO_2-saturated formation water. CorrosSci2016, 110, 123-133.

[13] A. Pfennig, R. Babler. Effect of CO_2 on the stability of steels with 1% and 13% Cr in saline water. CorrosSci2009, 51, 931-940.

[14] N. Mundhenk, P. Huttenloch, R. Babler, T. Kohl, H. Steger, R. Zorn. Electrochemical study of the corrosion of different alloys exposed to deaerated 80℃ geothermal brines containing CO_2. CorrosSci2014, 84, 180-188.

[15] D. J. Young. High Temperature Oxidation and Corrosion of Metals, second edition, Elsevier,

Netherlands, 2016.

[16] J. Yuan, A. Fu, C. Yin, X. Zhao, Y. Han, F. Li, L. Li, H. Zhang. Corrosion of N80 tubing steel in multi-thermal fluid used for enhancing the production of heavy oil, NACE – International Corrosion Conference Series, Corrosion Conference and Expo 2019, Nashville, TN, United states, March 24 – 28, 2019, Paper No. 13235.

[17] M. C. Biesinger, B. P. Payne, A. P. Grosvenor, L. W. M. Lau, A. R. Gerson, R. St. C. Smart. Resolving surface chemical states in XPS analysis of first row transition metals, oxides and hydroxides: Cr, Mn, Fe, Co and Ni. Appl Surf Sci2011, 257, 2717-2730.

[18] B. Wang, L. Xu, J. Zhu, H. Xiao, M. Lu. Observation and analysis ofpseudopassive film on 6.5% Cr steel in CO_2 corrosion environment. CorrosSci2016, 111, 711-719.

[19] J. Yuan, X. Wu, W. Wang, S. Zhu, F. Wang. Investigation on the enhanced oxidation of ferritic/martensitic steel P92 in pure steam. Materials2014, 7, 2772-2783.

[20] J. Yuan, W. Wang, H. Zhang, L. Zhu, S. Zhu, F. Wang. Investigation into the failure mechanism of chromia scale thermally grown on an austenitic stainless steel in pure steam. CorrosSci2016, 109, 36-42.

[21] L. Martinelli, C. Desgranges, F. Rouillard, K. Ginestar, M. Tabarant, K. Rousseau. Comparative oxidation behaviour of Fe – 9Cr steel in CO_2 and H_2O at 550℃: Detainled analysis of the inner oxide layer. CorrosSci2015, 100, 253-266.

[22] Y. Behnamian, A. Mostafaei, A. Kohandehghan, B. S. Amirkhiz, D. Serate, Y. Sun, S. Liu, E. Aghaie, Y. Zeng, M. Chmielus, W. Zheng, D. Guzonas, W. Chen, J. L. Luo. A comparative study of oxide scales grown on stainless steel and nickel-based superalloys in ultra-high temperature supercritical water at 800℃. CorrosSci2016, 106, 188-207.

[23] L. Xu, S. Guo, W. Chang, T. Chen, L. Hu, M. Lu. Corrosion of Cr bearing low alloy pipeline steel in CO2 environment at static and flowing conditions. Appl Surf Sci2013, 270, 395-404.

[24] D. J. Young, J. Zhang, C. Geers, M. Schutze. Recent advances in understanding metal dusting: A review. Mater Corros2011, 62, 7-28.

[25] Y. Behnamian, A. Mostafaei, A. Kohandehghan, B. S. Amirkhiz, J. Li, W. Zheng, D. Guzonas, M. Chmielus, W. Chen, J. L. Luo. Characterization of oxide layer and micro-crack initiation in alloy 316L stainless steel after 20000 h exposure to supercritical water at 500℃. Mater Charact2017, 131, 532-543.

[26] Z. Zhang, Z. F. Hu, L. F. Zhang, K. Chen, P. M. Singh. Effect of temperature and dissolved oxygen on stress corrosion cracking behavior of P92 ferritic – martensitic steel in supercritical water environment. J Nucl Mater2018, 498, 89-102.

本论文原发表于《Materials Research Express》2020年第7卷第10期。

Corrosion Behavior of Reduced-Graphene-Oxide-Modified Epoxy Coatings on N80 Steel in 10.0wt% NaCl Solution

Feng Chun[1,2]　Cao Yaqiong[1,2]　Zhu Lijuan[1,2]　Yu Zongxue[3]
Gao Guhui[4]　Song Yacong[1,2,]　Ge Hongjiang[5]　Liu Yaxu[1,2]

(1. CNPC Tubular Goods Research Institute;
2. State Key Laboratory for Performance and Structure Safety of Petroleum Tubular Goods and Equipment Materials;　3. Southwest Petroleum University;　4. TsinghuaRedbud Innovation Institute;
5. Oil Production Technology Research Institute of Petrochina Dagang Oilfield Company)

Abstract: This study is examined the effect of modified epoxy (EP) coatings with various contents of reduced graphene oxide(RGO) in high temperature and high salinity for long time on the corrosion resistance of N80 steel. The fracture surfaces of RGO-modified EP coatings were characterized by scanning electron microscopy (SEM). The corrosion resistance of the coatings after inmersion in 10.0wt% NaCl solution for 60 day sat 50℃ was characterized via electrochemical impedance spectroscopy (EIS) and Tafelcurves. It indicated that the addition of 1.0 wt% RGO nanosheets not only improve the strength and toughness of EP coatings, but also effectively decreased the size of pores by two to three times and the quantity of pores decreased by two magnitudes in the coatings than that of neat EP coatings after immersed in the 10.0 wt% NaCl for 60 days, which bring in excellent corrosion resistance.

Keywords: Reduced craphene oxide; Epoxy; High temperature; High salinity; Corrosion resistant

1 Introduction

Epoxy (EP) coatings have been widely used in the corrosion protection of tubing due to the excellent adhesion properties, low shrinkage low price, and outstanding chemical stability of epoxy[1-6]. However, with the development of ultra-deep oil fields, the problem of tubing corrosion is more severe due to the high-temperature and high-salinity exploitation environment[7,8]. In onshore oilfields with high sodium chloride content of the for mation brine, such as Pucheng oilfield, in which the Na$^+$ content is $(8.5 \sim 14.5) \times 10^4$ mg/L and the Cl$^-$ content is $(19.15 \sim 22.50) \times 10^4$ mg/L[9], the

---

Corresponding author: Feng Chun, fengchun003@cnpc.com.cn; Cao Yaqiong, caoyaqiong1992@163.com; Zhu Lijuan, zhulijuan1986@cnpc.com.cn.

corrosion of N80 tubing is extremely serious [10,11]. In these high temperature and high salinity exploitation environment, the EP coatingsis exposed its limitations: the high degree of crosslinking density makes it brittle, thereby reducing their fracture toughness[12,13], and they exhibit poor resistance to crack propagation[1]. Defects, such as pin holes, easily occur after curing[13]. Corrosive electrolytes, such as small molecules especially Cl⁻ in high – salinity and high – temperature environmens, will penetrate these cracks and pores to form corrosion products under the coatings. Therefore, the EP coatings can't provide long-term corrosion protection[14,15]. Fillers are typically added to improve the corrosion resistance of EP coatings[16-21].

In recent years, graphene has attracted extensive attention due to its substantial potential in improving the properties of resin-based materials and excellent physical properties. Graphene that is well-dispersed in the voids of the coatings can improve the compactness of the coatings, and the lamellar structure of graphene can effectively prevent the permeation of corrosive media, such as H_2O, O_2 and Cl⁻, thereby resulting in excellent corrosion resistance of the graphene – modified coatings[19-22]. The effects of graphene addition on the corrosion resistance of EP coatings in simulated seawater and normal atmospheric temperature environment have been widely reported[23-29]. Feng. [30] prepared graphene solid lubrication coating by using graphene water dispersion and water – based EP, and found thatgraphene solid lubrication coatingsshowed good corrosion resistanceafter 48 hours immersion in seawater with 3.5wt% NaCl aqueous solution at room tempareture. Wang Yuqiong et al. [31] added 0.5 wt% graphene dispersion to two – component waterborne EP to prepare EP coatings, and soaked in 3.5wt% NaCl simulated seawater solution for 48 hours at room temperature. The results show that the coatings has good corrosion resistance. However, the corrosion resistances of graphene modified EP coatings in high salinity and high temperature oil production environments are seldom reported.

In our previous study [32], RGO-reinforced EP composite coatings were prepared on N80 steel. These composite coatings showed improved adhesion, toughness, and corrosion resistance after immersion in simulated oil and gas production environments with high-temperature and high-salinity at 80℃ for 10 hours. The addition of 1.0 wt% RGO nanosheets effectively reduced the number and size of the pores in the as-prepared EP composite coatings. However, quantitative expressions and variation rules of the sizes of the pores in the EP composite coatings after corrosion testing with the contents of RGO have not been defined, and the long-term service performances of RGO reinforced EP composite coatings in oil and gas production environments remain unclear. The working temperature of most water injection tubings is approximately 50℃. However, the corrosion resistances of RGO – modified EP coatings in high – salinity and high – temperature oil production environments at 50℃ for long times are rarely reported.

Therefore, the corrosion resistance of RGO-modified EP coatings that are immersed in 10.0 wt% NaCl solution for 60 days at 50℃ was studied via EIS and potentiodynamic polarization curves in this study. In addition, the morphology and microstructure of the RGO was analyzed via FTIR and TEM. The fracture surfaces of RGO – modified EP coatings were characterized via SEM, and quantitative expressions and variation rules for the sizes of pores in the EP composite coatings as functions of the RGO content were defined.

2 Experimental

2.1 Preparation of modified EP coatings

N80 tubing samples (200mm×20mm×3mm) are selected as coating spraying substrates. The specimens were polished with silicon carbide paper and cleaned in an ultrasonic cleaner. The cleaning solution consisted of acetone and ethanol, and the steel substrates were reserved after drying. RGO was mixed into an organic solvent (0.8g/L), and the mixture was dispersed in an ultrasonic disperser. The prepared dispersions were added into EP coatings (solid content of 61.0 wt%) for magnetic stirring, and the stirring time was 15 min. The mixed slurries of RGO with various contents (0 wt%, 0.5 wt%, 1.0 wt%, 2.5 wt% and 4.0 wt%) were sprayed on the cleaned N80 tubing steel substrates, which were subsequently dried at 55 °C for 15 hours. The thickness of the coatings was approximately 300μm.

2.2 Characterization of RGO and RGO modified EP coatings

The morphology and microstructure of the RGO was investigated via FTIR (WQF-520, Beijing, China) and transmission electron microscopy (TEM, Tecnai G2 F20). The fracture surfaces of RGO-modified EP coatings after liquid nitrogen embrittlement were characterized via SEM (Inspect F50, FEI).

2.3 Electrochemical anti-corrosion test

Potentiodynamic polarization curves were plotted and electrochemical impedance spectroscopy (EIS) was conducted to evaluate the corrosion performance of coatings via potentiostat/galvanostat/ZRA (CS-350H, Wuhan Corrtest Instruments Co. Ltd., China). The frequency range was 10^5 Hz to 10^{-2} Hz, and the signal was a 10mV sinusoidal wave. Tafel analysis was conducted to measure the corrosion rates in the range of -0.25V to 0.25V (vs. OCP), and the scanning rate was 1 mV/s. A three-electrode system was adopted: a platinum electrode was used as the counter electrode, a saturated calomel electrode (SCE) was used as the reference electrode, and an N80 steel substrate with a coating was used as the working electrode. The working electrode was immersed in 10.0 wt% NaCl solution at 50°C. The impedance and polarization curves of the coatings were measured after 60 days of immersion. Then, Tafel fitting of the potentiodynamic polarization curves and equivalent electrical circuit (EEC) fitting of the EIS results were conducted using the ZSimpWin software.

3 Results and Discussion

3.1 Characterization of RGO

FTIR spectra of GO and RGO are presented in Fig. 1. The characteristic absorption bands of GO and RGO at 3447, 1741, 1625, 1200 and 1120 cm−1 correspond to −OH, C=O, C=C, C−O and C−O−O, respectively. For GO, many oxygen groups are distributed on the surface, and the carboxyl and carbonyl groups are distributed on the edges of the sheets[33]. In contrast to the groups of GO, the C=O, C−O and C−O−O groups in RGO almost disappeared. Therefore, the RGO that is used in this study has a high degree of graphitization.

The surface morphologies of RGO were identified via TEM. Fig. 2 presents a TEM image of dispersed RGO. The color of the RGO sheets is lighter; hence, the dispersion is satisfactory and there are no large aggregates. Due to the large surface energy of a single layer or several layers of RGO, curling and folding occur at the edges of the RGO sheet[34]. The thermodynamic stability of the two-dimensional structure is maintained by this curling and folding, which results in unique structural characteristics.

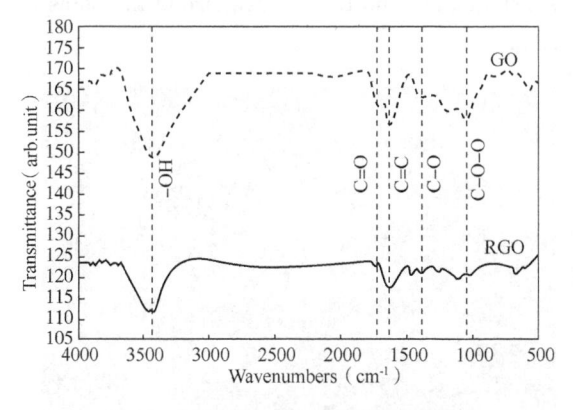

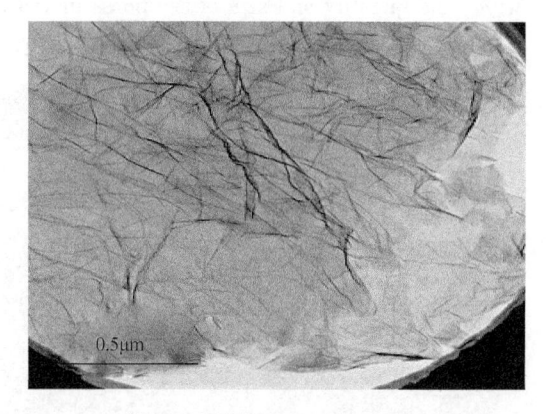

Fig. 1 FTIR spectra for GO and RGO

Fig. 2 TEM photograph of RGO

3.2 Characterization of RGO-modified EP coatings

In our previous work, we found that the addition of 1.0 wt% RGO nanosheets effectively reduced the numbers and sizes of the pores in the as-prepared EP composite coatings. To define the variation rules of the sizes of the pores in the EP composite coatings after long-term service, brittle fracture experiments of coatings with various RGO contents were conducted after immersing them at 50℃ in 10.0 wt% NaCl solution for 60 days. The fracture surfaces of 0 wt%, 0.5 wt%, 1.0 wt%, 2.5 wt%, and 4.0 wt% RGO-modified EP coatings are shown in Fig. 3. The cross-sectional surface of the 0 wt% RGO-modified EP coating shows a clean fracture surface, thereby implying a brittle fracture. However, in the ductile fracture morphology in Fig. 3[(b)-(e)], tearing ridges can be readily observed from the RGO-modified EP coatings, which show RGO that is warped without peeling off the resin matrix. Hence, the interfacial adhesion between the RGO and the EP was strong, and RGO can improve the strength and toughness of EP coatings. In the process of deformation, RGO, which is the stress concentration center, can cause the resin matrix around it to yield and absorb a substantial amount of energy. The interface between RGO and the EP is separated to form holes, which can passivate the cracks and prevent the generation of destructive cracks. Moreover, the specific surface area of RGO is large, which increases the contact area with the EP. If the material is under external force, it can produce more microcracks and absorb a substantial amount of stress. However, as the RGO content was increased to 4.0 wt%, the tearing ridge decreased significantly, which was likely due to the aggregation of the RGO nanosheets.

The pore sizes of the samples were randomly measured, and the top five sizes of pores for each sample are listed in Table 1. According to Table 1, the maximum pore size that was measured on the EP coatings with 0 wt% is 1305 nm, with an average size of 826.5 nm. The addition of 0.5 wt% RGO caused no significant change in the quantity or size of the pores [Fig. 3(a) and Fig. 3(b)]. However, when the RGO content was 1.0 wt%, the size of the pores decreased by two to three times and the quantity of the pores decreased by two orders of magnitude in the coatings compared to the neat EP coatings [Fig. 3(c)]. When the RGO contents were increased to 2.5 wt% and 4.0 wt%, the quantity and size of the pores increased substantially, and the average size of the pores in these coatings even exceeds that in the neat EP coating.

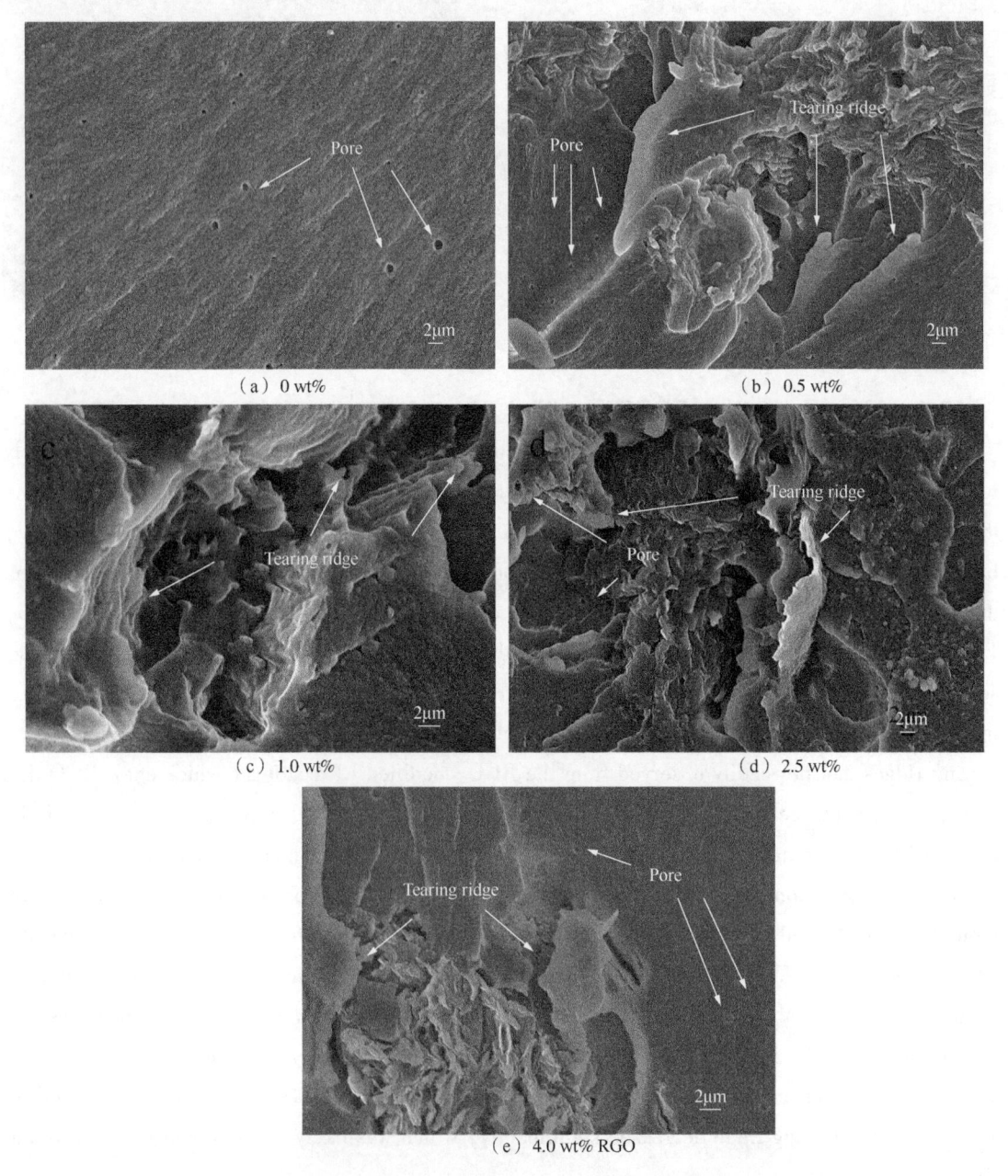

Fig. 3 SEM images of fractures in the EP coatings

Table 1 Pore size on the cross-sectional surfaces of coatings with various RGO contents

Content of RGO	Pore size (nm)					Average size (nm)
0 wt%	435.0	870.0	652.5	870.0	1305.0	826.5
0.5 wt%	431.0	431.0	862.0	1293.0	1115.0	826.4
1.0 wt%	652.5	261.0	211.0	196.0	356.0	335.3
2.5 wt%	862.0	1115.0	1115.0	1293.0	1293.0	1135.6
4.0 wt%	431.0	862.0	1293.0	1293.0	1293.0	1034.4

If the coating is subjected to external stress, large pore defects will lead to poor stress transfer and stress concentration, and corrosive media such as H_2O, O_2 and Cl^- will enter the coating more easily through the defects and reach the metal substrate surface, thereby forming severe corrosion, which will eventually cause the coating to fall off easily. Many studies have shown that the number of pores in the coating can be decreased by adding graphene into the EP. For example, nano oxide can be used to modify GO and blend with EP to form composite materials, which can fill pores in the coating[35-38]. Zhang[39] prepared a GO-modified EP coating, in which the GO modification effectively reduced the porosity of the coating. However, these studies use graphene to fill the pores, to decrease the number of pores, and the mechanism of graphene in the process of bubble generation is not discussed.

In this study, a possible mechanism is proposed: When RGO is dispersed by ultrasonic waves, the high temperature and high pressure of the generated bubbles through the cavitation of the rupture process promotes the uniform dispersion of RGO in EP coatings. Therefore, when the content of RGO is 0.5 wt%, many pores remain in the coatings. As the content of RGO increases, RGO will destroy large bubbles, thereby resulting in small bubbles. This facilitates bubble overflow; hence, the number of pores in the coatings will decrease. However, when the RGO content was increased to 2.5 wt% and 4.0 wt%, aggregation of the RGO nanosheets occurred, which significantly lowered the bubble-breaking effect, thereby resulting in the formation of pores of increased quantity and size in the coatings. Therefore, there is a balance between the pore reduction that is induced by the bubble-breaking effect of RGO nanosheets and the pores increase that is induced by the aggregation of the RGO nanosheets. This balance is realized when the optimal amount of RGO has been added into the coatings. The bubble-breaking effect and the mechanism of RGO will be investigated in our future work. Therefore, the addition of a suitable amount of RGO not only increases the toughness but also decreases the quantity and size of the pores of these coatings in high-temperature and high-salinity environments.

3.3 EIS analysis of RGO-modified EP coatings

EIS and potentiodynamic polarization curves were used to evaluate the corrosion performances of EP coatings with various contents of RGO. Many electrochemical studies on the coating have been conducted by immersing the coatings in 3.5% NaCl solution at room temperature in accordance with the marine simulation environment[23]. For example, GO was introduced into a polypyrrole (PPY) matrix, and PPY-GO composite coatings with various GO contents were electrodeposited in situ on 304 stainless steel (SS) bipolar plates to protect them from corrosion in an aggressive working environment[40]. N. N. Taheri[27] grafted GO particles onto zinc doped polyaniline (PANI), added it to the EP, and immersed it in 3.5% NaCl solution for electrochemical testing. To more closely approximate the conditions of the underground environment of onshore oil wells, prior to the test, the coatings were immersed in 10. 0 wt% NaCl solution at 50℃ for 60 days, and the results are presented in Fig. 4. The higher the impedance in the Nyquist plot, the better the anti-corrosion effect of the coatings. The addition of RGO increases the radius of the capacitance; hence, the addition of RGO effectively enhances the anticorrosion performance of the EP coating. The effect depends on the chemical stability of RGO. When RGO is added to the EP coating, a retardant layer can be formed, which prevents direct contact between the corrosive media and the substrate, and increases the tortuosity of the diffusion path. Moreover, the RGO is a nanometer-scale material, which can be used to fill the curing defects of the EP coating. Moreover, it is hydrophobic; hence, it can delay the pervasion of water and the corrosion. The TEM image in Section 3. 1 shows that there are many folds on the surface of RGO. When it is added to EP as a filler, the folds on the surface can have a larger contact area with the EP, thereby making the diffusion path of active media more tortuous and, thus, substantially improving the corrosion resistance of the coating. Therefore, according to the Nyquist plots and Bode diagrams in Fig. 4, with the increase of the RGO content, the radius of capacitance initially increases and subsequently decreases. The modified EP coating with 1. 0 wt% RGO shows excellent corrosion resistance, which is due to its high impedance and large radius of capacitance. However, when the content of RGO is high, agglomeration occurs in the coating, thereby resulting in many defects, such as pores in the EP coating, which provide channels for the diffusion of corrosive media; thus, the anti-corrosion performance of the coating is decreased.

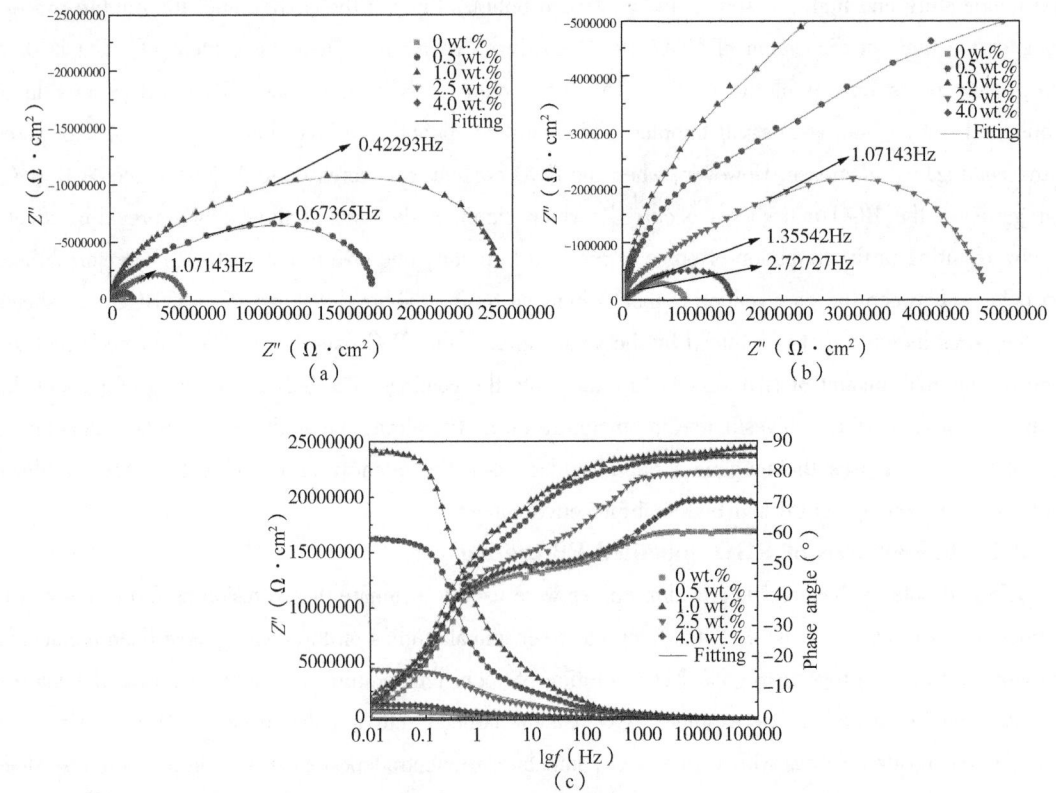

Fig. 4 (a) Nyquist plots, (b) locally enlarged Nyquist plots, and (c) Bode diagrams of the EP composite coatings after immersion in 10. 0 wt% NaCl solution at 50℃ for 60 days

To further characterize the anti-corrosion effect of the coatings, the impedance data are analyzed and fitted with the ZSimpWin simulation software. The equivalent circuit and the analysis results are presented in Fig. 5 and Table 2. The electronic components R_s, C_c, R_c, R_{ct} and CPE_{dl} in the equivalent circuit represent the electrolyte resistance, coating capacitance, coating resistance, charge transfer resistance and double layer capacitance, respectively. C_c represents the amount of corrosive media that penetrate into the coating, R_c can be used to represent the number and area of the pores on the coating surface, CPE_{dl} represents the failure area of the coating, and R_{ct} represents the resistance value of the charge transfer on the metal surface, which can directly reflect the corrosion rate of the interface between the coating and N80 steel.

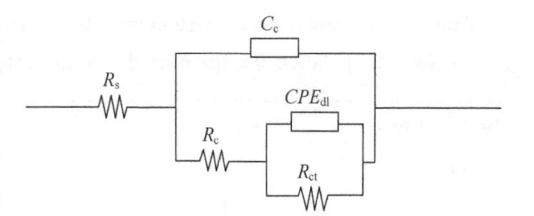

Fig. 5 Equivalent circuit of the sample

Table 2 EIS analysis of electrochemical parameters of the coating samples after immersion in 10.0 wt% NaCl solution for 60 days

Content of RGO	0 wt%	0.5 wt%	1.0 wt%	2.5 wt%	4.0 wt%
$R_{ct}(10^6 \Omega \cdot cm^2)$	0.11	2.81	10.16	1.02	0.18
$C_c(10^{-10} F \cdot cm^{-2})$	314.50	12.31	1.13	15.17	17.48
$R_c(10^5 \Omega \cdot cm^2)$	1.11	28.09	101.60	10.18	1.79
$CPE_{dl}(10^{-9} F \cdot cm^{-2})$	2.64	21290.00	1.67	36.87	77.35

As summarized above, $R_c(1.016 \times 10^7 \Omega \cdot cm^2)$ and $R_{ct}(1.016 \times 10^7 \Omega \cdot cm^2)$ attained their maximum values and $C_c(1.127 \times 10^{-10} F \cdot cm^{-2})$ and $CPE_{dl}(1.67 \times 10^{-9} F \cdot cm^{-2})$ reached their minimum values when the RGO content was up to 1.0 wt% in the modified EP coatings. Therefore, the optimal addition of RGO is 1.0 wt%. RGO can greatly improve the anti-corrosion performance of EP coatings. However, with the increase of graphene content, graphene will agglomerate, which will increase the defects of EP coatings and reduce the corrosion resistance.

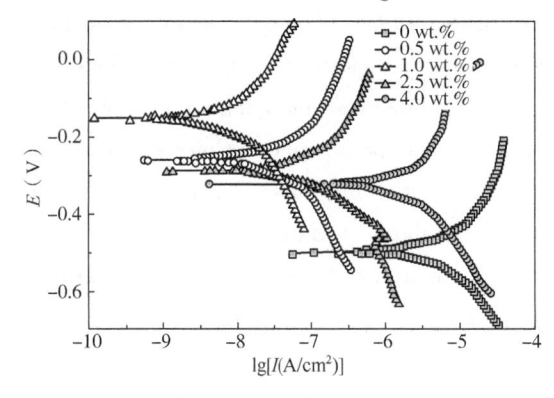

Fig. 6 Potentiodynamic polarization curves of the EP composite coatings after immersion in 10.0 wt% NaCl solution at 50 ℃ for 60 days

3.4 Tafel analysis of RGO-modified EP coatings

The potentiodynamic polarization curves are plotted in Fig. 6, and Table 3 lists the kinetic parameters that were calculated from the potentiodynamic polarization curves, which include I_{corr} and E_{corr}. The corrosion current (I_{corr}) decreased significantly and the corrosion potential (E_{corr}) exhibited a positive shift when graphene was added. I_{corr} initially decreased and subsequently increased with the increase of the RGO content, while E_{corr} initially exhibited a positive shift and subsequently exhibited a negative shift.

Table 3 Electrochemical parameters that were obtained from potentiodynamic polarization curves via Tafel extrapolation for the coated samples after immersion in 10. 0 wt% NaCl solution for 60 days

Content of RGO	0 wt%	0.5 wt%	1.0 wt%	2.5 wt%	4.0 wt%
I_{corr} ($\mu A/cm^2$)	9.190	0.074	0.009	0.705	5.415
E_{corr} (V)	-0.501	-0.262	-0.151	-0.287	-0.325

This is due to the small-size effect of RGO filling in the pores of the coating. The corrosive media can not penetrate the pore paths that are blocked by RGO t via natural convection mass transfer to the reaction area of the electrode interface corrosion. Due to the consumption of depolarization, the concentrations of H_2O, O_2 and Cl^- at the bottom of the pores can not be supplemented timely, thereby resulting in a concentration difference with the corrosive media in the solution of the coating surface at the top of the pore. Instead, Faradaic processes of the electrode reaction are controlled by the tangent diffusion of the corrosive media. The electrochemical polarization of the corrosion system is gradually weakened, and concentration polarization occurs.

Compared with neat EP coatings, the corrosion resistance of the coatings with RGO is substantially improved. The modified EP coatings with 1.0 wt% RGO had the lowest corrosion current density ($9.5172 \times 10^{-9} A/cm^2$), the highest corrosion potential ($-0.15107V$), and the maximum polarization resistance ($6.04 \times 10^9 \Omega \cdot cm^2$). Therefore, the optimum addition of RGO is 1.0 wt%, which is in accordance with the results of EIS.

3.5 Anti-corrosion mechanism

In this study, after immersion in 10.0 wt% NaCl solution for 60 days, the modified EP coatings with 1.0 wt% RGO still showed superior corrosion resistance to that of the neat EP coatings. Researchers proposed possible anti-corrosion mechanisms of graphene-modified coatings: (1) graphene can decrease the coating porosity[30-32]; (2) graphene causes a barrier effect[30-32]; (3) graphene can block the cathodic reaction between the coating and the metal interface[41]. As described in the above investigation, the addition of 1.0 wt% RGO nanosheets not only improved the strength and toughness of the coatings but also effectively decreased the size of the pores by two to three times and decreased the quantity of the pores by two orders of magnitude in the coatings compared to neat EP coatings. Therefore, the added RGO nanosheets effectively block the penetration of the corrosive media via the penetration effect and decrease the quantity and size of the pores in the coatings, which results in excellent anticorrosion performance. However, with the increase of the RGO content, RGO nanosheets will agglomerate, the balance between the pores reduction induced by the bubble-breaking effect of the RGO nanosheets and the pore increase that is induced by the aggregation of the RGO nanosheets will be broken, which will increase the severity of the defects in the EP coatings and decrease the corrosion resistance of the coatings.

4 Conclusions

(1) After immersion in 10.0 wt% NaCl solution at 50℃ for 60 days, the EP composite coatings with 1.0 wt% RGO nanosheets still showed improved strength and toughness, with pores that were two to three times smaller and two orders of magnitude less numerous compared to the neat

EP coatings.

(2) The modified EP coatings with 1.0 wt% RGO showed excellent corrosion resistance with the lowest corrosion current density (9.5172×10^{-9} A/cm^2), the highest corrosion potential (−0.15107V) and the maximum polarization resistance (6.04×10^9 Ω · cm^2) after immersion in the 10.0 wt% NaCl solution for 60 days at 50℃.

Acknowledgement

This work was supported by the National Natural Science Foundation of China (NO.51804335): Corrosion resistance and mechanism of graphene-modified epoxy coatings in a coupled oil and water multi-phase medium; Major science and technology projects of Petro China Co Ltd (2018E-11-06): Key technologies for improving the development level in complex fault block oilfields at the high water cut stage; and The National Key Research and Development Program of China(2019YFF0217504): Research and application of common technology in the national quality foundation-The complete process inspection and quality control technology of 13Cr / 110SS products in service.

References

[1] X. M. Shi, T. A. Nguyen, Z. Y. Suo, Y. J. Liu and R. Avci, Surface and Coatings Technology, 204 (2009) 237.

[2] I. Zaman, T. T. Phan, H. C. Kuan, Q. S. Meng, L. T. B. La, L. Luong, O. Youssf and J. Ma, Polymer, 52 (2011) 1603.

[3] S. Chatterjee, J. W. Wang, W. S. Kuo and N. H. Tai, Chemical Physics Letters, 531 (2012) 6.

[4] J. P. Pascault, R. J. J. Williams, Polymer Bulletin, 24 (1990) 115.

[5] D. Chmielewska, T. Sterzyński, and B. Dudziec, Journal of Applied Polymer Science, 131 (2014) 8444.

[6] S. Hichem, E. M. M. Lassaad, H. Guermazi, S. Agnel and A. Toureille, Journal of Alloys and Compounds, 477 (2009) 316.

[7] Y. G. Liu, Corrosion & Protection, 8 (2003) 361 (In Chinese).

[8] Y. Y. Ge, C. J. Han, Y. Q. Yuan, X. Y. Ma, S. F. Sheng and Y. C. Deng, Henan Petroleum, 19 (2005) 76 (In Chinese).

[9] G. P. He, Q. Y. Yang, Journal of Jianghan Petroleum University of Staff and Workers, 22 (2009) 56 (In Chinese).

[10] J. Z. Li, Inner Mongolia Petrochemical Industry, 20 (2013) 41 (In Chinese).

[11] A. Bisht, K. Dasgupta, and D. Lahiri, Journal of Applied Polymer Science, 135 (2018) 46101.

[12] A. Montazeri, J. Javadpour, A. Khavandi, A. Tcharkhtchi and A. Mohajeri, Materials & Design, 31 (2010) 4202.

[13] S. Chhetri, N. C. Adak, P. Samanta, P. K. Mallisetty, N. C. Murmu and T. Kuila, Journal of Applied Polymer Science, 135 (2018) 46124.

[14] H. Feng, X. D. Wang, and D. Z. Wu, Industrial & Engineering Chemistry Research, 52 (2013) 10160.

[15] R. J. Day, P. A. Lovell, and A. A. Wazzan, Composites Science and Technology, 61 (2001) 41.

[16] R. Bagheri, R. A. Pearson, Polymer, 41 (2000) 269.

[17] T. Kawaguchi, R. A. Pearson, Polymer, 15 (2003) 4239.

[18] S. H. Lee, D. R. Dreyer, J. An, A. Velamakanni, R. D. Piner, S. J. Park, Y. W. Zhu, S. O. Kim, C.

W. Bielawski and R. S. Ruoff, Macromolecular Rapid Communications, 31 (2010) 281.

[19] G. X. Wang, X. P. Shen, B. Wang, J. Yao and J. Park, Carbon, 47 (2009) 1359.

[20] Q. F. Jing, W. S. Liu, Y. Z. Pan, V. V. Silberschmidt, L. Li and Z. L. Dong, Materials & Design, 85 (2015) 808.

[21] T. Kuilla, S. Bhadra, D. H. Yao, N. H. Kim, S. Bose and J. H. Lee, Progress in Polymer Science, 35 (2010) 1350.

[22] M. Naderi, M. Hoseinabadi, M. Najafi, S. Motahari and M. Shokri, Journal of Applied Polymer Science, 135 (2018) 46201.

[23] H. H. Di, Z. X. Yu, Y. Ma, Y. Pan, H. Shi, L. Lv, F. Li, C. Wang, T. Long and Y. He, Polymers Advanced Technologies, 27 (2016) 915.

[24] Y. H. Tang, D. Xiang, D. H. Li, L. Wang, P. Wang and P. D. Han, Surface Technology, 47 (2018) 203.

[25] J. A. Q. Rentería, L. F. C. Ruiz, and J. R. R. Mendez, Carbon, 122 (2017) 266.

[26] Y. H. Wu, X. Y. Zhu, W. J. Zhao, Y. J. Wang, C. T. Wang and Q. J. Xue, Journal of Alloys and Compounds, 777 (2019) 135.

[27] N. N. Taheri, B. Ramezanzadeh, and M. Mahdavian, Journal of Alloys and Compounds, 800 (2019) 532.

[28] M. S. Selim, S. A. El-Saftya, N. A. Fatthallahd and M. A. Shenashen, Progress in Organic Coatings, 121 (2018) 160.

[29] F. D. Meng, T. Zhang, L. Liu, Y. Cui and F. H. Wang, Surface and Coatings Technology, 361 (2019) 188.

[30] C. Feng, L. J. Zhu, Y. Q. Cao, Y. Di, Z. X. Yu and G. H. Gao, International Journal of Electrochemical Science, 14 (2019) 1855.

[31] C. Feng, L. J. Zhu, Y. Q. Cao, Y. Di, Z. X. Yu and G. H. Gao, International Journal of Electrochemical Science, 13 (2018) 8827.

[32] L. J. Zhu, C. Feng, and Y. Q. Cao, Applied Surface Science, 493 (2019) 889.

[33] A. Dimiev, D. V. Kosynkin, L. B. Alemany, P. Chaguine and J. M. Tour, Journal of the American Chemical Society, 134 (2012) 2815.

[34] D. R. Nelson, T. Piran, and S. Weinberg, World Scientific, 2004, Singapore.

[35] Z. X. Yu, H. H. Di, Y. Ma, Y. He, L. Liang, L. Lv, X. Ran, Y. Pan and Z. Luo, Surface and Coatings Technology, 276 (2015) 471.

[36] Z. X. Yu, H. H. Di, Y. Ma, L. Lv, Y. Pan, C. L. Zhang, Y. He, Applied surface science, 351 (2015) 986.

[37] Y. Ma, H. H. Di, Z. X. Yu, L. Liang, L. Lv, Y. Pan, Y. Y. Zhang and D. Yin, Applied surface science, 360 (2016) 936.

[38] H. H. Di, Z. X. Yu, Y. Ma, F. Li, L. Lv, Y. Pan, Y. Lin, Y. Liu and Y. He, Journal of the Taiwan Institute of Chemical Engineers, 64 (2016) 244.

[39] Z. Y. Zhang, W. H. Zhang, D. S. Li, Y. Y. Sun, Z. Wang, C. L. Hou, L. Chen, Y. Cao and Y. Q. Liu, International journal of molecular sciences, 16 (2015) 2239.

[40] L. Jiang, J. A. Syed, H. B. Lu and X. K. Meng, Journal of Alloys and Compounds, 770 (2019) 35.

[41] S. S. Chen, L. Brown, M. Levendorf, W. W. Ca, S. Y. Ju, J. Edgeworth, X. S. Li, C. W. Magnuson, A. Velamakanni, R. D. Piner, J. Y. Kang, J. Park and R. S. Ruoff, ACS Nano, 5(2011)1321.

本论文原发表于《International Journal of Electrochemical Science》2020 年第 7 卷。

Experimental and Simulation Investigation on Failure Mechanism of a polyethylene Elbow Liner Used in an Oilfield Environment

Kong Lushi[1] Fan Xin[2] Ding Nan[3] Ding Han[1] Shao Xiaodong[1]
Li Houbu[1] Qi Dongtao[1] Liu Qingshan[3] Xu Yanyan[3] Ge Pengli[3]

(1. State Key Laboratory for Performance and Structure Safety of Petroleum Tubular Goods and Equipment Materials, Tubular Goods Research Institute, China National Petroleum Corporation; 2. Natural Food Macromolecule Research Center, School of Food and Biological Engineering, Shaanxi University of Science&Technology; 3. Key Laboratory of Enhanced Oil Recovery in Carbonate Fractured-vuggy Reservoirs, CNPC and SINOPEC Northwes Company of China Petroleum and Chemical Corporation)

Abstract: Internal collapse of a polyethylene (PE) liner in the steel elbow between furnace and manifold of second-stage separator occurred during the operation of the crude oil treatment system in a certain factory. To analyze the causes and prevent such cases from happening again, macroscopic and microscopic observation, FTIR, density and hardness measurements, thermogravimetric analysis (TGA), tensile test, and thermal stress simulation were conducted. The results show that the PE liner was swelled by the conveying medium, resulting in a decrease in yield strength and module. Moreover, high operating temperature leads to softening and expansion of the swelled liner pipe. Finally, the stress caused by the PE liner's inability to expand freely due to the external constraints of steel elbow increased dramatically and induced the liner pipe collapse inward, resulting in pipe blockage. Failure mechanism is further verified by computer simulation.

Keywords: Polyethylene; Elbow liner; Swelling; collapse

1 Introduction

In recent years, polymeric pipes are being considered as suitable candidates to substitute for metallic pipes in the transportation of oil and oil derivatives. Carbon steel is still the main material used to manufacture pipelines, but corrosion and a fairly high internal roughness are two main drawbacks of steel pipelines[1, 2]. Therefore, the substitution of old steel grids by new ones using polymers or steel tubes with an inner polymeric layer, instead of an all-steel pipe is a common trend, since polymers are corrosion resistant and can also be manufactured with very small surface

---

Corresponding author: Kong Lushi, kongls@cnpc.com.cn.

roughness[3]. Among the many available polymeric materials, polyethylene (PE) is a promising choice due to its good properties, availability, and cost[4-6].

PE pipes are mainly used in the protection of steel pipes in two fields: corrosion protection in the transportation of corrosive chemical reagents and repair of damaged steel pipes. [7] Trenchless interpenetration technology for repairing damaged pipelines was originally developed in Europe and North America for repairing water pipes. [8] During trenchless internal repairing, the diameter of PE pipe is first reduced by a special reducing tool, which reduces the cross-section area of the lining pipe by about 40%. The deformed PE pipe is pulled into the damaged pipe under the effect of traction. When PE liner pipe is in place, the deformed pipe will bounce back with the help of its memory characteristics or the effect of pressure and temperature, and the outer wall of the thermoplastic pipe will stick to the inner wall of the damaged steel pipe with an interference fit, forming a pipe-in-pipe structure. The joint will be connected by flange, ensuring good integrity and sealing of the pipe[9, 10]. The repaired steel pipe with PE liner combines the advantages of corrosion resistance of PE pipe and high strength and good pressure-bearing performance of the steel pipe. The conveying medium flows in the inner liner pipe without contact with the steel pipe, which prevents from corrosion of steel pipe and extends the service life of the steel pipe. The cost, using the trenchless interpenetration technology to repair the damaged pipelines, is only about 60% of the new pipeline comprehensive cost. In the engineering of the ground system construction, it can greatly shorten the construction period, achieve early start-up, and save the cost of investment[11].

Repairing steel pipe by trenchless interpenetration technology with thermoplastic plastics has drawn more and more attention. However, some problems have been exposed during the service life, such as collapse or rupture of the liner pipe[12-14]. In this work, an accident about the collapse of the elbow liner was reported. To avoid the recurrence of similar accidents, failure analysis was conducted on the polymeric liner pipe, and countermeasures were offered.

2 Background of the incident

The processing pipeline lined with a polymeric pipe in the crude oil treatment system station of a certain plant was put into operation in October 2013 with an operating pressure of 0.4 ~ 0.5 MPa. The conveying medium was crude oil, in which hydrogen sulfide content was less than 10%, carbon dioxide content was 4% ~ 6%, water content was 20% ~ 30%. The amount of daily treatment liquid was about 15000 tons. On May 15, 2019, a polymeric lined elbow located between the heating furnace and the manifold of the secondary separator was blocked. According to the record of the incident, the outlet pipeline temperature exceeded 90 ℃ when the failure occurred. Figure 1 shows the failed polymeric lined elbow. The inside diameter of the steel elbow is about 350 mm. The bend centerline radius and bend angle are 400 mm and 65°, respectively. The collapse appeared in the middle part of the polymeric liner, as shown and yellow-marked in Fig. 1(a) and (c). Besides, at one end of this elbow, the liner ruptured obviously between the liner body and the flanging zone.

3 Failure description

To conduct a complete visual examination, the polymeric liner was taken out from the steel

(a)　　　　　　　　　　　　(b)　　　　　　　　　　　　(c)

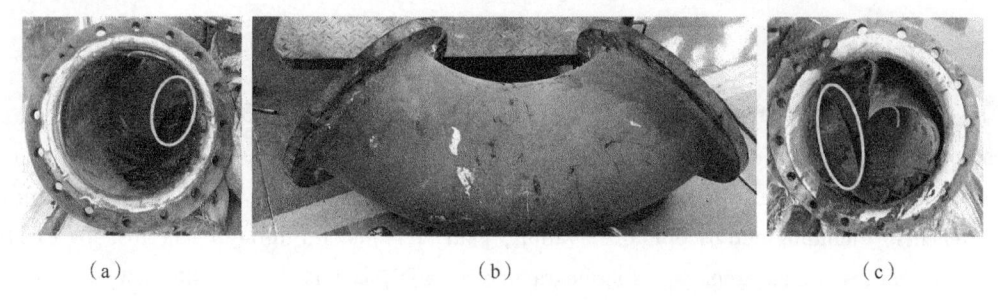

Fig. 1　Photographs of failed polymeric lined elbow

elbow. Fig. 2 shows the failed polymeric liner. It can be seen from Fig. 2(a)(b), and 2(c) that collapse occurred in three zones of the liner body, locating at the intrados side of the liner and the area of 90° and −90° from the intrados side, respectively. Among them, the most serious one happened in the area shown in Fig. 2(a) and the collapse runs almost through the liner body. In Fig. 2(d), the extrados side of the liner body remained intact. Besides, the three collapses are all close to the same one end of the liner at which more than half of the circle ruptured between the liner body and the flanging zone, as shown in Fig. 2(e). Fig. 2(f) shows that the other end of the liner had a crack at the inside edge of the flanging zone.

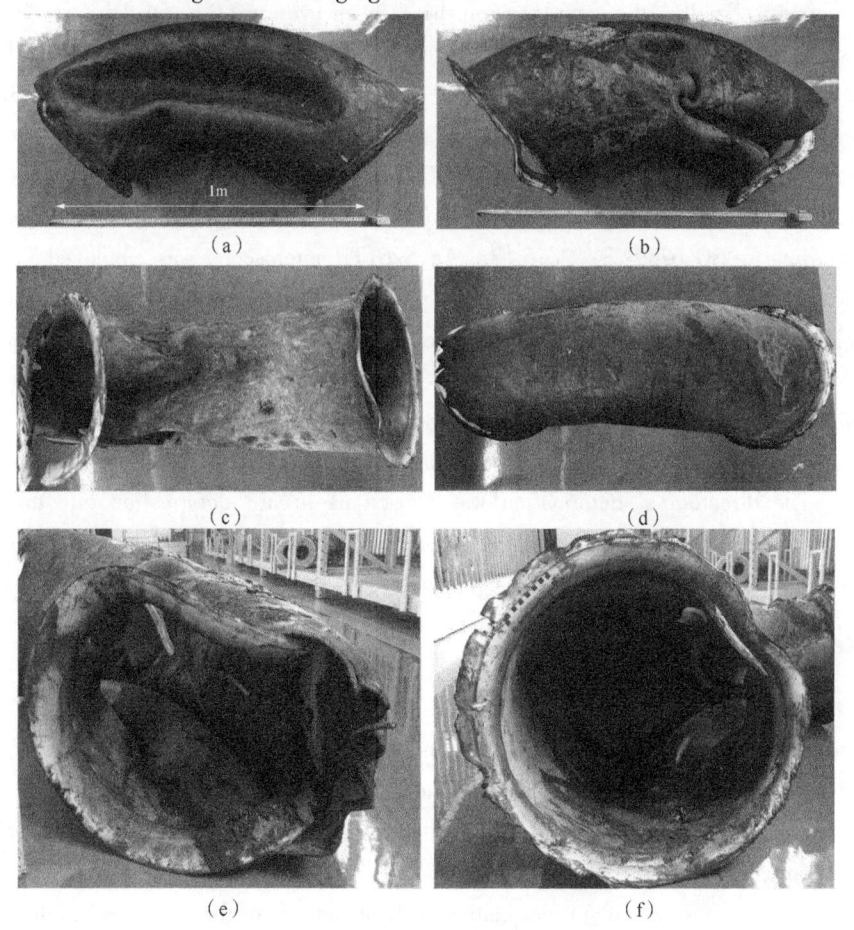

(a)　　　　　　　　　　　　(b)

(c)　　　　　　　　　　　　(d)

(e)　　　　　　　　　　　　(f)

Fig. 2　Photographs of failed polymeric liner

4 Methodology

In this work, the failure analysis contains three parts. First, based on the background information and failure description, the causes that might lead to the failures of the liner was proposed. Then, probable causes of the failure were systematically analyzed by kinds of measurements, including microscopic observation, Fourier Transform infrared spectroscopy (FTIR), density and hardness measurements, thermogravimetric analysis (TGA), tensile test, and Thermal stress simulation. Finally, a comprehensive analysis based on the above results was conducted and countermeasures are offered.

To analyze the causes of the failure, the samples used for chemical, physical, and mechanical tests were cut from the area close to the collapse zone and flanging zone, respectively, as shown and red-marked in Fig. 2(b) and 2(f). In theory, the material at the flanging zone was not in contact with the medium, which means that properties of the material are close to that of the liner material before service. So the sample obtained from the flanging zone was used as the reference sample. Before tests, all samples were firstly washed by kerosene to remove the oil pollution at the surface of the samples. Then, the samples were washed by ethanol and deionized water under ultrasonic treatment. Finally, the samples were dried in an oven at 60℃ for 4 h.

FTIR spectra of the samples via the attenuated total reflection model were collected by using a Nicolet iS 50 IR spectrometer. The density of the samples was measured by an ET-12SL electronic densitometer. The hardness of the samples was obtained by a TIME5410 Shore A durometer. Thermogravimetric analysis (TGA) was conducted from 40℃ to 650℃ at 30K/min at the N_2 atmosphere on a TGA-2 analyzer. The tensile tests were evaluated by an AGS-X10kN large stretch tensile testing machine according to ISO 6259-3: 2015[15]. An S261TR microscope was employed to observe the samples. Thermal stress simulation of elbow liner was performed by using Fusion 360 software. Polyethelene was selected as the liner material from the material database of the software. Because of the constraint of flange connection, the two end faces of the elbow liner were fixed. According to the record of the incident, the outlet pipeline temperature exceeded 90 ℃ when the failure occurred. Based on this information, the temperature of internal surface of the elbow liner was set as 90 ℃. The direction of deformation was chosen as inward deformation due to the external constraints of steel elbow. Based on 10% of the model size, The mesh was generated automatically by the software. There are 33575 nodes and 16399 elements in this model.

5 Results and discussions

5.1 Probable cause of the collapse

Based on the background information and the Failure description (Fig. 2), it can be predicted that the polymeric liner may suffer a swelling induced by conveying medium and thermal stress caused by the liner's inability to expand freely due to the external constraints of steel elbow under high working temperature. According to references[5, 6, 16], swelling can cause a severe decrease in the mechanical properties of polymeric materials. It means that the swelling could weaken the structural stability of this polymeric liner. Glock[17] derived an analytical expression relating buckling pressure with pipe geometry (w/D) valid within the elastic regime as shown in Eq. (1).

$$P_c = \frac{E}{1-v^2}\left(\frac{w}{D}\right)^{2.2} \tag{1}$$

Where P_c is the bucking collapse critical pressure expressed in MPa, E is the module of liner materials, v is the Poisson's ratio of liner materials, and w/D is the ratio of wall thickness and diameter.

According to Eq. (1), the decrease in E will result in a decrease in Pc, which means that the liner became easier to collapse. On the one hand, contact with a high-temperature medium can soften the materials, resulting in lower yield strength and module. On the other hand, at high temperature, thermal stress increased dramatically, posed a high risk of collapse of PE liner. Based on the above-mentioned analysis, it can be speculated that the collapse of the polymetric liner was caused by a combination of swelling and thermal stress. To verify this speculation, tests, including microscopic observation, FTIR spectra, density and hardness tests, TGA, tensile test, and thermal stress simulation, were conducted.

5.2 Microscopic observation

Fig. 3 shows the photographs of the surface and cross section of samples at the flanging and failed zone of the liner. It can be seen in Fig. 3(a) that the surface of the liner in the flanging zone is relatively smooth. The surface of the liner in the failed zone became rough and some pitting is distributed at the surface, as shown in Fig. 3(b). The color changed from white to black. Fig. 3(c) shows that the materials at the flanging zone are white and homogeneous, with flat edges and no cracks. Because of no contact with the conveying medium, there is no interaction between liner and medium and the liner material in the flanging zone remains in its original state. Fig. 3(d) shows that the color of the material at the failure zone changed significantly. The part near the upper and lower surfaces is black, the middle part is brown, and cracks appeared at the upper and lower surfaces, indicating that the conveying medium has infiltrated into and swelled the materials in the failure zone.

5.3 Density and hardness

The density and hardness of samples from the failed zone and flanging zone were measured respectively for comparison. The density of the sample from the ruptured zone (0.9377g/cm^3) is higher than that of the sample from the flanging place (0.9245g/cm^3). Compared with the sample from the flanging zone, the hardness of the sample from the ruptured zone shows an obvious decrease from 45.6 to 25.7. Based on the changes of color, density, and hardness, it can be inferred that the conveying medium with high density diffused into the liner, causing color change and density to increase. The medium in the liner may act as the plasticizer and soften the liner, causing hardness to decrease.

5.4 FTIR analysis

To confirm the chemical structure of the liner, samples from the ruptured zone and flanging place were characterized by FTIR. The FTIR spectra of the samples from the flanging zone and ruptured zone are shown in Fig. 4. The structure of PE can be identified from the characteristic peaks at 2915cm^{-1}, 2847cm^{-1}, 1462cm^{-1}, and 718cm^{-1}, corresponding to C-H stretching, C-H stretching, $-CH_2$ twisting and $-CH_2$ rocking, respectively. It implies that the chemical structure of the liner in the failed zone did not change obviously. Compared with the spectrum of the sample from the flanging zone, three new peaks at 1605cm^{-1}, 538cm^{-1}, and 450cm^{-1} appeared in the spectrum

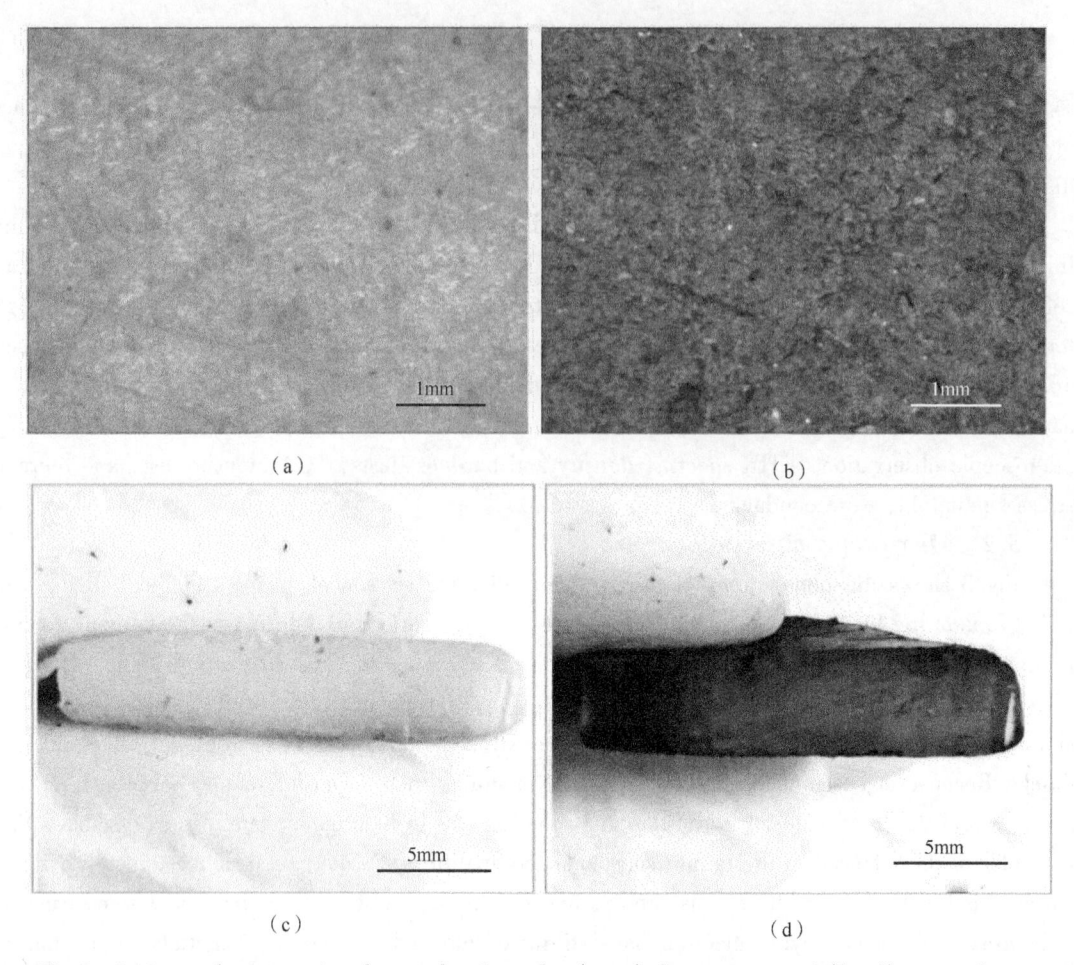

Fig. 3 Surface and cross section photographs of samples (a, c) flanging zone and (b, d) ruptured zone

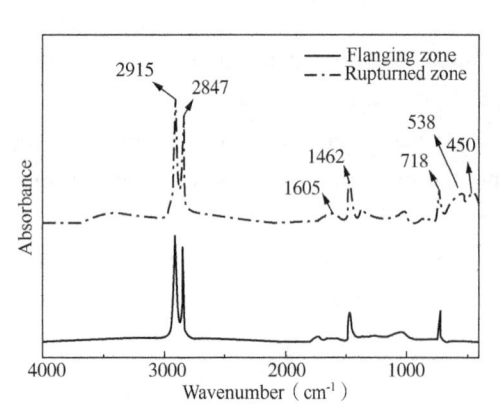

Fig. 4 FTIR spectra of the samples
(a) flanging place and (b) ruptured zone

of the sample from the ruptured zone. Considering the changes of color, density, and hardness, these three peaks could be attributed to characteristic absorption of the medium diffused into the matrix of the liner.

5.5 TGA analysis

To confirm the swelling of PE liner caused by the conveying medium, TGA was used to characterize the liner. TGA curves of the samples from the flanging place and ruptured zone are shown in Fig. 5. In Fig. 5(a), there is only one 100% mass loss step in the TGA curve, which could be attributed to the mass loss of thermal decomposition of PE in the flanging place. Being different from the TGA curve of PE in the flanging place, two mass loss steps appear in the TGA curve of PE in the failed zone [Fig. 5(b)]. The temperature of the first step is lower than the PE thermal decomposition temperature. This part of mass loss was

34.87%, which could be attributed to the evaporation of medium absorbed by the PE liner. The second step of mass loss (60.86%) belongs to the thermal decomposition of PE, being like the features of the TGA curve in Fig. 5(a). It should be noted that the residue is 4.27%, which may be attributed to some inorganic composition absorbed by PE during its service. The results of TGA confirm the swelling of PE by conveying the medium further. More importantly, the medium with a high weight percentage (34.87%) plays a key role in swelling and plasticizing of the PE liner. According to references[5, 6, 16], plastination can cause a severe decrease in mechanical properties of PE materials. It means that the plastination could weaken the structural stability of PE liner.

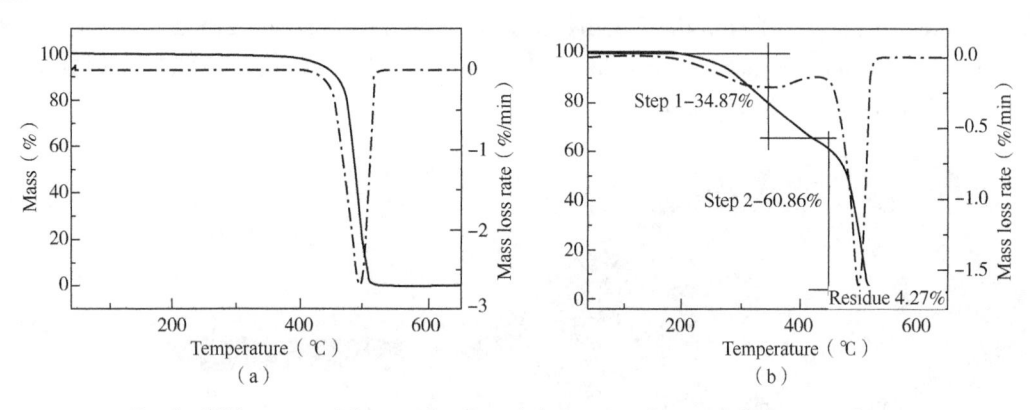

Fig. 5 TGA curves of the samples from (a) flanging place and (b) ruptured zone

5.6 Mechanical property of the PE liner

Fig. 6 shows the stress–strain curves of the sample in the failed zone and pristine PE. Compared with the pristine PE, yield strength, module, and breaking elongation of the sample in the failed zone exhibits a significant decrease. The yield strength of PE liner decreased from 20.4 MPa to 8.1 MPa and the module decreased from 822 MPa to 86.8 MPa. It verifies the plastination of medium absorbed by PE liner. In this case, the breaking elongation of the plasticized sample in the failed zone shows an obvious decrease instead of increase. The reason may be explained that pitting and cracks distributed at the surface, as shown in Fig. 3(b), will become stress concentrations during tensile, lead to the material break and low breaking elongation. It could be used to explain the rupture and crack of the liner in Fig. 2(e) and 2(f).

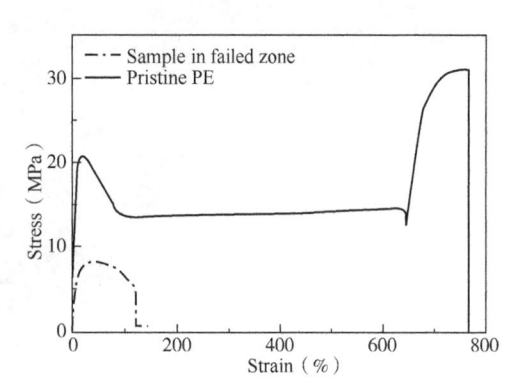

Fig. 6 Stress–strain curves of the sample in the failed zone and pristine PE

5.7 Thermal stress simulation

The results of FTIR, TGA, and tensile tests indicate that the structural stability of PE liner decreased significantly due to the swelling and plastination induced by the conveying medium exited in the PE matrix. Additionally, according to the incident records, the working temperature exceeded 90 ℃. On the one hand, contact with a high-temperature medium can soften the PE materials, resulting in lower yield strength and module. On the other hand, at high temperature, the stress caused by the PE liner 's inability to expand freely due to the external constraints of steel elbow increased dramatically and posed a high risk of collapse of PE liner. The results of the thermal stress simulation conducted by using Autodesk's Fusion 360 software confirmed the speculation. Fig. 7 shows the results of the thermal stress simulation at 90℃. Because of high stress, the intrados collapsed. The maximum stress is 10MPa, which is higher than that of yield strength of plasticized PE liner at room temperature. With increasing temperature, the yield strength and module of PE material will decrease[18]. At 90℃, the yield strength of PE is lower than that at room temperature. So, this high stress is likely to cause the yield or even break of PE liner. The position of maximum stress matches well with the ruptured zone shown in Fig. 2(a) and 2(b). Additionally, the stress in the transition part between flanging and interior is relatively high, which may cause the rupture of PE liner shown in Fig. 2(e).

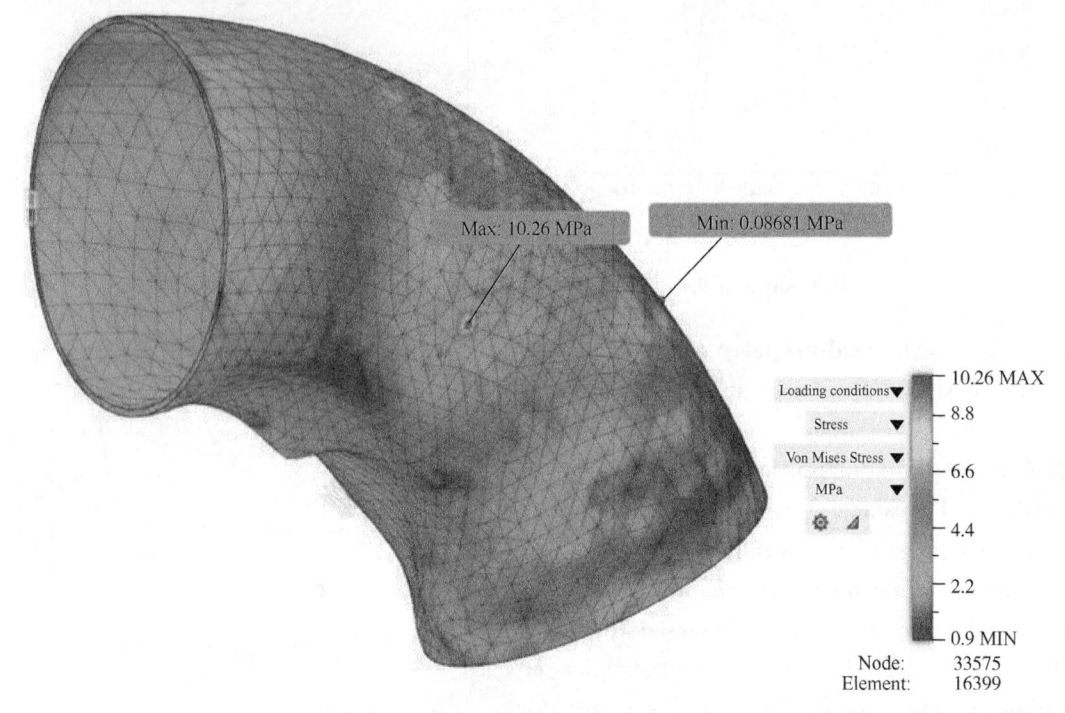

Fig. 7 Von Misses stress contours of deformed geometry obtained by FEM simulation
(The legend shows color references of Von Mises stress values) Color figure online

6 Conclusions and recommendations

6.1 Conclusions

Based on the background information of the failed PE lined elbow and results of tests, including macroscopic and microscopic observation, FTIR, density and hardness measurement, Thermogravimetric analysis (TGA), tensile test, and thermal stress simulation, it is deduced that the failure of this liner pipe was caused by a combination of swelling and heat. The intrinsic reason for the failure of the liner is that the PE liner was swelled by the conveying medium, resulting in a decrease of mechanical properties and poor structure stability. Meanwhile, high operating temperature (90 ℃) accelerated the decrease of the performance of the swelled liner pipe. At high temperature, the stress caused by the PE liner 's inability to expand freely due to the external constraints of steel elbow increased dramatically and induced the liner pipe collapse inward, resulting in pipe blockage.

6.2 Recommendations for failure prevention

According to working conditions, suitable materials and structure should be selected and designed to produce the liner pipe, such as high module materials and high w/D value to ensure the structural stability of liner. During its service, the working conditions, including the composition of medium, pressure, and temperature, should be continuously monitored to ensure that its change does not go beyond the capability of the material.

Acknowledgement

This work was supported by the SINOPEC Northwest Company of China Petroleum and Chemical Corporation (No. 34400007-19-ZC0607-0058), the Key Research and Development Program Shaanxi Province (No. 2018ZDXM - GY - 171), and Shaanxi Province Innovation Capability Foundation (No. 2019KJXX-092).

References

[1] A. H. U. Torres, J. R. M. d'Almeida, J. -P. Habas, Aging of HDPE Pipes Exposed to Diesel Lubricant, Polymer-Plastics Technology and Engineering 50(15) (2011) 1594-1599.

[2] L. Fan, F. Tang, S. T. Reis, G. Chen, M. L. Koenigstein, Corrosion Resistances of Steel Pipes Internally Coated with Enamel, Corrosion 73(11) (2017) 1335-1345.

[3] A. Habas-Ulloa, J. -R. M. D'Almeida, J. -P. Habas, Creep behavior of high density polyethylene after aging in contact with different oil derivates, Polymer Engineering & Science 50(11) (2010) 2122-2130.

[4] J. P. Manaia, F. A. Pires, A. M. P. de Jesus, S. Wu, Yield behaviour of high – density polyethylene: Experimental and numericalcharacterization, Engineering Failure Analysis 97 (2019) 331-353.

[5] M. Erdmann, M. Böhning, U. Niebergall, Physical and chemical effects of biodiesel storage on high-density polyethylene: Evidence of co-oxidation, Polymer Degradation and Stability 161 (2019) 139-149.

[6] W. Ghabeche, K. Chaoui, N. Zeghib, Mechanical properties and surface roughness assessment of outer and inner HDPE pipe layers after exposure to toluene methanol mixture, The International Journal of Advanced Manufacturing Technology 103(5-8) (2019) 2207-2225.

[7] E. Engle, Pipe Rehabilitation with Polyethylene Pipe Liners, Iowa. Dept. of Transportation, 2003.

[8] M. Kim, D. -G. Lee, H. Jeong, Y. -T. Lee, H. Jang, High Temperature Mechanical Properties of HK40-type Heat-resistant Cast Austenitic Stainless Steels, Journal of Materials Engineering and Performance 19 (2010) 700-704.

[9] A. Morris, T. Grafenauer, A. Ambler, City of Houston 30 – Inch Water Transmission Main Replaced by Compressed Fit HDPE Pipe Lining, Amer Soc Civil Engineers, New York, 2017.

[10] V. A. Orlov, I. O. Bogomolova, I. S. Gureeva, Trenchless renovation of worn-out pipelines through their prior destruction and dragging new polymer pipes in place of the old, Vestnik MGSU / Proceedings of Moscow State University of Civil Engineering (7) (2014) 101-109.

[11] Z. H. Deng, Y. S. Fu, P. Ning, The application case of high density polyethylene pipe linings for the renovation of water mains, in: M. J. Chu, X. G. Li, J. Z. Lu, X. M. Hou, X. Wang (Eds.), Progress in Civil Engineering, Pts 1-4, Elsevier Science Bv, Amsterdam, 2012, pp. 2424-2427.

[12] F. Rueda, A. Marquez, J. L. Otegui, P. M. Frontini, Buckling collapse of HDPE liners: Experimental set-up

and FEM simulations, Thin-Walled Structures 109 (2016) 103-112.
[13] S. R. Frost, A. M. Korsunsky, Y. Wu, A. G. Gibson, Collapse of polymer and composite liners constrained within tubular conduits, Plastics, Rubber and Composites 29(10) (2013) 566-572.
[14] F. Rueda, J. P. Torres, M. Machado, P. M. Frontini, J. L. Otegui, External pressure induced buckling collapse of high density polyethylene (HDPE) liners: FEM modeling and predictions. Thin-Walled Structures 96 (2015) 56-63.
[15] ISO 6259-3: 2015 Thermoplastics pipes-Determination of tensile properties-Part 3: Polyolefin pipes.
[16] W. Ghabeche, L. Alimi, K. Chaoui, Degradation of Plastic Pipe Surfaces in Contact with an Aggressive Acidic Environment. Energy Procedia 74 (2015) 351-364.
[17] D. Glock, Behavior of Liners for Rigid Pipeline Under External Water Pressure and Thermal Expansion, Der Stahlban (English translation) 7 (1997) 212-217.
[18] H. W. Mu Leijin, Tao Junlin, Experimental study on compressive property of polyethylene pipe material within wide temperature and wide strain rate range, New Building Materials 10 (2017) 132-138.

本论文原发表于《Journal of Failure Analysis and Prevention》2020 年第 20 卷第 6 期。

Failure Analysis on the Oxygen Corrosion of the Perforated Screens Used in a Gas Injection Huff and Puff Well

Fan Lei[1] Gao Yuan[1,2] Yuan Juntao[1] Yin Chenxian[1] Fu Anqing[1]
Zhao Mifeng[3] Li Yan[3] Suo Tao[4] Du Xiaoyi[4] Wang Yanhai[4]

(1. CNPC Tubular Goods Research Institute, State Key Laboratory for Performance and Structure Safety of Petroleum Tubular Goods and Equipment Materials; 2. School of Chemistry and Chemical Engineering, Xi'an University of Architeture and Technology; 3. Petro China Traim Oilfield Company; 4. Petrochina Changqing Petrochemical Company)

Abstract: The perforated screens of a gas injection huff and puff well broke down in an oilfield in western China. The physical and chemical properties of the failure – perforated screens met the requirements of relevant technical protocols. The injected gas was nitrogen produced by the membrane nitrogen method, which contains approximately 5% O_2. Therefore, a substantial amount of O_2 entered the well during the gas injection, and greatly promoteed the cathodic process of the electrochemical corrosion. High salinity formation water, especially with large amounts of Cl^-, further accelerated the corrosion. The significantoxygen corrosionunder the well finally led to the failure of the perforated screen.

Keywords: Gas injection and puff well; Nitrogen; Corrosion; Failure analysis; Oxygen corrosion

1 Introduction

Gas flooding is one of the most important means to improve the recovery of oil and gas. With exploitation, oilfields inevitably enter the secondary development stage wheregas flooding is widely applied. In this process, injected gases include hydrocarbon gas, flue gas, CO_2, N_2, and air[1,2]; however, the advantages of N_2 are evident[3-5]. For example, it can be widely sourced, is not restricted by region, and yields no pollution. In addition, the physical properties of N_2, including low viscosity, low thermal conductivity, low volatility (it is non-flammable and non-explosive), greater solubility in crude oil than in water, large compression coefficient, high expansion capacity, and large elastic energy, are excellent. Therefore, N_2 is asuperior source of injection gas.

A case of well failure in an oilfield in western China is an integrated opportunity for gas lift and

---

Corresponding author: Fan Lei, damifan198802@163.com.

injection. After six months of production, the nitrogen, which was produced by membrane nitrogen method, was injected into the well. After two months, aperforated screen fractured and the well broke down. H_2S or CO_2 could not be detected in this well. The depth of this well was 6600 m and the perforated screen was placed approximately 4500 m from the wellhead. The material of the perforated screen was C110 steel and its dimensions were 3 1/2in×6.45mm. The fracture of the failure screen was significantly corroded and damaged, therefore, fracture analysis could not be performed. Macroscopic examination, chemical composition analysis, mechanical properties testing, metallographic examination, scanning electron microscopy (SEM), X-ray diffraction (XRD), and electrochemical testing were used to comprehensively investigate the failure of the perforated screens.

2 Physical and chemical properties

The macroscopic morphology of the fracture of the failure-perforated screen is shown in Fig. 1. The hole of the perforated screen is observed around the fracture, which means that the fracture occurred near the perforation. The thickness of the failure-perforated screen near the fracture was measured by an ultrasonic thickness gauge and the results are shown in Fig. 2. The outer diameter near the fracture significantly decreases and necking occurs. Also, the thickness significantly decreases within 30mm from the fracture, reaching about 3.5mm at the end. The thickness of the failure screen (5.4~6.3mm) is slightly smaller than the standard API 5CT-2012 wall thickness requirement (6.45mm), which is mainly due to the uniform corrosion of the screen.

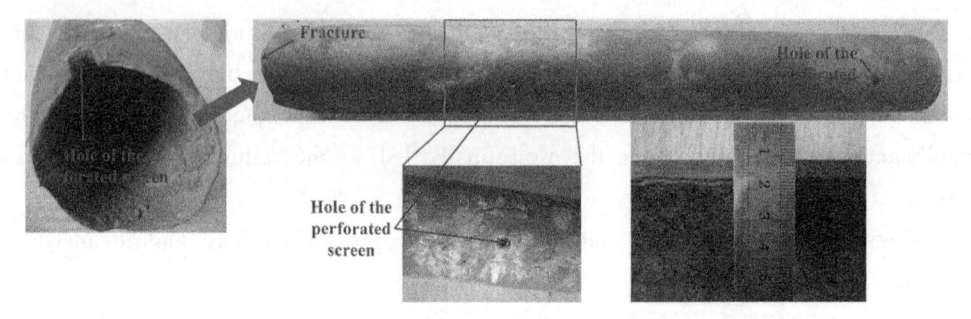

Fig. 1 Macroscopic morphology of the fractured perforated screens

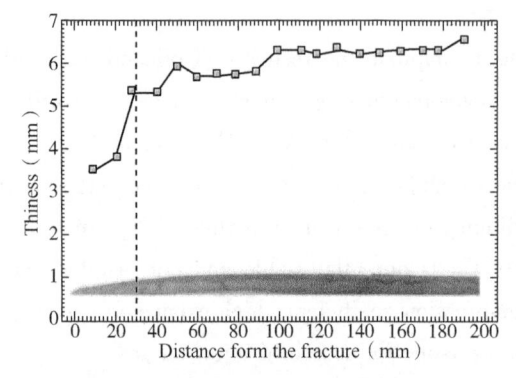

Fig. 2 Thickness curves of the failure perforated screens

The chemical composition of the perforated screen was analyzed by a direct reading spectrometer (ARL 4460). The results are summarized in Table 1 and indicate that the chemical composition of the screen meets the requirements of the standard API 5CT-2012.

For the measurement of the mechanical properties, the samples were machined from the fractured perforated screen by avoiding the punching and fracture. The mechanical properties were analyzed by a UTM5305 tensile tester, RB2002 type hardness tester, and PIT302D type impact tester. The temperature of the impact test is 0℃. The results are summarized in Table 2. The mechanical properties meet the requirements of the standard API 5CT-2012, except that the yield strength is lower than the standard.

Table 1 Chemical composition of the failure perforated screen (wt. %)

Element	C	Si	Mn	P	S	Cr	Mo	Ni	Nb	V	Ti	Cu	B	Al
Measured value	0.29	0.18	0.45	0.0069	0.0007	0.49	0.84	0.036	0.031	0.072	0.0022	0.037	0.0005	0.016
API 5CT-2012[6]	≤0.35	—	≤1.2	≤0.020	≤0.005	0.4~1.5	0.25~1.0	≤0.99	—	—	—	—	—	—

Table 2 The mechanical properties of the failure perforated screen

Property	Tensile property					Impact energy (J)	Rockwell hardness
	Size(mm×mm)	Tensile strength(MPa)	Yield strength(MPa)	Elongation(%)			
Measured value	19.1×50	803	625	17	50、34、35	18.8, 17.3, 17.2
API 5CT-2012	—	≥793	758~828	≥12	≥22.55	≤30

For the metallographic examination, the sample was machined around the fracture of the failure screen. The results show that the tissue is ferrite + pearlite. The deformation of the tissue could not be observed around the fracture. Slight decarburization at the outer surface [Fig. 3(b)] and some corrosion pits on the inner surface of the fracture [Fig. 3(c)] were observed.

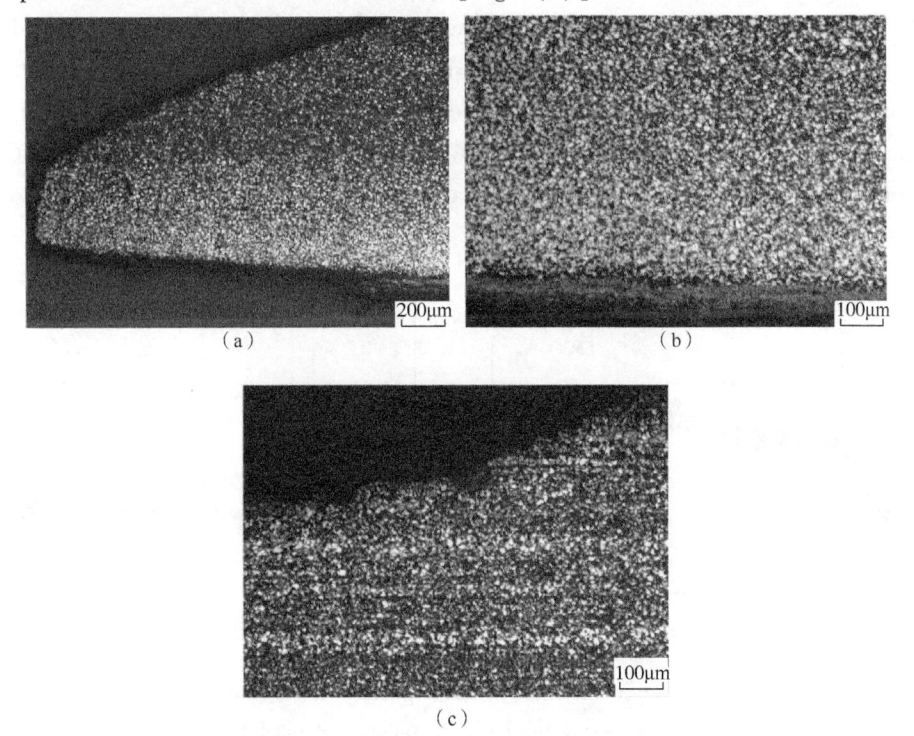

Fig. 3 Optical photomicrography of the fracture of the perforated screens
(a), outer surface (b), and inner surface(c)

3 Analysis of the corrosion product

The corrosion products are very abundant (Fig 1). The surface morphologies of the corrosion products around the fracture were analyzed by SEM. The results, presented in Fig. 4, showthat the even and dense corrosion products are on the outer surfacewhile the loose and porous corrosion products are on the inner surface with many holes and cracks.

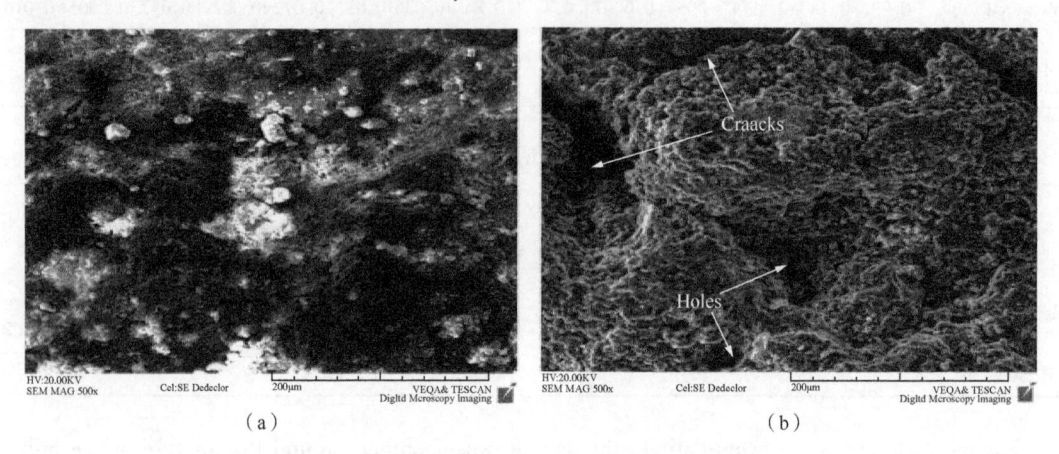

Fig. 4 Surface morphologies of the corrosion products on outer surface
(a) and the inner surface (b) around the fracture

The corrosion products of the inner surface and the outer surface of the failure-perforated screen were collected separately and the corrosion product powder was analyzed by XRD. The diffraction spectrum was analyzed by Jade 5 software. As shown in Fig. 5, the results indicate that the corrosion products on the outer surface of the screen are $Fe_8(O, OH)_{16}Cl_{1.3}$, $FeCO_3$, and Fe_3O_4. On the inner surface, the corrosion productsinclude Fe_3O_4, Fe_2O_3, $FeCO_3$, and $Fe_8(O, OH)_{16}Cl_{1.3}$. The crystal structure of $Fe_8(O, OH)_{16}Cl_{1.3}$ is similar to $\beta-FeOOH$ and its formation is related to chlorideions[6].

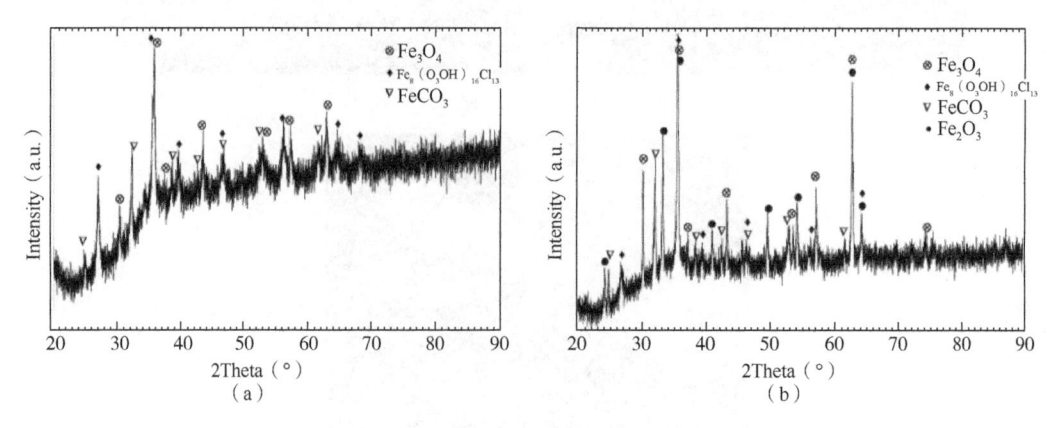

Fig. 5 XRD patterns of the corrosion products of the outer surface
(a) and inner surface (b) of the failure perforated screens

4 Electrochemical testing

4.1 Experimental method

Experiments were conducted at 1atm in a glass cell including a water bath to control the temperature. A typical three-electrode cell was used with an Ag/AgCl electrode (in saturated KCl solution) as the reference electrode, a large piece of platinum plate with a surface area of over 4cm^2 as the counter electrode, and the C110 steel from the failure screen as the working electrode. The specimens with an exposed surface of (10.0×10.0)mm^2 were machined from the failure screen and embedded in the epoxy resin. Only its cross-section contacted the electrolyte. Prior to each experiment, the exposed metal surface of each specimen was ground with wet silicon carbide paper to 1000 grade, degreased with acetone, cleansed with distilled water, and dried in compressed hot airflow.

The electrolyte was the simulating solution, which was prepared by using analytical reagent and distilled water. The ion content is shown in Table 3; the pH of the solution was adjusted using HCl to 6.2.

After theelectrolyte was heated to the set temperature, oxygen was removed by the purity N_2 for 30min. Then, 100% N_2, 5% O_2+95% N_2, 20% O_2+80% N_2, or100% O_2 was introduced into the electrolyte for 30min. Prior to the Tafel polarization tests, the samples were immersed in the test solution for 900 s before the beginning ofeach test to attain the steady open circuit potential (OCP). The Tafel polarization tests were carried out starting at −0.25 V (vs. OCP) and ending at currents exceeding 2mA at a potential scan rate of 0.33mV/s in a 1000 mL test solution. All potentials reported in this paper were measured with respect to the Ag/AgCl electrode (in saturated KCl solution). For better reproducibility, all electrochemical experiments were repeated more than three times.

Table 3 Chemical composition of the simulated solution

Ion species	Cl$^-$	HCO$_3^-$	Ca^{2+}	K$^+$+Na$^+$	SO$_4^{2-}$	Mg^{2+}
Content (mg/L)	8.79×10^4	2.35×10^2	8.27×10^3	4.687×10^4	336.2	487.1

4.2 Results

The OCP of the C110 steel was monitored when 100% N_2, 5% O_2+95% N_2, 20% O_2+80% N_2, and 100% O_2 were continuously introduced into the electrolyte at 25 ℃. The results, shown in Fig. 6, indicate that the OCP shifts towards a positive direction with the increasing O_2 pressure. The OCP of the C110 steel in the simulated solution containing 101kPa O_2 is close to −0.465 V$_{Ag/AgCl}$, which is 236 mV more positive than that measured in the absence of O_2.

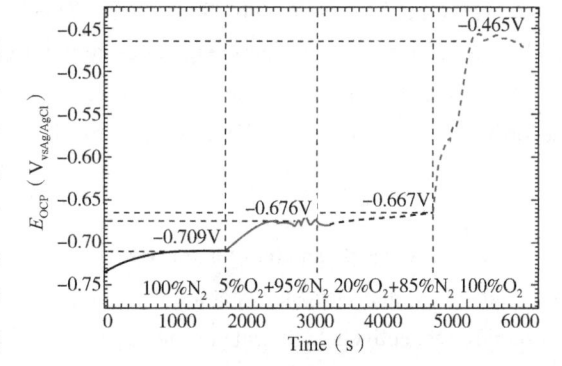

Fig. 6 Open circuit potential of the C110steel in the simulated solution continuously with 100%N_2, 5%O_2+95%N_2, 20%O_2+80%N_2, and 100%O_2 at 25℃

The Tafel polarization curves of the C110 steel in the simulating solutions in the absence and presence of different O_2 concentrations at 25 °C are shown in Fig. 7. A typical feature of these polarization curves is that the cathodic curve is remarkably accelerated by increasing O_2 concentration. However, the anodic curves were very similar, regardless of the O_2 concentration. As can be seen in Fig. 7, the anodic process of the C110 steel exhibited anodic iron dissolution.

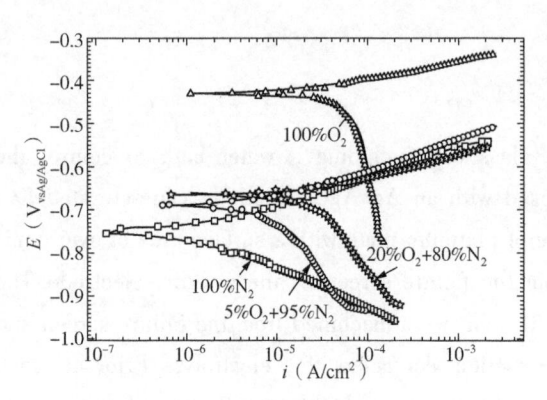

Fig. 7 Tafel polarization curves of C110 steel in 100%N_2(-□-), 5%O_2+95%N_2(-○-), 20%O_2+80%N_2(-☆-), or 100% O_2(-△-) at 25 °C

The determination of corrosion parameters (E_{corr}, R_p, b_a, b_c, and i_{corr}) could provide more information about the overall corrosion process. Jankowski and Juchniewicz[7] proposed a four-point method for determining the corrosion rate simply and accurately. In it, four current density values at four applied potentials, $E = \Delta E$, $-\Delta E$, $2\Delta E$, and $-2\Delta E$, on a polarization curve are used to determine the corrosion current density (i_{corr}) and harmonic mean of the Tafel constants (B) and each Tafel constant (b_a and b_c). The i_{corr}, b_a, and b_c are determined from the following equations[7-9]:

$$i_{corr} = \frac{I_1 I_{-1}}{\sqrt{I_2 I_{-2} - 4 I_1 I_{-1}}} \tag{1}$$

$$B = \frac{b_a b_c}{2.3(b_a + b_c)} = \frac{\Delta E}{\cosh^{-1}\left(\frac{I_2 I_{-2}}{2 I_1 I_{-1}} - 1\right)} \tag{2}$$

$$b_a = \frac{\Delta E}{\log\left\{\frac{I_1}{i_{corr}[1 - \exp(\Delta E/B)]}\right\}} \tag{3}$$

$$b_c = \frac{2.3 b_a B}{b_a - 2.3 B'} \tag{4}$$

where I_1, I_{-1}, I_2, and I_{-2} are the current densities at the four overpotentials ΔE, $-\Delta E$, $2\Delta E$, and $-2\Delta E$, respectively. These equations are derived from the Bulter-Volmer equation, which is valid in the potential range where reactions are controlled by the activation process[9]. Additionally, according to Stern and Geary[10], the polarization resistance (R_p) can be expressed as:

$$R_p = \frac{B}{i_{corr}} \tag{5}$$

The corrosion parameters for the C110 steel were determined with the four-point method. Table 4 summarizes the corrosion potentials (E_{corr}), Tafel slopes (b_a and b_c represent anodic and cathodic, respectively), harmonic mean of the Tafel constants (B), corrosion currents (i_{corr}), and calculated values of polarization resistance (R_p). The corrosion potential shifts towards a positive direction with the increasing pressure of O_2. For example, the corrosion potential of the C110 steel in the solution with 101kPa O_2 is close to $-0.430 V_{Ag/AgCl}$, which is 324mV more positive than that measured in the solution without O_2. The corrosion potential is the result of the two competitive reactions of the cathodic and anodic processes[9,11,12]. In this work, the anodic processes were not modified with various pressures of O_2; therefore, the cathodic process of the metalsurface was promoted with increasing O_2 concentration, which contributed to the positive shift of the corrosion potential.

Anodic Tafel slops, b_a (Table 4), in all environmentsare very similar (approximately 0.1V per decade), regardless of the concentration of O_2. This behavior indicates that the concentration of O_2 contributes similarly to the anodic process. However, theslopes of the cathodic branch, b_c, (Table 4) are considerably dependent on the concentration of O_2. This, along with greater values of these slopes compared with those from the anodic branch, indicate the complex nature of the reduction process. Additionally, the corrosion current density (i_{corr}) increases from 6.79×10^{-7} A·cm² to 4.15×10^{-5} A·cm² and the polarization resistance R_p values decrease with increasing O_2 concentration (Table 4). The results imply that the cathodic process is controlled by the concentration of O_2. The corrosion resistance of theC110 steel is deteriorated with increasing concentration of O_2.

The temperature increased with the depth of the well. To study the electrochemical corrosion behavior in the well, the Tafel polarization curves were measured at 85℃ in the absence and presence of O_2, and the results are shown in Fig. 9. The shape of the Tafel polarization curves and variation tendency of the cathodic and anodic processes were similarto that obtained at 25℃. The corrosion parameters were determined with the four-point method. The variation tendency of the corrosion parameters wassimilar to that obtained at 25℃. Compared to that at 25℃, the corrosion current density (i_{corr}) increased significantly at 85℃ in the absence and presence of O_2 (Table 4). This means that the corrosion accelerated with an increase in the well depth. Moreover, the corrosion was significantly accelerated by O_2 at 85℃ but the phenomenon seemed to be more noteworthy at 25℃.

Table 4　Corrosion parameters obtained fromthe Tafel polarization curves by four-point method

Temperature (℃)	Environment	Pressure of O_2(kPa)	E_{corr} (V)	i_{corr} (A/cm²)	b_a (V/dec)	b_c (V/dec)	B (V/dec)	R_p (Ω·cm⁻²)
25	100%N_2	0	-0.754	6.79×10^{-7}	0.118	0.0485	0.0149	2.50×10^4
25	5%O_2+95%N_2	5.1	-0.692	5.82×10^{-6}	0.105	0.0707	0.0184	3.58×10^3
25	20%O_2+80%N_2	20.3	-0.666	2.88×10^{-5}	0.074	0.211	0.0238	9.35×10^2
25	100%O_2	101	-0.430	4.15×10^{-5}	0.096	0.105	0.0218	5.97×10^2
85	100%N_2	0	-0.737	9.00×10^{-6}	0.102	0.103	0.0223	2.81×10^3
85	20%O_2+80%N_2	20.3	-0.700	2.39×10^{-5}	0.078	0.130	0.0211	9.99×10^2

5　Discussion

The physical and chemical properties of the failure perforated screens meet the requirements of API 5CT-2012.

The nitrogen was produced by the membrane method and wasapproximately 5% O_2. Therefore, a large amount of oxygen was also entered the well during the injection of nitrogen. The residual gas in the well contained 1% ~ 1.3% O_2 after some time. Therefore, oxygen corrosion occurred in the well. There are a lot of previous studies[13-21] shown that the oxygen corrosion is very different from the CO_2 or/and H_2S corrosion. Oxygen is a very strong oxidant[22], which directly participates the cathodic reduction reaction during the corrosion (Fig. 7, Fig. 8). The corrosion was accelerated as the increasing oxygen concentration (Table 4). The corrosion type of the C110 steel is changed from hydrogen separation in the absence of O_2 to the reaction of oxygen absorption in the presence of O_2. The temperature increased with the depth of the well and the corrosion is serious with the higher temperature[23-24]. Thus, the corrosion was accelerated as the well depth increased and the oxygen corrosion was severe in the down well. The corrosion product film formed

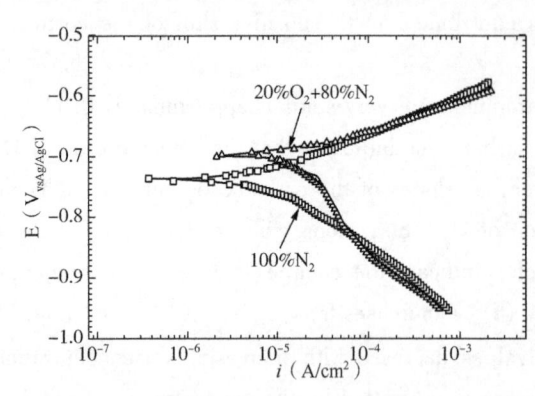

Fig. 8 Tafel polarization curves of the C110 steel in 100%N_2(-□-) or 20%O_2+80%N_2(-△-) at 85 ℃

was loose, porous (Fig. 4) during the oxygen corrosion is easily loosened from the steel surface. The cracks and holes were the channels of the short-circuiting diffusion. The corrosion medium diffused fast, further accelerating the corrosion. This means the corrosion product layer is poor inprotection performance for the substrate[24-25]. The formation waterhadhigh salinity, with a large amount of Cl^-, potentially causinglocal corrosion. Therefore, some corrosion pits were observed on the inner surface of the fracture [Fig. 3(c)]. The Cl^- also contributed to the formation of $Fe_8(O, OH)_{16}Cl_{1.3}$ (Fig. 5)[6].

Due to the aggravation of corrosion, the thickness and cross-section area of the perforated screen decrease. Furthermore, the tensile stress of the tube increases sharply. The perforation of the screens is the most serious corrosion area, which is where is the tube is most weak. Therefore, when the tensile stress of the screens exceeds its tensile strength, necking occurs (Fig. 2) and the perforated screen fractures at the perforation.

6 Conclusion

(1) The physical and chemical properties of the failure perforated screens meet the requirements of API 5CT-2012.

(2) The O_2 is brought into the well during the injection of nitrogen, resulting in severe oxygen corrosion. High salinity formation water, especially with a large amount of Cl^-, further accelerates the corrosion. The thickness of the perforated screen was decreased by the severe corrosion, resulting in fracture at the perforation.

(3) It is suggested that the oxygen content of the injected gas should be strictly controlled and that the corrosion inhibitor used in the oxygen-enriched environment should be applied to decrease the corrosion.

Acknowledgement

The project is supported by the Shanxi National Science Foundation (No. 2019JQ-213) and Basic Research and Strategic Reserve Technology Research Fund [project of China National Petroleum Corporation (2018Z-01)].

References

[1] Li Shilun, Guo Ping, Dai Lei, et al. Streathen gas injection for enhanced oil recovery [J]. Journal of southwest petroleum institute, 2000, 22: 41.

[2] Li Shilun, Zhou Shouxin, Du Jianfen et al. Review and prospects for the development of EOR by gas injection at home and abroad [J]. Petroleum geology and recovery efficiency, 2002, 9: 1

[3] Kuo, J. World's largest N2-generation plant starts up for cantarell reservoir pressure maintenance [J]. Oil Gas J, 2001, 99: 41

[4] Wang Jinan, Yue Lu, Yuan Guangjun, et al. Laboratory research on nitrogen drive [J], Petroleum exploration and development, 2004, 31: 119

[5] Wang Hao, Study of The Technology of Water Shut Off and Enhanced Oil Recovery by Nitrogen [D]. China University of Petroleum, 2015

[6] Hao Xianchao, Li Xiaogang, Xiao Kui, et al. Corrosion behaviors at the initial stage of Q235 steel in xisha atmosphere[J], Journal of Chinese society for corrosion and protection, 2009, 29: 465

[7] Jankowski J., Juchniewicz R.. A four-point method for corrosion rate determination[J]. Corrosion science, 1980, 20: 841

[8] Yi Y., Kim H., Park Y., et al. Effect of an inhibitor on the stress corrosion cracking behaviour of alloy 600 in a high-temperature caustic solution [J]. Corrosion, 2005, 61: 403

[9] Junwen Tang, Yawei Shao, Tao Zhang, et al. Corrosion behaviour of carbon steel in different concentrations of HCl solutions containing H_2S at 90 ℃ [J]. Corrosion science, 2011, 23: 1715

[10] Stern M., Geary A. L.. Electrochemical polarization: I. A theoretical analysis of the shape of polarization curves [J]. Journal of the electrochemical society, 1957, 104: 56.

[11] J. Liu, Y. Lin, X. Yong, X. Li, Study of cavitation corrosion behaviours and mechanism of carbon steel in neutral sodium chloride aqueous solution, Corrosion 61 (2005) 1061-1069.

[12] M. L. Doche, J. Y. Hihn, A. Mandroyan, R. Viennet, F. Touyeras, Influence of ultrasound power and frequency upon corrosion kinetics of zinc in saline media, Ultrason. Sonochem. 10 (2003) 357-362.

[13] Elgaddafi, R., Ahmed, R., Hassani, S., Shah, S.. Corrsion of C110 carbon steel inhigh-temperature aqueous environment with mixed hydrocarbon and CO_2 gas. J. Petrol. Sci. Eng., 2016, 146, 777-787.

[14] Elgaddafi, R., Ahmed, R., Shah, S.. Modeling and experimental studies on CO_2-H_2S corrosion of API carbon steels under high-pressure. J. Petrol. Sci. Eng. 2017, 156, 682-696.

[15] Elgaddafi, R., Ahmed, R., Shah, S.. Modeling CO_2-H_2S corrosion of tubular at elevated pressure and temperature. Res. J. Appl. Sci. Eng. Technol., 2016, 13, 510-524.

[16] Elgaddafi, R., Naidu, A., Ahmed, R., Shah, S., Hassani, S., Osisanya, S. O., Saasen, A.. Modeling and experimental study of CO_2 corrosion on carbon steel at elevated pressure and temperature. J. Petrol. Sci. Eng. 2015, 27, 1620-1629.

[17] Fassihi, M. R., Moore, R. G., Mehta, S. A., Ursenbach, M. G.. Safety considerations for high-pressure air injection into light-oil reservoirs and performance of the holt sand unit project. SPE Prod. Oper. 2016, 31, 197-206.

[18] Gao, S., Brown, B., Young, D., Nesic, S., Singer, M.. Formation mechanisms of iron oxide and iron sulfide at high temperature in aqueous H2S corrosion environment. J. Electrochem. Soc. 2018, 165, C171-C179.

[19] Gao, S., Brown, B., Young, D., Singer, M.. Formation of iron oxide and iron sulfide at high temperature and their effects on corrosion. Corrosion Sci. 2018, 135, 167-176.

[20] Gillham, T. H., Cerveny, B. W., Fornea, M. A., Bassiouni, Z.. Low cost IOR: an update on the W. Hackberry air injection project. In: SPE/DOE Improved Oil Recovery Symposium, 1998, Tulsa, OK.

[21] Gillham, T. H., Cerveny, B. W., Turek, E. A., Yannimaras, D. V.. Keys to increasing. production via air injection in Gulf Coast light oil reservoirs. In: SPE Annual Technical Conference and Exhibition, 1997, San Antonio, TX.

[22] Chen, L., Hu, J., Zhong, X., Zhang, Q., Zheng, Y., Zhang, Z., Zeng, D.. Corrosion behaviors of Q345R steel at the initial stage in an oxygen – containing aqueous environment: experiment and modeling. Materials, 2018, 11, 1462-1480.

[23] LIAO Guangzhi, YANG Huaijun, JIANG Youwei, REN Shaoran, LI Dangguo, WANG Liangang, WANG Zhengmao, WANG Bojun, LIU Weidong. Applicable scope of oxygen-reduced air flooding and the limit of oxygen content[J]. Petroleum Exploration and Development, 2018, 1: 111-117.

[24] Xiankang Zhong, Wenjun Lu, Huaijun Yang, Min Liu, Yang Zhang, Hongwei Liu, Junying Hu, Zhi Zhang, Dezhi Zeng. Oxygen corrosion of N80 steel under laboratory conditions simulating high pressure air injection: Analysis of corrosion products[J]. Journal of Petroleum Science and Engineering, 2019, 172, 162-170.

[25] Chen, M., Wang, H., Liu, Y., Ma, L., Wu, D., Wang, S.. Corrosion behavior study of oil casing steel on alternate injection air and foam liquid in air-foam flooding for enhance oil recovery. J. Petro. Sic. Eng. 2018, 165, 970-977.

本论文原发表于《Engineering Failure Analysis》2021 第 119 卷。

Investigation on Leakage Cause of Oil Pipeline in the West Oilfield of China

Liu Qiang[1] Yu Haoyu[2] Zhu Guochuan[1] Wang Pengbo[1] Song Shengyin[1]

(1. State Key Laboratory for Performance and Structural Safety of Petroleum Tubular Goods and Equipment Materials, CNPC Tubular Goods Research Institute; 2. School of Mechano-Electronic Engineering, Xidian University)

Abstract: Acrude oil pipeline used in an oilfield in the west oilfield of China leaked many times in a short time. Serious corrosion was found near the leaking position of the failed pipe, the wall thickness of pipeline also decreased near the failure section. To determine the reason for this, the failed pipeline was investigated and analysed by macroscopic analysis, chemical composition tests, metallurgical analysis, mechanical property analysis, scanning electronic microscopy (SEM) analysis, X-ray diffraction (XRD) analysis and corrosion simulation tests. Based on the systematic analysis, it can be concluded that the mechanical properties of the failed pipeline met the related standards. The crude oil conveyed in the pipeline contained a large amount of formation water; because the formation water was acidic (the pH value was 5.0~6.5) and had high salinity, the materials of the pipeline had a very high corrosion rate when exposed tothe crude oil, which was confirmed by the corrosion simulation test. At the same time, because the fluid flow changed suddenly in the pipeline lifting location, fouling deposition occurred inside the pipeline, which led to serious localized corrosion caused by thesmall anode/big cathode corrosion galvanic cell reaction, and the wall thickness decreased quickly in a short time. Finally, the pipeline leaked due to fluid erosion and the high speed impact of solid particles and grit in the crude oil. Improvement suggestions are put forward to the pipeline user at the end of this paper, such as improving the pipeline design, reducing the salinity in the conveying fluid, adding a corrosion inhibitor and replacing the pipeline lifting segment with an anticorrosive coating inside the tube; the crude oil pipeline did not leak again after followingthese suggestions.

Keywords: Pipeline; Leak failure; Localized corrosion; SEM; XRD; Corrosion simulation test

---

Corresponding author: Liu Qiang, liuqiang030@ cnpc. com. cn.

1 Introduction

In China's central and western oilfields, crude oil pipelines are prone to leakage and failure, often resulting in serious economic losses and environmental pollution. In recent years, the crude oil pipelines of Changqing oilfield in west of China had been leaking and failing frequently; one crude oil pipeline lifting segment used in the Changqing oilfield leaked many times in a short period. The specification of the pipeline in the lifting segment is 114mm×4.5mm, and the material is 20-steel (according to the Chinese GB/T 699—1999 standard[1]). The working pressure and output oil amount of this pipeline are 1.4 MPa and 13 m^3/h, respectively.

Many studies have been conducted to investigate similar leakage failures of oil and gas pipelines in the oilfield. Hu et al.[2] analysed the corrosion failure of an oil pipeline by chemical composition analysis, metallographic examination, mechanical testing and corrosion product analysis methods and found that under-scale corrosion was the direct cause of the failure. Ding et al.[3] studied the cause of gathering pipeline perforation in one western oil field through physical and chemical inspection as well as corrosion products analysis; the results showed that high CO_2 content in the pipes and a high chlorine ion environment caused local corrosion. Qin et al.[4] investigated the influences of various corrosion defects and the interaction of multiple corrosion defects on pipeline failure with full-scale pressure burst tests and the finite element method. Physical and chemical inspection, microstructure analysis, finite element analysis, corrosion product composition and structure analysis are the main methods to conduct this kind of pipeline failure analysis, but corrosion process research and simulation tests are rare. The lack of corrosion simulation tests under real conditions has a negative influence on the accurate analysis of the corrosion failure mechanism and the discovery of reasons for failure.

Fig. 1 Operation leak morphology of crude oil pipeline

To determine the reason for the failure of this pipeline lifting segment, the failed part of the pipeline lifting segment was dug out and examined by a nondestructive test. The leakage position was at the bottom of pipeline near the corner of the lifting segment, as shown in Fig. 1. An obvious wall thickness decrease was also found near the failure position, as shown in Fig. 2. To determine the cause of the leak failure in the pipeline, macroscopic analysis, tensile tests, flattening tests, metallurgical analysis and chemical composition tests were used for the pipeline. Material physical-chemical examination was conducted, and corrosion products near the leak position were investigated by X-ray diffraction (XRD) analysis and scanning electronic microscopy (SEM) analysis. Finally, corrosion simulation tests were designed for this study to simulate the corrosion process of the pipeline for the analysis of leakage causes.

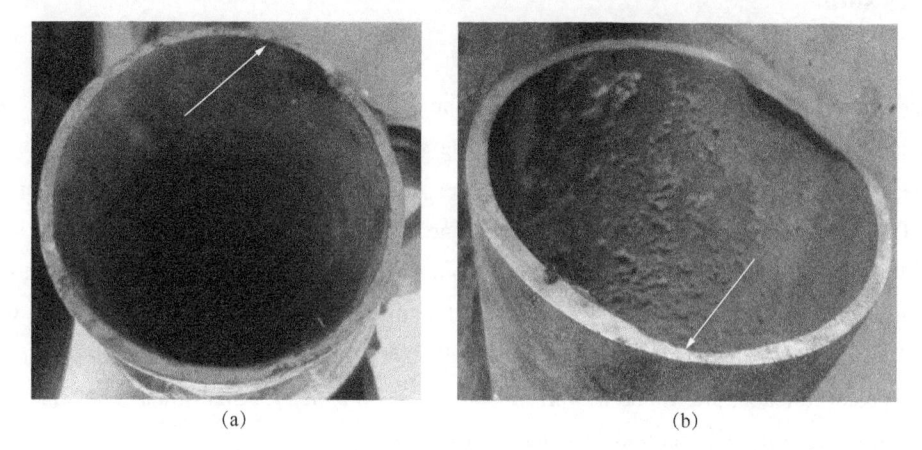

Fig. 2 Wall thickness decreasemorphology of crude oil pipeline(arrow position)
(a)position 1, (b) position 2

2 Experiments

The chemical composition sample of the failed pipeline was taken from the longitudinal pipe and analysed by a Baird Spectrovac 2000 direct reading spectrometer and a LECO CS-444 IR Carbon-Sulfur Spectrometer. To prepare specimens for metallurgical analysis, the samples were cut near the leak position after being ground and mechanically polished with diamond pastes of 6 μm and 1 μm. Specimens were etched and observed by MEF3A and MEF4M metallographic microscopes.

The mechanical properties of the pipeline were determined by tensile testing and flattening testing. Tensile test samples were cut along the pipe longitudinal direction with a section gauge of 50 mm×11.2mm, and the tests were carried out by MTS810-15 machines according to the GB/T 228.1-2010 standard[5]. The tensile properties were determined at room temperature. The leak surface morphology and corrosion products were analysed by visual examination, scanning electronic microscopy (SEM) with a TESCAN-VEGA II system and measurements with an OXFORD-INCA350 Energy Dispersive Spectrometer (EDS). The compositions of the corrosion products were examined by X-ray diffraction with a D8 Advance instrument.

To simulate the corrosion process of the pipeline, corrosion simulation tests were carried out in a high-temperature autoclave made by Cortest Co. Ltd. The hanging corrosion samples with dimensions of 50mm×10mm×3mm were prepared from a longitudinally failed pipeline and polished by silicon carbide papers of 240 to 1200 grit; the surface roughness was less than 1.6μm. Corrosion samples were loaded into the autoclave before the corrosion test, and high purity nitrogen was injected into the autoclave for more than 10 h for deoxygenation; then the samples were loaded, and the autoclave was sealed. The high purity nitrogen was continuously injected for deoxygenation, and the medium gas was injected. The test pressure was set to 1.8MPa, and the autoclave was heated to 40℃, which is the same as the pipeline working temperature. When the temperature in the autoclave was raised to the required temperature, the test timing was started. Corrosion simulation tests were carried out for 120h. The samples were weighed with a balance before and after testing.

3 Results

3.1 Visual examination

Macroscopic examination was carried out on the failed pipeline. As shown in Fig. 3, the leak failure occurred at the bottom of the pipeline lifting segment, which was 40cm above the welding seam. The wall thickness inspection showed that the wall thickness of the pipeline bottom was not uniform. The AB section wall thickness of the pipeline bottom 20cm away from the welding seam was normal (4.5cm). In the BC section, 20~40cm above the welding seam, the wall thickness of the pipeline gradually became thinner at the bottom; the thickest portion of the BC section was only 1.79mm. The wall thickness of the CD section, which was 20~140cm above the welding seam, gradually became thicker; the thickest wall of the CD section was 3.75cm thick. The wall thickness of the pipeline above the CD section was normal (4.5cm).

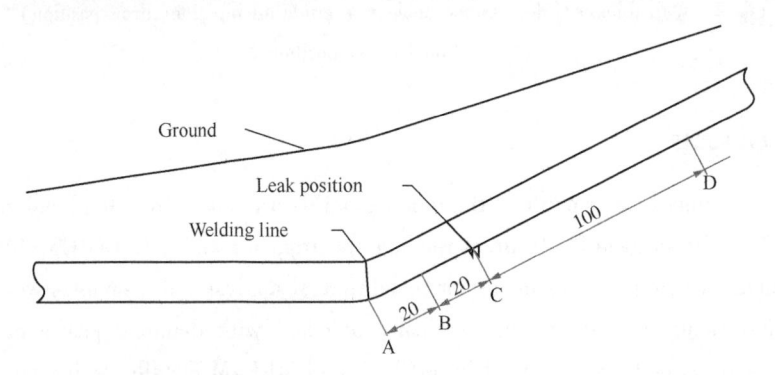

Fig. 3 Measurement Position of external diameter of the burst steam injection pipe

The external diameters of the failed pipeline were also measured. The results showed that there was no outer diameter distortion in the pipeline; the leak position morphology of the pipeline is shown in Fig. 4.

3.2 Chemical composition

The chemical composition tests of the failed pipeline were carried out, and the results are shown in Tab. 1. The pipeline materials were qualified and complied according to the requirements of national standard GB/T 8163—1999[6].

Fig. 4 Leak position morphology of failure crude oil pipeline

Tab. 1 Chemical composition results of leak failure pipeline (wt., %)

Sample	Elements								
	C	Si	Mn	P	S	Cr	Mo	Ni	Cu
2#	0.19	0.23	0.43	0.022	0.0085	0.014	—	0.0078	0.010
GB/T 8163—2008 requirement	0.17~0.23	0.17~0.37	0.35~0.65	≤0.030	≤0.030	≤0.25	—	≤0.30	≤0.25

3.3 Microstructure examination

The microstructure of failure pipeline material is ferrite and pearlite, as shown in Fig. 5. According to standard GB/T 6394—2002[7], the grain fineness of ferrite and pearlite is 10 grade. The non-metallic inclusion grade are A0.5, B0.5 and D1.0. It is indicated that metallographic test results were qualified and accorded with the standard GB/T 8163—1999 requirement.

The inner surface of leak failure pipeline was analyzed by metallographic microscope, it can be seen that there were several corrosion pits at the bottom of pipeline BC section, no crack was found around the leak position, the microstructure of the material under the corrosion pits is still ferrite and pearlite, no microstructure deformation was occurred, as shown in Fig. 6.

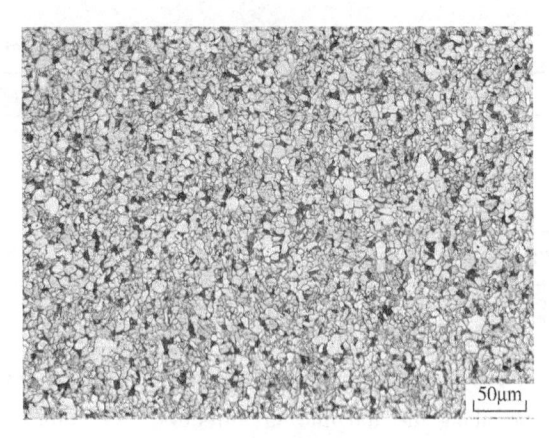

Fig. 5 Microstructure of the material of the crude oil pipeline with leak failure

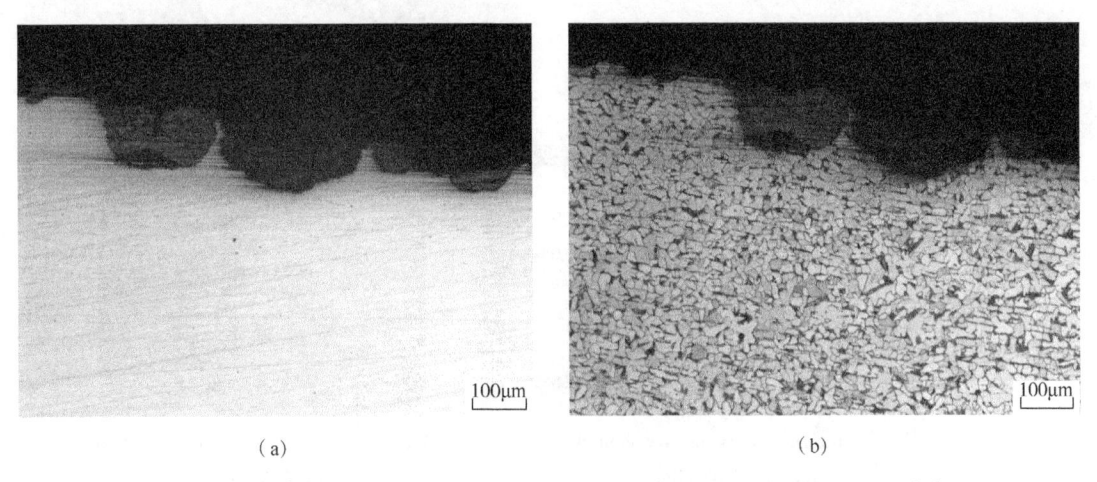

Fig. 6 Inner surface corrosion pitsmorphology(a) and microstructure of pipeline(b)

3.4 Mechanical properties

The tensile and flattening properties of the failed pipeline at room temperature are shown in Tab. 2 and 3, respectively. The mechanical properties of the burst pipe met the requirement of standard GB/T 8163—1999.

Tab. 2 Results of tensile test

Items	Ultimate tensile strength(MPa)	Yield tensile strength(MPa)	Elongation(%)
Results	448	330	22
	486	334	29
	464	309	36
Average	466	324	29
GB/T 8163—2008 requirement	410~530	≥245	≥20

Tab. 3 Results of flattening test

Sample		Plate distance	Results
Item	Diameter(mm)		
pipe	φ114×100	D/3(38mm)	NO crack
GB/T 8163—2008 Requirement		D/3(38mm)	NO crack

3.5 Corrosion products XRD analysis

The leaking section (BC section) of the failed pipeline was selected and opened in the middle plane along the longitudinal direction. The internal corrosion morphology of the failed pipeline is shown in Fig. 7. The inner surface of the pipeline has corrosion products attached; significant wall thinning at the bottom of the tube and corrosion perforation of local areas were both observed. After the corrosion products were cleaned through physical methods, many corrosion pits and grooves were visible around the leak position (Fig. 8).

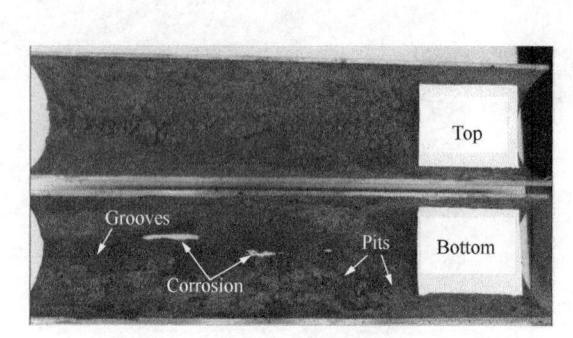

Fig. 7 Internal corrosion morphology of the failed pipeline

Fig. 8 Corrosion pits and grooves in the failed pipe

Corrosion products collected from the inner surface of the pipeline were examined by X-ray diffraction (XRD) analysis, as shown in Fig. 9. The XRD results for the corrosion products are shown in Fig. 10. The results revealed that these corrosion products were composed of ferric oxide (Fe_2O_3), alkali-type ferrous oxide ($FeO(OH)$), silicon dioxide (SiO_2) and a small amount of ferrous salt carbonate ($FeCO_3$). The ferric oxide is produced by the pipeline material reacting with oxygen dissolved in the oil-water mixture, while $FeO(OH)$ was the reaction product of oxygen in the atmosphere of the pipeline operation and/or oxygen dissolved in the liquid. $FeCO_3$ was the corrosion product of the pipeline material and CO_2 dissolved in the liquid. SiO_2 in the corrosion products was derived from impurities in the oil-water mixture pipe adhered to the pipeline wall.

Fig. 9　Corrosion products from the failed pipeline 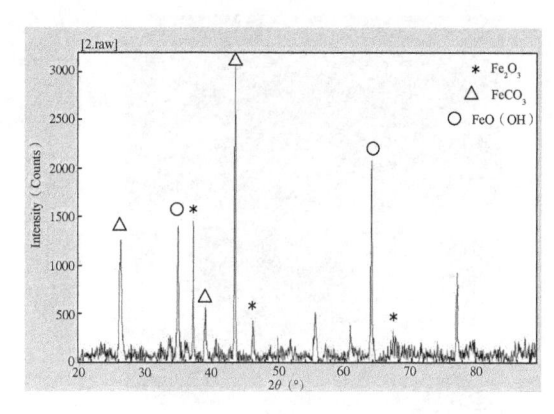　　　Fig. 10　XRD analysis results of the corrosion products

3.6　SEM and EDS analysis

The inner corroded surface of the failed pipeline was further examined by SEM and EDS, as shown in Fig. 11. The results revealed that the inner surface of the pipeline presented an obvious corrosion morphology; many etched pits and grooves were found near the leakage. The EDS analysis revealed that there were obvious thick corrosion products covering the inner surface of the pipeline, and the major elements of the corrosion products were Fe, O, C and Si. Based on the analysis of the atomic ratio, the corrosion products may be Fe_2O_3 and $FeCO_3$. Small amounts of S, Ca, Si and other elements were also found in the inner surface, as shown in Fig. 12. According to the information provided, sulfates were contained in the oil-water mixture, so the products may be calcium sulfate salt and silica, which are consistent with the XRD analysis.

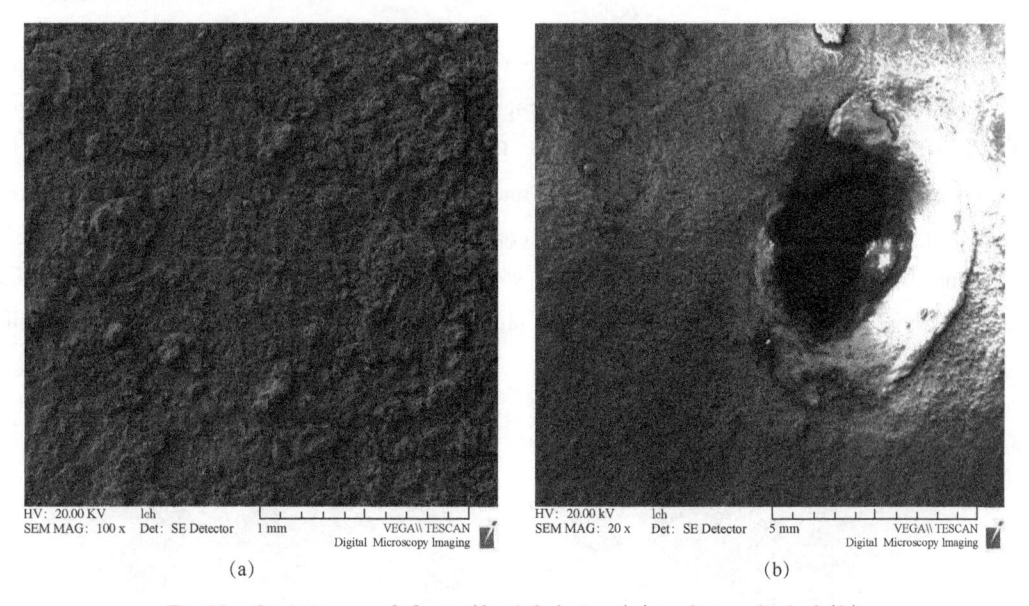

Fig. 11　Corrosion morphology of the failed pipe (a) and near the leak (b)

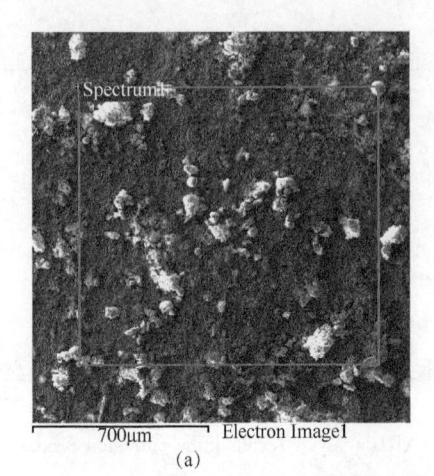

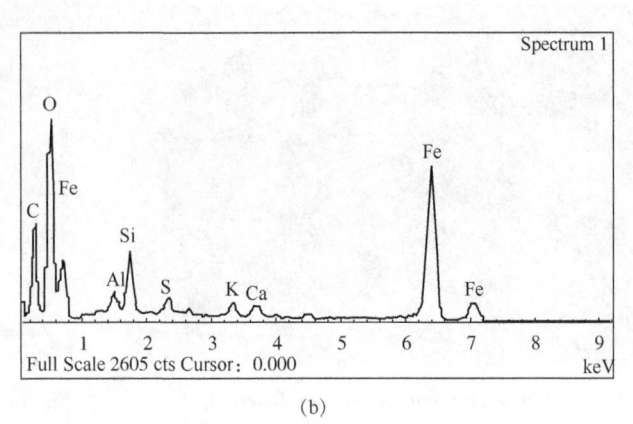

Fig. 12 EDS analysis results(b) and the failed pipe corrosion products(a)

3.7 Corrosion simulation tests

The diameter and weight of the samples were measured before and after the corrosion simulation tests, as shown in Fig. 13. The corrosion rate of the pipeline material was calculated according to the NACE RP0775-99 standard[8], as shown in Tab. 4. The results showed that under the field conditions of the working medium, temperature and pressure, the samples obtained after corrosion simulation tests presented apparent corrosion characteristics according to the corrosion degree classification rules in NACE standard RP0775-99, as shown in Tab. 5. The corrosion rate of the pipeline materials in this working condition was calculated to be 0.174357mm/a, which is classified as a serious corrosion level.

Fig. 13 Sample surface morphology of the failed pipe before (a) and after (b) the corrosion simulation test

Tab. 4 Results of the samples corrosion rates calculation

Items		Weight before test (g)	Weight after test (g)	Weight loss (g)	The sample surface area (mm^2)	corrosion rates (mm/a)
samples	4-1	11.7577	11.7298	0.0279	14.2909	0.181551
	4-2	11.6316	11.609	0.0226	14.22141	0.147781
	4-3	11.5057	11.4762	0.0295	14.15972	0.193741
Average corrosion rates						0.174357

Tab. 5 Classification of the corrosion rates in NACE RP-0775-99

Classification	Corrosion rate(mm/a)
Slight corrosion	<0.025
Moderate corrosion	0.025~0.125
Serious corrosion	0.126~0.254
Extremely Serious corrosion	>0.254

4 Discussion

The results obtained indicate that the chemical composition, mechanical properties and microstructure of the failed pipelinewere all in accordance with the relevant technical requirements of the standards. Macromorphological observation of the pipeline indicates that the outer surface of the pipeline is intact with no obvious corrosion. This illustrates that the outer soil corrosion of the pipeline was very slight, so this failure was caused by internal corrosion perforation of the pipeline from the inside to the outside.

The leaking section of the pipeline was opened longitudinally, showing that there were thinning areas all around the lower part of the pipeline above the corner; this is a typical local corrosion phenomenon. There was a mass of corrosion products covering the surface of the corrosion and perforation site; these corrosion products were thicker and easily peeled off the pipeline. The inner structure of the corrosion products was loose and porous. There were many corrosion pits and grooves under the corrosion product, so local corrosion very easily occurred under these loose corrosion products.

The SEM and XRD analysis results revealed that the corrosion products were composed of ferric oxide, alkali-type ferrous oxide, silicon dioxide, a small amount of ferrous salt carbonate and calcium sulfate. The ferric oxide was produced by the pipeline material reacting with oxygen dissolved in the oil-water mixture. FeO(OH) is a reaction product from the corrosion products reacting with oxygen in the atmosphere of the pipeline operation and/or oxygen dissolved in the liquid[9]. $FeCO_3$ is obtained by Fe^{2+} reacting with CO_3^{2-} generated from CO_2 dissolved in the liquid[10-12]. Silicon dioxide and calcium sulfate werederived from mineral salts and gravel conveyed with the oil-water mixture. From the analysis results of the oil-water mixture conveyed in the pipeline, the liquid contained a large amount of formation water, which was high in Cl^-, Ca^{2+} and salinity. Previous studies have reported that the existence of Ca^{2+} ions could significantly improve the ionic strength of the liquid, the electrical conductivity of the medium and the scaling tendency, which accelerated the local corrosion of the pipeline[13]. A large number of Cl⁻ ions penetrated into the pores and defects of the corrosion product film, which covered the steel surface and resulted in a microrupture of the corrosion product film. The nucleation of pitting corrosion also occurred. As a result of the constant accumulation of Cl⁻ ions, the pitting corrosion of the pipeline was also accelerated under the action of the occluded cell[14-17]. Because the formation water was acidic (the pH value was 5.0~6.5) and had high salinity, the oil-water mixture conveyed in the pipeline was strongly corrosive, which was the internal cause of corrosion and perforation of the pipeline[18].

According to the information provided by the oil field, corrosion perforations of the pipeline could only be found in positions with a large drop height. Moreover, the leakages often occurred approximately 200 ~ 400 mm from the upstream welding seam of pipeline elbows, as shown in Fig. 1. The velocity of the transmission fluid decreased at these climbing sections of the pipeline, and then the oil-water separation, sediment deposition and deposition of corrosive media took place in these positions. The corrosion product film and $CaCO_3$ incrustation easily formed beneath the sediment. These corrosion product films and incrustations did not completely prevent corrosion of the pipeline due to the poor uniformity and compactness of the film, which is composed of corrosion products and incrustation. Some weak spots existing in the film became anodes for the corrosion reaction, and the residual surface protected by the film became the cathode for the corrosion reaction. It is common knowledge that a large cathode coupled with a small anode has an important effect on acceleration of corrosion[19]. Localized corrosion is very serious near these weak spots; corrosive pitting can form on the inner surface and reduce the wall thickness of the pipeline. In the worst cases, the corrosive pitting will lead to leakage of the pipeline[15].

At the same time, the bottom of the pipeline was covered by scale and corrosion products, which were scoured by many small solid particles and gravel, resulting in the corrosion product film being destroyed and corrosion occurring at the bottom of the pipeline[20]. With increased accumulation of a high concentration of Cl⁻ ions in the corrosion pits, the dissolution rate of anodic Fe in the corrosion pits was accelerated, so the wall thickness of the pipeline continuously decreased over a short time. The pipeline leaked because the remaining wall thickness was not enough to bear the pipeline working pressure.

To prevent such leak failures, the crude oil pipeline lifting segment should be upgraded with an internal anti-corrosion coating. A corrosion inhibitor should also be added to the conveyed liquid to decrease scaling and reduce the corrosion rate. Deoxygenation should be carried out regularly in the pipeline in order to reduce the salinity of the conveyed crude oil, especially the contents of Cl⁻ and Ca^{2+} ions, which could decrease the corrosion rate. At the same time, the design of the pipeline structure should be optimized when the pipeline crosses ravines; a sudden large height deviation in the pipeline should be avoided in the pipeline design, and large radius elbows are proposed for use in the corners of the pipeline. Regular pigging operation is also recommended. The crude oil pipeline in this oilfield did not leak again after following these suggestions.

5 Conclusions

The chemical composition, mechanical properties and microstructure of thefailed pipeline are all in accordance with the relevanttechnical requirements of the standards. The corrosion rate of the pipeline in the working conditions is 0.174357 mm/a, which is a serious corrosion level. The crude oil conveyed in the pipeline contained a large amount of formation water and was very corrosive. Because the fluid flow changed suddenly at the pipeline lifting location, the fouling deposition occurred at the inner bottom of the pipeline, which resulted in serious localized corrosion caused by the small anode/big cathode corrosion galvanic cell reaction. The wall thickness of the pipeline decreased significantly in a short time. Finally, the pipeline lifting segment leaked due to fluid erosion and the high speed impact of solid particles and grit in the crude oil.

Acknowledgment

The author thanks the financial supports from the Essential Research and Strategic Reserve Technology Research Fund Program of China national petroleum corporation with Grand No. 2015Z-05, and the Shaanxi Innovation Capability Support Program with Grand No. 2018KJXX-046.

References

[1] GB/T 699-1999 Quality Carbon Structure Steel[S]. National Standard of the People Republic of China.
[2] J.-GHu, H.-j Luo, Z.-h Zhang, et. al, Analysis on corrosion failure of an oil pipeline in Changqing oilfield [J], Corrosion & Protection, 2018, 39(12): 962-970.
[3] H. Ding, G.-f Lin, Y. Long, et. al, Failure Analysis of Ground Pipeline in the Western Oil Field [J], Petroleum Tubular Goods & Instruments, 2017, 3(1): 56-63.
[4] P.-c Qin, C.-b Xiong, Z. LI, et. al, Failure Pressure Assessment of Submarine Pipelines Considering theEffects of Multiple Corrosion Defects Interaction [J], Surface Technology, 2020, 49(1): 237-244.
[5] GB/T 228.1-2010 Metal materials-Tensile testing-Part 1: Method of test at room temperature [S]. National Standard of the People Republic of China.
[6] GB/T 8163-1999 Seamless Steel Tubes for Liquid Service[S]. National Standard of the People Republic of China.
[7] GB/T 6394-2002, Metal-methods for estimating the average grain size [S]. National Standard of the People Republic of China.
[8] NACE Standard RP0775-99 Preparation, Installation, Analysis, and Interpretation of Corrosion Coupons in Oilfield Operations [S]. NACE International, 1999.
[9] Y.-P. Nin. Study of the Oil Transportation Pipeline Corrosion and Corrosion Rate Prediction[J]. Pipeline Technique and Equipment. 2011, 32(5): 45-47.
[10] Nesic S. Key Issues Related to Modeling of Internal Corrosion of Oil and Gas Pipelines A Review[J]. Corrosion Science, 2007, 49(12): 4308-4338.
[11] Ogundele G.I., White W.E., Some Observation on Corrosion of Carbon Steel in Aqueous Environment Containing Carbon Dioxide[J]. Corrosion, 1986, 42(2): 71~77.
[12] G.-x ZHAO, H.-x LV. Analysis of the Reason on Tubing Corrosion Failure[J]. Journal of Materials Engineering, 2010, 3: 51-55.
[13] C.-z ZHANG. Metal corrosion and protection[M]. Metallurgical industry press, 2000. 105.

[14] Fierro G, Ingo G M, Mancla Fi. XPS-investigation on the corrosion behavior of 13Cr martensitic stainless steel in $CO_2-H_2S-Cl^-$ environment[J]. Corrosion, 1989 (10): 814.

[15] W. -j ZHOU, D. GUO, Y. ZHANG, H. -p ZHU, Comprehensive Research to the Corrosion Failure Behavior of X52 Pipeline Steel in Acidic Corrosive Medium[J]. Pipeline, 2009, 23(9): 13-18.

[16] L. -p WAN, Y. -f MENG, C. -x WANG, L. YANG, Mechanism of Corrosion and Scale Deposit of Tubes in Western Oilfield[J]. Journal of Chinese Society for Corrosion and Protection, 2007, 27(8): 247-251.

[17] Z. -c LIAO, B. -q LIN, J. -c ZHANG, The corrosion and protection of oil and gas well[J]. Storage, Transportation & Preservation of Commodities, 2008, 30(6): 115- 118.

[18] Bassam G. N. M., MounaA., Mohammed H. M., et. al. Inspection of internal erosion-corrosion of elbow pipe in the desalination station[J]. Engineering Failure Analysis, 2019, 102(8): 293-302.

[19] A. Colombo, L. Oldani, S. P. Trasatti, Corrosion failure analysis of galvanized steel pipes in a closed water cooling system[J]. Engineering Failure Analysis, 2018, 84(2): 46-58.

[20] S. Arielya, A. Khentovb, Erosion corrosion of pump impeller of cyclic cooling water system[J]. Engineering Failure Analysis, 2006, 13(6): 925-932.

本论文原发表于《Engineering Failure Analysis》2020 年第 113 卷。

Comparative Study on Hydrogen Embrittlement Susceptibility in Heat-Affected Zone of TP321 Stainless Steel

Xu Xiuqing[1]   Niu Jing[2]   Li Chengzheng[3]
Huang Hangjuan[4]   Yin Chengxian[1]

(1. State Key Laboratory of Performance and Structural Safety for Petroleum Tubular Goods and Equipment Materials, TGRI, CNPC; 2. Xi′an Jiaotong University; 3. Oilfield Development Division of PetroChina Changqing Oilfield Company; 4. No. 6 Gas Plant of PetroChina Changqing Oilfield Company)

Abstract: TP321 stainless steel is widely used in hydrogenation refining pipes owing to its excellent performance of creep resistance and high-temperature resistance. In this study, the thermal simulation tests were carried out for the welding heat-affected zone (HAZ) of TP321 stainless steel at temperatures of 1300℃, 1100℃, and 850℃. Slow strain tensile tests were conducted under the condition of electrolytic hydrogen charging (EHC) and the metallographic microstructure of the cracks as well as the morphology of the fractures were analyzed in detail. The result shows that hydrogen could change the fracture mode of the tensile specimen. Hydrogen significantly decreased the plastic deformation capability of HAZ and the reduction of area after the fracture decreases by 58%, 41%, and 45% at 1300℃, 1100℃, and 850℃, respectively. The existence of δ ferrite was considered to be the main reason for the aggravation of hydrogen-induced plasticity loss.

Keywords: TP321 stainless steel; thermal simulation test; heat-affected zone; electrolytic hydrogen charging; crack; plastic deformation

1 Introduction

Refining hydrogenation pipes are generally manufactured using austenitic stainless steel materials with excellent high-temperature mechanical properties because they need endure a high-temperature, high-pressure and hydrogen environment. The most commonly used refining hydrogenation pipe is TP321 stainless steel which are usually connected by welding during the manufacturing of pipes. However, severe safety accidents occur from time to time in actual operations because of frequent cracking of welded parts in the equipment. This is a major potential safety hazard in the production of refining hydrogenation pipes.

Corresponding author: Xu Xiuqing, xuxiuqing@cnpc.com.cn.

Several studies have been reported on the hydrogen service performance of welding joints of austenitic stainless steels. These studies mainly focuses on two aspects: (1) the hydrogen service performance of welding joints of chrome-molybdenum steels in hydrogenation reactor materials[1-3] and (2) the hydrogen-induced damage of austenitic stainless steel body or welded metal[4-7]. After electrolytic hydrogen charging (EHC) by electrochemical methods, Michalska et al.[8] investigated the hydrogen damage behavior of super austenitic stainless steel 904L in seawater by a slow strain tensile test. The results showed that the performance of 904L stainless steel clearly degraded at a hydrogen current of 20mA. Murakami et al.[9] studied the hydrogen embrittlement mechanism of austenitic stainless steels as microscopic ductile fracture resulting from hydrogen concentration at crack tips leading to hydrogen-enhanced slip. Li et al.[10] studied the hydrogen corrosion of 304 stainless steel after the service for 20000h at 450℃ under a hydrogen pressure of 5MPa and hydrogen corrosion was attributed to the existence of $Cr_{23}C_6$ carbide. The relevant literature about hydrogen corrosion of austenitic stainless steel welding joints has not been searched at present.

In this study, thermal simulation tests were performed to obtain different microstructures of the heat-affected zone (HAZ) in 321 stainless steel, and the effect of hydrogen on the microstructure and performance of welding HAZs in 321 stainless steel was evaluated by dynamic EHC tests. The cracking mechanism of welding HAZ in 321 hydrogenation pipe was elucidated. This study has very important practical significance in preventing safety accidents in refining enterprises.

2 Materials and methods

The experimental material was TP321 stainless steel. To investigate the microstructure and performance of different welding HAZs in 321 stainless steel, a Gleeble 3800 thermal simulation testing machine was used to enlarge the welding HAZs in the specimens. Considering the characteristics of welding HAZ of austenitic stainless steel, the peak temperatures of welding were selected as 850℃, 1100℃ and 1300℃. For austenitic stainless steels, 850℃ is the highest sensitization temperature in the range 450~850℃ and can be used to simulate the effect of short-stay at the sensitization temperature on the microstructure and performance of 321 stainless steel. A higher temperature of 1100℃ can homogenize the alloy elements within austenite, which is the representative temperature of solid-solution treatment of austenitic stainless steels. The highest temperature is selected as 1300℃ because the effect of high temperature on the grain size of 321 stainless steel was mainly considered. The resistance heating and jet cooling modes were used in the thermal simulation tests, and the related experimental parameters are shown in Table 1.

Table 1 HAZ thermal cycling parameters determined by thermal simulation tests

E(kJ/cm)	W(℃/s)	T(℃)	$T_{8/5}$(s)	t_H(s)	
10.6	130	1300	11.44	1100℃	900℃
				14.20	21.39
		1100	11.44	1000℃	900℃
				17.24	21.39
		850	11.44	800℃	700℃
				27.22	35.82

Note: E = weld heat input, W = rate of heating, T = peak temperature, $T_{8/5}$ = cooling rate, t_H = residence time of high temperature.

Because hydrogen is only distributed near the specimen surface layer after the static EHC of tensile specimens, dynamic EHC and slow strain tensile tests (using INSTRON 1195) were used to study the hydrogen embrittlement sensitivity of welding HAZs in 321 stainless steel. The specimens after thermal simulation tests at 1300℃, 1100℃, and 850℃ were machined into tensile specimens according to the ASTM A370 standard. The tensile specimens after machining were ultrasonically cleaned with alcohol, blow dried, and placed in a drying oven for use. The electrolyte for EHC test was a mixture of 0.5M H_2SO_4 + 1.85 M $Na_4P_2O_7$, and deionized water was used to dilute the solutions. The test rate was 0.1 mm/min and the hydrogen current was 30mA. To guarantee the reliability of test data, three parallel specimens were measured in each group in the slow strain tensile tests.

The macroscopic morphology and microstructure of slow strain tensile specimens before and after dynamic EHC were observed using a Nikon ECLIPSE MA200 optical microscope. The fracture and banded structure morphologies of slow strain tensile specimens were analyzed using a VEGA II XMUINCA scanning electron microscope (SEM). The composition of the banded structure was analyzed using an energy-dispersive X-ray spectrometer (EDS).

3 Experimental results and discussion

3.1 Effect of hydrogen on fracture characteristics of different hazs in 321 stainless steel

Fig. 1 shows the fracture positions and morphologies of tensile specimens for different HAZs before and after EHC at 850℃. The 321 stainless steel specimensafter different thermal cycles show no significant difference in the fracture morphologies after slow strain tensile tests without EHC. A clear necking phenomenon was observed in the three specimens, and these specimens all fractured in the end, as shown in Fig. 1(a). No clear crack was observed on the specimen surface, but a deformation trace was visible. The tensile specimens after EHC showed that many cracks were vertical to the tensile direction on the specimen surface and no clear necking phenomenon was observed in the fractures of specimens after EHC as seen in Fig. 1(b). The fracture was ragged and the fracture surface had a certain angle to the tensile axis. These characteristics show a striking contrast with those of tensile fracture of specimens without EHC. The fracture mode of specimen changed fundamentally from a microvoid coalescence fracture to a tensile fracture similar to that of a notched tensile specimen. The final fracture of the specimens under EHC is directly related to these cracks.

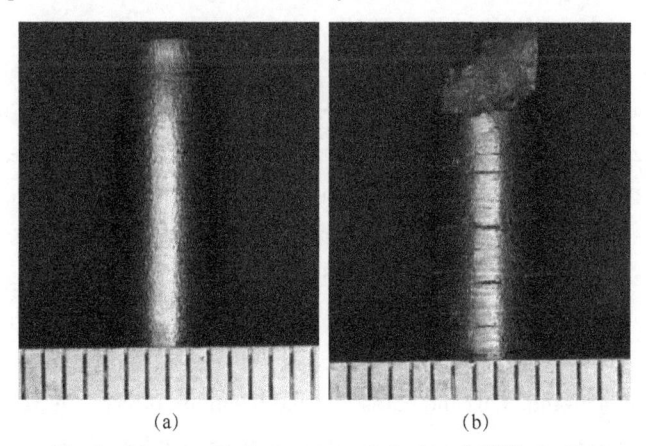

Fig. 1 Macroscopic fracture morphologies of HAZs in 321 stainless steel (a) before EHC and (b) after EHC (T = 850 ℃)

3.2 Effect of hydrogen on tensile performance of different HAZs in 321 stainless steel

Table 2 shows the effect of hydrogen on the tensile performance of 321 stainless steel base material and HAZ. EHC led to a significant decrease from 730MPa to 660MPa in the tensile strength of 321 stainless steel base material. For the HAZs, the tensile strength at 850℃ slightly decreased, and that at 1100℃ slightly increased after EHC. The two tensile strengths after EHC are both higher than that of the base material. In contrast, the tensile strength at 1300℃ significantly decreased after EHC even 40 MPa lower than that of the base material. These results indicate that different welding HAZs of 321 stainless steel have different hydrogen sensitivities. Hence, it is particularly important to control the welding parameters and thus reduce the area of high-temperature HAZ during the welding for improving the bearing capacity of welding positions of 321 stainless steel.

Table 2 Effect of hydrogen on tensile performance of HAZ in 321 stainless steel

$T(℃)$	Hydrogen State	Tensile strength (MPa)	Elongation A(%)	Reduction of area Z(%)	Hydrogen embrittlement sensitivity coefficient IA(%)	Hydrogen embrittlement sensitivity coefficient IZ(%)
1300	Without EHC	669.7	63.6	70.3	45.9	56.1
	EHC	618.0	34.4	29.4		
1100	Without EHC	684.3	62.3	72.6	24.6	42.6
	EHC	692.0	47.0	42.8		
850	Without EHC	686.3	63.3	74.0	28.7	45.5
	EHC	674.3	45.1	40.4		
Base material	Without EHC	730.3	41.7	70.6	-7.7	60.6
	EHC	660.3	44.9	27.8		

The percentage elongation after the fracture of different regions of HAZ after EHC has a certain difference, i.e., the percentage elongation at 850℃ after EHC is comparable to that of base material. It is slightly higher at 1100℃ than that of base material, whereas the percentage elongation of HAZ at 1300℃ is much lower than that of base material. However, by comparing the percentage elongation of base material and HAZ without EHC, it was found that the percentage elongation of base material after EHC is higher than that before EHC. The main reasons for this abnormal phenomenon have probably two aspects. On one hand, dynamic EHC has a slight effect on the percentage elongation after the fracture of 321 stainless steel. On the other hand, this phenomenon is related to the abundant surface cracking in the parallel section of specimens after EHC during the slow strain tensile tests. Unlike the base material, dynamic hydrogen has a relatively large effect on the percentage elongation after the fracture of different regions of HAZ. After EHC, the percentage elongation of HAZs at 850℃, 1100℃, and 1300℃ decreased by about 18%, 16%, and 30%, respectively.

The percentage reduction of area of all the regions of HAZ in 321 stainless steel after EHC is higher than that of base material. This indicates that the plasticity of HAZ in 321 stainless steel under a hydrogen service environment is better than that of base material. Compared with the situation without EHC, the percentage reduction of area of welding HAZ and base material both significantly decreased under the condition of dynamic EHC. This result indicates that the welding HAZs of 321 stainless steel and base material are both very sensitive to hydrogen.

3.3 Fracture mechanism analysis of welding HAZ in 321 stainless steel in hydrogen environment

Fig. 2 shows the low-power fracture morphology of 850℃ HAZs before and after EHC. Under dynamic EHC, the fracture surface of fractured specimen shows relatively large undulations and is mainly composed of flat regions near the surface and an oblique plane with a certain angle to the tensile axis [Fig. 2(b)]. As shown in Fig. 2(a), a typical cup fracture morphology is observed without EHC, which is significantly different from that of under EHC. These results indicate that the fracture mode of 321 stainless steel after EHC has essential differences from that of 321 stainless steel without EHC. The cracks gradually propagate to the interior of specimens with a continuous increase in load, finally leading to an unstable fracture of the specimens under the state of EHC. Hence, it is clearly that the role of hydrogen that leads to a change in the tensile failure mode of 321 stainless steel.

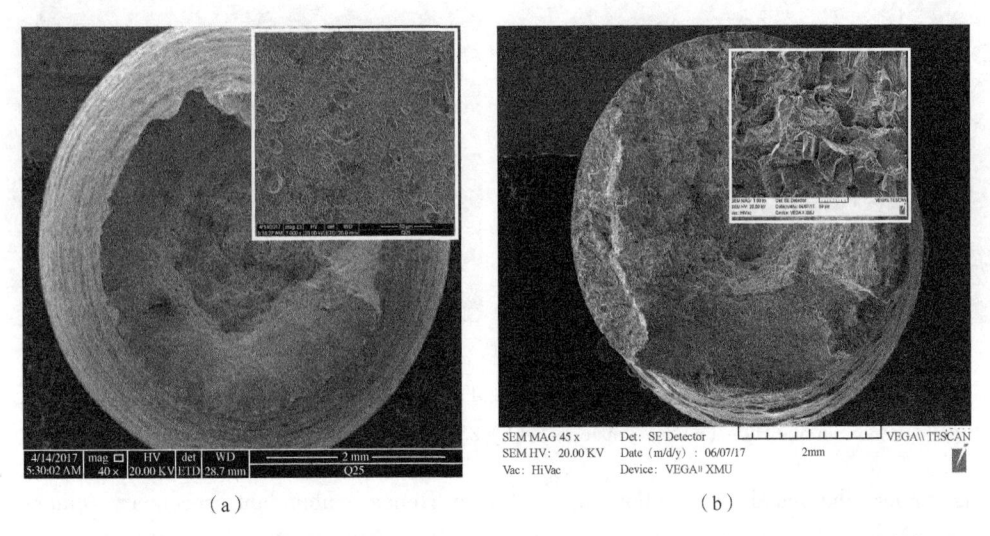

(a)　　　　　　　　　　　　　　(b)

Fig. 2　Fracture surface morphologies of 850 ℃ HAZ of 321 stainless steel (a) before EHC and (b) after EHC

Fig. 3 shows the metallographic images of cracks in 321 stainless steel base material and different HAZs under EHC. The crack orientation is perpendicular to the axial direction ofspecimens and a certain bifurcation phenomenon was observed at the crack tips. The tensile stress field of the crack tip can aggravate hydrogen diffusion, migration and accumulation. Finally, the hydrogen-induced cracking occurs before the principal crack arrives. The principal crack plays an interconnection role under the action of tensile stress.

Unlike the crack initiation and propagation characteristics of 321 stainless steel base material, it was found that the cracks in HAZ initiate from the relatively large plastic deformation zone on specimen surface. The crack propagation and arrest may be associated with the banded structure inside the specimens, as shown in Fig. 3(c) and (d). Especially at 1300 ℃, the banded structure is very clear [Fig. 3(d)] and the intergranular cracking morphology of the principal crack was also observed in addition to the intergranular fracture cracking caused by the banded structure. This is mainly because the effective segmentation effect of banded structure on austenitic microstructure results in complex stress states at the junction of austenite grain boundaries, inducing intergranular

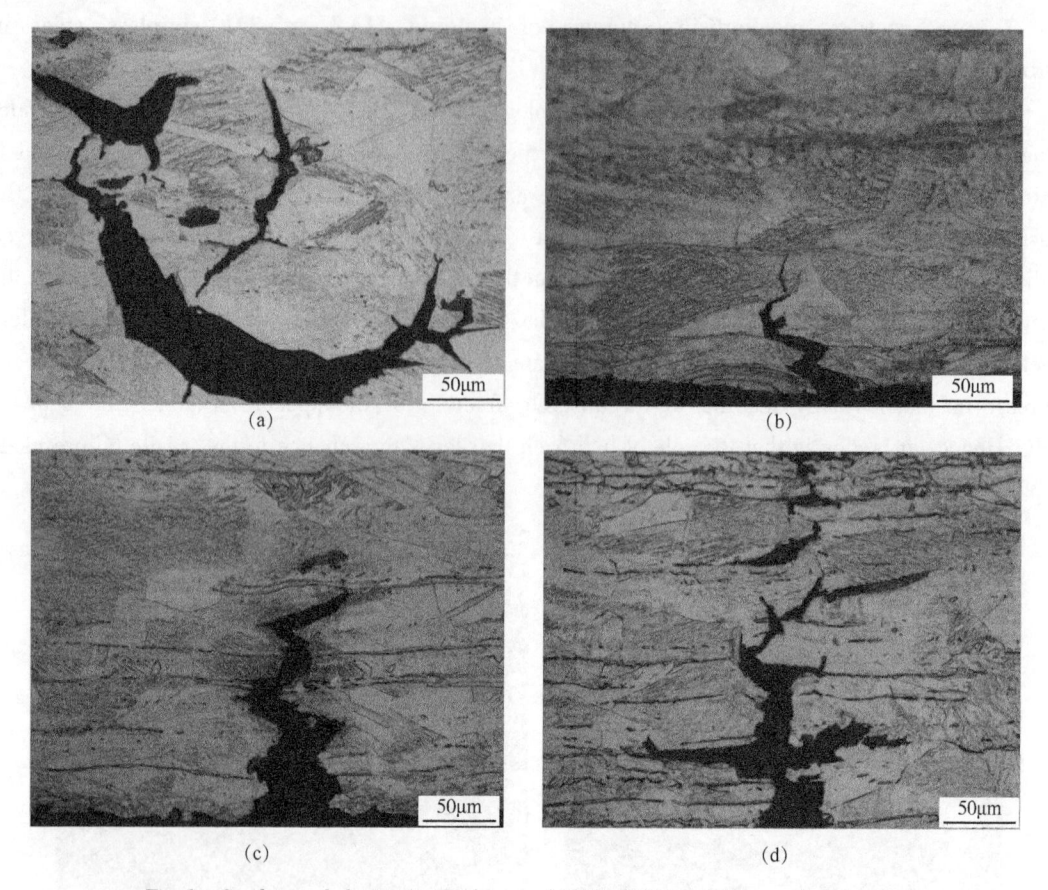

Fig. 3 Crack morphologies for the base and HAZ of 321 stainless steel under the condition of EHC: (a) base material, (b) 850 ℃, (c) 1100 ℃ and (d) 1300℃

cracking under the combined action of hydrogen. Hence, abundant secondary cracks and intergranular fractures were observed on the fracture surface [Fig. 3(d)]. In addition, in the HAZ at 1300 ℃, the transgranular cracks showed a propagation morphology along the intracrystalline slip band in the austenitic microstructure. This is probably related to hydrogen diffusion and accumulation during the sliding deformation inside austenitic grains.

Therefore, it can be deduced that the banded structure in the welding HAZ of 321 stainless steel is one of the main reasons that lead to the differences in hydrogen embrittlement sensitivity between the HAZ and base material. The HAZs at different peak temperatures have different densities of banded structure, leading to the differences in hydrogen embrittlement sensitivity of different HAZs.

Fig. 4 shows the SEM morphology of banded structure, and the EDS analysis results are shown in Table 3. The carbon content (C in

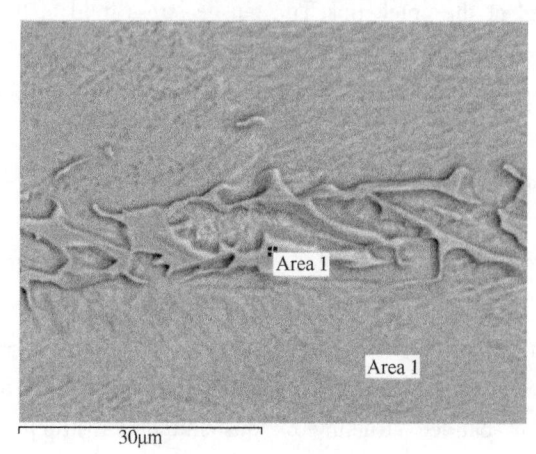

Fig. 4 SEM morphology of banded structure

Table 3) in different selected areas has no difference. The contents of Cr and Si in the bone-shaped structure (Selected Area 1) is higher than those in the austenitic structure (Selected Area 2), but the contents of Ni and Mn in Selected Areas 1 is lower than those in Selected Area 2. Cr and Si are ferritizing elements, whereas Ni and Mn are austenitizing elements. Thus, it can be further confirmed that the striped and bone-shaped structures are both δ ferrite structure.

Table 3 EDS analysis results of different HAZs

Elements	Area 1		Area 2	
	wt. %	at. %	wt. %	at. %
C K	0.06	0.26	0.06	0.25
Si K	0.87	1.69	0.76	1.48
Cr K	23.28	24.39	17.40	18.35
Mn K	—	—	0.94	0.94
Fe K	70.56	68.81	71.67	70.40
Ni K	5.23	4.85	9.19	8.58

The content of δ ferrite structure in different HAZs of 321 stainless steel has aclear difference (Fig. 3) which gradually increased with the increase in peak temperature. In the EHC process, the diffusion coefficient and solubility of hydrogen element in δ ferrite with a body-centered cubic structure and austenite with a face-centered cubic structure are different. This difference probably causes hydrogen accumulation on the interface of δ ferrite and austenite, further weakening the interface bonding force. As a result, in the slow strain tensile process under the state of EHC, the interface of δ ferrite and austenite is prone to form lamellar cracking under the action of shear stress. This is also the reason why the mechanical properties of HAZ at 1300 ℃ degrade significantly.

4 Conclusions

The hydrogen embrittlement sensitivity of different welding HAZs in 321 stainless steel was analyzed through dynamic EHC and slow strain tensile tests by thermal simulation. And the reasons for the hydrogen embrittlement sensitivity difference of 321 stainless steel were elucidated from microstructure analysis, SEM analysis of fracture surface and metallographic analysis of the cracks. The main conclusions are as follows。

(1) After EHC, the fracture mode of the specimen under a slow strain tensile test changed fundamentally, from a microvoid coalescence fracture to a tensile fracture similar to that of a notched tensile specimen. The fracture mode on the specimen surface and near the specimen surface was intergranular plus transgranular cracking, and that in the center was a dimple fracture.

(2) Embrittlement due to the increase of temperature can be seen by reduction of total elongation and also the reduction of area.

(3) The existence of δ ferrite is the main reason for the decrease in hydrogen-induced plasticity index.

Acknowledgement

The authors acknowledge the financial support of the Innovation Capability Support Program of Shaanxi No. 2019TD-038 and CNPC Project of Science Research and Technology Development No. 2017D-2307.

References

[1] X. G. Li. Review and prospect of hydrogen corrosion research at high temperature and high pressure, Petrochemical Corros. & Prot., 3 (2000) 5-7. (in Chinese).

[2] J. Koukal, M. Sondel, D. Schwarz. Correlation of creep properties of simulated and real weld joints in modified 9%Cr steels, Welding in the World, 54 (2010) 27-34.

[3] R. G. Rizvanov, A. M. Fairushin, D. V. Karetnikov. Research of vibration treatment effect on the stress-strain properties and fracture strength of pipe welding joints of 15Cr5Mo creep-resisting steel, Oil & Gas Business, 1 (2013) 369.

[4] D. Hardie, J. Xu, E. A. Charles, et al. Hydrogen Embrittlement of stainless steel overlay materials for hydrogenators, Corros. Sci., 46 (2004) 3089-3100.

[5] L. Zhang, M. Wen, M. Imade, et al. Hydrogen embrittlement of austenitic stainless steels at low temperatures, Environ. Assist. Fract., (2006) 1003-1004.

[6] C. Pan, Y. J. Su, W. Y. Chu, et al. Hydrogen embrittlement of weld metal of austenitic stainless steels, Corros. Sci., 9 (2002) 1983-1993.

[7] C. L. Briant. Hydrogen assisted cracking of sensitized 304 stainless steel, Metallurgical Transactions A, 5 (1978) 731-733.

[8] J. Michalska, B. Chmiela, J. Łabanowski, et al. Hydrogen damage in super austenitic 904L stainless steels, J. Mater. Eng. Perform., 8 (2014) 2760-2765.

[9] Y. Murakami, T. Kanezaki, Y. Mine, et al. Hydrogen embrittlement mechanism in fatigue of austenitic stainless steels, Metall. Mater. Trans., 6 (2008) 1327.

[10] X. G. Li, H. Chen, Z. M. Yao, et al. Hydrogen attack on austenitic steel 304 under high temperature and high pressure, Acta. Metall. Sin., 6 (1993) 374-378.

本论文原发表于《Materials Science Forum》2020 年第 993 卷。

External Stress Corrosion Cracking Risk Factors of High Grade Pipeline Steel

Zhu Lixia[1,2]   Luo Jinheng[1,2]   Wu Gang [1,2]   Li Lifeng[1,2]   Li Lei[1,2]   Wu Yi[3]

(1. CNPC Tubular Goods Research Institute; 2. State Key Laboratory for Performance and Petroleum Tubular Goods and Equipment Materials; 3. Petrochina West Pipeline Company)

Abstract: The influence weights of various sensitive factors, such as steel grade, applied potential, temperature and soil environment, on stress corrosion cracking behavior of high-grade pipeline steel were studied by means of orthogonal test at mixed level and slow strain rate test. The stress corrosion behavior of X80 pipeline steel in near neutral and high pH simulated soil solution under overprotection potential was observed by SEM. The results showed that the applied potential has the greatest weight on stress corrosion index of ISSRT, followed by temperature, steel grade and soil environment. X80 pipeline steel exhibits high stress corrosion sensitivity in near-neutral and high pH simulated soil solutions at applied potential of -1500mV, and its SCC mechanism is hydrogen embrittlement (HE). This study can provide guidance for practical engineering application of pipeline steel.

Keywords: Pipeline steel; Stress corrosion cracking; Risk factor; Weight; Applied potential

1 Introduction

The rapid growth of China's oil and gas energy demand has led to the rapid development of high strength steel for oil and gas pipelines. In recent years, China has vigorously developed high strength pipeline steel, and X70 and X80 pipeline steel have become the major steels of the oil and gas pipeline in China[1,2]. The stress corrosion cracking of pipeline steel is a sort of phenomenon of brittle fracture under low stress, which happened under combined action of stress and corrosion environment. Due to its unpredictability and serious destruction, it has become one of the main factors for the safety and integrity of buried pipelines[3,4]. Three conditions for external stress corrosion cracking are stress, the sensitive pipeline materials and the specific corrosion environment[5-7]. Accordingly, the risk factors affecting the external stress corrosion cracking of pipeline steel can be divided into three categories, among which the material risk factors mainly refer to the grade of steel, the environmental risk factors include soil, temperature, coating, cathodic

---

Corresponding author: Zhu lixia, zhulx@cnpc.com.cn.

potential etc., and the stress risk factors mainly include circumferential stress, axial stress and residual stress.

A lot of studies have done on the risk factors of external stress corrosion cracking of pipeline steel, particularly on the synergistic effect of single or multiple factors on stress corrosion behavior. The influence of single factor indicated that: (1) The SCC sensitivity of X70 - X100 pipeline steel is significantly affected by the applied potential, and the SCC mechanism varies with the applied potential[8-14]. (2) Temperature has a great and complex influence on SCC behavior of pipeline steel. The sensitivity of SCC increases with the temperature increase[15,16]. (3) In terms of soil effects, the SCC mechanism of pipeline steel in high pH value solution is hydrogen embrittlement[17,18]. But, in near neutral pH solution, it is a combined mechanism of anodic dissolution (AD) and hydrogen embrittlement (HE), which is related to applied potential[9,12,13]. (4) In terms of mechanical factors, Zhiying Wang et al. studied the influence of stress ratio, loading frequency and external transverse load on the SCC crack initiation of X70 pipeline steel and believe that low stress ratio promoted the cracks initiative of stress corrosion cracking of pipeline steel[19]. Studies on the synergy of multiple factor mainly focus on the interaction of two factors on stress corrosion. Yang Fu et al.[20] studied the influence of temperature and applied potential on X70 pipeline steel stress corrosion behavior in simulated soil solution. They indicated that the effect of potential variation is dominant in the interaction test of temperature, and applied potential, and the variation trend of stress corrosion susceptibility at different temperatures is consistent.

A tpresent, there are few studies involved three or more risk on stress corrosion of pipeline steel. Liang Zhang et al.[21] studied the influencing factors of stress corrosion cracking of high strength pipeline steel in near neutral pH soil. It is founded that in soil with near neutral pH value, the higher the steel grade, more negative the external potential and the lower the fluctuation stress ratio led to the stronger sensitive of pipeline steel stress corrosion cracking. However, the interaction of these factors and the influence weight of each factor were not considered. Fei Xie et al.[22] studied the influence of temperature, pH value and dissolved oxygen content on X80 steel in korla soil simulation solution. They found that the main and secondary order factors were dissolved oxygen content, pH value and temperature, but the effects of material factors, potential and anion in environmental factors were not considered.

Based on hybrid levels orthogonal experiment method, this study discusses how the different material and environmental factors affect the stress corrosion cracking behavior of pipeline steel. Four risk factors, steel grade, applied potential, temperature and soil properties, are involved, and the influence weights of each risk factor are analyzed, which in order to provide some references for the research of external stress corrosion cracking of high grade pipeline and provide guidance for practical engineering application of pipeline steel.

2 Experimental procedure

2.1 Experimental material

Experimental materials are X70 and X80 spiral seam submerged arc welded steel pipes. The specifications are $\phi 1016mm \times 14.6mm$ and $\phi 1219mm \times 18.4mm$. The chemical compositions are shown in Table 1. The microstructure was acicular ferrite, as shown in Fig. 1. The experimental environment used in the experiment was based on the actual soil in the western region of the actual crossing area of the first and second line of the West-East Gas Pipeline. According to the analysis results of physical and chemical properties of collected soil, pure chemical and deionized water were used to prepare near-neutral and high-pH value soil simulation solution. The specific components of soil simulation solutions are shown in Table 2.

Table 1 Chemical composition of pipeline steel[wt%]

Steel grade	C	Si	Mn	P	S	Cr	Mo	Ni	Nb	V	Ti	Cu	B	Al
X70	0.045	0.19	1.64	0.012	0.0023	0.23	0.13	0.0060	0.088	0.0050	0.011	0.0097	0.0003	0.033
X80	0.074	0.21	1.86	0.0099	0.0024	0.27	0.27	0.054	0.079	0.030	0.018	0.058	0.0003	0.037

Table 2 Chemical composition of Soil simulation solution

Soil solution	pH Value	Cl$^-$(%)	SO$_4^{2-}$(%)	HCO$_3^-$(%)
Near-neutral soil solution	7.28	0.336	0.155	0.028
High pH value soil solution	8.63	0.451	0.612	0.061

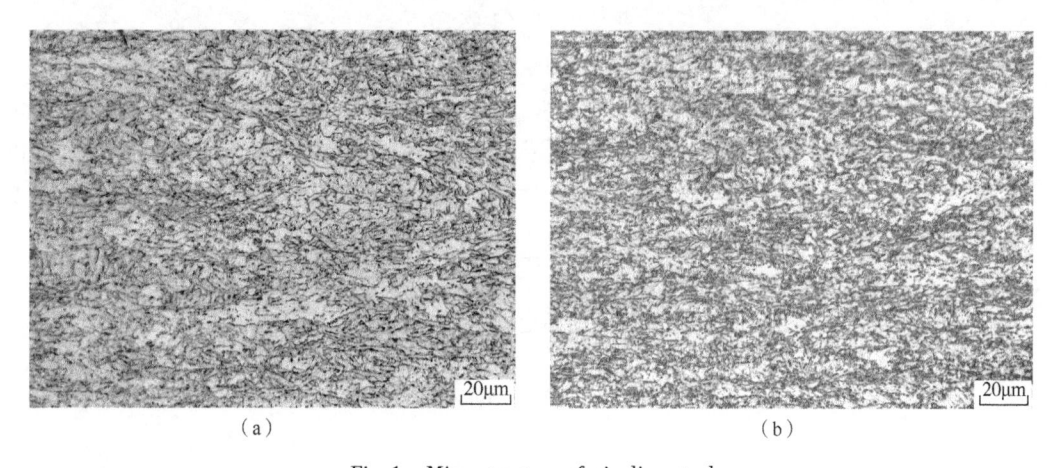

(a) (b)

Fig. 1 Microstructure of pipeline steel

2.2 Experimental method

Using mixed level orthogonal table to design the slow strain rate tensile test and analysis the weight of the influencing factors. The design of the orthogonal table of L8($4^1 \times 2^3$) is shown in Table 3, in which there are 4 levels of cathode potential: -500mV, -850mV, -1000mV and -1500mV, and two levels of temperature: 10℃, 40℃. There are two kinds of soil: near-natural soil solution and high-pH value soil solution and two steel grades: X70 and X80.

Table 3 Factors and levels of orthogonal design

No.	CathodepPotential	Temperature	Soil environment	Steel grades
1	1	1	1	1
2	1	2	2	2

Continued

No.	CathodepPotential	Temperature	Soil environment	Steel grades
3	2	1	1	2
4	2	2	2	1
5	3	1	2	1
6	3	2	1	2
7	4	1	2	2
8	4	2	1	1

The slow strain rate tensile test was carried out on MFDL100 slow strain rate stress corrosion test machine. The tensile samples were cut along the circular direction of the steel pipe, and the sample size is shown in Fig. 2. Before the tensile test, the sample was polished by $150^{\#} \sim 800^{\#}$ metallographic sandpaper, degreased with acetone and cleaned with anhydrous ethanol. Before the experiment, the solution was deoxidized with 99.9% nitrogen for 1h. The sample was stretched under the strain rate of 1×10^{-6} s^{-1} in different cathode potential conditions. The test temperatures were 10℃ and 40℃. The fracture morphology and secondary cracks morphology were observed by JSM-6390A scanning electron microscope. The test result was evaluated by ISSRT stress corrosion index.

$$I_{SSRT} = 1 - \frac{\sigma_{fw}(1+\delta_{fw})}{\sigma_{fA}(1+\delta_{fA})} \quad (1)$$

where, σ_{fw} is the breaking strength of the sample in the environmental medium, MPa, δ_{fw} is the breaking strength of the sample in the air, MPa, δ_{fw} is the breaking elongation of the sample in the environmental medium, %, δ_{fA} is the breaking elongation of the sample in air, %.

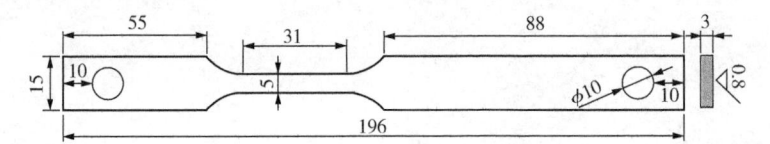

Fig. 2 Sample schematic diagram of SSRT

3 Results and discussion

3.1 Weight analysis of influencing factors

The orthogonal test results are shown in Table 4. Ⅰ, Ⅱ, Ⅲ, Ⅳ in the table are the sum of the stress corrosion index of each factor in different levels. K_1, K_2, K_3, K_4 are the average values of Ⅰ, Ⅱ, Ⅲ, Ⅳ, respectively. R is the range, which represents the dispersion degree of date and reflects the influence degree of various factors and levels on the stress corrosion index. From the R value in Table 4, it can be seen that: $R_{\text{Cathode Potential}} > R_{\text{temperature}} > R_{\text{steel grade}} > R_{\text{soil environment}}$. Therefore, cathode potential is the most important factor affecting the stress corrosion index, following by temperature and steel grade. Soil environment has no obvious influence on the stress corrosion index. In addition, it can be seen from Table 4 that the higher the temperature and grade, the higher the stress corrosion susceptibility of pipeline steel. The stress corrosion susceptibility of pipeline steel in high pH soil solution is similar to that in near neutral soil solution. The direct analysis of average value of each level of loading potential can be seen in Fig. 3. The stress corrosion sensitivity of pipeline steel is the highest when the loading potential is −1500mV.

Table 4 Analysis of the experimental results

No.	Cathode potential	Temperature	Soil environment	Steel grades	Issrt
1	−500	10	near neutral pH	X70	0.05
2	−500	40	high pH	X80	0.064
3	−850	10	near neutral pH	X80	−0.038
4	−850	40	high pH	X70	0.037
5	−1000	10	high pH	X70	0.006
6	−1000	40	near neutral pH	X80	0.104
7	−1500	10	high pH	X80	0.084
8	−1500	40	near neutral pH	X70	0.07
I	0.114	0.102	0.186	0.163	
II	−0.001	0.275	0.191	0.214	
III	0.11				
IV	0.154				
K_1	0.057	0.0255	0.0465	0.04075	
K_2	−0.0005	0.06875	0.04775	0.0535	
K_3	0.055				
K_4	0.077				
R	0.0775	0.04325	0.00125	0.01275	

3.2 Stress corrosion behavior of X80 steel under overprotective potential

According to the orthogonal test results in Table 4, X80 pipeline steel has strong stress corrosion sensitivity both in high pH value and near neutral pH value soil simulation solution, when the cathodic protective potential was −1500mV and the temperature is 40 ℃. SEM was used to observe fracture of X80 slow strain rate tensile sample in two kinds of simulation solution when temperature is 40℃ and cathodic protective potential was 1500mV, and the fracture was compared with that in air.

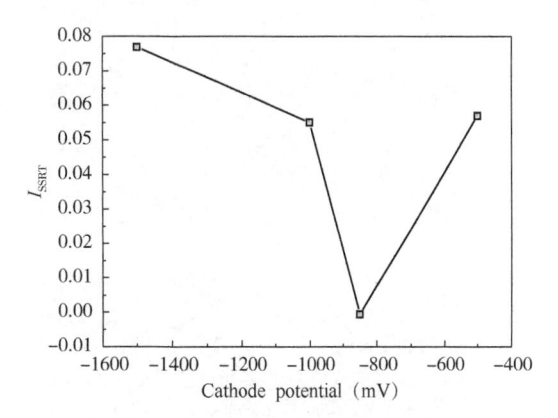

Fig. 3 Visual analysis diagrams of I_{SSRT} affected by applied potential

Fig. 4(a)−(f) shows the SEM morphology of the SSRT fracture, and its side in the air. The fracture is a typical ductile. The main fracture showed equiaxial dimple with similar size, with holes at its bottom and no cleavage morphology. There are no secondary cracks and no stress corrosion cracking in the side morphology. Fig. 4(c)−(d) showed the SEM morphology of SSRT fracture and the side of X80 pipeline steel in high−PH value soil simulation solutions at temperature of 40℃ and the cathodic protective potential of −1500mV. It can be seen that under the negative potential of −1500mV, the main fracture morphology of X80 steel shows cleavage characteristic, which is brittle fracture. A large number of short and deep secondary micro cracks, which is perpendicular to the tensile direction can be observed on the side of the fracture. And many adjacent secondary cracks combined to be long cracks showing higher stress corrosion sensitivity. Fig. 4(e)-(f) shows the SEM morphology of SSRT fracture and side of X80 pipeline steel in near-neutral soil simulation solution at a temperature of 40℃, cathodic protective potential of −1500mV. The whole micro fracture morphology is mainly composed of cleavage surface, with small flat dimples, brittle fracture is dominant. There were secondary micro cracks on the side of fracture, which shows obvious SCC sensitivity.

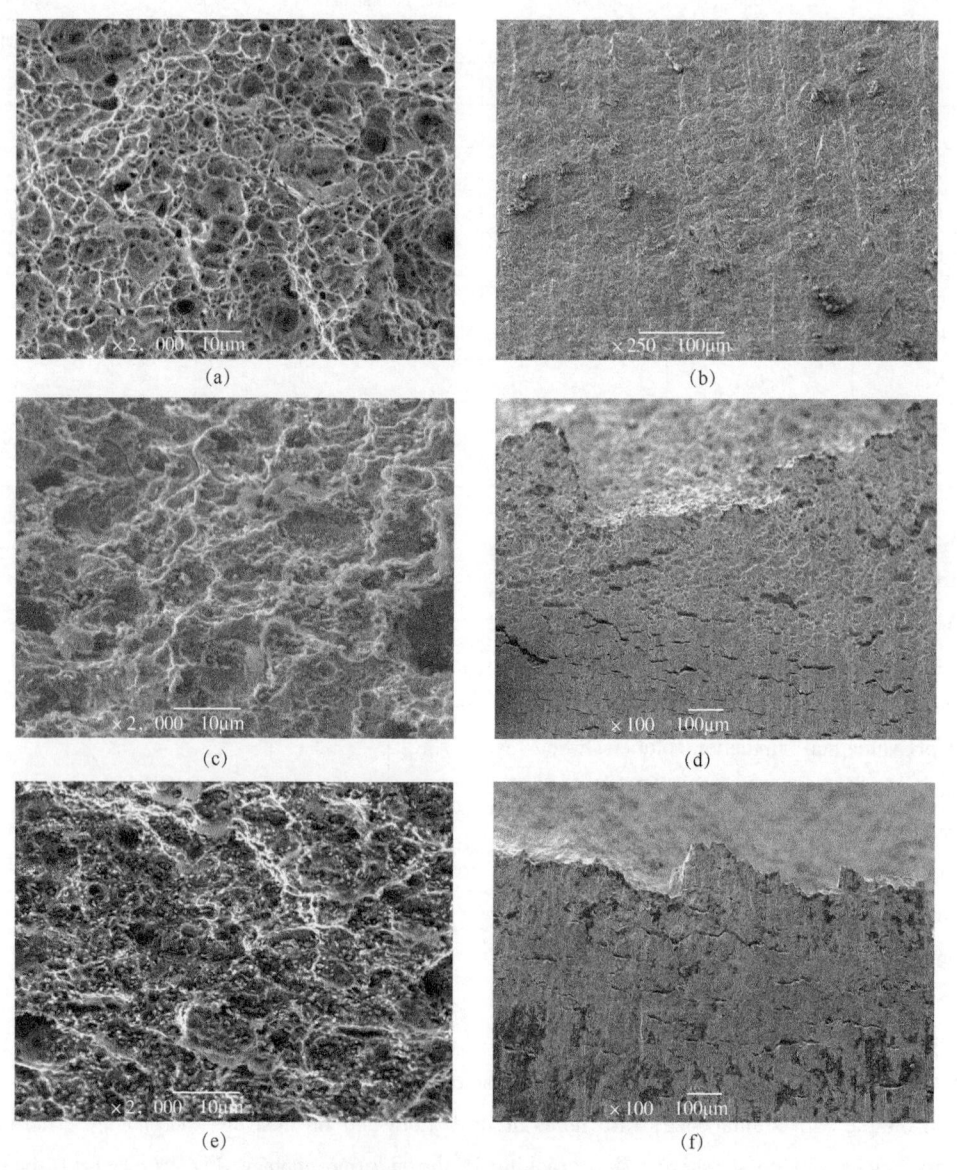

Fig. 4 Micro fractographies of X80 base metal under different conditions, (a, b) in air, (c, d) −1500mV, high pH solution, (e, f) −1500mV, near neutral solution

According to the above analysis, under the overprotective potential of −1500mV, X80 pipeline steel shows obvious hydrogen − induced cracking in both high − pH value and near neutral soil solution. This is because the higher cathodes current promotes the hydrogen evolution vacation on the steel surface, which allows more hydrogen to enter into the steel. And then hydrogen induced cracking happened and caused the low stress brittle fracture.

4 Conclusions

(1) Among the risk factors affecting the SCC sensitivity of pipeline steel, the weight of external potential on the stress corrosion sensitivity ISSRT is the largest, followed by temperature and steel grade. The effect of soil environment on stress corrosion sensitivity is not obvious.

(2) Under protective potential of −1500mV, X80 pipeline steel shows obvious hydrogen induced cracking in both high pH value and near neutral pH value soil solution. And its SCC mechanism is hydrogen embrittle (HE) mechanism.

References

[1] Feng Yaorong, Huo Chunyong, Ji Lingkang, etc. Progress and prospects of research and applications of high grade pipeline steels & steel pipes in China [J]. Petroleum Science Bulletin, 2016, 1(01): 143-153.

[2] Liu Zhiyong, Li Xiaogang, Du Cuiwei. Key factors affecting corrosion of high rength pipeline steels in soli enviroments[C]. Langfang: Proceeding of CIPC2011, 2011: 24-29.

[3] Liang P, Li X G, Du C W, et al. Stress corrosion cracking of X80 pipeline steel in simulated alkaline soil solution [J]. Materials and Design, 2009, 30(5) : 1712-1717.

[4] Z. Y. Liu, X. G. Li, Y. F. Cheng. In-situ characterization of the electrochemistry of grain and grain boundary of an X70 steel in a near-neutral pH solution [J]. Electrochem Commun, 2010, 12 (7) : 936-938.

[5] Zhang Liang, Li Xiaogang, Du Cuiwei, etc. Progress in study of factors affecting stress corrosion cracking of pipeline steels[J]. Corrosion science and protection technology, 2009, 21(1): 62-65.

[6] R. L. Eadie, K. E. Szklarz, R. L. Sutherby. Corrosion fatigue and near-neutral pH stress corrosion cracking of pipeline steel and the effect of hydrogen sulfide [J]. Corrosion, 2005, 61: 2.

[7] Ming Chun Zhao, K. Yang, Strengthening and improvement of sulfide stress cracking resistance in acicular ferrite pipeline steels by nano-sized carbonitrides[J]. Scr. Mater, 2005, 52 (9): 881-886.

[8] Zhang Liang, Li Xiaogang, Du Cuiwei, etc. Effect of applied potentials on stress corrosion cracking of X70 pipeline steel in simulated Ku'erle soil solution [J]. Journal of chinese society for corrosion and protection, 2009, 29(5): 353-359.

[9] Zhao Xinwei, Zhang Guangli, Zhang Liang, etc. Influence of applied potential on stress corrosion cracking behavior of X80 pipeline steel in near-neutral pH soil environment [J]. Oil & Gas Storage and Transportation, 2014, 33(11): 1152-1158.

[10] Liu Zhiyong, Wang Changpeng, Du Cuiwei, etc. Effect of applied potentials on stress corrosion cracking of X80 pipeling steel in simulated YingTan soil solution[J]. Acta metallurgica sinica, 2011, 47(11): 1434-1439.

[11] Wang Dan, Xie Fei, Wu Ming, etc. Effect of cathode potentials on stress corrosion behavior of X80 pipeline steel in simulated alkaline soil solution[J]. Journal of central south University (Science and Technology), 2014, 45(9): 2985-2992.

[12] Yuan Hongzhong, Liu Zhiyong, Li Xiaogang, etc. Influence of applied potential on the stress corrosion behavior of X90 pipeline steel and its weld joint in simulated solution of near neutral soil environment[J]. Acta

Metallurgica Sinica, 2017, 53(7): 797-807.

[13] Luo Jinheng, Luo Sheji, Li Lifeng, etc. Stress corrosion cracking behavior of X90 pipeline steel and its weld joint at different applied potentials in near-neutral solutions[J]. Natural Gas Industry, 2018, 38(8): 96-102.

[14] Jia Yizheng, Li Hui, Hu Nannan, etc. Effect of applied cathode potential on the behavior of near-neutral pH SCC of X100 pipeline steel[J]. Journal of SiChuan University (engineering Science Edition), 2013, 45(4): 186-191.

[15] Guo Hao, Li Guangfu, Cai Xun, etc. Stress corrosion cracking behavior of X70 pipeline steel in near-neutral pH solutions at different temperatures[J]. Acta Metallurgica Sinica, 2004, 40(9): 967-971.

[16] Ye Cundong, Kong Dejun, Zhang Lei. Effects of temperature on stress corrosion of X70 pipeline steel in solution with oxygen[J]. Journal of central South University (Science and Technology), 2015, 46(7): 2432-2438.

[17] Zhang Guoliang. Stress Corrosion Cracking of pipeline steels in soil enviroments[D]. ShangHai: Shanghai Institute of Materials, 2007.

[18] Fan Lin, Liu Zhiyong, Du Cuiwei, etc. Relationship between high pH stress corrosion cracking mechanisms and applied potentials [J]. Acta Metallurgica Sinica, 2013, 49(6): 689-698.

[19] Wang Zhiying, Wang Jianqiu, Han Enhou, etc. Effect of mechanical factors on SCC initiation of pipeline steel [J]. Journal of Chinese society for corrosion and protection, 2008, 28(5): 282-286.

[20] Fu Yang, Wu Ming, Xie Fei, etc. Effect of temperature and applied potential on stress corrosion of X70 pipeline steel in simulated soil solution[J]. Heat treatment of metals, 2015, 40(8): 183-187.

[21] Zhang Liang, Jia Haidong, Yang Yonghe, etc. Stress Corrosion Cracking Influence Factors of High Strength Pipeline Steel in Simulated Near-neutral pH Soil Environment [J]. Welded pipe, 2016, 39(10): 19-23.

[22] Xie Fei. Corrosion behavor of X80 pipeline steel and its welds in soil simulated solution in KURLER [D]. Beijing: China University of Petroleum, 2013.

本论文原发表于《Materials Science Forum》2020 年第 993 卷。

Analysis of Corrosion Behavior on External Surface of 110S Tubing

Han Yan[1] Li Chengzheng[2] Zhang Huali[3] Li Yufei[3] Zhu Dajiang[3]

(1. CNPC Tubular Goods Research Institute, State Key Laboratory for Performance and Structure Safety of Petroleum Tubular Goods and Equipment Materials;
2. Oilfield Development Division, Changqing Oilfield Branch of Petrochina;
3. PetroChina Southwest Oil and Gas Field Company Engineering Technology Research Institute)

Abstract: The failure analysis of 110S tubing during acidizing process was addressed. Results showed that serious pitting corrosion occurred on the outer wall of tubing, and there was no obvious pitting on the inner wall. The maximum pitting depth on the outer wall was 1019 μm. According to the results of simulation corrosion test, needle-shaped pitting appeared on the sample surface in the test without inhibitor, the maximum depth of pitting was 158 μm; and no pitting was found on the sample surface in the test within 1.5% TG501 inhibitor; the original pitting were deepened after spent acid test, and the sample with no pitting at the beginning also showed deep pitting corrosion after 96 hours spent acid test. It was indicated that the spent acid accelerated the development of pitting significantly. The external surface corrosion of the 110S tubing was caused by the chemical reaction between the high-concentration acidifying liquid and the outer wall of the tubing. There is a gap between the tubing and coupling threaded connection, which caused the acid solution entered into the thread position, and hence the severe corrosion of the thread and pin end of the tubing happened, the joint strength was continuously reduced with corrosion development till the tripping of the coupling, and then the lower string dropped. Some suggestions were proposed for avoiding or slowing down this kind of failure based on this study.

Keywords: 110S tubing; Acidizing; Spent acid; Pitting corrosion; Inhibitor; Threaded connection

1 Introduction

Acid-fracture technology is the most effective oil extraction process to increase production of carbonatite. It expands, extends, presses and communicates cracks through the chemical dissolution of acid and the hydraulic action when acid is squeezed into the formation. The oil and gas seepage channel with high

---

Corresponding author: Han Yan, hanyan@cnpc.com.cn.

circulation capacity can effectively stabilize the oil production and ensure the stability during the oil exploitation process [1]. Therefore, acid-fracture technology is a very important and necessary construction technique in the oil exploitation process [1-6]. However, the acid solution will cause various degree corrosion to casing, tubing and construction equipment during the acidification process. Although the corrosion of the wellbore can be alleviated by adding corrosion inhibitor to the acid solution, there is still much oil casing corrosion failure occurs caused by acidification [7-11]. The long residence time of fresh acid and residual acid in the string, the high temperature in the downhole, and the failure of the corrosion inhibitor to play an effective role are the main causes of casing failure [10,11]. In this paper, the corrosion failure of external wall of 110S tubing after acid-fracture process was systematically analyzed. The reason of failure was verified by the simulation experiment, while the relevant preventive measures and suggestions were proposed, in order to prevent or slowdown the corrosion failures that may occur during oilfield acidification design and construction.

2 Sample preparation and experimental methods

The tubing was made of 110S steel with the specifications of $\phi 73$ mm × 5.5 mm. The well was approximately 3472 m deep. It was found that the packer could not be sealed tightly during the acid-fracture construction, and the tubing string was found to be tripped at 1894 m after string lifting, which resulted in the 158 lower tubing falling into the well, and the fish was tubing collar. After salvage, there were 62 tubing with seriously corroded on outer wall, included 29 tubing which not-falling into the well and 33 tubing which falling into the well. Some of the tubing also had obvious pitting corrosion on pin thread and sealing face. The severely corroded tubing was located near the tripping position of tubing string, and there was no significant corrosion was found elsewhere.

To study the cause of the external corrosion, the following series of tests were conducted: (1) visual examination of corrosion characteristics; (2) chemical composition analysis; (3) metallographic structural characterization; (4) scanning electron microscopy (SEM) observation of corrosion morphology and energy-dispersive spectrometry (EDS) analysis of corrosion products; (5) simulated condition corrosion test.

In order to clarify the corrosion resistance of the 110S tubing under acidification solution, simulated corrosion tests were carried out under three conditions. The specific test conditions are shown in Table 1. Coupon specimens with the length 40 mm, width 10 mm and thickness 3 mm, were used.

The acidification solution and duration time in simulated test were determined based on the on-site acid-fracture process. The test temperature was similar with the temperature at tripping position. Since the inhibitor used in the field cannot be obtained, 1.5% TG201 inhibitor was used in fresh acid with inhibitor test (Test No. 2 in Table 1). The spent acid was fetched from site.

Table 1 Test condition under simulated environmental

Test No.	Temperature (℃)	Solution	Duration time (h)
1	80	20% HCl	4
2	80	20% HCl + inhibitor (1.5% TG201)	4
3*	80	Spent acid	96

* Note: The specimens used in test No. 3 are the specimens used in the test No. 1 and test No. 2.

3 Results and discussion

3.1 Visual examination

The outer diameter of the corroded tubing is 73 mm, the wall thickness is 5~6 mm, and the outer wall presents a dense pitting morphology. The corrosion products are rust and black colors, as shown in Fig. 1(a). The thread of the tubing is also seriously corroded. There are many corrosion pits on the outer wall, inner wall and end face of the thread, as shown in Fig. 1(b) and Fig. 1(c). The thread was damaged too severely to connect with coupling.

(a) overall morphology

(b) external thread (c) end face of the thread

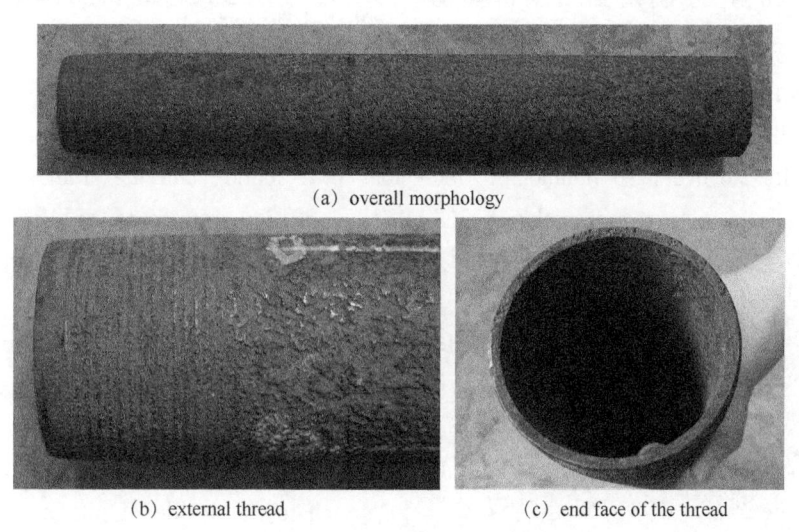

Fig. 1 Corrosion morphology of the tubing

Low magnification observation in Fig. 2 shows that the corrosion pits on the outer wall of the tubing are mostly circular shaped, and two or more corrosion pits are connected and superimposed to aggravate corrosion. Sampling was carried out at the severe pitting corrosion area. The maximum depth of the corrosion pit on samples fetched from outer surface and thread end were 1019 μm and 936 μm separately, as shown in Fig. 2.

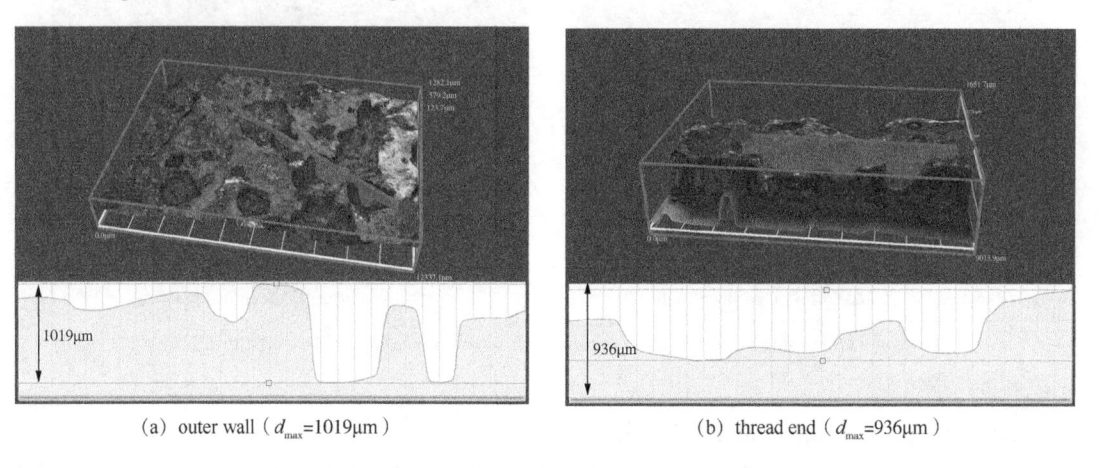

(a) outer wall (d_{max}=1019μm) (b) thread end (d_{max}=936μm)

Fig. 2 Characteristic and maximum depth of pitting

3.2 Chemical composition analysis

The chemical composition of the 110S tubing is shown in Table 2. The elements content accorded with the requirement of SY/T 6857.1—2012 standard [12].

Table 2 Chemical composition of the tubing (wt. %)

Elements	C	Si	Mn	P	S	Cr	Mo	Ni	Nb	V	Ti	Cu
110S tubing	0.26	0.22	0.48	0.0080	0.0030	1.01	0.45	0.014	0.0012	0.0052	0.003	0.024
SY/T 6857.1—2012	≤0.35	≤0.35	≤1.00	≤0.010	≤0.005	0.40~1.20	0.15~1.00	≤0.15	≤0.040	≤0.050	≤0.040	≤0.15

3.3 Metallographic structural characterization

The metallographic structural of the 110S tubing is shown in Fig. 3, and the microstructure is tempered sorbite. No oversized nonmetallic inclusions were found, and the grain size of tubing is ASTM 9.0 grade, which is smaller than SY/T 6857.1—2012 requirement: ASTM 7.0 grade.

The characteristics of localized corrosion are shown in Fig. 4. The typical corrosion pits on the tubing surface are in round or elliptical, which are all defined as the primary pits, as shown in Fig. 4 (a). The gray corrosion products, as marked by arrow in Fig. 4 (a), attach to the inside of pits. In addition, some new pits continue to stack along the depth direction of the bottoms of primary pits, as shown in Fig. 4(b).

Fig. 3 Microstructure of the tubing

(a) microstructure and corrosion products in the pitting　　(b) characteristic of corrosion pitting

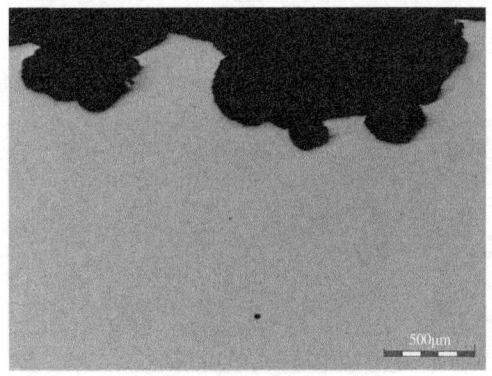

Fig. 4 Morphology and the microstructure of pitting on outer-surface of the tubing

3.4 SEM observation and EDS analysis

The micro-morphology and energy spectrum analysis specimen were taken from the pipe body and the thread end. It was found that various degree of corrosion occurred in all parts of the tubing. Among them, pitting corrosion mainly occurred on the outer wall and the thread area, the inner wall of the tubing undergoes slight uniform corrosion, as shown in Fig. 5. The outer wall of the tubing was covered with circular corrosion pitting, and the corrosion product film exhibits loose and uneven characteristic, which has no protection against further corrosion.

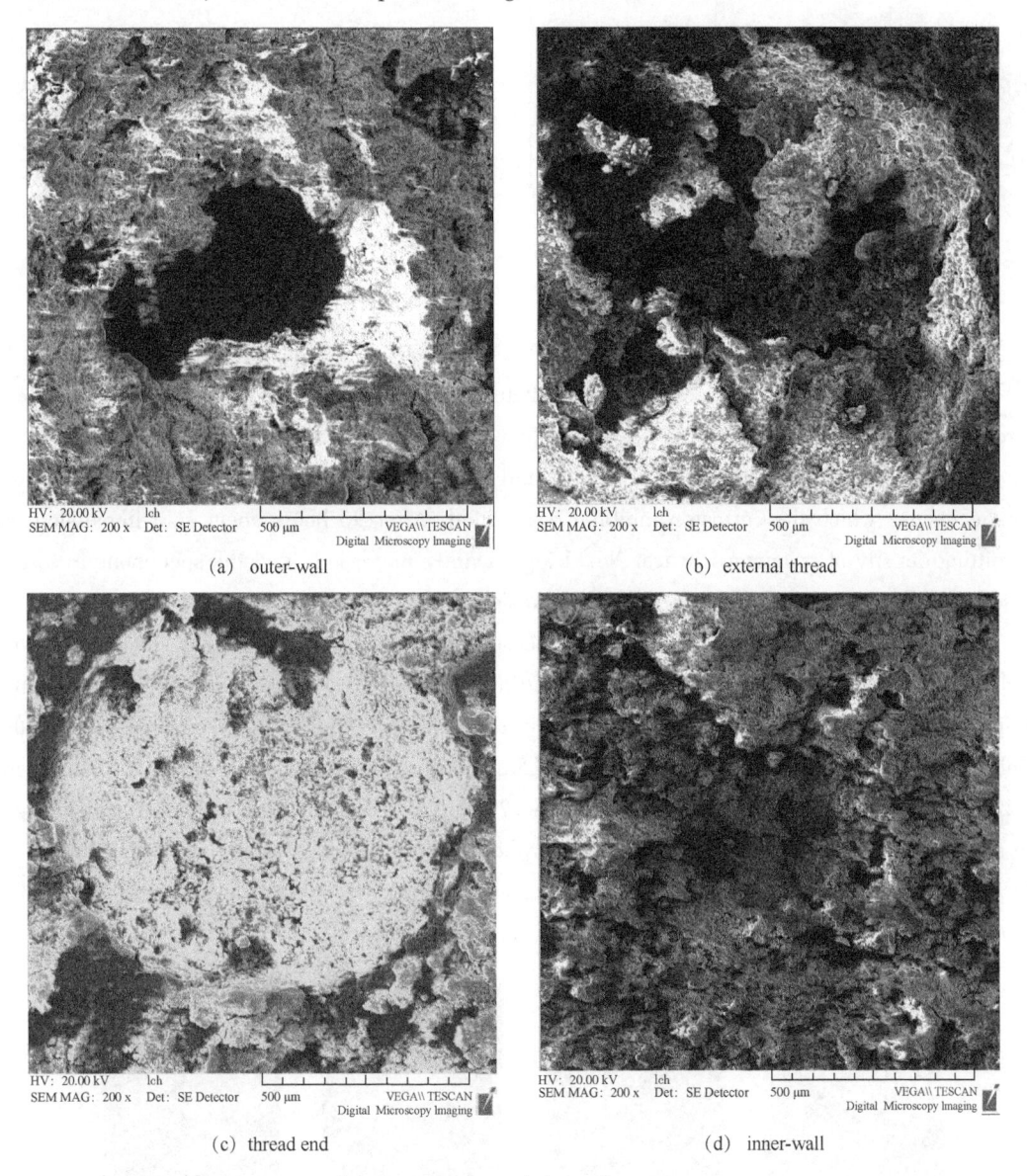

(a) outer-wall

(b) external thread

(c) thread end

(d) inner-wall

Fig. 5 Micro-morphology of corrosion

The EDS results show that the corrosion products mainly contain Fe, C, O and S elements, and the Fe and Cr content in corrosion products in outer wall is significantly higher than that in inner wall, while the content of C and O elements in corrosion products in inner wall is higher than that in outer wall, as shown in Fig. 6. This suggest that the outer wall corrosion pit exhibits the composition of the metal matrix, while the inner wall corrosion product is mainly CO_2 corrosion product.

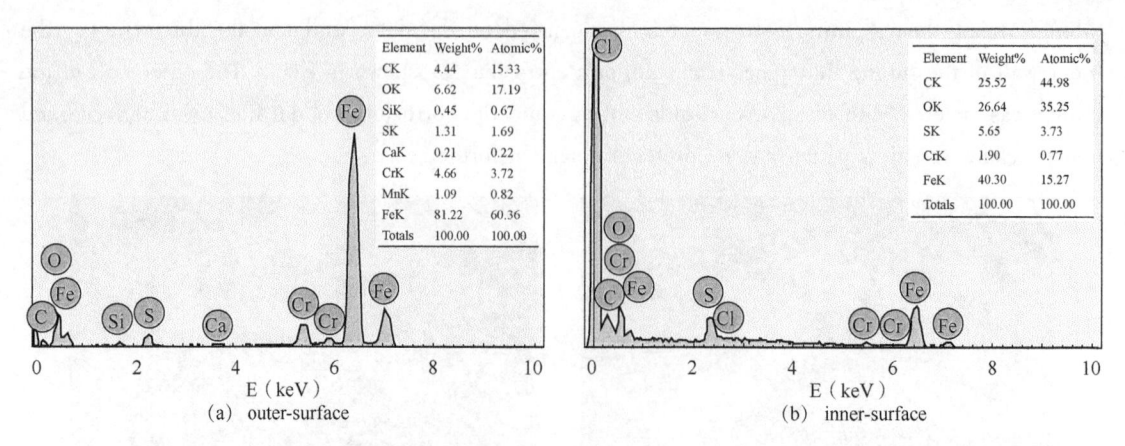

Fig. 6 Energy spectrum of the corrosion products in the pitting

3.5 Simulated condition corrosion test

Photographs of the specimens after simulated test are shown in Fig. 7, and the micro-morphology was shown in Fig. 8. It was found that the surface of the specimens were covered with pin-like pitting after 20% HCl fresh acid test, and no obvious pitting was observed on the surface of the specimens, which the corrosion inhibitor was added. After 96 hours spent acid immersion test, the pitting density of specimens in test No. 1 significantly increased, and the specimens in test No. 2, which had no obvious pitting and pitting corrosion.

Ten pitting pits were selected for depth measurement randomly, and the result was shown in Table 3. After the spent acid test, the original pitting pits were deepened, and the specimens with no pitting pits at the beginning also showed deep pitting corrosion. After 4 hours' fresh acid test without inhibitor, the maximum depth of the pitting is 158μm, and the maximum depth of the pitting after 96 hours' spent acid experiment are 226μm and 178.2 μm, respectively. The pitting morphology of the specimens is consistent with the pitting morphology of the failure tubing. It is distributed in a plurality of circular pit-like shapes, and has traces of superposition and development in the depth direction. The profiles of the pitting were shown in Fig. 9 to Fig. 10.

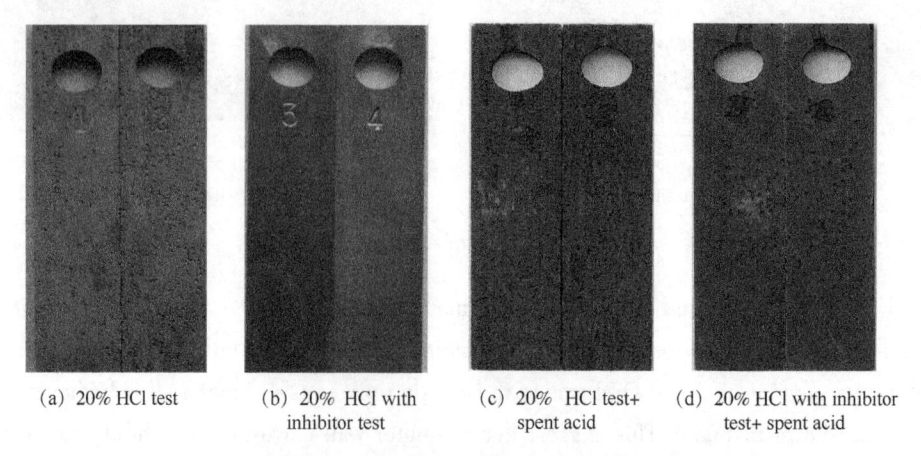

Fig. 7 Photographs of specimens after simulated test

(a) 20% HCl

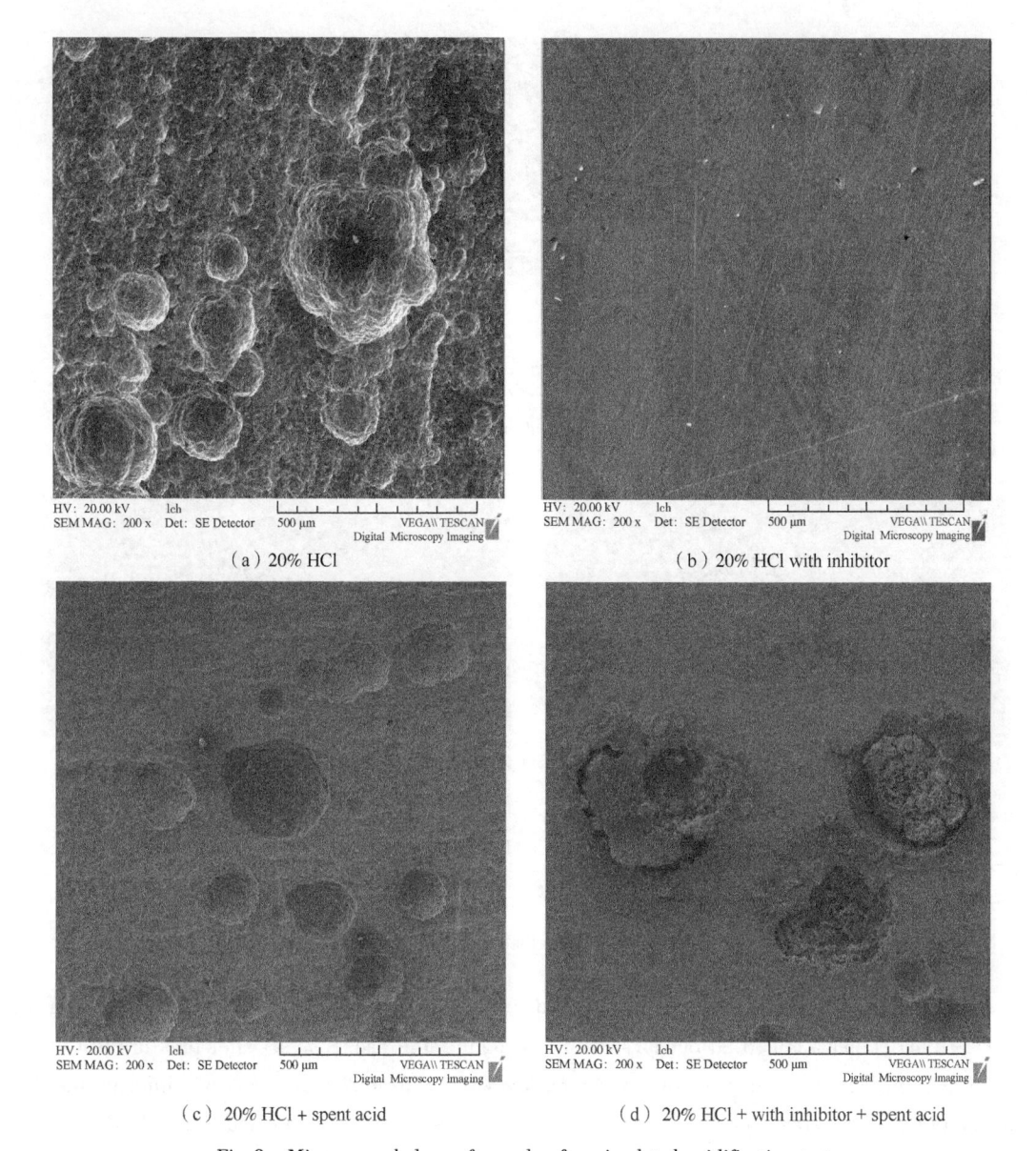

(b) 20% HCl with inhibitor

(c) 20% HCl + spent acid

(d) 20% HCl + with inhibitor + spent acid

Fig. 8　Micro-morphology of sample after simulated acidification test

Table 3　Results of the pitting depth after simulated corrosion test(μm)

Test sample		\multicolumn{10}{c}{Points}	Average
		1	2	3	4	5	6	7	8	9	10	
Test No. 1	1-1#	147.7	132.8	139	107	132	134	151	137	151	158	138.95
	1-2#	122	128.8	140	125	122	128	131	110	126	131	126.38
Test No. 3	3-1#	226	93	93	142	224	118	153	159	124	124	145.6
	3-2#	171.6	127.1	115	125.4	141	157	135	90	153	113	132.81
	3-3#	139.4	57	114.2	149	58	161.8	49	93	54	116	99.14
	3-4#	143.4	57	131	85	73	178.2	74	156.4	63	90	105.1

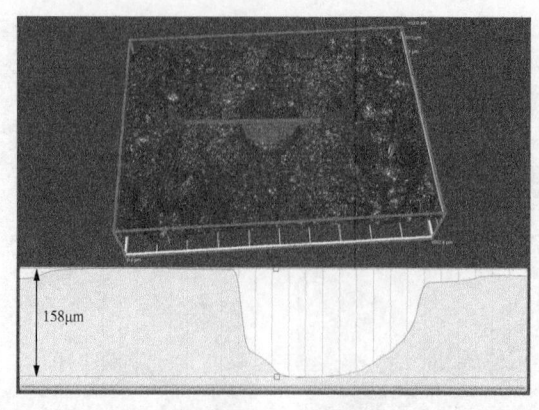

Fig. 9 profile of the pitting after 20% HCl ($d_{max} = 158\mu m$)

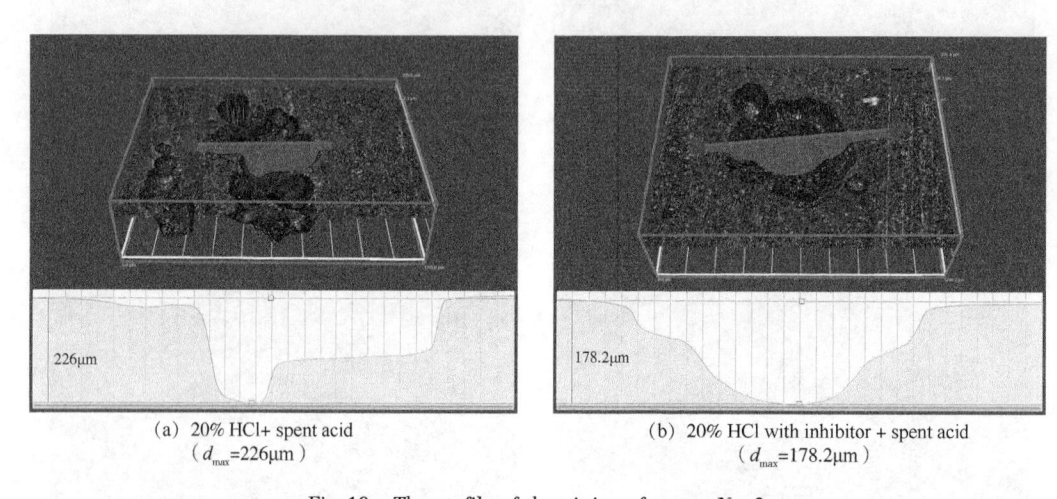

(a) 20% HCl+ spent acid
(d_{max}=226μm)

(b) 20% HCl with inhibitor + spent acid
(d_{max}=178.2μm)

Fig. 10 The profile of the pitting after test No. 3

According to the construction record, theacidizing process as follows: Firstly, acidification of the 3414~3416m section was carried out for 4 hours. Secondly, the packer is set after pressure test of the column to 68 MPa, the sealing pressure of the packer was 19 MPa. 204.8m^3 acid was injected into the tubing with 61MPa average pressure, and then 27m^3 common acid was injected with 54MPa average pressure, the bursting pressure is 63.5MPa, the pump pressure is 36MPa when stop, the highest applied pressure is 64MPa, the balance pressure is between 14MPa to 26MPa. Subsequently, the fracturing pump car casts the ball and to raise the pressure of column, but it was found that the packer cannot be sealed tightly and the upper section acidified fracturing cannot be implemented.

Therefore, the process of spray, drain, kill well, and wash off with clean water were executed until the water quality of the import and export is consistent, and then salvage. The upper 198 tubing, have been subjected to acidification for about 4 days, and the lower 158 tubing have been salvaged out 9 days after acidification. From the observation of more than 300 tubing, it was found that only 62 tubing corroded seriously on their outer surface, and the corrosion range is about 300m above and below the tripping position. Some of the threads and sealing surfaces were corroded heavily, and no serious corrosion was found on other tubing. It can be seen from the corrosion morphology of the corroded tubing that the corrosion pits are deep circular hole, and there are several small corrosion pits at the bottom of the large corrosion pit, which are superimposed pitting, which is consistent with the acid corrosion characteristics [9, 10], and the pits has the same shape with the corrosion pit after the simulated corrosion test in the laboratory. The pitting corrosion of the tubing is caused by the corrosion of high concentration acid.

Based on the serious corrosion at the external thread of the tubing, it can be judged thatat field end of the tubing there exist a gap between the buckle and the coupling. The injection of acid with high pressure caused the acid solution penetrate into the thread connection gap, then severe localized corrosion occurred at this area due to the poor fluidity at the threaded joints. As the thread corrosion strengthen, the acid solution entered into the casing annulus and caused corrosion on the outer wall of the tubing near the leakage location. At last, the coupling tripping as the continuously deteriorate of the sealing performance of threaded, and the lower tubing string dropped eventually.

Hydrochloric acid is an acid commonly used in acid-fracture processes, but it will cause severe corrosion to the tubing string. Therefore, a certain corrosion inhibitor must be added in acid-fracture processes to reduce the corrosion. The rate of uniform corrosion caused by acid can reach extreme high (>100mm/a) in absence of corrosion inhibitors or inhibitor do not play an effective role, and it will increase exponentially with increasing temperature and acid concentration[13]. SHI Zhi-ying et al.[14] shown that, the corrosion inhibitor in the acid is absorbed by the rock clay and minerals as the acid squeezes into the formation, so the corrosive of spent acid from the formation will increase higher. Considered the long processing cycle of spent acid (3~7 days), it will bring more severe corrosion than fresh acid with inhibitor. FU An-qing's studies[15] shown that, P110 steel showed different pitting degrees after the acid and spent acid tests, and the pitting density of the sample in the spent acid increased significantly. When steel material is corroded in the electrolyte solution, the anode process is the dissolution of Fe as shown in Eq (1), and the cathode process is different due to the change of the medium. The hydrogen depolarization reaction of the cathode is prone to occur in the reflux liquid with pH value of 1.5, as shown in the Eq (2). The reaction rate increases with the concentration increase of the H^+. The corrosion rate of P110 tubing steel in the reflux fluid can reach 47mm/a [16].

Anode process $\qquad$ $Fe \longrightarrow Fe^{2+} + 2e$ $\qquad$ (1)

Cathode process $\qquad$ $2H^+ + 2e \longrightarrow H_2$ $\qquad$ (2)

The corroded tubing undergoes 4 hours acidification time for fresh acid with inhibitor and 4 days in spent acid before lifted from the well, it can be inferred that the spent acid is more likely to cause serious corrosion on the outer surface of the 110S tubing.

4 Conclusions and suggestions

The chemical composition and microstructure of the 110S tubing meet the requirements of SY/T 6857.1—2012 standard. According to the results of macroscopic and microscopic morphology observation, the pitting corrosion on the outer wall is circular and shows the characteristics of superimposed corrosion, which is consistent with the acid corrosion characteristics. Combined with the results of simulated corrosion test, the pitting corrosion of the tubing is caused by the corrosion of high concentration acid. The pitting morphology of specimens in simulated test is consistent with the pitting morphology of the failure tubing. The depth of pitting is deepened after the spent acid immerse.

There is a gap between the tubing and coupling threaded connection, which caused the acid medium entered into the thread position, it brings severe corrosion of the thread and pin end of the tubing, and the joint strength continuously reduced with the development of corrosion till the tripping of the coupling, then the lower string dropped.

In order to avoid this kind of failure from happening, construction management on-site should be strengthened, and to ensure the reasonable torque we mustcarry out the torque according to the requirements of the technical specification strictly. Meanwhile, it is necessary to verify the effectiveness of the corrosion inhibitor before use.

Acknowledgement

This work was supported by the National Key R&D Program of China (2017YFB0304905), Key project of shanxi key research and development plan (2018ZDXM-GY-171) and Major science and technology project of CNPC (2016E-0606).

References

[1] Matjaž Finšgar, Jennifer Jackson. Application of corrosion inhibitors for steels in acidic media for the oil and gas industry: A review [J]. Corrosion Science, 2014, 86: 17 – 41.

[2] Luo Jing-qi. Acidizing fundamentals [M]. Beijing: petroleum industry press, 1983.

[3] M. U. Shafiq, H. B. Mahmud, MOHSEN Ghasemi. Integrated mineral analysis of sandstone and dolomite formations using different chelating agents during matrix acidizing[J]. Petroleum, 2018, in press: 1-10.

[4] A. Q. Fu, Y. R. Feng, R. Cai, et al. Downhole corrosion behavior of Ni-W coated carbon steel in spent acid & formation water and its application in full-scale tubing[J]. Engineering Failure Analysis, 2016, 66: 566-576.

[5] R. Abdollahi, S. R. Shadizadeh. Effect of acid additives on anticorrosive property of henna in regular mud acid [J]. Scientia Iranica Transactions C: Chemistry and Chemical Engineering, 2012, 19: 1665-1671.

[6] Lv Zhi-feng, Zhan Feng-tao, Wang Xiao-na, et al. Synthesis of ethyl chloroacetate quinolinium salt derivatives and evaluation of its acidification inhibition corrosion properties[J]. Materials protection, 2018, 51 (5): 25-29.

[7] X. W. Lei, Y. R. Feng, A. Q. Fu, et al. Investigation of stress corrosion cracking behavior of super 13Cr tubing by full-scale tubular goods corrosion test system[J]. Engineering Failure Analysis, 2015, 50: 62-70.

[8] StefanBachu, Theresa L. Watson. Review of failures for wells used for CO_2 and acid gas injection in Alberta, Canada[J]. Energy Procedia, 2009, 1(1): 3531-3537.

[9] Xie Junfeng, FU An-qing, Qin Hong-de, et al. Influence of Surface Imperfection on Corrosion Behavior of 13Cr Tubing in Gas Well Acidizing Process [J]. Surface technology, 2018, 47(6): 51-56.

[10] Li Yan, Xie Jun-feng, Chang Ze-liang, et al. Reasons for Tube Corrosion in a Gas Well Due to Acidizing in Tarim Oilfield [J]. Corrosion and protection, 2016, 37(10): 861-864.

[11] Zhanc Shuang-shuang, Zhao Guo-xian, Lv Xiang-hong. Corrosion Behavior of TP140 in Simulated Oilfield Fresh Acid and Spent Acid Environments [J]. Corrosion and protection, 2014, 35(1): 28-32.

[12] SY/T 6857.1-2012, Petroleum and natural gas Industries – OCTG used for special environment–Part 1: Recommended practice on selection of casing and tubing of carbon and low alloy steels for use in sour service

[S]. Beijing: Beijing: petroleum industry press, 2012.
[13] E. Barmatov, J. Geddes, T. Hughes, et al. Research on corrosion inhibitors for acid stimulation[C], NACE, 2012, C2012-0001573.
[14] Shi Zhi-ying, TiaN Zhen-yu, Chen Li. The study of spent acids' corrosiveness and its control[J]. Oil drilling technology, 1999, 6(3): 52-53.
[15] Fu An-qing, Geng Li-yuan, Li Guang, et al. Corrosionfailure analysis of an oil tube used in a western oilfield [J]. Corrosion and protection, 2013, 34(7): 645-648.
[16] Liu Zongzhao, Wang Yu, Wang Jun. Research on corrosion behavior of release sewage of acidizing in the exploration of LD10-1 oilfield[J]. Total corrosion control, 2013, 27(6): 35-39.

本论文原发表于《Materials Science Forum》2020年第993卷。

Failure Analysis and Solution to Bimetallic Lined Pipe

Li Fagen[1]　Li Xunji[2]　Li Weiwei[1]　Wang Fushan[2]　Li Xianming[2]

(1. CNPC Tubular Goods Research Institute, State Key Laboratory of Performance and Structural Safety for Petroleum Tubular Goods and Equipment Materials;
2. PetroChina Tarim Oilfield Company)

Abstract: Owing to low cost, excellent pressure and corrosion resistance, bimetallic lined pipes were regarded as one of the most important methods to resolve corrosion of traditional steel pipes used for oilfield. Nowadays, the pipes were widely used in the projects of oil and gas gathering and transportation. However, there were some failure cases in succession in recent years. In this paper, the failure causes were excavated from multiple perspectives, and on this basis further countermeasures were put forward. Firstly, three typical failure accidents, including CRA layer collapse, CRA layer corrosion and weld joint failure, were listed throughout the whole life cycle from product ordering, construction technology to later operation. Secondly, failure analysis was carried out from five aspects: product quality, test technology, welding process, standard specification and application threshold, and a serial of comprehensive views were proposed. (1) Manufacturing period: Water seepage and tightness between CRA layer and backing pipe could not be effectively monitored. The ratio of collapse test was low so that proposed relevant risk could not be eliminated. (2) Welding period: the process has high risk failure in theory and lack of process assessment and construction acceptance standards. Potential danger could not be effectively assessed, and weld quality could not be guaranteed. (3) Application period: The CRA application range remained unclear, and bimetallic lined pipes were used in the environment beyond the threshold sometimes. The paper also summarized further improvement measures and research directions about these five aspects, including process quality, inspect technique, welding process, standard specification and application range. Finally, solution suggestions were proposed for the whole chain from manufacturers, testing institutions, construction units to oilfield users.

Keywords: Bimetallic lined pipe; CRA layer collapse; CRA layer corrosion; Weld joint failure; Failure analysis

Corresponding author: Li Fagen, lifg@ cnpc. com. cn.

1 Introduction

Owing to low cost, excellent pressure and corrosion resistance, bimetallic lined pipes were regarded as one of the most important methods to resolve corrosion of traditional steel pipes used for oilfield. Nowadays, the pipes were widely used in the projects of oil and gas gathering and transportation. Preliminary statistics showed that there were more than 2500 km long bimetallic lined pipes used in oilfields in China[1-4].

Bimetallic lined pipes could be equally allocated pressure to backing pipe and corrosion to CRA layer. Material combinations dramatically reduced the cost. Furthermore, cost effectiveness increased as lengths extended, and operational expenses were lower than the corrosion inhibitor in a long period. Regarding corrosion environment, bimetallic lined pipe selected suitable CRA as the layer, totally owned the performance like CRA materials. Comparing with corrosion inhibitors, this solution avoided the risk resulted in the complicated management process, and it guaranteed operational safety. Due to special structure with two layers, there were higher requirements in manufacturing technology, performance testing, welding construction and field application about bimetallic lined pipe than the single pipe. Therefore, despite the development of manufacturing technology and application technology for many years, there were some failure cases, including CRA layer collapse, CRA layer corrosion, welding joint failure in succession in recent years. Frequent failure seriously disturbed the normal production order in oil fields[5-10].

At present, the scholars have done a lot of work to analyze the failure causes of bimetallic lined pipes, but the understanding was still lack of systematic, and no comprehensive solution had been put forward[5-13]. Based on that, this paper would focus on typical failure cases, analyze failure reasons from five aspects (product quality, inspect technology, welding process, standard specification and application range), and then put forward five solution suggestions.

2 Typical failure problems

2.1 CRA layer collapse

The CRA layer was affixed or tightly fitted to the external pipe full length by expansion, or some other means. There was inevitable gap between CRA layer and backing pipe. Therefore, the CRA layer might collapse once it was subjected to external pressure. In practice, CRA Layer collapse shown in Fig.1 had been one of the main failure modes. For example, a wide range of collapse with the highest proportion even reaching 28% happened in the early stage of external anticorrosion in one project, while in the later stage of operation in another project more than 250 similar failure cases were found.

2.2 CRA layer corrosion

The corrosion resistance was one of the key factors to determine service life. However, transportation environment in oilfields was complex and severe, and even residual acid was sometimes mixed into the oil/gas medium. It was important to ensure CRA material used in appropriate conditions. Once the operation was improper, the risk of pipe corrosion was high. Fig.2 showed that corrosion case about 316L liner pipe occurred.

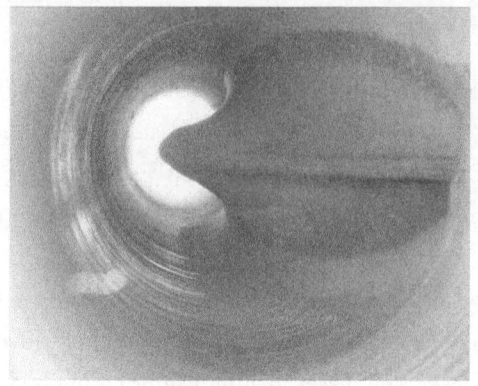

Fig. 1 Failure morphology of CRA layer collapse   Fig. 2 Failure morphology of CRA layer corrosion

2.3 Weld joint failure

The welded joint structure and process about 316L lined pipes were complex so that failure cases occurred frequently. The structure and process were often consisted of five steps: seal weld, root weld, transition weld, filler weld and cap weld. Root pass and transition weld pass were welded by 309Mo or 309MoL electrode, while carbon steel electrode was often selected to weld fill pass and cap pass in China. However, the first success rate of welding operation was low. Moreover, welding quality was not easy to guarantee. Weld joint failure, such as the crack and corrosion shown in Fig. 3 had taken place for many times.

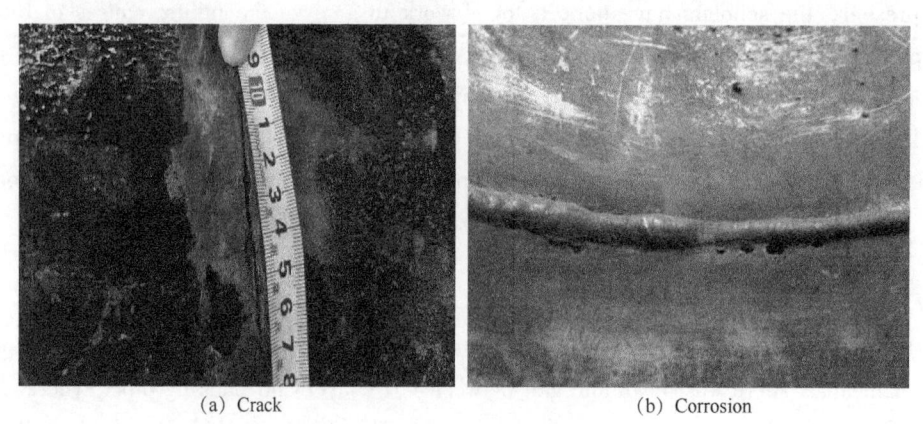

(a) Crack   (b) Corrosion

Fig. 3 Failure morphology of weld joint

3 Failure factor analysis

3.1 Product quality

CRA layer collapse was essentially one style of stiffness instability. in bimetallic lined pipes, there must be an instability threshold whose calculation formula was given in the literature[14]. Once external pressure was more than CRA's instability threshold, layer collapse would happen. As a matter of fact, not only 316L's stiffness was very low, but also diameter–thickness ratio and tightness that effected instability threshold were also difficult to control. The instability threshold about 316L liner pipes was not high, and it was easy to collapse. Additionally, impurities, such as water and air between backing pipe and CRA layer, could not be completely removed during early manufacture stage. External pressure was introduced while impurities were heated during external anticorrosive stage. Similarly, once oil and gas medium was entered into the gap between backing pipe and CRA layer due to weld corrosion and other reasons, stress load would also be introduced at the moment of shutdown during operation stage. Once external pressure went beyond 316L's instability threshold, the collapse phenomenon would occur.

3.2 Inspect technique

The quality control of bimetallic lined pipes depended not only on the standard requirements and manufacturing process, but also on a series of inspect techniques. However, there were still many problems to resolve about inspect technology, such as how to monitor tightness in the manufacturing stage and variation of corrosion resistance elements in the seal weld stage. If tightness could be accurately tested and controlled by real-time, product performance would be more reliable. In recent years, although monitoring modal parameters had been proposed to control the tightness by the relationship between modal parameters and tightness timely, the practical results need still to be proved by further application [15]. Similarly, there was also lack of on-line monitoring methods for the seal weld quality. If appropriate test could be used to find out the change of corrosion resistance elements, and then weld current was dynamically adjusted and the corrosion resistance might not turn worse.

3.3 Welding process

The early weld process was not mature and it was closely related to frequent weld joints failures. Previous process would easily result in weld defects in the region of seal weld, promote crack initiation and rapid propagation both in seal weld and weld joint. Weld defects became the failure source, while the high hardness martensite area was transformed into the crack propagation pass further, and then cracks passed through weld joint. Thermal stress between backing pipe and CRA layer promoted weld defects formation. The martensite would form during the welding process by carbon steel electrode under stainless steel transition pass. Additionally, previous process also resulted in weld corrosion. Once weld current was slightly larger during seal weld stage, local ablation was more serious when CRA was needed to repeat heating. The joint welding by the cored wire without gas protective measures would be oxidized and the protective effect couldn't meet the demand. Therefore, the two together led to poor corrosion resistance in the heat affected zone and promoted to weld corrosion [16].

3.4 Standard specification

API Spec 5LD standard was used to control the product quality, but its performance requirement was too loose to meet the application requirements. The pipe end sizes could not make sure CRA layer to be effective welded. In addition, the standard also has the problem of insufficient testing items. The tightness mentioned above not only had no acceptance index, but also had no test mean to prevent CRA layer collapse. In weld quality control, weld joints must meet both the strength requirement of backing pipe and the corrosion resistance of CRA layer. Nowadays, the assessment and construction specification still remained unsolved, and the welding quality was difficult to guarantee. The existing standards, which mainly aimed at carbon steel or stainless steel, were not obviously suitable for bimetallic lined pipes.

3.5 Application range

Corrosion resistance of CRA was limited, and each material had its usage threshold in severe corrosion conditions. The corrosion resistance range about 316L was given[17,18]. However, the oilfield environment was too complex to take into account all of effect factors, and there was one consensus that 316L could not be used in the low pH value condition, high chloride ion and high temperature environment. In the process of production allocation, it was easy to overlook CRA's threshold by users. For example, one pipe was used in 90℃ environment for a long time despite its design temperature was 60℃. In addition, high corrosion media, such as residual acid, was mixed in the pipe mentioned above without any protective measures. It was obvious that service condition beyond the threshold would lead to CRA corrosion.

4 Application recommendations

Following suggestion were address for manufacturers, testing institutions, construction units and oilfield users.

(1) The manufacturer should improve the manufacturing process to reduce or eliminate the problem of CRA Layer Collapse. A new process might be tried to solve the problem [19]. Meanwhile, structure optimization, such as improving dimension accuracy, has been considered to reduce the collapse probability. Finally, manufacturers could also make efforts to develop cheap bimetallic clad pipes and remove the trouble of CRA layer collapse.

(2) Corrosion environment applicability and collapse sensitivity, which were not mentioned in API Spec 5LD standard, were the vital factors of service life about bimetallic lined pipes. It was necessary to develop relevant test technology including dynamic monitoring tightness, non-destructive testing weld joint etc.

(3) The new process that the whole joint was welded by CRA materials was not only widely used abroad but also has been applied in China[20]. The new process had more simple operation and more reliable quality than previous process mentioned above in China. Later large-scale promotion was carried out about the new process. It was suggested that large-scale welding application should be carried out around the new process in the future.

(4) It was very urgent to compile technical specifications applicable to the site and welding process qualification and construction specifications. At present, the main work was to optimize or form the evaluation methods and acceptance requirements based on the key issues of tightness, dimensions, and collapse of pipes, mechanical properties and corrosion resistance of welded joints, and so forth.

(5) It was suggested that design units, oil fields, manufacturers and relevant research departments worked together to screen corrosion factors, and clarify the scope of materials applicability further. Finally, bimetallic lined pipes were used in the suitable environment so that it could reduce unnecessary corrosion damage.

Acknowledgement

The authors acknowledge the financial support of the Key Research and Development Program of Shaanxi Province (No. 2019KJXX-091, No. 2018ZDXM-GY-171).

References

[1] LI Fagen, WEI Bin, SHAO Xiaodong, et. al. Bimetal Lined/Clad Pipe Used in Highly Corrosive Oil and Gas Fields. Oil & Gas Storage and Transportation, 2010, 32 (12): 92-96.

[2] Russell D, Wilhelm S M. Analysis of Bimetallic Pipe for Sour Service. SPE Production Engineering, 1991, 6 (3): 291-296.

[3] ROMMERSKIRCHEN I. New Progress Caps 10 Years of Work with Bubi Pipes. World Oil, 2005, 226(7): 69-70.

[4] Fu, G. H. Technical Solutions Analysis of CO_2 Corrosion Problems in Xushen Gas Field. J. Oil-Gasfield Surface Engineering, (2008) 66-67.

[5] PAN Xu, ZHOU Yongliang, FENG Zhigang, et al. Problems and Suggestions on Collapse of Double Metal Composite Pipe Lining. Petroleum Engineering Construction, 2017, 43(1): 57-59.

[6] GUO Chongxiao, JIANG Qinrong, ZHANG Yanfei, et al. Stress Corrosion Cracking Failure Analysis on Bimetal Composite Pipe Lining Layer. Welded Pipe and Tube, 2016, 39(02): 33-38.

[7] CHEN Hao, GU Yuanguo, JIANG Shengfei. Failure Reasons for 20G/316L Double Metal Composite Pipe. Corrosion & Protection, 2015, 36(12): 1194-1197.

[8] Fu A. Q, Kuang X. R, Han Y. Failure Analysis of Girth weld Cracking of Mechanically Lined Pipe sed in Gas field Gathering System. Engineering Failure Analysis, 2016, (68): 64-75.

[9] CHEN Haiyun, CAO Zhixi. Influence of Heat Load to the Reisdue Pressure of Bimetal Composite Pipe. Journal of Plasticity Enginnering, 2007, 14(2): 86-89.

[10] WEI Fan, JIANG Yi, WU Ze, et al. Mechanism Analysis and Testing Research on the Buckling of the Bimetal Lined Pipe. Natural Gas and Oil, 2017, 35(5): 06-11.

[11] LIN Yuan, KYRIAKIDS Stelios. Wrinking Failure of Lined Pipe under Bending. 32nd International Conference on Ocean, Offshore and Arctic Engineering, Nantes, France, 2013: 1: 7.

[12] CHANG Zeliang, JIN Wei, CHEN Bo, et al. Effect of Welding Process on Pitting Potential of Welded Joints of 316L Lining Composite Pipe. Total Corrosion Control, 2017, 30(11): 18-22.

[13] VASILIKIS D, Karamanos S A. Mechanical Behavior and Wrinkling of Lined Pipes. International Journal of Solids and Structures, 2012, (49): 3432-3446.

[14] DANIEL Vasilikis, Spyros A. Karamanos. Mechanics of Confined Thin-Walled Cylinders Subjected to External Pressure. Applied Mechanics Reviews, 2014, 66: 010801-1: 010801-15.

[15] WEI Fan, ZHANG Yanfei, GUO, et al. Inspection and Control of Bonding Strength for Mechanical Composite Pipe, Welded Pipe and Tube, 2015, 38(2): 32-36.

[16] LI Fagen, MENG Fanyin, GUO Lin, et al. Analysis of Welding Technology about Bimetal-Lined Pipe. Welded Pipe and Tube, 2014, 37(6): 40-43.

[17] Li Weiwei, Liu Yaxu, Xu Xiaofeng, et al. One girth weld process of bimetallic lined pipe, China, ZL201310202717. X. 2013-10-02.

[18] LI Ke, SHI Daiyan, et al. Application Boundary Conditions for 316L Austenite Stainless Steel Used in Gas Field Containing CO_2. Materials for Mechanical Engineering, 2012, 36(11): 26-28.

[19] Bruce D Craig. Selection Guidelines for Corrosion Resistant Alloys in the Oil and Gas Industry. https://stainless-steel-world.net/pdf/10073.pdf.

[20] YANG Gang. Welding Technology and Quality Control about X65/316L Bimetallic Lined Pipe. Welding Technology, 2012, 41(12): 56-57.

本论文原发表于《Materials Science Forum》2020 年第 993 卷。

外加电位对 X80 管线钢在轮南土壤模拟溶液中应力腐蚀行为的影响

朱丽霞[1,2] 贾海东[3] 罗金恒[1] 李丽锋[1] 金 剑[3] 武 刚[1] 胥聪敏[4]

(1. 中国石油集团石油管工程技术研究院·石油管材及装备材料服役行为与结构安全国家重点实验室；2. 西安理工大学材料学院；
3. 中石油管道有限责任公司西部分公司；
4. 西安石油大学材料科学与工程学院)

摘 要：采用慢应变速率拉伸(SSRT)实验、SEM 观察和动电位极化曲线测量等方法，研究了外加电位对 X80 管线钢母材及焊缝在轮南土壤模拟溶液中的应力腐蚀开裂(SCC)行为。结果表明，X80 钢母材及焊缝在轮南土壤模拟溶液中均具有一定的应力腐蚀敏感性。在同一外加电位下，X80 钢焊缝的 SCC 敏感性高于母材的；X80 钢 SCC 敏感性及开裂机理受外加电位影响显著，在-500mV 外加阳极电位下，X80 钢的 SCC 机理为裂尖阳极溶解—膜破裂机制，在-800mV 阴极电位以下(-850mV、-1000mV 和-1500mV)，氢脆作用在 SCC 过程中的影响明显增强，阴极析氢反应促进了钢的氢致开裂，导致 X80 钢 SCC 敏感性显著增加。

关键词：X80 管线钢；应力腐蚀开裂；外加电位；土壤环境；腐蚀机理

近年来随着我国石油及能源工业的快速发展，埋地管线里程越来越长，油气管道建设稳步推进，高压、大管径、高钢级管线钢是石油和天然气输送管道发展的必然趋势[1-3]。我国 70%的石油和 99%的天然气运输全部依赖埋地管道进行输送，管道运输关系着经济命脉，同时也关系到公共安全[4-7]。截至 2018 年，我国长输管线已达 1.3×10^5 km，预计到 2025 年将完成 2.4×10^5 km 的铺设，其中西气东输一线、二线、三线和中俄东线为代表的高钢级管道就有 4×10^4 km[8-10]。国家重大工程"西气东输"工程是目前世界上 X80 管线钢用量最大、铺设里程最长的管线工程[3,11]，该工程几乎途经我国全部地形、地貌和气象单元，这些因素对管线钢的长周期安全运行将带来极大影响。因此，迫切需要对油气输送管道外腐蚀实施控制，尤其应该开展 X80 管线钢在我国西气东输工程沿线各种典型土壤环境下的服役安全性研究和数据积累工作。

土壤介质引起的应力腐蚀(SCC)是长输管道服役过程中最大的安全隐患之一[12,13]。随着服役时间的增加，埋地管道普遍存在外部涂层破损和剥离缺陷，在外加电位和土壤介质的共同作用下，将会发生不同 pH 值土壤环境下的 SCC，导致高强管线钢存在严重的 SCC 风险[14]。目前，国内外学者已经对管线钢在高 pH 值含高浓度 CO_3^{2-}/HCO_3^- 的涂层下滞留液

基金项目：国家重点研发计划(2016YFC0801204)和陕西省自然科学基金(2019JQ-937)。
作者简介：朱丽霞，女，1980 年生，博士生，高级工程师，研究方向为石油管材性能与服役安全，E-mail：zhulx@cnpc.com.cn。

(pH=8.0~12.5)和近中性 pH 值(pH=5.5~8.0)模拟溶液中的 SCC 进行了大量的研究[15-18]，还有一些国内学者研究了管线钢在我国实际土壤模拟溶液中的 SCC 敏感性[19-21]。轮南作为"西气东输"工程的起点，这里蒸发强烈，土壤次生盐渍化严重，以粗砂为主且含盐量较高，是我国西部地区典型的内陆盐土，对材料的腐蚀作用很大。然而，目前尚未系统开展 X80 管线钢在我国西部盐渍土壤环境下的研究工作。

本文以我国新疆轮南土壤模拟溶液为实验介质，研究了外加电位对 X80 钢母材及焊缝在轮南土壤中 SCC 行为与敏感性的影响，探究其腐蚀机理和规律，为管线的运行和管理提供理论及数据支持。

1 实验方法

实验所用试样是从 X80 螺旋缝埋弧焊管上线切割而来，X80 钢的具体化学成分(质量分数,%)为：C 0.047，Mn 1.81，Si 0.19，P 0.01，S 0.0021，Cr 0.35，Mo 0.11，Nb 0.066，Ni 0.14，V 0.003，Ti 0.015，Cu 0.17，Fe 余量。X80 钢在室温下的力学性能为：屈服强度 604MPa，抗拉强度 727MPa。断后伸长率为 38%。

选取新疆轮南地区土壤环境为模拟研究介质，依据轮南土壤的主要理化数据配制的模拟溶液成分(质量分数,%)为：Cl^- 0.336，SO_4^{2-} 0.155，HCO_3^- 0.028，溶液 pH 值为 7.28，用分析纯 NaCl，Na_2SO_4、$NaHCO_3$ 及去离子水配制，整个实验过程中，持续向溶液中通入 95% N_2+5% CO_2(体积分数)以维持近中性 pH 值环境，实验温度为 10℃。

采用极化曲线测试辅助分析不同外加电位下的 SCC 机理。采用 M2273 电化学综合测试系统进行极化曲线的测试，测试采用三电极体系，工作电极为 X80 钢试样，参比电极为饱和甘汞电极，辅助电极为 Pt 片，电位测量范围：$-500mV$(vs 自腐蚀电位 E_{corr})~ $+1.6V$，扫描速率采用快速扫描(50mV/s)与慢速扫描(0.5mV/s)两种，裂纹尖端的电化学行为适合用快扫极化曲线模拟；非裂尖区域的电化学行为适合用慢扫极化曲线模拟[22,23]。

使用 MFDL100 型慢应变速率应力腐蚀试验机进行慢应变速率拉伸试验(SSRT)，拉伸应变速率是 $1×10^{-6}s^{-1}$。依据 GB/T 15970 制作试样，具体的试样尺寸与形状如图 1 所示。为了确保试样主受力方向在拉伸时与实际受力方向一致，X80 母材试样是沿着实际管道的环向进行取材的；X80 焊接试样从螺旋缝埋弧焊管上截取，其中焊缝区处于焊接试样标距中间位置。每个试样测试面经 SiC 水磨砂纸逐级打磨至 1000#，然后使用丙酮除油、蒸馏水清洗、无水酒精脱水，冷风吹干后放入干燥器中备用。外加电位的 SSRT 实验施加的电位分别为：$-1500mV$，$-1000mV$，$-850mV$ 和 $-500mV$，均相对于饱和甘汞电极(SCE)。实验温度为 10℃。采用扫描电子显微镜(SEM，JSM-6390A)观察试样断口横截面及侧面的形貌。

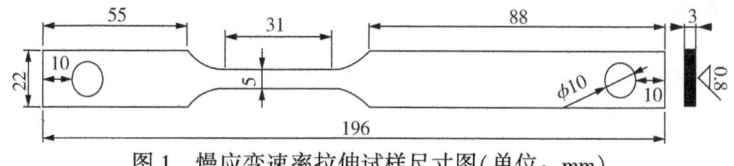

图 1 慢应变速率拉伸试样尺寸图(单位：mm)

SCC 敏感性可使用断面收缩率损失系数 I_ψ 来表征，其表达式如下：

$$I_\psi = (1-\psi/\psi_0)×100\% \tag{1}$$

式中：ψ 和 ψ_0 分别表示试样在腐蚀介质与空气中的断面收缩率。通常情况下，材料—

介质体系的 SCC 敏感性随 I_ψ 增大而增强。

2 结果与讨论

2.1 SSRT 拉伸实验

图 2 是 X80 管线钢母材及焊缝在空气中与轮南土壤模拟溶液中不同外加电位下的 SSRT 曲线。可见，X80 管线钢焊缝的 SSRT 试样均在焊接热影响区发生断裂。除了 X80 钢母材在 −850mV 电位下的延伸率高于空气中的，其他电位下的延伸率均明显低于空气中的，并且对比焊缝和母材的 SSRT 曲线可知，焊缝区域的延伸率较母材低，表明 X80 钢母材和焊缝在轮南土壤模拟溶液中具有一定的 SCC 敏感性。

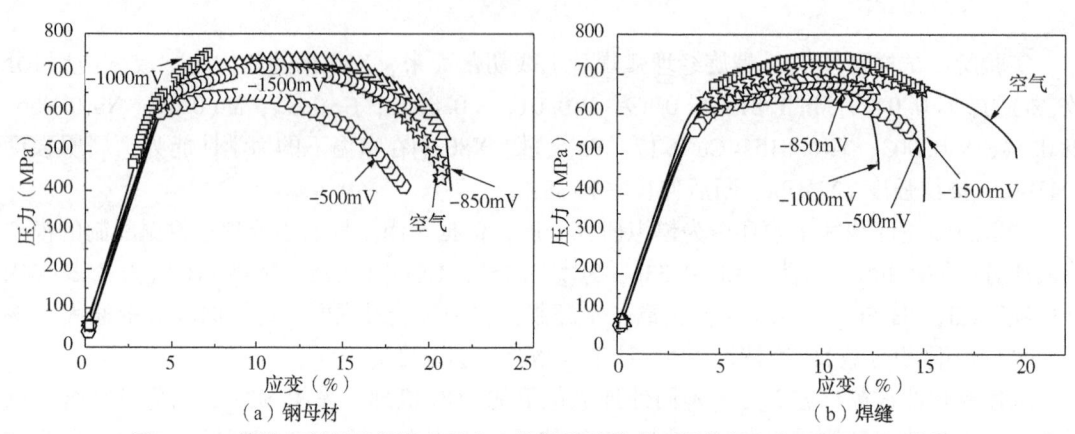

（a）钢母材

（b）焊缝

图 2 X80 管线钢母材及焊缝在轮南土壤模拟溶液中不同外加电位下的 SSRT 曲线

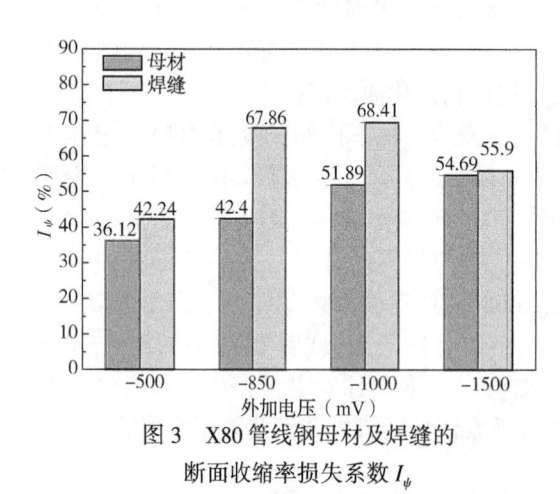

图 3 X80 管线钢母材及焊缝的断面收缩率损失系数 I_ψ

根据拉断后试样的测量数据，计算得到的 X80 管线钢母材及焊缝在轮南土壤模拟溶液中的 I_ψ 如图 3 所示。可见，X80 钢母材及焊缝基本上都具有较明显的 SCC 敏感性；且 X80 钢焊缝在不同电位下的 SCC 敏感性指标 I_ψ 均高于母材的，说明 X80 钢焊缝的 SCC 敏感性高于母材，这可能与焊缝组织相变和冶金反应有关；X80 钢母材的 I_ψ 随外加电位的负移而逐渐增大，说明外加电位的降低可使 X80 母材的 SCC 敏感性增加；X80 钢焊缝的 I_ψ 随外加电位的负移先增加后降低，在 −1000mV 时达到最高，其 SCC 敏感性排序为 $I_\psi(-500\text{mV}) < I_\psi(-1500\text{mV}) < I_\psi(-850\text{mV}) < I_\psi(-1000\text{mV})$，说明外加电位保护并不能有效抑制 X80 钢的 SCC。

2.2 断口及裂纹形貌观察

为了进一步研究不同外加电位对 X80 钢母材和焊缝 SCC 敏感性的影响，通过 SEM 观察了 X80 钢在不同电位下断口正面与侧面的微观形貌，如图 4 和图 5 所示。X80 钢宏观断口在空气中出现了显著的颈缩现象，从图 4(a) 和图 4(b) 可以看出，其微观断口表现为等轴韧窝与韧窝间微孔洞相间而生，表现为典型的韧窝微孔型的韧性断裂特征，这是由于 X80 钢在空气中的慢拉伸过程中产生了显著的塑性变形，当应力高于钢的屈服极限 σ_s 后，材料内部缺陷在相界、晶界、亚晶界和缺陷等部位形成位错塞积群，在应力集中处形成微孔洞，这些微孔洞随形变增加而相互吞并变大，最后导致颈缩和断裂的发生。X80钢母材和焊缝在不同电位下的宏观断口均发生了一定程度的颈缩，但其颈缩比例远低于空气中的，断口微观形貌主要由浅平小韧窝、微孔洞、撕裂棱和准解理面组成，表现为韧—脆混合断裂特征。在-500mV电位下，X80钢母材和焊缝的断口形貌主要以浅平小韧窝为主，但韧窝特征不如空气中的明显，且断口中间区域出现了一些准解理小刻面，表明此时X80钢已表现出一定的SCC敏感性；在-850~-1500mV电位下，X80钢母材和焊缝的断口形貌以准解理和解理特征为主，在脆性区之间存在少量的扁平小韧窝形貌，表明随外加阴保电位的负移，X80钢SCC敏感性进一步增加。断口形貌特征与SCC敏感性测试结果相一致。

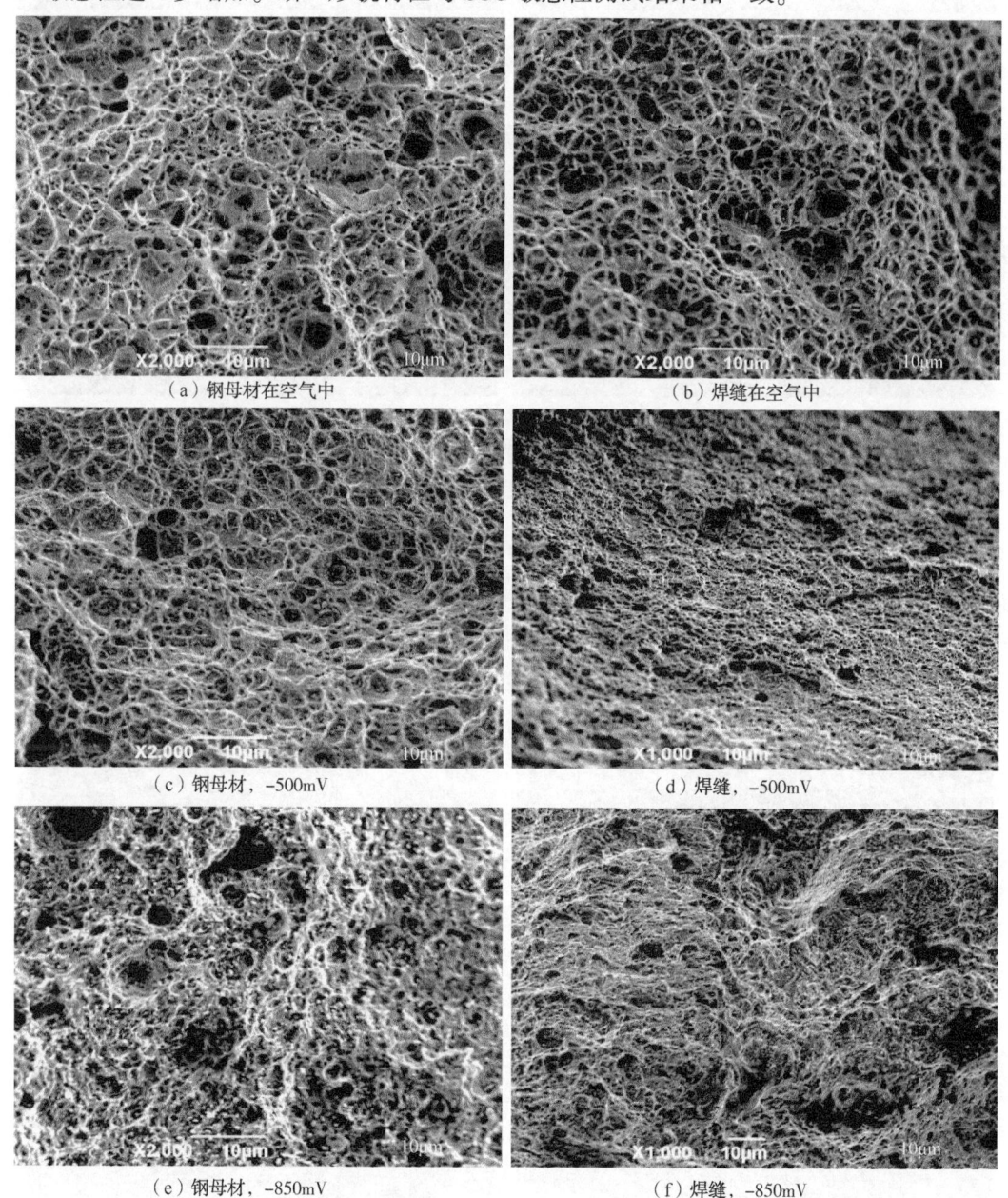

（a）钢母材在空气中　　　　　　　　　　　（b）焊缝在空气中

（c）钢母材，-500mV　　　　　　　　　　　（d）焊缝，-500mV

（e）钢母材，-850mV　　　　　　　　　　　（f）焊缝，-850mV

图4　X80管线钢母材及焊缝在空气中和不同外加电位下的断口SEM形貌

（g）钢母材，-1000mV

（h）焊缝，-1000mV

（i）钢母材，-1500mV

（j）焊缝，-1500mV

图4 X80管线钢母材及焊缝在空气中和不同外加电位下的断口SEM形貌（续）

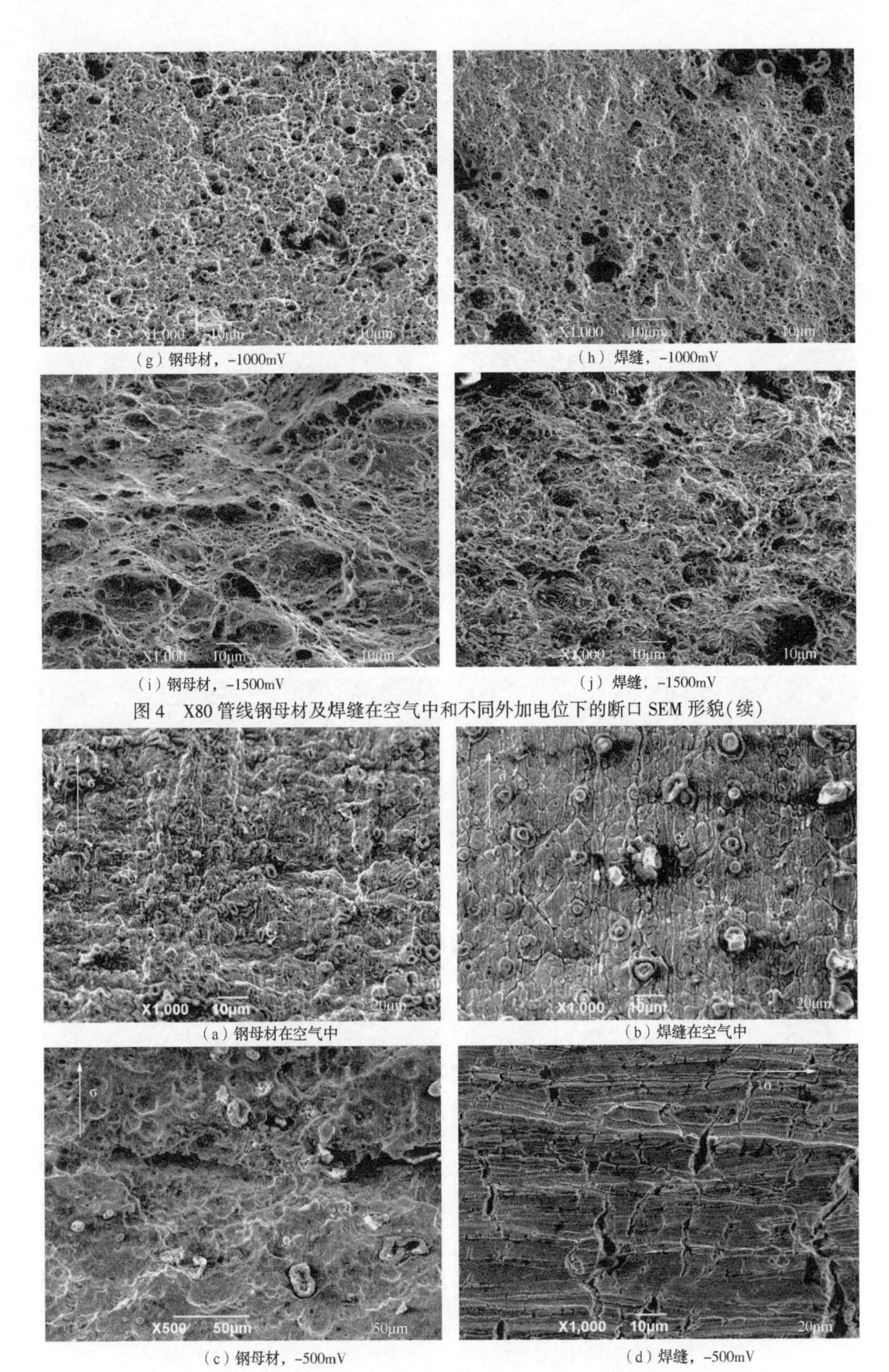

（a）钢母材在空气中

（b）焊缝在空气中

（c）钢母材，-500mV

（d）焊缝，-500mV

图5 X80管线钢母材及焊缝在空气中和不同外加电位下的断口侧面SEM形貌

(e) 钢母材, -850mV (f) 焊缝, -850mV

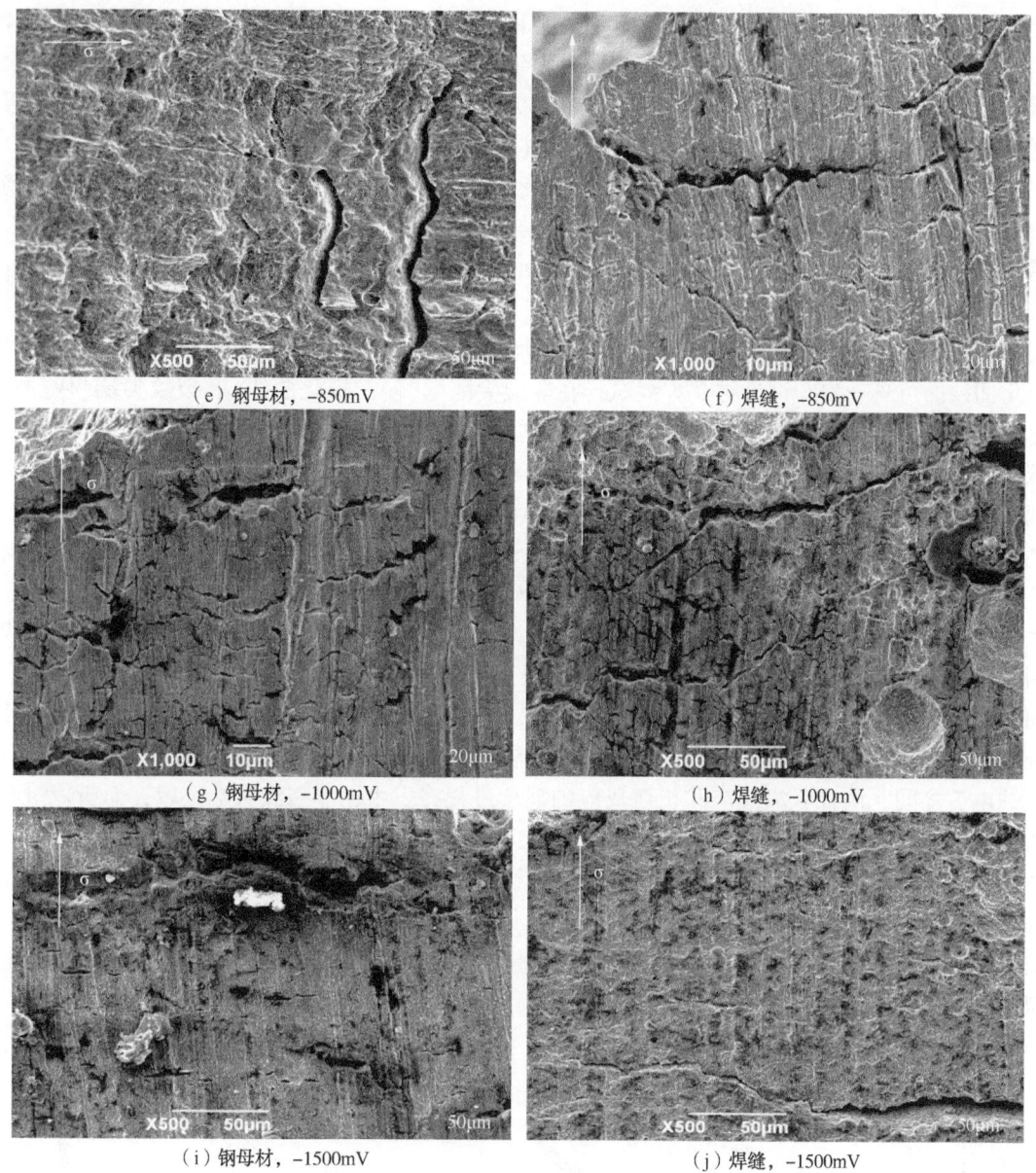

(g) 钢母材, -1000mV (h) 焊缝, -1000mV

(i) 钢母材, -1500mV (j) 焊缝, -1500mV

图 5　X80 管线钢母材及焊缝在空气中和不同外加电位下的断口侧面 SEM 形貌(续)

一般认为,在腐蚀性介质中拉伸试样断口侧面存在微裂纹(二次裂纹),则表明该材料对 SCC 敏感[24]。图 5 是 X80 钢在空气中和不同外加电位下的断口侧面 SEM 形貌图。可见,X80 钢母材与焊缝在空气中的 SSRT 断口侧面均发生了显著的塑性变形,并无二次裂纹产生,因此不具有 SCC 敏感性[图 5(a)和图 5(b)]。在不同外加电位条件下,X80 钢母材和焊缝断口侧面仅出现了少量的塑性变形,且侧表面上均出现了不同程度的二次裂纹,这些二次裂纹基本上均与拉伸载荷方向垂直。在-500mV 电位下,X80 钢母材断口侧面二次裂纹较少,但出现大量溃疡状的点蚀坑,表明此种状态下 SCC 的萌生受阳极溶解(AD)过程的影响较大,X80 钢焊缝断口侧面也出现了一些小的腐蚀坑,同时出现了大量细小的裂纹,表明 X80 钢在-500mV 电位下具有一定程度的 SCC 敏感性[图 5(c)和图 5(d)]。在-850mV 电位下,X80 钢母材和焊缝断口侧面腐蚀程度较轻微,但二次裂纹的长度和密度明显增加,表明此种情况下X80钢的SCC敏感性进一步增加[图5(e)和图5(f)];在-1000mV和-1500mV电位下,X80钢母材和焊缝断口侧面的二次裂纹长度/密度进一步增加,与-850mV时相比,该电位下试样表面的AD过程进一步被抑制,析氢作用增强,表明析氢作用可促进X80钢母材及焊缝的SCC过程,增加SCC敏感性[图5(g)至图5(j)]。

由图5可知,X80钢母材与焊缝的二次裂纹均沿直线方向扩展,与管线钢在近中性环境中的SCC特征相一致。拉伸时裂纹尖端发射的位错可形成位错反塞积群,在裂纹尖端和位错反塞积群之间形成无位错区,SCC裂纹可在该处不连续形核并扩展[21]。

2.3 极化曲线测量

图6是X80钢母材及焊缝在轮南土壤模拟溶液中分别进行快/慢速率扫描得到的电化学极化曲线。可见,X80钢母材及焊缝试样在非裂尖区域(慢扫极化曲线)和裂纹尖端(快扫极化曲线)的电化学行为相似,X80钢母材及焊缝在非裂尖区域一直处于活化状态,说明该区域的阴极和阳极均表现为活化控制特征;而X80钢母材及焊缝在裂尖区域的阳极曲线中出现了轻微的活化—钝化转变区和稳定钝化区,说明裂尖区域的阳极首先受活性溶解过程控制、接着发生了轻微钝化、最后钝化膜破裂导致裂尖进一步溶解,其开裂机理为裂尖阳极溶解—膜破裂机理;根据快扫极化曲线和慢扫极化曲线零电流电位的差异可将外加电位分为三个区域:(1)在慢扫极化曲线的自腐蚀电位以上,快扫与慢扫的极化曲线均为阳极曲线,表明SCC裂纹萌生(即非裂尖区域的电化学过程)受阳极过程控制,其SCC机制为AD过程,而裂纹扩展(即裂尖区域的电化学过程)的SCC机制为裂尖AD-膜破裂过程。从该电化学特征判断,-500mV测试条件处于AD-膜破裂机制电位区域。(2)在快扫与慢扫极化曲线的自腐蚀电位之间,阴极吸氧反应和析氢反应的混合过程将发生在非裂纹尖端区域,而析氢过程将促进SCC微裂纹的成核与扩展,进一步强化非裂尖区域的氢脆(HE)机制,而裂尖区域产生的非稳态AD过程直接促进了裂尖的阳极溶解与扩展,说明该区域的SCC机制为AD+HE的混合机制[23]。(3)在快扫极化曲线自腐蚀电位(大约-800mV)以下,裂尖与非裂尖区域的电化学过程均为阴极析氢过程,表明SCC机制以HE过程为主,X80钢的HE敏感性在此电位范围内显著增强[23]。从该电化学特征判断,-850mV,-1000mV和-1500mV这3种测试条件均处于HE机制电位区域。

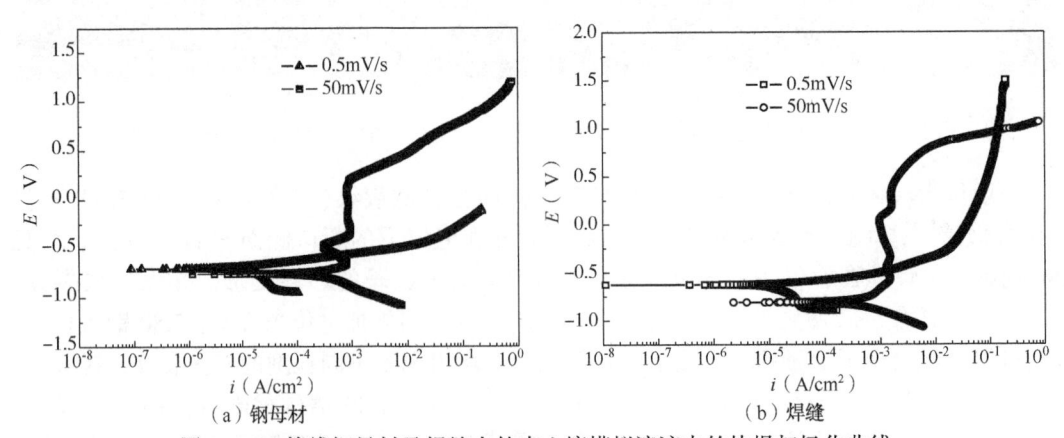

(a) 钢母材　　(b) 焊缝

图6　X80管线钢母材及焊缝在轮南土壤模拟溶液中的快慢扫极化曲线

2.4 分析讨论

通常将长输管线埋在地下1.5m以下的土层中,该土层属于低O_2或无O_2的服役环境,故在本次实验过程中,通过向模拟溶液通入高纯 N_2 来进行除 O_2 处理,这样就可以忽略 O_2 的去极化过程,因此 X80 钢表面可能会发生如下电极反应[25,26]:

阳极反应: $\quad\quad\quad\quad\quad\quad Fe \longrightarrow Fe^{2+} + 2e \quad\quad\quad\quad\quad\quad$ (2)

阳极次级反应: $\quad\quad\quad\quad Fe^{2+} + CO_3^{2-} \longrightarrow FeCO_3 \quad\quad\quad\quad$ (3)

阴极反应: $\quad\quad\quad\quad\quad\quad 2H^+ + 2e \longrightarrow H_2 \quad\quad\quad\quad\quad\quad$ (4)

$\quad\quad\quad\quad\quad\quad\quad 2H_2CO_3 + 2e \longrightarrow 2HCO_3^- + H_2 \quad\quad\quad\quad$ (5)

$\quad\quad\quad\quad\quad\quad\quad 2HCO_3^- + 2e \longrightarrow 2CO_3^{2-} + H_2 \quad\quad\quad\quad$ (6)

在外加阳极电位-500mV下,因阳极溶解作用导致 X80 钢表面出现了大量的点蚀坑(图5b)。这些蚀坑底部由于受到应力集中作用而成为潜在的裂纹形核处,进而影响 X80 钢的 SCC 过程[25]。然而,由于阳极溶解作用致使萌生的裂纹被溶解掉而不能产生有效的扩展,因此,-500mV 电位在一定程度上又降低了 X80 钢 SCC 敏感性。由图6可知,X80 钢在-850mV,-1000mV 和-1500mV 这3种阴极电位条件下电化学行为均受阴极析氢过程控制,因此,当外加阴极电位负移,X80 钢表面析氢反应逐渐加强。高的阴极电流使进入钢中的 H 增加,从而降低了钢的韧性,进而促进氢致开裂(HIC)的发生。可见,在-800mV 电位以下,阴极反应产生的 H 增多,X80 钢的 SCC 敏感性增大。

此外,SCC 扩展由初始扩展阶段与快速扩展阶段组成,同时还受到应力与电化学反应的联合作用[23,27]。在裂纹扩展初期,点蚀坑成为裂纹形核源,此时的裂纹扩展缓慢,主要受 AD 与应力的共同控制;裂纹扩展到一定尺寸后,逐渐进入快速扩展期,此时主要受 HE 和应力的共同作用,呈现出脆性断裂特征[28](图4和图5)。由图3至图5还可看出,X80 钢焊缝的 SCC 敏感性高于母材,随着外加电位由阳极电位负移到阴极电位,裂纹扩展得更快。

3 结论

(1) X80 管线钢及其焊缝在轮南土壤模拟溶液中表现出了一定的 SCC 敏感性。在同一外加电位下,X80 钢焊缝的 SCC 敏感性高于母材的,这可能与焊缝组织相变和冶金反应有关。

(2) 外加电位对 X80 管线钢在轮南土壤模拟溶液中的 SCC 敏感性与开裂机理具有显著影响。在-500mV 阳极电位范围内,X80 钢的 SCC 机理为裂尖 AD-膜破裂机制,在-800mV 电位以下(-850mV,-1000mV 和-1500mV),由于 HE 作用在 SCC 过程中产生的影响更大,阴极析氢反应会促进钢的氢致开裂,X80 钢 SCC 敏感性显著增加。

参 考 文 献

[1] Yan M C, Xu J, Yu L B, et al. EIS analysis on stress corrosion initiation of pipeline steel under disbonded coating in near-neutral pH simulated soil electrolyte[J]. Corros Sci., 2016, 110: 23.

[2] Liu Z Y, Li X G, Du C W, et al. Effect of inclusions on initiation of stress corrosion cracks in X70 pipeline steel in an acidic soil environment[J]. Corro. Sci., 2009, 51: 895-900.

[3] 李鹤林,吉玲康,田伟. 西气东输一、二线管道工程的几项重大技术进步[J]. 天然气工业,2010,30(4):1.

[4] Mohd M H, Paik J K. Investigation of the corrosion progress characteristics of offshore subsea oil well tubes [J]. Corrosion Science, 2013, 67: 130.

[5] 杜燕飞．华建敏：十二五末我国长输油气管道总里程超10万公里[DB/OL]．(2010-09-25)．

[6] Dang D N, Lanarde L, Jeannin M, et al. Influence of soil moisture on the residual corrosion rates of buried carbon steel structures under cathodic protection[J]. Electrochim. Acta, 2015, 176：1410.

[7] Caleyo F, Valor A, Alfonso L, et al. Bayesian analysis of external corrosion data of non–piggable underground pipelines[J]. Corrosion Science, 2015, 90：33.

[8] 人民网．油气管道智能化建设待提速[N]．中国能源报，2019-03-29(19)．

[9] 王小强，王保群，王博，等．我国长输天然气管道现状及发展趋势[J]．石油规划设计，2018，29(5)：1．

[10] 中国经济网．2025年我国油气管网规模将达到24万公里[DB/OL]．(2017-07-12)．

[11] 毕宗岳，刘海璋，牛辉．西气东输二线用X80管材及其焊接工艺[J]．焊接，2011，(11)：47．

[12] Yan M C, Sun C, Xu J, et al. Role of Fe oxides in corrosion of pipeline steel in a red clay soil[J]. Corros. Sci., 2014, 80：309.

[13] Fang B Y, Atrens A, Wang J Q, et al. Review of stress corrosion cracking of pipeline steels in "low" and "high" pH solutions[J]. J. Mater Sci., 2003, 38：127.

[14] Chen X, Li X G, Du C W, et al. Effect of cathodic protection on corrosion of pipeline steel under disbonded coating[J]. Corros. Sci., 2009, 51：2242.

[15] Javidi M, Horeh S B. Investigating the mechanism of stress corrosion cracking in near-neutral and high pH environments for API 5L X52 steel[J]. Corros. Sci., 2014, 80：213.

[16] 杨东平，胥聪敏，罗金恒，等．0.8设计系数用X80管线钢在近中性pH溶液中的应力腐蚀开裂行为[J]．材料工程，2015，43：89．

[17] Cui Z Y, Liu Z Y, Wang L W, et al. Effect of plastic deformation on the electrochemical and stress corrosion cracking behavior of X70 steel in near-neutral pH environment[J]. Mater. Sci. Eng., 2016, A677：259.

[18] 胥聪敏，罗金恒，周勇，等．SRB对X100管线钢在西北盐渍土壤中应力腐蚀开裂行为的影响[J]．材料热处理学报，2016，37(5)：82．

[19] 罗金恒，胥聪敏，杨东平．SRB作用下X100管线钢在酸性土壤环境中的应力腐蚀开裂行为[J]．中国腐蚀与防护学报，2016，36：321．

[20] 刘智勇，王长朋，杜翠薇，等．外加电位对X80管线钢在鹰潭土壤模拟溶液中应力腐蚀行为的影响[J]．金属学报，2011，47：1434．

[21] Asher S L, Leis B N, Colwell J, et al. Investigating a mechanism for transgranular stress corrosion cracking on buried pipelines in near-neutral pH environments[J]. Corrosion, 2007, 63：932.

[22] 苑洪钟，刘智勇，李晓刚，等．外加电位对X90钢及其焊缝在近中性土壤模拟溶液中应力腐蚀行为的影响[J]．金属学报，2017，53：797．

[23] 褚武扬．断裂与环境断裂[M]．北京：科学出版社，2000：109．

[24] Egbewande A, Chen W X, Eadie R, et al. Transgranular crack growth in the pipeline steels exposed to near-neutral pH soil aqueous solutions: Discontinuous crack growth mechanism[J]. Corros Sci, 2014, 83：343.

[25] Cheng Y F, Niu L. Mechanism for hydrogen evolution reaction on pipeline steel in near-neutral pH solution[J]. Electrochem Commun, 2007, 9：558.

[26] Marshakov A I, Ignatenko V E, Bogdanov R I, et al. Effect of electrolyte composition on crack growth rate in pipeline steel[J]. Corros. Sci., 2014, 83：209.

[27] Barbalat M, Lanarde L, Caron D, et al. Electrochemical study of the corrosion rate of carbon steel in soil: Evolution with time and determination of residual corrosion rates under cathodic protection[J]. Corros. Sci., 2012, 55：246.

本论文原发表于《中国腐蚀与防护学报》2020年第40卷第4期。

页岩气输送用转角弯头内腐蚀减薄原因分析

朱丽霞[1,2]　罗金恒[1]　李丽锋[1]　刘　畅[3]　孙明楠[3]

(1. 中国石油集团石油管工程技术研究院·石油管材及装备材料服役行为与结构安全国家重点实验室；2. 西安理工大学材料科学与工程学院；3. 西南油气田分公司安全环保与技术监督研究院)

摘　要：针对某页岩气输送平台用转角弯头发生严重内腐蚀减薄的问题开展原因分析，明确腐蚀类型及机理，指导防腐处理，提高转角弯头的服役安全性。以页岩气输送用转角弯头为研究对象，针对转角弯头内腐蚀减薄行为开展基础研究，通过宏观观察及尺寸测量分析内腐蚀的腐蚀形貌及分布，并进行理化检测、微观观察、物相分析，探究腐蚀产物，综合分析转角弯头内腐蚀减薄的原因。宏观分析发现，转角弯头内壁外弧侧与中性区过渡区域有壁厚发生突变而产生的腐蚀台阶，最大壁厚减薄率达 63.4%。电子显微形貌与金相分析表明，弯头内壁的腐蚀坑呈纵深发展，逐层剥离，腐蚀产物疏松、形貌多样，且可观察到细菌形貌。腐蚀产物的能谱及 XRD 分析发现，管体内壁的腐蚀产物主要是 FeS、Fe_2O_3、$FeCO_3$ 等，内腐蚀可能与 CO_2、H_2S、SRB 等有关。弯头腐蚀减薄是硫酸盐还原菌(SRB)-CO_2 腐蚀协同作用的结果，SRB 的存在对 CO_2 腐蚀起催化作用，此外，Cl^- 对腐蚀产物膜的破坏和弯头外弧侧的冲刷加速了腐蚀作用。建议确定 SRB 细菌来源，以便有效投放杀菌剂，同时对管线内壁定期进行清理，避免菌落长期附着于管体内壁。此外，建议添加多级气液分离装置，严格控制气相中的含水量。

关键词：转角弯头；腐蚀减薄；硫酸盐还原菌；内腐蚀；协同作用

集输管道中的腐蚀失效一直是影响国内外整个油气传输工艺的主要问题，不仅直接影响着管道的运行成本[1]，而且在油气工业中具有潜在的灾难性后果和巨大的经济损失[2-6]。国外研究发现，CO_2 腐蚀和微生物腐蚀在油气管道的腐蚀失效中起着重要的作用，Yaro[7] 和 Papavinasam[8] 等人对管道 CO_2 腐蚀进行了广泛的研究和建模，采用失重法、动电位极化法和腐蚀表面表征法研究了低碳钢在含 CO_2 气田水中的腐蚀速率，发现腐蚀速率随温度、醋酸浓度和旋转速度的增加而增加，随溶液 pH 值的增加而降低。同时对于管线钢在硫酸盐还原菌(SRB)生物膜下的腐蚀研究发现，该生物膜在模拟的 CO_3^{2-} 饱和油田采出水培养基中形成，预培养较长时间的生物膜增加了钢的腐蚀速率，SRB 促进了钢的局部腐蚀，SRB 生物膜一旦停用，将不再影响钢的腐蚀[9,10]。我国学者针对油气田集输管道的研究发现，腐蚀是导致

基金项目：国家重点研发计划项目(2016YFC0801204)；中国石油天然气股份有限公司重大科技专项(2016E-0610)。

作者简介：朱丽霞(1980—)，女，博士研究生，高级工程师，主要研究方向为油气管道的性能及服役安全。邮箱：zhulx@cnpc.com.cn。

集输管道失效的首要原因[11-14]，其中，内腐蚀引起的事故往往具有突发性和隐蔽性[15]，后果相对严重。对集输管道的内腐蚀影响因素进行研究，探讨不同影响因素下的腐蚀机理，对于提高管道的防腐能力，保证管道的服役安全具有十分重要的意义。

以上研究中对于输气管道的腐蚀机理已经进行了深入探究，但是对于页岩气输送平台用转角弯头发生严重内腐蚀减薄的问题研究很少。通过上述研究可以发现，转角弯头内腐蚀减薄行为与CO_2和微生物还原菌有较大的相关性，要深入了解页岩气输送平台用转角弯头发生严重内腐蚀减薄的原因，需要对腐蚀形貌和腐蚀产物进行研究。我国四川某页岩气集气平台频繁发生管道侧漏及内腐蚀减薄失效事故，弯头位于集气平台的站内竖管转角连接处，规格为 φ76mm×7.62mm，材质为 L245N。输送介质为页岩气，气相中含的有害气体为 CO_2，不含 O_2 和 H_2S。此外，页岩气中含有少量水分，水相成分中的阳离子主要为 K^+、Na^+、Ca^{2+} 以及少量的 Mg^{2+}，阴离子主要为 Cl^-、SO_4^{2-}、HCO_3^-，pH 值为 6.5。本文针对一起典型的转角弯头内腐蚀减薄失效，进行宏观分析及尺寸测量、理化检验、微观形貌分析、物相分析等，旨在探明该转角弯头腐蚀减薄的原因，并针对性地提出防腐措施，确保管线服役安全。

1 试验

试验材料为该页岩气集气平台发生腐蚀减薄的转角弯头，按如图1所示位置，采用超声波测厚仪对其进行壁厚测量，然后将弯头试样沿 0°~180°轴线剖开（图2），对其内壁进行宏观观察，并结合输送介质组分对弯头的腐蚀特征进行初步判断。利用扫描电子显微镜及能谱分析仪对内壁腐蚀产物进行形貌观察和微区成分分析，并取内壁腐蚀产物研磨成粉末进行 XRD 分析。

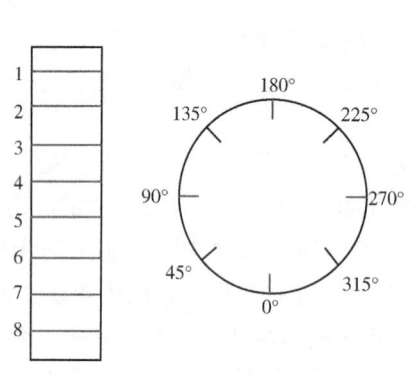

图 1 壁厚测量位置    图 2 弯头内壁腐蚀形貌

2 结果及分析

2.1 宏观分析

弯头 3 个截面的壁厚测量结果见表1，可以看出，弯头135°~225°附近壁厚明显减薄，最大减薄率为63.4%。壁厚减薄区平均壁厚为3.96mm，结合服役年限可粗略计算出平均腐蚀速率为1.20mm/a。

表1 弯头几何尺寸测量结果

位置	0°(内弧侧)	45°	90°(中性区内弧侧)	135°	180°(外弧侧)	225°	270°(中性区外弧侧)	315°
位置1	6.84	6.48	6.38	5.49	4.29	5.74	6.38	7.18
位置2	7.06	6.24	6.41	3.20	4.35	3.46	6.49	7.31
位置3	6.73	6.01	6.36	2.83	3.65	2.60	6.08	6.89
平均值	6.88	6.24	6.38	3.84	4.10	3.93	6.32	7.13

宏观分析显示，弯头内壁产生了明显的腐蚀坑，腐蚀产物可见分层特征，外层腐蚀产物为黄褐色，内层腐蚀产物呈墨绿色，如图3所示。此外，外弧区域腐蚀坑较密，局部区域腐蚀坑连成一片，形成长条状腐蚀坑，且呈线状分布。外弧侧与中性区过渡处可见腐蚀坑有台阶状特征，使得壁厚在此区域呈现突变，形貌如图4所示。

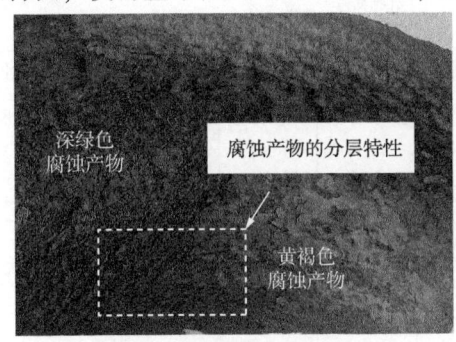

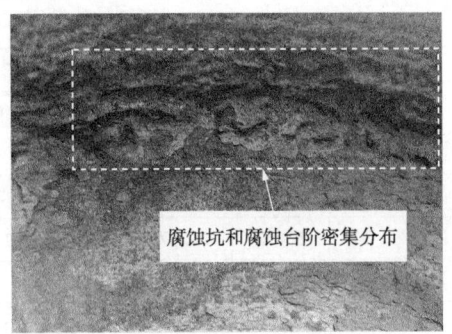

图3 腐蚀产物分层特征　　图4 腐蚀坑密集分布及腐蚀台阶

2.2 腐蚀形貌及成分分析

用扫描电镜观察内表面的腐蚀坑形貌，如图5所示。腐蚀坑纵深发展，腐蚀产物形态疏松，具有逐层剥离的特点。腐蚀坑截面形貌如图6所示，内表面腐蚀坑底腐蚀产物较厚，且连续分布。试样内表面可观察到多种腐蚀产物形貌，如图7所示。表层腐蚀产物按形貌分成Ⅰ、Ⅱ、Ⅲ类，三类腐蚀产物的主要元素含量见表2。除了Fe、C、O、S等元素以外，腐蚀产物Ⅰ和Ⅲ还存在有Cl、Ca等元素，与水相中的化学成分吻合。腐蚀坑底的腐蚀产物存在杆状物体，其特征如图8所示，是细菌的典型形貌。

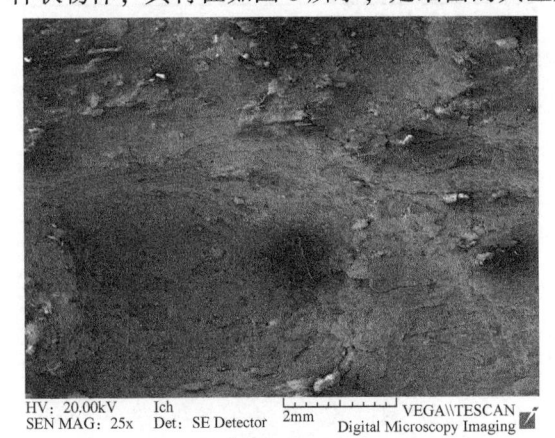

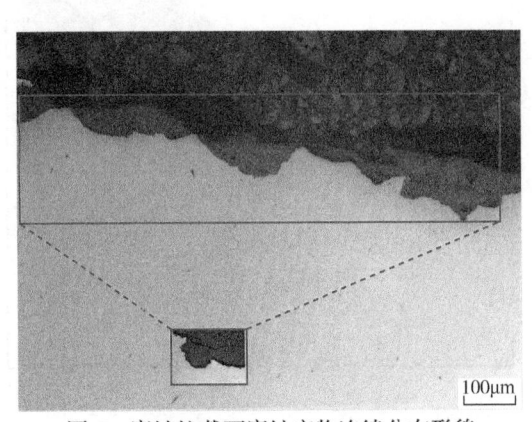

图5 腐蚀坑纵深发展形貌　　图6 腐蚀坑截面腐蚀产物连续分布形貌

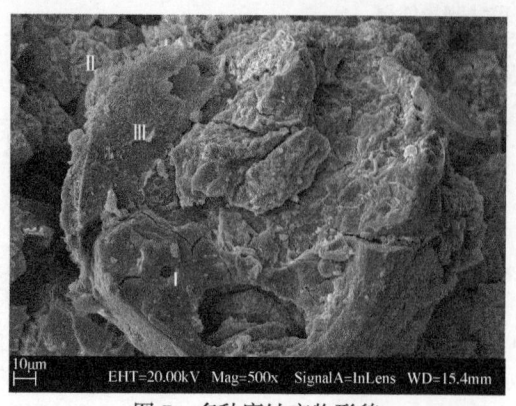

图7 多种腐蚀产物形貌

图8 腐蚀坑底短杆状微生物形貌

表2 各类腐蚀产物能谱分析结果

元素	腐蚀产物Ⅰ 质量分数(%)	腐蚀产物Ⅰ 原子分数(%)	腐蚀产物Ⅱ 质量分数(%)	腐蚀产物Ⅱ 原子分数(%)	腐蚀产物Ⅲ 质量分数(%)	腐蚀产物Ⅲ 原子分数(%)	内表面腐蚀产物 质量分数(%)	内表面腐蚀产物 原子分数(%)	管底腐蚀产物 质量分数(%)	管底腐蚀产物 原子分数(%)
C	1.82	4.41	12.13	28.92	7.34	22.62	4.65	13.44	3.07	6.34
O	33.24	60.50	29.07	52.01	12.18	28.20	24.70	53.63	45.36	70.29
S	0.76	0.79	9.35	8.35	1.10	1.27	8.31	9.00	1.07	0.83
Fe	61.17	31.89	1.39	0.71	65.63	43.53	22.10	13.75	49.80	22.11
Ca	0.75	0.54	—	—	0.41	0.38	—	—	0.70	0.43
Cl	2.27	1.87			0.52	0.54	—	—		
Ba			48.05	10.01	12.82	3.46	40.24	10.18		

为了进一步分析腐蚀产物的构成,分别对表层及腐蚀坑底部的腐蚀产物进行能谱分析(图9,表2)。内表面腐蚀产物的主要元素为 C、O、S、Ba、Fe,含量最高的为 Ba、Fe、C、O 四种元素,推测可能的腐蚀产物为铁的氧化物及 $FeCO_3$、$BaSO_4$。从腐蚀产物推断腐蚀类型可能存在二氧化碳腐蚀和氧腐蚀,但由于输送介质中未发现 Ba,推测其来源于钻井液。

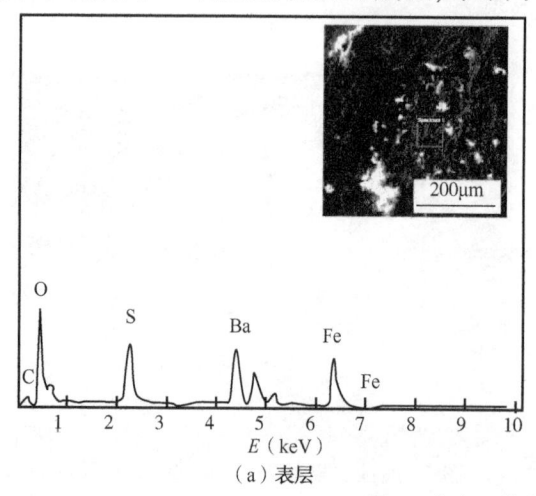

(a)表层

(b)腐蚀坑底部

图9 内表面能谱分析位置及谱图

2.3 腐蚀产物 XRD 分析

能谱分析确定腐蚀产物中含有 Fe、C、O、S 等元素,为了进一步确定腐蚀产物的组成,刮取管体内壁腐蚀产物(图10),研磨均匀后进行 X 射线衍射分析,结果如图11所示。结果表明,管体内壁的腐蚀产物主要是 FeS、Fe_2O_3、$FeCO_3$、SiO_2、$BaSO_4$ 等。从衍射强度可知,$FeCO_3$、Fe_2O_3、$BaSO_4$ 含量较高。一般来说,$FeCO_3$ 腐蚀产物是 CO_2 腐蚀的典型腐蚀产物[16-17],Fe_2O_3 是氧腐蚀的典型腐蚀产物[18-19],$BaSO_4$ 来源于钻井液。由于输送介质中不含氧气,因此可以排除氧腐蚀可能,Fe_2O_3 可能来源于暴露在空气条件下产生的氧腐蚀。另外,XRD 图中的硫化物与能谱中的 S 元素相对应,也说明腐蚀产物的形成与 H_2S 有关。由此推断,腐蚀可能与 CO_2、H_2S 等有关。

图10 内壁腐蚀产物

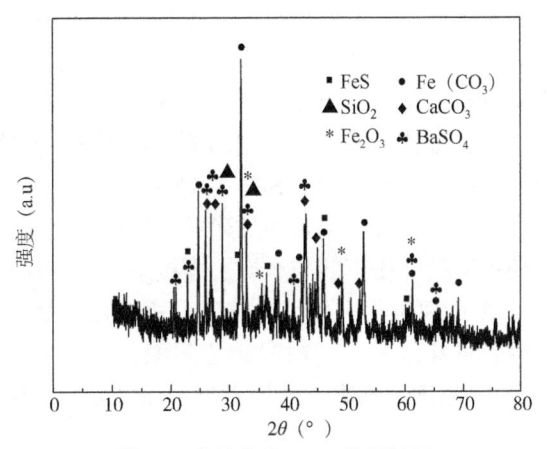

图11 腐蚀产物 XRD 分析结果

2.4 腐蚀原因综合分析

2.4.1 原因分析

宏观及微观分析结果表明,转角弯头内壁均发生不同程度的腐蚀减薄,外弧侧腐蚀坑密集分布,壁厚明显减薄,最大减薄率为63.4%,且外弧侧与中性区过渡处可见明显的壁厚突变。腐蚀产物的能谱及 XRD 分析表明,内腐蚀可能与二氧化碳、硫化氢等有关。以下结合腐蚀产物分层形貌、输送介质成分等对腐蚀原因进行进一步分析。

(1) CO_2 腐蚀。输送介质中,CO_2 的摩尔分数约为1.4%。输送压力为3.5~6.3MPa,可计算出 CO_2 分压约为0.049~0.088MPa。研究表明[20-22],CO_2 腐蚀速率和 CO_2 分压成正比,可分为以下三种情况:CO_2 分压<0.021MPa,无腐蚀,不需防 CO_2 腐蚀;CO_2 分压为0.021~0.21MPa,应考虑防 CO_2 腐蚀;CO_2 分压>0.21MPa,CO_2 腐蚀严重,需采用特殊防腐措施。由此可见,弯头内的输送介质中 CO_2 分压在 CO_2 腐蚀的范围,CO_2 分压越高,腐蚀反应速率越大。

(2) 微生物腐蚀。XRD 检测发现,腐蚀产物中含有 FeS,由于输送介质中不含 H_2S,所以判断 H_2S 是在服役过程中产生的。SEM 观察表明,弯头内表面腐蚀坑底的腐蚀产物中存在短杆状细菌,是硫酸盐还原菌(SRB)的典型形貌。为了进一步确定 SRB 的形成,使用稀释的0.005mol/L 盐酸清洗腐蚀产物,发现内表面的元素 S 主要以 SO_4^{2-} 的形式存在,少量的以 S^{2-} 形式存在。由于输送介质中不含 S^{2-},由 SRB 消耗 SO_4^{2-},从而将硫酸盐转化为 S^{2-},并

促进酸性环境中的 H_2S 和 HS^- 形成。HS^- 在溶液中与铁离子反应，生成含铁硫化物，然后沉积在转角弯头内表面。含铁硫化物能有效地传输电子，因此，硫化物可以增强内表面与腐蚀产物之间的电子转移，从而加速弯头内表面的腐蚀。此外，H_2S 的离子化还会导致溶液酸化，进一步增强 SRB 的数量和活性，从而加快转角弯头内表面的腐蚀过程。文献资料显示[23]，硫酸盐还原菌（SRB）能够将硫酸根离子还原成 H_2S，有文献中给出 SRB 的繁殖条件：pH 为 6~9，温度为 30~50℃，有机物为养料，无氧环境，低流速等[24-26]。根据现场资料可知，输送介质水相 pH 值为 6.5，输送温度平均值在 30℃，无氧气。可以看出，管线的输送环境满足 SRB 细菌的繁殖条件。对平台水相中的细菌含量进行测量，发现取样点的硫酸盐还原菌数量达到 14000 个，已严重超标。SRB 将硫酸根离子还原成 S^{2-}，进而生成 H_2S，所以腐蚀产物中的 S 元素来源于硫酸盐还原菌。

（3）冲刷腐蚀。壁厚检测结果显示，弯头的外弧侧及周边 135°~225°附近壁厚明显减薄，使得外弧侧与中性区的壁厚呈现突变。在实际工况条件下，页岩气经管体向输送平台用转角弯头流动过程中，流体流速可以达到 8m/s 左右。较高流速导致流体在转角弯头处产生明显的湍流，湍流产生较大的湍动能，对转角弯头产生明显的冲蚀作用，导致外弧侧与中性区的壁厚减薄。可见，除了 SRB 腐蚀为主因之外，弯头外弧侧还受到输送介质的冲刷，切向作用力阻碍金属表面保护膜的形成，并破坏已形成的保护膜。保护膜被剥落后，露出新的金属，使得腐蚀介质不断与金属表面接触，进入腐蚀—冲蚀循环，导致腐蚀加速。

（4）Cl^- 的腐蚀促进作用。能谱分析显示，腐蚀产物中含有 Cl 元素。由于 Cl^- 具有极强的穿透性，可穿过金属腐蚀产物膜间隙，形成点蚀坑。随着介质内的 Cl^- 不断向蚀坑内迁移，坑内 Cl^- 浓度远远高于金属平坦表面区域，基体中的 Fe 不断腐蚀溶解，使点蚀坑迅速扩展，对局部腐蚀起到促进作用。

2.4.2 腐蚀机理分析

在 pH 值大于 6 的含 SRB-CO_2 体系中，初期主要表现为 CO_2 腐蚀，其阴极反应为[27]

$$2H_2O + 2e \longrightarrow 2OH^- + H_2$$

阳极反应为

$$Fe \longrightarrow Fe^{2+} + 2e$$

$$Fe^{2+} + CO_3^{2-} \longrightarrow FeCO_3$$

$$Fe^{2+} + 2HCO_3^- \longrightarrow Fe(HCO_3)_2$$

$$Fe(HCO_3)_2 \longrightarrow FeCO_3 + CO_2 + H_2O$$

当 SRB 数量达到一定量，SRB 将硫酸盐作为有机物异化时的电子受体，并在代谢活动中产生硫化物，其阴极反应如下：

$$SO_4^{2-} + 9H^+ + 8e \longrightarrow HS^- + 4H_2O$$

代谢产生的 HS^- 与溶液中的 Fe^{2+} 反应生成 FeS。此外，研究还表明，在 CO_2 环境中，SRB 的存在会加速阴极去极化反应[28]，并使腐蚀介质的 pH 值降低，导致介质的腐蚀性增强，对 CO_2 腐蚀起催化作用。

页岩气输送平台用转角弯头发生严重内腐蚀减薄行为是硫酸盐还原菌（SRB）-CO_2 腐蚀协同作用的结果。在腐蚀阶段，由 SRB 消耗 SO_4^{2-}，从而将硫酸盐转化为 S^{2-}，并促进酸性环

境中的 H_2S 和 HS^- 形成。HS^- 在溶液中与铁离子反应，生成含铁硫化物，然后沉积在转角弯头内表面，增强了内表面与腐蚀产物之间的电子转移，从而加速弯头内表面的腐蚀。此外，H_2S 的离子化还会导致溶液酸化进一步增强 SRB 的数量和活性，从而加快转角弯头内表面的腐蚀过程。H_2S 不但参与 CO_2 腐蚀阴极过程，而且对 CO_2 腐蚀产物也有显著的影响，H_2S 的含量随着 SRB 的数量和活性增加而增大，从而进一步影响 CO_2 腐蚀过程和产物。CO_2 分压的提高又影响 SRB 的生物活性，并加速阴极去极化反应，导致腐蚀介质酸化，从而增强介质的腐蚀性，对 CO_2 腐蚀起催化作用，从而加快腐蚀速率，形成 SRB-CO_2 腐蚀协同作用的结果。此外，Cl^- 对腐蚀产物膜的破坏和弯头外弧侧的冲刷加速了腐蚀作用，最终导致转角弯头发生腐蚀穿孔，管件泄漏。

3 结论

弯头腐蚀减薄是 SRB-CO_2 腐蚀协同作用的结果，SRB 的存在对 CO_2 腐蚀起催化作用。此外，Cl^- 对腐蚀产物膜的破坏及弯头外弧侧的冲刷作用加速了腐蚀作用。

综合以上分析结果，为防止弯头内腐蚀减薄失效的再次发生，建议采取如下措施：确定 SRB 细菌来源，以便有效投放杀菌剂；在管线内壁涂覆防腐层，并添加缓蚀剂，降低 SRB 腐蚀的作用；对管线内壁定期进行清理工作，避免菌落长期附着于管道内壁；建议添加多级气液分离装置，严格控制气相中的含水量。

参 考 文 献

[1] 聂永臣，何敏，苏继祖. 油气集输管道内腐蚀及内防腐[J]. 油气田地面工程，2015，34(1)：83-84.

[2] Zhang G A, Zeng Y, Guo X P, et al. Electrochemical corrosion behavior of carbon steel under dynamic high pressure H_2S/CO_2 environment[J]. Corrosion science, 2012, 65: 37-47.

[3] Zeng L, Zhang G A, Guo X P. Erosion-corrosion at different locations of X65 carbon steel elbow[J]. Corrosion science, 2014, 85: 318-330.

[4] Gaber M A F M, Zewail T M, Amine N K. Evaluation of erosion corrosion in liquid-solid and liquid-gas via experimental analysis inside 90° copper elbow[J]. Journal of failure analysis and prevention, 2016, 16(3): 410-416.

[5] Velazquez J C, Cruzramirez J C, Valor A, et al. Modeling localized corrosion of pipeline steels in oilfield produced water environments[J]. Engineering failure analysis, 2017, 79(9): 216-231.

[6] 赵毅，许艳艳，朱原原，等. 油气集输管道内防腐技术应用进展[J]. 装备环境工程，2018，15(6)：53-58.

[7] Yaro A S, Abdul-Khalik K R, KHADOM A A. Effect of CO_2 corrosion behavior of mild steel in oilfield produced water[J]. Journal of loss prevention in the process industries, 2015, 38: 24-38.

[8] Papavinasam S. Corrosion control in the oil and gas industry[J]. Corrosion control in the oil & gas industry, 2013, 42: 116-129.

[9] Liu H, Gu T, Zhang G, et al. Corrosion of X80 pipeline steel under sulfate-reducing bacterium biofilms in simulated CO_2-saturated oilfield produced water with carbon source starvation[J]. Corrosion science, 2018, 52: 136-149.

[10] LIU H, FU C, GU T, et al. Corrosion behavior of carbon steel in the presence of sulfate reducing bacteria and iron oxidizing bacteria cultured in oilfield produced water[J]. Corrosion science, 2015, 100: 484-495.

[11] 岳明，汪运储. 页岩气井下油管和地面集输管道腐蚀原因及防护措施[J]. 钻采工艺，2018，41(5)：125-127.

[12] 王婷, 王新, 李在蓉, 等. 国内外长输油气管道失效对比[J]. 油气储运, 2017, 36(11): 1258-1264.
[13] 王海秋, 张昌兴, 李双林, 等. 油气管道腐蚀失效概率统计与预测模型[J]. 油气田地面工程, 2007, 26(4): 14.
[14] 王冰, 刘晓娟, 熊哲, 等. 某天然气管道内腐蚀原因及防控措施[J]. 表面技术, 2018, 47(6): 89-94.
[15] 石仁委, 郝毅, 宁华东. 管道失效的腐蚀因素分析[J]. 全面腐蚀控制, 2014, 28(11): 16-21.
[16] 孙成, 王佳, 贾思洋. CO_2腐蚀及控制研究进展[J]. 腐蚀科学与防护技术, 2007, 19(5): 350-353.
[17] 王凤平, 李晓刚, 杜元龙. 油气开发中的 CO_2 腐蚀[J]. 腐蚀科学与防护技术, 2002, 14(4): 223-226.
[18] 韩霞. 郑408块火烧驱油注气井腐蚀原因分析及对策[J]. 腐蚀科学与防护技术, 2010, 22(3): 247-250.
[19] 王磊, 胡锐, 王新虎, 等. S135钻杆钢在钻井液中的氧腐蚀行为[J]. 石油机械, 2006, 34(10): 1-4.
[20] 杨涛, 杨桦, 王凤江, 等. 含 CO_2 气井防腐工艺技术[J]. 天然气工业, 2007, 27(11): 116-118.
[21] Mansoori H, Young D, Brown B, et al. Influence of calcium and magnesium ions on CO_2 corrosion of carbon steel in oil and gas production systems—A review[J]. Journal of natural gas science and engineering, 2018(8): 20-25.
[22] Lim T H, Wang E R. Effects of temperature and partial pressure of CO_2/O_2 on corrosion behaviour of stainless-steel in molten Li/Na carbonate salt[J]. Fuel and energy abstract, 2001, 42(1): 31-32.
[23] 敬加强, 刘黎, 谢俊峰, 等. 输油管道腐蚀垢样中硫酸盐还原菌对Q235钢腐蚀行为的影响[J]. 腐蚀与防护, 2018, 39(1): 6-16.
[24] 陈悟, 汪文俊, 向福, 等. 腐蚀生物膜垢中硫酸盐还原菌的系统进化分析[J]. 微生物学通报, 2008, 35(2): 161-165.
[25] 张小里, 陈志昕, 刘海洪, 等. 环境因素对硫酸盐还原菌生长的影响[J]. 中国腐蚀与防护学报, 2000, 20(4): 224-229.
[26] Xu C M, Zhang Y H, Cheng G X, et al. Corrosion behavior of 316L stainless steel in the combination action of sulfate-reducing and iron-oxidizing bacterias[J]. Transactions of materials and heat treatment, 2006, 110(4): 104-108.
[27] 范梅梅. 二氧化碳存在条件下SRB在管道内腐蚀状况研究[D]. 武汉: 华中科技大学, 2011.
[28] 刘玉秀. 硫酸盐还原菌(SRB)对碳钢腐蚀行为影响的研究[D]. 大连: 大连理工大学, 2002.

本论文原发表于《表面技术》2020年第49卷第8期。

16Mn 管线钢的焊缝表面冲蚀机理研究

武 刚 李德君 罗金恒 白 强 李丽锋 朱丽霞

(中国石油集团石油管工程技术研究院·石油管材及
装备材料服役行为与结构安全国家重点实验室)

摘 要：在管道工程中，16Mn 管线钢异径接头—管体焊缝(简称焊缝)表面冲蚀行为是引发失效的主要原因之一，通过研究焊缝余高和管体流体作用，以此来探究焊缝表面冲蚀机理。以 16Mn 管线钢为研究对象，针对焊缝区表面冲蚀行为开展基础研究，通过电化学和腐蚀模拟实验，研究了流体初期焊缝区表面腐蚀行为，并通过流体模拟实验，研究了焊缝余高和流体速度对焊缝区冲蚀过程的影响，揭示了管线钢焊缝的冲蚀机理。在腐蚀模拟实验中，焊缝、接头母体和管体的开路电位分别为 −0.717V、−0.686V、−0.687V，焊缝区发生电化学腐蚀的倾向在模拟腐蚀液中最严重，焊缝的自腐蚀电流密度为 $7.9\mu A/cm^2$，母材的自腐蚀电流密度为 $3.2\mu A/cm^2$，焊缝的电化学腐蚀倾向性更大，焊缝区金属腐蚀速率最大，在焊缝表面形成了疏松的 FeO 产物层。在流体模拟实验中，流体在焊缝余高作用下形成了湍流，流速的增加也提高了湍动能，流速为 15m/s 和 30m/s 时，焊缝凹槽的深度分别为 3mm 和 8mm，焊缝凹槽相差 5mm。湍动能在焊缝余高的 FeO 腐蚀产物表面产生了变形磨损和切削效应，使得焊缝表面疏松的 FeO 产物层脱落，加速了腐蚀过程，最终形成了冲蚀凹陷区。16Mn 管线钢焊缝的冲蚀行为是腐蚀和流体冲蚀共同作用的结果，可分为初期的腐蚀和流体冲蚀两个阶段，形成了腐蚀与冲蚀循环交替过程。焊缝余高和流体速度对冲蚀影响较大，内焊缝余高和流体速度的增加将导致余高处的湍动能急剧增加，加速焊缝金属腐蚀产物层剥离，导致焊缝表面冲蚀。研究结果可以为管道失效行为和安全服役设计提供理论基础。

关键词：16Mn 管线钢；焊缝余高；流体模拟；冲蚀；腐蚀；变形磨损；切削效应

管道运输作为五大运输方式之一，在天然气运输中发挥着巨大作用，而腐蚀是影响管道寿命的重要因素之一。输气管道焊缝在工作过程中会受到冲蚀等作用，最终导致管体失效，从而造成巨大的安全隐患。

目前，针对管道的冲蚀失效问题开展了多项研究，Velázquez 等[1]研究了 X60 管线钢在油田采出水环境中的局部腐蚀，在不同的实验条件和周期下进行了浸没实验，建立了局部腐

基金项目：国家"十三五"国家重点研发计划(2017YFC0805804)；陕西省创新人才推进计划-青年科技新星项目(2017KJXX-06)。

作者简介：武刚(1985—)，男，硕士，工程师，主要研究方向为油气管道及储存设施完整性。E-mail：wugang010@cnpc.com.cn。

蚀缺陷最大深度的时间演化模型。Paul等[2]研究了X42管线钢的冲蚀行为及机理,在较低的速度下观察到塑性变形以及腐蚀颗粒在靶材表面的嵌入和犁削,且犁耕机理以较高的速度和较长的实验时间为主。Zheng[3]和Ahmed[4]等分别研究了管道金属在液固冲击条件下的腐蚀与冲蚀作用,发现冲蚀与腐蚀之间具有较强的相关性。林楠[5]进行了输气管道中颗粒属性及流场作用对冲蚀磨损的影响研究,认为输气管道的减薄及穿孔破坏主要是由管道内固体颗粒冲蚀磨损造成的。偶国富[6]及李增亮等[7]基于有限元方法的颗粒冲蚀数值计算模型,建立了流固耦合数理模型,对异径管冲蚀失效进行流固耦合数值模拟,研究了冲蚀因素对冲蚀速率及表面变形特征的影响规律。

目前,学者们已经深入探究了输气管道的腐蚀或冲蚀的单独机理,对于液固两相的腐蚀与冲蚀共同作用也有了新的发现,但是对于输气管道中气固两相的冲蚀和腐蚀共同作用机理的研究很少。通过上述研究可以发现,冲蚀过程与焊缝区的腐蚀行为和流体的状态有较大的相关性,要深入了解16Mn管线钢的焊缝冲蚀过程及失效机理,需要对焊缝区的腐蚀和流体冲蚀行为进行研究。本文通过电化学和腐蚀模拟实验,研究了流体初期的16Mn管线钢(某油气田作业公司提供)焊缝区腐蚀行为,并通过流体模拟实验,研究了其冲蚀过程,明确了焊缝失效的冲蚀过程机理,建立了冲蚀失效模型。研究结果可以为16Mn管线钢焊缝的冲蚀行为和防护理论提供科学依据。

1 实验

1.1 实验材料

本文的研究对象包括16Mn管线钢管体焊缝(简称焊缝)、接头母体和管体[图1(a)]。接头母材和管体的材质都为16Mn,其中主要研究区域为焊缝[图1(b)],研究对象化学成分见表1。

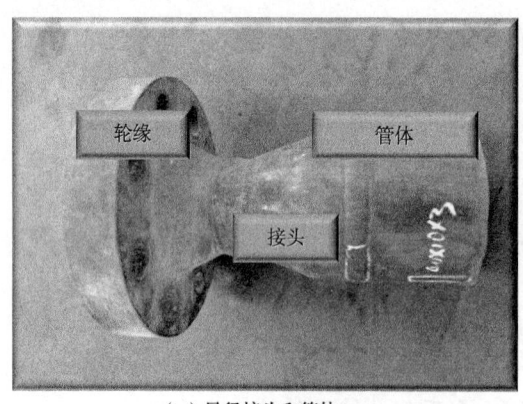

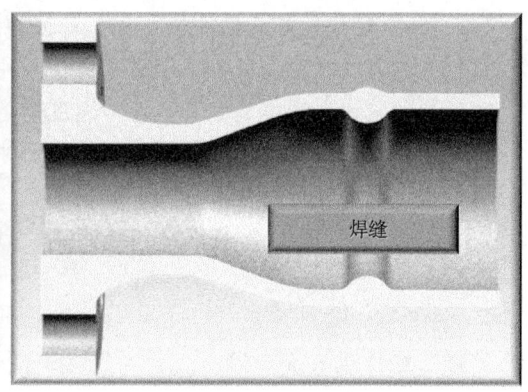

(a)异径接头和管体　　　　　　　　(b)焊缝示意图

图1　16Mn管线钢异径接头—管体焊缝

表1　管件和焊缝金属化学成分检测结果　　　　单位:%(质量分数)

材料	C	Si	Mn	P	S	Cr	Mo	Ni	Nb
异径接头	0.180	0.42	1.57	0.0280	0.0076	0.024	<0.005	0.0075	<0.005
管体	0.160	0.29	1.39	0.0130	0.0058	0.026	<0.005	0.0300	<0.005
焊缝	0.089	0.37	1.40	0.0083	0.0055	0.046	<0.005	0.5200	<0.005

1.2 电化学和腐蚀模拟实验

管件材料与焊缝填充金属间可能因为化学成分差异而发生电偶腐蚀，因此本文采用电化学和腐蚀模拟实验，研究母材与焊缝金属的腐蚀特性。分别在接头母体、接头焊缝、管体上取样，加工为$\phi 15mm \times 3mm$的电化学试样，将电化学试样背面焊接铜导线，用环氧树脂将试样包裹，依次用240#、400#、600#、800#、1000#、1200#、1500#和2000#砂纸打磨并抛光成镜面，然后用无水乙醇和去离子水冲洗，烘干后备用。电化学实验仪器为CS370型电化学工作站，工作电极为待测材料，辅助电极为石墨电极，参比电极为Ag/AgCl/饱和KCl电极。对16Mn管线钢的接头母体、接头焊缝和管体进行开路电位和极化曲线测试。测试使用的腐蚀溶液根据某油气田作业公司提供的水质分析结果，在实验室内使用分析纯NaCl、NaHCO$_3$和Na$_2$SO$_4$进行配制，溶液中Cl$^-$、HCO$_3^-$、SO$_4^{2-}$质量浓度分别为1666.18mg/L、650.98mg/L、126.67mg/L，测量待测金属的开路电位和极化曲线，测试范围为$-1.2 \sim 0.3V$，扫描速率为1mV/s。

为了进一步研究焊缝的现场腐蚀情况，采用高温高压釜模拟现场的高压腐蚀工况。腐蚀挂片沿管件纵向截取，焊缝位于腐蚀挂片的中间，腐蚀挂片的尺寸为50mm×10mm×3mm。腐蚀溶液根据某油气田作业公司提供的水质分析结果，在实验室内进行配制，为了加速实验过程，将二氧化碳分压设定为1MPa，液体流速为1m/s，温度为50℃，压力为8MPa，实验周期为168h。取1组（3件）腐蚀挂片试样置于液相中浸泡90min。

1.3 组织和成分检测方法

分别在接头母体、接头焊缝、管体上切取纵向常规金相试样，依次用240#、400#、600#、800#、1000#、1200#、1500#和2000#砂纸打磨并抛光成镜面，然后用无水乙醇和去离子水冲洗，烘干后在4%硝酸乙醇溶液中腐蚀5~8s，用BX51M光学显微镜观察试样金相组织。经过腐蚀实验后，将腐蚀挂片取出，用酒精清洗并干燥，采用扫描显微镜Hitachi S4800(SEM)观察试样表面形貌。采用D8先进X射线衍射(XRD)技术，在20°~80°范围内、0.05(°)/s条件下，利用铜-钾α射线衍射仪对腐蚀挂片进行物相鉴定，扫描速度为1步/s。

1.4 流态模拟实验和流速模拟实验

在实际的工况条件下，造成管线钢焊缝失效的原因有很多（冲蚀、焊缝组织和成分差异都会对焊缝失效产生影响），本文主要对冲蚀的影响进行了研究。考虑到焊缝余高和流体作用是造成冲蚀的主要原因，本文进行了流态模拟和流速模拟实验。为了探究焊缝余高对流体流态产生的影响，选择标准的$k-\varepsilon$模型作为求解计算的湍流模型，利用ANSYS软件进行了流体模拟实验。因为管道输送的天然气中还包含有微量沙粒、水及凝析油等，输送流体属于多相流体，模拟过程过于复杂，所以本文对多相流进行了简化处理；因为输送流体中液相含量较少，液相对管壁的冲刷剪切作用很低，所以管内流体可以简化看成是气体和微量砂粒的两相流体；因为颗粒细小而且含量较少，所以在模拟时可以将管内流体简化为单相流体。为了进一步探究不同流体流速对焊缝失效的影响，同样采用标准的$k-\varepsilon$模型作为求解计算的湍流模型进行流速模拟实验。

2 结果和讨论

2.1 焊缝腐蚀行为研究

16Mn管线钢接头母体、焊缝和管体的极化曲线如图2(a)所示，焊缝区的开路电位为$-0.717V$，接头母体的开路电位为$-0.686V$，管体的开路电位为$-0.687V$。其中焊缝区开路电位最低，所以焊缝区发生电化学腐蚀的倾向在模拟腐蚀液中最严重。在极化曲线中，焊缝的自腐蚀电流密度为 $7.9\mu A/cm^2$，母材的自腐蚀电流密度为 $3.2\mu A/cm^2$，焊缝的自腐蚀电流密度大于母材，表明焊缝的腐蚀敏感性大于母材，电化学腐蚀倾向性更大，在腐蚀介质中的腐蚀速率最大。同时，接头母体金相组织[图2(b)]和管体金相组织[图2(c)]为多边形铁素体(PF)+粒状贝氏体($B_{粒}$)，焊缝金相组织为"PF+P+$B_{粒}$"[图2(d)]，均无明显组织缺陷，保证了电化学和腐蚀模拟实验的相对准确性。在之前研究中也发现类似的结果。Zhang等人[8]采用电化学研究了X65管线钢的腐蚀和局部腐蚀行为，提出了局部腐蚀模型，并发现裸钢和焊缝区之间存在电偶效应，焊缝金属在腐蚀介质中的腐蚀速率最大。

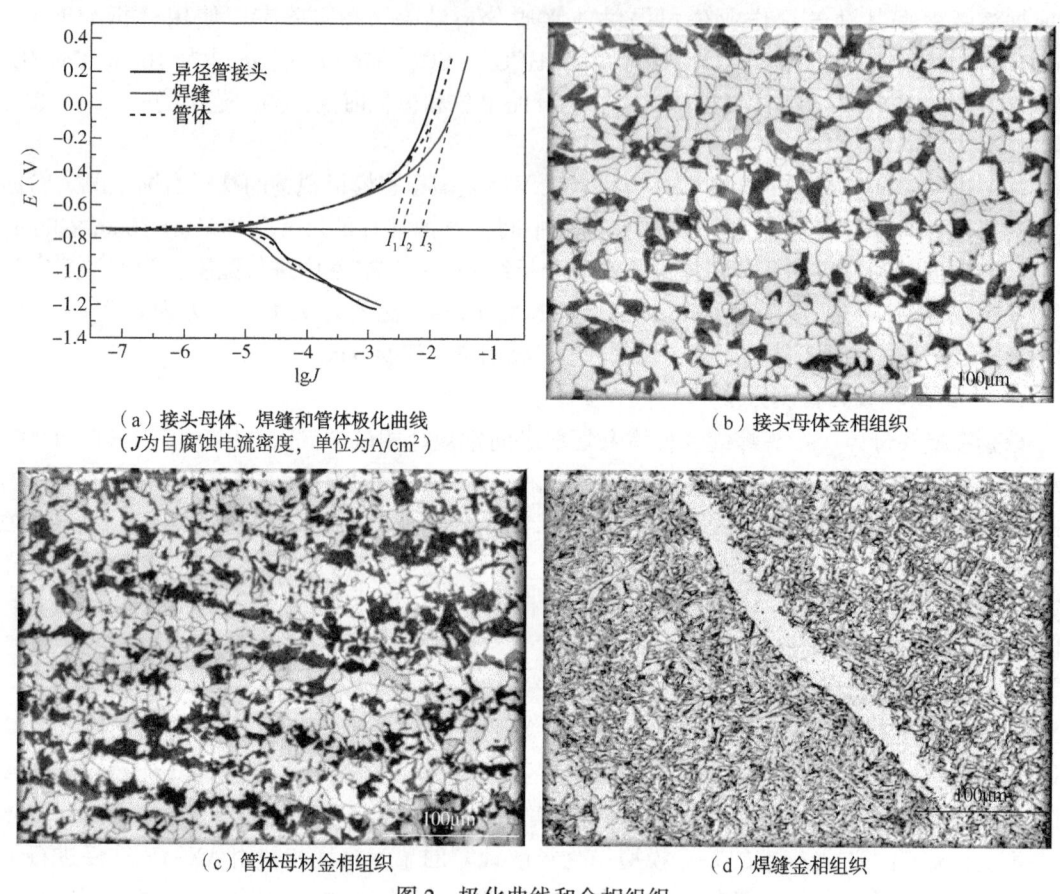

(a) 接头母体、焊缝和管体极化曲线
（J为自腐蚀电流密度，单位为A/cm^2）

(b) 接头母体金相组织

(c) 管体母材金相组织

(d) 焊缝金相组织

图2　极化曲线和金相组织

静态浸泡腐蚀可以模拟流体在初期静止条件下对母材和焊缝的腐蚀过程，腐蚀挂片取样方位如图3(a)所示。其中，从液相中腐蚀挂片的腐蚀形貌图[图3(b)]可以看出，腐蚀挂片腐蚀严重，呈均匀腐蚀形貌，焊缝区域的腐蚀产物层呈多孔疏松状，并且氧化膜(FeO)厚度异常薄，未经脱层处理就已经开裂脱落。利用扫描电子显微镜对焊缝与母材区的腐蚀产物膜层形貌进行观察，管体母材的FeO腐蚀产物层[图3(c)]相对致密，致密的氧化层可以有效阻隔腐蚀介质和基体进一步接触，具有一定的保护作用；然而焊缝处的FeO腐蚀产物层[图3(d)]多孔疏松，焊缝区更容易被腐蚀介质腐蚀。由X射线衍射图谱[图3(e)]可以看出，内壁上检测到的物相主要为Fe，同时含有一定量的Fe_3C和少量的FeO，可以观察到X射线衍射图谱[图3(e)]中FeO峰较弱，主要因为焊缝区内壁上残留的腐蚀产物膜很薄，使得X

射线在极短的时间内穿透 FeO 腐蚀产物层，打到基体金属，获得了以基体金属为主的 Fe 和 Fe$_3$C 相。

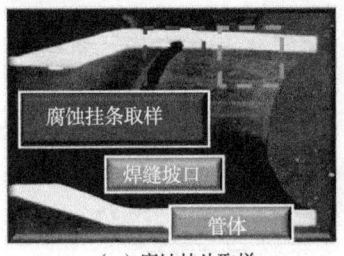

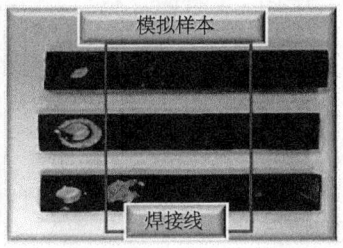

(a) 腐蚀挂片取样　　(b) 液相中腐蚀挂片的腐蚀形貌图　　(c) 管体母材腐蚀产物层形貌

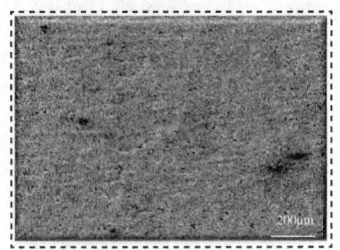

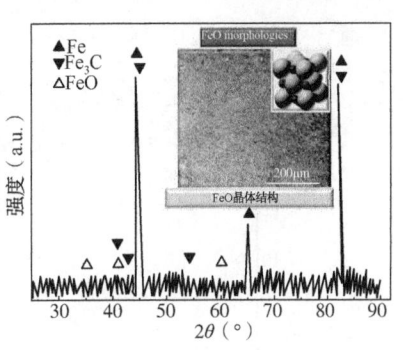

(d) 焊缝腐蚀产物层形貌　　(e) 焊缝腐蚀产物XRD谱线

图 3　腐蚀模拟的实验结果

母材区致密均匀的膜层可以有效抵抗冲蚀剥离，而焊缝区腐蚀产物层因为多孔疏松，在流体作用下不断被冲蚀剥离，使得新鲜的基体金属不断暴露出来而被腐蚀，冲蚀的存在也使腐蚀产物层对层下金属的保护作用大大降低，从而加快了腐蚀速度，加剧了焊缝的腐蚀失效。在其他研究中也发现类似的结果。Bilmes[9]和温建萍[10]等分别进行了16Mn钢的动态全浸实验和焊缝金属的显微组织研究，发现母材的腐蚀产物膜更致密，焊缝的腐蚀产物膜比较疏松，焊缝区更容易被介质腐蚀。

2.2　流态和流速模拟行为研究

管件内壁连续平滑时的湍动能[图4(a)]在接头焊缝出口端最大，但是紧挨管壁处的湍动能则明显降低。当管体内壁存在内焊缝余高导致管体内壁不连续光滑时，在内焊缝余高处会产生湍流，并且截面突变处会发生流体回流现象，管件内壁焊缝余高处的湍动能[图4(b)]最大，湍流最为剧烈，紧挨焊缝余高处的湍动能则明显降低。在其他研究中也发现类似的结果。Hung 等[11]认为湍动能(k)可以利用湍流强度进行估算，其计算公式为

$$k = \frac{3}{2}(uI)^2 \tag{1}$$

式中：u 为平均速度，m/s；I 为湍流强度，MPa。采用直接数值求解方程研究了后向台阶上的湍流流动，并发现由于流体通道的内径在焊缝处发生变化，导致焊缝余高处的流体流速发生了不规则变化，产生了湍流，湍流的存在会产生湍动能[12]，其中焊缝余高处的湍动能最大。

不同内焊缝余高条件下的湍动能[图4(c)]差别较大，内焊缝余高为10mm时的最大

湍动能是2mm时的两倍，内焊缝余高越大，余高处的湍动能越大，湍流越剧烈，对焊缝余高的冲蚀破坏越严重，并且当焊缝余高被破坏导致轮廓不完整时，余高处的湍动能急剧增加，最大湍动能是焊缝余高为10mm时的2倍，是2mm时的4倍。湍动能的存在会对焊缝余高产生冲蚀作用，湍流强度越大，湍动能越大，对焊缝余高产生的冲蚀作用越大。

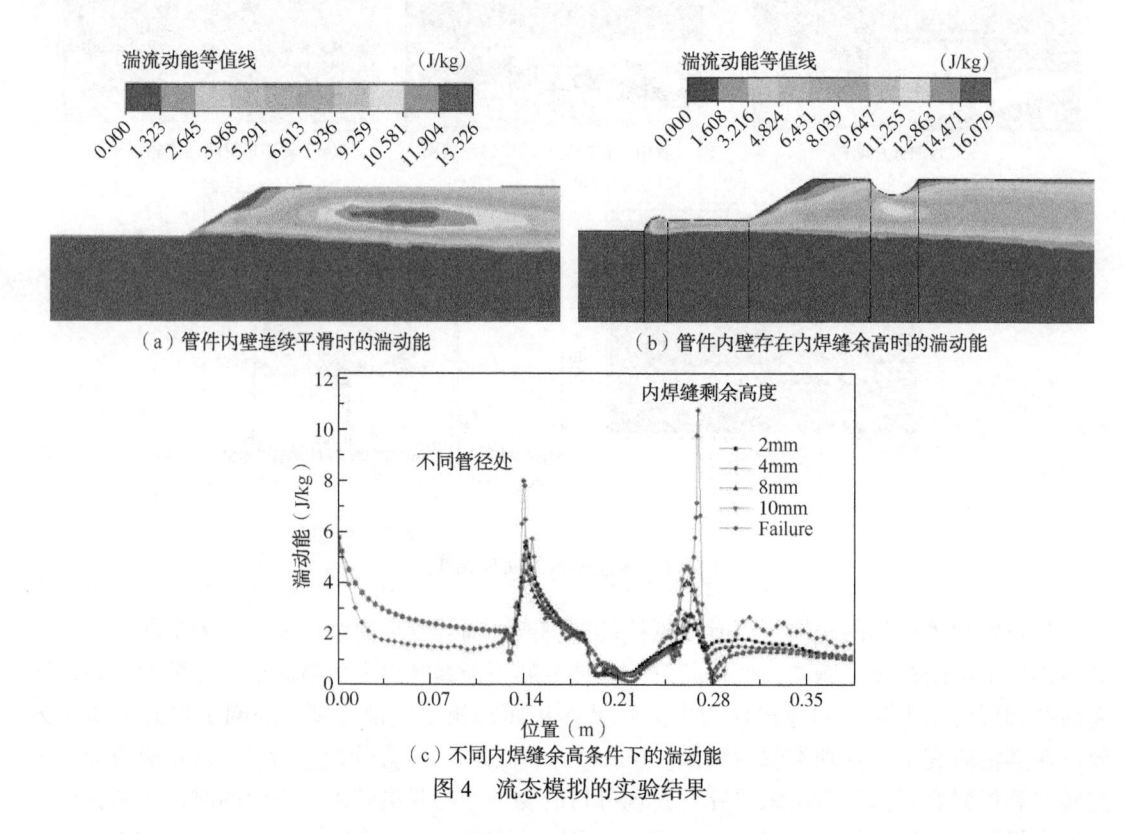

（a）管件内壁连续平滑时的湍动能

（b）管件内壁存在内焊缝余高时的湍动能

（c）不同内焊缝余高条件下的湍动能

图4 流态模拟的实验结果

此外，研究报道发现流速是影响冲蚀行为的主要因素之一[12]。本研究中发现流体经接头焊缝向管体流动，当管体内壁光滑连续时，由于流体通道的内径在接头焊缝处发生变化，在接头小径出口端有流体回流现象[图5（a）]；当存在内焊缝余高，导致管体内壁不光滑连续时，在内焊缝余高处会产生明显的湍流，并且截面突变处还会发生流体回流现象[图5（b）]。不同流速条件下焊缝的湍动能[图5（c）]差别也很大，内焊缝余高在流速为30m/s时的最大湍动能是15m/s时的四倍，随着流体速度的升高，内焊缝余高处的湍动能单调增加，尤其是当流速超过15m/s时，湍动能增加趋势显著，导致流体对焊缝余高的冲蚀作用显著增加。在管道增压过程中，流体速度越大，湍动能越大，湍动能对焊缝余高产生的冲蚀作用就越大。在高文祥等人[13]的研究中也发现类似的结果。

在相同的环境参数和时间下，采用流速实验研究了速度分别为15m/s、30m/s时的焊缝冲蚀情况。焊缝区冲蚀形貌如图6所示，可以发现随着冲蚀速度的变大，冲蚀行为加重。经过测量发现，流速为15m/s时，焊缝凹槽深度为3mm[图6（a）]；流速为30m/s时，焊缝凹槽深度为8mm[图6（b）]，焊缝凹槽相差5mm。其研究结果与模拟结果吻合，说明随着流体速度的增大，内焊缝余高处的湍动能单调增加，流体对焊缝余高的冲蚀作用变大。

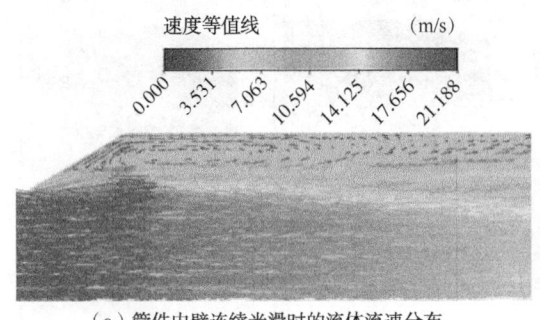

(a)管件内壁连续光滑时的流体流速分布

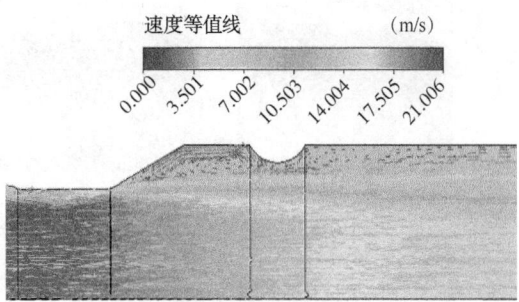

(b)管件内壁存在内焊缝余高时的流体流速分布

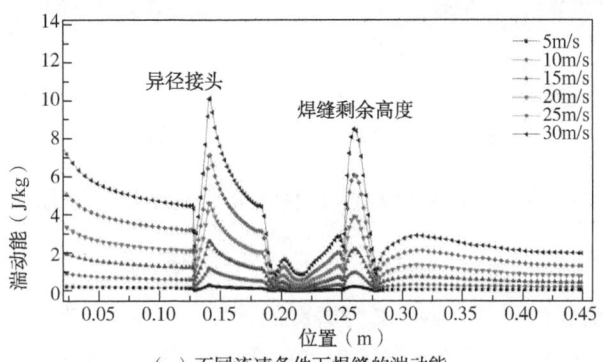

(c)不同流速条件下焊缝的湍动能

图5 流体流速模拟的实验结果

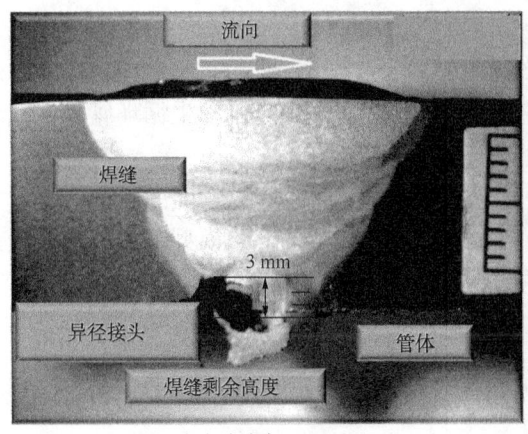

(a)流速为15m/s

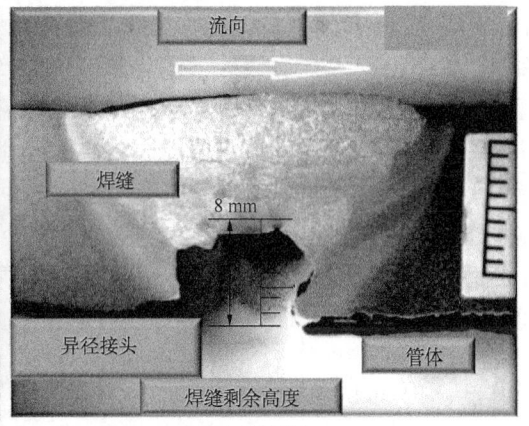

(b)流速为30m/s

图6 不同流速下内焊缝余高的冲蚀情况

2.3 冲蚀机理研究

针对冲蚀过程和机理的研究发现[13-15]，焊缝失效主要是腐蚀和冲蚀的作用，其失效过程和冲蚀机理分别如图7所示。在焊缝失效初期，焊缝金属主要以静态的腐蚀为主(图7中第一步)，氧化产生了FeO附着在焊缝余高处，主要发生的电化学腐蚀过程为

$$Fe^{2+} + O^{2-} \longrightarrow FeO \tag{2}$$

其中主要的腐蚀产物为FeO，表面形貌疏松多孔。随着流速的增加，进入了冲蚀阶段(图7中第二步Ⅰ&Ⅱ)，FeO产物层在湍流作用下易产生磨损和切削作用。Farhat[14]研究了冲蚀角和冲蚀速度对冲蚀的影响，认为当球粒冲击内焊缝余高平面时，可能会产生弹性或塑

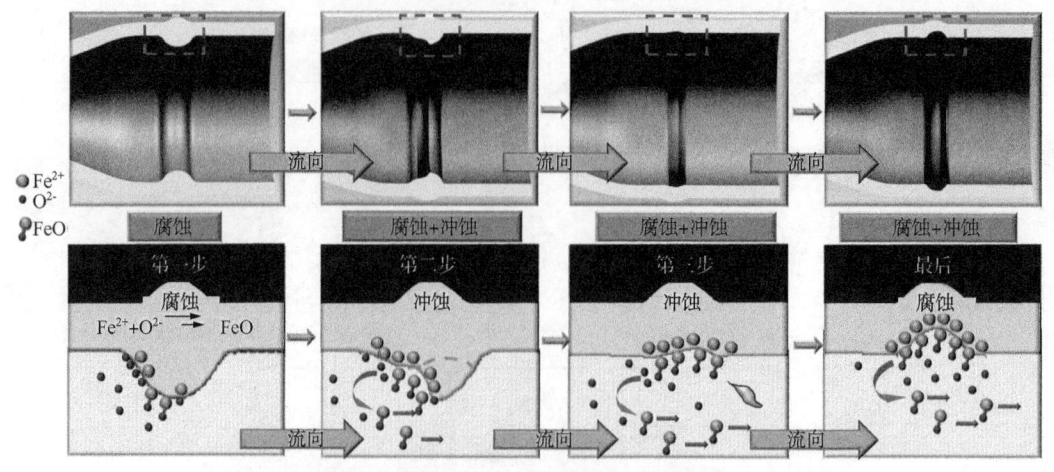

图 7 内焊缝余高冲蚀过程机理示意图

性变形,但这取决于冲击力是否达到材料的屈服强度。其中变形磨损量 W_D 为

$$W_D = M(V\sin\alpha - K)^2/2\varepsilon \tag{3}$$

式中：M 为冲击磨粒的质量,kg；V 为磨粒的速度,m/s；α 为冲击角,(°)。焊缝区在冲蚀初期(图 7 中第二步Ⅰ),磨粒镶嵌于焊缝的表面,此时焊缝的冲蚀磨损量较小,将产生负磨损,此阶段称为孕育期,经过一段时间后,当焊缝磨损量大大超过镶嵌量时,转变为正磨损。随着冲蚀时间的持续,焊缝的磨损量也稳定增加,这一阶段为稳定冲蚀期(图 7 中第二步Ⅱ),磨粒造成的磨损称为一次磨损。由于腐蚀作用的存在,固体粒子在冲击焊缝表面时,往往会使 FeO 腐蚀产物产生破碎,并产生大量碎片,这些碎片能造成前期冲击时焊缝表面所形成的挤出唇或翻皮的去除,未被腐蚀的金属就会暴露出来,发生循环腐蚀生成腐蚀 FeO 氧化层,这时固体粒子的冲蚀磨损机理主要是延性切削机理。Wang 等人[15]利用有限元模型,以延性材料为例,模拟了冲击角、冲击速度和颗粒侵蚀对焊缝表面的影响,认为固体粒子以一定的速度和角度侵入焊缝表面,会对材料产生切削作用。该模型假设一颗质量为 m 的磨粒,以一定速度 v 和冲击角 α 冲击到焊缝的表面,由理论分析可得出材料的磨损体积为

$$V = K\frac{mv^2}{p}f(\alpha) \tag{4}$$

式中：p 为焊缝表面的流动应力,MPa；K 为常数。焊缝的磨损体积与流体的质量和流体速度的平方(即流体的动能)成正比,与焊缝的流动应力成反比,其结果验证了随着流体速度的升高,内焊缝余高处的湍动能单调增加,尤其是当流速超过 15m/s 时,流体对焊缝余高的冲蚀作用显著增加。随着内焊缝余高的变化,流体的腐蚀作用也随之变化。在冲蚀过程中,内焊缝余高的形貌从刚开始的圆弧状变为最后的凹槽状(图 7),在内焊缝余高冲蚀阶段初期(图 7 中第一步),焊缝金属与流体相接触产生腐蚀,形成多孔疏松的 FeO 产物层,内焊缝余高呈圆弧状且高度最大,焊缝余高会影响流体的流态,使流体在焊缝余高处产生湍流,湍流的冲蚀作用使氧化层易发生剥落,因为焊缝余高最大,此时受冲蚀直接影响的焊缝表面积最大,使得冲蚀起主导作用(图 7 中第二步Ⅰ),冲蚀作用较强,同时流体中携带的固体粒子也会对氧化层产生冲蚀磨损,氧化层被剥落后露出新的焊缝金属,进入腐蚀—冲蚀

循环。随着时间的累积，焊缝金属因为腐蚀和冲蚀影响作用，外形随之发生变化，形成凹槽状（图7中第二步Ⅱ），凹槽的形成更进一步加剧了凹槽底部的冲蚀，湍流和回流将在凹槽内壁产生，凹槽内壁将持续受到冲刷，FeO产物层不断被冲蚀，使得新鲜的金属被腐蚀，焊缝处的壁厚不断减薄，凹槽将不断变大，一旦凹槽的深度达到一定值时，凹槽底部流体速度将显著降低，凹槽内积聚的腐蚀介质继续腐蚀焊缝金属，腐蚀作用取代冲蚀成为控制因素（图7中最后），最终在腐蚀作用下，局部区域发生腐蚀穿孔，导致管件泄漏。

3 结论

（1）焊缝区开路电位最低，在模拟腐蚀液中发生电化学腐蚀的倾向较大，同时，焊缝区腐蚀电流密度最大，在腐蚀介质中的腐蚀速率较大。另外，内焊缝余高越大，余高处的流体湍动能越大，焊缝余高受到的冲蚀破坏越严重，当余高轮廓被破坏时，余高处的湍动能急剧增加。随着流体速度变快，流体对焊缝余高的冲蚀作用增强。焊缝区比母材的耐蚀性差，焊缝区腐蚀产物层在流体冲蚀作用下，更加容易剥离。

（2）管线钢焊缝的冲蚀行为是腐蚀和流体冲蚀共同作用的结果，可分为初期腐蚀和流体冲蚀两个阶段：在焊缝失效初期，焊缝区主要以静态的腐蚀为主，在焊缝表面形成了疏松多孔的腐蚀产物层；在流体冲蚀阶段，冲蚀起主导作用，冲蚀使得焊缝表面的腐蚀产物层脱落，形成了腐蚀—冲蚀交替过程，最终形成了冲蚀凹陷区。针对焊缝组织和成分差异引起的静态和动态腐蚀作用，将在后期进行相关研究。

参 考 文 献

[1] Velazquez JC, Cruzramirez JC, Valor A, et al. Modeling localized corrosion of pipeline steels in oilfield produced water environments[J]. Engineering failure analysis, 2017, 79(9)：216-231.

[2] Okonkwo P C, Shakoor R A, Ahmed E, et al. Erosive wear performance of API X42 pipeline steel[J]. Engineering failure analysis, 2016, 60：86-95.

[3] Zheng Z B, Zheng Y G. Erosion-enhanced corrosion of stainless steel and carbon steel measured electrochemically under liquid and slurry impingement[J]. Corrosion science, 2016, 102：259-268.

[4] Islam M A, FarhaT Z N, Ahmed E M, et al. Erosion enhanced corrosion and corrosion enhanced erosion of API X-70 pipeline steel[J]. Wear, 2013, 302(1-2)：1592-1601.

[5] 林楠. 输气管道中颗粒属性及流场作用对冲蚀磨损的影响研究[D]. 北京：北京交通大学, 2017.

[6] 偶国富, 许根富, 朱祖超, 等. 弯管冲蚀失效流固耦合机理及数值模拟[J]. 机械工程学报, 2009, 45(11)：119-124.

[7] 李增亮, 杜明超, 董祥伟, 等. 固体颗粒冲蚀磨损模型的建立及有限元分析[J]. 计算机仿真, 2018, 35(6)：275-281.

[8] Zhang G A, Cheng Y F. Localized corrosion of carbon steel in a CO_2-saturated oilfield formation water[J]. Electrochimica acta, 2011, 56(3)：1676-1685.

[9] Bilmes P D, Llorentel C L, Méndez C M, et al. Microstructure, heat treatment and pitting corrosion of 13CrNiMo plate and weld metals[J]. Corrosion science, 2009, 51(4)：876-881.

[10] 温建萍, 李博明, 温涛, 等. 油田回注污水对常用管线钢的腐蚀性[J]. 腐蚀科学与防护技术, 2006(1)：28-31.

[11] Le H, Moin P, Kim J, et al. Direct numerical simulation of turbulent flow over a backward-facingstep[J]. Journal of fluid mechanics, 1997, 54：349-374.

[12] Wyngaard J C, Cote O R. The budgets of turbulent kinetic energy and temperature variance in the atmospheric

surface layer[J]. Journal of the atmospheric sciences, 1971, 28(2): 190-201.
[13] 高文祥, 王治国, 曹银萍, 等. 超级 13Cr 钢在液固两相流体中的冲蚀实验研究[J]. 科学技术与工程, 2014, 14(31): 179-182.
[14] Islam M A, Farhat Z N. Effect of impact angle and velocity on erosion of API X42 pipeline steel under high abrasive feed rate[J]. Wear, 2014, 311(1-2): 180-190.
[15] Finnie I. Erosion of surfaces by solidparticles[J]. Wear, 1960, 3(2): 87-103.
[16] Wang H, Yu Y, Yu J, et al. Numerical simulation of the erosion of pipe bends considering fluid-induced stress and surface scar evolution[J]. Wear, 2019, 440: 203043.

本论文原发表于《表面技术》2020 年第 49 卷第 3 期。

酸洗钝化对 316L/L415 双金属机械复合管环焊缝耐蚀性的影响

宋成立[1] 王福善[2] 冯 泉[2] 范 磊[1] 白真权[1] 方 艳[1]

(1. 中国石油集团石油管工程技术研究院·石油管材及装备材料服役行为与结构安全国家重点实验室；
2. 中国石油塔里木油田分公司)

摘 要：针对 316L/L415 双金属机械复合管环焊缝内腐蚀刺漏的问题，配制酸洗钝化液($25\% HNO_3+5\% HF+70\% H_2O$)对其进行酸洗钝化，并利用形貌观察、电化学试验、点腐蚀试验等对酸洗钝化前后焊接试样的耐蚀性进行对比分析。结果表明：酸洗钝化后的试样表面变得平整、光亮、洁净；酸洗钝化后，焊缝和热影响区试样表面的耐蚀性元素 Cr、Ni 含量均有明显增加，但母材的 Cr、Ni 含量变化很小；酸洗钝化后的母材、焊缝及热影响区试样的自腐蚀电位均发生正移，腐蚀率更低。酸洗钝化可以在环焊缝表面生成一层新且更致密的钝化膜，明显提高 316L/L415 复合管环焊缝的耐蚀性。

关键词：双金属复合管；环焊缝；腐蚀；酸洗钝化；耐蚀性

随着油气开采的深入，综合含水不断升高，部分油气田还含有 H_2S 和 CO_2 等酸性气体，导致管道内腐蚀问题越来越严峻[1-2]。当前，碳钢材质管道已无法适应这种腐蚀环境，而碳钢+缓蚀剂、不锈钢管、非金属管的使用成本又较高，因此，双金属复合管的应用逐步扩大[3-4]。目前应用最多的是碳钢基材内衬 316L 不锈钢这种复合管，但由于现场组对焊接存在质量问题，导致双金属复合管在使用过程中仍然存在较为严重的腐蚀问题，主要表现为环焊缝的腐蚀刺漏[5-6]。研究大量双金属复合管环焊缝失效案例可知，其环焊缝腐蚀刺漏往往始于焊缝热影响区的点蚀，逐渐发展至穿透 316L 衬管，最终导致基管穿孔刺漏[7-8]。而有关研究表明，不锈钢在焊接过程中金属表面容易产生贫铬、氧化等问题，导致金属表面电化学性质不均匀而使其耐蚀性下降，这正是焊缝热影响区易产生点蚀的原因[9-10]。

鉴于此，为提高焊缝热影响区的耐蚀性，本工作对双金属复合管的环焊缝进行酸洗钝化，并对酸洗钝化前后 316L 环焊缝的耐蚀性差异进行相关腐蚀试验评价，以期为双金属复合管的现场实际应用提供科学依据。

1 试验

1.1 试样

试样来自现场焊接的 316L 双金属机械复合管，材质为 L415(基管)+316L(衬管)，尺寸

基金项目：中国石油天然气股份有限公司科学研究与技术开发项目(2017D-1608)。
作者简介：宋成立(1989—)，工程师，硕士，主要从事石油管材腐蚀与防护及失效分析，029-81887624，songcl@cnpc.comc.cn。

为φ508mm×(14.2+2.5)mm,制造标准为SY/T 6623—2012《内覆或衬里耐腐蚀合金复合钢管规范》。采用手工焊接组对,焊接方法为钨极氩弧焊+焊条电弧焊,焊接材料见表1。

表1 焊接材料

焊缝	焊材型号	焊材牌号	验收标准	直径(mm)
封焊	R309LT1-5	ATS-F309L	AWS A5.22—2012	2.2
根焊	R316LT1-5	ATS-F316L	AWS A5.22—2012	2.2
过渡焊	ER309LMo	ATS-309MoL	AWS A5.9—2012	2.4
填充盖面焊	E309MoL-15	AES-309MoLZ	GB/T 983-1995	2.6

1.2 试验方法

1.2.1 理化检验

采用ARL 4460直读光谱仪,对316L复合管内衬层进行化学成分分析;采用MEF4M金相显微镜及图像分析系统,对316L复合管内衬层环焊缝进行显微组织分析。

1.2.2 酸洗钝化试验

加工规格为50mm×30mm×$t$mm(t为试样壁厚)的焊缝接头试样,采用丙酮对其进行超声波清洗,随后采用树脂和硅胶进行密封。并配制25%(质量分数,下同)HNO_3+5% HF+70% H_2O的酸洗钝化溶液,用塑料刷将其均匀涂抹到试样的表面,并保持试样表面湿润,待30min后用蒸馏水冲洗干净至表面pH值为7。再配制1g$K_3[Fe(CN_6)]$+3ml HNO_3+100ml H_2O的溶液,用滤纸浸渍该溶液,贴附于酸洗后的试样表面,30s后观察滤纸是否变色,以检验酸洗钝化后是否还有铁离子污染(称为蓝点试验)。

1.2.3 扫描电镜观察

采用TESCAN VEGA Ⅱ扫描电子显微镜及其附带的XFORD INCA350能谱分析仪分别对酸洗钝化前后的母材、焊缝及热影响区的形貌和化学成分进行分析。

1.2.4 电化学试验

分别在酸洗钝化前后的焊缝、热影响区及母材上取电化学试样(5mm×5mm),试片用导线焊接后采用环氧树脂镶嵌,并采用SiC砂纸(100-800号)逐级打磨,辅助电极为Pt电极,参比电极为饱和KCl甘汞电极,试验温度为(20±1)℃,试验溶液为3.5%NaCl水溶液。

1.2.5 点腐蚀试验

依据GB/T 17897—2016《金属和合金的腐蚀不锈钢三氯化铁点腐蚀试验方法》,配制0.16%(质量分数,下同)HCl+6% $FeCl_3$+93.84% H_2O的溶液,对酸洗钝化前后的试样进行35℃条件下的浸泡点腐蚀试验,待24h后对试样表面的点蚀坑形貌进行观察,并计算腐蚀率。

2 结果及讨论

2.1 理化检验结果

由表2可见:316L内衬层的化学成分符合SY/T 6623—2012标准要求。由图1可见:母材、焊缝及热影响区的内衬层金相组织均为A+少量α相,未见其他异常。

表2 内衬层的化学成分检测结果   单位:%

项目	w_C	w_{Si}	w_{Mn}	w_P	w_S	w_{Cr}	w_{Mo}	w_{Ni}
内衬层(316L)	0.024	0.5	1.19	0.036	0.0012	17.7	2.01	10
SY/T 6623—2012	≤0.035	≤1.00	≤2.00	≤0.045	≤0.030	16.0~18.0	2.0~3.0	10.0~14.0

(a）母材　　　　　　　　　　（b）焊缝　　　　　　　　　　（c）热影响区

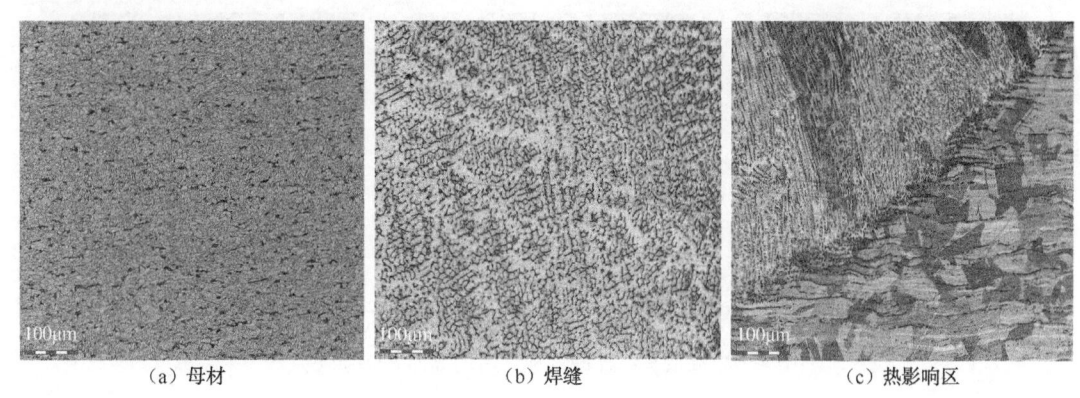

图 1　焊接接头试样的显微组织

2.2　酸洗钝化试验结果

酸洗钝化后的焊缝试样表面变得更加光亮、洁净（图2），对其进行蓝点试验后发现滤纸未变色，表明酸洗钝化是合格的且不存在铁离子污染。

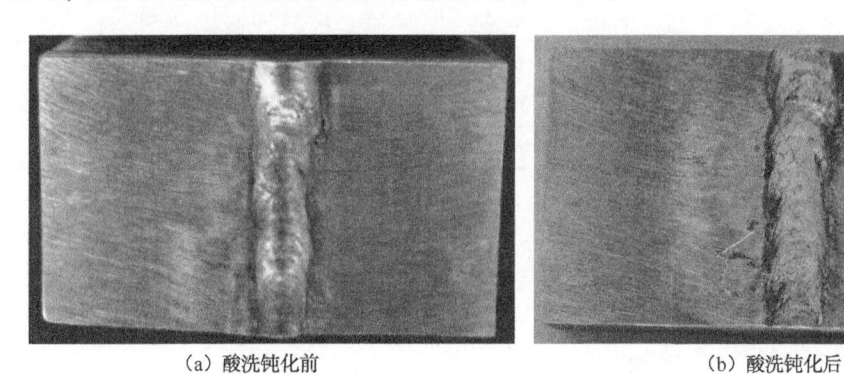

(a）酸洗钝化前　　　　　　　　　　　　　　　　（b）酸洗钝化后

图 2　焊缝经酸洗钝化试验前后的宏观形貌

2.3　扫描电镜及能谱分析结果

由图3和图4可见：试样表面的原始划痕经酸洗钝化后变得平整、洁净且划痕消失。由表3可知，酸洗钝化后焊缝和热影响区的 Cr、Ni 含量均明显增加，但母材的 Cr、Ni 含量变化很小。

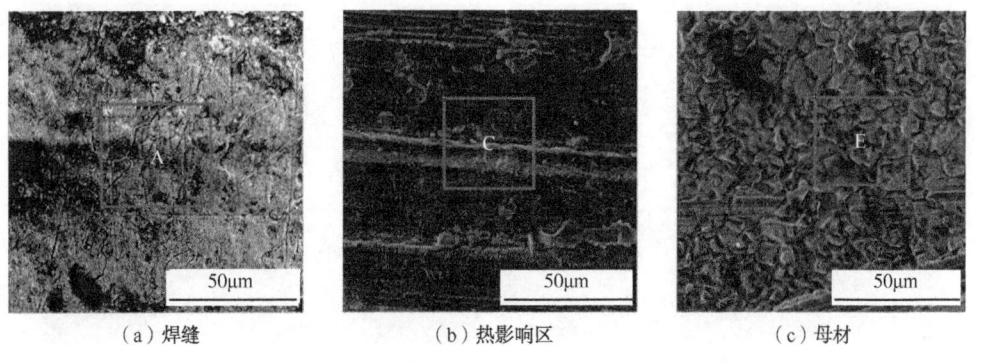

(a）焊缝　　　　　　　　　　（b）热影响区　　　　　　　　　　（c）母材

图 3　焊接接头试样酸洗钝化前的表面 SEM 形貌

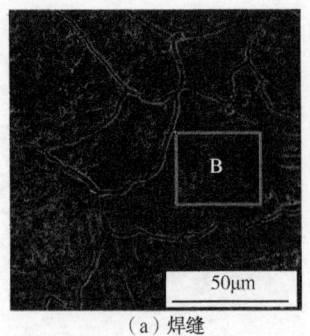

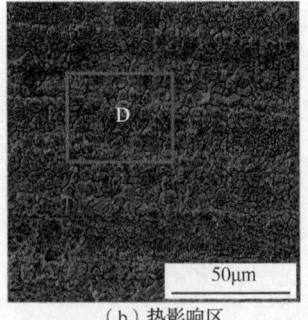

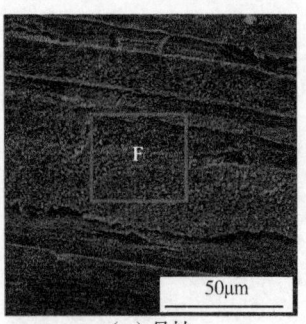

(a) 焊缝　　　　　　　　　　(b) 热影响区　　　　　　　　　(c) 母材

图4　焊接接头试样酸洗钝化后的表面SEM形貌

表3　试样的EDS分析结果　　　　　　　　　　　　　　　　　　　　　%

区域	w_C	w_O	w_{Cr}	w_{Fe}	w_{Ni}	w_{Cu}	w_{Mo}	w_{Hg}
A	6.95	7.51	14.88	45.18	7.28	5.36	7.64	5.20
B	—	1.12	18.35	66.69	12.20	—	1.64	—
C	5.68	13.75	15.76	45.40	7.53	4.90	6.98	—
D	—	1.13	18.91	69.51	8.85	—	1.61	—
E	1.83	2.11	18.42	66.68	9.18	—	1.77	—
F	—	1.08	18.28	69.28	9.18	—	2.18	—

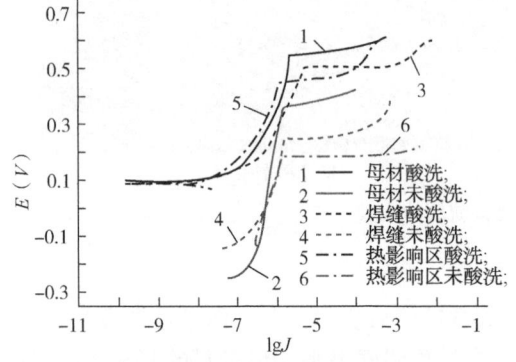

2.4　电化学试验结果

如图5所示，经酸洗钝化后，母材、焊缝及热影响区试样的自腐蚀电位均发生了正移。

2.5　点腐蚀试验

如图6所示，酸洗钝化后的试样表面点蚀坑数量更少，计算可知焊接接头试样酸洗钝化前后的腐蚀率分别为299.17 g/(m²·h)和36.39 g/(m²·h)，可见酸洗钝化后试样的腐蚀率更低。

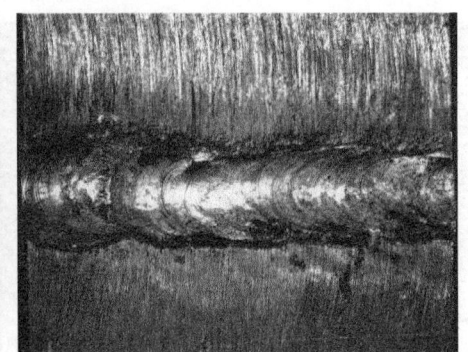

图5　酸洗钝化前后试样的动电位极化曲线
（J为自腐蚀电流密度，单位为A/cm²）

1——母材酸洗；2——母材未酸洗；3——焊缝酸洗；4——焊缝未酸洗；5——热影响区酸洗；6——热影响区未酸洗

(a) 酸洗钝化前　　　　　　　　　(b) 酸洗钝化后

图6　酸洗钝化前后试样的点蚀形貌

2.6 讨论

由以上检测和试验结果可知,316L 双金属复合管环焊缝的化学成分符合相关标准要求,且金相组织未见异常,这表明试验材料的性能是合格的。

焊接接头试样酸洗钝化后,表面变得更加光亮,主要是由于酸洗过程可以除去残存的油污,并且与酸洗液作用后形成新的钝化膜,这层膜独立存在,通常是氧和金属的化合物[11,12],它起着与腐蚀介质完全隔开的作用,钝化膜越致密,耐蚀性越强[13]。同时,SEM 结果表明表面粗糙、有原始划痕且附着杂质的试样经酸洗钝化后变得平整、洁净且划痕消失。由此可知,酸洗钝化清除了金属表面的划痕和附着杂质。试样表面的能谱分析(表3)结果显示:(1)酸洗钝化后的主要元素是 O、Cr、Fe、Ni、Mo,表明钝化膜的主要组成元素 Cr、Fe、Ni、Mo 是以氧化物形式存在的[14];(2)酸洗钝化后焊缝和热影响区的 Cr、Ni 含量均明显增加,但母材的 Cr、Ni 含量变化很小,而且酸洗钝化后的焊缝、热影响区和母材三个区域同种元素的含量变得更加接近,表明酸洗钝化过程缓解了环焊缝焊接时造成的焊缝及热影响区表面耐蚀性元素(Cr、Ni)含量下降的问题,并在焊缝及热影响区形成了新的钝化膜,将使其耐蚀性得到提高。

同时,酸洗钝化后的母材、焊缝和热影响区的点蚀电位均比未酸洗钝化的高,自腐蚀电位均发生了正移,这说明酸洗钝化提高了316L复合管环焊缝的耐蚀性[15]。点蚀试验结果表明,未酸洗钝化试样的腐蚀速率均比酸洗钝化试样的高,而且点蚀坑数量也更多,与电化学试验结果一致,再次验证了酸洗钝化提高了环焊缝的耐蚀性。

3 结论

利用 25% HNO_3 +5% HF+70% H_2O 酸洗钝化液对 316L 双金属复合管环焊缝表面进行 30min 酸洗钝化可以清除焊缝、母材及热影响区表面的划痕和附着杂质,使其表面变得更加平整、光亮、洁净,而且在其表面生成更多的钝化膜,使其自腐蚀电位发生正移。酸洗钝化可以明显提高 316L 复合管环焊缝的耐蚀性。

参 考 文 献

[1] 肖雯雯,宋成立,白真权,等. 油田地面集输管道腐蚀穿孔风险分析[J]. 油气田地面工程,2017,36(4):81-85.

[2] 叶帆,高秋英. 凝析气田单井集输管道内腐蚀特征及防腐技术[J]. 天然气工业,2010,30(4):96-101.

[3] 李雪,朱庆杰,周宁,等. 油气管道腐蚀与防护研究进展[J]. 表面技术,2017,46(12):206-217.

[4] 李发根,魏斌,邵晓东,等. 双金属复合管技术经济性分析[J]. 腐蚀科学与防护技术,2011,23(1):86-88.

[5] Fu A Q, Kuang X R, Han Y, el al. Failure analysis of girth weld cracking of mechanically lined pipe used in gasfield gathering system[J]. Engineering Failure Analysis,2016,68:64-75.

[6] 张立君,张燕飞,郭崇晓. 2205 双相不锈钢双金属复合管焊接工艺研究[J]. 焊管,2009,32(4):30-34.

[7] 赵为民. 金属复合管生产技术综述[J]. 焊管,2003,26(3):10-14.

[8] 曾德智,杜清松,谷坛,等. 双金属复合管防腐技术研究进展[J]. 油气田地面工程,2008,27(12):64-65.

[9] 丁晗,李先明,安超,等. L245/316L 双金属复合管的失效分析[J]. 腐蚀与防护,2018,39(2):

157-162.

[10] 李磊，邝献任，姬蕊，等.某油田316L/L360NB 机械式双金属复合管失效行为及原因分析[J].表面技术，2018，47(6)：224-231.

[11] 王玲玲，丁毅，马立群，等.316L 不锈钢酸洗工艺及耐点蚀性能研究[J].压力容器，2006(8)：21-23.

[12] 夏浩，周栋，丁毅，等.304不锈钢环保型酸洗钝化工艺及其性能研究[J].表面技术，2009，38(4)：47-49.

[13] Geng S N, Sun J S, Guo L Y. Effect of sandblasting and subsequent acid pickling and passivation on the microstructure and corrosion behavior of 316L stainless steel[J]. Materials & Design, 2015, 88：1-7.

[14] 应红，李岩，刘飞华，等.不锈钢现场钝化技术[J].材料保护，2014，47(12)：11-12.

[15] Wallinder D, Pana J, Leygraf & A C, et al. Electrochemical investigation of pickled and polished 304L stainless steel tubes [J]. Corrosion Science, 2000, 42：1457-1469.

本论文原发表于《腐蚀与防护》2020 年第 41 卷第 11 期。

干湿交替砂土环境下 X80 管线钢的腐蚀行为研究

李丽锋[1,2]　李 超[3]　罗金恒[1]　罗立辉[4]　昝聪敏[4]

(1. 中国石油集团石油管工程技术研究院·石油管材及装备材料服役行为与结构安全国家重点实验室；2. 中国石油大学(华东)储运与建筑工程学院；
3. 中石油管道有限责任公司西部分公司；
4. 西安石油大学材料科学与工程学院)

摘 要：针对管道在土壤干湿交替下的外腐蚀风险因素，采用腐蚀失重实验、电子显微分析、电化学测试等方法，研究了 X80 钢在砂土饱和水和干湿交替两种环境下的腐蚀行为。结果表明：X80 在干湿交替土壤环境下腐蚀速率达 0.33mm/a，属极严重腐蚀，是在饱和水土壤中的 2~3 倍，这主要是由于在干湿交替下其腐蚀产物作为氧化剂参与反应使得阴极电流变大，加速了阳极 Fe 的溶解。电化学测试表明：在同等土壤腐蚀环境下，焊缝比母材更易被腐蚀，焊缝试样自腐蚀电流更高，腐蚀速率更大。

关键词：X80 管线钢；干湿交替；土壤；腐蚀；焊缝；母材

随着我国油气消费量和进口量的快速增长，截止到 2018 年我国已建成油气长输管道 1.3×10^{5}km，其中以西气东输一线、二线、三线、陕京四线、中俄东线为代表的高钢级、大口径管道已达 4×10^{4}km，建设规模与需求仍在逐年提升。这些管道为我国长距离油气输送发挥重要作用的同时，在服役过程中，仍承受着腐蚀、地质灾害、第三方破坏等风险因素的威胁，因此，如何预测与预防管道风险，避免恶性事故发生，保障管道安全运行，是当前重大安全需求[1,2]。土壤外腐蚀是管道失效的主要风险因素之一，而含水率是影响土壤外腐蚀严重程度的重要因素，含水量的多少决定了腐蚀过程中的电导率和氧的扩散速度，从而影响材料腐蚀速度。土壤含水率对管线钢腐蚀行为已有大量的研究[3-6]，主要集中在两个方面：一是在真实土壤中通过添加不同含水率来研究土壤腐蚀情况；二是在土壤模拟溶液中通过间隔浸泡的干湿交替来研究管线钢的腐蚀情况。

在含水率影响方面，陈旭等[7-8]研究了含水量对我国典型的滨海盐渍土壤和鹰潭酸性土壤对管线钢腐蚀的影响，发现 X70 管线钢在滨海盐渍土壤中水质量分数为 20% 和 25% 时发生局部腐蚀，当含水量高于 30% 时试样会发生均匀腐蚀，主要是当含水量达到一定值时试样表面会形成连续液膜，而水中溶解氧有限不易形成氧化膜，造成腐蚀进程减缓；在研究酸性土壤时发现管线钢的腐蚀速率与含水率存在一定的线性关系，当土壤水饱和时腐蚀速率也达到最大，腐蚀形态由局部腐蚀逐渐发展到均匀全面腐蚀，在土壤含水量为 13% 时，阴极

基金项目：陕西省创新人才推进计划—青年科技新星项目(2018KJXX-064)资助。
作者简介：李丽锋(1983—)，高级工程师，硕士，主要研究方向为油气管道及储运设施风险评估、检测评价及失效分析，电话：029-81887682，E-mail：lilifeng004@cnpc.com.cn。

反应为吸氧反应，随着含水量的增加 H^+ 和氧共同参与阴极反应，土壤含水率与 pH 值共同作用影响管线钢的电化学腐蚀行为。

在干湿交替影响方面，任呈强等[9]研究了 X80 钢在 24h 内不同间浸时间下的腐蚀情况，所采用的方法是首先将待测试样全浸在腐蚀液中，腐蚀浸泡到既定时间，取出样品并悬挂在空气中到一个间浸结束，每个间浸循环为 24h。结果显示，当干/湿交替时间比为 9/15h 时腐蚀速率最高，是全浸腐蚀速率的 1.8 倍，相比于全浸试样，间浸试样的腐蚀产物水氧充足，因此，Fe^{2+} 氧化生成更加稳定的 Fe^{3+}，产物以 FeOOH 为主，而 FeOOH 又可消耗阳极产物 Fe^{2+} 生成 Fe_3O_4，与原本的阴极氧水解反应共同作用加速试样腐蚀；刘建国等[10]研究发现在干湿交替腐蚀阴极反应中氧控制特征基本消失，腐蚀以原产物还原的电荷控制。钢材的干湿交替腐蚀电流是自身的腐蚀电流和腐蚀产物氧化还原电流的叠加，这使得钢模拟干湿交替的腐蚀速率远大于全浸区腐蚀速率。

上述两个方面在一定程度上表征了土壤含水率对管线钢腐蚀的影响，前者在土壤中以几种恒定的含水量研究其对管线钢的腐蚀情况，未考虑土壤环境中含水量动态变化的特点，后者模拟了干湿交替的土壤环境，但通过在配置的土壤模拟溶液中间隔浸泡试样来实现干湿交替工况，且干湿交替时间过短，与管线钢服役的实际土壤环境存在差异，且未能考虑实际土壤中各因素对腐蚀的交互影响。为此，本工作选用实际土壤，通过添加去离子水保证土壤含水率的方式，研究了 X80 管线钢在干湿交替砂土环境下的腐蚀行为，并与水饱和土壤环境下进行了对比，以深化管线钢土壤腐蚀的认识，并为含外腐蚀缺陷管道剩余寿命预测和外腐蚀风险评估提供技术支撑。

1 试验材料与方法

1.1 试验材料

试验用土样取自陕北靖边，土壤类型为砂土，pH 值为 7.86，饱和含水率为 23%，Cl^- 浓度为 0.522%，SO_4^{2-} 浓度为 0.638%，HCO_3^- 浓度为 0.045%，试验前经自然风干后并 120 目筛子过滤后备用。试验用管材为西二线 X80 螺旋焊管，化学成分见表 1。采用线切割在母材纵向和螺旋焊缝上分别取样，试样类型包括片状和圆形试样，规格分别为 40mm×10mm×3mm 和 ϕ10mm×3mm。片状试样用于失重试验、微观形貌观察及腐蚀产物分析，试验前采用 200～1000 号水磨砂纸依次打磨，并经蒸馏水冲洗，丙酮除油，无水酒精脱脂后吹干；圆形试样用于电化学试验，顶面焊接通路导线，用环氧树脂和固化剂混合密封非工作面。

表 1 X80 管线钢的化学成分(质量分数)

元素	C	Mn	Si	S	P	Cr	Mo
w(%)	0.042	1.860	0.240	0.002	0.010	0.025	0.230
元素	Nb	Ni	V	Ti	Cu	Fe	
w(%)	0.060	0.180	0.004	0.015	0.170	余量	

1.2 试验方法

将干燥后的土壤置于恒温水浴锅内，并将管材试样埋入 5cm 下的土壤中进行水饱和环境和干湿交替环境下腐蚀试验。水饱和腐蚀试验时每天定时称重，并添加去离子水保证土壤的水饱和性；干湿交替腐蚀试验时每天定时称重，待水含量蒸发减重 80% 左右添加去离子水至饱和，循环往复直到试验结束。试验温度为 35℃ 恒温，试验过程中保持自然通风状态，试验方案见表 2。达到埋入周期后，将试样从土壤中取出，用软毛刷蘸取除锈液(500mL 盐酸+500mL 去离子水+3.5g 六次甲基四胺）彻底清除试样表面锈迹，然后用蒸馏水冲洗，酒精脱水，冷风吹干后放入干燥器，试样充分干燥后经分析天平称量，采用失重法计算试样腐蚀速率。采用 JSM-6390A 扫描电镜观察腐蚀后试样表面形貌，并进行能谱分析。

表 2 干湿交替试验方案

试验编号	材质	土壤	含水量	埋入周期(d)
1	X80 母材	砂土	水饱和	20、40、60、90
2	X80 母材	砂土	干湿交替	20、40、60、90
3	X80 焊缝	砂土	水饱和	20、40、60、90
4	X80 焊缝	砂土	干湿交替	20、40、60、90

采用 CS350 电化学工作站，将管材、饱和甘汞和铂片分别作为工作电极、参比电极和辅助电极，采用土壤模拟溶液作为导电溶液，测量腐蚀时间分别为 0d、3d、10d 的管材试样极化曲线，扫描速度为 0.5mV/s。依据 Tafel 曲线外推法对不同腐蚀状况的试样计算自腐蚀电流密度 J_{corr}，分析其变化规律。

2 结果与分析

2.1 水饱和和干湿交替砂土环境下 X80 管材的腐蚀速率

图 1 是水饱和和干湿交替砂土环境下 X80 母材及焊缝的腐蚀速率结果。如图 1 所示，对于 X80 母材，在饱和含水率土壤中的腐蚀速率为 0.12mm/a，基本属于中度腐蚀，随着埋入土壤时间的增大，腐蚀速率虽略有减缓但随后又恢复，这可能与前期生成的腐蚀产物阻碍了氧气及水分的扩散有关，而随着腐蚀时间的增加，靖边土样中较高含量的腐蚀性离子破坏腐蚀产物膜使得腐蚀速率加快；而在干湿交替的环境下，试样在埋入土壤 20d，腐蚀速率最高可以达到 0.3325mm/a，属于极严重腐蚀程度，在 40d、60d、90d 的时间里，虽然腐蚀速率有明显降低，但仍然属于严重腐蚀的程度。对于 X80 焊缝试样，在水饱和土壤中腐蚀较母材更为严重，基本属于严重腐蚀程度，腐蚀速率在 40d 时达到最大，随着埋入土壤时间进一步增大，腐蚀速率也呈现出减缓态势；在干湿交替的环境下，焊缝试样的腐蚀也都属于严重腐蚀，腐蚀速率在 20d 时能达到 0.3159mm/a，属于极严重腐蚀程度。

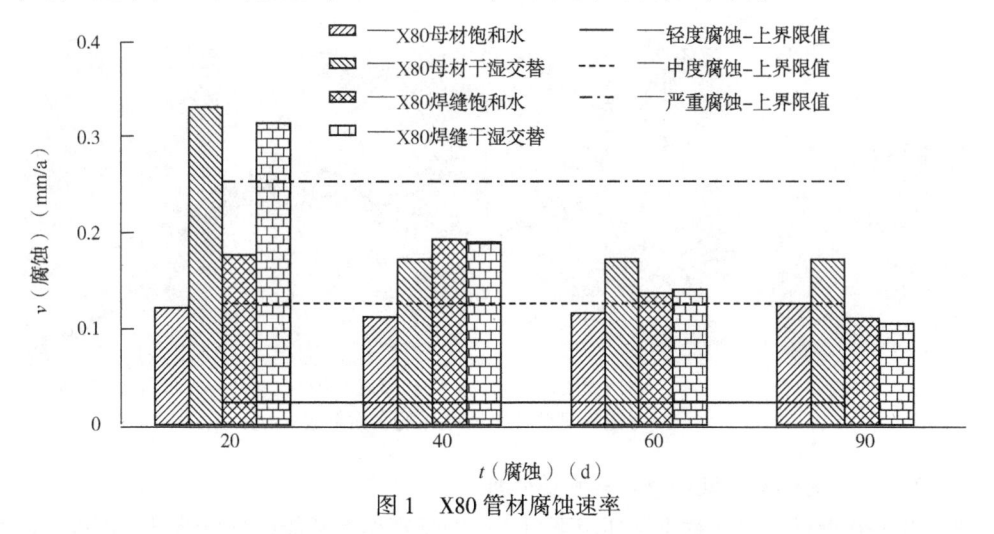

图 1 X80 管材腐蚀速率

2.2 腐蚀试样表面形貌与能谱分析

2.2.1 腐蚀试样表面宏观形貌

图2为试验周期20d后X80试样表面宏观形貌,可以看出,无论是母材还是焊缝试样,在埋入20d后表面都已被砂砾胶结覆盖,也可以很明显地看出,水饱和土壤中的试样砂砾层较为致密,砂砾层由内侧致密的细沙和外侧的粗砂组成,胶结非常紧密[图2(a)和图2(c)];干湿交替环境下的试样胶结砂砾层空隙粗大,用小刀轻微刮掉砂砾胶结层,可发现砂砾层分为两部分,内层靠近试样基体的是细小沙粒,呈蜂窝状附着在黑色锈层上,黏附力非常大,外层是较为粗大的砂砾,砂砾间黏结非常紧密。试样基体存在较为严重的沟垄状腐蚀形貌,同时,也能明显看出干湿交替环境下的试样砂砾层颜色较深一些,这应该是干湿交替环境下氧气能更加完整的形成稳定的 Fe_3O_4 的黑褐色产物[图2(b)和图2(d)]。随着时间的增加,试样表面的砂砾层越加致密厚重,这可能是腐蚀速率降低的直接原因。

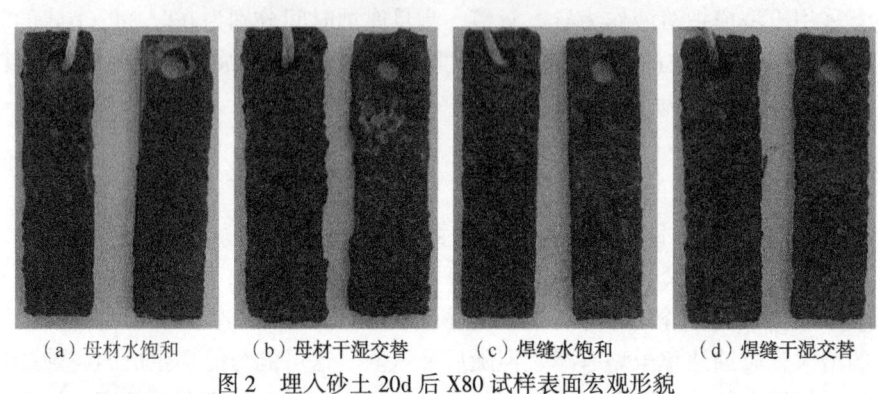

(a)母材水饱和　　(b)母材干湿交替　　(c)焊缝水饱和　　(d)焊缝干湿交替
图2　埋入砂土20d后X80试样表面宏观形貌

图3为去除腐蚀产物后试样的表面形貌,由图3可以看出:埋入土壤中20d的试样去除腐蚀产物后,水饱和土壤中母材试样有轻微的全面腐蚀,但基体机械加工划痕痕迹仍然可见[图3(a)],焊缝试样除了全面腐蚀外,基体局部还存在腐蚀坑[图3(c)];在干湿交替土壤情况下,母材试样基体已被完全覆盖,存在大量细小的腐蚀坑,加工痕迹已经基本消失[图3(b)],而焊缝试样腐蚀更为严重,不仅基体表面存在着大量腐蚀坑,并且腐蚀坑呈溃疡状态,基体表面已经完全被破坏[图3(d)]。

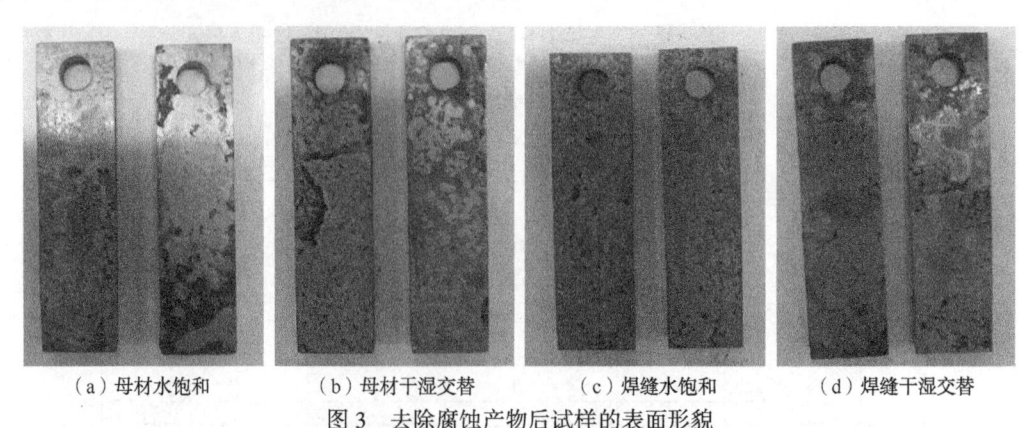

(a)母材水饱和　　(b)母材干湿交替　　(c)焊缝水饱和　　(d)焊缝干湿交替
图3　去除腐蚀产物后试样的表面形貌

2.2.2 腐蚀试样表面微观形貌与能谱分析

图4为X80母材干湿交替土壤环境下20d后的微观形貌及能谱分析结果。可以看出腐蚀

产物在试样表面呈多层状分布,内层有较为致密的腐蚀膜存在[图4(a)],外侧的产物层已经完全龟裂,一部分已经滑落,部分完全龟裂的产物层附着在试样表面,裂隙呈晶界状分布[图4(b)],无法抵挡水分及腐蚀性离子的进入,这与失重试验的结果相互吻合。能谱分析结果表明,产物以铁的氧化物为主,还有相当的硫化物存在,这可能是土壤中存在的微生物腐蚀代谢的产物[图4(d)、图4(e)]。

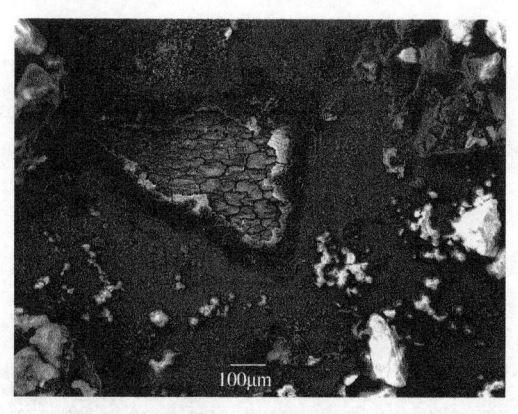

(a)腐蚀产物形貌

(b)外层腐蚀产物

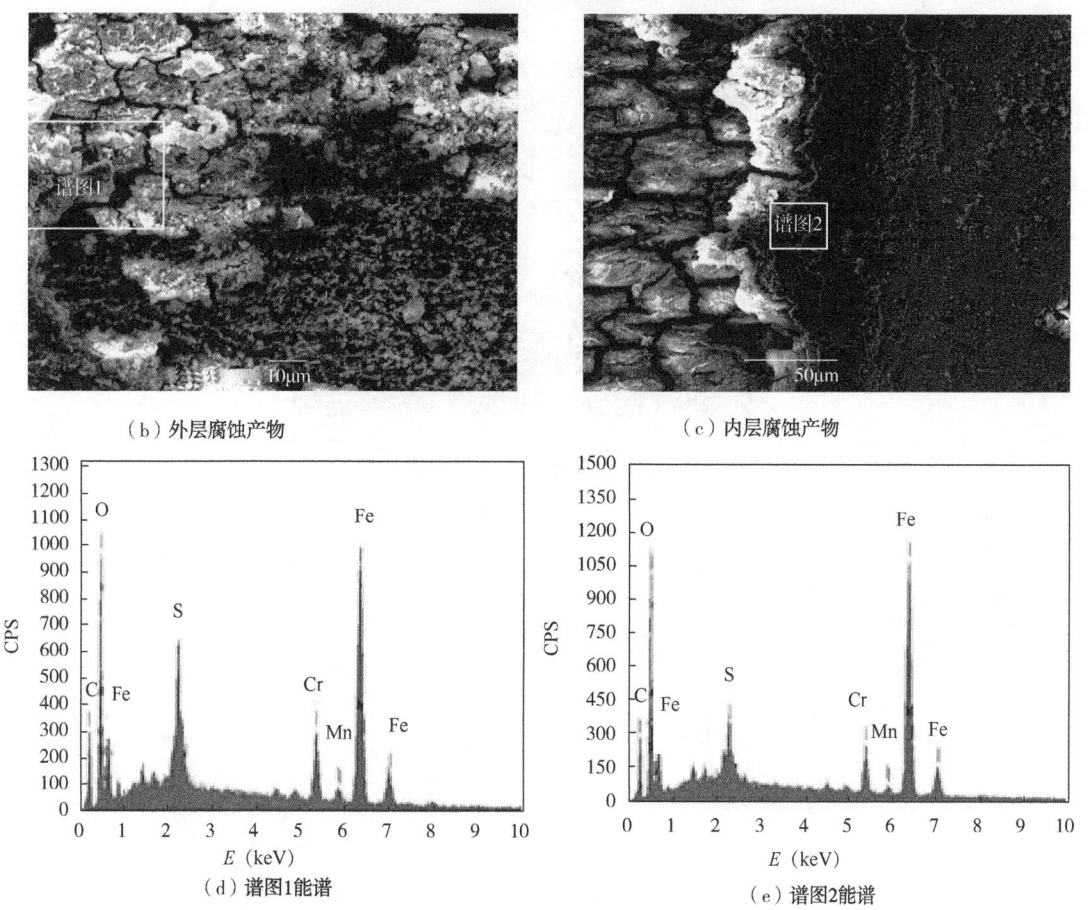

(c)内层腐蚀产物

(d)谱图1能谱

(e)谱图2能谱

图4 20d 干湿交替土壤腐蚀环境下 X80 母材表面的 SEM 形貌及 EDS 谱

图5是水饱和土壤环境下 X80 焊缝在 20d 的 SEM 及 EDS 图,可以看出,腐蚀产物呈内

外两层分布[图5(a)]，内层腐蚀产物膜并不完整，腐蚀形貌多样，这可能是焊缝试样基体活化程度不同导致的，外层腐蚀产物是在内层腐蚀产物膜层上局部向外发展而成，呈团簇状半球形扩展，其延伸孔隙较大，并且靠近内层腐蚀产物的膜层存在龟裂，裂隙呈放射状向外扩展。能谱分析结果表明[图5(c)、图5(d)]，内层膜状腐蚀产物以铁的氧化物为主，外层腐蚀产物主要为铁的氧化物，且呈 FeOOH 的片状或棉花状形貌特征[10]，并存在大量的 Cl⁻吸附。

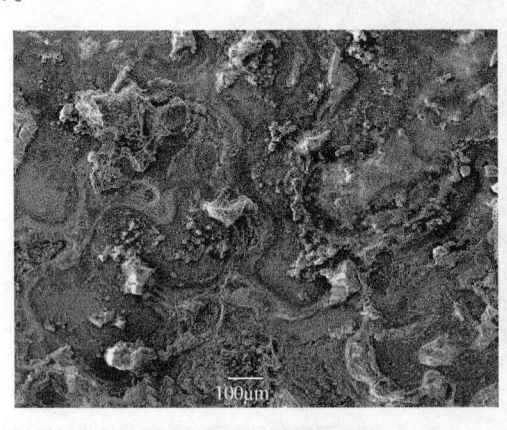

(a) 腐蚀产物形貌　　　　　　　(b) 外层腐蚀产物

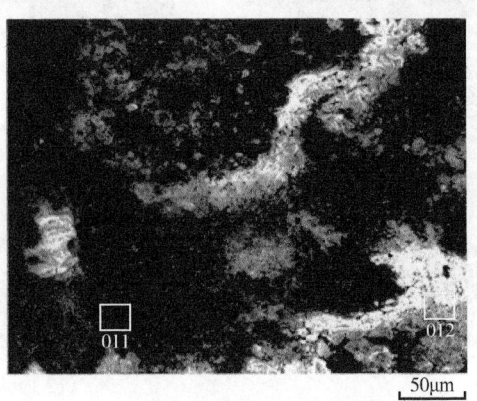

(c) 内、外层腐蚀产物能谱分析位置

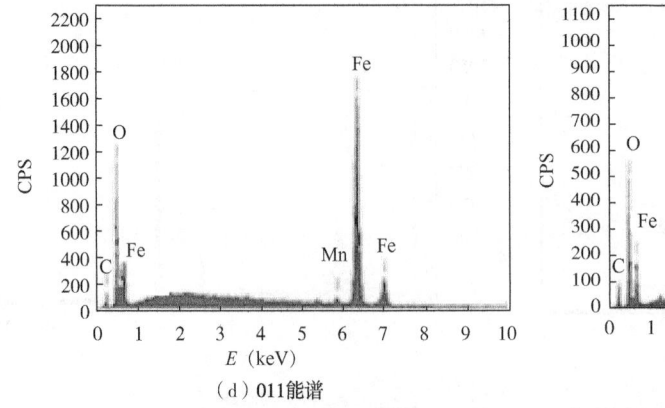

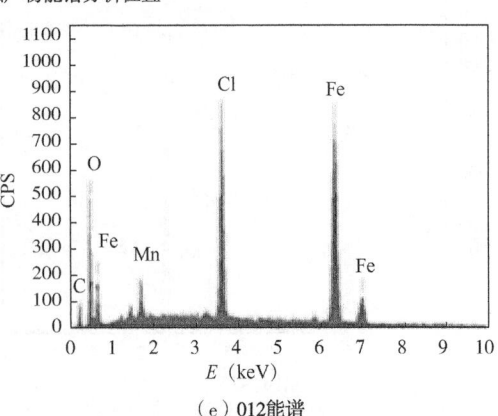

(d) 011能谱　　　　　　　(e) 012能谱

图 5　20d 水饱和土壤腐蚀环境下 X80 焊缝表面 SEM 形貌及 EDS 谱

图 6 是 X80 焊缝在干湿交替土壤中 20d 的 SEM 形貌及 EDS 谱。

· 420 ·

可以看出，试样基体表面存在厚重的氧化产物层，产物层存在粗大的裂隙，呈龟裂状覆盖在试样基体表面[图6(a)、图6(b)]，产物层可分为两层，靠近试样基体的内层膜状产物层厚重，外层分布有取向各异的层片状形貌产物，沿试样表面发展成片状聚集，腐蚀产物整体并不致密[图6(c)]，内侧的腐蚀产物膜层已经龟裂，裂隙沿六边形方向延伸，外层的片层状产物缝隙最大。能谱分析结果表明，腐蚀产物都以铁的氧化物为主，还存在少量的硫化物。

(a) 腐蚀产物形貌

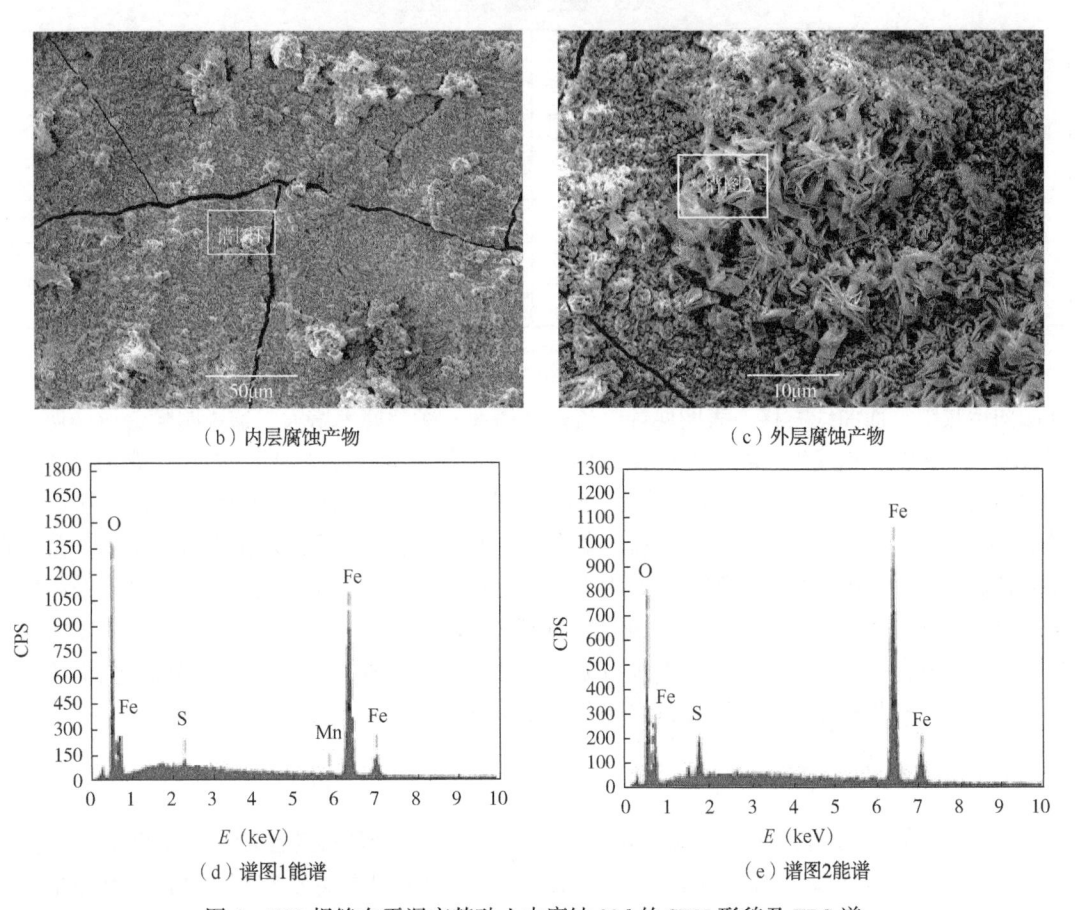

(b) 内层腐蚀产物

(c) 外层腐蚀产物

(d) 谱图1能谱

(e) 谱图2能谱

图6 X80焊缝在干湿交替砂土中腐蚀20d的SEM形貌及EDS谱

2.3 电化学极化曲线分析

图7是X80母材及焊缝试样在砂土模拟溶液中的极化曲线,表3为对应的极化曲线拟合结果。

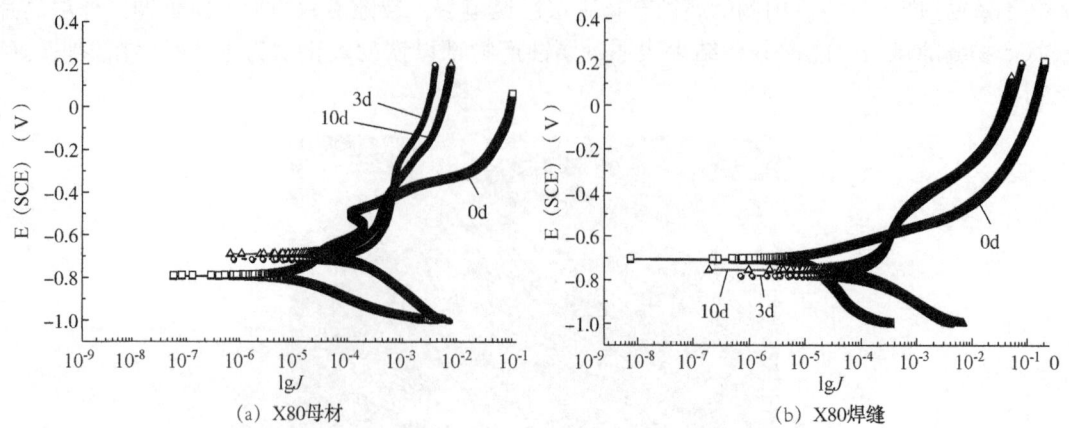

(a) X80母材

(b) X80焊缝

图7 砂土模拟溶液中不同表面腐蚀状态的X80钢极化曲线

(J为自腐蚀电流密度,单位为A/cm^2)

表3 极化曲线拟合数据

试 样	$t(d)$	$b_a(mV)$	$b_c(mV)$	$J_{corr}(A \cdot cm^{-2})$	$E_{corr}(mV)$
X80母材	0	179.69	99.908	$9.503×10^{-6}$	−794
	3	1562.3	289.18	$4.638×10^{-4}$	−714
	10	990.69	294.26	$2.861×10^{-4}$	−690
X80焊缝	0	78.45	307.30	$1.296×10^{-5}$	−707
	3	1469.8	219.29	$3.023×10^{-4}$	−781
	10	740.94	209.6	$2.189×10^{-4}$	−755

可以看出,X80母材在砂土溶液中腐蚀3d后,曲线上移,试样自腐蚀电位升高,说明腐蚀倾向性减缓,而曲线右移,表明腐蚀电流密度增大,腐蚀速率增大,这说明随着试样腐蚀时间的增加,试样表面腐蚀产物对基体保护作用不大,当继续浸泡时,曲线右移,腐蚀电流密度增幅不大,这说明随着腐蚀时间的延长,后期试样基体表面的腐蚀产物阻碍了进一步的腐蚀[图7(a)]。同时可以看出,X80母材试样在3d,10d情况下,阳极区曲线斜率很大,其阳极tafel常数均高于阴极,反应以阳极控制为主,说明其腐蚀电阻增大,基体外部的锈层及板结的土壤对基体保护接近于钝化。X80焊缝试样同样随着浸泡腐蚀时间的延长,曲线持续右移下移,说明其腐蚀倾向性持续增大,腐蚀电流密度增大,速率增大,这可能与焊缝表层活化无法生成完整膜状产物有关,X80焊缝在溶液中一直处于腐蚀加速状态[图7(b)]。

3 腐蚀机理分析

根据土壤成分分析,在土壤饱和含水情况下,试样所处环境与土壤模拟溶液类似,使得土壤中离子的迁移不受土壤因素的干扰,因此,该腐蚀过程主要分为两个过程,首先是动力学控制阶段,金属的腐蚀主要受氧传输的控制,即水中溶解氧向金属表面的扩散控制,在腐蚀过程中,金属的溶解会形成表面锈层,这时候的锈层非常薄并且疏松多孔,不会产生明显的电阻极化，腐蚀量与浸泡时间呈正比，由于锈层没有保护作用，氧很容易扩散到达金属界面参与阴极还原的去极化，进入到金属界面的氧越多，腐蚀就越快，腐蚀速度就越高。其次是氧扩散控制阶段，随着腐蚀过程的进行，锈层逐渐地增厚并且分层，易钝化金属表面形成致密的锈层逐渐阻止溶解氧向金属界面的传送，产生明显的阴极极化，表面锈层的不断增厚使腐蚀速率也逐渐降低，而由于土壤中的无机阴离子如 Cl^-、SO_4^{2-} 的含量都较为丰富，会发生近程力引起的非库仑力吸附，阴离子水化程度低，能够击穿试样表层的腐蚀产物膜层使得腐蚀继续较快进行[11,12]。根据土壤分析表明，靖边土壤中腐蚀性离子高，因此其具有强的腐蚀性，这与失重试验结论相吻合。

干湿交替情况下腐蚀可分干湿两个阶段，首先当在湿润条件下，土壤水含量充足，含氧量下降，水溶解土壤中的无极阴离子如 Cl^-、SO_4^{2-} 等，这些离子靠近试样基体发生特征吸附，在金属表面膜薄弱处排挤掉试样表层晶格中的 Fe，使得 Fe^{2+} 游离于基体外部，反应式见下式，表现为阳极区 Fe 失去电子；

$$Fe \longrightarrow Fe^{2+}+2e \tag{1}$$

由于试验用土壤 pH 值均呈弱碱性，因此自由 H^+ 含量有限，基本不会发生的 H^+ 在电极表面溶液层中的还原反应，所以在阴极发生吸氧反应，氧分子发生还原反应，见式(2)至式(5)。同时，由于土壤中均有较高含量的 Cl^- 和 SO_4^{2-}，在阴极区也可发生反应见式(6)、式(7)[13]。

$$2H_2O+O_2+4e \longrightarrow 4OH^- \tag{2}$$

$$Fe(OH)_2+O_2 \longrightarrow 2FeOOH \tag{3}$$

$$2Fe(OH)_3+nH_2O \longrightarrow Fe_2O_3 \cdot nH_2O+3H_2O \tag{4}$$

$$Fe_2O_3 \cdot nH_2O \longrightarrow Fe_2O_3+nH_2O \tag{5}$$

$$SO_4^{2-}+4H_2O+8e \longrightarrow S^{2-}+8OH^- \tag{6}$$

$$Fe^{2+}+S^{2-} \longrightarrow FeS \tag{7}$$

其次当土壤水分逐渐蒸发过程中，含水率下降土壤透气性增强，作为腐蚀反应动力的氧含量增加，腐蚀产物表层，水分和氧气供应充足，因此，Fe^{2+} 会生成 Fe^{3+}，产物以 FeOOH 为主；该阶段中阳极反应仍然为 Fe 基体的腐蚀，且前期湿润阶段基体周围固结的土壤板结层能保留一定的水分，土壤中腐蚀性离子可以通过锈层孔洞的电解质溶液到达基体，而在阴极区除了发生吸氧反应外，前期的腐蚀产物 FeOOH 作为氧化剂，发生氧化还原反应转化为 Fe_3O_4，阴极区发生的反应导致阴极电流变大，从而大大加速了腐蚀的发生。

而在相同条件下，焊缝的耐蚀性更差，无论是失重试验还是电镜观察，焊缝试样由于焊接热影响，基体活化重结晶，与母材试样相比，焊缝区域各部位因加热和冷却的条件不同，导致基体电位差异，再加上焊接后更加活化的晶界，使得焊缝试样的腐蚀倾向性更大。外部离子也更易侵入基体造成腐蚀，腐蚀倾向性更加明显。

4 结论

（1）相比水饱和土壤环境，X80 管线钢在干湿交替土壤环境下腐蚀更为严重，是在水饱和土壤中的 2~3 倍，这主要是由于在干湿交替下其腐蚀产物 FeOOH 作为氧化剂参与反应使得阴极电流变大，加速了阳极 Fe 的溶解。

（2）在同等土壤腐蚀环境下，焊缝比母材更易被腐蚀，焊缝试样自腐蚀电流更高，腐蚀速率越大。

参 考 文 献

[1] 杨峰平，卓海森，罗金恒，等．油气输送管失效案例与原因分析[J]．石油管材与仪器，2015，1(3)：63-66．

[2] 王婷，王新，李在蓉，等．国内外长输油气管道失效对比[J]．油气储运，2017，36(11)：1258-1264．

[3] 郭建永，胡军，李庆达，等．国内外油气管道腐蚀与防护的研究进展[J]．材料保护，2017，50(6)：83-87．

[4] AYDIN H, NELSON T W. Microstructure and mechanical properties of hard zone in friction stir welded X80 pipeline steel relative to different heat input [J]. Materials Science & Engineering A, 2013, 586(6): 313-322.

[5] 李雪，朱庆杰，周宁，等．油气管道腐蚀与防护研究进展[J]．表面技术，2017，46(12)：206-217．

[6] 刘建国，李言涛，侯保荣．海洋浪溅区钢铁腐蚀与防护进展[J]．腐蚀与防护，2012，33(10)：833-836．

[7] 陈旭，杜翠薇，李晓刚，等．含水量对 X70 钢在大港滨海盐渍土壤中腐蚀行为的影响[J]．北京科技大学学报，2008，30(7)：730-734．

[8] 陈旭，杜翠薇，李晓刚，等．含水率对 X70 钢在鹰潭酸性土壤中腐蚀行为的影响[J]．石油化工高等学校学报，2007，20(1)：55-58．

[9] 任呈强，李丽，王煦，等．管线钢在干湿交替环境下的腐蚀[J]．腐蚀与防护，2011，33(4)：272-275．

[10] 刘建国，李言涛，侯保荣．N80 钢模拟全浸区和干湿交替试样的腐蚀行为[J]．腐蚀与防护，2012，33(11)：925-927．

[11] 胥聪敏，张璇，罗立辉．海滨盐碱土壤中硫酸盐还原菌对 X100 管线钢腐蚀行为的影响[J]．材料保护，2017，50(6)：40-43．

[12] 陈旭，吴明，何川，等．外加电位对钢及其焊缝在库尔勒土壤模拟溶液中行为的影响[J]．金属学报，2010，46(8)：951-958．

[13] 石志强，张秀云，王彦芳，等．SO_4^{-2} 对 X100 管线钢在盐渍性溶液中点蚀行为的影响[J]．中国石油大学学报(自然科学版)，2016，40(1)：128-133．

本论文原发表于《Material Protection》2020 年第 53 卷第 3 期。

一种乙烯裂解炉管高温损伤评估的新方法

徐秀清[1,2]　吕运容[3,4]　尹成先[1,2]　李伟明[3,4]

(1. 中国石油集团石油管工程技术研究院；
2. 石油管材及装备材料服役行为与结构安全国家重点实验室；
3. 广东省石化装备故障诊断重点实验室；4. 广东石油化工学院)

摘　要：渗碳和蠕变是引发乙烯裂解炉管失效的最主要损伤机理。按照渗碳体形态特征和蠕变孔洞的状态裂解炉管的高温损伤可分为7个阶段，在这7个阶段中，微观上晶内Cr/Ni含量发生明显且持续的变化，这种变化将导致材料的磁特征参数——矫顽力发生变化，主要表现为在高温损伤早期矫顽力曲线呈现"U"形特征，而材料进入损伤末期阶段则呈现出倒"U"形曲线，因此可以用矫顽力作为磁特征参数，用于识别乙烯裂解炉管所处的损伤状态，从而评估炉管的损伤状态等级。

关键词：渗碳；蠕变；矫顽力；饱和磁化强度；损伤状态评估

乙烯作为重要的化工原料，是目前为止用量最大的化学品之一[1]。目前，中国已成为世界第二大乙烯生产国[2]，并且仍处于一个迅猛发展的时期[3]。就工艺而言，乙烯的生产工艺主要通过裂解石油烃来获得乙烯产物[4]，其中绝大部分是采用管式炉裂解技术来实现的[5]。而作为乙烯裂解炉的最核心部件，裂解炉管的花费占到乙烯装置总投资的10%，其失效有可能引起炉内燃烧等安全事故，不仅会造成重大的经济损失，而且可能造成人员伤亡[6]。虽然裂解炉管的设计使用寿命一般在10×10^4h以上，但是寿命期内的失效事故时有发生，其中主要原因为渗碳[7]和蠕变[8]。目前，国内外研究主要集中在损伤机理的定性研究上[9]，其研究表明，对于离心铸造的HP型高温炉管来说，在室温下呈骨架状的基体组织，会在渗碳最终发展为网状组织，并在晶界碳化物附近出现蠕变孔洞和微裂纹[10]。通过实验证明，渗碳层越厚，炉管热膨胀系数越低，而热膨胀系数的差异又会使炉管力学性能持续恶化[11]。一般认为，这类损伤的深度由表面到心部，只是损伤程度有所不同[12]。

1 乙烯裂解炉管的损伤状态划分方法

一般认为，由于乙烯裂解炉管工作温度往往在800℃甚至更高，会在炉管内表面形成高碳势环境，有利于渗碳机理的形成，且蠕变孔洞与渗碳晶界有关，所以可以综合渗碳体形态特征和蠕变孔洞的状态对高温损伤的严重程度进行分类和划分等级[13]。本文根据电子扫描显微镜和能谱分析结果，将炉管从原始状态到失效之间分为7个状态，即基本完好阶段、晶

基金项目：国家重点研发计划资助项目(编号：2017YFF0210406)。
作者简介：徐秀清(1981—)，女，山东日照人，博士，高级工程师，研究领域为炼化装置腐蚀与防护，已发表文章20余篇。

内二次渗碳体析出阶段、骨架状晶界开始粗化溶解阶段、蠕变早期萌生阶段、骨架状晶界全面溶解阶段、失效临界状态阶段和失效状态阶段。按照晶界形态、晶内二次渗碳体、晶内 Cr/Ni 元素分布、蠕变孔洞形态、晶内 Cr/Ni 含量 5 个特征描述各阶段的基本特征。

1.1 基本完好阶段

晶界形态的扫描电镜结果如图 1 所示,为骨架和块状形态,且晶界纤细而完整。从能谱的面扫描结果来看,Ni 元素为均匀分布,Cr 元素仅在晶界处有少量点状集中,也基本为均匀分布状态,如图 2 所示。总结图 1 和图 2 以及能谱元素分析结果,本阶段材料状态描述见表 1。该阶段没有任何蠕变孔洞存在,二次渗碳体也几乎不会发生,Cr 和 Ni 元素的集中分布区域,其含量较平均值仅有微量增加。

图 1 基本完整阶段的 SEM 图像

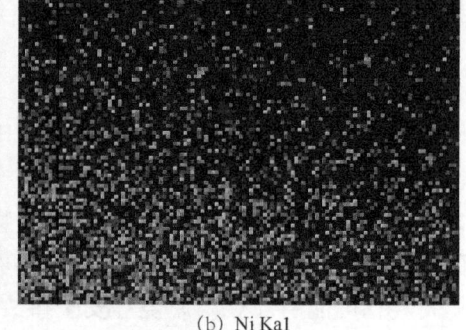

(a) Cr Kα1　　(b) Ni Kα1

图 2 基本完整阶段的 Cr/Ni 分布状态

表 1 基本完整阶段的材料微观特征描述

特征参量	特征描述
晶界形态	骨架和块状形态,且纤细、完整
晶内二次渗碳体	几乎无任何析出的二次渗碳体
晶内 Cr/Ni 元素分布	Cr 含量符合标准要求,且分布基本均匀,仅在晶界有稍微集中的分布
蠕变孔洞	无蠕变孔洞
晶内 Cr/Ni 含量	Cr 含量(质量分数)较扫描范围内的平均值低 2%~2.5%,Ni 含量(质量分数)较扫描范围内的平均值高 1.5% 以下

1.2 晶内二次渗碳体析出阶段

本阶段晶界形态如图 3 所示,整体仍为骨架形态,但晶界有所粗化。从能谱的面扫描结果来看,Ni 元素较前一阶段没有明显变化,而 Cr 元素则由于晶界粗化而呈现出向链状变化的趋势,整体仍为均匀分布,如图 4 所示。结合能谱元素分析结果,本阶段材料状态描述见表 2。该阶段仍没有任何蠕变孔洞出现,开始有细小颗粒状二次渗碳体从晶内析出,Cr 和 Ni 元素的集中程度较上一阶段有小幅度增加。

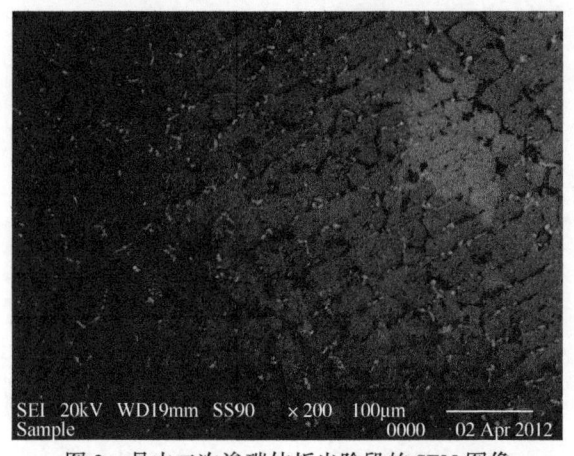

图3 晶内二次渗碳体析出阶段的SEM图像

(a) Cr Ka1　　　　　　　　　　　　(b) Ni Ka1

图4 晶内二次渗碳体析出阶段的Cr/Ni分布状态

表2 晶内二次渗碳体析出阶段的材料微观特征描述

特征参量	特征描述
晶界形态	仍为骨架和块状形态,且较之前有所粗化,但仍然完整
晶内二次渗碳体	少量点状细小二次渗碳体析出
晶内Cr/Ni元素分布	Cr含量无明显变化,且分布基本均匀,仅在晶界有稍微集中的分布
蠕变孔洞	无蠕变孔洞
晶内Cr/Ni含量	Cr含量(质量分数)较扫描范围内的平均值低2.5%~4%,Ni含量(质量分数)较扫描范围内的平均值高1.5%~2.0%

1.3 骨架状晶界开始粗化溶解阶段

本阶段晶界形态较基本完好节点变化比较明显,如图5所示,开始出现明显的不连续晶界,且晶界粗化越加明显。从能谱的面扫描结果来看,Ni元素分布图中开始出现少量非连续的"黑洞",而Cr元素在晶界的分布形态随着晶界明显粗化而转化为链状分布,如图6所示。结合能谱元素成分分析结果,本阶段基本状态描述见表3。该阶段仍没有任何蠕变孔洞出现,但细小颗粒状二次渗碳体在晶内大量析出且呈现弥散状分布,Cr和Ni元素的集中程度较前一阶段有明显增加,元素偏析量增加近1倍左右。

图5 骨架状晶界开始粗化溶解的SEM图像

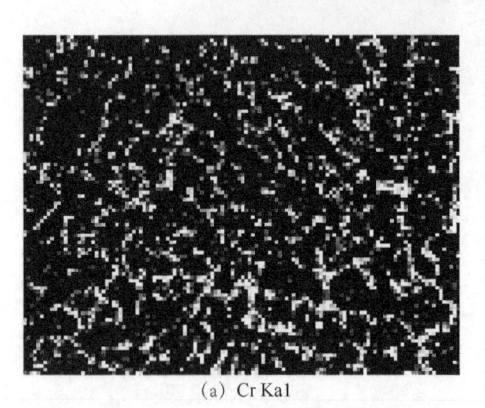

(a) Cr Kα1　　　　　　　　　　　(b) Ni Kα1

图6 骨架状晶界开始粗化溶解的Cr/Ni分布状态

表3 骨架状晶界开始粗化溶解阶段的材料微观特征描述

特征参量	特征描述
晶界形态	骨架状晶界开始溶解且开始明显变粗，出现断续的不连续晶界
晶内二次渗碳体	弥散大量点状细小二次渗碳体析出，且数量和密度逐渐增加
晶内Cr/Ni元素分布	Cr含量较最初状态微量减少，差别不大，但晶界有明显的集中分布，同时在晶界上开始出现相对孤立的贫Ni"黑洞"
蠕变孔洞	无蠕变孔洞
晶内Cr/Ni含量	Cr含量(质量分数)较扫描范围内的平均值低6%~8%，Ni含量(质量分数)较扫描范围内的平均值高3%~4%

1.4 蠕变早期萌生阶段

本阶段晶界形态如图7所示，骨架状晶界大部分溶解为块状不连续晶界，且晶界进一步粗化。从能谱的面扫描结果来看，Ni元素分布图中在晶界位置出现连续分布的"黑洞"，且"黑洞"位置基本与晶界上Cr元素的集中分布位置相对应，如图8所示。再结合能谱元素成分分析结果，本阶段基本状态描述见表4。该阶段蠕变孔洞开始以独立孔洞的分布形态出现，细小颗粒状二次渗碳体开始合并和粗化，Cr和Ni元素的集中程度较前一阶段并没有明显的变化，这与如图8所示的元素集中分布范围扩大但集中程度并未有明显提高的情况相

· 428 ·

吻合。

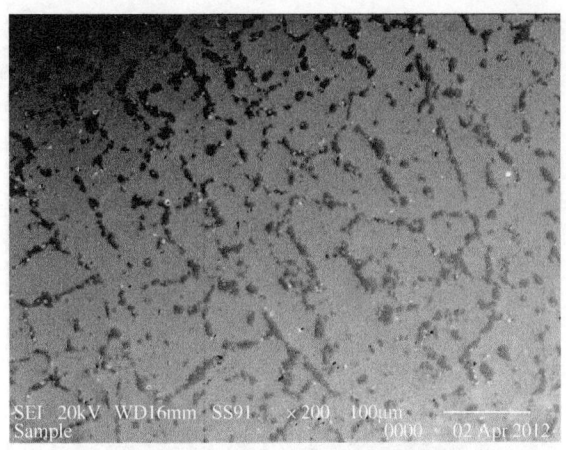

图 7 蠕变早期萌生的 SEM 图像

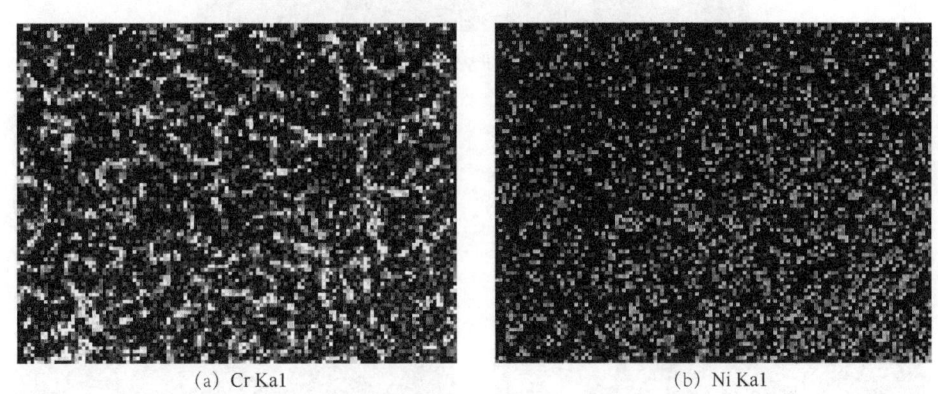

(a) Cr Ka1   (b) Ni Ka1

图 8 蠕变早期萌生的 Cr/Ni 分布状态

表 4 蠕变早期萌生阶段的材料微观特征描述

特征参量	特征描述
晶界形态	骨架状晶界大部分已发生溶解，且粗化明显，整体呈现为由块状渗碳体组成的链状晶界
晶内二次渗碳体	仍然弥散大量碳化物，但是细小碳化物明显出现合并和粗化
晶内 Cr/Ni 元素分布	Cr 含量较最初状态有所减少，减少百分量保持在个位数以内，且晶界 Cr 元素集中程度进一步增加，同时在晶界上出现连续的贫 Ni "黑洞"
蠕变孔洞	部分晶界边缘可以看到独立的蠕变孔洞，但孔洞并未连接成链状
晶内 Cr/Ni 含量	Cr 含量(质量分数)较扫描范围内的平均值低 7.5%～8.5%，Ni 含量(质量分数)较扫描范围内的平均值高 4.0%～5.0%

1.5 骨架状晶界全面溶解阶段

本阶段晶界形态如图 9 所示。骨架状晶界溶解并粗化为断续条块状晶界，且晶界溶解的渗碳体为晶内二次渗碳体的长大提供重要元素，导致 Cr 集中分布区域从晶界向晶内扩散，如图 10(a)所示。表现为 Cr 元素分布图上出现大量弥散分布的白色斑点，相应地，Ni 元素分布图中的"黑洞"也从晶界向晶内扩散，并在晶内出现大量且弥散分布的代表 Ni 元素高度集中的浅绿色亮斑。再结合能谱元素成分分析结果，本阶段基本状态描述见表 5。该阶段蠕变孔洞开始出现少量连续孔洞,细小颗粒状二次渗碳体团聚粗化为大块团状,Cr 和 Ni 元素的集中程度较前一阶段明显升高,升高幅度接近 1 倍左右,这与如图 10 所示的元素集中分布情况相吻合。

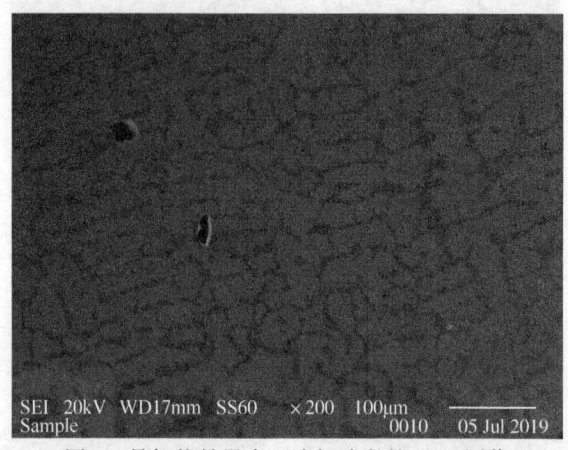

图 9 骨架状晶界全面溶解阶段的 SEM 图像

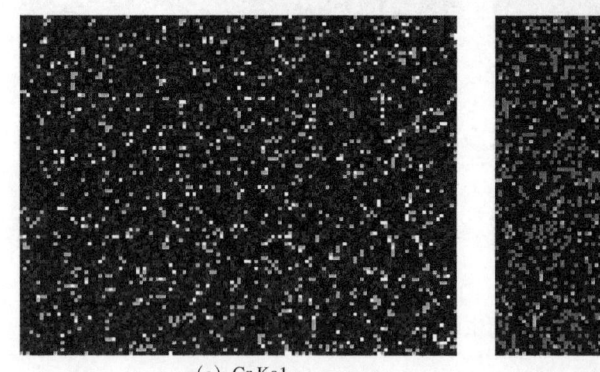

(a) Cr Ka1　　(b) Ni Ka1
图 10 骨架状晶界全面溶解阶段的 Cr/Ni 分布状态

表 5 骨架状晶界全面溶解阶段的材料微观特征描述

特征参量	特征描述
晶界形态	大量溶解并粗化,晶界表现为断续的条块状晶界
晶内二次渗碳体	二次渗碳体相互融合,继续粗化,并团聚为大块的团状粗化二次渗碳体
晶内 Cr/Ni 元素分布	Cr 含量进一步有所减少,且减少百分量达到双位数,但其含量仍为双位数,晶界 Cr 元素集中程度进一步增加,同时在晶界和晶内出现大量且连续的贫 Ni "黑洞"
蠕变孔洞	大量独立的蠕变孔洞,开始出现少量连续蠕变孔洞
晶内 Cr/Ni 含量	Cr 含量(质量分数)较扫描范围内的平均值低 8.0% ~ 15.0%,Ni 含量(质量分数)较扫描范围内的平均值高 5.0% ~ 13.0%

1.6 失效临界状态阶段

本阶段晶界形态如图 11 所示。由于残余未溶解的粗化晶界与长大的二次渗碳体相互连接,晶界转变为间距更加密集的网状晶界,导致 Cr 高度集中分布于新晶界上,如图 12(a)所示。表现为新境界在 Cr 元素分布图上呈现出白色网状分布,晶内呈现 Cr"黑洞";相应地,Ni 元素分布图中的"黑洞"延网状晶界连续分布,并在晶内出现连续分布的代表 Ni 元素高度集中的浅绿色团块。结合能谱元素成分分析结果,本阶段基本状态描述见表6。该阶段蠕变孔洞出现大量连续孔洞集中于新的网状晶界上,且孔径扩大,Cr和Ni元素的集中程度较前一阶段又进一步明显升高,偏差幅值均达到20%左右,这与如图12所示的元素块状集中分布的情况相吻合。

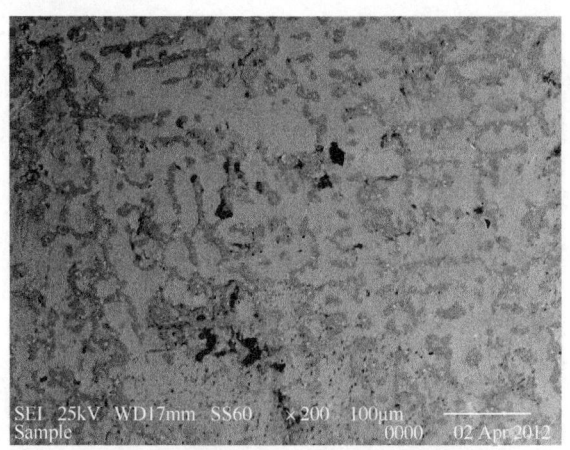

图11 失效临界状态阶段的SEM图像

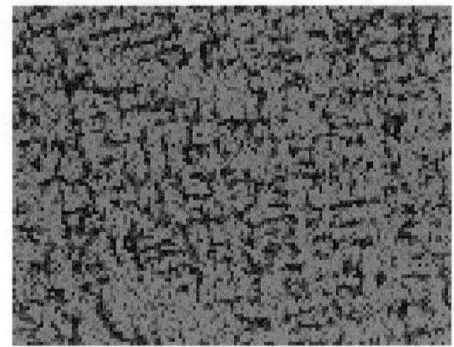

(a) Cr Ka1　　　　(b) Ni Ka1

图12 失效临界状态阶段的Cr/Ni分布状态

表6 失效临界状态阶段的材料微观特征描述

特 征 参 量	特 征 描 述
晶界形态	残留粗化晶界和晶内二次渗碳体重新连接形成粗化且网格密度更大(较原始骨架状晶界的网格密度)的网状晶界
晶内二次渗碳体	二次渗碳体进一步相互溶解,团聚,且数量大量减少
晶内Cr/Ni元素分布	Cr含量明显减少至个位数,晶界Cr元素集中程度进一步增加,导致晶内出现明显的贫Cr"黑洞",同时在晶界上出现呈条块状、网格状的贫Ni"黑洞"
蠕变孔洞	大量连续蠕变孔洞出现在晶界上(孔径小于$10\mu m$)
晶内Cr/Ni含量	Cr含量(质量分数)较扫描范围内的平均值15.0%~22.0%,Ni含量(质量分数)较扫描范围内的平均值高12.0%~18.5%

1.7 失效状态阶段

本阶段晶界形态如图13所示。网状晶界持续粗化,导致Cr在晶界上的集中程度进一步增加,晶界边缘出现大量连续蠕变孔洞,且孔径扩大,如图14(a)所示,Cr元素向晶界集中，而 Ni 元素向晶内集中的程度越加严重。通过能谱元素成分结果分析，本阶段基本状态描述见表 7。该阶段 Cr 和 Ni 元素的集中程度较失效临界状态阶段进一步升高，但增幅较失效临界状态阶段有所降低，可以认为在失效临界状态阶段是材料失效前损伤进入末期加速发展期的临界阶段。

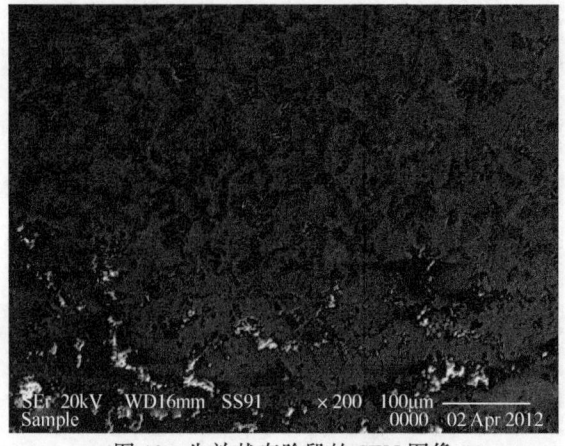

图 13 失效状态阶段的 SEM 图像

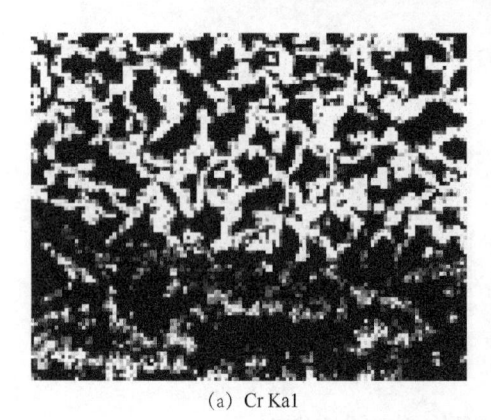

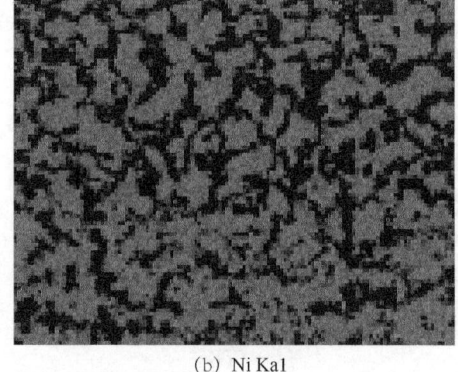

(a) Cr Kα1 (b) Ni Kα1

图 14 失效状态阶段的 Cr/Ni 分布状态

表 7 失效状态阶段的材料微观特征描述

特征参量	特征描述
晶界形态	高密度的网状晶界进一步粗化，但较之前的状态变化不明显
晶内二次渗碳体	二次渗碳体进一步团聚且粗化加重，且数量持续减少
晶内 Cr/Ni 元素分布	Cr 含量继续减少，晶界 Cr 元素集中程度进一步增加，导致晶内出现块状的贫 Cr"黑洞"，同时在晶界上出现网格化的贫 Ni"黑洞"
蠕变孔洞	大量连续蠕变孔洞出现在晶界上，且几乎布满所有晶界，甚至晶内也出现连续孔洞，(孔径小于 10μm)，这阶段较前一阶段最明显的变化为大部分孔洞连接成链状，甚至出现微裂纹
晶内 Cr/Ni 含量	Cr 含量(质量分数)较扫描范围内的平均值低 20.0%~26.5%，Ni 含量(质量分数)较扫描范围内的平均值高 17.0%~21.0%

2 乙烯裂解炉管各损伤状态的磁特征信号

从表1至表7可知，晶内Cr/Ni含量是唯一随着损伤程度增加而连续变化的特征参量，主要表现为高温损伤程度越深，晶内的Cr含量越低而Ni含量越高。就此可以认为，一个试样的高温损伤程度可以用晶内Cr含量或Ni含量与试样平均Cr含量或Ni含量之差加以表征。

另外，从高温合金的磁特征来说，材料中的Ni元素属于铁磁性材料，但是Cr元素和奥氏体相铁，即γ-Fe相，均为顺磁性，因此在各元素均匀分布状态下，含量占优的Cr元素和γ-Fe相严重削弱了Ni的铁磁性表现，使得材料整体表现出顺磁性。但是，随着高温损伤过程的不断深入，晶内的C和Cr元素会以渗碳体$Cr_{23}C_7$和Cr_7C_2的形式向晶界聚集，导致晶内脱C和脱Cr。晶内脱C使晶内的γ-Fe转变为铁磁性的α-Fe，脱Cr则导致晶内Ni元素含量增加，最终因晶内铁磁性的Ni和α-Fe含量相对增加而使材料的磁特性由顺磁性转化为铁磁性，在极端情况下，晶内的Cr含量接近0的"黑洞"处可以形成接近强磁性坡莫合金的Ni/Fe比例。

综上所述，晶内Ni/Cr比即可以表征材料高温损伤严重程度，也可以表示材料的磁特性变化程度，从而建立材料高温损伤严重程度与材料磁特征信号之间的定量关联。因此，本文以采用晶内Ni/Cr比与材料Ni/Cr比均值(含晶内和晶界在内)之差作为高温损伤过程微观特征变量，研究高温损伤程度与材料磁特征信号之间的定量关系和变化规律。

以矫顽力作为磁特征变量，针对7个损伤阶段的不同试样进行测量，并以晶内Ni/Cr比与材料Ni/Cr比均值(含晶内和晶界在内)之差作为横轴坐标，建立起高温损伤全寿命周期内矫顽力的变化规律曲线，如图15所示。其中，A至F分别表示对损伤阶段A到损伤阶段F对应的数据点。

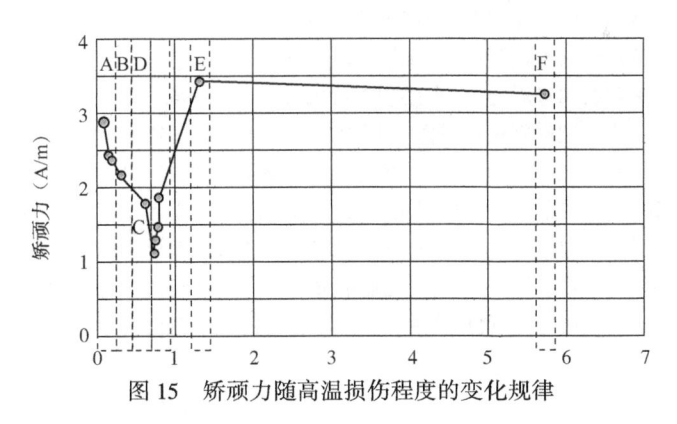

图15 矫顽力随高温损伤程度的变化规律

3 乙烯裂解炉管全寿命周期磁特征信号讨论

由图15可知，矫顽力随高温损伤严重程度并不是一个单调变化关系，而是前期呈现出一个"U"形曲线，而末期呈现倒"U"形曲线。通过观察图1至图14以及表1至表7可以发现A、B、C三个阶段的矫顽力之所以下降，主要是由于渗碳体不断形成导致晶内脱碳脱铬，使晶内饱和磁化强度增大，而晶内仅有少量二次渗碳体，且没有形成密集分布，不足以抵消饱和磁化强度增大对矫顽力的降低作用，从而导致这三个阶段矫顽力不断降低。在D、E阶段，晶内大量形成密集的二次渗碳体，由于这些二次渗碳体的直径均较小，会产生钉扎效应，大量的钉扎效应作用导致这2个阶段矫顽力开始回升。当高温损伤进入F阶段，晶内Cr含量降低到个位数，Ni/Fe比例接近强磁性坡莫合金，饱和磁化强度大幅增加，同时二次渗碳体大量融合，蠕变孔洞也开始大量形成，并开始长大或链接成链状（蠕变孔洞成核）。此时，杂质的反磁化核成核效应代替钉扎效应成为主导效应，导致进入失效前的临界阶段矫顽力开始从最高点下降，但该阶段更加密集的网状晶界已经形成，且晶界和蠕变孔洞仍有持续长大的趋势，对矫顽力的下降速度起到了抑制作用。最终，在该阶段矫顽力虽然开始回落，但回落速度较最初的A、B、C三个阶段要低很多。

4 结论

乙烯裂解炉管，按照渗碳体形态特征和蠕变孔洞的状态，从原始状态到失效之间大致可分为7个状态，即基本完好阶段、晶内二次渗碳体析出阶段、骨架状晶界开始粗化溶解阶段、蠕变早期萌生阶段、骨架状晶界全面溶解阶段、失效临界状态阶段和失效状态阶段，各状态之间可按照晶界形态、晶内二次渗碳体、晶内Cr/Ni元素分布、蠕变孔洞形态、晶内Cr/Ni含量5个特征参量加以区分。其中，晶内Cr/Ni元素在微观层面表征炉管的损伤程度，并且晶内Cr/Ni元素之比对材料的磁特性有显著影响，可以利用矫顽力等磁特征参数来测量高温损伤的这种微观变化，从而判断炉管高温损伤所处的阶段。其矫顽力特征曲线在高温损伤早期呈现为"U"形曲线，而在高温损伤末期则呈现倒"U"形曲线，这种特征可以用于识别高温损伤的发展阶段，从而实现乙烯裂解炉管的损伤状态评估。

参 考 文 献

[1] 刘方涛. 我国乙烯工业现状及发展前景[J]. 化学工业, 2010, 28(1)：1-4.
[2] 钱伯章. 中国乙烯工业市场和原料分析[J]. 中外能源, 2011, 16(6)：62-73.
[3] 王红秋, 郑轶丹. 我国乙烯工业强劲增势未改[J]. 中国石化, 2019(1)：27-30.
[4] 王可, 张洪林. 当代乙烯技术进展[J]. 当代化工, 2006(2)：117-120.
[5] 王国清, 曾清泉. 裂解技术进展[J]. 化工进展, 2002(2)：92-96.
[6] 曹菊勇. 多种工况下乙烯裂解管蠕变损伤模拟[D]. 天津大学, 2012.
[7] 李琦. HPM裂解炉管的渗碳损伤研究[D]. 大连：大连理工大学, 2014.
[8] 郑显伟. 我国乙烯裂解炉辐射炉管的使用状况[J]. 压力容器, 2013, 30(5)：45-52.
[9] 黄雷. 裂解炉管长期高温组织损伤研究[D]. 大庆：东北石油大学, 2011.
[10] 金沛斌. 高温复杂环境对乙烯裂解炉管性能影响分析[D]. 徐州：中国矿业大学, 2017.
[11] 谭家隆, 李德俊, 于永泗, 等. 奥氏体耐热钢抗渗碳性能的研究[J]. 大连理工大学学报, 1988(4)：37-44.
[12] 牟卫萍. 测定20Cr钢渗碳层深度方法的比较[J]. 实验室研究与探索, 2001(1)：48-49.
[13] 许颖恒. 乙烯裂解炉管的蠕变损伤失效和寿命评级[J]. 石化技术, 2015, 22(11)：26-28.

本论文原发表于《机电工程技术》2020年第49卷第6期。

炼化企业常压塔顶露点温度计算及缓蚀剂性能研究

范 磊[1]　刘宏铭[2]　杜笑怡[2]　赵儒盼[2]　王 峰[2]　王延海[2]

(1. 中国石油集团石油管工程技术研究院·石油管材及装备材料服役行为与结构安全国家重点实验室；
2. 中国石油长庆石化公司)

摘　要： 常减压装置塔顶低温 $HCl-H_2S-H_2O$ 腐蚀是石油炼化行业普遍存在的腐蚀问题，注中和缓蚀剂和注水是有效控制腐蚀的主要手段，这些注剂的注入位置是保证其防腐效果的关键因素；为了精准定位注剂的注入位置，必须准确判定塔顶介质中水的露点温度；采用 Aspen plus 软件建模，并且根据现场实际生产参数优化模型，模拟计算得到正常负荷条件下常压塔顶的水露点温度为 98.2℃。

关键词： 露点温度；缓蚀剂；中和剂；常压塔顶；常减压装置

常减压蒸馏是原油在石油炼化过程中的第一道工序，是整个石油炼化的基础[1]。根据 API 571—2011《炼油厂设备损伤机理》，常减压装置涉及的损伤类型一共有20种之多，常压塔顶系统就有8种。常减压装置的常压塔顶系统的低温腐蚀一直是困扰国内外炼厂的重要问题，制约炼化企业长周期运行的瓶颈。根据常压塔顶物性分析可知，塔顶油水混合物中水的酸性偏低，甚至 pH 值达到 1～1.3[2]，可能存在严重的液态酸性水腐蚀。特别是在露点位置，HCl 含量更高，甚至有报道称在露点位置 HCl 在液态水中的含量(质量分数)可达 1%～3%[2-4]，碳钢在该酸性环境将发生快速的露点腐蚀，露点腐蚀是饱和蒸汽因冷却而凝结成液体对材料造成的腐蚀[5]，露点位置的腐蚀容易造成点蚀[6]。目前，针对常压塔顶腐蚀，炼厂工艺防腐为主，材料防腐为辅。最常见的防腐方法为"一脱三注"，即电脱盐、注水、注缓蚀剂、注中和剂[3]。其中，注缓蚀剂、注水和注中和剂的最佳位置为露点位置，因此确定塔顶酸性水凝结温度，并且根据凝结温度评价中和缓蚀剂，对于制定常压塔顶防腐方案、缓蚀剂等注剂的筛选和确定加注位置至关重要。

根据某炼厂目前加工的原油性质和常压系统流程，结合产品实测性质、现场操作条件与物料平衡数据，运用化工流程设计模拟软件 Aspen plus，建立模拟模型。经过不断调整运行参数，优化运行结果，并与实测数据相互印证，使 Aspen plus 软件运行结果尽量接近实际运行工况，获得常压塔关键部位操作参数，最终根据常压塔顶油气产品的冷凝曲线，获得常压塔顶的露点温度。模拟常压塔顶露点腐蚀，对现场再用中和剂和缓蚀剂进行评价。

1　露点温度计算

1.1　模型建立

Aspen plus 软件是基于稳态化工模拟、优化、灵敏度分析和经济评价的大型化工流程模拟软件，拥有一套完整的单元操作模块，是全世界公认的标准大型化工流程模拟软件，深受广大化工厂工程师的喜爱，也是各化工设计单位广泛使用的流程模拟软件。

结合该炼厂实际工艺过程中初馏塔、常压塔进料、侧线抽出板等关键塔板位置，建立如图1所示的常压蒸馏系统的Aspen模拟工艺流程，主要包括初馏塔和常压塔两部分。原油经过电脱盐后作为初馏塔原料进入初馏塔。具体的工艺流程为初馏塔顶油气经冷凝冷却后进入初顶回流罐；初顶回流罐冷凝油经初顶回流泵抽出一部分升压打入初馏塔顶做回流，一部分作为石脑油出装置。初馏塔侧线油经增压泵升压后并入常一中返塔线，进入常压塔。初底油抽出并升压后，经两路换热并经常压炉加热后进入常压塔第四层，作为常压塔进料。常压塔蒸馏后，常压塔顶油气经过冷凝冷却后进入常顶回流罐，液相经常顶回流泵抽出升压后一部分打回常压塔顶做热回流，另一部分进入常顶回流油冷却器冷却作为石脑油并入常顶石脑油出装置线。未冷凝油气由常顶回流罐顶进入常顶油后气冷系统经再冷却后进入常顶产品罐，常顶产品罐的液相经常顶产品泵升压后作为石脑油出装置。常压塔从上到下侧线依次馏出的常一线、常二线、常三线、常四线经换热器换热冷却后送出装置。常压塔设有一个顶回流，三个中段回流（常顶、常一中、常二中），常底油经泵抽出，进入下一个蒸馏过程。

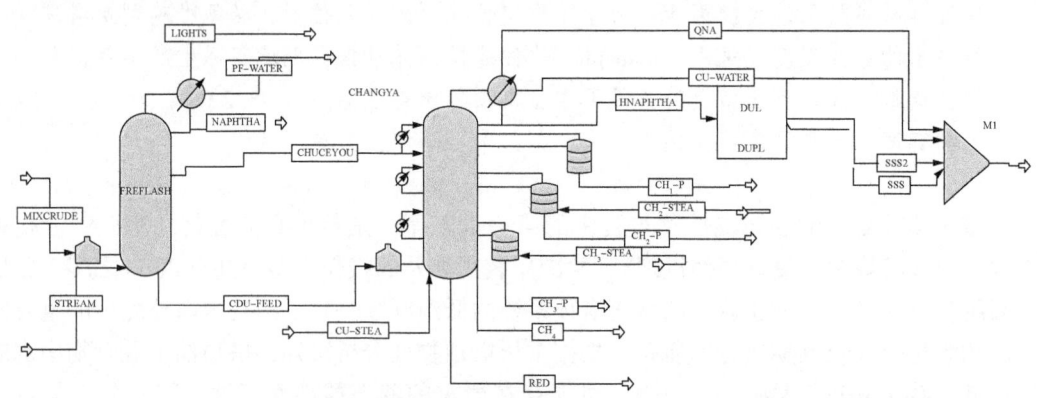

图1 某炼厂常压蒸馏系统的Aspen模拟工艺流程图

1.2 模拟的露点温度

根据原油的组分性质、实沸点蒸馏数据、API度和某炼厂常减压装置的生产运行数据，建立模拟工况。对比现场运行参数、物料平衡数据，结合塔顶产品与各侧线产品质量分析报告，调节模拟参数，最终得到与实际工况相符的常压塔系统模型和工况。模拟得到的各项参数误差均在5%以内，因此认为建立的模型符合实际工况，模拟得到的结果能真实地反映实际情况。

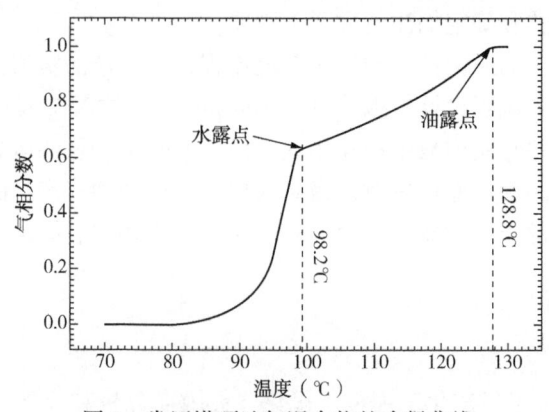

图2 常压塔顶油气混合物的冷凝曲线

当处理量为 13000t/d 时，模拟计算得到的常压塔顶油气混合物在常压塔操作压力（58kPa）下的冷凝曲线如图 2 所示。可见，常压塔顶油气混合物在 128.8℃和 98.2℃分别出现拐点。128.8℃为油的露点温度，98.2℃为水的露点温度。

2 缓蚀剂防腐技术研究

2.1 实验材料及方法

材料为常压塔顶常用管线钢 20#碳钢。试样线切割加工为 50mm×10mm×3mm，经 240 号至 800 号砂纸逐级打磨，然后用丙酮和酒精除油清洗并用冷风吹干待用，采用游标卡尺测量其实际长、宽、高，试样腐蚀前后均采用精度为 0.1mg 的电子天平称量。

模拟溶液为分析纯浓盐酸和分析纯九水合硫化钠配置的 1000mg/L HCl+1000mg/L H_2S 溶液。实验采用油浴加热，实验温度为露点温度（98.2℃±0.5℃）。中和剂和缓蚀剂均为某炼厂现场用药剂。加入中和剂时，缓慢滴加至溶液 pH 值为 6.5~7，缓蚀剂加量为 100mg/L。为了保证数据的准确性，实验均采用 3 个试样作为平行实验，腐蚀速率取平均值。

腐蚀后用失重法计算平均腐蚀速率，计算公式如下：

$$V = \frac{8.76 \times 10^4 \times (m_0 - m)}{S \cdot t \cdot \rho} \quad (1)$$

式中：V 为腐蚀速率，mm/a；m_0 为实验前的质量，g；m 为实验后的质量，g；S 为试样表面积，cm^2；t 为实验时间，h；ρ 为材料密度，g/cm^3。

根据腐蚀速率计算缓蚀率，计算公式如下：

$$\eta = \frac{V^o - V}{V^o} \times 100\% \quad (2)$$

式中：V^o 为不添加药剂时的腐蚀速率，mm/a；V 为添加药剂时的腐蚀速率，mm/a。

2.2 结果与讨论

如图 3 所示为 20#碳钢在不同环境中的腐蚀速率。空白条件下，20#碳钢的腐蚀速率达到 21.45mm/a，HCl 和 H_2S 导致非常严重的腐蚀，这主要是因为 HCl 环境为强酸，H^+ 作为强去极化剂，快速与碳钢反应，形成的腐蚀产物为可溶的 $FeCl_2$，没有保护性能。溶液中的 Cl^- 易变形，产生离子极化，具有较高的极性和穿透性能，从而，在金属表面上有较高的吸附率，强烈地促进了电子交换反应强，因此虽然 Cl^- 没有直接参加反应，但也能明显加速腐蚀速率[7]。

$$Fe+HCl \longrightarrow FeCl_2+H_2 \quad (3)$$

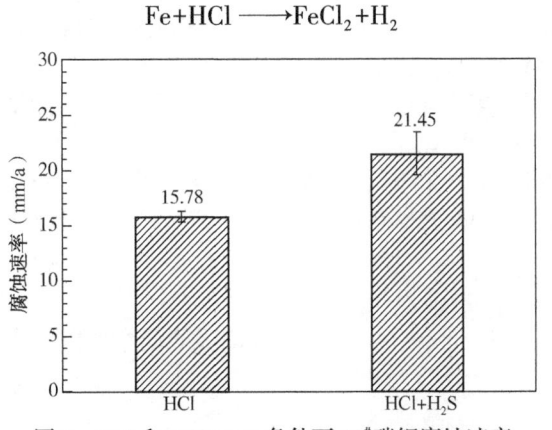

图 3 HCl 和 HCl+H_2S 条件下 20#碳钢腐蚀速率

H₂S一方面提供阴极反应所需的H⁺，加速阴极过程[8]。另一方面H₂S两级解离产生S²⁻，当阳极反应生成的Fe²⁺与H₂S解离产生的S²⁻相遇时生成Fe_xS_y，消耗Fe²⁺，加速腐蚀阳极过程。因此H₂S的存在，进一步加快腐蚀的发生。造成常压塔顶管线腐蚀非常严重，现场生产过程中必须采取高效的防腐技术。

$$H_2S \longrightarrow HS^- + H^+ \longrightarrow S^{2-} + 2H^+ \tag{4}$$

中和剂调节溶液pH值或者加入缓蚀剂均能有效降低腐蚀速率，缓蚀率分别为90.5%和81.1%。中和剂为有机胺，能与溶液中的强去极化剂H⁺反应，抑制阴极反应过程，从而降低腐蚀速率。现场用缓蚀剂为季铵盐类缓蚀剂，季铵盐类缓蚀剂分子中含有N，可以与基体的Fe形成配位键，在试样表面形成吸附膜，阻挡腐蚀的发生，降低腐蚀速率。当中和剂和缓蚀剂两者共同作用时，腐蚀速率非常低，缓蚀率达到95.9mm/a。因此在单独的中和剂调节溶液pH值或者加入缓蚀剂条件下绝对腐蚀速率仍然较高，不能满足炼化企业长周期运行的要求，而需要采用中和剂与缓蚀剂配套使用的方法(图4)。

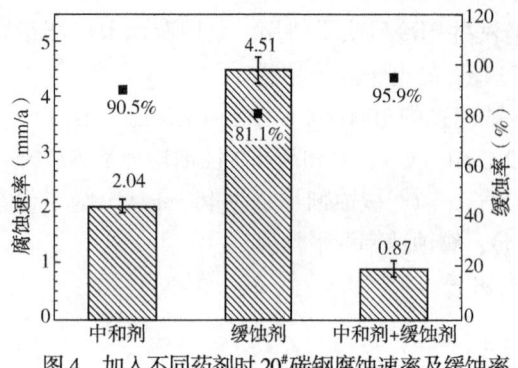

图4 加入不同药剂时20#碳钢腐蚀速率及缓蚀率

$$R_3N + H^+ \longrightarrow R_3NH^+ \tag{5}$$

3 结论

（1）采用Aspen软件建立的模型合理，能真实反映某炼厂常减压系统的实际工况。
（2）常压塔顶油气混合物的水露点温度为98.2℃。
（3）中和剂与缓蚀剂配套使用时缓蚀率达到95.6%，能有效减缓常压塔顶腐蚀，有利于常减压装置的长周期运行。

参 考 文 献

[1] 庞娅静. 常压塔顶系统腐蚀及防护技术应用研究[J]. 石油化工应用，2016，35(1)：98-100.
[2] 李志平. 常减压装置的腐蚀与应对措施[J]. 安全技术，2007，7(9)：15-17.
[3] 陈洋. 常减压装置塔顶系统腐蚀与控制技术现状[J]. 全面腐蚀控制，2011，25(8)：10-13.
[4] 汪东汉. 常减压蒸馏装置设备腐蚀典型事例与防护[J]. 石油化工腐蚀与防护，2004，21(5)：10-15.
[5] 许适群. 关于露点腐蚀及用钢的综述[J]. 石油化工腐蚀与防护，2000，174(1)：1-4.
[6] 偶国富，赵露露，裴克梅，等. 温度和pH值对20#钢盐酸露点腐蚀行为的影响[J]. 工程科学学报，2018，40(9)：1099-1107.
[7] 李春杰. 盐酸腐蚀低碳钢的动力学研究[J]. 大气石油学院学报，1999，32(3)：28-30.
[8] 郭金彪，李艳. HCl-H₂S-H₂O体系中HCl与H₂S对20碳钢腐蚀的交互作用[J]. 材料保护，2013，46(5)：2，64-66.

本论文原发表于《化学工程与装备》2020年第2期。

连续纤维复合材料环形试样蠕变行为研究

张冬娜[1,2] 邵晓东[1,2] 蔡雪华[1,2] 王 航[1,2] 丁 楠[1,2]

(1. 石油管材及装备材料服役行为与结构安全国家重点实验室;
2. 中国石油集团石油管工程技术研究院)

摘 要：对连续玻璃纤维复合材料进行了拉伸蠕变试验研究,为了模拟复合材料在压力容器中的受力状态并减少夹具加持力对试样的影响,采用环形复合材料试样拉伸蠕变试验方法。对复合材料环形试样的拉伸强度及不同应力等级下的拉伸蠕变性能进行了研究,并基于时间—应力等效原理,通过双对数法拟合出压力容器50年使用寿命时复合材料的最大蠕变应力,为复合材料压力容器的设计提供支持。并基于时间—应力等效原理,通过双对数法拟合出压力容器50年使用寿命时复合材料的最大蠕变应力应低于其拉伸强度的44.4%。

关键词：蠕变；复合材料；拉伸；寿命

连续纤维复合材料通常由连续纤维(包括玻璃纤维、碳纤维、芳纶纤维等)与树脂复合而成,结合了纤维高强度、高模量及树脂材料耐腐蚀的特点,广泛应用于各类压力容器与压力管道的制造中,如各种金属内胆缠绕气体、玻璃钢管、柔性管等,其中金属内胆缠绕气瓶通过复合材料的增强,不仅减轻了质量,还获得了更好的安全性能,具有明显的节能、减排意义[1]。玻璃钢管和柔性管具有耐腐蚀、质量轻、综合成本低等优点,在油气输送中的应用也越来越广泛[2]。

为了保证压力容器在使用年限内的安全运行,对复合材料黏弹性行为的研究尤为重要,压力容器在运行时复合材料一直保持承载状态,因此研究复合材料的蠕变行为对非金属压力容器的使用寿命及承压能力有非常重要的意义。蠕变现象是在一定温度和低于材料断裂强度的恒定外力作用下,材料的形变随时间增加而逐渐增大的现象。针对复合材料的蠕变特性及长期性能的预测,国内外开展了相应的研究,通常土木工程要求50年的使用寿命,聚合物复合材料管道一般要求的设计使用寿命也为50年[3],目前对于埋地的玻璃纤维增强复合材料(glass reinforced polymer composite, GRP),要求进行10000h以上的蠕变试验[4]。为了保证汽车在使用年限(通常15年)中结构的安全,Ren[5]对复合材料的蠕变应力加上由汽车振动、加速、减速、转弯等产生的循环载荷进行了研究。Goertzen和Nakada[6,7]等研究了树脂基复合材料的拉伸及弯曲蠕变行为,并用时—温等效原理对复合材料的长期蠕变性能进行了预测。

为了在较短的周期内提供非金属压力容器及管道用复合材料基础设计数据,本文针对目前非金属压力容器最常使用的玻璃纤维增强热固性树脂进行了蠕变行为的研究,由于复合材

基金项目：中国石油天然气集团公司科学研究与技术开发项目"非金属复合油气管关键技术研究"(2016B-3001)。

作者简介：张冬娜,zhangdna@cnpc.com.cn。

料在服役状态下主要受到拉伸应力的作用,因此模拟材料的受力状态,使用环形试样进行拉伸蠕变试验方法,并对材料的使用寿命进行了预测。

1 实验部分

1.1 主要原料

玻璃纤维,158B-2400tex,欧文斯科宁公司。
环氧树脂,E51,南通星辰合成材料有限公司。
固化剂甲基四氢苯酐,HN-2200,日立化成株式会社。
促进剂,DMP-30,广州市三昌化工有限公司。

1.2 主要设备及仪器

复合材料制样机,E42-1,上海扶夷机电工程有限公司。
材料试验机,Instron 8804,美国英斯特朗公司。

1.3 样品制备

使用复合材料制样机加工复合材料环形试样,固化条件为90℃下2h后升温至130℃固化3h;复合材料中纤维沿环向分布,通过测试得出复合材料试样的树脂含量为27.5%~28%,试样尺寸如GB/T 1458—2000中要求,试样厚度分别为1.5mm及3.0mm,其中1.5mm厚的环形试样用于拉伸强度测试,3.0mm的试样用于拉伸蠕变性能测试。

1.4 蠕变试验方法

目前复合材料拉伸蠕变试验可参考的标准是ASTM D2990-17,在本标准中,要求将复合材料加工成薄板进行拉伸蠕变测试。国内拉伸蠕变试验的标准有GB/T 2039—2012《金属材料单轴拉伸蠕变试验方法》和GB/T 11546.1《蠕变性能的测定 第1部分:拉伸蠕变》,其中GB/T 2039—2012适用于金属材料,GB/T 11546.1适用于硬质和半硬质的非增强、填充和纤维增强的塑料材料,目前没有针对纤维增强树脂基复合材料的拉伸蠕变试验国家标准。

通过对容器和管道结构及受力的分析,使用复合材料环试样进行蠕变行为的测试,试样尺寸参考GB/T 1458—2008《纤维缠绕增强塑料环形试样力学性能试验方法》中的要求,即内径为150mm,宽度为6mm。设计了复合材料环形试样蠕变试验的夹具,如图1、图2所示。在进行蠕变试验时,将环形试样装入2个如图1所示的夹具中,试验机的夹头加持加持杆。

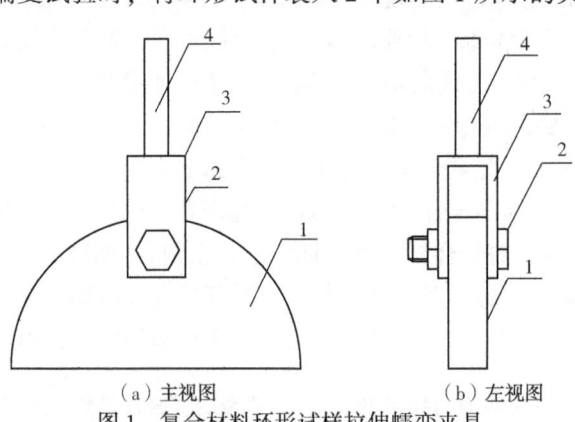

图1 复合材料环形试样拉伸蠕变夹具
1—盘体;2—螺栓;3—支架;4—与试验机连接的加持杆

图2 环形试样蠕变试验状态

使用环形试样进行拉伸蠕变测试，一方面更接近复合材料在压力容器服役时的工作状态，另一方面蠕变试验机夹具直接加持的是加持杆，减少了夹具加持力对复合材料试样的影响。

1.5 试验方案

首先按照 GB/T 1458—2008 进行复合材料环形试样的拉伸强度测试，拉伸速率为 5mm/min，拉伸强度即为材料的极限抗拉强度(ultimate tensile strength，UTS)，之后分别以复合材料 UTS 的 90%、80%、70% 和 60% 进行蠕变试验，记录材料发生蠕变破坏的时间。拉伸试验及蠕变试验均在室温下进行，图3为进行蠕变试验中的试样状态，图3为蠕变试验结束时破坏的试样状态

2 结果与讨论

2.1 UTS 测试

对复合材料环形试样进行拉伸强度测试，具体结果见表1，在复合材料中连续纤维起到主要的承载作用，树脂的作为是保护纤维及传递应力，平均值为 985.1MPa，标准差为 20.16，变异系数为 2.05%，拉伸强度的分散性较小，试样均匀性好。为了确保测试结果对工程应用的指导意义，保证压力容器运行的安全可靠性，尽量选取测试结果中较高的强度值作为蠕变应力的基准，即复合材料的 UTS 为 1000MPa。

图3 蠕变试样破坏的环形试样

表1 复合材料环形试样 UTS 测试结果

样品编号	UTS/MPa	样品编号	UTS/MPa
1#	998	6#	942
2#	1000	7#	1000
3#	988	8#	1000
4#	989	9#	976
5#	999	10#	959

2.2 拉伸蠕变性能测试

复合材料的黏弹性由于纤维的存在比聚合物更为复杂，连续纤维的加入使材料表现出了各向异性，同时纤维和基体的蠕变特性不同。在复合材料中，蠕变主要包括以下几个方面：(1)纤维蠕变：尽管试验表明，室温下玻璃纤维的蠕变可以忽略，但在高温和高应力作用下，尤其是在腐蚀介质中，纤维材料还具有一定程度的蠕变现象。(2)纤维逐渐拉直：在复合材料的加工过程中，可能会存在纤维不完全伸直的现象，在应力作用随着基体蠕变逐渐被拉直。(3)基体蠕变：基体蠕变是复合材料蠕变的主要来源，但复合材料中的纤维不仅提高了材料的强度，同时对树脂基体的蠕变有很大的抑制作用。对于本文研究的室温复合材料的蠕变行为，主要的蠕变变形来源于纤维的逐渐拉直和基体的蠕变。

在室温下，分别以复合材料 UTS 的 90%、80%、70% 和 60% 进行蠕变试验，图4为蠕变应力为 UTS 的 60% 时的蠕变应变—时间曲线，从曲线上可以看出，测试的复合材料蠕变分为3个阶段，即瞬间蠕变、稳态蠕变及非稳定蠕变阶段。在瞬间蠕变阶段，蠕变应变迅速增加，这部分变形主要来源于树脂基体高分子链内键长和键角发生的瞬间变化，之后蠕变应变

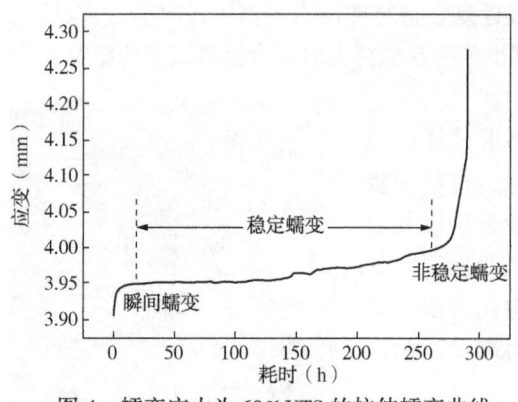

图4 蠕变应力为60%UTS的拉伸蠕变曲线

速率逐渐减小,进入稳态蠕变阶段。对于此阶段通常用式(1)表达,其中,0<n<1:

$$\varepsilon = Ct^{-n} \quad (1)$$

式中:t 为时间;C 为材料常数。

稳态蠕变阶段蠕变速率稳定,这一阶段持续时间较长,通常与应力水平有关,这一阶段蠕变变形主要是聚合物链段对拉伸应力响应而产生的变形,部分纤维在此阶段会发生破坏[8],此阶段可用一次线性表达式描述,见下式:

$$\varepsilon = \varepsilon_0 + vt \quad (2)$$

式中:ε_0 为瞬间蠕变阶段的应变量;v 为稳态蠕变阶段的蠕变速率;t 为时间。

第三阶段为非稳定蠕变阶段,材料由于内部的损伤逐渐累积发展,蠕变速率不断增加至试样破坏。在整个蠕变过程中,纤维伴随着树脂基体发生形变,主要是未伸展的部分充分伸展,如果材料的应力水平较低,蠕变曲线将不会出现非稳定蠕变阶段,即材料不会发生蠕变破坏[9],该阶段尚无权威的表达式来描述。

在进行复合材料结构设计时,需考虑材料的使用寿命,目前大多使用时—温等效原理[10]对复合材料的寿命进行预测。本文通过在室温下对不同应力水平拉伸蠕变性能的测试,对复合材料的使用寿命进行预测。

2.3 复合材料使用寿命预测

时—温等效原理常用来预测复合材料的使用寿命,较低温度下分子运动所需松弛时间长,聚合物分子运动对外力的响应需要较长时间才能达到宏观某一物理量,若升高温度,则分子运动加速,运动所需的松弛时间减小,就可以在较短的时间内达到其力学相应。为了缩短试验时间,常提高温度进行试验,并根据时—温等效原理进行寿命预测。与时间—温度等效原理类似,还存在时间—应力等效原理[11,12]。聚合物在一定外界温度下,达到某一种力学性能所需要的时间和应力之间也存在着等效关系,即增大外加应力与延长时间对聚合物的运动和黏弹性行为产生的作用是相等的,应力和时间等效性的研究对预测聚合物长期的蠕变性能和使用寿命具有重要的意义[13]。

在室温下,分别以复合材料UTS的90%、80%、70%和60%进行蠕变试验,记录各种应力水平下的蠕变破坏时间,见表2。蠕变破坏时间随蠕变应力的增大而降低,更高应力下损伤在材料内积累更快,加速了材料的破坏。

表2 不同蠕变应力下的蠕变破坏时间

应力水平(%)UTS	蠕变破坏时间(h)		蠕变破坏时间平均值(h)
60	220.65	279.43	250.04
70	80.00	65.50	72.75
80	1.05	2.35	1.70
90	0.12	0.12	0.12

为了对复合材料的长期性能进行评价，利用获得的试验数据，通过双对数线性回归的方法获得材料服役的长期性能数据，该方法参考 ASTM D3681 中附录 A 的内容，在长期性能试验中，有研究使用此种方法研究了玻璃纤维复合材料在腐蚀条件下 50 年后剩余强度[14]，类似的方法还被用来预测复合管的使用寿命[15]。通过蠕变试验曲线可知，复合材料的蠕变破坏时间主要是第二阶段即蠕变稳定阶段的时长，见式(2)，第二阶段的表达式为直线方程，可使用双对数拟合的方法。参考 ASTM D 3681 中的方法，x 轴为 $\lg t$，其中 t 为时间，单位为小时(h)，y 轴为 $\lg V$，其中 V 为 UTS 百分比，使用蠕变破坏时间的平均值进行拟合。

对不同蠕变应力下蠕变破坏时间的数据点进行线性拟合，如图 5 所示，拟合的直线见下式，$R^2 = 0.82656$。

$$\lg V = 1.92342 - 0.04899 \lg t \quad (3)$$

其中通过拟合的曲线计算复合材料在 50 年时对应的 UTS 百分比值，得出对应的 UST 百分比为 44.4%。因此在进行管道设计时，要保证复合材料在 50 年的服役年限内不发生蠕变破坏，其环向应力最大应低于其 UTS 的 44.4%。影响蠕变破坏时间长度的主要是蠕变稳定阶段的时长，在这个阶段随着蠕变时间的增加，材料的应变逐渐变大，蠕变应力更高时，应变增加更快，在非常短的时间内就进入了蠕变非稳定阶段。根据时间—温度等效及时间-应力等效原理，更高的蠕变应力相当于试样在更高的温度下进行蠕变，链段热运动能更高，各个运动单元的松弛时间缩短，分子链的伸展和滑移运动加剧，聚合物的蠕变黏弹性力学行为更显著，因此在更短的时间内发生蠕变破坏。但是当应力水平足够低时，试样在蠕变过程中的稳定蠕变阶段会足够长，甚至不出现蠕变非稳定阶段。

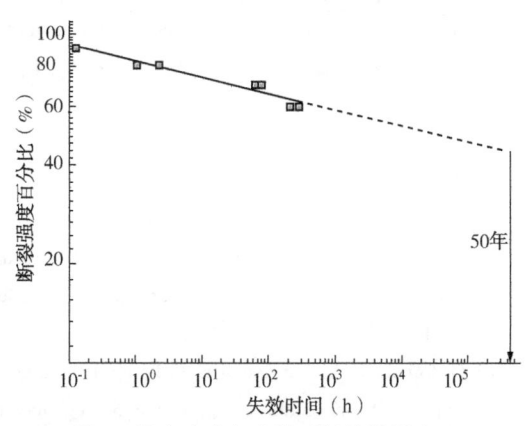

图 5 蠕变应力与失效时间线性拟合

3 结论

（1）采用复合材料环形试样进行拉伸蠕变测试，根据时间—应力等效原理，分别以不同的拉伸应力进行蠕变试验，记录每种应力状态下的蠕变破坏时间，通过对测试结果的拟合，得出复合材料 50 年的应力极限为 UST 的 44.4%。

（2）复合材料的蠕变破坏时间主要取决于蠕变稳定阶段的时长，更高应力状态下聚合物的黏弹性行为更明显，试样以更快的速度发展到蠕变非稳定状态，因此复合材料在服役时，最高的应力状态应根据蠕变试验结果进行设计，保证在运行期间的安全。

（3）由于埋地输水非金属复合管的使用环境和温度较为稳定，因此本测试结果可为此类管道的设计提供支持，对于高温及腐蚀环境下复合材料的长期性能还需进行进一步的研究。

参 考 文 献

[1] 成志刚. 大容积钢质内胆环向缠绕气瓶标准介绍[J]. 压力容器，2013(7)：47-55.

[2] 张冠军，齐国权，戚东涛. 非金属及复合材料在石油管领域应用现状及前景[J]. 石油科技论坛，2017(2)：26-31.

[3] Pellowe S. Fiberglass pipe design manual[S]. Wash-ington: American water Works Association, 2005.
[4] Guedes R M. Creep and fatigue lifetime prediction of polymer matrix composites based on simple cumulative damage laws[J]. Composite Part A, 2008, 39(11): 1716-1725.
[5] Ren W, Brinkman C R. Creep and creep-rupture behavior of a continuous strand, swirl mat reinforced polymer composite in automotive environments [C]//Preceddings of International Composites Expo. Nashviles, Tennessee, Jan, 1998: 19-21.
[6] Goertzen W K, Kessler M R. Creep behavior of carbon fiber/epoxy matrix composite[J]. Materials Science and Engineering: A, 2006, 421: 217-225.
[7] Nakada M, Miyano Y, Cai H, et al. Prediction of long-term viscoelastic behavior of amorphous resin based on the time-temperature superposition principle[J]. Mechanics of Time-Dependent Material, 2011, 15(3): 309-316.
[8] 刘鹏飞, 赵启林, 王景全. 树脂基复合材料蠕变性能研究进展[J]. 玻璃钢/复合材料, 2013(3): 109-117.
[9] 沈观林. 复合材料力学[M]. 北京: 清华大学出版社, 2006: 294-297.
[10] Willliams M L, Landel R F, Ferry J D. The temperature dependence of relaxation mechanisms in amorphous polymers and other glass-forming liquids[J]. Journal of the American Chemical Society, 1955, 77(14): 3701-3707.
[11] O'Connel P A, McKenna G B. Large deformation response of polycarbonate: time-temperature, time-aging time, and time-strain superposition[J]. Polymer Engineering and Science, 1997, 37(9): 1485-1495.
[12] Luo W B, Yang T Q, An Q. Time-temperature-stress equivalence and its application to nonlinear viscoelastic material[J]. Acta Mechanica Solida Sinica, 2001, 14(3): 195-199.
[13] 贾玉. 碳纳米管/热塑性聚合物复合材料的制备及蠕变行为研究[C]. 合肥: 中国科学技术大学, 2012.
[14] SPOO K. Measuring corrosion of glass fibers using an abbreviated technique [C]//Corrosion. 2014, San Antonio, USA, No. 3901.
[15] Higuchi Y, Tanishita T, Yamada K, et al. Development of the ring creep testing method for composite pipes [C]//Proceeding of the 17th Plastic Pipes Conference. Sep, 22-24, 2014, Chicago, USA.

本论文原发表于《中国塑料》2020年第34卷第1期。

非金属智能连续管拉伸层力学特性研究

丁楠[1,2] 李厚补[1,2] 古兴隆[3] 朱永凯[3] 吴燕[4]

(1. 石油管材及装备材料服役行为与结构安全国家重点实验室；
2. 中国石油集团石油管工程技术研究院；3. 南京航空航天大学；
4. 新疆油田公司工程技术研究院)

摘 要：复合材料连续管是连续管的重要发展方向之一。在非金属智能连续管内植入辅助线缆，会对拉伸层结构的力学特性造成影响，必须通过合理的结构设计来保证拉伸层结构的力学性能。鉴于此，针对非金属智能连续管理结构中的拉伸器，基于ABAQUS软件开展拉伸层典型结构有限元建模，采用数值仿真分析其力学特性，并讨论了拉伸层相关参数对典型结构力学特性的影响规律。分析结果表明：在拉伸层结构中，增加光纤信号线对其力学性能影响较小；增大增强带宽度能够明显提高其刚度特性；当缠绕角度小于50°时，增大缠绕角度将降低其刚度特性，当缠绕角度大于50°时，继续增大缠绕角度对拉伸层刚度特性的影响十分有限。研究结果可以为非金属智能连续管的设计和使用提供技术借鉴。

关键词：复合材料；连续管；力学特性；拉伸层；有限元模型；弹性模量；增强带；信号线

连续管钻井技术在提高油田采收率和降低成本方面具有一定优势[1]，目前已经在美国和加拿大等国家得到普遍应用[2,3]。连续管的发展经历了高强度低合金碳钢、高强度低合金调质钢、钛合金及复合材料等4个阶段。目前国外已研制出强度高、柔韧性好和耐蚀性能优良的复合材料连续管，并在制造过程中实现了动力和信号线缆的内置，有效提高了连续管施工的测控精度[4,5]。同金属连续管相比，复合材料连续管的主要优势在于力学性能好、耐腐蚀、质量轻、材料和结构的可设计性强、辅助性功能易于实现等。因此，复合材料连续管能够用于具有特殊需求的石油工程作业中，是连续管的重要发展方向之一[3]。国内主要使用的是金属连续管，复合材料连续管的制造和应用还处于起步阶段，同国外的技术水平相比存在较大差距[6,7]。

根据石油工业用连续管的性能要求和使用环境，本文设计了多层非黏结型复合材料智能连续管，并对其拉伸层的力学特性进行了研究。

非金属智能连续管结构层的材料、制作工艺和功能等各不相同。其管内植入辅助线缆，

基金项目：中国石油天然气集团公司基础研究和战略储备技术研究基金项目"非金属复合材料智能连续油管研究"[2017D5008(2018Z-02)]。

作者简介：丁楠，高级工程师，生于1981年，2008年毕业于北京航空航天大学材料物理与化学专业，获硕士学位，现从事油田用非金属管材研究工作。地址：(710077)陕西省西安市。E-mail：dingnan@cnpc.com.cn。

必然会对拉伸层结构的力学特性造成影响，尤其是结构的刚度特性，这可能导致连续管在拉伸载荷下变形过大，同时线缆也会与拉伸层一起在载荷作用下发生变形，过大的变形量会降低信号的传输效率。因此必须通过合理的结构设计来保证拉伸层结构的力学性能[8-10]。本文针对非金属智能连续管结构中的拉伸层，采用有限元分析方法，基于商用 ABAQUS 软件，开展拉伸层典型结构力学特性数值仿真分析，研究了不同工况下拉伸层典型结构的力学特性，并分析了相关参数对拉伸层结构力学性能的影响规律。

1 复合材料连续管结构及拉伸层结构建模

1.1 复合材料连续管结构

复合材料智能连续管结构如图 1 所示。其内衬层采用热塑性塑料经模具挤出成型，该层直接与输送介质接触，可起密封防渗、耐腐蚀的作用；环向层由高强度连续纤维和树脂制成，通过缠绕工艺成型，主要用于增加内衬层强度并承担部分内压载荷；骨架层由高强度连续纤维和树脂制成，通过缠绕工艺成型，承担外压径向载荷，保证弯曲均匀；拉伸层分为内、外两层，由高强度连续纤维与树脂制成的增强带彼此分离，通过缠绕工艺成型，主要用于承担轴向拉伸载荷；辅助性线缆内置在外拉伸层中，包括动力缆和信号线，用于向井下作业工具传输电力和控制信号；保护层由热塑性塑料经模具挤出成型，位于管体最外侧，用于保护管体内部结构。

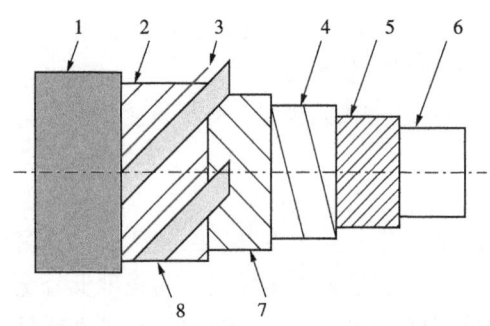

图 1 复合材料智能连续管设计原型结构示意图
1—保护层；2—外拉伸层；3—信号线；4—骨架层；
5—环向层；6—内衬层；7—内拉伸层；8—动力缆

该复合材料连续管的设计特点在于主体结构采用了全非金属材质，辅助线缆赋予了管体"智能化"，可称其为"非金属智能连续管"。

为了保证受力平衡，拉伸层分为内、外两层，两层的材质和结构形式相同，但两层的缠绕方向相反，所以两层的力学行为基本相同。由于辅助线缆内置在外拉伸层中，所以以外拉伸层为基础进行研究。构建典型结构：壁厚为 4.0mm；增强带宽度为 9.8mm，缠绕角度为 30°，共计 15 根，分 3 组，每组 5 根；动力缆宽度为 11.3mm，缠绕角度为 30°，共计 3 根，沿管周向均匀分布在 3 组增强带之间；采用开槽预埋的方式将 3 条信号线嵌入复合材料增强带中，线槽深度 1.0mm，宽度 3.0mm。

1.2 动力缆有限元模型

动力缆为三相输电线缆，沿拉伸层环向均匀分布，材质为金属铜，分析中将其视为均匀各向同性材料，其弹性模量为 115GPa，泊松比为 0.36。采用各向同性本构关系描述其力学特性，采用扫掠网格方法对结构进行网格划分，单元类型为 8 节点减缩积分三维体单元，单元编号 C3D8R，单元数量为 8352。动力缆有限元模型如图 2 所示。

图 2 动力缆有限元模型

1.3 复合材料增强带有限元模型

外拉伸层包含15条复合材料增强带，均分为3组，沿拉伸层环向均匀分布，将其视为均匀宏观各向异性变形体，采用三维实体建模。分析中对信号线缆结构进行简化，将3条信号线合并为1条，在增强带表面开槽，用于铺设光纤信号线。

增强带为单向纤维增强复合材料，增强纤维为Tex9600，基体为8201树脂，分析中采用横观各向同性本构关系描述其力学特性，采用composite layup工具进行逐层铺放，Tex9600/8201复合材料单层厚度为2mm，其基本性能参数见表1。强度及损伤演化参数见表2。表1及表2中E_{11}为纵向弹性模量，E_{22}和E_{33}为横向弹性模量，G_{12}和G_{13}分别为纵、横向剪切模量，G_{23}为横向剪切模量，μ_{12}和μ_{13}分别为纵、横向泊松比，μ_{23}为面内泊松比，X_T、X_C分别为纵向拉伸和压缩强度，Y_T、Y_C分别为横向拉伸和压缩强度，S_X、S_Y为剪切强度。以非金属智能连续管截面圆心为坐标原点，轴向为Z轴，径向为R轴，环向为θ轴建立柱坐标，采用离散坐标系定义增强带材料方向，R轴为堆叠方向，Z轴为主方向。

表1 Tex9600/8201单层复合材料基本性能参数

E_{11}(GPa)	E_{22}(GPa)	E_{33}(GPa)	μ_{12}	μ_{13}	μ_{23}	G_{12}(GPa)	G_{13}(GPa)	G_{23}(GPa)
51.432	5	5	0.26	0.26	0.3	2.5	2.5	1.923

表2 Tex9600/8201单层复合材料强度性能参数

X_T (MPa)	X_C (MPa)	Y_T (MPa)	Y_C (MPa)	S_X (MPa)	S_Y (MPa)	I型断裂韧度 (J/m²)	II型断裂韧度 (J/m²)
950	820	60	70	45	45	404.4	2309.7

采用扫掠网格方法对结构进行网格划分，网格扫掠方向沿管截面径向R方向。考虑到非金属智能连续管内径与外径大小的差异，沿R方向采用渐进式网格划分方式。增强带有限元模型如图3所示。

图3 增强带有限元模型

1.4 光纤信号线有限元模型

外拉伸层中铺放有3条光纤信号线，建模中对其进行归并简化处理，视其为宏观各向异性体，采用横观各向同性本构关系描述其力学特性，采用composite layup工具进行截面属性设置，其基本性能参数见表3。强度及损伤演化参数见表4。光纤信号线局部铺层方向定义与增强带定义相同。采用扫掠网格方法对其进行网格划分，网格扫掠方向沿管截面径向R方向，单元尺度为2mm。光纤信号线有限元模型如图4所示。

表3 光纤信号线材料基本性能参数

E_{11}(GPa)	E_{22}(GPa)	E_{33}(GPa)	μ_{12}	μ_{13}	μ_{23}	G_{12}(GPa)	G_{13}(GPa)	G_{23}(GPa)
68.337	5	5	0.26	0.26	0.3	2.5	2.5	1.923

2 拉伸层力学特性分析

2.1 轴向载荷工况

图 10 给出了 27kN 拉伸载荷下外拉伸层典型结构应力云图。从图 10 可以看出，嵌埋光纤信号线对外拉伸层应力分布的影响很小，并未改变外拉伸层整体应力分布规律。

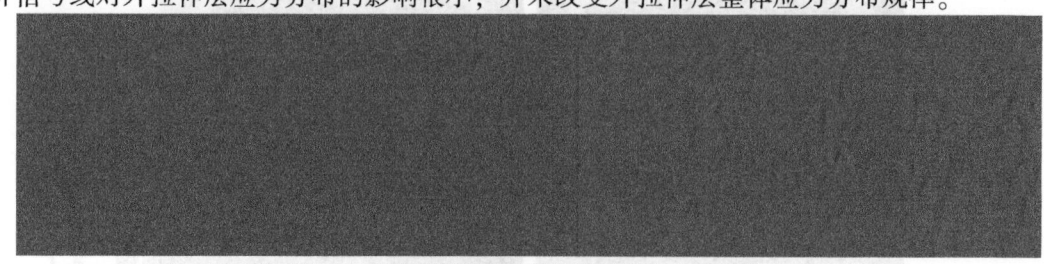

图 10 拉伸载荷下外拉伸层典型结构应力云图

图 11 给出了 27kN 拉伸载荷下拉伸层典型结构增强带、树脂区和辅助线缆的应力云图。从图 11 可看出：拉伸载荷下，增强带承受了主要的载荷，其应力水平最大；动力缆应力水平很低，对拉伸层承载能力影响较小；树脂区轴向拉伸应力水平很低，表明拉伸载荷下增强带间的相互挤压作用较小；光纤信号线应力水平低，对开槽区影响有限，增强带上的线槽区应力集中并不明显，表明开槽对增强带的影响不大。

图 11 拉伸载荷下增强带、动力缆、树脂区和光纤信号线的应力云图

根据数值计算结果，提取拉伸层顶面加载点处的位移和载荷，计算拉伸层弹性模量 E_L：

$$E_L = \frac{\mathrm{d}\sigma}{\mathrm{d}\varepsilon} = \frac{\dfrac{4\Delta F}{\pi(D_2^2 - D_1^2)}}{\dfrac{\Delta u}{L}} \quad (1)$$

式中：F 为载荷；u 为位移；D_1 和 D_2 分别为拉伸层内径和外径；L 为外拉伸层典型结构长度。

通过计算，外拉伸层典型结构的弹性模量为 25.385GPa。

2.2 内压载荷工况

图 12 给出了内压载荷下外拉伸层典型结构增强带、树脂区和辅助线缆的应力云图。

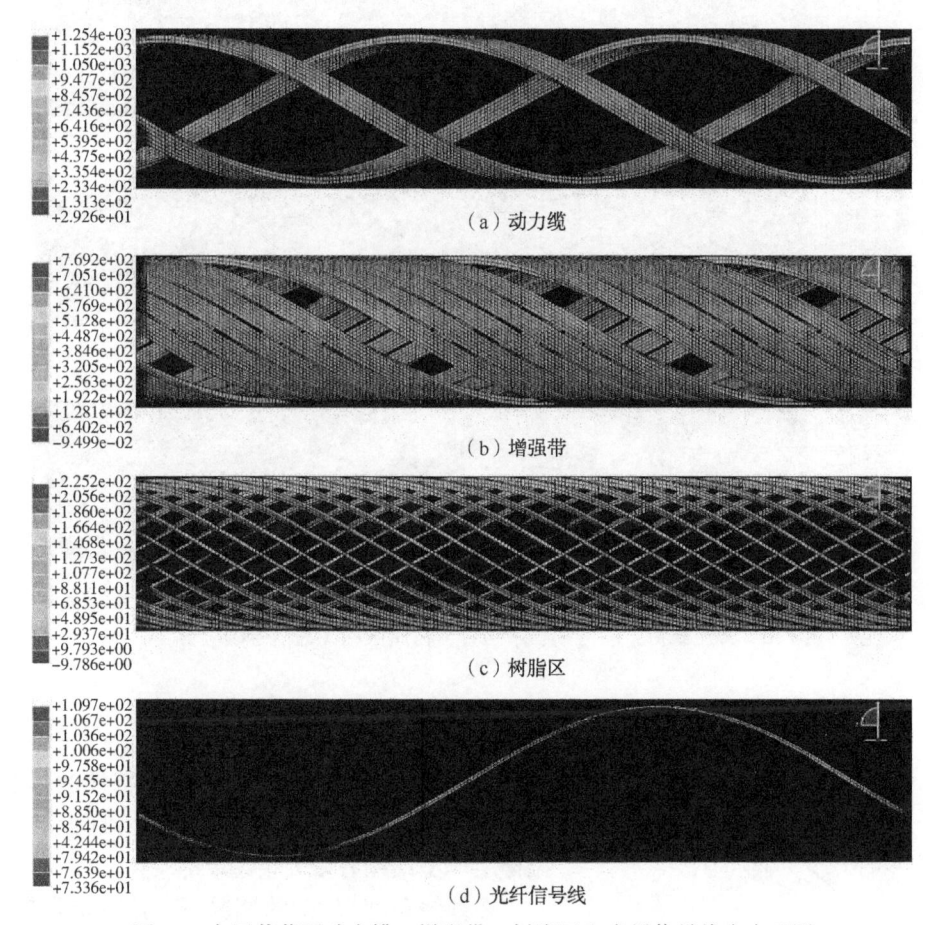

（a）动力缆

（b）增强带

（c）树脂区

（d）光纤信号线

图 12 内压载荷下动力缆、增强带、树脂区和光纤信号线应力云图

从图 12 可看出：内压载荷下，增强带承受了主要的载荷，线槽区应力集中现象明显，应力水平较高，25MPa 内压载荷下，线槽区应力 500MPa 左右，小于增强带强度指标；信号线缆应力比其承受轴向载荷时有所增大，但总体应力水平仍然很低，对拉伸层承载能力影响较小；树脂区中，仅与开槽区相连的部分应力较高，其他部位应力分布无变化。

2.3 外压载荷工况

图 13 给出了外压载荷下外拉伸层典型结构增强带、树脂区和辅助线缆的应力云图。从图

13可看出：外压载荷下，增强带承受了主要的载荷，线槽区出现应力集中现象，应力水平较高，7MPa外压载荷下，线槽区应力180MPa左右，小于增强带强度指标；辅助线缆应力较内压载荷时有所增大，但总体应力水平仍然很低，对拉伸层承载能力影响较小；树脂区中，仅与开槽区相连的部分应力较高，且内表面应力高于外表面，其他部位应力分布无变化。

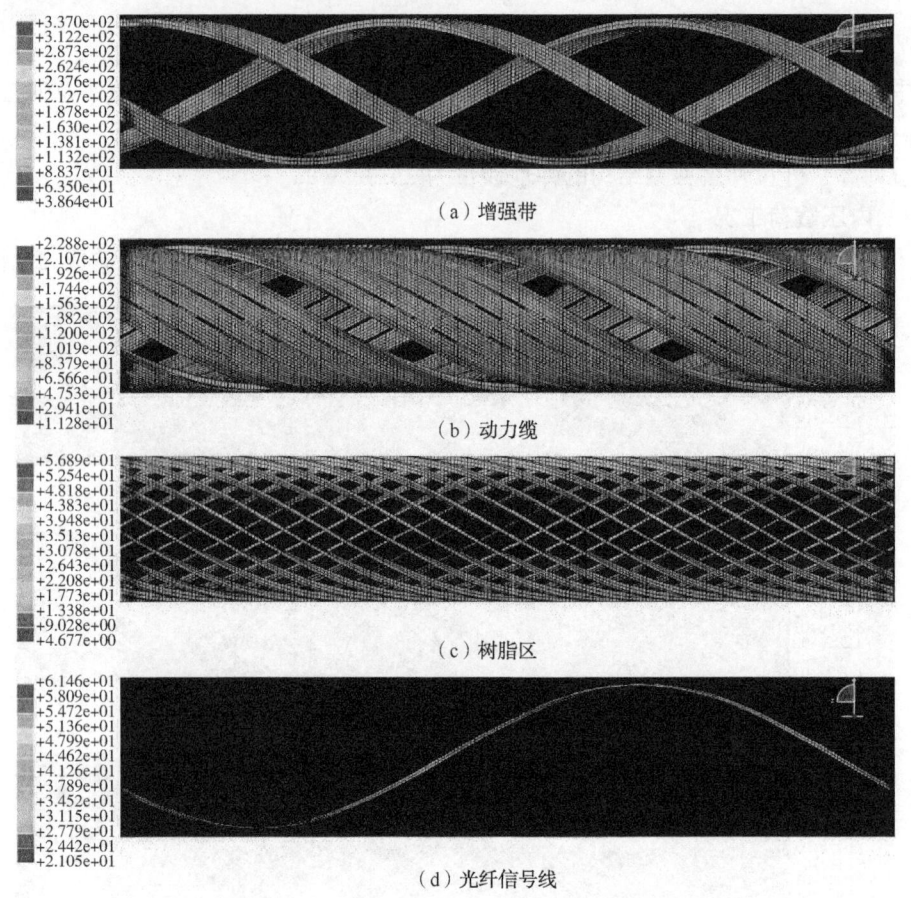

（a）增强带

（b）动力缆

（c）树脂区

（d）光纤信号线

图13 外压载荷下动力缆、增强带、树脂区和光纤信号线应力云图

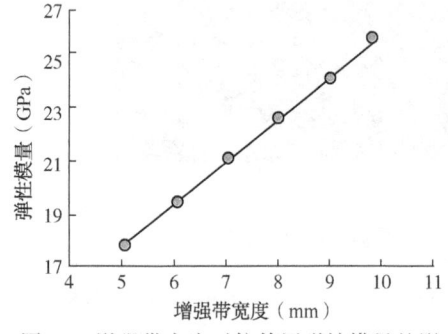

图14 增强带宽度对拉伸层弹性模量的影响

3 参数分析

为研究结构参数对拉伸层力学特性的影响规律，建立了不同增强带宽度、缠绕角度和信号线宽度的拉伸层典型结构有限元模型。

在缠绕角度为30°的情况下，选取增强带宽度为5.0mm、6.0mm、7.0mm、8.0mm、9.0mm和9.8mm，分别建立拉伸层典型结构有限元模型，计算不同增强带宽度下拉伸层典型结构的轴向弹性模量，结果如图14所示。从图14可看出：随着增强带宽度逐渐增大，拉伸层轴向弹性模量逐渐增大，增强带宽度与弹性模量近似呈线性关系。

保持增强带宽度为9.4mm，选取15°、30°、40°、50°、60°、70°和80°的缠绕角度，分别建立拉伸层典型结构有限元模型，计算不同缠绕角度下拉伸层典型结构的轴向弹性模量，结果如图15所示。从图15可以看出：随着缠绕角度逐渐增大，拉伸层轴向弹性模量逐渐减小，缠绕角度在趋近于80°时，弹性模量趋近于最小值5GPa；随着缠绕角度减小，弹性模量急剧增大。

选取2mm、3mm、4mm、5mm、6mm、7mm和8mm的光纤信号线宽度，分别建立拉伸层典型结构有限元模型，计算不同信号线宽度下拉伸层典型结构的轴向弹性模量，结果如图16所示。从图16可以看出，随着信号线宽度逐渐增大，拉伸层轴向弹性模量逐渐减小，信号线宽度与弹性模量呈线性关系。

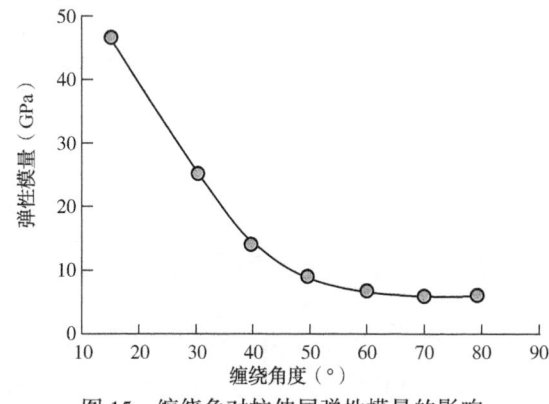

图15 缠绕角对拉伸层弹性模量的影响

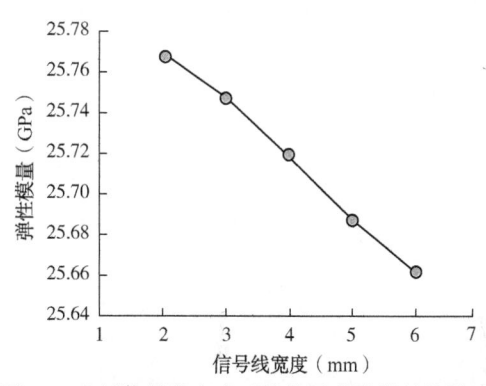

图16 光纤信号线宽度对拉伸层弹性模量的影响

4 结论

（1）非金属智能连续管在拉伸层结构中增加光纤信号线对其力学性能影响较小；增大增强带宽度能够明显提高其刚度特性。

（2）当缠绕角度小于50°时，增大缠绕角度将降低其刚度特性；当缠绕角度大于50°时，继续增大缠绕角度，对拉伸层刚度特性的影响将十分有限。

参 考 文 献

[1] 于京阁，董怀荣，江正清，等．复合材料连续管设计技术研究[J]．石油钻探技术，2011，39(3)：114-117．

[2] 陈立人．国外连续管材料技术及其新进展[J]．石油机械，2006，34(9)：127-130，137．

[3] Perry K F. Microhole coiled tubing drilling：a low cost reservoir access technology[J]．Journal of Energy Resources Technology，2009，131(1)：41-49．

[4] Newman K. Coiled tubing technology continues its rapid growth[J]．World Oil，1998，219(1)：64-70．

[5] 陈树杰，赵薇，刘依强，等．国外连续油管技术最新研究进展[J]．国外石油机械，2010，26(11)：44-50．

[6] 韩秀明．浅谈连续油管技术的现状与展望[J]．石油管材与仪器，2015，1(2)：1-3，6．

[7] 刘成．吐哈油田连续油管技术的应用[J]．石油矿场机械，2001，30(3)：45-47．

[8] 张洪伟．连续油管力学分析[D]．青岛：中国石油大学(华东)，2010．

[9] 张辛，徐兴平，王雷，等．复合材料连续管设计研究[J]．石油机械，2012，40(11)：19-22．

[10] 赵广慧，梁政．连续油管力学性能研究进展[J]．钻采工艺，2008，31(4)：97-101．

本论文原发表于《石油机械》2020年第48卷第11期。

四、其他

A Strain Rate Dependent Fracture Model of 7050 Aluminum Alloy

Cao Jun[1]　Li Fuguo[2]　Ma Weifeng[1]　Wang Ke[1]
Ren Junjie[1]　Nie Hailiang[1]　Dang Wei[1]

(1. Tubular Goods Research Institute of CNPC;
2. State Key Laboratory of Solidification Processing, School of Materials Science and Engineering, Northwestern Polytechnical University)

Abstract: The purpose of this research is to predict fracture loci and fracture forming limit diagrams (FFLDs) considering strain rate for aluminum alloy 7050 – T7451. A fracture model coupled Johnson – Cook plasticity model was proposed to investigate its strain rate effect. Furthermore, a hybrid experimental–numerical method was carried out to verify the strain rate–dependent fracture model by using fracture points of uniaxial tension, notched tension, flat–grooved tension, and pure shear specimens. The results show that the fracture points are in accordance with the fracture loci and FFLDs under different strain rates. The increasing strain rate decreases the FFLDs of aluminum alloy 7050–T7451. The difference of force–displacement responses under different strain rates is larger for notched tension and pure shear conditions.

Keywords: Strain rate; Fracture forming limit diagrams; Fracture criterion; Stress state; Simulation

1 Introduction

Forming limit diagrams (FLD) have been widely used to predict the formability in the sheet metal forming processes, and it can be constructed by experiments such as Marciniak cup tests[1]. Fracture forming limit diagrams (FFLDs) present forming limits in the space of (ε_1, ε_2) from the uniaxial compression to the balanced biaxial tension[2]. To fully investigate and understand the FFLDs of structural part under various strain rates, the research of mechanical behavior of materials under various loading paths becomes significant.

Some experimental studies have revealed that the mechanical behaviors of materials, including initial yield stress, plastic deformation evolution and even the final fracture strain, present apparent difference under various strain rates[3-5]. Strain rate effect on necking behavior drew more attention because of its effect on local deformation[6].

Corresponding author: Cao Jun, caojun@ cnpc. com. cn.

Johnson−Cook (J−C) plasticity model represents the flow stress as a function of the equivalent plastic strain, strain rate, and temperature, and it has been widely used[7,8]. Swift model[9] is a classic constitutive equation to describe the stress−strain behavior of metallic alloy, it is a power-law equation, which can be coupled in the J−C plasticity model. The J−C plasticity model could be used to predict the viscoplastic response of materials when it is up to very high strains[10-12]. Xiao et al.[13] studied the fracture behavior of the ZK60 alloy by using modified J−C model by considering the effects of strain, stress state, and strain rate. A cryomilled nanostructured 5083 aluminum alloy exhibits the higher ductility at lower strain rates because of a diffusion−mediated stress relaxation mechanism[14].

The effect of strain rate on forming limit diagram (FLD) is different, Roth and Mohr[15] reported the increasing of strain rate increases the fracture loci for TRIP780 and DP590, while Lee et al. found[16] FLD of AZ31 alloy sheet was worse on higher strain rate and Mirfalah−Nasiri et al.[17] found that increasing of strain rate decreases the FLD of AA3104 sheet metal. In addition, Guo et al.[18] found that higher strain rate results in the decreasing of elongation to fracture for Al−Zn−Mg−Cu alloy and Dong et al.[19] also found failure strain decreases with an increase in the strain rate of U−5.5Nb alloy. These results indicate the effect of strain rate on FLD is sensitive to materials. Khan and Liu[20] developed an isotropic ductile criterion by including strain rate and temperature dependences. The experimental results showed that in quasi−static loading region, the ductile fracture of the alloy was shown to have negligible strain rate dependence.

The purpose of this research is to predict fracture loci and FFLDs considering strain rate for aluminum alloy (AA) 7050−T7451. To investigate the effect of strain rate on the fracture loci and FFLDs of AA 7050−T7451, a strain rate dependent fracture model is proposed. The specimen geometries used in the analysis is to present strain evolution under various strain paths in the fracture loci or FFLDs. The experiments and simulations of uniaxial tension, notched tension, flat−grooved tension and pure shear specimens on three strain rates, $0.001s^{-1}$, $0.01s^{-1}$ and $0.1s^{-1}$ were carried out to analyze the relation between strain rate and FFLDs. The results show that the strain histories come from verified finite element (FE) model agree with theoretical predictive fracture loci and FFLDs considering strain rate.

2 Materials and methods

2.1 Material and experiment

Uniaxial tension, notched tension, flat−grooved tension and pure shear experimentsof AA 7050−T7451 were carried out on Instron 3382 at $0.001s^{-1}$, $0.01s^{-1}$ and $0.1s^{-1}$ which are quasi-static ($\leqslant 0.01s^{-1}$) and low−impact strain rates ($\leqslant 50s^{-1}$)[21]. The sizes and shapes of the four kinds of specimens are shown in the Figure 1(a). In addition, the relative displacements for these experiments were measured using digital image correlation (DIC) technique which was explained in the section 2.3. Prior to experiments with DIC technique, random speckle needs to be evenly distributed to a thin white layer on the specimen, as shown in Figure 1(b). The images for DIC are obtained from the high-speed cameras.

The constitutive parameters of Swift model were determined by fitting the true stress−strain curve of AA7050 – T7451 at 0.001s^{-1}, so K, n, and ε_0 were determined as 675 MPa, 0.07642, -0.00114, respectively. The parameters of strain rate – dependent fracture model are calculated according to the uniaxial tension and pure shear experiments[22], as listed in Table 1 (the parameters are defined in section 2.4).

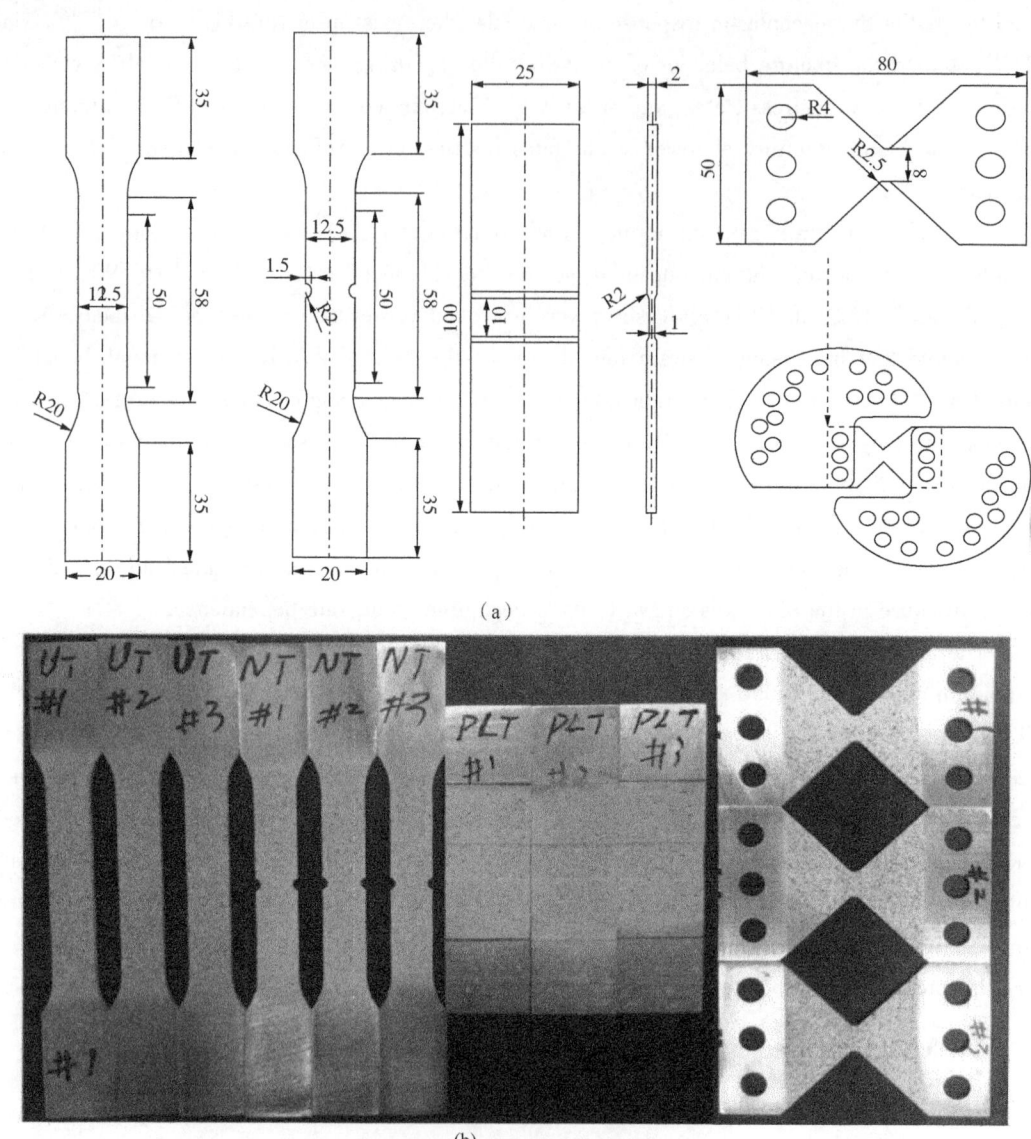

Figure 1 The (a) size and shape of uniaxial tension, notched tension, flat-grooved, and pure shear specimens and (b) actual four kinds of specimens

Table 1 The parameters of strain rate dependent fracture model

Strain rate(s^{-1})	α	β($R_\sigma>0$)	β($R_\sigma<0$)	τ_0
0.001	0.1434	1	0.5	383
0.01	0.1564	1	0.5	401
0.1	0.1624	1	0.5	407

2.2 Numerical simulation

The numerical simulations of uniaxial tension, notched tension, flat-grooved tension and pure shear specimens were conducted using commercial software ABAQUS 6.14 (SIMULIA, Wakeison France) The J-C plasticity model was employed through user subroutine. Since the symmetry of the tensile specimens, a quarter of uniaxial tension, notched tension and flat-grooved tension specimens were applied in the 3D FE models. The eight-node hexahedra elements with reduced integration (C3D8R) were used and the meshes near the notches and grooves were refined.

2.3 Digital image correlation method

Digital Imaging Correlation (DIC) is a technique which may prove to be ideally suited for the study of material deformation in real-world applications, as it has the potential to become a cheap, simple yet accurate solution[23]. DIC method was applied to obtain the local strain evolution and displacement of uniaxial tension, notched tension, flat-grooved tension, and pure shear specimens. The surfaces of four kinds of specimens were sprayed with a thin layer of white paint. Then some black speckles were evenly painted on the surfaces of the specimens. The DIC frame rate was set to 1 fps, 10 fps, and 100 fps for the strain rates $0.001s^{-1}$, $0.01s^{-1}$, and $0.1s^{-1}$. The optical measurements were done using the XITUDIC_ VS system with 6-megapixel resolution of cameras. To reduce noise, the strain filter used a (3×3) square elements and to calculate the strain value of center point. The (3×3) square elements [$(3\times3\times13)$ pixel] was chosen to calculate the local strain.

2.4 Theoretical method

The J-C plasticity model took the effect of strain hardening, strain rate and temperature into account, respectively. It is widely used in the numerical analysis under dynamic loading and high temperature condition. The constitutive equation of J-C plasticity model is expressed as

$$f(\bar{\varepsilon}_p, \dot{\bar{\varepsilon}}_p, T) = f_\varepsilon(\bar{\varepsilon}_p) f_{\dot{\varepsilon}}(\dot{\bar{\varepsilon}}_p) f_T(T) \tag{1}$$

The Swift hardening model is applied in the J-C plasticity model. It is expressed as

$$f_\varepsilon(\bar{\varepsilon}_p) = K(\bar{\varepsilon}_p + \varepsilon_0)^n \tag{2}$$

where K is strength coefficient, $\bar{\varepsilon}_p$ is equivalent plastic strain, ε_0 is offset strain, n is strain hardening index. The function in terms of strain rate is presented as

$$f_{\dot{\varepsilon}}(\dot{\bar{\varepsilon}}_p) = \begin{cases} 1, & \dot{\bar{\varepsilon}}_p < \dot{\bar{\varepsilon}}_0 \\ 1 + C\ln\left(\dfrac{\dot{\bar{\varepsilon}}_p}{\dot{\bar{\varepsilon}}_0}\right), & \dot{\bar{\varepsilon}}_p > \dot{\bar{\varepsilon}}_0 \end{cases} \tag{3}$$

where the reference strain rate $\dot{\bar{\varepsilon}}_p$ is taken as 0.001 under quasi-static loading condition and C is sensitivity coefficient of strain rate which can be fitted. In this study, the effect of temperature is not considered, and all the experiments were carried out at the room temperature, therefore, $f_T(T) = 1$ in this study. The parameter C is determined by fitting the relationship between $\sigma/K(0.002 + \varepsilon_0)^n$ and $\dot{\bar{\varepsilon}}_p/\dot{\bar{\varepsilon}}_0$, and the C is determined as 0.03529.

A modified elliptical fracture criterion was proposed based on the strain energy density[22]. It is expressed as

$$\tau^2 + \alpha^2\beta\sigma^2 \mathrm{sgn}(\sigma) = \tau_0^2 \quad (4)$$

where α is intrinsic parameter which is τ_0/σ_0. τ_0 is critical shear stress, σ_0 is critical normal stress, β is external parameter, $\mathrm{sgn}(\sigma)$ is symbolic function which is 1 when normal stress $\sigma > 0$ and is -1 when normal stress $\sigma < 0$.

Based on the derivation process of the modified elliptical fracture criterion in[22], equivalent stress can be expressed as:

$$\bar{\sigma}^2 = \frac{(1-\alpha^2\beta^*)\tau_0^2}{\alpha^2\beta^*\left(R_\sigma - \dfrac{L}{3\sqrt{L^2+3}}\right)^2 + (1-\alpha^2\beta^*)\left(\dfrac{1}{\sqrt{L^2+3}}\right)^2} \quad (5)$$

where R_σ is stress triaxiality, L is Lode parameter, β^* is β when $\sigma > 0$, β^* is $-\beta$ when $\sigma < 0$. In order to obtain the expression of equivalent plastic strain $\bar{\varepsilon}_p$ in the space of $(R_\sigma, L, \bar{\varepsilon}_p)$, the J–C plasticity model need to be substituted into the Equation. (5). Therefore, the J–C plasticity model is coupled in the modified elliptical fracture criterion in the expression of $\bar{\varepsilon}_p$. So $\bar{\varepsilon}_p$ is expressed as:

$$\bar{\varepsilon}_p = \begin{cases} \left[\dfrac{(1-\alpha^2)\tau_0^2}{\alpha^2 K^2\left(R_\sigma - \dfrac{L}{3\sqrt{L^2+3}}\right)^2 + K^2(1-\alpha^2)\left(\dfrac{1}{\sqrt{L^2+3}}\right)^2}\right]^{1/2n} - \varepsilon_0, & \dot{\varepsilon}_P < \dot{\varepsilon}_0 \\[2em] \left(\dfrac{(1-\alpha^2)\tau_0^2}{\alpha^2 K^2\left[1+\mathrm{Cln}\dfrac{\dot{\varepsilon}_P}{\dot{\varepsilon}_0}\right]^2\left(R_\sigma - \dfrac{L}{3\sqrt{L^2+3}}\right)^2 + K^2\left[1+\mathrm{Cln}\dfrac{\dot{\varepsilon}_P}{\dot{\varepsilon}_0}\right]^2(1-\alpha^2)\left(\dfrac{1}{\sqrt{L^2+3}}\right)^2}\right)^{1/2n} - \varepsilon_0, & \dot{\varepsilon}_P > \dot{\varepsilon}_0 \end{cases} \quad (6)$$

FFLD and fracture loci in strain spaces with consideration of strain rate were constructed based on the Equation. (6) and the assumption of proportional straining, and the plotting method were illustrated in[22].

3 Results

Figures. 2–5 show force–displacement responses and local equivalent plastic strain versus displacement between experiments and simulations at the strain rates of $0.001\mathrm{s}^{-1}$, $0.01\mathrm{s}^{-1}$, and $0.1\mathrm{s}^{-1}$ for uniaxial tension, notched tension, flat-grooved and pure shear specimens, additionally, the contour maps between DIC technique and simulation near fracture point are inserted in the figures. The comparisons between experiments and simulations present good agreement, which validate the correctness of FE models. In the Figure 2, the ultimate tensile strength (UTS) increases with the increase of strain rate, this reason is discussed in the next paragraph. In addition, the local strain at lowest strain rate are the largest. In the Figure 3, the effect of strain rate on fracture displacement for notched tension specimen is remarkable and the fracture displacement at highest strain rate is lowest. In the Figure 4, the failure region under plane strain tension is concentrated in the center and the effect of strain rate does not largely affect the force and displacement. In the Figure 5, the necking regions of pure shear specimens are largely influenced by strain rate, and the higher strain rate leads to the smaller necking degree. The stress triaxiality of flat-grooved process is largest and that of pure shear process is lowest. The stress triaxiality is the ratio of mean stress to equivalent stress, which indicates that the ratio of volume deformation to shape deformation. Therefore, the process of dislocation motion for pure shear is largest since the ability of shape deformation is largest, and then the strain rate has larger effect on the pure shear deformation. Since stress concentration exists in the notched tension process, the strain rate only affects the hardening behavior in the root of notch. The plastic deformation at the root of notch proceeds more adequately at lowest strain rate. So, the fracture displacement of notched tension specimen is largest at the lowest strain rate.

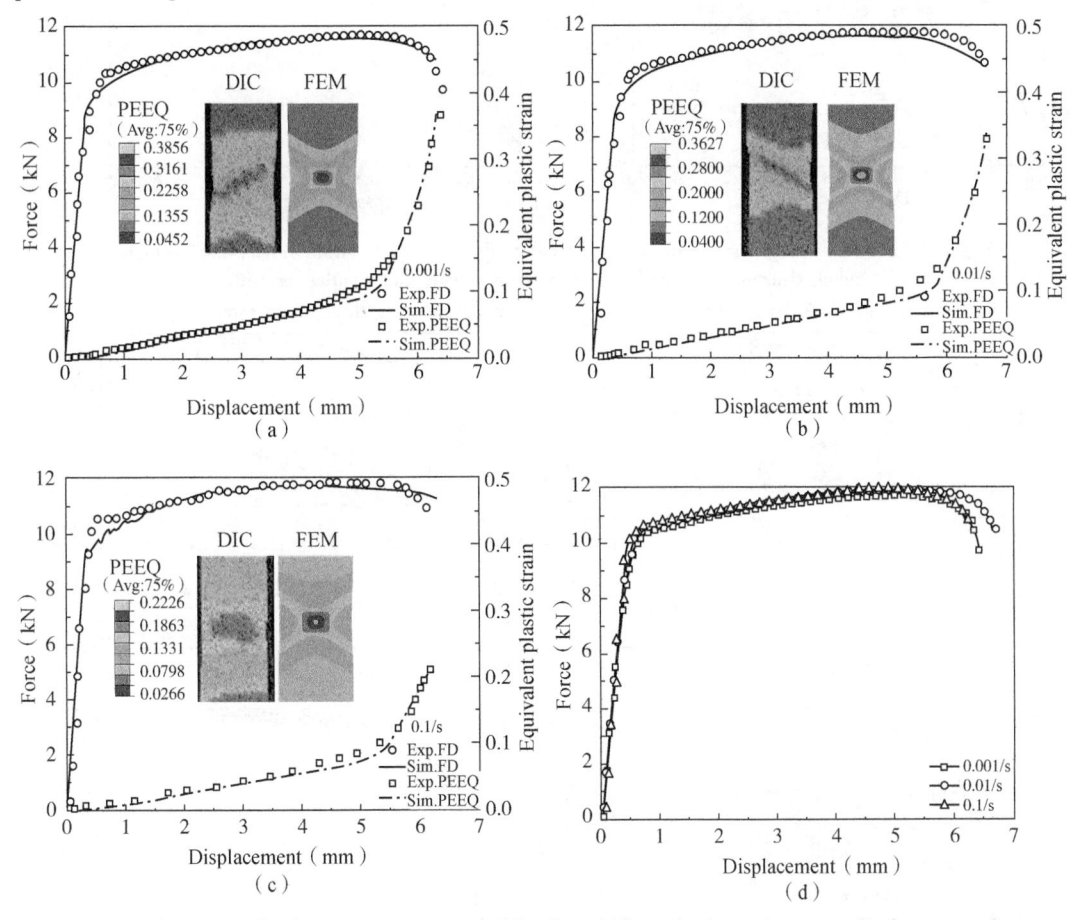

Figure 2 Force-displacement responses and local equivalent plastic strain versus displacement of uniaxial tension specimens between experiments and simulations with (a) 0.001 s^{-1}, (b) 0.01 s^{-1}, (c) 0.1 s^{-1}, and (d) three strain rates. (PEEQ is abbreviation of equivalent plastic strain).

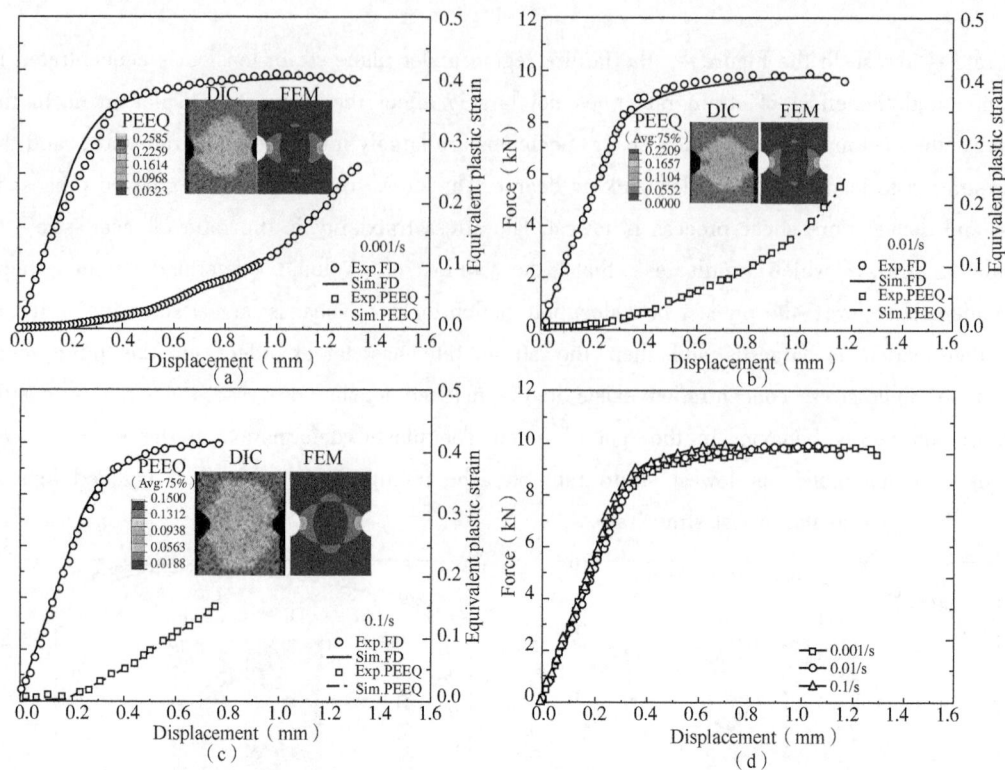

Figure 3 Force-displacement responses and local equivalent plastic strain versus displacement of notched tension specimens between experiments and simulations with
(a) 0.001s^{-1}, (b) 0.01s^{-1}, (c) 0.1s^{-1}, and (d) three strain rates

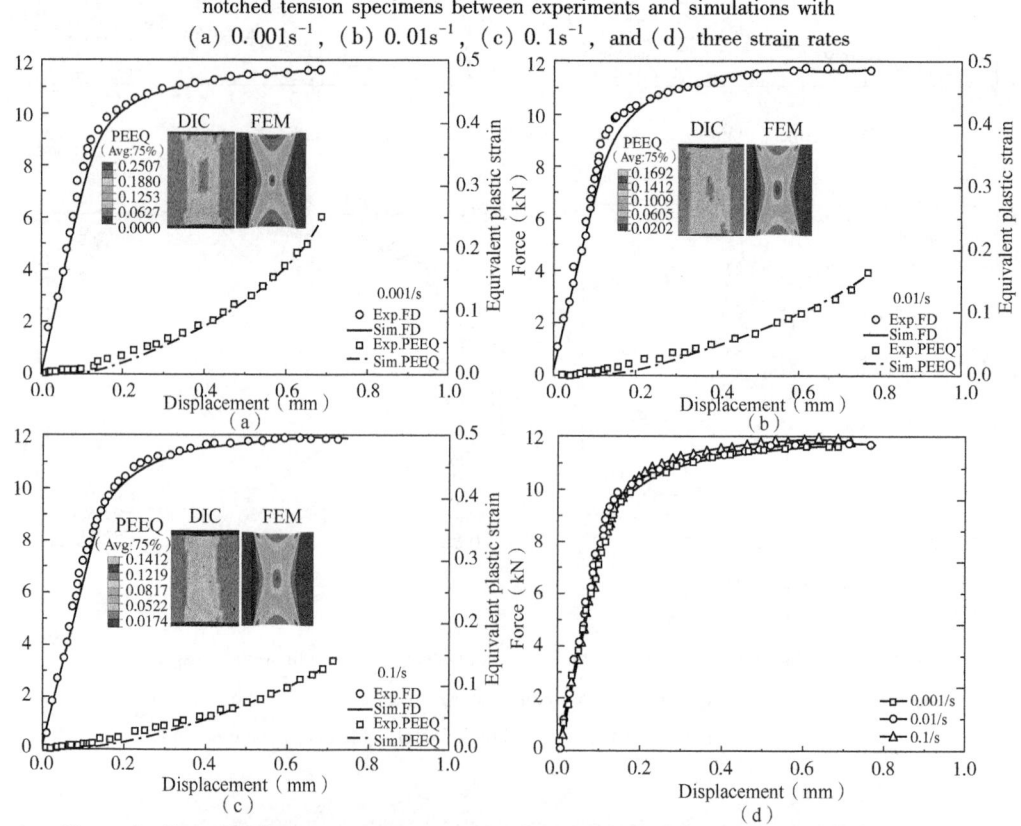

Figure 4 Force-displacement responses and local equivalent plastic strain versus displacement of flat-grooved specimens between experiments and simulations with
(a) 0.001 s^{-1}, (b) 0.01 s^{-1}, (c) 0.1 s^{-1}, and (d) three strain rates

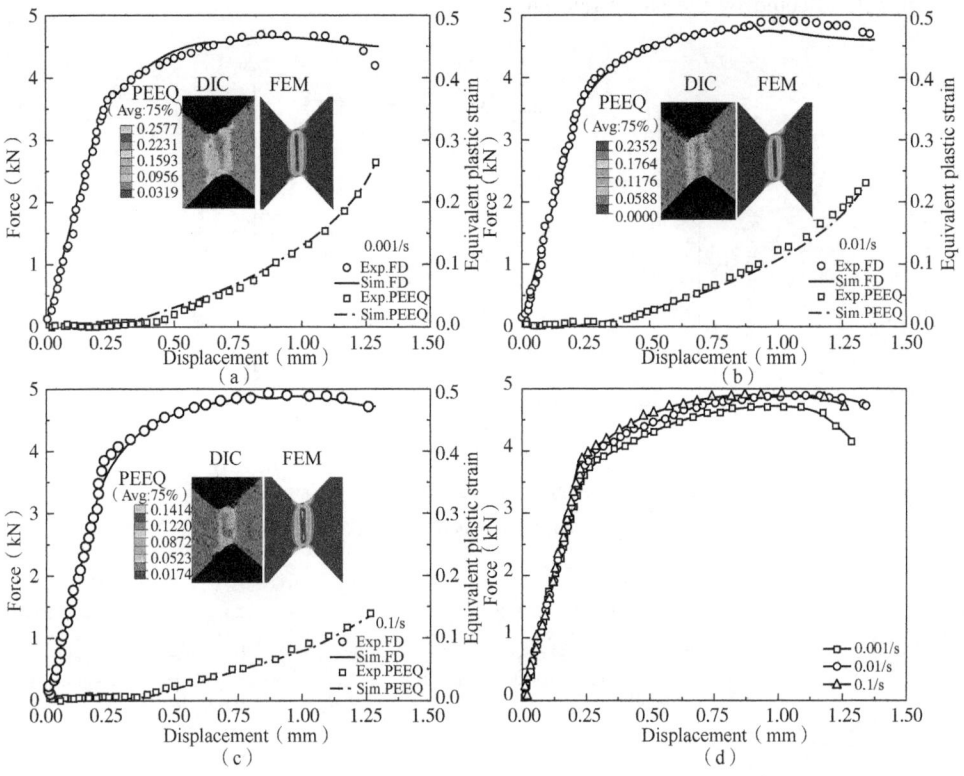

Figure 5 Force-displacement responses and local equivalent plastic strain versus displacement of
pure shear specimens between experiments and simulations with
(a) $0.001s^{-1}$, (b) $0.01s^{-1}$, (c) $0.1s^{-1}$, and (d) three strain rates

Figure 6 shows the experimental yield strengths (YS) and UTS of AA7050 - T7451 at $0.001s^{-1}$, $0.01s^{-1}$ and $0.1s^{-1}$ of strain rates based on Figure 2. It reflects that the effect of strain rate on YS is larger than that on UTS. Since the mechanisms of yielding and hardening saturation process are different, the two processes are initial and final stage of plastic deformation. In the initial stage, only one slip system may be at work and the effect of work hardening is not obvious. However, in the final stage of plastic deformation, two or more intersecting slip surfaces slide along a slip direction, the effect of work hardening gradually decreases. Therefore, the strain rate effect has large influence at yielding stage, while it plays a less role at saturate work hardening stage.

Figure 7 shows FFLDs at the three strain rates ($0.001s^{-1}$, $0.01s^{-1}$ and $0.1s^{-1}$) and the loading histories with experimental fracture strain for the four kinds of specimens. The four fracture points are in good accordance with the FFLDs at three strain rates. It indicates that the forming ability decreases with the increase of strain rate. For instance, the major strain of uniaxial tension specimen at $0.001s^{-1}$ of strain rate is ~0.4 while it reduces to ~0.2 at $0.1s^{-1}$ and the necking degree is not obvious. It indicates that the high strain rate improves the brittleness of AA 7050 - T7451. From the variation of loading paths for the four kinds of specimens, the loading path becomes more and more straight which indicates that the degree of local deformation decreases with the increasing of strain rate. Hence, the fracture strain is largely affected by the local deformation which is largely affected by the strain rate effect.

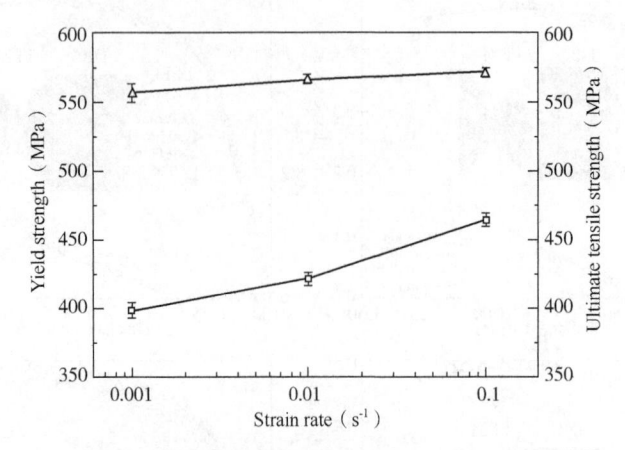

Figure 6　The effect of strain rate on yield strength and ultimate tensile strength of AA7050-T7451

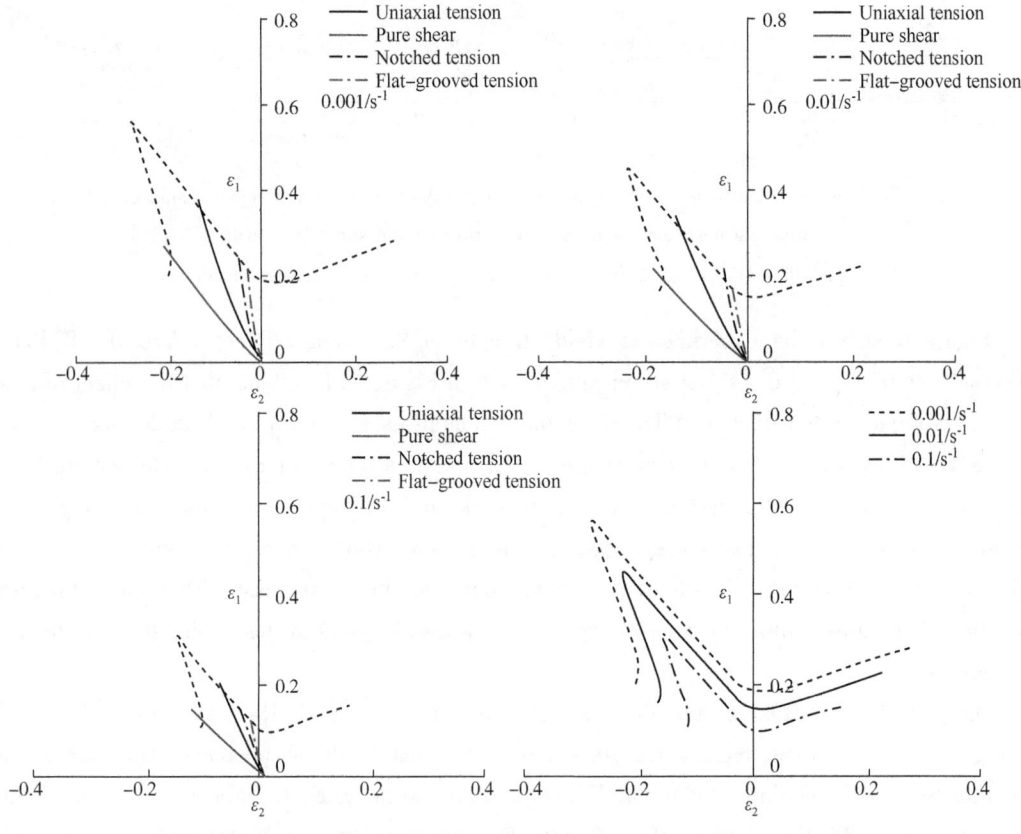

Figure 7　Fracture forming limit diagram(FFLDs) and the loading histories of experimental strain for uniaxial tension, notched tension, flat-grooved tension and pure shear specimen with (a) 0.001s^{-1}, (b) 0.01s^{-1}, (c) 0.1s^{-1}, and (d) three strain rates

Figure 8 shows the fracture loci in the space of (R_σ, $\bar{\varepsilon}_p$) and strain histories versus stress triaxiality at three strain rates (0.001s^{-1}, 0.01s^{-1} and 0.1s^{-1}) for the four kinds of specimens. The comparisons between fracture loci and fracture points at 0.001s^{-1}, 0.01s^{-1} and 0.1s^{-1} are consistent. The strain is obtained from DIC measurement and the stress triaxiality is obtained from FE models. The stress triaxiality of uniaxial tension, notched tension, and flat tension deformation are concentrated in $0.3 \sim 0.5$ and the stress triaxiality increases with the process of plastic deformation. In addition, the stress triaxiality of pure shear specimen is concentrated in $0 \sim 0.1$. These changes of stress triaxiality reflect the degree of local deformation under various strain rates. It also indicates that the extent of strain rate effect is larger for pure shear and uniaxial tension processes. Meanwhile, the difference between largest and lowest equivalent plastic strain gradually decreases with the increase of strain rate because of the lack of sliding time at higher strain rate.

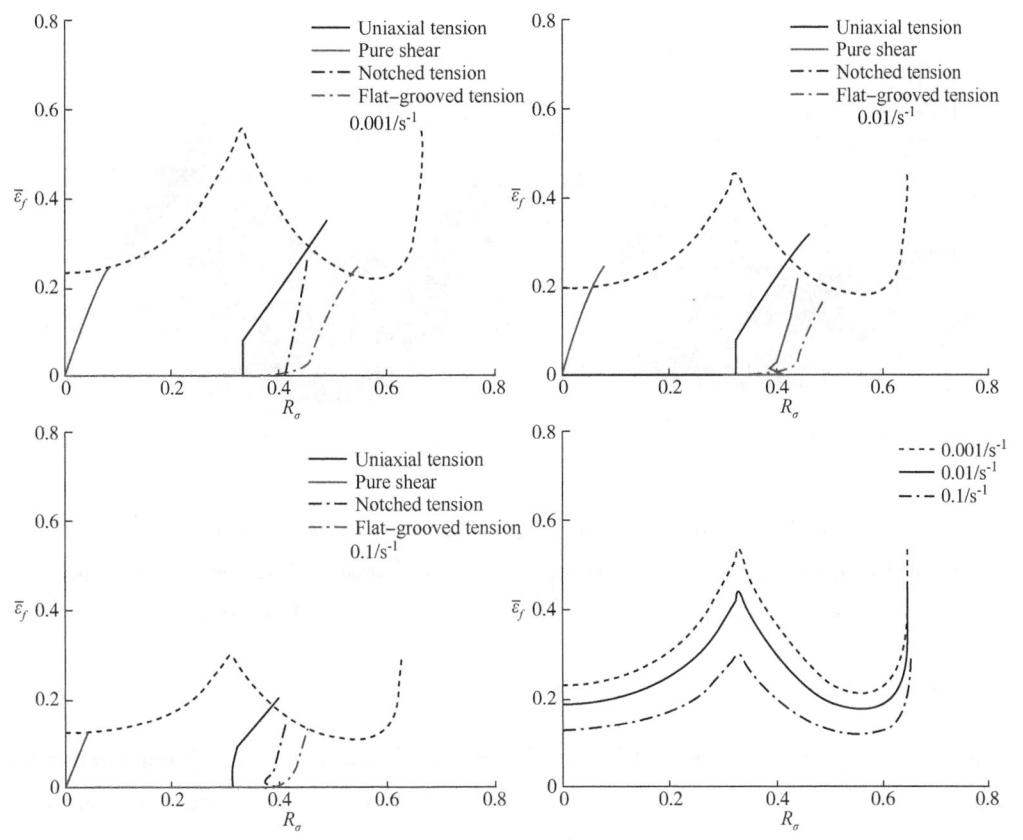

Figure 8 Fracture loci in the space of (R_σ, $\bar{\varepsilon}_p$) and the loading histories of experimental strain for uniaxial tension, notched tension, flat-grooved tension, and pure shear specimen with (a) 0.001 s^{-1}, (b) 0.01 s^{-1}, (c) 0.1 s^{-1}, and (d) three strain rates

Figure 9 shows the fracture surfaces in the space of (R_σ, L, $\bar{\varepsilon}_p$) at the three strain rates and the loading histories of the four kinds of specimens. The loading histories with stress triaxiality and Lode parameter of the four kinds of specimens reflect the evolution of stress states and strain states. The four fracture points agree with the fracture surfaces at the three strain rates (0.001 s^{-1}, 0.01 s^{-1} and 0.1 s^{-1}).

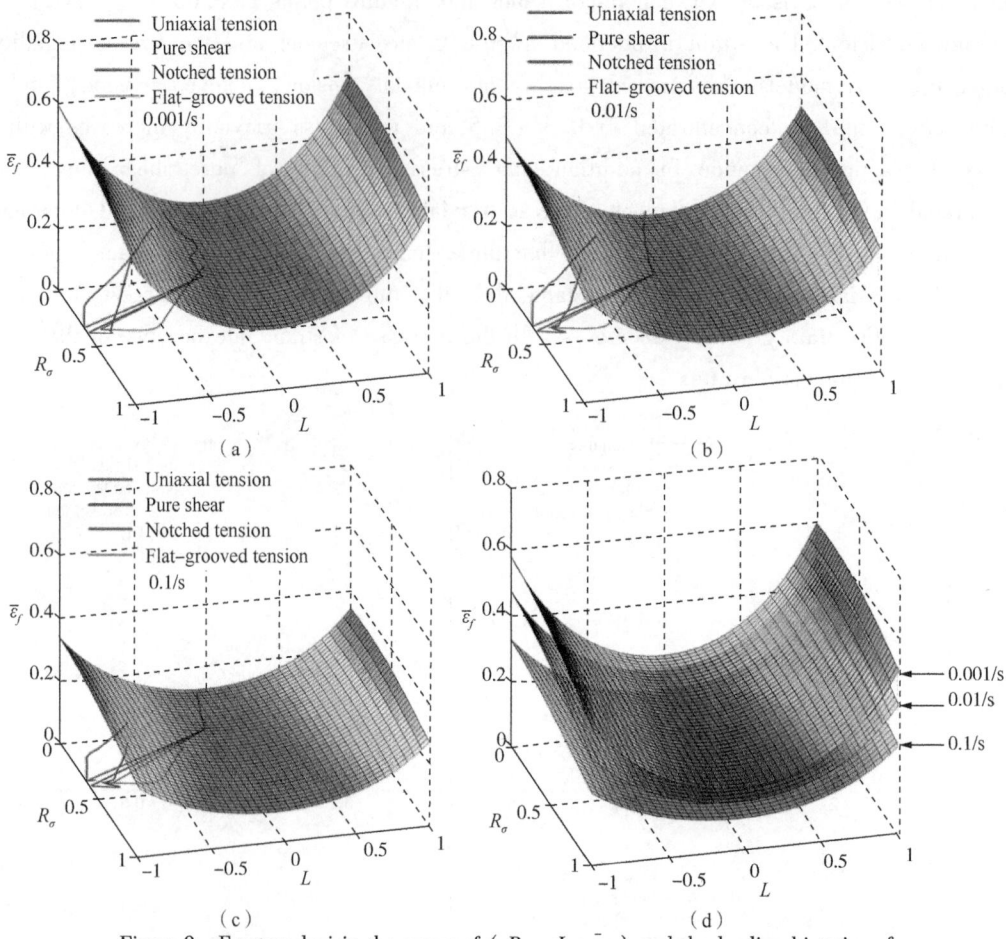

Figure 9 Fracture loci in the space of (R_σ, L, $\bar{\varepsilon}_p$) and the loading histories of experimental strain for uniaxial tension, notched tension, flat-grooved tension, and pure shear specimen with (a) 0.001s^{-1}, (b) 0.01s^{-1}, (c) 0.1s^{-1}, and (d) three strain rates

4 Discussion

From Figure 9, the differences of fracture loci at $L=-1$ and $L=0$ with different strain rates are different. The equivalent plastic strain is largest at $L=-1$ and $L=1$, and the strain rate effect is larger at these two strain states (normalized uniaxial tension and normalized uniaxial compression state)[24]. In order to explain this issue, the distortional strain energy density[22] is introduced:

$$(\mathrm{d}W/\mathrm{d}V)_d = (1+\nu)[(\sigma_1-\sigma_2)^2 + (\sigma_2-\sigma_3)^2 + (\sigma_1-\sigma_3)^2]/6E \tag{7}$$

Since Lode parameter factually is representation of the effect of second principal stress σ_2, the maximum value of $(\mathrm{d}W/\mathrm{d}V)_d$ is determined when σ_2 is $(\sigma_1+\sigma_3)/2$ and $L = (2\sigma_2-\sigma_1-\sigma_3)/(\sigma_1-\sigma_3)$. So the $(\mathrm{d}W/\mathrm{d}V)_d$ is minimum value when $L=0$, and $(\mathrm{d}W/\mathrm{d}V)_d$ is larger when $L=1$ and $L=-1$. $(\mathrm{d}W/\mathrm{d}V)_d$ is representation of the ability of shape deformation, hence, the effect of strain rate on fracture loci is larger when $L=-1$ and $L=1$. It is because that the strain rate can largely affect the dislocation movement.

Since the strain rate of 0.001 s^{-1}, 0.01 s^{-1}, and 0.1 s^{-1} are quasi-static and low impact loading conditions[21], the dynamic effects are not significant for tension and shear behavior. The effect of strain rate on failure strain mainly depends on the responses of necking behavior. Within the range of quasi-static and low impact loading conditions, with the increase of strain rate, the resistance of dislocation movement in materials increases, which leads to dislocation accumulation in local areas of materials, thus the yield stress and tensile strength of materials slightly increases, and decreasing the uniform elongation significantly, leading to the early fracture of materials, which reduces the fracture strain of the materials. However, in the range of high strain rate, damage and adiabatic temperature rise may occur during plastic deformation, resulting in different degrees of reduction of deformation resistance, and increase of fracture strain[15].

5 Conclusions

Based on J-C plasticity model and a modified elliptical fracture criterion, a strain rate-dependent fracture criterion was proposed to predict fracture loci in strain spaces and FFLD of AA 7050-T7451. Experiments and simulations of uniaxial tension, notched tension, flat-grooved, and pure shear specimens were carried out to analyze the effect of strain rate on fracture loci and FFLD. The conclusions can be drawn as follows.

(1) The FFLDs and fracture loci decrease with the increase of strain rate under 0.001 s^{-1}, 0.01 s^{-1}, and 0.1 s^{-1}. These results indicate that the ductility under different strain paths reduces as increasing the strain rate. The effect of strain rate on YS is larger than that of UTS of AA7050-T7451. It reflects that the strain rate effect has large influence at initiate stage of plastic deformation, while it plays a lesser role at the saturate work hardening stage.

(2) The UTS increases with the increase of strain rate, and the local strain at lowest strain rate is largest for uniaxial tension specimen. The effect of strain rate on fracture displacement for notched tension specimen is remarkable and the fracture displacement at highest strain rate is lowest. The necking regions are influenced by strain rate under pure shear loading and higher strain rate leads to smaller necking degree.

(3) The FFLDs, fracture loci in the space of (R_σ, $\bar{\varepsilon}_p$) and the fracture loci in the space of (R_σ, L, $\bar{\varepsilon}_p$) at three strain rates are consistent with the fracture points of experimental strain for uniaxial tension, notched tension, flat-grooved tension and pure shear specimens. It verifies the correctness of the strain rate-dependent fracture model of AA 7050-T7451.

Acknowledgement

This work was supported by the National Science Foundation for Young Scientists of China under Grant (No. 51904332), the National Key R&D Program of China under Grant (No. 2016YFC0802101), and the State Key Laboratory of Solidification Processing (NWPU) of China under Grant (No. 130-QP-2015).

References

[1] Marciniak, Z. A.; Kuczyński, K.; Pokora, T. Influence of the plastic properties of a material on the forming

limit diagram for sheet metal in tension. International Journal of Mechanical Sciences 15, 789-800.

[2] Lou, Y.; Huh, H.; Lim, S.; Pack, K. New ductile fracture criterion for prediction of fracture forming limit diagrams of sheet metals. International Journal of Solids & Structures 49, 3605-3615.

[3] Hopperstad, O. S.; Børvik, T.; Langseth, M.; Labibes, K.; Albertini, C. On the influence of stress triaxiality and strain rate on the behaviour of a structural steel. Part I. Experiments. European Journal of Mechanics 2003, 22, 1-13.

[4] Clausen, A. H.; Børvik, T.; Hopperstad, O. S.; Benallal, A. Flow and fracture characteristics of aluminium alloy AA5083-H116 as function of strain rate, temperature and triaxiality. Materials Science & Engineering A 2004, 364, 260-272.

[5] Liu, Y. J.; Sun, Q. A dynamic ductile fracture model on the effects of pressure, Lode angle and strain rate. Materials Science & Engineering A 2014, 589, 262-270.

[6] Sato, K.; Yu, Q.; Hiramoto, J.; Urabe, T.; Yoshitake, A. A method to investigate strain rate effects on necking and fracture behaviors of advanced high-strength steels using digital imaging strain analysis. International Journal of Impact Engineering 2015, 75, 11-26.

[7] Qiao, J. W.; Chu, M. Y.; Cheng, L.; Ye, H. Y.; Yang, H. J.; Ma, S. G.; Wang, Z. H. Plastic flows of in-situ metallic glass matrix composites upon dynamic loading. Materials Letters 119, 92-95.

[8] Farahani, H. K.; Ketabchi, M.; Zangeneh, S. Determination of Johnson-Cook Plasticity Model Parameters for Inconel718. J Mater Eng. Perform 2017, 26, 5284-5293.

[9] Swift, H. W. Plastic instability under plane stress. Journal of the Mechanics and Physics of Solids 1952, 1, 1-18.

[10] Verleysen, P.; Peirs, J.; Van Slycken, J.; Faes, K.; Duchene, L. Effect of strain rate on the forming behaviour of sheet metals. Journal of Materials Processing Technology 2011, 211, 1457-1464.

[11] Erice, B.; Gálvez, F.; Cendón, D. A.; Sánchez-Gálvez, V. Flow and fracture behaviour of FV535 steel at different triaxialities, strain rates and temperatures. Engineering Fracture Mechanics 2012, 79, 1-17.

[12] Kajberg, J.; Sundin, K.-G. Material characterisation using high-temperature Split Hopkinson pressure bar. Journal of Materials Processing Technology 2013, 213, 522-531.

[13] Yue, X.; Qi, T.; Hu, Y.; Jian, P.; Luo, W. Flow and Fracture Study for ZK60 Alloy at Dynamic Strain Rates and Different Loading States. Materials Science & Engineering A 2018, 724, 208-219.

[14] Han; B., Q.; Huang; J., Y.; Zhu; Y., T.; Lavernia; E., J. Effect of strain rate on the ductility of a nanostructured aluminum alloy. Scripta Materialia 2006, 54, 1175-1180.

[15] Roth, C. C.; Mohr, D. Effect of strain rate on ductile fracture initiation in advanced high strength steel sheets: Experiments and modeling. International Journal of Plasticity 2014, 56, 19-44.

[16] Lee, Y.; Kwon, Y.; Kang, S.; Kim, S.; Lee, J. Forming limit of AZ31 alloy sheet and strain rate on warm sheet metal forming. Journal of materials processing technology 2008, 201, 431-435.

[17] Mirfalah-Nasiri, S.; Basti, A.; Hashemi, R. Forming limit curves analysis of aluminum alloy considering the through-thickness normal stress, anisotropic yield functions and strain rate. International Journal of Mechanical Sciences 2016, 117, 93-101.

[18] Yue, G.; Zhou, M.; Sun, X.; Long, Q.; Li, L.; Xie, Y.; Liu, Z.; Di, W.; Yang, L.; Tong, W. Effects of Temperature and Strain Rate on the Fracture Behaviors of an Al-Zn-Mg-Cu Alloy. Materials 2018, 11, 1233.

[19] Dong, C.; Chen, X.; Rong, M.; Tang, Q.; Su, B.; Wang, Z.; Zhang, X.; Meng, D. Effect of strain rate on fracture of U-5.5Nb alloy. Mater. Sci. Technol. 2018, 34, 12-19.

[20] Khan, A. S.; Liu, H. Strain rate and temperature dependent fracture criteria for isotropic and anisotropic metals. International Journal of Plasticity 2012, 37, 1-15.

[21] Safari, M.; Azodi, H. D. An investigation into the effect of strain rate on forming limit diagram using ductile fracture criteria. Meccanica 2012, 47, 1391-1399.

[22] Cao, J.; Li, F.; Ma, X.; Sun, Z. A modified elliptical fracture criterion to predict fracture forming limit diagrams for sheet metals. Journal of Materials Processing Technology 2018, 252, 116-127.

[23] Mccormick, N.; Lord, J. Digital Image Correlation. Materials Today 2010, 13, 52-54.

[24] Lou, Y.; Yoon, J. W.; Huh, H. Modeling of shear ductile fracture considering a changeable cut-off value for stress triaxiality. International Journal of Plasticity 2014, 54, 56-80.

本论文原发表于《Metals》2020年第10卷第3期。

Effects of Pre-Oxidation on the Corrosion Behavior of Pure Ti under Coexistence of Solid NaCl Deposit and Humid Oxygen at 600℃: the Diffusion of Chlorine

Fan Lei[1,3]　Liu Li[2]　Lv Yuhai[4]　Wang Hao[4]
Fu Anqing[1]　Yuan Juntao[1]　Li Ying[3]　Wang Fuhui[2]　Yin Chengxian[1]

(1. State Key Laboratory for Performance and Structure Safety of Petroleum Tubular Goods and Equipment Materials, CNPC Tubular Goods Research Institute;
2. Key Laboratory for Anisotropy and Texture of Materials (MoE), School of Materials Science and Engineering, Northeastern University;
3. Institute of Metal Research, Chinese Academy of Sciences;
4. NO. 1 Gas Production Plant of Changqing Oilfield Company)

Abstract: The effect of pre-oxidation on the corrosion behavior of pure Ti covered with a solid NaCl deposit in the humid O_2 flow at 600℃ is studied. The oxide scale, formed by pre-oxidation, protects the substrate from the NaCl induced corrosion during the initial stage. However, the corrosion of the pre-oxidized sample is severely accelerated by solid NaCl after an incubation period. The chlorine, generated from the decomposition of solid NaCl, diffuses into the oxide/substrate interface as ions during the incubation period, which was observed by ToF-SIMS. The chlorine at the oxide/substrate interface induces the fast corrosion after the incubation period although the pre-oxidation scale is complete and compact.

1 Introduction

Compressor blades of airplanes and ships suffer severe corrosion in marine environment, which is mainly due to the fact that marine air contains abundant salts (especially NaCl) and water vapor. At the temperature of 300~600℃ where compressor blades operate, NaCl appears in solid state and gets deposited on the metal surface. The corrosion of several metals and alloys of compressor blades gets accelerated under a synergistic effect of the solid NaCl deposit and humid oxygen.

Extensive works have been done on the corrosion behavior of several metals and alloys of

Corresponding author: Liu Li, liuli@ mail. neu. edu. cn.

compressor blades. The authors in paper[1-11] demonstrated that the corrosion behavior of pure Fe, pure Cr, Fe-Cr alloy, Ti alloy and Ni alloy at 600℃ under a deposit of solid NaCl deteriorated sharply. Similarly, the authors investigated the influence of KCl on the oxidation of the 304-type (Fe18Cr10Ni) austenitic stainless steel at 600℃ in 5% O_2 and in 5% O_2 + 40% H_2O in paper[12-17], and the results showed that small additions of potassium chloride strongly accelerated high temperature corrosion. The fast corrosion is mainly due to a chemical reaction of the solid NaCl/KCl deposit with oxides. This reaction destroys the protective oxide scale on the surface and forms a porous non-protective scale (i. e. , the scale has numerous holes and voids[1,6,8,9,15,17]). It is also reported in these investigations that the key factor for the fast corrosion is the chlorine, HCl or Cl_2, which cyclically reacts with the substrate at the oxide/substrate interface. The gaseous molecule of chlorine, which is formed during the chemical reaction of solid NaCl and oxide[1,8,18], diffuses inward quickly through macro defects (like cracks and holes) in the scale of corrosion products. The chlorine then reacts cyclically with the substrate to generate volatile products (for example $CrCl_3$) which break the protective scale [1,3,6,8].

In our previous work[11], we observe that the scale of corrosion products of Ti60 alloys can be divided into an outer the corrosion layer and an inner corrosion layer after exposure at 600℃ in the environment of NaCl + H_2O + O_2. The chlorine exists in a Ti-Cl bond which is formed by replacing O with Cl within lattices of Ti oxides of the inner corrosion layer. We think the chlorine can diffuse inward as ion. In this case, Cl^- comes from the decomposition of solid NaCl and the residual sodium reacts with oxides on the surface to form metallic acid salts (like $Na_2CrO_4^1$ and $Na_4Ti_5O_{12}^{10}$). In order to make clear the corrosion mechanism, we need to first figure out the state of the chlorine and study its diffusion in the scale of corrosion products, especially during the destruction process of the scale of protective oxide, like TiO_2 and Cr_2O_3.

When pure Ti is oxidized in the pure O_2 condition at 600℃, a compact and even oxide scale forms on the surface[10]. This oxide scale can simulate the passive film to protect the substance from the corrosive environment. When the pre-oxidized pure Ti is exposed in the NaCl + H_2O + O_2 environment, this oxide scale can slow down the destruction process of the protective passive film such that the behavior of the solid NaCl can be investigated in detail. Thus, to clarify the micro-mechanism under the specific environment of NaCl + H_2O + O_2, we need to study the effects of pre-oxidation on the corrosion behavior of pure Ti.

In this paper, we investigate the effects of pre-oxidation on the corrosion behavior of pure Ti underneath a solid NaCl deposit in humid O_2 flow at 600℃, using scanning electron microscope (SEM) equipped with an energy dispersive spectrometer (EDS), X-ray diffraction (XRD) and time of flight-secondary ion mass spectrometry (ToF-SIMS). After examining the detailed properties of the corrosion products and the diffusion of chlorine in the oxide scale, we discuss possible micro acceleration mechanisms of the chlorine based on experiment results.

2 Result

2.1 Corrosion kinetics

The mass gain curves of pure Ti under different conditions are shown in Fig. 1. For the pre-oxidized samples, the mass gain under the condition of H_2O+O_2 (about 0.17mg/cm^2) is slightly larger than that under the condition of O_2 (about 0.09mg/cm^2). In fact, under both the above two conditions the corrosion is minor. In the initial stage (incubation period), the mass gain is slight when the pre-oxidized pure Ti samples under NaCl deposit. In contrast, the mass gain increases rapidly after this incubation period. After 20h exposure in NaCl+H_2O+O_2 or NaCl+O_2, the mass gains are about 1.8mg/cm^2 and 1.4mg/cm^2, respectively, which are about one order of magnitude larger than that in the absence of the solid NaCl deposit. Thus, we reach the conclusion that the solid NaCl accelerates the corrosion of the pre-oxidized pure Ti in O_2 or humid O_2.

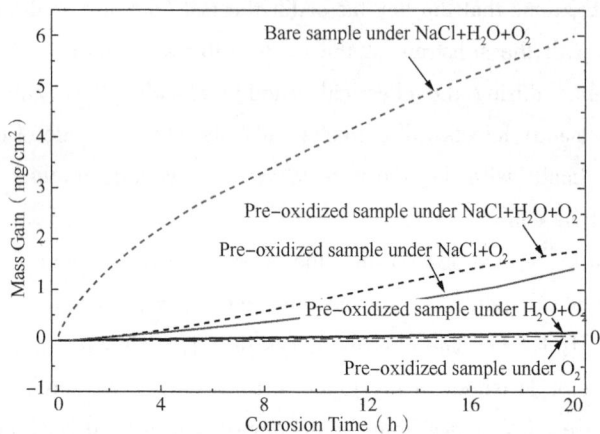

Fig. 1 The mass gain curves for the corrosion at 600℃ of the bare Ti under NaCl + H_2O + O_2 (red line) and the pre-oxidized Ti under O_2 (cyan line), H_2O + O_2 (black line), NaCl + O_2 (green line) and NaCl + H_2O + O_2 (purple line)

For the bare samples, the mass gain increases dramatically during the whole exposure in NaCl+H_2O+O_2. After exposed for 20h, the mass gain is about 6.0mg/cm^2, which is increased by 3.5 times compared with the pre-oxidized samples. Therefore, we can conclude that the pre-oxidation improves the corrosion resistance of pure Ti coated with a solid NaCl deposit in humid O_2 at 600℃, especially in the incubation period.

2.2 Phase composition and microstructure of the pre-oxidized sample

For the pure Ti samples after pre-oxidation in O_2 flow for 20h, the surface and the cross-sectional morphologies are shown in Fig. 2. As can be observed in Fig. 2(a)(b), a continuous, compact and even oxide scale is formed on the surface. The thickness of the oxide scale is about 1μm [Fig. 2(b)]. We show the results of EDS analysis on the oxide scale in Fig 2(c). As can be seen, it consists of Ti and O. Through XRD analysis (which is not shown here), the phase composition of the pre-oxidized sample surface is only TiO_2.

2.3 Phase composition and microstructure of the hot corrosion products forming on the bare samples

For the bare samples after 20h exposure in NaCl+H_2O+O_2, the surface and the cross-sectional morphologies are shown in Fig. 3. As can be observed, the scale of corrosion products contains two parts: one with a compact structure, and the other with a grid structure part. The corrosion products with a compact structure [marked in Fig. 3(a-b)] are very complete, uniform and compact. The

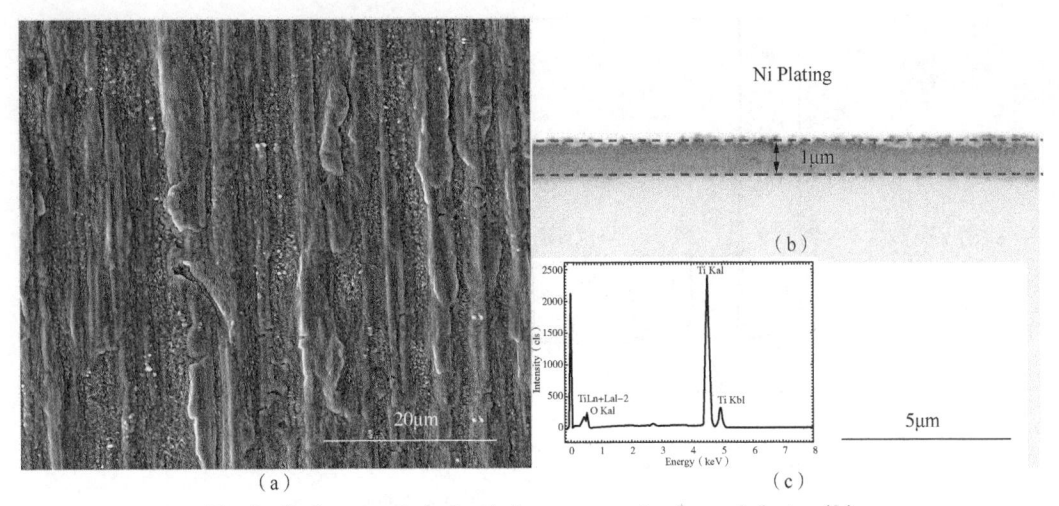

Fig. 2 Surface morphologies (a), cross-sectional morphologies (b) and EDS patterns (c) of pure Ti oxidized in dry O_2 at 600℃ for 20h

corrosion products with a grid structure [marked in Fig. 3(a-b)] contain lots of holes [filled with Ni plating as shown in Fig. 3(b)], which penetrate the corrosion scale. These through-holes are the fast diffusion path for the corrosion medium. Below the corrosion scale, a loose layer, containing some holes (close pores) and cracks, is formed within the substrate due to the outward diffusion of Ti. The thickness of the corrosion product scale and the loose layer is about 45μm and 15μm, respectively.

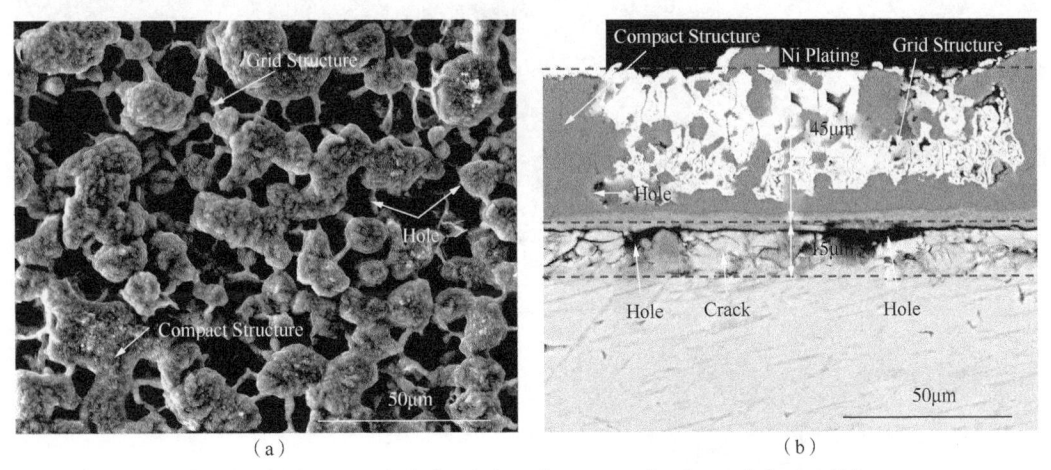

Fig. 3 Surface morphologies (a) and cross-sectional morphologies (b) of the bare samples after corrosion in NaCl + H_2O+O_2 at 600℃ for 20h

In Fig. 4, we present the XRD patterns of the bare samples exposed in NaCl+H_2O+O_2 up to 20h. The results show that the corrosion products mainly consist of TiO_2 and some $Na_4Ti_5O_{12}$, with the residual NaCl.

2.4 Phase composition and microstructure of the hot corrosion products forming on the pre-oxidized samples

For the pre-oxidized pure Ti samples after 100h exposure in dry O_2 flow, the surface and the

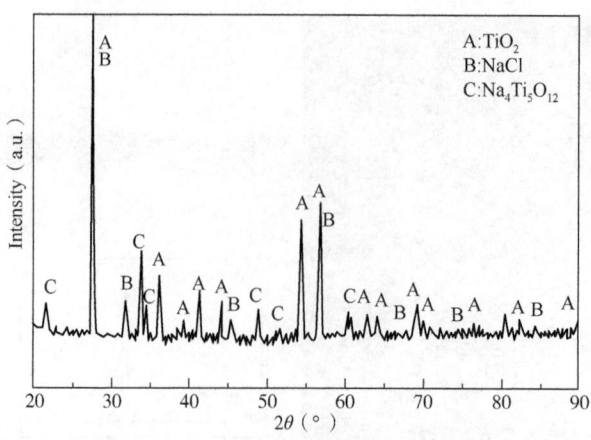

Fig. 4 X-ray diffraction patterns of the bare samples exposed in
NaCl + H$_2$O + O$_2$ up to 20 h at 600℃

cross-sectional morphologies are shown in Fig. 5. As can be seen in Fig. 5(a)(b), a very thin, compact and continuous scale is formed on the pre-oxidized samples after 100 h exposure in dry O$_2$ flow. From Fig. 5(c), we can see that the thickness of the oxide scale is about 2.5μm.

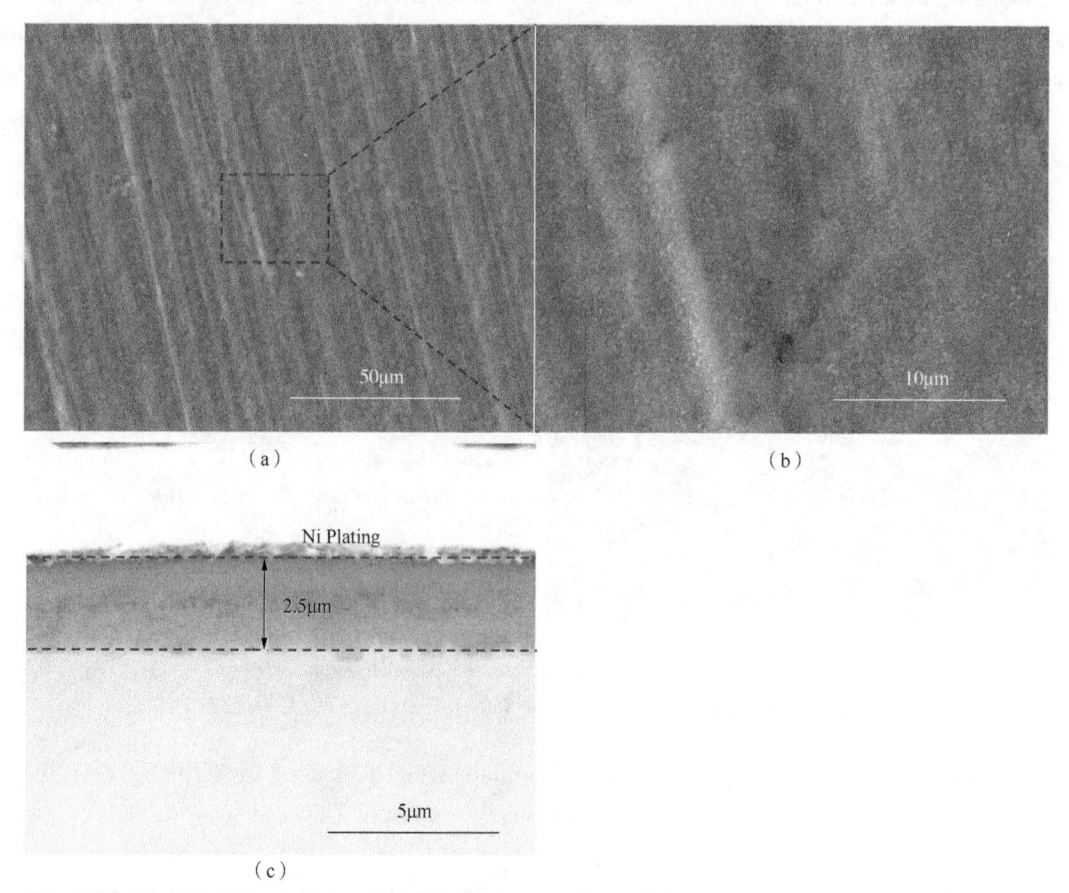

Fig. 5 Surface morphologies (a), the high magnification image (b) and cross-sectional morphologies (c) of the pre-oxidized samples after oxidized in dry O$_2$ at 600℃ for 100h

For the pre-oxidized samples after 100 h exposure in humid O_2 flow, the surface and the cross-sectional morphologies are presented in Fig. 6. As can be observed from Fig. 6(a)(b), a very thin, compact and continuous scale which contains both granulated corrosion products and compact products was formed on the pre-oxidized samples. From Fig. 6(c), we can see that the oxide scale is divided into a loose outer layer and a compact inner layer, with a thickness of about 1.2μm and 1μm, respectively.

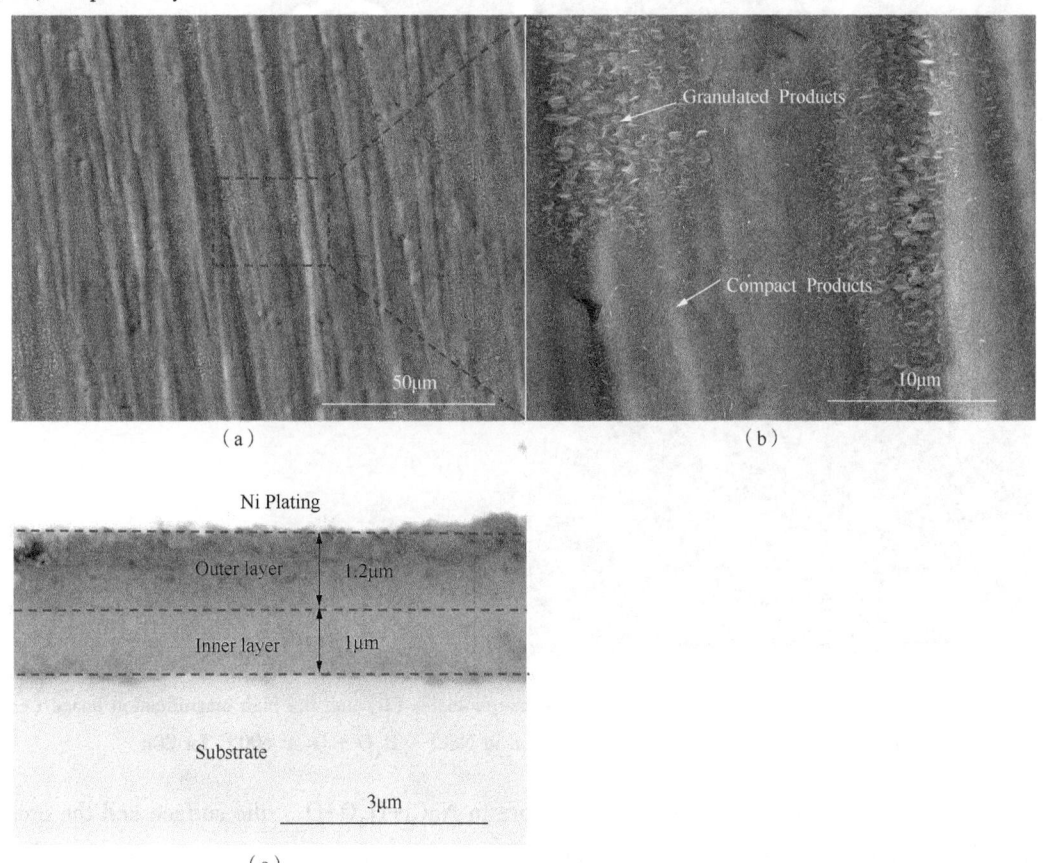

Fig. 6 Surface morphologies (a), the high magnification image (b) and cross-sectional morphologies (c) of the pre-oxidized samples after oxidized in humid O_2 at 600℃ for 100h

For the pre-oxidized samples after 20h exposure in NaCl+O_2, the surface and the cross-sectional morphologies are displayed in Fig. 7. As can be seen in Fig. 7(a), the corrosion products on the surface are in clusters. As can be observed in Fig. 7(b)(c), the corrosion product scale can be divided into an outer layer and an inner layer. The outer layer is filled with clustered corrosion products, and the inner layer whose thickness is about 5μm contains loose corrosion products.

For the pre-oxidized samples after 20h exposure in NaCl+H_2O+O_2, the surface and the cross-sectional morphologies are depicted in Fig. 8. As can be observed in Fig. 8(a)(b), corrosion products have a very thin, compact and continuous scale, with some blind holes forming on the surface where we can also observe some defects (like pore). Furthermore, the scale can be divided into a loose inner corrosion layer and a compact outer corrosion layer. As shown in Fig. 8(c), the thickness of the outer corrosion layer and the inner corrosion layer is 8μm and 10μm, respectively.

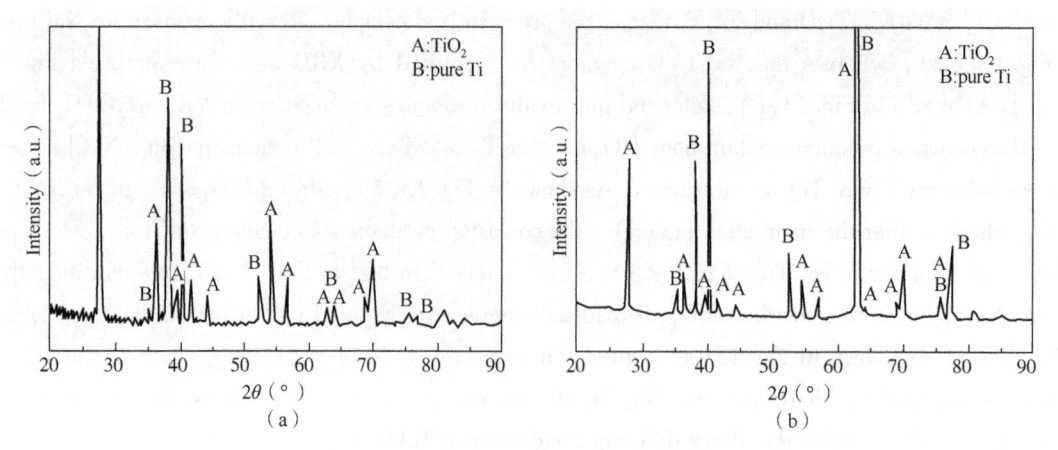

Fig. 11 X-ray diffraction patterns of the pre-oxidized samples after exposed in dry O_2 flow (a) and in humid O_2 flow (b) for 100h

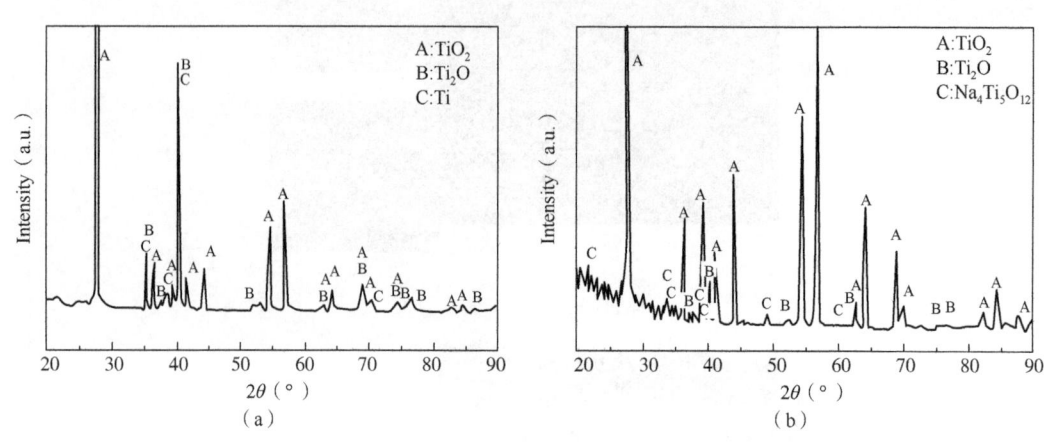

Fig. 12 X-ray diffraction patterns of the pre-oxidized samples after exposed in NaCl + O_2 for 5h (a) and for 20h (b)

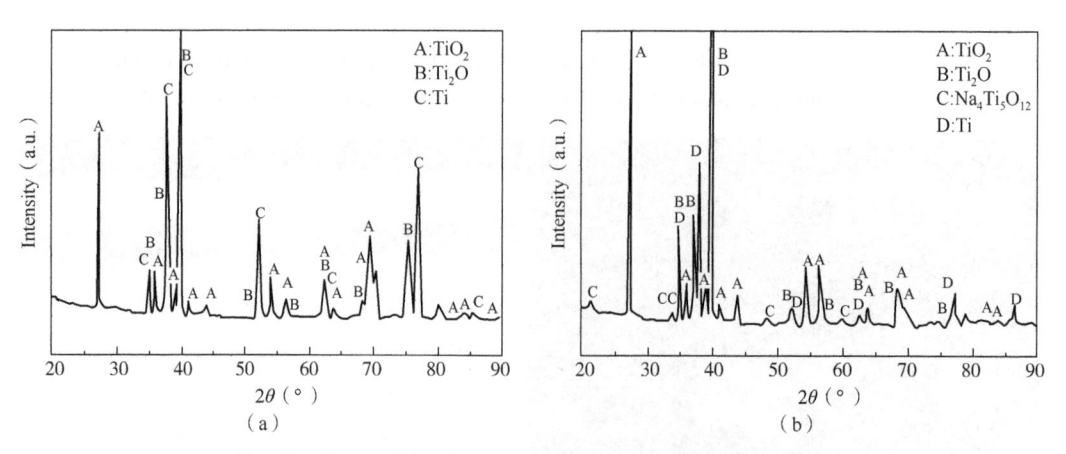

Fig. 13 X-ray diffraction patterns of the pre-oxidized samples after exposed in NaCl +H_2O+ O_2 for 1h (a), 5h (b) and 20h(c)

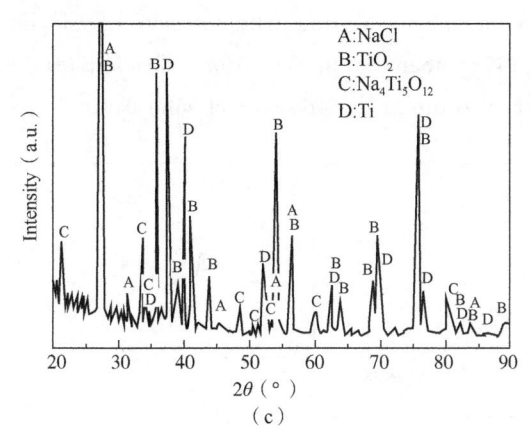

Fig. 13 X-ray diffraction patterns of the pre-oxidized samples after expose dn NaCl +H$_2$O+ O$_2$ for 1h (a), 5h (b) and 20h(c)(Continue)

Table 1 Corrosion products forming on the bare samples and the pre-oxidized samples after hot corrosion

Samples	Environment	Corrosion Time			
		1h	5h	20h	100h
Bare Samples	NaCl+H$_2$O+O$_2$	—	—	TiO$_2$, Na$_4$Ti$_5$O$_{12}$	—
Pre-oxidized Samples	NaCl+H$_2$O+O$_2$	TiO$_2$, Ti$_2$O	TiO$_2$, Ti$_2$O, Na$_4$Ti$_5$O$_{12}$	TiO$_2$, Ti$_2$O, Na$_4$Ti$_5$O$_{12}$	—
	NaCl +O$_2$	—	TiO$_2$, Ti$_2$O	TiO$_2$, Ti$_2$O, Na$_4$Ti$_5$O$_{12}$	—
	H$_2$O+O$_2$	—	—	—	TiO$_2$
	O$_2$	—	—	—	TiO$_2$

2.5 Time of flight-secondary ion mass spectrometry, ToF-SIMS

Fig. 14 shows the depth profiles for the negative ions from the pre-oxidized samples after 1 h and 5h exposure in NaCl+H$_2$O+O$_2$. The O$_2^-$ ion signal in Fig. 14 is applied here to indicate the distribution of oxygen in the scale and to identify the interface between the corrosion products scale and the substrate. In Fig. 14(a), the O$_2^-$ ion exhibits two levels of concentration, and the transition region is around the interface of the corrosion product scale and the substrate. Therefore, the average of logarithmic value is pinpointed as the interface as shown in Fig. 14(a). Similarly, this interface is also identified in Fig. 14(b).

In Fig. 14(a), we can also observe that after 1 h exposure in NaCl+H$_2$O+O$_2$, Cl$^-$ ions first enrich on the surface of the pre-oxidized samples (which was caused by the residual NaCl) and then decrease rapidly in the corrosion product scale. Afterwards, Cl$^-$ is enriched at the oxide/metal interface, which indicates that the chlorine penetrates the oxide and accumulates at the oxide/metal interface. In Fig. 14(b), after the pre-oxidized samples are exposed in NaCl+H$_2$O+O$_2$ for 5 h, a similar distribution of Cl$^-$ ion in the corrosion product scale was observed. Particularly, the Cl$^-$ ion is still enriched when the O$_2^-$ ion decreases around the oxide/metal interface, which implies that some chlorine diffuses into the metal substrate. The intensity of Cl⁻ ion around the oxide/metal interface is stronger in Fig. 14(b) than that in Fig. 14(a). This implies that more chlorine penetrates into the oxide after a longer exposure in the presence of solid NaCl.

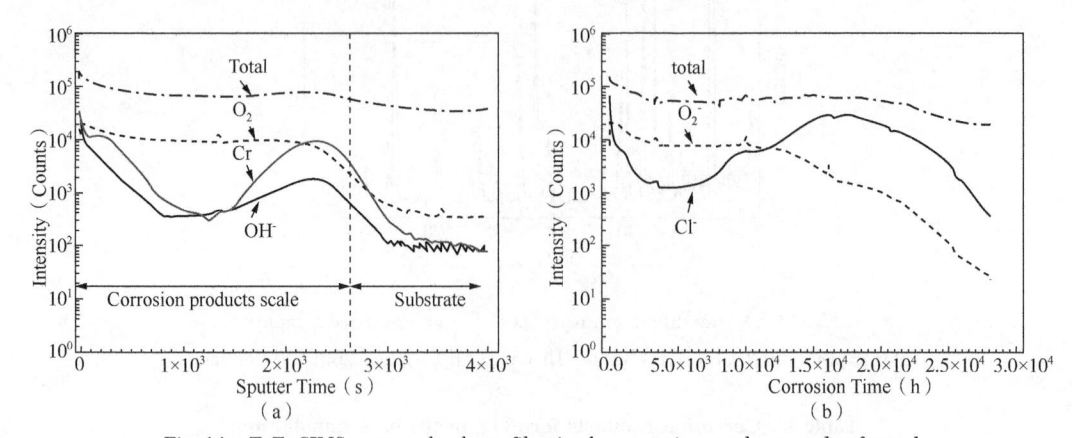

Fig. 14 ToF-SIMS sputter depth profiles in the corrosion products scales formed on the pre-oxidized samples under a NaCl deposit in an atmosphere of humid O_2 at 600℃ for 1h (a) and 5 (b)

The OH⁻ signal in Fig. 14 is applied here to indicate the distribution of H in the scale. From Fig. 14a, the distribution of the OH⁻ is similar with Cl⁻. First, the OH⁻ ion enriches on the surface and then decrease rapidly in the corrosion product scale. This is the information of water vapor. Afterwards, OH⁻ is enriched at the oxide/metal interface where Cl⁻ is accumulated, which indicates that the H penetrates the oxide along with the chlorine.

3 Discussion

3.1 The acceleration of NaCl to the corrosion of pure Ti

As shown in Fig. 2, after pure Ti samples were oxidized in the pure O_2 flow at 600℃ for 20h, a compact TiO_2 scale with a thickness of about 1μm was formed on the surface, which is caused by the following reaction between the metal substrate and O_2:

$$Ti + O_2 = TiO_2 \quad (1)$$

In contrary, the mass gain of pure Ti under a solid NaCl deposit layer ($NaCl + H_2O + O_2$) increased quickly and the mass gain is about 6.0mg/cm² after 20h exposure (Fig. 1), which is much larger than that in the absence of solid NaCl (about 0.3mg/cm²[10]). The resulting scale consists of plentiful and porous corrosion products (Fig. 3) and its thickness is about 45μm. The above phenomena indicate that pure Ti suffered severe corrosion when exposed in $NaCl + H_2O + O_2$. This severe corrosion has been discussed in our previous work[10] and the main reason is that the protective TiO_2 scale cannot be formed due to the occurrence of a series of chemical reactions, and the corrosion becomes active. In the previous mechanism, NaCl first destroys the protective scale as follow:

$$5TiO_2 + 4NaCl + 2H_2O = Na_4Ti_5O_{12} + 4HCl \quad (2)$$

$$6TiO_2 + 4NaCl = Na_4Ti_5O_{12} + TiCl_4 \quad (3)$$

The $TiCl_4$ reacts with H_2O to form HCl:

$$TiCl_4 + 2H_2O = TiO_2 + 4HCl \quad (4)$$

Then, HCl reacts with the substrate cyclically:

$$2HCl + Ti = TiCl_2 + H_2 \quad (5)$$

$$4HCl + Ti = TiCl_4 + 2H_2 \quad (6)$$

$$2TiCl_2 + O_2 + 2H_2O = 2TiO_2 + 4HCl \quad (7)$$

$$2H_2 + O_2 = 2H_2O \quad (8)$$

These reactions are thermodynamically spontaneous due to the negative $\Delta G°$ at 600℃, as presented in Table 2.

Table 2 Standard Gibbs free energy changes of reactions at 600℃

Reaction	$\Delta G°$ * (kJ/mol)	Reaction	$\Delta G°$ * (kJ/mol)
$Ti + O_2 \longrightarrow TiO_2$	-785	$Ti + 2HCl \longrightarrow TiCl_2 + H_2$	-174
$4NaCl + 6TiO_2 \longrightarrow Na_4Ti_5O_{12} + TiCl_4$	-65	$Ti + 4HCl \longrightarrow TiCl_4 + 2H_2$	-256
$4NaCl + 5TiO_2 + 2H_2O \longrightarrow Na_4Ti_5O_{12} + 4HCl$	-195	$2TiCl_2 + 2H_2O + O_2 \longrightarrow 2TiO_2 + 4HCl$	-905
$TiCl_4 + 2H_2O \longrightarrow TiO_2 + 4HCl$	-130	$2H_2 + O_2 \longrightarrow 2H_2O$	-399

* The $G°$ value of $Na_4Ti_5O_{12}$ was calculated by the first principles. And others were calculated by the HSC Chemistry.

As shown in Fig. 3(b), in the corrosion product scale there are holes which provide rapid diffusion channels for corrosive species (for example, oxygen, chlorine and water vapor). Meanwhile, a corrosion scale is formed by the outward diffusion of Ti which results in a 15μm loose layer that is observed under the corrosion products. Hence, the corrosion product scale is non-protective and the corrosion of pure Ti is greatly accelerated by solid NaCl in an $O_2 + H_2O$ environment at 600℃, which is similar with the observation for several Ti alloy[11,18-20].

3.2 The protection of pre-oxidation scale to the pure Ti

After pre-oxidation in O_2 flow for 20h, a continuous, compact and even oxide scale is formed on the surface (Fig. 2). The pre-oxidation scale is only TiO_2 [Fig. 2(c)] and its thickness is about 1μm [Fig. 2(b)]. Since this scale can protect the substrate, the mass gain of the pre-oxidized samples is small in the dry O_2 or humid O_2 (Fig. 1).

When pre-oxidized samples are exposed in the presence of solid NaCl (NaCl+ O_2 and NaCl+ H_2O+O_2), the mass gain of the pre-oxidized samples is smaller than that of the bare samples (Fig. 1), especially in the incubation period (the mass gain increases slowly). The resulting scale on the pre-oxidized samples (which is compact as shown in Fig. 7 and Fig. 8) is thinner than that on the bare samples (which is porous as shown in Fig. 3). Thus, the pre-oxidation can protect the pure Ti samples from the corrosion under NaCl deposit. This is mainly attributed to the compact TiO_2 scale formed on the surface in the pre-oxidation (Fig. 2). During the pre-oxidation, the thickness of the formed TiO_2 scale is about 1μm, and the pre-oxidation scale is larger thicker than the passive film forming on the bare sample. Thus, the TiO$_2$ scale can be a barrier layer for the diffusion of corrosive species (for example, oxygen, water vapor and HCl) and reduce the corrosion to a certain extent, especially in the incubation period.

3.3 Thedestruction of pre-oxidation scale

When exposed in humid O$_2$, the pre-oxidized samples suffer a slightly more serious oxidation than when exposed in dry O$_2$ (Fig. 1). This is because H$_2$O tends to dissociate at the defects on TiO$_2$/(110) planes into free H atoms and OH$^-$ groups[21-23] when the pre-oxidized samples is exposed in humid O$_2$. The papers[21,22] reported that along the crystal channels in c-axis direction, the diffusion rate of hydrogen is at least one order of magnitude larger than that in the perpendicular direction of c-axis. Thus, the generated hydrogen atoms are more prone to diffuse into TiO$_2$ through the former one. This dissolution of hydrogen in TiO$_2$ will form hydrogen defects, thus the hydrogen can increase the concentration of crystal defects and increase the outward diffusion of Ti ions[23]. This fast diffusion in the scale causes a larger corrosion rate and hence the hole forms in the oxide scale (Fig. 6), which make the corrosion product scale thicker and more porous [Fig. 6 (c)]. Thus, the water vapor slightly reduces the protection of the pre-oxidation scale by forming the H defects.

From Fig. 1, we can observe the following: (1) after the incubation period, the mass gain of the pre-oxidized samples increases more rapidly when exposed in the presence of the solid NaCl (NaCl+O$_2$ or NaCl+H$_2$O+O$_2$) than when exposed in the absence of the solid NaCl (O$_2$ or H$_2$O+O$_2$); (2) after 20 h exposure, the pre-oxidized samples has a larger mass gain under the condition of NaCl +O$_2$ or NaCl+H$_2$O+O$_2$ than under the condition of O$_2$ or H$_2$O+O$_2$. Additionally, we can observe that the corrosion product scale of the pre-oxidized samples is thicker when exposed in the presence of the solid NaCl (Fig. 7 and Fig. 8) than when exposed in the absence of the solid NaCl (Fig. 5 and Fig. 6). Thus, we can claim that the corrosion of the pre-oxidized samples is accelerated by solid NaCl. The fast corrosion and the formation of Na$_4$Ti$_5$O$_{12}$ on the surface [Fig. 9(c-d)] indicate that the "protective oxide" formed in the pre-oxidation is not completely inert to solid NaCl.

3.4 Thediffusion of the chlorine in the pre-oxidation scale

The evidence from XRD and SEM/EDX show that Na$_4$Ti$_5$O$_{12}$ formed on the surface [see Fig. 4 and Fig. 9 (c-d)]. This indicated that the solid NaCl reacted with the TiO$_2$ scale. Similar observations have been reported for Ti alloy using the same experimental conditions[18]. In that case, it was argued that the fast corrosion in an O$_2$ + H$_2$O environment is triggered by titanate formation. Thus, in an O$_2$ + H$_2$O + NaCl environment, the formation of titanate destroys the protective scale formed on the surface, and then the gathered chlorine circularly reacts with the substrate. Thus, for the corrosion of pure Ti/Ti alloy in the NaCl + H$_2$O + O$_2$ environment, the chlorine is the main factor.

For the pre-oxidized samples, a thick TiO$_2$ scale is formed on the surface; thus, it takes more time to destroy the protective TiO$_2$ scale by Na$_4$Ti$_5$O$_{12}$ formation and the mass gain changes little during the incubation period (Fig. 1). Furthermore, after the sodium forms Na$_4$Ti$_5$O$_{12}$, the gathered chlorine diffuses inward to the metal/oxide interface [Fig. 14(a)]. Shu et al.[18] report that the chlorine permeates into the corrosion product scale as molecule, like HCl or Cl_2. For the pre-oxidized samples, a compact TiO_2 formed on the surface protects the substrate from the inward diffusion of chlorine and the fast corrosion. Nevertheless, during the incubation period, the chlorine also diffuses into the metal/oxide interface [Fig. 14(a)] while the TiO_2 scale is complete (Fig. 9). This implies that the chlorine cannot be molecule. In our previous work, the chlorine forms Ti-Cl bond in the inner corrosion layer by the dopant Cl replacing O in the Ti oxides lattice when Ti60 alloy was exposed in the NaCl + H_2O + O_2. On the other hand, the diffusion of ion in the TiO_2 is fast and some NaCl is split into ions (Na^+ and Cl^-) at 600℃. Hence, we think the chlorine permeates into the corrosion product scale as ion (Cl^-).

As shown in Fig. 14b, when the pre-oxidized samples suffer from severe corrosion in NaCl + H_2O + O_2, the chlorine is also observed around the metal/oxide interface by SIMS. Additionally, the intensity of chlorine is still large when the intensity of oxygen decreases to a relatively small value. As aforementioned, the substrate locates at the places where the intensity of oxygen is small. Hence, we can infer that the chlorine diffuses into the substrate and directly binds with the metal of the substrate during the fast corrosion. The gathered compound diffuses outward, resulting in the loss of metal in the substrate. After 20h exposure, a mass of the metal reacts with the chlorine and diffuses outward, and the inner corrosion layer is formed under the outer corrosion layer [Fig. 8 (b)].

4 Conclusion

During the pre-oxidation, a compact and thick TiO_2 scale forms on the surface of pure Ti. This TiO_2 scale is the barrier layer for the diffusion of corrosive species (for example, oxygen, chlorine, and water vapor), which can protect the substrate from fast corrosion. However, the "protective oxide" is not completely inert to solid NaCl and the corrosion rate of the pre-oxidized sample is also greatly accelerated by solid NaCl after an incubation period. During the incubation period, the chlorine, generated in the decomposition of solid NaCl at 600℃, diffuses into the oxide/substrate interface as ions to enhance the corrosion rate, and then the residual sodium reacts with the oxide on the surface. After the incubation period, the chlorine diffuses into the substrate and directly reacts with the substrate, resulting in the fast corrosion.

4.1 Materials preparation

The material used in this study was pure Ti and the samples were cut into the size of 10mm × 15mm × 2.5mm. Prior to experiments, all samples were mechanically grinded with 800# SiC paper, ultrasonically degreased in alcohol for about 20min, and dried in the air. The bare samples were then exposed in the pure O_2 flow at 600℃ for 20h for the pre-oxidization. The surfaces of preheated samples were covered with a layer of NaCl deposit by repeatedly brushing and drying with saturated NaCl solution[1,6,8,9,18,24], until about (4 ± 0.2) mg/cm^2 solid NaCl was deposited on.

4.2 Corrosion experiments

The corrosion testswere carried out in a thermo-balance[1,6,8,9,24]. The continuous mass gain during the corrosion experiment was obtained with a thermo-gravimetric analysis (TGA).

The pure O_2 was bubbled into the distilled water with a glass bubbler (when the inner diameter of the tube was about 3.2cm, the flow rate of pure O_2 was about 140mL/min in this study) to produce the test atmosphere (humid O_2). According to the relationship between the vapor pressure of water and its temperature, we precisely set up the temperature of the distilled water in the glass bubbler to control the amount of the water vapor. In this study, the temperature of the distilled water was about 70℃, producing about 30.8 vol. % water vapor. In order to avoid the water vapor condensing inside the thermo-balance, a counter flow of pure N_2 was passed through the thermo-balance, whose flow rate was about 400mL/min. After the furnace reached 600℃ and the gas flows of humid O_2 as well as pure N_2 stabilized, the samples were quickly lowered into the constant temperature zone of the furnace tube.

In this study, the corrosion experiments were respectively carried out under the following four conditions: the first condition was with a solid NaCl deposit layer in a humid O_2 flow at 600℃ (denoted as $NaCl+H_2O+O_2$); the second condition was carried out in a humid O_2 flow at 600℃ (denoted as H_2O+O_2); the third condition was with a solid NaCl deposit layer in a dry O_2 flow at 600℃ (denoted as $NaCl+O_2$); and the fourth condition was carried out in a dry O_2 flow at 600℃ (denoted as O_2). We present the explicit environmental parameters in Table 3.

Table 3 Experimental parameters

Environment	Mass of NaCl (mg/cm^2)	Water pressure (kPa)	Flow rate of O_2 (mL/min)	Temperature (℃)
$NaCl+H_2O+O_2$	4.0	31	140	600
$NaCl+O_2$	4.0	0	140	600
H_2O+O_2	0	31	140	600
O_2	0	31	140	600

4.3 Morphologies and chemical composition analysis

The surface morphologies and the cross-sectional morphologies of corrosion products scale were collected by SEM-EDS. The chemical composition of corrosion products was identified by XRD.

To protect the oxide scale from fracture and spall during the metallographic preparation, the corroded samples were wrapped into a thin nickel foil by electroless plating and were embedded into the epoxy resin. Then, the corroded samples were grinded to 3000 grit with SiC paper. Finally, the corroded samples were polished with diamond paste. The samples were washed with distilled water to remove the residual solid NaCl, and then dried in air, before surface investigation by SEM. Under the condition of H_2O+O_2 and O_2, the mass gains of the pre-oxidized sample were very small (Fig. 1) and the corrosion was minor. The surface and the cross-sectional morphologies after a longer corrosion time (100h) were investigated.

4.4 Time of flight-secondary ion mass spectrometry, ToF-SIMS

The pre-oxidized samples were first exposed in either $NaCl+H_2O+O_2$ or $NaCl+O_2$. Then, they are ultrasonically washed with distilled water to remove the residual solid NaCl. After the pre-oxidized samples were further dried in the air, they were analyzed using a ToF-SIMS5 instrument (ION-TOF GmbH), which allowed parallel mass registration with high sensitivity and high mass resolution. A cesium liquid-metal ion (LMI) gun at 20 keV beam energy was used for spatially resolved ToF-SIMS analysis.

Acknowledgement

The investigation is supported by the National Key Research and Development Program of China (2017YFF0210406), the Shanxi National Science Foundation (2019JQ-213) and the Basic Research and Strategic Reserve Technology Research Fund (project of China National Petroleum Corporation) (2018Z-01).

References

[1] Shu, Y., Wang, F. & Wu, W. Corrosion behavior of pure Cr with a solid NaCl deposit in O_2 plus water vapor. Oxidation of Metals 54, 457-471, doi: 10.1023/a: 1004690518225 (2000).

[2] Liu, L., Li, Y., Zeng, C. & Wang, F. Electrochemical impedance spectroscopy (EIS) studies of the corrosion of pure Fe and Cr at 600℃ under solid NaCl deposit in water vapor. Electrochimica Acta 51, 4736-4743, doi: 10.1016/j.electacta.2006.01.033 (2006).

[3] Tang, Y., Liu, L., Li, Y. & Wang, F. The electrochemical corrosion mechanisms of pure Cr with NaCl deposit in water vapor at 600℃. Journal of The Electrochemical Society 158, C237, doi: 10.1149/1.3596167 (2011).

[4] Tang, Y., Liu, L., Li, Y. & Wang, F. Evidence for the occurrence of electrochemical reactions and their interaction with chemical reactions during the corrosion of pure Fe with solid NaCl deposit in water vapor at 600℃. Electrochemistry Communications 12, 191-193, doi: 10.1016/j.elecom.2009.11.021 (2010).

[5] Tang, Y., Liu, L., Fan, L., Li, Y. & Wang, F. The corrosion behavior of pure iron under solid Na_2SO_4 deposit in wet oxygen flow at 500℃. Materials 7, 6144-6157, doi: 10.3390/ma7096144 (2014).

[6] Wang, F. & Shu, Y. Influence of Cr content on the corrosion of Fe-Cr alloys: The synergistic effect of NaCl and water vapor. Oxidation of Metals 59, 201-214, doi: Doi 10.1023/A: 1023083309041 (2003).

[7] Pujilaksono, B. et al. Oxidation of binary FeCr alloys (Fe-2.25Cr, Fe-10Cr, Fe-18Cr and Fe-25Cr) in O_2 and in O_2+H_2O environment at 600℃. Oxidation of Metals 75, 183-207, doi: 10.1007/s11085-010-9229-z (2011).

[8] Shu, Y., Wang, F. & Wu, W. Synergistic effect of NaCl and water vapor on the corrosion of 1Cr-11Ni-2W-2Mo-V steel at 500-700℃. Oxidation of Metals 51, 97-110, doi: 10.1023/a: 1018854202982 (1999).

[9] Wang, F., Geng, S. & Zhu, S. Corrosion behavior of a sputtered K38G nanocrystalline coating with a solid NaCl deposit in wet oxygen at 600 to 700℃. Oxidation of Metals 58, 185-195, doi: 10.1023/a: 1016072726338 (2002).

[10] Fan, L. et al. Corrosion Behavior of Pure Ti under a Solid NaCl Deposit in a Wet Oxygen Flow at 600 degrees C. Metals 6, 11, doi: 10.3390/met6040072 (2016).

[11] Fan, L. et al. Corrosion Behavior of Ti60 Alloy under a Solid NaCl Deposit in Wet Oxygen Flow at 600 degrees C. Scientific Reports 6, doi: 10.1038/srep29019 (2016).

[12] Pettersson, J., Asteman, H., Svensson, J. E. & Johansson, L. G. KCl Induced corrosion of a 304-type austenitic stainless steel at 600℃; The role of potassium. Oxidation of Metals 64, 23-41, doi: 10.1007/s11085-005-5704-3 (2005).

[13] Pettersson, C., Pettersson, J., Asteman, H., Svensson, J. E. & Johansson, L. G. KCl-induced high temperature corrosion of the austenitic Fe-Cr-Ni alloys 304L and Sanicro 28 at 600℃. Corrosion Science 48, 1368-1378, doi: 10.1016/j.corsci.2005.05.018 (2006).

[14] Pettersson, C., Johansson, L. G. & Svensson, J. E. The influence of small amounts of KCl(s) on the initial stages of the corrosion of alloy sanicro 28 at 600℃. Oxidation of Metals 70, 241–256, doi: 10.1007/s11085-008-9118-x (2008).

[15] Jonsson, T. et al. The influence of KCl on the corrosion of an austenitic stainless steel (304L) in oxidizing humid conditions at 600℃: A microstructural study. Oxidation of Metals 72, 213–239, doi: 10.1007/s11085-009-9156-z (2009).

[16] Pettersson, J., Svensson, J. E. & Johansson, L. G. KCl-Induced corrosion of a 304-type austenitic atainless ateel in O_2 and in $O_2 + H_2O$ environment: The influence of temperature. Oxidation of Metals 72, 159–177, doi: 10.1007/s11085-009-9153-2 (2009).

[17] Jonsson, T., Folkeson, N., Svensson, J. E., Johansson, L. G. & Halvarsson, M. An ESEM in situ investigation of initial stages of the KCl induced high temperature corrosion of a Fe-2.25Cr-1Mo steel at 400 degrees C. Corrosion Science 53, 2233–2246, doi: 10.1016/j.corsci.2011.03.007 (2011).

[18] Shu, Y., Wang, F. & Wu, W. Corrosion behavior of Ti60 alloy coated with a solid NaCl deposit in O_2 plus water vapor at 500–700℃. Oxidation of Metals 52, 463–473, doi: 10.1023/a: 1018864216554 (1999).

[19] Dumas, P. & Stjohn, C. Nacl-Induced accelerated oxidation of a titanium alloy. Oxidation of Metals 10, 127–134, doi: Doi 10.1007/Bf00614242 (1976).

[20] Yao, Z. & Marck, M. NaCl - induced hot corrosion of a titanium aluminide alloy. Materials Science and Engineering: A 192–193, 994–1000, doi: 10.1016/0921-5093(95)03345-9 (1995).

[21] Johnson, O. W., Paek, S. H. & Deford, J. W. DIFFUSION OF H AND D IN TIO2 - SUPPRESSION OF INTERNAL FIELDS BY ISOTOPE - EXCHANGE. Journal of Applied Physics 46, 1026–1033, doi: 10.1063/1.322206 (1975).

[22] Fowler, J. D., Chandra, D., Elleman, T. S., Payne, A. W. & Verghese, K. T DIFFUSION IN AL2O3 AND BEO/D BEO. Journal of the American Ceramic Society 60, 155–161, doi: 10.1111/j.1151-2916.1977.tb15493.x (1977).

[23] Douglass, D. L., Kofstad, P., Rahmel, A. & Wood, G. C. International workshop on high–temperature corrosion. Oxidation of Metals 45, 529–620, doi: 10.1007/bf01046850 (1996).

[24] Wang, C., Jiang, F. & Wang, F. Corrosion inhibition of 304 stainless steel by nano - sized Ti/silicone coatings in an environment containing NaCl and Water Vapor at 400–600℃. Oxidation of Metals 62, 1–13, doi: 10.1023/B: OXID.0000038782.77162.5b (2004).

本论文原发表于《Scientific Reports》2020年第10卷。

Failure Analysis of Crankshaft of Fracturing Pump

Wang Hang[1,2] Yang Shangyu[1,2] Han Lihong[1,2]
Fan Heng[3] Jiang Qingfeng[4]

(1. Tubular Goods Research Institute of CNPC; 2. State Key Laboratory of Performance and Structural Safety for Petroleum Tubular-Goods and Equipment Materials;
3. School of Electronic Engineering, Xi'an Shiyou University;
4. Sichuan Baoshi Machinery Special Vehicle Co., Ltd)

Abstract: Factory fracturing is characterized by high pressure, large discharge capacity and long-timeservice for shale gas well, then unite of fracturing pump operates under more severe condition with complex load and corrosion. Failure analysis was performed for crankshaft via theoretical calculation, finite element modelling and fractographic observation in fracturing pump with model QWS-2800. The results indicated that fatigue fracture was the dominant failure mechanism, with evidence of beach and striation marks and multiple fatigue cracks. The fracture plane had nearly 45° inclination with respect to the shaft axis. The main cause of insufficient fatigue strength was due to absence of surface hardening treatment. Fillet radius was a main factor for fatigue crack initiation, the fatigue crack mostly located in the thread root of radial oilhole. Finite element analysis results indicated that the stress increased remarkably as fillet radius decreased. Then stress concentration became much higher with the transition of thread root from smooth to sharp fillet. It provides some suggestions to prevent the crankshaft rupture in terms of fatigue strength and thread structure.

Keywords: Crankshaft; Fillet radius; Factory fracturing; Fatigue; Crack growth

1 Introduction

Fracturing pump is one of key equipment during fracturing operation with large scale, which mainly enhances both of pressure nd discharge capacity for fracturing liquid. Meanwhile, as a main component, the crankshaft often run with harmonic torsion combined with cyclic bending stress due to the loads of the cylinder pressure [1]. The fracture of crankshaft was reported [1-6] frequently for crankshaft in diesel engine, the tendency to increase of engine power can be observed. Both the high power and large torsional moment of modern diesel cause a significantly increase of operational

---

Corresponding authors: Wang Hang, wanghang008@ cnpc. com. cn; Han Lihong, hanlihong@ cnpc. com. cn; Fan Heng, fan_h@ xsyu. edu. cn.

stresses in the crankshaft. It often results in a decline in fatigue life of engine components. In addition, failure analysis was conducted for crankshaft in compressor as well [7-12], which undergoes complex load, such as periodic gas pressure, inertia force and torsional vibration.

Both staged fracturing and factory fracturing operation were considered to be successful experience for exploitation of shale gas in USA. In comparison with conventional mode, fracturing operation performed in the form of both multi-layers and multi-stages. Moreover, well group was fracturing with intersection during factory fracturing. Therefore, the typical work condition was characterized by large discharge capacity, large amount of liquid and sand, high pressure and long-time service for shale gas well. Currently, Sichuan and Chongqing area became the main block for increase reserve and production in China. The parameter of fracturing operation was much higher than that in North American regions, such as operation pressure in excess of 105MPa, discharge capacity with 16m^3/min, and time of duration more than a week. As a result, these conditions accelerated damage for unite of fracturing equipment. However, a large number of research works mainly focused on pump housing [13-15], little on crankshaft of fracturing pump.

The objective of this paper is to uncover the failure mechanism of fractured crankshaft, as well as specific improvement suggestion. It is expected to provide technology support for safety and durable reliability of fracturing equipment during factory fracturing.

2 Engineering background

The accident of fracturing equipment happened in shale gas well, located in Sichuan and Chongqing area. Firstly, the tailstock of fracturing truck shaken sharply, emitted black smoke and then fired. Subsequently, another four fracturing trucks were ignited, as shown in Fig. 1.

Fig. 1 Fire site of fracturing truck during operation for a shale gas well in Sichuan-Chongqing area

The failure analysis indicated that the 1st crank pin of crankshaft fractured in the fracturing pump, of which one end close to the 1st crank web located on the side of the transmission gearbox, another next to 2nd crank web situated on the side of power transmission, as shown in Fig. 2. On a closely examination, connecting rod stuck between crankshaft and power end box, as shown in left top corner Fig. 2(b). The macrograph of integral failed crankshaft was shown in Fig. 3. Operational record exhibited that this fracturing pump with model of QWS-2800 put into use since September

2015, cumulative running time of 2214.0h, operation layer of 750 ones. The operating pressure ranged from 86MPa to 120MPa.

Fig. 2 Macroscopic appearance of fractured crankshaft:
(a) one side of transmission gearbox, (b) another of power box

Fig. 3 The macrograph of the integral fractured crankshaft

3 Strength check

3.1 Static strength check

According to design parameter, static strength was checked for crankshaft using theoretical calculation method. Crankshaft undergoes both torque (T) and bending moment (P) under operation, as shown in Fig. 4.

The formulation of safety factor as follows:

$$S = \frac{S_\sigma \times S_\tau}{\sqrt{S_\sigma^2 + S_\tau^2}} \quad (1)$$

Of which, S_σ -Safety factor of blending load; S_τ -Safety factor for torque load.

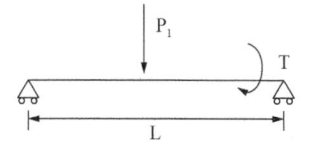

Fig. 4 The schematic diagram of loads in crankshaft

Under condition of blending load, the formulation of safety factor as follows:

$$S_\sigma = \frac{\sigma_s}{\sigma_w} = \frac{\sigma_s}{(M/Z)} \qquad (2)$$

Of which, σ_s - Yield strength, MPa; σ_w - Blending stress, MPa; M - Load of blending moment, N·m; Z - Section coefficient of bending resistance of dangerous cross section.

Under condition of torque load, the formulation of safety factor as follows:

$$S_\tau = \frac{\tau_s}{\tau} = \frac{\tau_s}{(T/Z_p)} \qquad (3)$$

Of which, τ_s - Torsional strength, MPa; τ - Distorting stress, MPa; T - Torque load, N·m; Z_p - Section coefficient of torsion resistance of dangerous cross section.

$$S_\sigma = 4.76, \quad S_\tau = 2.10$$
$$S = 1.925$$

According to empirical value [16], the recommendation safety factor is 1.7~2.2. It can be seen that the safety factor of static strength meet the requirement for fatigued crankshaft.

3.2 Fatigue strength check

Fatigue strength depends on multiple factors, such as surface quality, stress concentration, structural size and fatigue limit. Fatigue strength was checked in terms of dangerous section in crankshaft. The formulation of safety factor for fatigue strength as follows:

$$S = \frac{S_\sigma S_\tau}{\sqrt{S_\sigma^2 S_\tau^2}} \geqslant S_p \qquad (4)$$

Of which, S_σ - Safety factor of bending moment load; S_τ - Safety factor of torque load; S_p - allowable safety factor by industry recommendation.

Under condition of bending moment load, the formulation of safety factor as follows:

$$S_\sigma = \frac{\sigma_{-1}}{\dfrac{K_\sigma}{\beta \varepsilon_\sigma}\sigma_a + \phi_a \sigma_m} \qquad (5)$$

Of which, σ_{-1} - Bending fatigue limit under symmetric cyclic stress, MPa; K_σ - Effective stress concentration coefficient under bending load; β - The factor of surface quality; ε_σ - Dimension influence factor under bending; φ_σ - Conversion coefficient of average stress under bending load; σ_a - Stress amplitude under bending load, MPa.

σ_m - Average stress value under bending load, MPa; The calculation result as follows,

$$S_\sigma = 1.013$$

Under condition of torque load, the formulation of safety factor under torque load as follows:

$$S_\tau = \frac{\tau_{-1}}{\dfrac{K_\tau}{\beta \varepsilon_t}\tau_a + \varphi_a \tau_m} \qquad (6)$$

Of which, τ_{-1} - Torsional fatigue limit under symmetric cyclic stress, MPa; K_τ - Effective stress concentration coefficient under torsion load; ε_τ - Dimension influence factor under torsion; φ_τ - Conversion coefficient of average stress under torsion load; τ_a - Stress amplitude under torsion load, MPa; τ_m - Average stress value under torsion load, MPa.

The relative calculation result as follows:

$$S_\tau = 0.715$$

Based on the both results of S_σ and S_τ, the safety factor of dangerous section as follows:

$$S = 0.584$$

Therefore, the safety factor of fatigue strength is much less than the allowable one (1.30 ~ 1.50)[16].

As for surface quality (β), the factor is determined to be 1.0 without surface hardening treatment. Meanwhile, this factor can be increased to the range of 1.7 ~ 2.1 with surface nitrided layer. The metallographic examination revealed that there is no hardened layer in the surface of failed crank pin [Fig.5(a)], but the depth of the hardened layer is about 10μm for connecting rod [Fig.5(b)], accompanying with evidence of micro-hardness measurement. These results demonstrated that the failed crankshaft operated without surface hardening treatment.

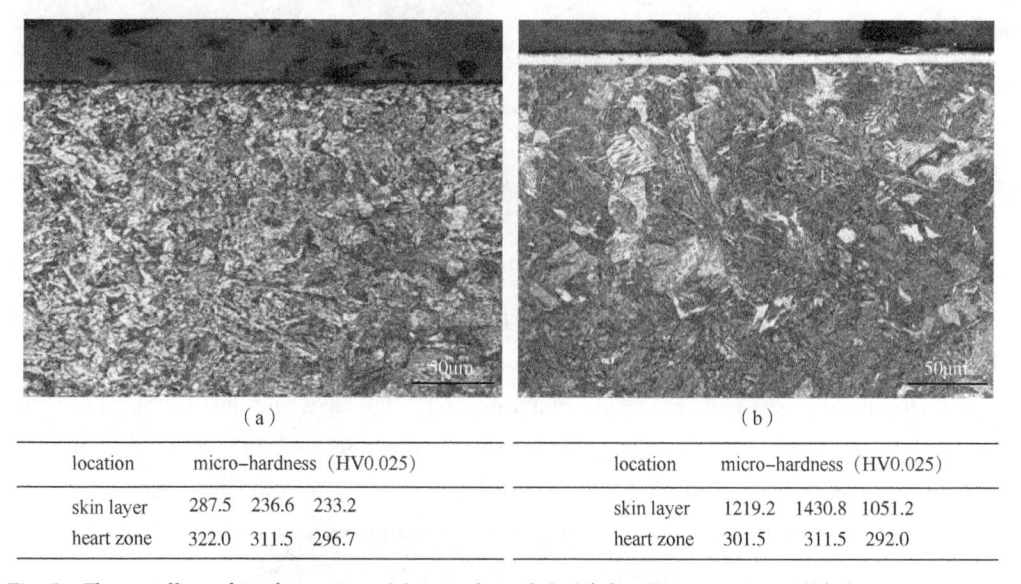

location	micro-hardness (HV0.025)	location	micro-hardness (HV0.025)
skin layer	287.5 236.6 233.2	skin layer	1219.2 1430.8 1051.2
heart zone	322.0 311.5 296.7	heart zone	301.5 311.5 292.0

Fig. 5 The metallographic observation of fractured crankshaft(a) and connecting rod(b) along cross section

4 Fracture analysis

4.1 Load characteristic

The schematic diagram was showed in Fig. 6, including a spline, five segments of crank pin, and six segments of crank web. These serial number were marked by arrow in Fig. 6. FEM (Finite element model) was established for the integral crankshaft, using their structural parameters. Load boundary conditions were applied in finite element mesh respectively, including torque (T) and bending moment (P). The stress distribution was characterized by steady state method for crankshaft with different segments.

The related results indicated that the fillet region undertook the higher stress between the crank pin and web, as shown in Fig. 7(a). The maximum equivalence stress reached to 369MPa in the region close to the 1st crank web. While the stress was 211MPa in the body. Thus, the stress concentration factor (K) was determined to be 1.75. As for the region of radial oil-hole, the stress equaled to 379MPa, the stress concentration factor(K) increased to 1.79, as shown in Fig. 7(b).

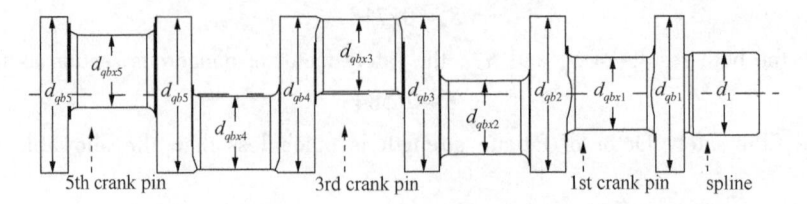

Fig. 6 Schematic diagram of integral crankshaft structure

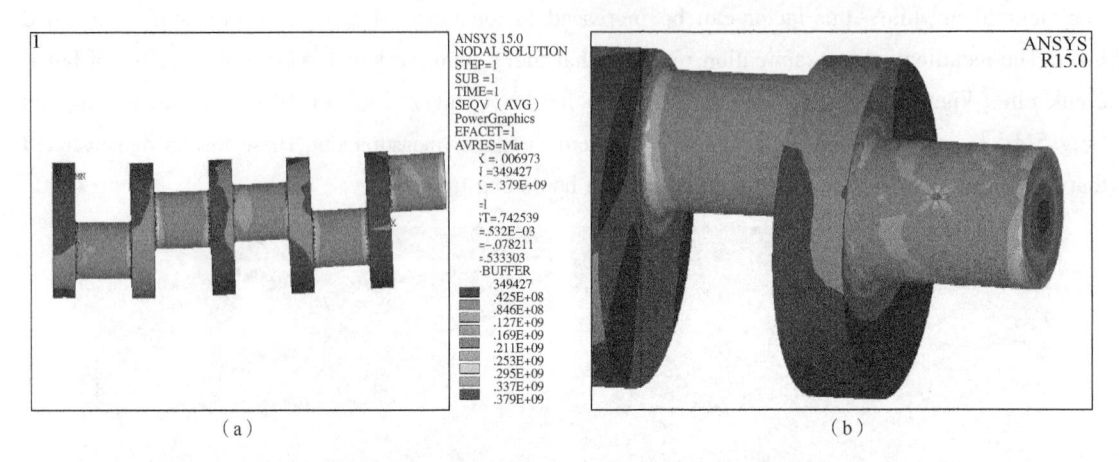

Fig. 7 Stress distribution of crankshaft: (a) integral structure, (b) the first crank pin

The profile of section was shown in Fig. 8 for the failed crankshaft. The fracture plane has about 45° inclination with respect to the shaft axis. The left section close to the fillet region of the 2nd crank web, the middle part penetration the radial oil hole, the right one next to the fillet region of the 1st crank web. It can be seen that the fracture surface located in the section with stress concentration, along with the radial oil-hole and the two-crank pin-web fillet regions.

Fig. 8 Macroscopic morphology of fracture in ruptured crankshaft

4.2 Fractograph

The macrograph of integral fracture surface was shown in Fig. 9 in failed crankshaft. There are a large number of striation marks in the right area of the radial oil-hole, which is a typical feature of fatigue failure. On a closely examination, multiple fatigue cracks were observed at the thread root around the radial oil-hole, as shown by arrow in Fig. 10(a). Meanwhile, clear striation marks with tight spacing presented in the stable crack propagation region, as shown in Fig. 10(b). The fracture process was examined using scan electron microscopy (SEM). These results further confirmed that fatigue crack located in the thread root, as shown in Fig. 11(a). Beach mark appeared at the region of stable crack propagation, implying the crack growth in the form of trans-granular mode, as shown in Fig. 11(b). It is worth noting that a step-like pattern with radial morphology presented close to fillet region, accompanying with a second micro-crack, as shown by arrow in Fig. 12. These results indicated that the fillet region was another site for fatigue crack formation.

Fig. 9 Macrograph of fracture surface in 1st crank pin of failed crankshaft

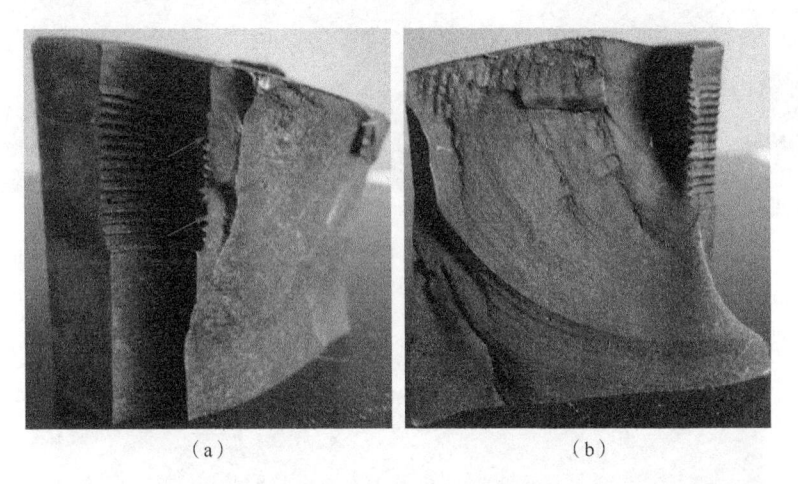

Fig. 10 Fracture morphology in radial oil-hole: (a) fatigue source, (b) crack propagation

4.3 Damage feature

The longitudinal section of the thread in the radial oil-hole was shown in Fig. 13. The thread almost sticked or deformed seriously. Meanwhile, the body of part thread collapsed or fractured. On a closely examination, there were apparent differences with respect to the morphology feature in these thread roots. For example, some with sharp fillet (i.e., the fillet radius is zero), other with

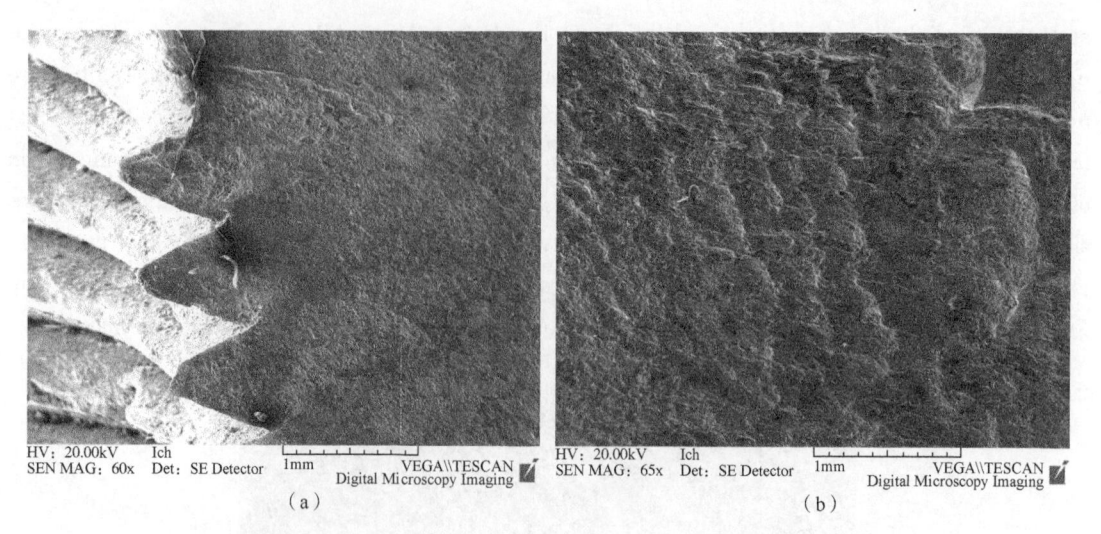

Fig. 11 Microscopic fracture morphology in thread root: (a) fatigue origin, (b) crack propagation region

smooth or fillet one, as shown in dotted line in Fig. 14. Generally speaking, the sharp fillet could be attributed to the unreasonable design or machine. Meanwhile, both of thread sticking and fracture possibly resulted from the cross threading due to improper installation of screw-in oil control plug.

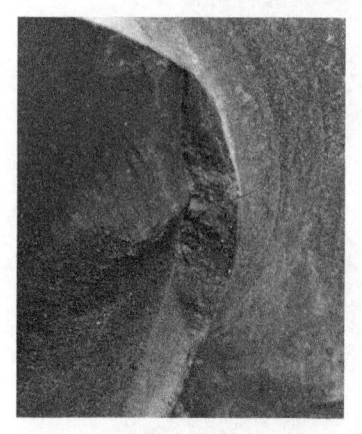

Fig. 12 The local magnification morphology of fracture surface close to crankpin-web fillet region

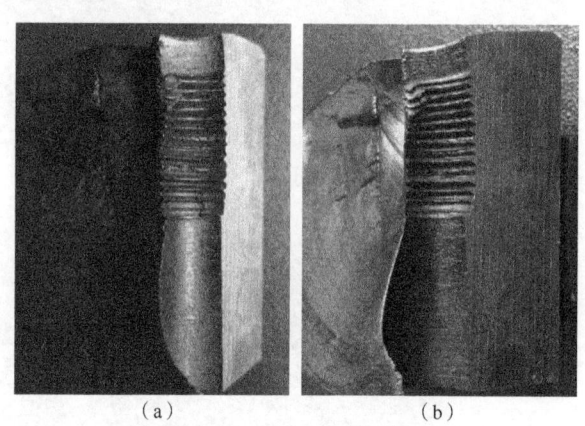

Fig. 13 Morphology feature of thread in radial oil-hole in the first segment eccentric turning

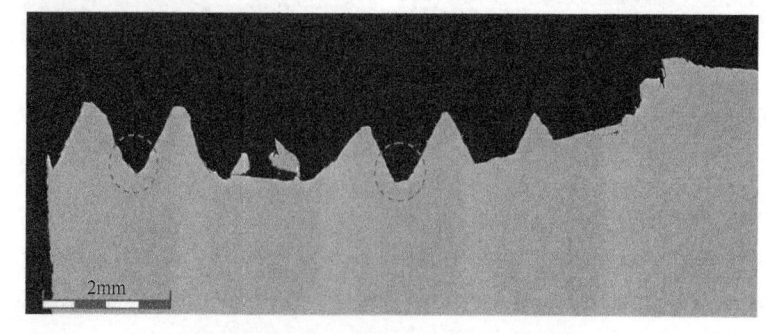

Fig. 14 Morphology of thread in radial oil hole in the 1st crank pin along longitudinal section

The optical microscope (OM) observation revealed that micro-crack presented at the thread root with sharp fillet, as shown in Fig. 15(a). The metallographic examination confirmed that the trace of plastic deformation occurred in both the tip and two sides of crack, as shown in Fig. 15(b). It can be speculated that much higher stress was possible to be undertook at thread root with sharp fillet, then stress concentration occurred, which consequently resulted in plastic deformation initially, followed by initiation of crack and then crack propagation.

The feature of morphology was shown in Fig. 16 for thread root with smooth fillet. The body cracked or ruptured, and even separated from the matrix, as shown in Fig. 16(a). The metallographic observation revealed that the trace of plastic deformation occurred around two sides of thread, as shown in Fig. 16(b). It was worth noting that no micro-crack was observed in the thread root with smooth fillet, which indicated that the fatigue crack was not formed in this region.

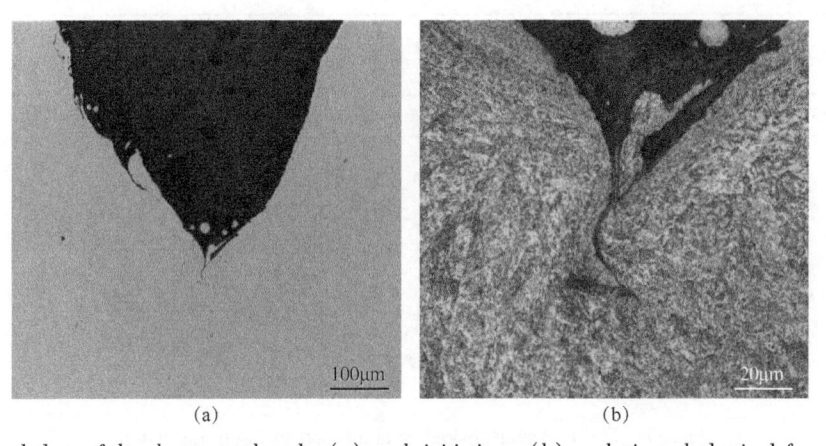

Fig. 15 Morphology of the sharp root thread: (a) crack initiation, (b) crack tip and plastic deformation region

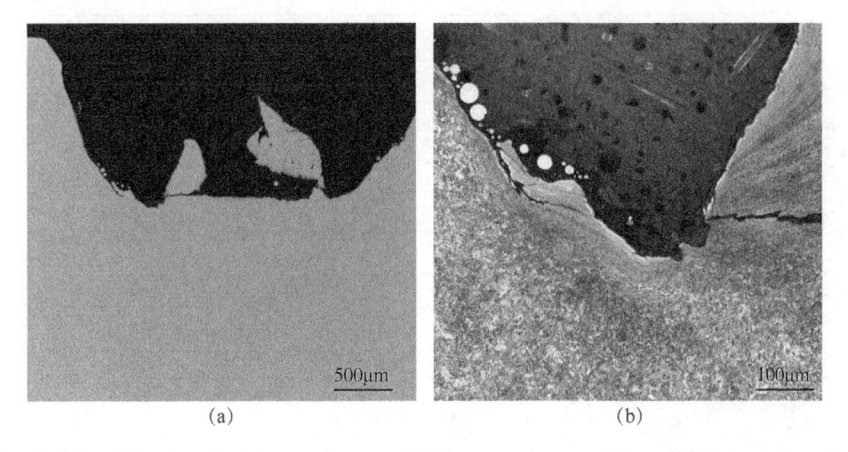

Fig. 16 Morphology of the thread with smooth root: (a) damage characteristic, (b) deformation microstructure

5 Finite element analysis

According to operation parameter, the torque load is 272, 720N · m, bending moment is 92, 535N · m. The material property listed in Table 1, Its mark was 4340 steel with America brand, the chemical composition consists of C 0.36 (wt,%), Si 0.30, Mn 0.60, P 0.0069, S 0.002, Cr 1.65, Mo 0.25, Ni 1.37, Cu 0.12, Al 0.02, Fe balance. Finite element model of crankshaft was shown in Fig. 17, its left side was fixed, the right side subjected to both torque (T_t) and bending moment (P_b). Finite element analytical results revealed that thread root undertook much higher load in both sharp and smooth fillet, as shown in Fig. 18 and Fig. 19, respectively. The relationship between fillet radius and stress was established, as shown in Fig. 20. The curve revealed that the load of thread root directly depended on the fillet radius. The maximum equivalence stress decreased significantly as fillet radius increased. Particularly, as fillet radius was 0.05, the stress was 1355.5MPa. Meanwhile, when fillet radius increased to 0.4 mm, the stress declined to 804.2MPa. These results exhibited that stress concentration can be released significantly when filletradius increased reasonably. Therefore, fatigue property can be improved by means of optimization of fillet radius.

Table 1 Property of material

Item	Material mark	Density (g/cm³)	Elastic modul (GPa)	Poisson ratio	Yield strength (MPa)	Impact toughness (J)	Elongation (%)
Para	4340	7.8	206	0.28	777.0	73.0	19.4

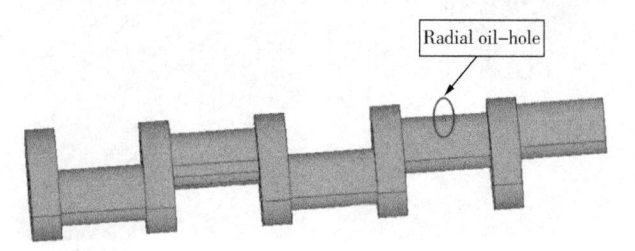

Fig. 17 Finite element model of crankshaft in fracturing pump

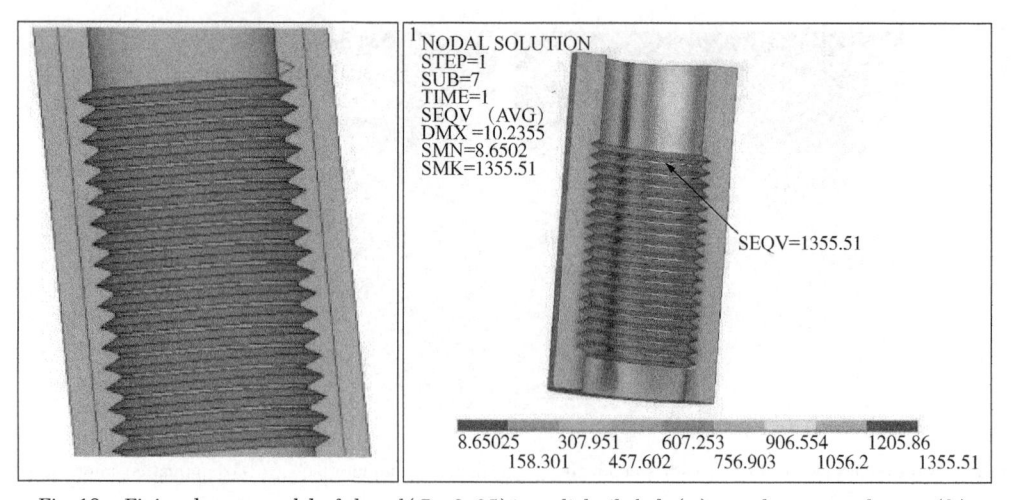

Fig. 18 Finite element model of thread ($R = 0.05$) in radial oil-hole(a), and stress nephogram(b)

6 Analysis of the cause of failure

The typical feature of fracture surface revealed that the fatigue fracture was the dominant failure mechanism [5,6], which included striation and beach marks at stage of stable fatigue crack growth. These results accorded with the characteristic of loads in crankshaft during fracturing

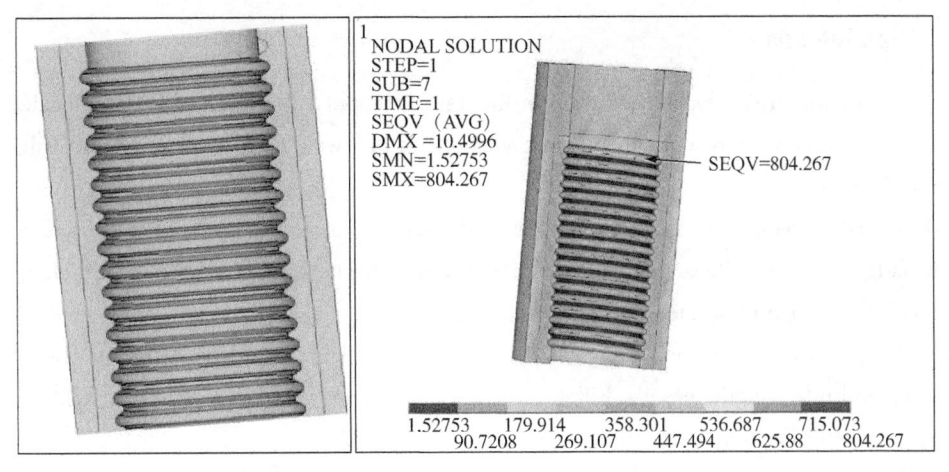

Fig. 19 Finite element model of thread($R=0.4$) in radial oil-hole(a), and stress nephogram(b)

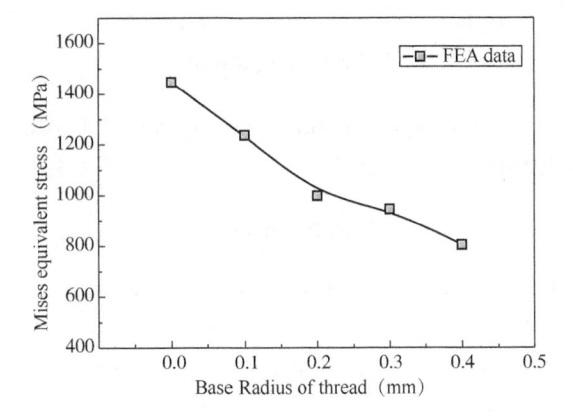

Fig. 20 The curve of fillet radius vs the maximum equivalence stress of thread in radial oil-hole

operation, such as harmonic torsion in combination with cyclic bending stress and high-frequency vibration.

Fatigue strength was regarded as one of important parameters for fatigue property, which closely associated with surface quality, stress concentration, structural size and fatigue limit. There is no hardened layer in the surface of ruptured 1st crank pin. According to literature [16], the factor of surface quality is only 1.0 without surface hardening treatment. On the contrary, this factor can be increased to the range of 1.7~2.1. Therefore, the cause of premature failure was insufficient fatigue strength due to absence of surface hardening treatment.

Fatigue origin mostly located in the thread root of radial oil-hole, as well as the crank pin-web fillet region. These features can be explained on the basis of FEA results. The maximum equivalence stress presented in these area, compared with that in the region of crank shaft body, their stress concentration factors consequently reached to 1.75 and 1.79 respectively, which accelerated the initiation of fatigue crack. Particularly, fillet radius was a main factor for fatigue crack initiation, the thread root with sharp fillet became a site for fatigue crack formation.

7 Conclusions

(1) The fractured crankshaft occurred in the 1st crank pin, the fracture surface located in the about 45° inclination section with stress concentration, along with both crank pin-web fillet region and radial oil-hole.

(2) Strength check demonstrated that the static strength meet the requirement of safety factor, while the fatigue strength does not satisfy the requirement. The main cause is insufficient fatigue strength duo to absence of surface hardness treatment.

(3) Fatigue fracture is dominant failure mechanism, with evidence of striation and beach marks, and multiple fatigue cracks. Fillet radius was a main factor for fatigue crack initiation, fatigue crack mostly located in the thread root of radial oil-hole.

(4) The fracture failure can be prevented from increasing in fatigue strength and optimization of fillet radius of thread root.

Acknowledgement

The authors gratefully acknowledge the financial support of CNPC science and technology development project (No. 2019B-4013).

References

[1] M. Fonte, V. Infante, M. Freitas, L. Reis, Failure mode analysis of two diesel engine crankshaft, Procedia Struct. Integrity 1(2016)313-318.

[2] L. Witek, M. Sikora, F. Stachowicz, T. Trzepiecinski, Stress and failure analysis of the crankshaft of diesel engine, Eng. Fail. Anal. 82(2017)703-712.

[3] V. Infante, J. M. Silva, M. A. R. Silvestre, R. Baptista, Failure of a crankshaft of an aeroengine: a contribution for an accident investigation, Eng. Fail. Anal. 35(2013)286-293.

[4] F. Jiménez Espadafor, J. Becerra Villanueva, M. Torres García, Analysis of a diesel generator crankshaft failure, Eng. Fail. Anal. 16(2009)2333-2341.

[5] Z. W. Yu, X. L. Xu, Failure analysis of a diesel engine crankshaft, Eng. Fail. Anal. 12(2005)487-495.

[6] R. K. Pandey, Failure of diesel engine crankshafts, Eng. Fail. Anal. 10(2003)165-175.

[7] J. A. Becerra, F. J. Jimenez, M. Torres, D. T. Sanchez, E. Carvajal, Failure analysis of reciprocating compressor crankshafts, Eng. Fail. Anal. 18(2011)735-746.

[8] G. Y. Yu, W. S. Xiao, J. Liu, H. M. Wang, H. Q. Zhou, Optimization of crankshaft on ocean high-power compressor, China Petro Mach. 44(2016)62-67.

[9] X. Q. Huang, X. L. Wang, X. H. Shen, F. Xiao, Effect of the shape of railway wheel plate on its stresses and fatigue evaluation, Eng. Fail. Anal. 97(2019)718-726.

[10] M. Masoumi, A. Sinatora, H. Goldenstein, Role of microstructure and crystallographic orientation in fatigue crack failure analysis of a heavy haul railway rail, Eng. Fail. Anal. 96(2019)320-329.

[11] A. AI-Juboori, D. Wexler, H. Li, H. Zhu, C. Lu, A. McCusker, J. McLeod, S. Pannil, Z. Wang, Squat formation and the occurrence of two distinct classes of white etching layer on the surface of rail steel, I, J. Fatigue 104(2017)52-60.

[12] Y. D. Li, C. B. Liu, N. Xu, X. F. Wu, W. M. Guo, J. B. Shi, A failure study of the railway rail serviced for heavy cargo trains, Case Stud. Eng. Fail. Anal. 1(2013)243-248.

[13] S. B. Jiang, Z. Q. Li, B. B. Jia, S. G. Hu, W. W. Yang, Structural innovations for Valve housing using in fracturing at unconventional oil gas fields, Drill Produ. Tech. 40(2017)72-74.

[14] C. H. Wang, C. C. Wang, G. Y. Zhang, Fatigue resistance solution for high-pressure pump housing, China Petro Mach. 44(2016)71-76.

[15] J. Q. Hu, Q. L. Wang, L. B. Zhang, H. T. Wang, K. W. Li, W. W. He, Fatigue failure evolution of fracturing pump under changing pressure condition, China Petro Mach. 45(2017)67-73.

[16] A. X. Yin, et al., Handbook of Mechanical Design, fifth ed., Chemical Industry Press, China, 2008.

本论文原发表于《Engineering Failure Analysis》2020 年第 109 卷。

Failure Analysis of a Sucker Rod Fracture in an Oilfield

Ding Han[1,2] Zhang Aibo[1] Qi Dongtao[2] Li Houbu[2] Ge Pengli[3]
Qi Guoquan[1,2] Ding Nan[2] Bai Zhenquan[2] Fan Lei[2]

(1. Department of Applied Chemistry, School of Science, Northwestern Polytechnical University;
2. Tubular Goods Research Institute, China National Petroleum Corporation&State Key Laboratory for Performanceand Structure Safety of Petroleum Tubular Goods and Equipment Materials;
3. Northwest Oil Field Branch Company Petroleum Engineering Technology Research Institute)

Abstract: The sucker rod was fracture after 729 days' service in an oilfield. The fracture position is located at the welding seam that is connecting the sucker rod's upsetting end and the polished rod end. The failure causes were analyzed by the nondestructive test(NDT) of penetration detection, direct-reading spectrometer, tensile strength test machine, impact test machine, vickers hardness tester, optical microscope (OM), macroscopic fracture morphology, scanning electron microscopy (SEM) in this paper. The process of the fracture is under cyclic loading and the stress concentration, the sucker rod'soutside weld was brittle cracking in the first fracture at the beginning. Then the crack in first facture is passed to the inside welding residual, which is corresponding to the source area of the second fracture. Finally, the crack in the second fracture propagation along the weld seam and connected with the first fracture surface lead the rod broke completely. The causes of sucker rod breaking for the sucker rod did not meet the requirements of the HL grade rod material and the friction welding defects.

Keywords: Sucker rod; Friction welding; Fracture; Fatigue.

1 Introduction

A history of human exploitation of energy is the history of human civilization. Since thousands of years ago, man began to make use of fire, which enabled man to survive and develop in the harsh natural environment. Britain has a plenty of coal resources, the invention of steam engine prompted the British to develop and utilize a large number of coal resources, thus promoting the first industrial revolution and making a qualitative leap in human development. After a hundred years of development of modern industry, petroleum has replaced coal as the blood of modern industry, and the petroleum industry is like a huge heart, pumping continuous fresh blood for global industry.

Corresponding authors: Zhang Aibo, zhab 2003@ nwpu. edu. cn, Qi Dongtao, qidt@ cnpc. com. cn.

In the process of oil production, we will inevitably encounter all kinds of failure problems, and our job is to find out the cause of failure and avoid the recurrence of such problems. China's most of oilfield is located in the hinterland of the desert or gobi, where the natural environment is harsh and the temperature difference between day and night is huge. Many scholars have studied the failure cases of the oilfield. A. Q. Fu was studied the downhole corrosion behavior of NiW coated carbon steel in spent acid and formation water and girth weld cracking of mechanically lined pipe in the northwest oilfield in 2016[1,2], Y. Long studied the 13Cr valve cage of tubing pump failure in an oilfield in 2018[3], H. Ding analyzed the connecting rod for oil pumping unit in China western oilfield[4]. Production and transportation pipelines and equipment used in oil and gas fields can easily fail under harsh service conditions or due to material quality. D. L. Duan studied the failure of sucker rod coupling or tubing in sucker rod pumping system[5], J. An studied the effect of boronizing from solid phase on the tensile mechanical properties and corrosion and wearresistances of steel AISI 8620[6].

In the oilfield, after 729 days' of service, the sucker rod failed to break and failure. The fracture position is located at the welding seam which connecting the upsetting end of the sucker rod with the polished rod, and its macroscopic morphology was shown in Fig. 1. Polished rod model is $\phi 38mm \times 11mm$, the grade is HL, material made of 35CrMoand welding methods was friction welding. In order to find out the reasons of the rod's breaking, the macro and micro morphology of the sample and the physical and chemical properties of the material were tested and analyzed, and the failure reasons of the sample were analyzed comprehensively.

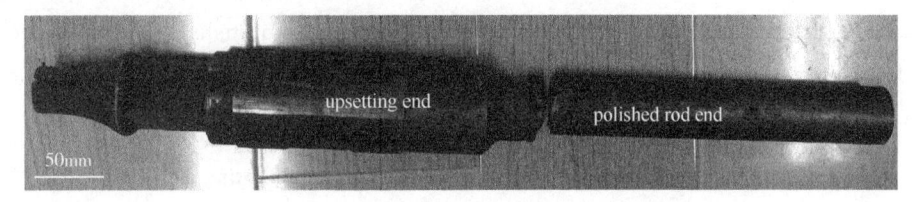

Fig. 1 Fracture of the sucker rod visual appearance

2 Experimental

In order to find out the causes of the friction welding seamfracture of a sucker rod, a series of materials characterization was carried out on the fracture sucker rod. The nondestructive test(NDT) of the II C-d method adopted, penetration detection was carried out near the fracture. The chemical compositions were detected by ARL4460 photoelectric direct reading spectrometer. TheUTM-5305 Mechanical testing machine was used to test the tensile strength and yield strength of the fracture sucker rod each side on room temperature. The PIT752D-2(300J) impact machine was used to test the impact energy of the polished rod end and upsetting end at room temperature respectively. TheKB30BVZ-FA vickers hardness testing machine was used to test the HV_{10} hardness of the base material, weld and the heat affected zone (HAZ). Besides, the OLS-4100 laser confocal microscope was used to observe the microstructure of the base material, weld and the HAZ. Finally, the fracture surface and crack growth of the fracture were analyzed by macroscopic morphology and the scanning electron microscope(SEM).

3 Results and discussion

3.1 Nondestructive testing

According to NB/T 47013.5—2015 standard[7], the results of penetration detection near the fractureshowed in Fig. 2, that no cracks were found near the fracture surfaces of the sucker rod except the fracture position.

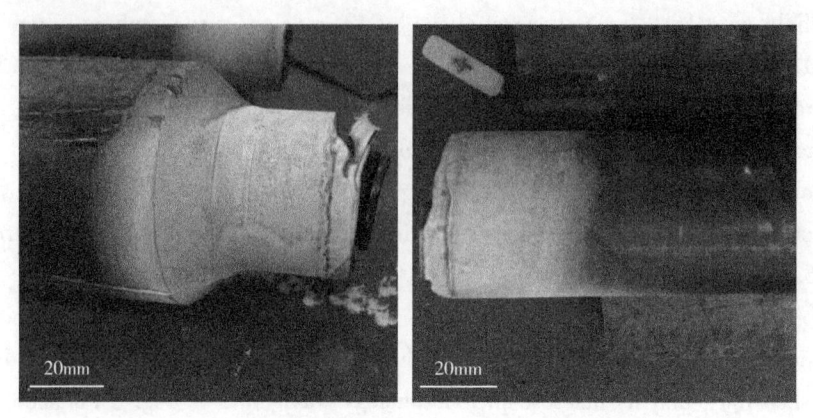

Fig. 2 Penetration detectionfor the sucker rod

3.2 Chemical composition

Spectral analysis for material composition of the sucker rod end and the upsetting end are shown in Table 1. The results of chemical composition analysis showthat the chemical composition of the sucker rod end material has some deviation from the standard requirements, mainly because the content of C element is lower than the requirements of GB/T 26075—2010 standard[8] for 35CrMo low-alloy steel.

Tab. 1 Chemical composition of the rod end and the upsetting end (wt. %)

Element	C	Si	Mn	P	S	Cr	Mo	Ni	Cu
GB/T 26075—2010	0.32~0.40	0.17~0.37	0.50~0.70	≤0.025	≤0.025	0.80~1.10	0.15~0.25	≤0.30	≤0.20
rod end Measured	0.28	0.23	0.64	<0.01	<0.008	0.98	0.17	0.019	<0.02
upsetting end Measured	0.38	0.23	0.61	0.022	0.0098	0.88	0.18	0.028	0.067

3.3 Mechanics property analysis

The tensile test results of the sucker rod end and upsetting end are shown in Table 2. The tensile strength of the sample on the rod end at room temperature was higher than the upper limit of SY/T 5029—2013 standard[9] for HL-grade rod, and the tensile strength and yield strength of the upsetting end sample were lower than the minimum requirements of the standard. Compared the stress-strain curves in Fig. 3, there is a vast difference in mechanical properties between the rod end and the upsetting end obviously.

The impact energy test results of the sucker rod end and the upsetting end are shown in Table 3. As to the test results, the impact performance of the sucker rod end and the upsetting end meets the requirements of SY/T 5029—2013 standard for HL-level sucker rod.

Table 2 Tensile test results at room temperature

Sample	Size(mm)	Tensile Strength(MPa)	Yield Strength(MPa)	Elongation(%)
rod end	φ6.25×25	1232	1176	17
upsetting end		1223	1145	16
		1219	1145	17
		742	632	26
		747	612	22
		749	647	20
SY/T 5029—2013		965~1195	≥793	≥10

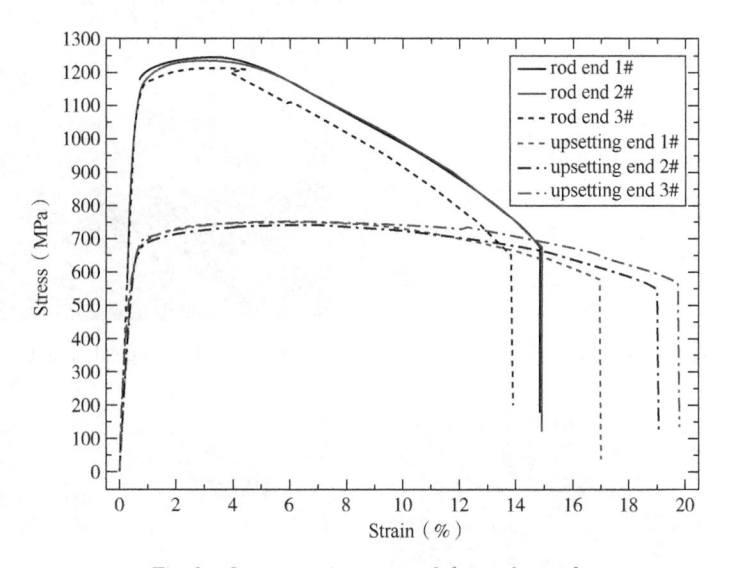

Fig. 3 Stress-strain curves of the sucker rod

Table 3 Impact test result(J)

Sample	Size(mm)	Notch	Temperature(℃)	Measured	SY/T 5029—2013
rod end	10×10×55	U	20	140	≥60J
upsetting end				140	
				147	
				144	
				136	
				143	

The HV_{10} hardness test results are shown in Table 4. As the vickers hardness test results distribution in Fig. 4, we can clearly see the hardness of the weld seam is much higher than that of the base material and the HAZ.

Table 4 Vickers hardness test results(HV_{10})

Measuring position	HAZ	Weld	Base metal
Measured	280, 256	386	286, 281

3.4 Metallographic analysis

The fracture sample was cut and prepared for the metallurgical examination. Metallurgical structure of the weld is high hardness tempered martensite as shown in Fig. 5, that's also explains why the hardness of weld is much higher thanthat of the HAZ and base metal. The structure of the HAZ is tempered sorbite with grain size of 9.5, besides the metallographic structure has the characteristics of flow line deformation as shown in Fig. 6. The base metal of the rod body is tempered sorbite with grain size of 8.0as shown in Fig. 7.

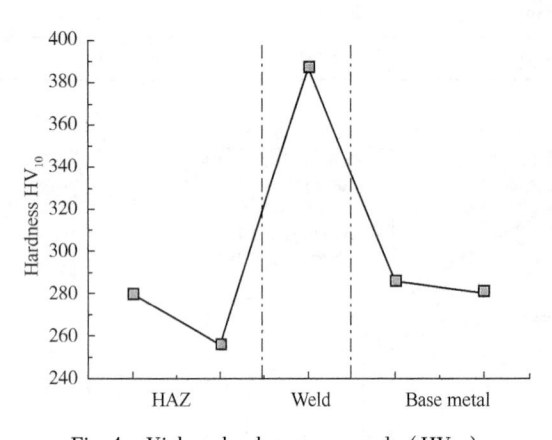

Fig. 4　Vickers hardness test results(HV_{10})

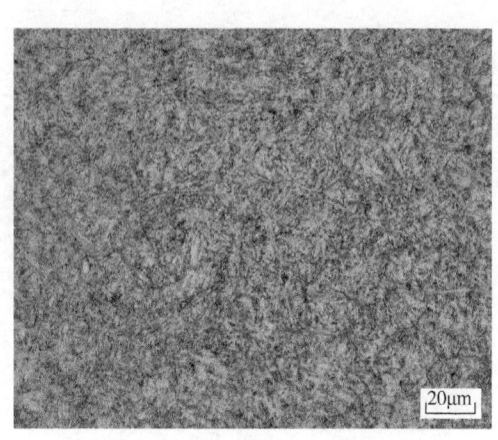

Fig. 5　Metallurgical structure of the weld

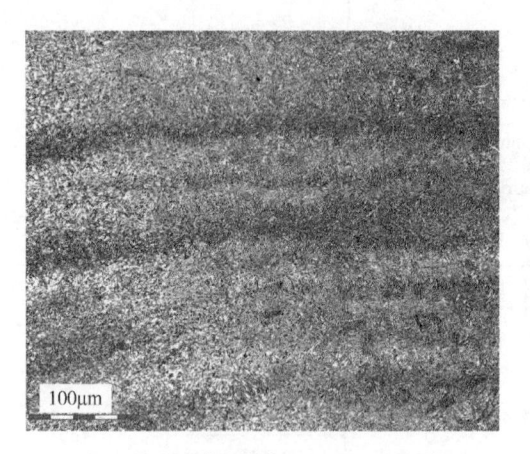

Fig. 6　Metallurgical structure of HAZ

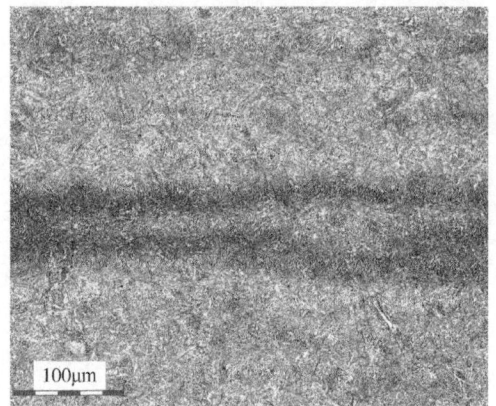

Fig. 7　Metallurgical structure of base metal

3.5 Fractography analysis

Fig. 8 shows the side morphology of the sucker rod fracture. It can be seen that the fracture cracks laterally on both sides of the weld seam. Clearly, we observe two fracture surface on different plane, besidesthe residual height of the weld is 0.5mm higher than the rod body. The sucker rod fractured on both sides of the weld, crack growth along the two cross section one after another. The crack in the first fracture starts from the outside of the sucker rod, and stoppedwhen the crack extends to its 1/2 circle. The first fracture'scrack end corresponds to the the second fracture'ssource area. After the the second fracture'scrack continues to expand thoroughly, the sucker rod breaks completely.

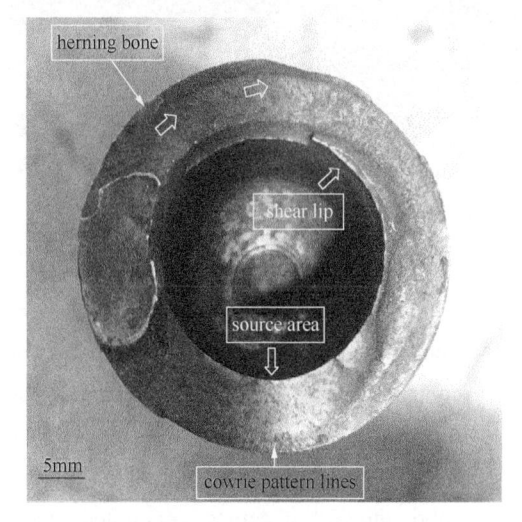

Fig. 9 shows the macroscopic morphology of the second fracture surface. It can be seen that the second fracture surface is divided into two characteristic sections: the flat section and the shear lip section. The flat zone occupies about 1/3 of the circumference of the fracture and has brittle cracking characteristics. The cowrie pattern lines points to the source area where inner side of the weld seam. The herring bone pattern points to the direction of the crack source along the rod circle. The shear lip area occupies about 2/3 of the circumference of the fracture, which is the instantaneous fracture area formed by the final fracture of the rod body. Brittle cracking occurred in the source region of the second fracture after the first fracture's crack extended to its end. The second fracture's crack continuous growth lead the sucker rod fractured completely.

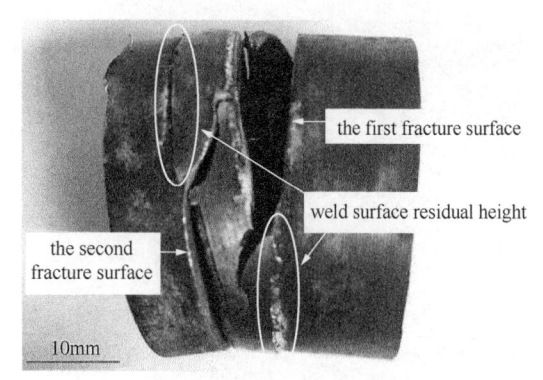

Fig. 8 Macroscopic morphology of fracture side

Fig. 10 shows the axial section of the fracture on the sucker rod. The fracture surface is located at the joint of the residual height of the inner weld seam and the base material. The crack originates from the inner surface of the interface between the weld seam and the base material, which is consistent with the cowrie pattern lines of the fracture surface points direction in Fig. 8, and the residual height of the weld inner side is curled and itas high as 4mm.

Fig. 9 The macroscopic morphology of the second fracture surface

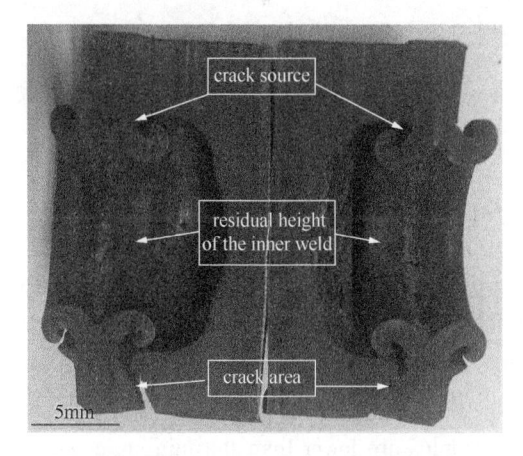

Fig. 10 The macroscopic morphology of the fracture was cut along the axial direction

The fracture source area SEM photo shown as in Fig. 11, the surface is relatively flat. Fig. 12 and Fig 13 show the micromorphology of the fracture propagation zone. A large number of fatigue striations can be seen at the micro level of the fracture in Fig. 12, extrusion marks were shown in Fig. 13 as the white arrow pointed, indicating that the fracture has fatigue propagation after repeated folding and extrusion.

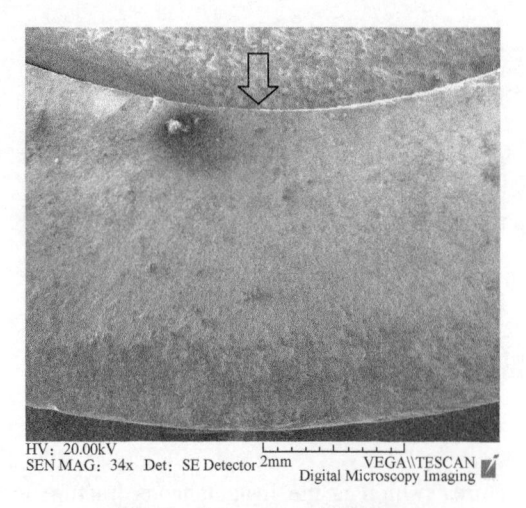

Fig. 11　Source area SEM photo 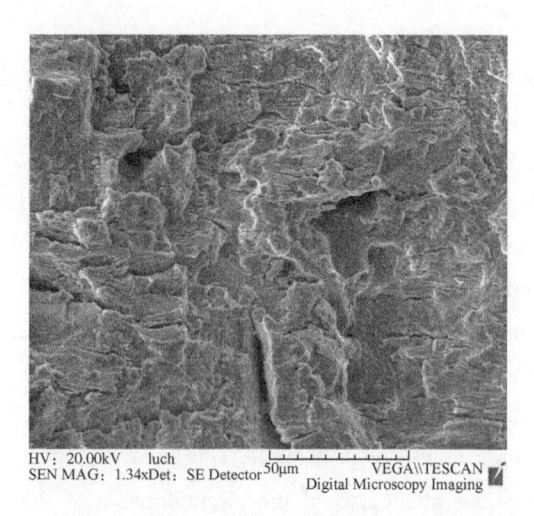　　　　　　　Fig. 12　Propagation area fatigue striations

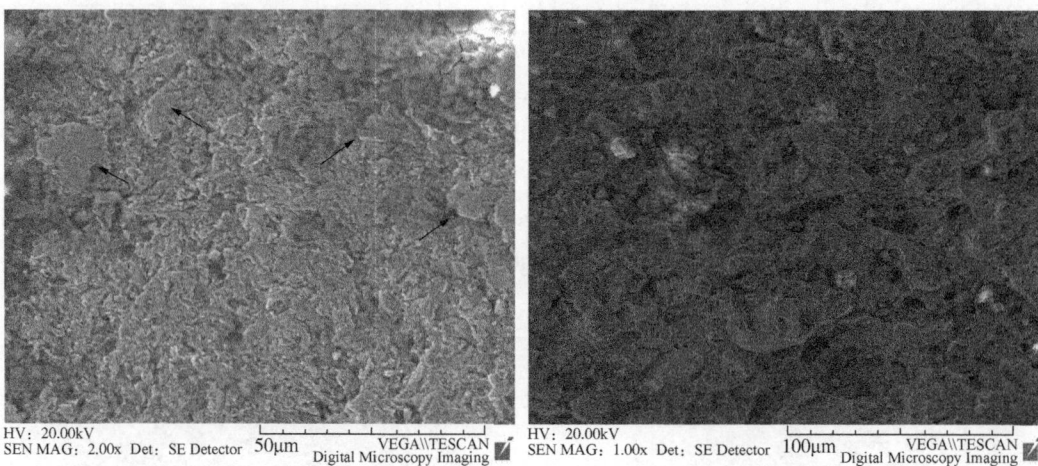

Fig. 13　Propagation area extrusion trace

4　Root cause analysis

Based on the test results and analysis above, the chemical composition of the sample has some deviation from the standard requirements, the metallographic structure of the weld is high hardness tempered martensite as shown in Fig. 5, the tensile properties at room temperature do not meet the standard requirements of SY/T 5029—2013, the tensile strength and yield strength of the upsetting end samples are lower than the minimum requirement of the standard. It can be concluded that the poor mechanical properties of the material is one of the main reasons for the fracture of the rod.

The main reasons and type of the rod's fracture can be determined by combining the macroscopic morphology and service stress of the sample. First of all, because the rod is a welding structure member, it is prone to dimensional change at the welding seam. According to the macroscopic morphology of the fracture in Fig. 10 above, the weld reinforcement of the inner and outer sides weld reaches 4mm and 0.5mm respectively, and the stress concentration in this part was caused by dimensional change change. Secondly, in the mechanical test, the hardness value of the weld seam is significantly higher than the base material and the heat affected area, because of high hardness tempered martensite was formed. So the weld became more prone to crack than the base material. Finally, during the service period, the sucker rod is subjected to alternating load. When such cyclic load exists for a long time, it is easy to cause fatigue damage of the structure. Such fatigue damage will often give priority to the formation of fatigue cracks in the material dimension where is not uniform or the dimensional change site. From the macroscopic morphology of the fracture in Fig. 8-10, it can be seen that the crack originated from the weld residual height part, and it expanded laterally along the lateral residual height edge until complete fracture. The fatigue striation and extrusion marks of the fracture can be clearly seen from the fracture micromorphology in Fig. 12 and Fig. 13. In addition, the metallographic analysis also shown that the microstructure of the heat-affected zone of the weld has the characteristics of streamline deformation in Fig. 6. It can be inferred that, under the action of cyclic loading, fatigue and fracture occurred in the high weld residual part after repeated folding and extrusion, which eventually led the weld of the sucker rod to fatigue fracture.

5 Conclusions and recommendations

5.1 Conclusions

(1) The reason for the fracture of the sucker rod is fatigue fracture. The fracture process is as follows: under cyclic loading, the residual height of the sucker rod in the inner side of the weld forms a crack source because of stress concentration at first, then the crack propagates along the heat affected zone, after further extends along the residual height of the outer weld lead the sucker rod complete fracture.

(2) The content of C in the chemical composition of polished rod end material is lower than the requirement of GB/T 26075—2010 standard for 35CrMo low alloy steel, and its room temperature tensile property does not meet the requirement of SY/T 5029—2013 standard.

5.2 Recommendations

(1) Improving the structure of sucker rod, especially the excess weld height in the welded part, needs to be effectively handled.

(2) Improve welding quality, control welding temperature, nondestructive testing of weld seam after welding, etc.

(3) The quality of sucker rod products should be effectively controlled, and materials that meet the standards and specifications should be selected.

References

[1] A Q Fu, Y R Feng, R Cai, et al. Downhole corrosion behavior of NiW coated carbon steel in spent acid & formation water and its application in full-scale tubing Engineering Failure Analysis, Volume 66, August 2016, Pages 566-576.

[2] Fu A Q, Kuang X R, Han Y, et al. Failure analysis of girth weld cracking of mechanically lined pipe used in gasfield gathering system. Engineering Failure Analysis, 2016, 68: 64-75.

[3] Y. Long, G. Wu, A. Q. FuFailure analysis of the 13Cr valve cage of tubing pump used in an oilfieldEngineering

Failure Analysis, Volume 93, November 2018, Pages 330-339.
[4] H. Ding, J. F. Xie, Z. Q. BaiFracture analysis of a connecting rod for oil pumping unit in China western oilfieldEngineering Failure Analysis, Volume 105, November 2019, Pages 313-320.
[5] D. L. Duan, Z. Geng, S. L. Jiang, S. Li Failure mechanism of sucker rod coupling Engineering Failure Analysis 36(2014)166-172.
[6] J. An, C. Li, Z. Wen A study of boronizing of steel AISI 8620 for sucker rods Metal Science and Heat Treatment, Vol. 53, Nos. 11-12, March, 2012.
[7] NB/T 47013.5—2015, Nondestructive testing of pressure equipments, Part 5: Penetant testing.
[8] GB/T 26075—2010, Steel bars for sucker rods, 2010.
[9] SY/T 5029—2013, Sucker rods, 2013.

本论文原发表于《Engineering Failure Analysis》2020 年第 109 卷。

Stress Analysis of Large Crude Oil Storage Tank Subjected to Harmonic Settlement

Zhang Shuxin[1] Liu Xiaolong[2] Luo Jinheng[1] Sun Bingbing[2]
Wu Gang[1] Jiang Jinxu[3] Gao Qi[2] Liu Xiaoben[3]

(1. Tubular Goods Research Institute, China National Petroleum Corporation & State Key Laboratory for Performance and Structure Safety of Petroleum Tubular Goods and Equipment Materials; 2. Petrochina West Pipeline Company; 3. National Engineering Laboratory for Pipeline Safety/MOE Key Laboratory of Petroleum Engineering/Beijing Key Laboratory of Urban Oil and Gas Distribution Technology, China University of Petroleum-Beijing)

Abstract: Differential settlement has a significant effect on the safe operation of tanks. In order to investigate the stress response of crude oil storage tank subjected to harmonic settlement. A numerical simulation model of the steel storage tank was developed in this study. Based on a validated finite element model, parametric analysis was conducted based on the main factors. The results show that the stress at the top of the tank wall is symmetrically distributed along the circumference of the tank. The stress result increases linearly with the increasing harmonic amplitude. The axial stress increases firstly, and then decreases for large wave number. Meanwhile, The stress value increases with the increment of the height-to-radius ratio, and decreases with the increment of the radius-to-thickness ratio. Especially, when the height-to-radius ratio ranging from 1.0 to 1.5 or the ratio r/t is more than 1500, the tendency of stress variation is slighter. This study could be referenced instrength or safety assessment of crude oil storage tank subjected to harmonic settlement.

1 Introduction

With the rapid growth of oil demand, the need of storage devices has increased. As important storage equipment, large scale oil tanks have been widely used in recent decades. Meanwhile, the steel tanks are usually constructed on coastal areas where large differential settlement may easily occur. And the failure behaviour caused by foundation settlement is an urgent problem. In order to ensure the safe operation of storage tanks, investigation on mechanical response of oil storage tank subjected to settlement is important and necessary.

Corresponding author: Liu Xiaoben, xiaobenliu@cup.edu.cn

In recent years, extensive experimental and numerical research has been conducted to analyze the performance of large scale steel tanks subjected to differential settlement. Gong et al.[1-4] adopted numerical simulation methods to buckling strength of the fixed-roof tank subjected to harmonic settlement. Based on the symmetrical finite element model of the tank shell, Ahmed et al.[5] investigated the effect of tank wall thickness on the buckling mode and the critical buckling settlement displacement. Shi et al.[6,7] studied the effect of wave number, harmonic amplitude and liquid level on the radial deformation of the tank wall. Full-scale model of tank was developed to simulate stress state caused by differential settlement. Li et al.[8] studied the deformation and stress response of large oil storage tanks subjected to foundation settlement. Chen et al.[9,10] investigated the deformation behavior of steel tank wall under uneven foundation settlement, and proposed solution for predicting deformation of steel tank under differential settlement. Shang et al.[11] studied the stress distribution of storage tanks under harmonic settlement, and Fourier series was adopted to represent the measured settlement value. The strength evaluation of the tank wall and bottom was carried out. Yang[12] investigated structural response of thin-walled cylindrical shells under uneven settlement, based on the result of experimental research. He[13] developed the finite element method to obtain the deformation result and stress distribution of steel tanks subjected to uneven settlement.

As it can be seen from there view of previous literature, investigations on the mechanical response of large scale steel tank subjected to differential settlement. In this stage, parametric analysis was conducted based on the main factors that influence the buckling strength of storage tank. While the stress and deformation analysis of tanks is performed by 2D finite element method. Numerical simulation results can not accurately reflect the true service status of the storage tank. Meanwhile, influence factors of tank's stress response were not sufficiently investigated. In this study, stress results of large crude oil storage tank was simulated by 3D model. The influence of the harmonic amplitude, wave number, height-to-radius ratio and radius-to-thickness ratio on the stress response of tank was studied.

2 Finite element model

2.1 Geometric model

In this study, a full 3D numerical simulation model of crude oil storage tank was developed. The radius and height of the model are $R = 40$m, $h = 22$m, respectively. The variations of wall thickness were considered. The shell and bottom plates of the tank were modeled by S4R. The elements modelling the shell-to-bottom fillet welds and tank foundation are C3D8R. Based on the results of a preliminary mesh sensitivity analysis, the finite element model was divided into 70671 elements as shown in Figure 1.

(a) storage tank model   (b) Section in tank wall

Figure1 Finite element model of tank

2.2 Material properties

The elastic modulus, the density and the Poisson ratio of the tank material are 206 GPa, 0.3 and 7850kg/m³, respectively. The yield strength of Q235-B, 12MnNiVR and 16MnR adopted in the model are 235MPa, 490MPa and 345MPa. In order to estimate the mechanical response of steel tank subjected to measured differential settlement. The material nonlinearity was considered in the model. As shown in Figure 2, the well-recognized Ramberg-Osgood model was adopted to regress the true stress-strain curve of the tank materials.

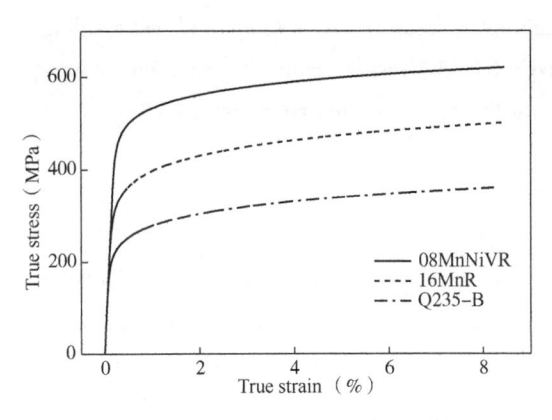

Figure 2 Stress-strain relationship for tank materials

2.3 Boundary conditions

The own-weight of the steel tank was calculated using the gravity constant g = 9.81m/s. Meanwhile, the effect of the hydrostatic pressure should be considered. The liquid pressure value decreases with the increasing height. The effect of the harmonic settlement was investigated by imposing axial displacement load to the circumference of the foundation, while the radial displacement and the circumferential displacement are constrained. The settlement value can be can be indicated as formula 1. The friction coefficient between tank foundation and bottom plate is 0.2 according to verification analysis.

$$u = u_n \cos(n\theta + \varphi_n) \quad (1)$$

Where u_n stands for harmonic amplitude, n represents wave number, φ_n is phase angle.

2.4 FE model verification

The FE model established in this study is validated by a stress result based on field tests[6]. The liquid level of the storage tank is set to 19.76m. The axial stress of the tank wall is plotted against the distance from bottom plate in Figure 3(a). It can be observed from the figure that the trends of the numerical simulation results of axial stress in this study and field test result are basically identical. The maximum of axial stress is mainly located at the shell-to-bottom Fillet

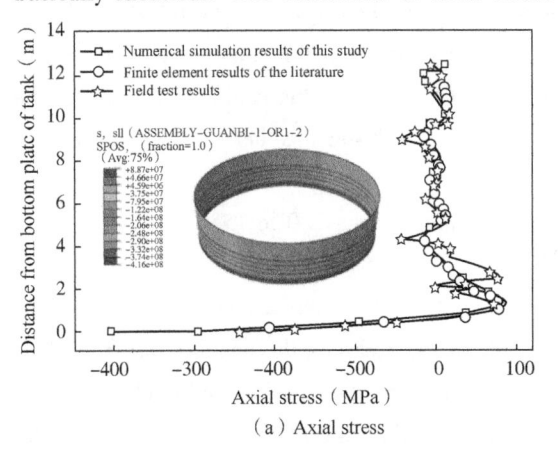

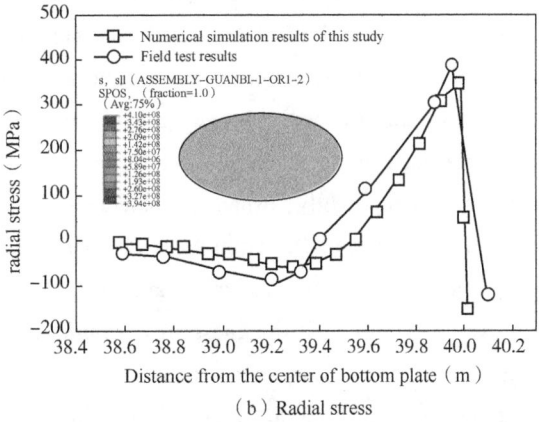

(a) Axial stress

(b) Radial stress

Figure 3 Comparative analysis of numerical simulation results and existing research results

Welds. As can be seen from the figure 3(b), the simulation result of radial stress and test result are highly similar. Comparing the finite element analysis (FEA) results with the existing research results, we can find that the maximum relative error of the model in this paper is 9.15%.

3 Parametricstudy

3.1 Effect of the harmonic amplitude(u)

Harmonic amplitude has a significant effect on the stress response of the tank subjected to harmonic settlement. As the amplitude increases, a large radial displacement, buckling of the shell, and even the failure of the tank may occur. In order to investigate the stress result of tank under differential settlement, the harmonic amplitude varies from 20mm to 100mm, and the Wave number is set as 3. Axial stress and hoop stress are calculated. The stress response result at the top of the tank wall are plotted against the harmonic amplitude in Figure 4. As observed in the figures, the stress is symmetrically distributed along the circumference of the tank. The axial stress and hoop stress presents a similar trend as follow: The stress result increases linearly with the increasing harmonic amplitude.

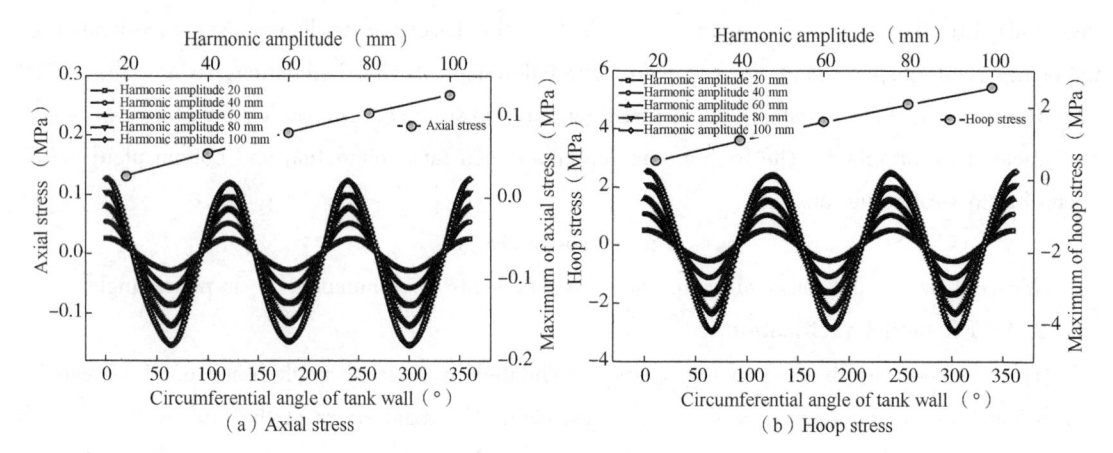

Figure 4 The stress response results versus circumferential angle of tank wall for various harmonic amplitude

3.2 Effect of the height-to-radius ratio(h/r)

The height-to-radius ratio(h/r) of large crude oil storage tanks varies. As the ratio h/r of tank increases, the stability of the shell will significantly change under the foundation settlement. In order to obtain the effect of the h/r on the stress response of tank wall, the axial stress results at the top of the tank wall with different height-to-radius ratio are displayed together in Figure 5(a). It can be clearly noticed that the axial stress is monotonically increasing. The value of axial stress increases more slightly, when the height-to-radius ratio ranging from 1.0 to 1.5.

3.3 Effect of the radius-to-thickness ratio(r/t)

The wall thickness design of large scale storage tank follows the equal strength design criterion. The value of tank wall thickness in the numerical model varies. Keeping the average wall thickness constant at 18mm, the stress response of tank wall are obtained when radius-to-thickness ratios ranging from 500 to 2000. Figure 5(b) illustrates the effect of the ratio r/t on the axial stress. The plots in the figure show that the axial stress decreases significantly as the radius-to-thickness ratio increases. Meanwhile, as for large ratio (e.g. the ratio r/t is more than 1500), the stress decreases slightly with increasing the ratio. The anti-deformation capacity of tank wall with larger radius-to-thickness ratio will be weaken.

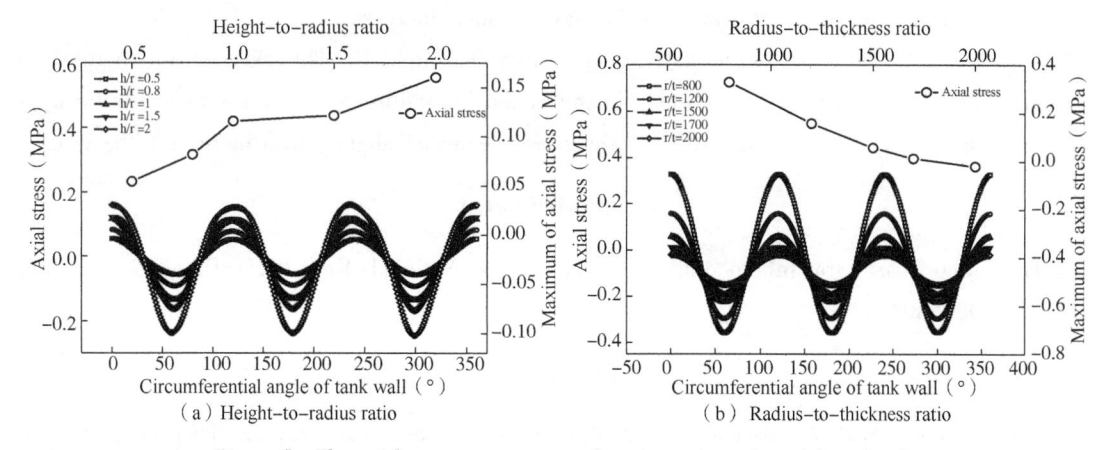

Figure 5 The axial stress response results versus circumferential angle of tank wall for various tank dimensions

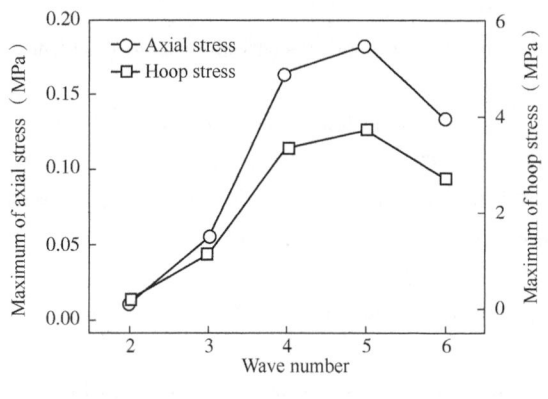

Figure 6 The stress response results versus wave number

3.4 Effect of the wave number (n)

As the most dangerous settlement type, differential settlement can be simplified as the harmonic forms. The harmonic settlement with wave number varying from 2 to 6 can more accurately reflect the real state of the foundation settlement. In this section, effect of the wave number was investigated by varying n from 2 to 6 while keeping the harmonic amplitude constant at 40mm. Figure 6 present axial stress of tank wall versus wave number. It clearly shows that the wave number has a significant effect on the stress response. The axial stress increases firstly, and then decreases for wave number n=6. This is due to the separation between the tank bottom plate and foundation becomes more obvious, as the wave number increases. The effect of foundation settlement on the tank wall stress is weakened.

4 Conclusions

A comprehensive investigation on stress response of large steel storage tank subjected to harmonic settlement has been performed throughout this paper. A numerical model was established by nonlinear finite element software ABAQUS. Based on the numerical model verified by existing research results, parametric analyses were conducted to derive how the parameters influences the mechanical response of large scale tank. Some remarkable conclusions can be drawn as follows.

(1) The stress value is symmetrically distributed along the circumference direction of the tank. The axial stress and hoop stress presents a similar trend. The stress result increases linearly with the

increasing harmonic settlement amplitude.

(2) The axial stress increases firstly, and then decreases for large wave number (e. g. when the wave number is more than 6). This may be due to the separation between the tank bottom plate and foundation becomes more obvious, as the wave number increases.

(3) When height – to – radius ratio ranges from 0.5 to 2.0. The axial stress increases s monotonically. The axial stress decreases with the increasing radius-to-thickness ratio. As for large ratio (e.g. the ratio r/t is more than 1500), the stress decreases slightly with increasing the ratio.

Acknowledgment

The authors are grateful to the fund support of National Key R&D Program of China (2017YFC0805804).

References

[1] Gong J G, Cao Q S(2017). Buckling strength of cylindrical steel tanks under measured differential settlement: Harmonic components needed for consideration and its effects[J]. Thin-Walled Structures, 119: 345-355.

[2] Zhao Y, Wang Z, Cao Q S, et al. (2013). Buckling behavior of floating-roof steel tanks under measured differential settlement. Thin-Walled structures.

[3] Chen Z P, Fan H G, Cheng, & Jian, et al. (2018). Buckling of cylindrical shells with measured settlement under axial compression. Thin-Walled Structures.

[4] Gong, J, Tao J, Zhao J, Zeng S, & Jin T. (2013). Buckling analysis of open top tanks subjected to harmonic settlement. Thin-Walled Structures, 63(FEB.), 37-43.

[5] Ahmed Shamel Fahmy, Amr Mohamed Khalil(2016). Wall thickness variation effect on tank's shape behaviour under critical harmonic settlement, Alexandria Engineering Journal, Volume 55, Issue 4, Pages 3205-3209, ISSN 1110-0168.

[6] Shi L(2016). Research on the Strength and Stability of Large Crude Oil Tanks[D], China University of Petroleum(Beijing).

[7] Shi L, Shuai J, Xu K, et al. (2014) Assessment of large-scale oil tanks foundation settlement based on FEA model and API 653[J]. China Safety Science Journal, 24(03): 114-119.

[8] Li W, Song W, SongJ L (2016). Deformation and stress analysis of large storage tanks under different settlement modes[J], Spatial structure, 22(04): 78-84.

[9] Chen, Y F, Ma, S, Dong S H. (2020). Deformation of Large Steel Tank under Uneven Foundation Settlement. IOP Conference Series: Earth and Environmental Science.526.012229.10.1088/1755-1315/526/1/012229.

[10] Zhao Y T Lou F Y Chen Y F et al. (2019) Contrastive Analysis on Tank's Behavior Model Under Foundation Settlement[J]. Industrial safety and environmental protection, 04: 30-33.

[11] Shang Guan F Y(2017). Stress Analysis of Large tank under Foundation Settlement[D]. Southwest Petroleum University.

[12] Yang Y. (2011) Experimental research on thin-walled cylindrical shells under uneven settlement [D]. Zhejiang University.

[13] He G F(2012). Deformation and stress analysis of large-scale storage tank under local foundation subsidence [J]. Special Structures, 2012(02)27-31.

本论文原发表于《Earth and Environmental Science》2020 年第 585 卷。

Study on Remaining Oil Distribution of Single Sand Body in Yan 10 Reservoir in Zhenbei Area, Ordos Basin

Wang Shuai[1] Luo Jinheng[1] Li Sen[2] Deng Liangguang[2] Han Binhu[2]

(1. State Key Laboratory of Performance and Structural Safety for Petroleum Tubular Goods and Equipment Materials, CNPC Tubular Goods Research Institute;
2. No. 11 Oil Production Plant, Changqing Oilfeild Company)

Abstract: In yan 10 reservoir of Jurassic in Zhenbei oilfield, there are many small laminations vertically, many casing broken wells, and the water cut rises rapidly. It is necessary to carry out the research on the distribution law of remaining oil in single sand body, so as to effectively guide the development of potential tapping measures. The logging electrical standard of single sand body division is established, the vertical and horizontal contact relationship of single sand body is clarified, and the influence of different contact patterns of single sand body on water breakthrough is studied. The geological model which can accurately reflect the spatial distribution of single sand body is established, and the numerical simulation research is carried out to clarify the distribution characteristics of remaining oil under different single sand body superposition patterns. The vertical superposition pattern of single sand body in yan 10 reservoir in Zhenbei area is mainly separated and isolated, and the horizontal contact relationship is mainly divided into bay and embankment contact. According to the influence of different single sand body styles on water content rise, the water content rise speed of single channel is the fastest, followed by side cutting and docking contact. In terms of the distribution law of remaining oil, the vertical separation type and isolated type of single sand body are rich in remaining oil, the horizontal up cutting type and docking type of single sand body are rich in remaining oil, and the single sand body is rich in remaining oil in the area with imperfect injection production horizon. The distribution rule of remaining oil under the control of single sand body is clear, which can effectively guide the development of potential exploration measures. At the same time, it can be used for reference in fine single sand body remaining oil characterization of similar reservoirs.

Keywords: Single sand body; Contact relationship; water drive; Remaining oil

---

Corresponding author: Wang Shuai, wangshuaib@cnpc.com.cn.

1 Geological survey

Zhenbei oilfield is located in the southwest of Ordos Basin, and its structure is located in the slope belt of Northern Shaanxi. The slope is a west inclined gentle monocline with a dip angle of only 0.5°~0.8°. There is a nose like structure in some parts. Yan10 reservoir in Zhenbei area is a very low permeability layered reservoir, with an average porosity of 12.8% and an average permeability of 3.30mD. The reservoir has poor physical properties and strong heterogeneity. In the process of waterflood development, there are some problems such as one-way breakthrough and low efficiency of waterflood, which lead to rapid water cut rise and low recovery degree, which seriously affect the development effect of oilfield.

1.1 Sedimentary characteristics

Based on the observation and description of core system of cored well, single well facies analysis is carried out in combination with geological logging, well logging curve, core analysis and test data, and through systematic research on color, sedimentary cycle, grain size characteristics, sedimentary structure, rock maturity, etc. yan10 oil reservoir group in Zhenbei area is determined as discernible river deposit, which is mainly composed of flood and sandy channel, side beach and core beach Facies, flood plain and other sedimentary microfacies.

1.2 Reservoir characteristics

1.2.1 Petrological characteristics

The lithology of yan10 reservoir group in Zhenbei area is mainly lithic arkose and feldspathic lithic sandstone with medium component maturity. The clastic components are mainly quartz, feldspar and rock debris, and the rock debris are generally medium and low-grade metamorphic rocks. The content of feldspar, quartz and rock debris is 18.1%, 55.1% and 26.8% respectively. The fillings are mainly hydromica (6.1%), siliceous (3.9%), ferridolomite (3.0%), kaolinite (2.3%), and the average content is 19.6%.

1.2.2 Pore structure characteristics

According to the statistics of casting thin section test data, the average face rate of yan10 reservoir in Zhenbei area is 5.3%. The main pore types are intergranular pore, followed by feldspar dissolution pore, intergranular pore and debris dissolution pore. According to the pore structure parameters, the displacement pressure of yan10 reservoir is 0.22MPa, the average throat radius is 0.84μm, the mercury removal efficiency is 36.2%, and the separation coefficient is 2.34. Image pore analysis shows that the average coordination number of yan10 reservoir in Zhenbei area is 0.32, the average pore diameter is 65.7μm, the average pore throat ratio is 3.0, and the average throat width is 5.6μm. The results show that the pore assemblage type of yan10 sandstone reservoir is mainly pore micro throat type[1].

2 Division of single sand body

2.1 Identification mark of single sand body logging

According to the rock electric calibration of core wells, the logging facies characteristics of single sand body identification marks, such as argillaceous interbed, calcareous interbed and physical interbed, are summarized. The characteristics of mud interlayer recognition are: SP return, GR rise, AC high and RT low. The recognition features of calcareous interlayer are slight return of SP, obvious decrease of GR, low value of AC and obvious increase of RT. The recognition characteristics of the physical interlayer are that the amplitude of natural unit(SP) and natural gamma(GR) is slightly abnormal, and the amplitude is related to the shale content. If there is calcium, the acoustic time difference(AC) shows a low value, and the resistivity(RT) shows a high value. On this basis, the yan10 reservoir group is subdivided into six stages of single sand bodies: yan10-1-1, yan10-1-2, yan10-2-1, yan10-2-2, yan10-3-1 and yan10-3-2.

2.2 Characteristics of single sand body superimposed longitudinally

Combined with the results of single well facies analysis, through in-depth dissection of the study area, the stacking patterns of some single sand bodies in the study area along the vertical and along the direction of material source are statistically analyzed, and the vertical and lateral stacking patterns of single sand bodies in the study area and their identification marks are summarized[2,3].

According to the statistics of the distribution frequency of the vertical stacking relation of the single sand bodies of yan10 sand formation in Zhenbei area, the vertical stacking relation of the single sand bodies of yan10-1-1~yan10-1-2, yan10-1-2~yan10-2-1, yan10-2-1~yan10-2-2 and yan10-2-2~yan10-3-1 is mainly separated type, accounting for 58.57%, 62.72%, 65.82% and 57.76% respectively, and the second is isolated type, only in the local area is superposition type and shearing type contact. The vertical stacking relationship of yan10-3-1~yan10-3-2 single sand body is mainly isolated type, accounting for 57.69%, followed by separation type, accounting for 35.58%, only in local area, superposition type and cutting superposition type contact(Fig. 1).

(a) superposition type　　　　(b) cutting superposition type　　　　(c) separated type

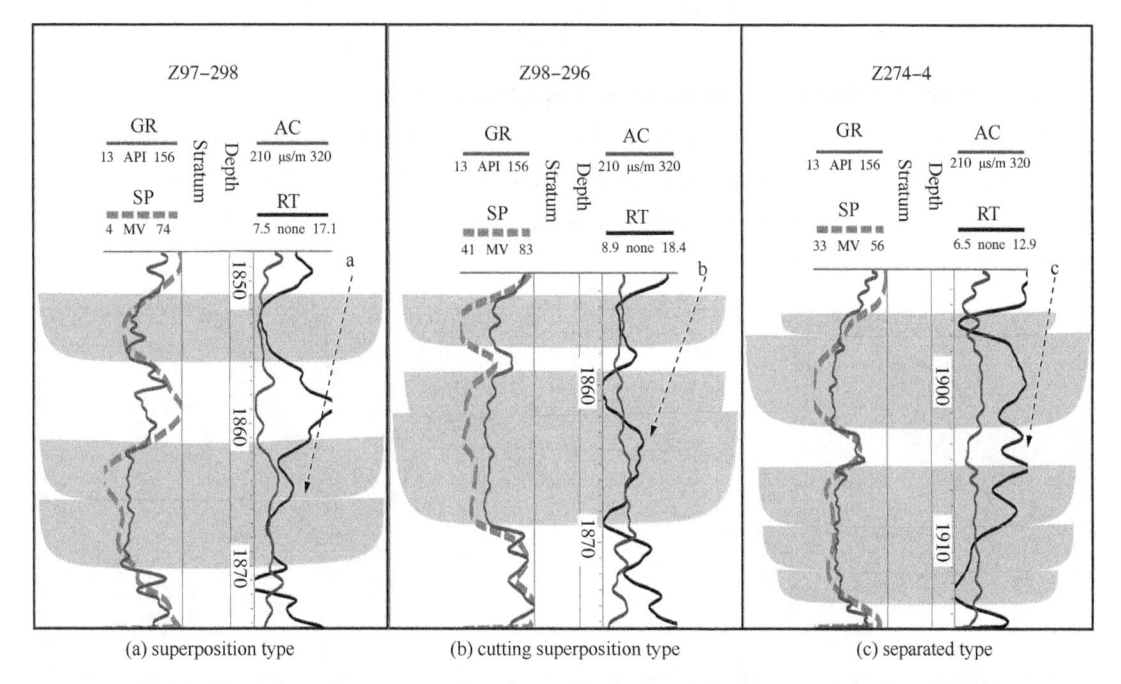

Fig. 1　Vertical stacking pattern of single sand body of yan 10 reservoir in Zhenbei Oilfield

2.3 Lateral contact characteristics of single sand body

According to the statistics of the distribution frequency of the single sand body lateral contact patterns of yan10 sand formation in Zhenbei area, the main lateral contact patterns of yan10-1-1, yan10-1-2 and yan10-3-2 sand bodies are bay type contact, accounting for 46.07%, 44.91% and 51.25% respectively; the main lateral contact patterns of yan10-2-1 and yan10-3-1 sand bodies are docking type and side cut contact, accounting for 34.36% and 31.67% respectively The main lateral contact pattern of 10-2-2 single sand body is embankment contact, accounting for 32.44% (Fig. 2).

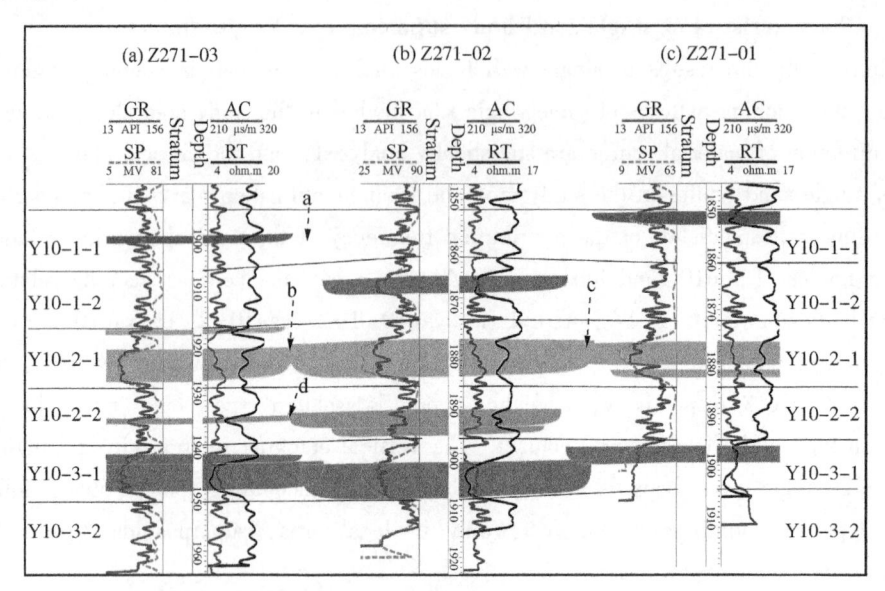

Fig. 2 Single sand body lateral contact pattern of yan10 reservoir in Zhenbei Oilfield
a: bay type contact; b: docking type; c: side cut type; d: embankment contact

2.4 Plane distribution characteristics of single sand body

It can be seen from the single sand body distribution map of each small layer of yan10-2-1 and yan10-2-2 in Z277 area of Zhenbei oilfield that there are 14~18 river channels from northeast to southwest in yan10-2-1 and yan10-2-2. The width of single sand channel is 150~350m and the thickness of sand is 4~7m (Fig. 3).

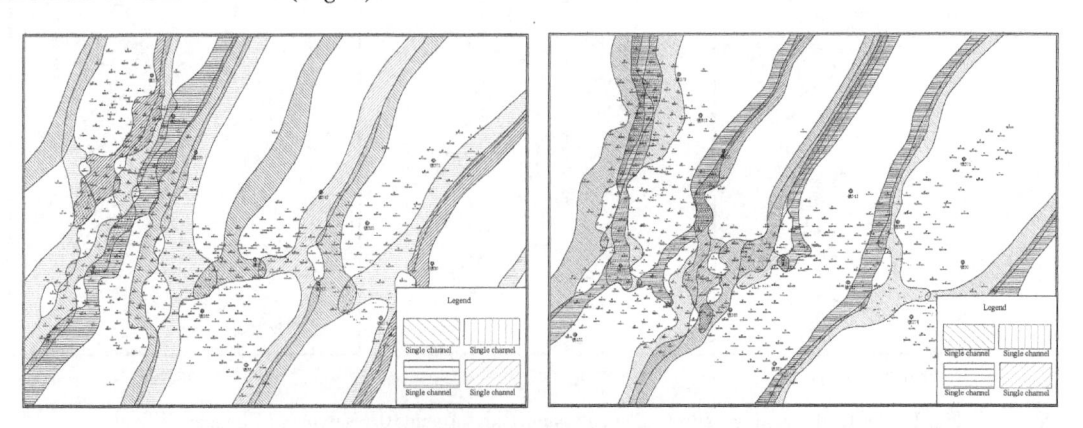

Fig. 3 Distribution Of Single Sand Body In Yan10-2-1 and Yan10-2-2 Of The Study Area

3 Effective water breakthrough characteristics of single sand body

3.1 Effective characteristics of single sand body

The water breakthrough range, direction and characteristics of each single sand body are different. It can be seen from the histogram of correlation analysis between single sand body lateral contact pattern and effective water breakthrough characteristics that water is easy to see in a single channel, followed by side cutting and docking lateral contact(Fig. 4).

3.2 Water breakthrough characteristics of single sand body

The water breakthrough range, direction and characteristics of each single sand body are different. It can be seen from the histogram of correlation analysis between single sand body lateral contact pattern and effective water breakthrough characteristics that water is easy to see in a single channel, followed by side cutting and docking lateral contact[4](Fig. 5).

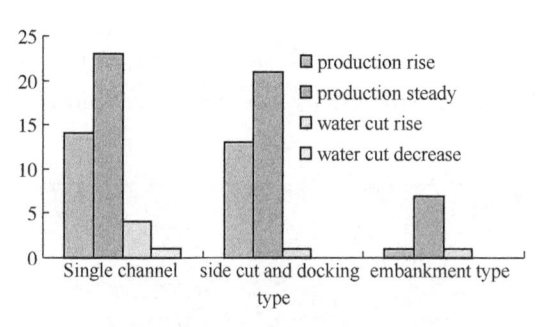

Fig. 4 Correlation Histogram Between Side Contact Pattern And Effective Characteristics Of Yan10 Single Sand Body In Zhenbei Area

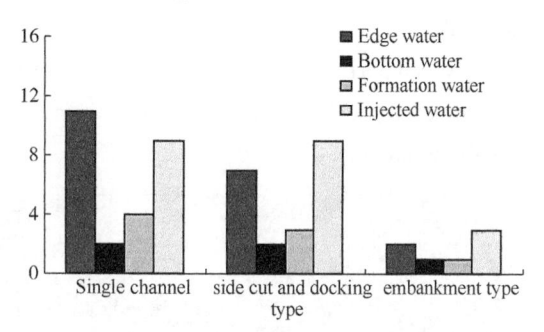

Fig. 5 Correlation histogram between side contact pattern and water breakthrough characteristics of yan10 single sand body in Zhenbei area

4 Distribution characteristics of remaining oil in single sand body

Based on the study of single sand body division and water drive law, the remaining oil distribution characteristics of single sand body are analyzed by numerical simulation method. Three dimensional geological modeling uses manual drawing of single sand body map constraint to establish lithofacies model, which reflects the lateral contact relationship of single sand body[5]; vertical single sand body contact relationship is realized by reserving the interlayer identified by logging. On this basis, the reservoir numerical simulation is carried out to study the influence of single sand body on the distribution characteristics of remaining oil[6-8].

4.1 Residual oil distribution of single sand body with different stacking patterns

Vertically separated type and isolated type are rich in remaining oil. Superposition type and cutting superposition type have been perforated and have good physical properties, and the water flooded area is relatively strong, with less remaining oil. As can be seen in Fig. 6, the superposition pattern of yan10-1-2 and yan10-2-1 single sand bodies in Z275-4 is separate type. Yan10-2-1 single sand body is not perforated and the remaining oil is rich. The superposition style of yan10-2-1 and yan10-2-2 single sand bodies is superposition type. Yan10-2-2 single sand bodies have been perforated, and the next physical properties are better than water flooding.

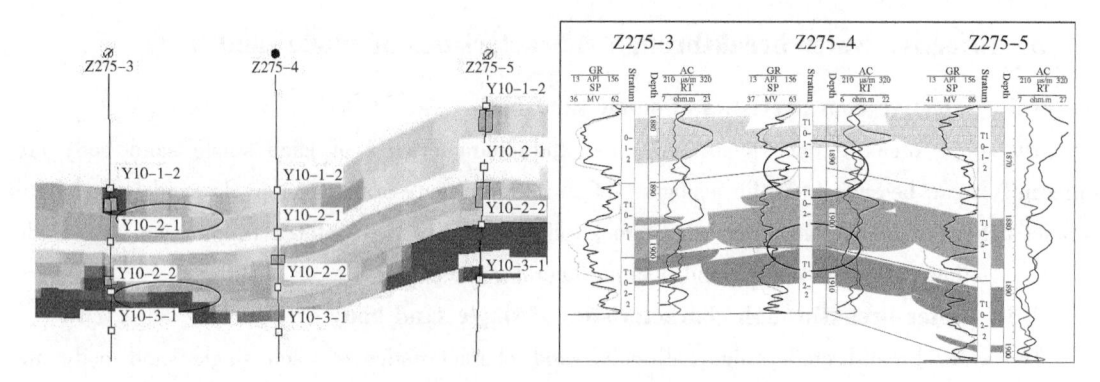

Fig. 6 Vertical superposition pattern of Z275-3 well Z275-5 single sand body and distribution comparison of remaining oil

4.2 Distribution characteristics of residual oil in single sand body lateral contact pattern

By comparing the lateral contact pattern of single sand body with the distribution rule of remaining oil, it is considered that the single sand body is enriched by side cutting and docking remaining oil(Fig. 7).

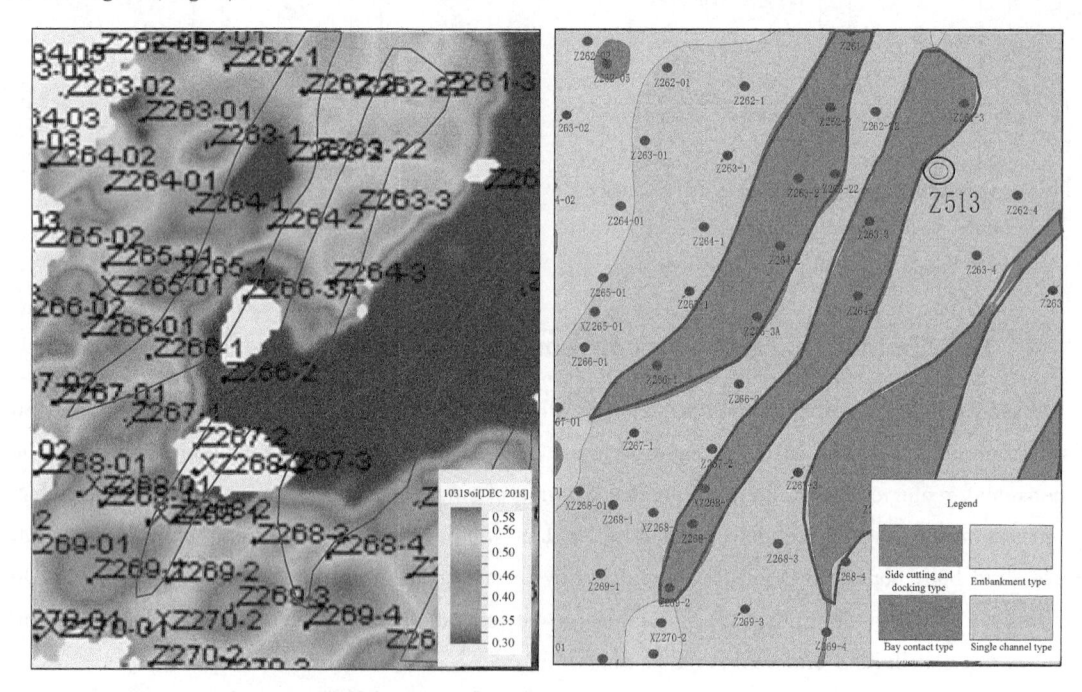

Fig. 7 Comparison of lateral contact pattern and remaining oil distribution of yan10-2-1 single sand body in yan10 reservoir of Zhenbei Oilfield

4.3 The distribution characteristics of residual oil in single sand body with imperfect injection production horizon

Because the injection and production of local single sand body are not corresponding, there is a situation of injection or production or injection, and the remaining oil in this area is relatively rich. It can be seen from the remaining oil profile of well Z275-3~well Z275-6 that the remaining oil of well Z275-4 and well Z275-6 with or without injection and production is enriched in small layer yan10-1-2. It can be seen from the remaining oil profile of well Z273-01 ~ well Z274-2 that the remaining oil of well Z273-01 and well Z274-2 with or without injection is enriched in small layer yan 10-2-1. It is necessary to carry out hole mending measures to improve the injection production system[9-10] (Fig. 8).

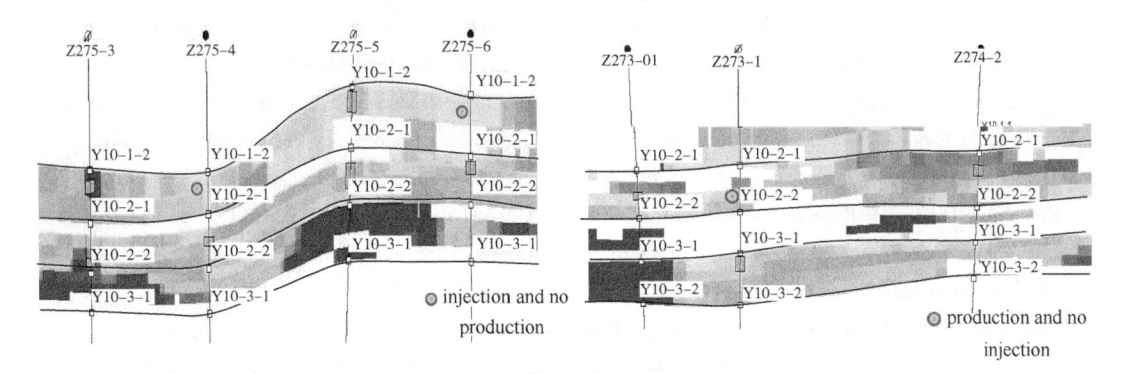

Fig. 8 Residual oil profile of single sand body injection and production imperfect area

5 Conclusion

(1) In yan10 reservoir of Zhenbei area, the single sand body is mainly composed of argillaceous interlayer, calcareous interlayer and physical interlayer, and the lateral contact of single sand body is mainly composed of distributary bay and embankment contact, followed by single channel distribution, and a few of them are side cut and docking contact. The vertical stacking pattern is mainly of separation type and isolation type, and the superposition type is located in the center and intersection of the river channel. The channel width of single sand body is between 150~350m.

(2) In the single channel, the main production is rising and stable, and the second is side cutting and docking contact. It is easy to see water breakthrough in a single channel, and then it is easy to see water breakthrough in side cutting and docking contact.

(3) Vertically separated type and isolated type are rich in remaining oil. Superposition type and cutting superposition type have been perforated and have good physical properties, and the water flooded area is relatively strong, with less remaining oil. By comparing the lateral contact pattern, the single sand body is enriched by side cutting and docking remaining oil. And the remaining oil is rich in the area with imperfect injection production horizon of single sand body.

Acknowledgement

The project is supported by Major National Science And Technology Projects (Number 2016ZX05016-003).

References

[1] GAO Shusheng, HU Zhiming, LIU Huaxun, et al. Microscopic pore characteristics of different lithological reservoirs. Acta Petroei Sinica 37(2), 248-256(2016).

[2] Liu Yuming, Hou Jiagen, Wang Lianmin, et al. Architecture analysis of braided river reservoir. Journal of China University of Petroleum: Natural Science Edition 33(1), 7-17(2009).

[3] Bai Zhenqiang, Study on the 3D architecture geological modeling of braided fluvial sandbody. Journal of Southwest Petroleum University: Science and Technology Edition 32(6), 21-24(2010).

[4] REN Xiaojuan, QU Zhihao, SHI Chengen, et al. Micro-flowing behaviors of water displacing oil in low permeability weak oil-wet reservoirs, Xifeng oilfield. Journal of Northwest University(Natural Science Edition) 35(6), 766-770(2005).

[5] TONG Kaijun, LIU Huiqing, ZHANG Yingchun, et al. Three diminsional physical modeling of waterflooding in metamorphic fractured reservoir. Petroleum Exploration and Development 42(4), 538-544(2015).

[6] YU Chunlei, MI Lidong, WANG Chuan, et al. Percolation characteristics investigation of microscopic remaining oil in water flooding reservoir with ultra-high water cut. Fault-Block Oil & Gas Field 23(5), 592-594(2016).

[7] HE Qiaolin, GUAN Zengwu, ZUO Xiaojun, et al. Waterflooding Experiment and Micro-Distribution of Remaining Oil in JIV Reservoir in Lunnan-2 Oilfields. Xinjiang Petroleum Geology 38(1), 81-84(2017).

[8] Li Junjian, Effects of microscopic pore structure heterogeneity on the distribution and morphology of remaining oil. Petroleum Exploration And Development 45(6), 1043-1052(2018).

[9] Dong Lifei, YUE Xiang'an, SU Qun, et al. Distribution of remaining oil by water flooding in heterogeneous reservoirs and indoor simulation study for its potential tapping. Oil Drilling & Production Technology 37(6), 63-66(2015).

[10] Feng Congjun, Bao Zhidong, Yang Ling et al. Reservoir architecture and remaining oil distribution of deltaic front underwater distributary channel. Petroleum Exploration And Development 41(3), 323-329(2014).

本论文原发表于 The 2019 Internation Field Exploration and Development Conference。

第二篇　成果篇

第三集　流果辞

一、省部级科技奖励

2020年获得省部级科技奖励见表1。

表1　2020年获得省部级科技奖励概览

序号	成果名称	授奖部门	获奖等级	获奖类型
1	OD1422mm X80管线钢管研制及应用技术	中国石油天然气集团有限公司	特等奖	科技进步奖
2	复杂工况油气井管柱腐蚀控制技术及工程应用	陕西省人民政府	一等奖	科技进步奖
3	双金属复合管产品研发及工业化应用	中国腐蚀与防护学会	一等奖	科技进步奖
4	苛刻环境油气钻采用管材选用关键技术研究及应用	中国腐蚀与防护学会	一等奖	科技进步奖
5	西部油气田集输管线内腐蚀控制技术及工程应用	中国石油天然气集团有限公司	一等奖	科技进步奖
6	SY/T 6859—2012 油气输送管道风险评价导则	中国石油天然气集团有限公司	一等奖	优秀标准奖
7	特殊性能钻采设备和管材应用及评价关键技术	中国石油天然气集团有限公司	二等奖	科技进步奖
8	石油工业用高性能膨胀管及其性能评价技术	陕西省人民政府	二等奖	科技进步奖

1. OD1422mm X80管线钢管研制及应用技术

随着国民经济发展和能源战略的实施，以及受土地、环保、建设与运营等因素制约，发展年输气量近 $400×10^8m^3$ 大输量天然气管道工程迫在眉睫。管材钢级的提高、输送压力和钢管口径增大等技术手段，是降低天然气长距离输送成本的有效途径。其中OD1422mm、12MPa、X80天然气管道方案的经济输量范围为 $(320～440)×10^8m^3/a$，最大输气量可达 $500×10^8m^3/a$。对于直径1422mm，厚壁X80管材，我国尚没有开发和应用的经验。2012年中国石油设立"第三代大输量天然气管道工程关键技术"重大专项，系统开展高钢级、大口径管线钢的研制，经过科研攻关，取得了三项重大创新成果。

（1）提出了OD1422mm X80管材化学成分要求，研究确定了管材的关键技术指标和试验方法，形成了管材系列技术标准。

① 研究设计了OD1422mm X80管线钢的化学成分。通过降低C含量、控制Nb含量0.05%~0.08%（技术指标对比见表2），严格控制微合金元素波动范围，使管线钢具有较低的碳当量，经反复工业化试验验证，可有效兼顾管线钢管的强韧性和焊接性。

表2　技术指标对比表

标准		C含量(%)	Si含量(%)	Mn含量(%)	Mo含量(%)	Ni含量(%)	Cu含量(%)	Cr含量(%)	Nb+V+Ti含量(%)
中俄东线标准	要求值	≤0.09	≤0.42	≤1.85	≤0.35	≤0.50	≤0.30	≤0.45	≤0.15
	直缝推荐值	≤0.07	≤0.30	≤1.80	0.08~0.30	0.10~0.30	≤0.30	≤0.30	0.095~0.135
	螺旋推荐值	≤0.07	≤0.30	≤1.80	0.12~0.27	0.15~0.25	≤0.30	0.15~0.30	0.105~0.135
西二线/三线标准	要求值	≤0.09	≤0.42	≤1.85	≤0.35	≤0.50	≤0.30	≤0.45	≤0.15
中国石油通用标准(CDP)	要求值	≤0.12	≤0.45	≤1.85	≤0.5	≤1.0	≤0.50	≤0.50	≤0.15
API 5L X80	要求值	≤0.12	≤0.45	≤1.85	≤0.5	≤1.0	≤0.50	≤0.50	≤0.15

② 通过对OD1422mm钢管的包申格效应、韧脆转变行为、各向异性研究，确定了OD1422mm X80钢管的强韧性指标，提出了拉伸、DWTT试验取样位置和方法。

③ 国内首次针对OD1422mm X80螺旋缝焊管和直缝焊管建立了管材/板材化学成分、启裂和止裂韧性等关键技术指标，形成了系列技术标准，用于指导国内钢厂、管厂进行管材开发和试制。

（2）攻克了OD1422mm X80管材、弯管、管件的制造工艺，研制出OD1422mm X80焊管、感应加热弯管等系列产品，形成了性能测试与表征技术。

① 开发了螺旋焊管成型包角角度显示仪和周长自动测量专用工装、大径厚比直缝钢管的成型对中定位装置和JCOE成型专用模具，解决了大径厚比钢管尺寸控制精度难题。

② 开发了适合X80钢级OD1422mm×21.4mm规格螺旋埋弧焊管低应力成型工艺技术和OD1422mm厚壁直缝钢管的残余应力控制技术，显著降低了钢管的残余应力，保证了钢管服役的安全性。

③ 通过开发新型焊丝、焊剂，优化坡口形式、焊接参数，形成了OD1422mm X80螺旋/直缝埋弧焊管高速、高强韧性埋弧焊接技术。

④ 开发了OD1422mm X80钢级感应加热弯管专用工装，包括专用前夹具、加热线圈等。通过优化煨制工艺，增加扶正辊轮等，开发出了OD1422mmX80钢级感应加热弯管椭圆度控制技术。

⑤ 开发了OD1422mm、X80板材、焊管、弯管及管件产品。开展了板材、焊管、弯管、管件的单炉、小批量试制及质量评价和研究，制定了OD1422mm X80板材、焊管及配套弯管、管件的综合评价方案，形成了管材性能测试和表征技术。

（3）修正了高压输气管道的止裂韧性预测模型及方法，发展了高压输气管线止裂预测技术。

① 分析了BTC、HLP、Sumitomo模型对OD1422mm X80管道止裂韧性计算的适用性，首次引入土壤因子，修正了高压输气管道止裂韧性BTC中管材流变应力、能量释放率与断裂韧性之间的关系。

② 完善了高压输气管道的止裂韧性预测模型及方法，提高了BTC方法的计算精度，发展了高压输气管线止裂预测技术。

③ 结合现有X80气体爆破试验数据库，提出了OD1422mm X80管道的止裂韧性指标。

④ 针对管道具体的服役条件，通过断裂力学、气体减压波特性，制定了OD1422mm X80高压输气管道断裂控制方案，解决了高钢级、大口径管道安全应用的核心问题。

本项目获发明专利15件，实用新型专利9件，编制发布产品和试验技术标准14部，发表论文21篇，累计供应研制规格的钢管近$80×10^4$t，租赁及销售研制施工装备97台套，近2年获得直接经济效益25.5939亿元。该研究成果填补了国内OD1422mm X80管材开发应用技术空白，成果已应用于中俄东线等国家重大管道建设，应用效果证明研究成果完全符合大输量管道建设工程需求，未来将具有广阔的推广应用价值。该成果刷新了我国高压大口径天然气管道建设记录，提升了高强度大口径油气管道建设技术水平，推动了我国钢铁冶金、材料加工、机电学科和油气输送管道领域的技术进步。

2. 复杂工况油气井管柱腐蚀控制技术及工程应用

油气井管柱作为油气生产的唯一通道，是油气勘探开发和安全生产的基础。我国每年消耗油气井管柱$(300～350)×10^4$t，耗资$(250～300)$亿元。随着我国油气勘探开发向"深、海、低、非"发展，油气井管柱服役工况环境日益复杂苛刻，失效事故频发，据统计，我国60%以上的油气井管柱失效是因腐蚀造成的。我国油气井主要有两大类复杂苛刻工况，一类是一次开发井工况，主要表现为高温（井底温度最高达200℃）、高压（地层压力最高达138MPa）、高矿化度地层水（Cl⁻含量高达160000mg/L）、高含CO_2/H_2S（含量超过15%）；另一类是二次增产井工况，随着我国大部分油气田开发进入中后期，大规模应用酸化压裂、CO_2驱、高温蒸汽驱、空气泡沫驱等增产工艺引入新的高腐蚀工况。复杂苛刻工况造成油气井管柱腐蚀失效，严重影响井筒的完整性和油气安全高效生产，国外复杂苛刻工况油气井也面临类似问题，如美国墨西哥湾高温高压气井、加拿大稠油井、英国和挪威北海油井、东南亚海上油气井等，属于世界性难题。本项目从复杂工况油气井管柱腐蚀失效难题入手，联合重点油气田和优势高校开展多学科联合攻关，阐明了油气井全生命周期内的多场耦合作用下的管柱腐蚀失效机制，建立了跨尺度腐蚀评价及选材技术，形成了一体化腐蚀控制技术体系，支撑我国重点油气田高效勘探开发和安全生产。

（1）揭示了油气井全生命周期内"材料（组织）—力学（载荷）—化学（环境）—工艺（增产作业）"多场耦合作用下的管柱腐蚀失效机制。

多环境场连续作用下点蚀失效机制（图1）：创新提出了基于油气井全生命周期服役环境（"鲜酸→残酸→凝析水→地层水"）的点蚀诱发及生长机制，主要表现为高酸度促使钝化膜减薄（鲜酸酸化阶段）、高浓度Cl⁻攻击钝化膜薄弱部位并诱发点蚀萌生（残酸返排阶段）、高分压CO_2阳极溶解形成闭塞环境引发点蚀扩展（凝析水阶段）、高分压CO_2和高浓度Cl⁻双重作用加速点蚀生长（地层水阶段）。揭示了13Cr不锈钢管柱点蚀发展过程的遗传效应，主要表现为点蚀坑内部微环境的改变滞后于外部环境的变化，使得前序腐蚀环境得以遗传并诱导点蚀自催化生长，加速点蚀进程。探索了微观组织对13Cr不锈钢亚稳态/稳态点蚀行为的影响，揭示了逆变奥氏体抑制界面贫铬从而提高耐点蚀性能的微观机制。多尺度结构界面失稳诱发应力腐蚀开裂失效机制（图2）：基于马氏体组织复杂多尺度结构界面（原奥氏体晶界、域界、块界、板条界），揭示了界面失稳对13Cr不锈钢管柱内（酸性环境）/外（碱性环境）表面应力腐蚀开裂诱发机制。阐明了13Cr不锈钢管柱在大排量酸液环境中出现的"阳极溶解"向"氢致开裂"的转变机制，即管柱表面缓蚀剂吸附膜局部破损导致"大阴极—小阳极"引发点蚀萌生，蚀坑内"低pH-高应力"微环境进一步诱发裂纹并沿原奥氏体晶界扩展。首次发现了13Cr不锈钢管柱在碱性环境中的应力腐蚀开裂行为，揭示了缓慢"阳极溶解"开裂机制，即马氏体多尺度结构界面发生铬元素的贫化，导致界面耐蚀性显著降低，在高应力下易发生裂纹萌生及扩展。基于上述理论，形成了对马氏体不锈钢在酸性及碱性溶液体系中应力腐蚀开裂敏感性的新认识，明确了13Cr不锈钢管柱在复杂苛刻工况环境中的应用边界条件。

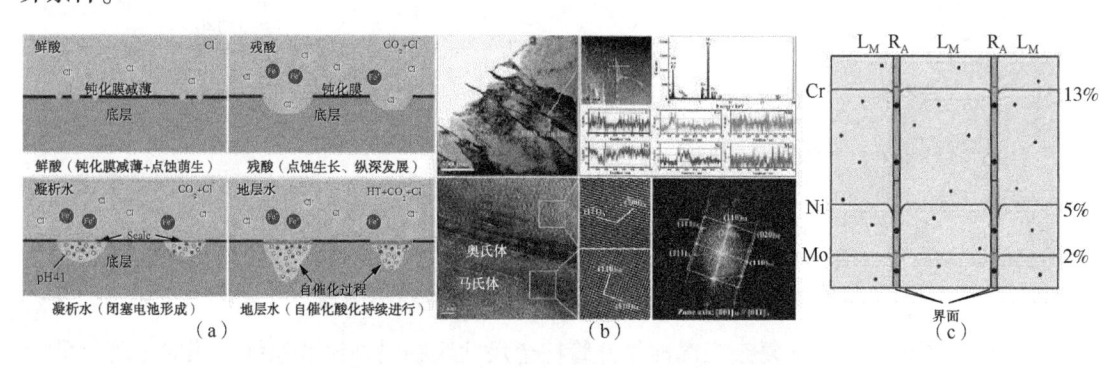

图1 多环境场连续作用诱导点蚀失效机制

(a)

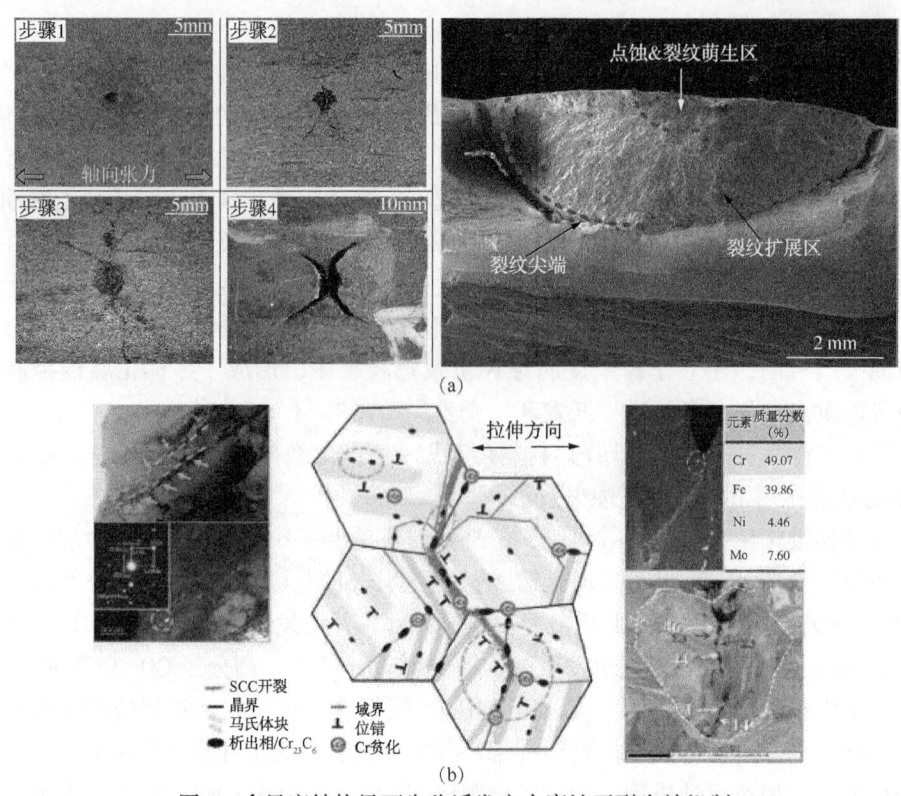

(b)

图 2 多尺度结构界面失稳诱发应力腐蚀开裂失效机制

电偶/应力促进作用下的缝隙腐蚀失效机制(图3)：针对油气井管柱连接结构(如螺纹连接、管柱与井下工具接触等)存在的缝隙—电偶—应力等多因素耦合作用下的腐蚀失效，发现了异种金属、同种金属、金属—非金属在油田产出水中的缝隙腐蚀敏感性规律。揭示了非金属材料在扭矩作用下的轻微变形诱发闭塞电池形成从而加速缝隙腐蚀的失效机制(13Cr-PTFE 的加速幅度约为140%，TP140-PTFE 的加速幅度约为21%)；揭示了异种金属连接中电偶加速缝隙腐蚀的失效机制(13Cr-TP140 的加速幅度约为70%，13Cr-G3 的加速幅度约为210%)；阐明了腐蚀产物在缝隙口沉积导致传质阻滞进而形成金属离子浓差电池的缝隙腐蚀机制；揭示了应力对缝隙腐蚀的倍增式加速效应(缝隙对腐蚀的加速幅度约为800%，应力对腐蚀的加速幅度约为170%，而缝隙与应力的综合加速幅度约为530%)。

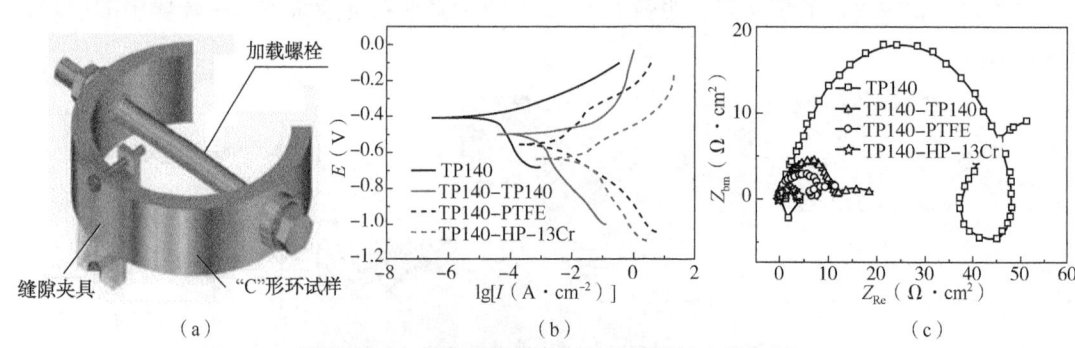

图 3 电偶/应力促进作用下的缝隙腐蚀失效机制

(2) 研发了国际上首套复杂工况油气井管柱全尺寸实物腐蚀试验系统，建立了跨尺度的腐蚀评价及选材技术。

基于油气井全生命周期的管柱腐蚀评价技术(图4)：针对油气井全生命周期的服役工况，建立了"动态多环境"腐蚀试验评价技术，用于表征"油气增产→生产工况"对管柱腐蚀过程的影响，形成了基于油气井全生命周期的管柱腐蚀选材技术标准，支撑了复杂工况油气井管柱的腐蚀选材及评价。全尺寸实物管柱应力腐蚀试验评价技术(图5)：针对传统小试样腐蚀实验方法因尺寸小、无管柱结构特征、无内压和应力等不足，自主设计研发了集环境、载荷、结构于一体的全尺寸油气井管柱应力腐蚀试验系统，实现了100MPa内压、1000t拉伸载荷、200℃高温、油气水多相腐蚀介质等复杂工况的多参量实验模拟。试验可获得管柱的极限承载能力、连接结构失效模式、表面处理工艺可靠性等关键信息。突破了油套管在应力作用下腐蚀和密封耦合作用试验瓶颈，解决了应力腐蚀试验中的结构和尺寸效应难题。为了进一步让室内实验结果接近现场，建立了油气井管柱"小尺寸试验筛选(Screening Test)→全尺寸试验适用性评价(Fitness for Service)→现场试验段验证(Validation)"相结合的跨尺度试验评价技术体系(图6)，实现了复杂工况油气井全生命周期内"多环境""多尺度""多因素"的综合评价。该技术先后在长庆、塔里木、大庆等油气田的CCUS驱采井、高温高压气井、高含硫气井的管材适用性、连接结构选型及优化、涂层/镀层工艺可靠性等评价研究中得到应用。

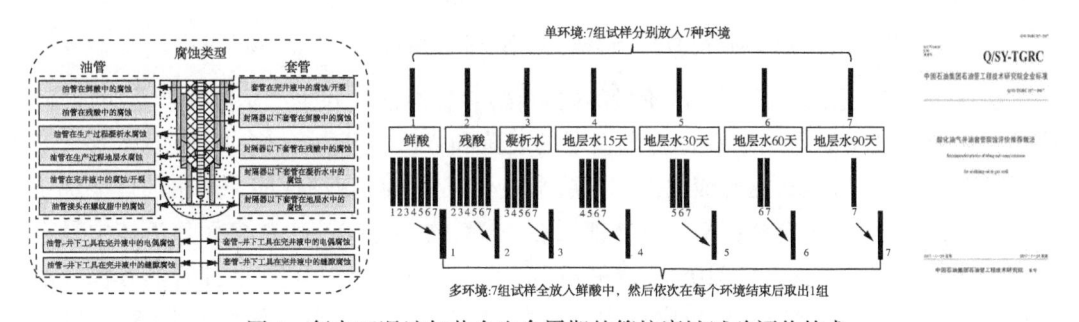

图4 复杂工况油气井全生命周期的管柱腐蚀试验评价技术

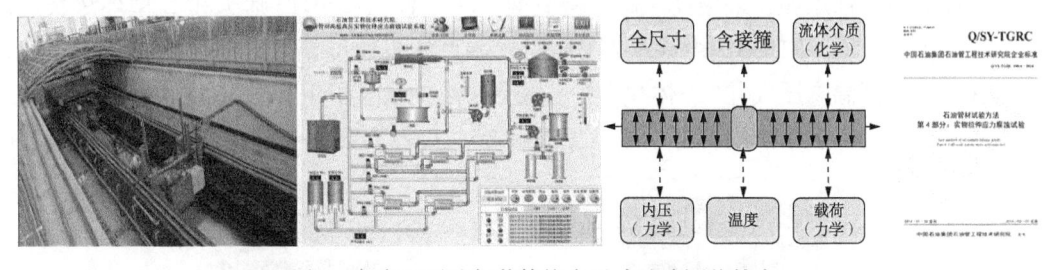

图5 复杂工况油气井管柱全尺寸试验评价技术

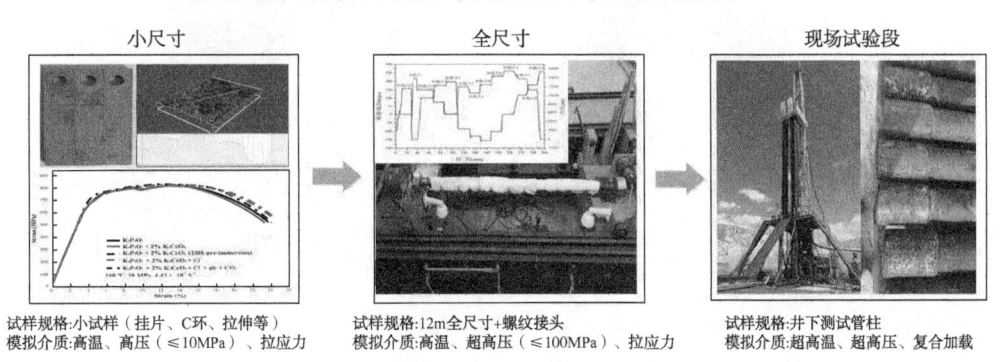

图6 复杂工况油气井管柱的跨尺度试验评价技术

（3）建立了以"高温酸化缓蚀剂技术""合金镀层技术""甲酸盐完井液防腐技术"为核心的复杂工况油气井管柱"一体化"腐蚀控制技术体系。高温酸化缓蚀剂技术（适用于增产改造阶段的内防腐）（图7）：针对高温酸化压裂增产工艺中的酸液腐蚀问题，建立了"空间多分子层多吸附中心"酸化缓蚀剂设计模型，以喹啉季铵盐化合物和曼尼希碱化合物为空间多分子层架构，植入多种金属离子（Sb^{2+}、Cu^+、Ca^{2+}及Al^{3+}）进行复配，形成系列高温酸化缓蚀剂配方和产品（TG201、TG201-II、TG202）。其中TG201-II缓蚀剂产品被认定为中国石油天然气集团公司自主创新重要产品。第三方检测机构的检测结果表明，在180℃的酸化液中，TG202缓蚀剂的加入将13Cr不锈钢的腐蚀速率控制在$18g/(cm^2·h)$，远优于SY/T 5405—1996标准的一级指标$[70g/(cm^2·h)]$。TG系列酸化缓蚀剂已经量产，并在高温酸化压裂增产井中广泛使用。

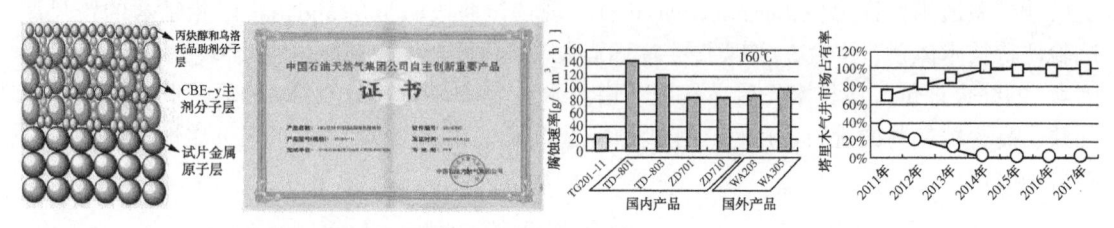

图7 高温酸化缓蚀剂的设计模型、产品证书、性能指标以及市场占有率

合金镀层技术（适用于长期生产阶段的内防腐）：针对油气井生产阶段的严苛腐蚀环境，发明了机械研磨技术制备纳米镍基合金镀层的装置和方法，开发了Ni-W镀层和Ni-Sn-P镀层。其中，Ni-W镀层在返排酸化液和地层水中具有优异的耐蚀性能，其耐蚀性能优于马氏体不锈钢，接近于G3合金；Ni-Sn-P镀层在240℃的$CO_2-O_2-H_2O$多元热流体中具有优异的耐蚀性能，其耐蚀性能与310S不锈钢相当。全尺寸实物Ni-W镀层油管具有优异的抗黏扣和密封性能。甲酸盐完井液防腐技术（适用于长期生产阶段的外防腐）（图8）：针对油套环空管柱应力腐蚀难题，采用高温高压慢应变速率拉伸法（SSRT）筛选和C型环法长周期验证，探究了"$K_2CO_3-KHCO_3$-甲酸盐"缓冲配方体系的服役可靠性，首次提出了以该体系替代磷酸盐作为高温高压气井"一体化"作业用完井液，建立了甲酸盐完井液质量控制和现场应用技术规范，出版了国内首部甲酸盐完井液技术专著，建立了完井液"优选—质控—应

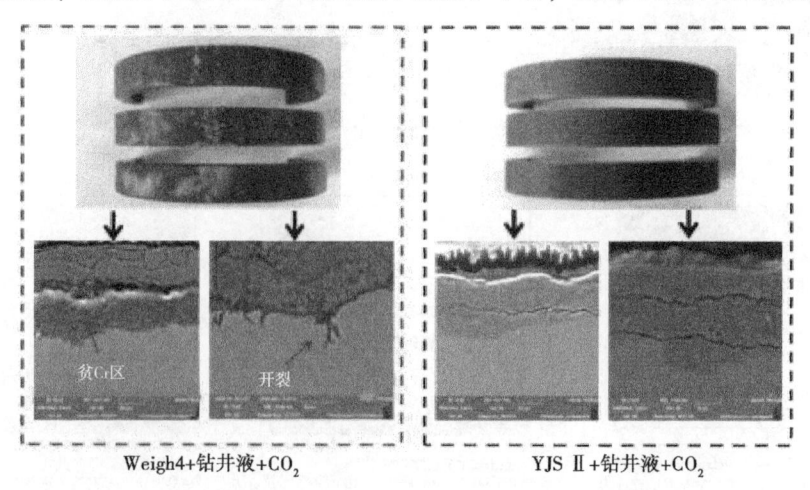

图8 甲酸盐完井液"优选—质控—应用"环空防腐技术体系

图 8　甲酸盐完井液"优选—质控—应用"环空防腐技术体系(续)

用"环空防腐技术体系。甲酸盐完井液于2017年起在塔里木高温高压区块推广应用85口井，新井A环空压力异常比例由50%降到0，至今未出现管柱腐蚀失效，全油田高温高压气井完整性提高至79%，高于墨西哥湾井完整性55%的国际先进水平。

3. 西部油气田集输管线内腐蚀控制技术及工程应用

随着塔里木("三高"油气田)和长庆("三低"油气田)等西部重点油气资源的深入开发，复杂苛刻采出介质导致集输管道腐蚀泄漏频发，严重影响油气正常生产和有油田经济效益。特别是自2015年新的"两法"实施以来，油气管道腐蚀泄漏不仅是经济问题，更是安全和环境问题。为了适应安全环保新形势和新要求，本项目针对油气田复杂苛刻环境中油气管道腐蚀泄漏难题，突破关键技术瓶颈，最终形成油气田管道"腐蚀失效识别→实验研究方法→防腐技术产品→腐蚀检测技术→现场防腐治理技术"体系，大大降低油田地面集输管道腐蚀穿孔频次，对保障我国"西气东输"主要气源安全输送具有重要战略意义。本项目取得了五项主要创新成果。

(1) 揭示了油气管道在复杂工况环境中"材料—化学—工艺—力学—电学"多因素耦合腐蚀失效机理。基于油气生产工艺，立足管道材料本构特性，结合油气管道的化学环境和受力状态以及电场干扰，系统研究了油气田地面集输系统用低碳钢、双金属复合管、2205双相不锈钢、内涂层管等管道的失效退化机理，建立了含不同缺陷管道的腐蚀预测模型和剩余寿命评价方法。

(2) 自主研发了全尺寸油气管道腐蚀实验系统，建立了油气管道"小试样筛选→全尺寸适用性评价→现场试验段验证"腐蚀评价技术。创新研发了集环境、应力、流态、结构于一体的全尺寸油气管道腐蚀实验系统，可实现极端工况多参量腐蚀实验模拟。为了进一步接近现场，建立了油气管道现场试验段橇装装置。最终形成了油气管道"小试样—全尺寸—现场试验段"的腐蚀评价技术体系(图9)。该技术先后在长庆、塔里木、大庆等油田管道评价研究中得到应用。

(3) 攻克了"三高一低"复杂苛刻环境油气管道的缓蚀难题，自主研发了桐油松香咪唑啉曼尼西碱系列缓蚀剂产品及配套技术。首创提出了"多分子层多吸附中心缓蚀模型"，采

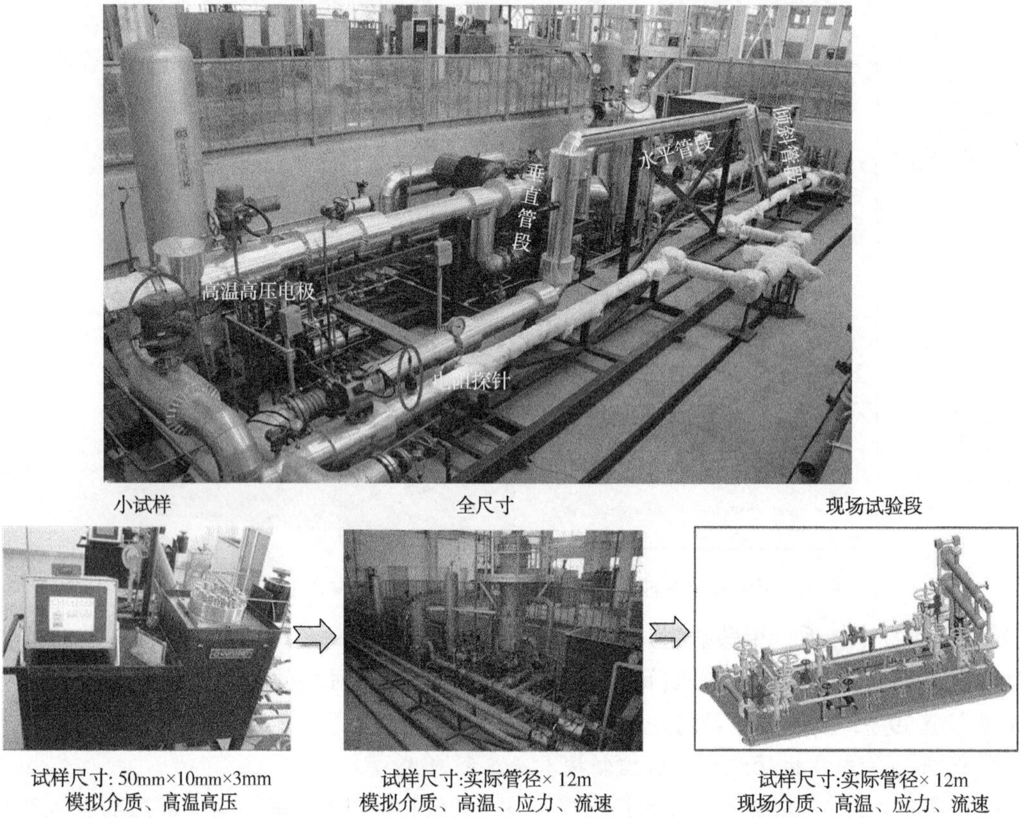

图 9 油气管道"小试样—全尺寸—现场试验段"腐蚀评价技术体系

用量子化学计算优化了缓蚀分子自组装结构,借助在线红外和量热合成技术精准合成了以桐油松香咪唑啉曼尼西碱为主剂的预膜、高 H_2S、高 CO_2、含氧等系列集输缓蚀剂产品(图10),配套开发了缓蚀剂智能加注技术,建立了缓蚀技术"研究—生产—应用—服务"完整性体系。在塔里木油田碳钢管线推广应用 5000 多千米,管道腐蚀穿孔率如图 11 所示。

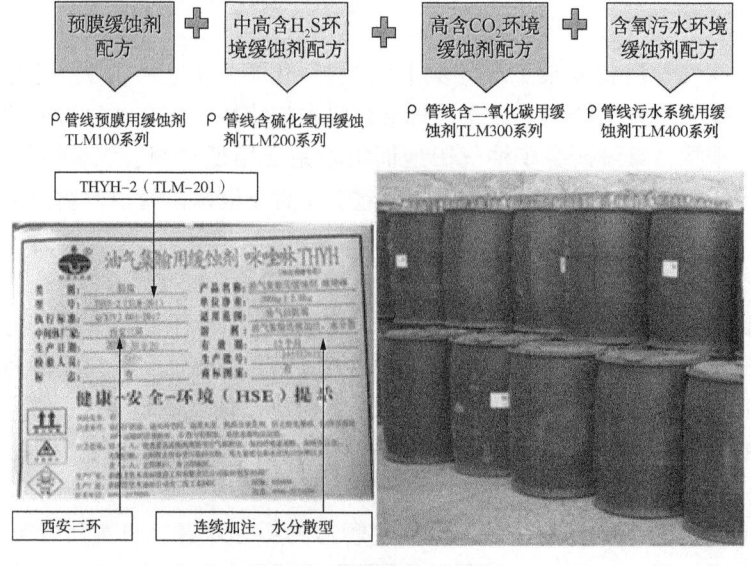

图 10 集输缓蚀剂产品

· 532 ·

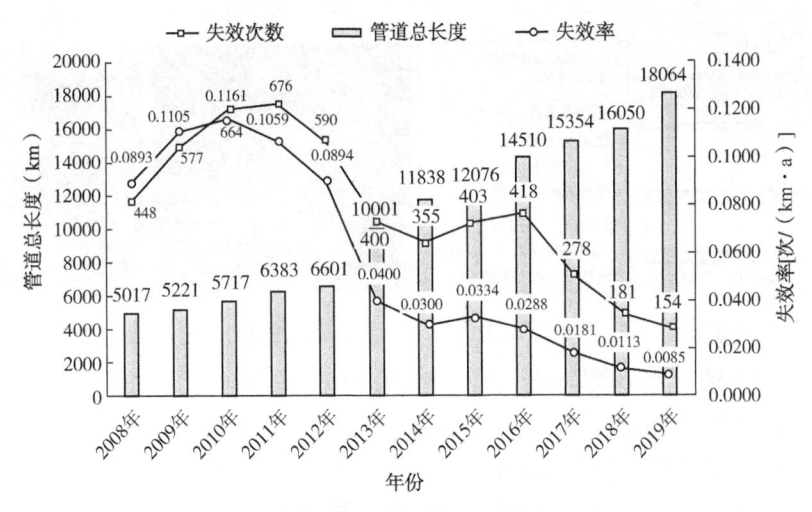

图 11 管道长度、腐蚀穿孔率随时间的变化

（4）攻克了小口径油气管道内腐蚀检测技术。为了解决油气田小口径管道（<DN200mm）内腐蚀检测的世界性技术难题，首创研发了小口径管道内窥检测、内防腐层测厚、电磁涡流检测为核心的检测技术，实现了机器人化、计算机远程控制及无线传输功能，填补了国内外技术空白。实验室和现场结果表明检测准确率达90%以上，在长庆油田示范应用260km（图12）。

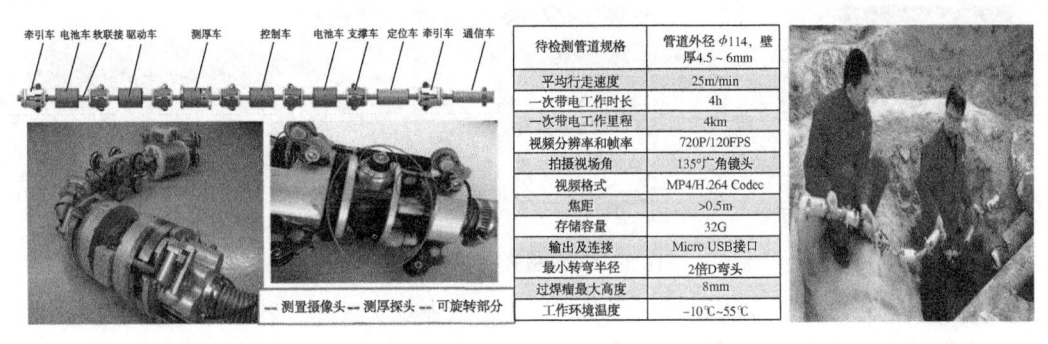

图 12 小口径油气管道内腐蚀检测技术示意图

（5）创新研发了油气田在役管道"在线风送挤涂"和"聚乙烯内衬"防腐修复工艺及标准规范。针对油气田在役管道开挖更换难且成本高的难题，建立了油气田地面系统在役管道"在线风送挤涂"和"聚乙烯内衬"防腐修复工艺及标准规范。以上两种技术在长庆油田累计应用超过11000千米，成功实现了油气田小口径在役油气管道的防腐修复与治理。长庆油田管道泄漏频次由2016年的0.18次/(km·a)下降至2019年的0.065次/(km·a)，年腐蚀泄漏频次减少70%（图13）。

本项目获授权专利18项（发明专利8项）、制修定标准9项（国/行标4项）、发表论文24篇（SCI/EI12篇）、专著3部。自2016年以来，累计产生经济效益10.3872亿元。项目总体成果达到国际先进水平，其中抗CO_2/H_2S缓蚀剂产品、全尺寸油气管道腐蚀实验系统及小口径管道内腐蚀检测技术领先于国内外同类技术和产品。该项目成果已在塔里木油田和长庆油田得到成功推广应用，为我国油气集输管道腐蚀研究和防腐技术应用提供了重要技术支持，推动了行业技术进步和产业发展。

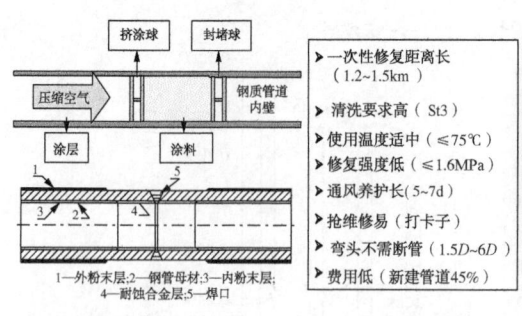

风送挤涂原理及连接技术　　技术主要参数　　在线修复后管道内表面　　现场施工后检测

图13 "在线风送挤涂"和"聚乙烯内衬"防腐修复工艺示意图

4. 石油工业用高性能膨胀管及其性能评价技术

该项目率先提出基于应变诱发塑性机制的膨胀管管材设计理念,即利用相变诱发塑性(Transformation induced plasticity,TRIP)机制和孪晶诱发塑性(Twinning induced plasticity,TWIP)机制显著提高管材的加工硬化能力和均匀延伸率,以金属塑性变形导致的应变(加工)硬化为突破口,提高低屈服管材的强度,有效解决易胀与高强度之间的突出矛盾。综合利用微合金化技术、炉外精炼技术、径向锻造技术、先进管材成形与热处理技术实现对管材几何尺寸精度、微观组织与性能的精准控制,如图14所示,开发出系列化高性能膨胀管管材。

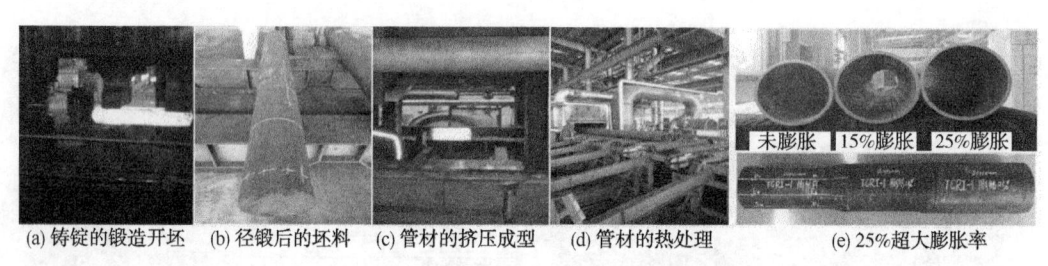

(a) 铸锭的锻造开坯　(b) 径锻后的坯料　(c) 管材的挤压成型　(d) 管材的热处理　(e) 25%超大膨胀率

图14 高性能膨胀管制管主要工艺流程与效果

在获得管材高尺寸精度的同时,微观组织内引入亚稳奥氏体,其在塑性变形过程中将持续转变为马氏体或形成机械孪晶,通过TRIP和TWIP机制获得持续稳定的高加工硬化率提高管材的均匀塑性变形能力和胀后强度,如图15和图16所示,满足高性能膨胀管对管材高均匀延伸率与加工硬化能力的要求,并且管材低的原始屈服强度满足了其易膨胀的要求。开发的新型系列膨胀管管材均匀延伸率大于20%,最高均匀延伸率可达50%以上,经测试其内径膨胀率达25.75%,可满足"单一井径"技术对膨胀管超大变形率的要求。胀前管材屈服强度约350MPa,经过约15%膨胀,管材的屈服强度可迅速提高至80ksi,管材综合性能超过美国亿万奇公司的水平,达到国际领先水平。

项目设计开发了两种膨胀管专用特殊螺纹接头,并形成相应的配套加工技术,其中单钩形端面金属自密封膨胀管螺纹接头外螺纹端和内螺纹端具有负角度的钩形内外螺纹,拧接后紧密咬合,接触应力分布均匀,具有较强的抗拉强度。单钩形螺纹的圆柱形金属密封面和橡胶密封圈共同密封,提高了膨胀管螺纹接头的密封可靠性,该结构具膨胀后密封性良好,加工成本低等优点。双钩形膨胀管螺纹采用变导向面偏角和螺纹锥度设计使得膨胀管接头具有双钩内外螺纹紧密啮合,消除了应力集中,具有更高的强度,且采用多重金属—金属、金属—橡胶—金属密封设计,具有膨胀后接头强度分布均匀与密封强度高的特点。

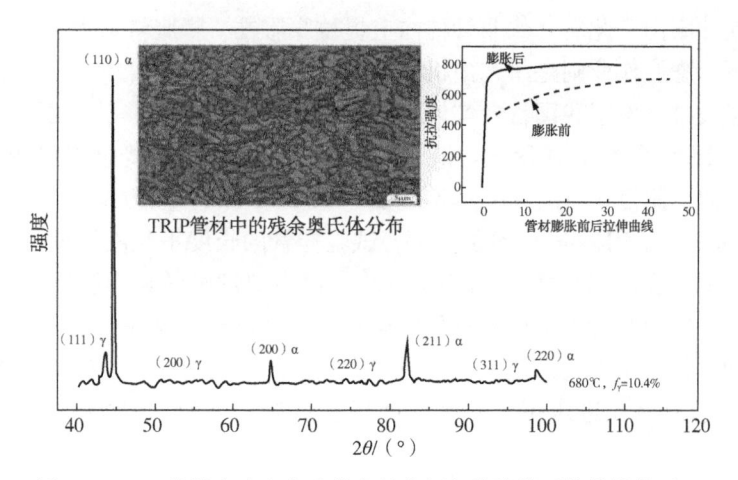

图 15　TRIP 管材中残余奥氏体定量分析与膨胀前后拉伸性能对比

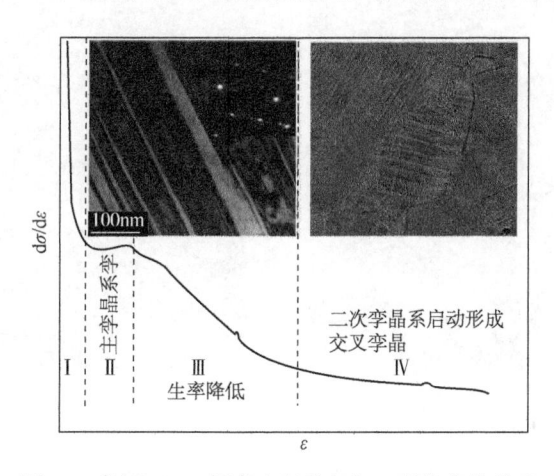

图 16　高锰 TWIP 钢孪生行为与加工硬化率的关系

基于螺纹结构设计形成一套完整的膨胀管特殊螺纹加工与检验技术，确保了螺纹质量，如图 17 所示。本项目设计的膨胀管螺纹性能达到国外同类产品技术水平。

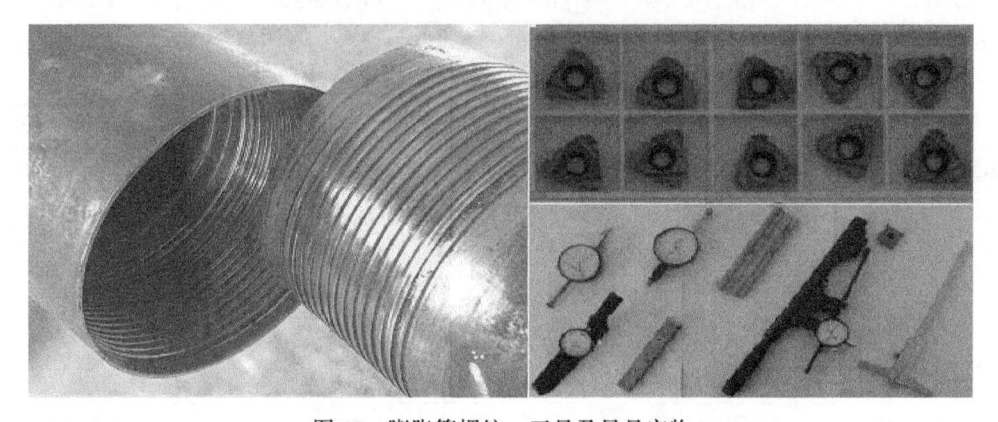

图 17　膨胀管螺纹、刀具及量具实物

膨胀工艺性能是膨胀管首要的服役性能，膨胀力的大小是衡量管材膨胀工艺性能的重要参数，很多失败案例都是由于膨胀压力过高而导致的。在现场进行膨胀作业时，首先通过水压实现膨胀锥的运动，将钢管的内径扩开。当膨胀锥通过预设在膨胀管底端的锚定橡胶后会将膨胀管固定在井眼内。此后，作业管柱向上提拉膨胀锥，膨胀将由水压与机械拉力共同完成。特殊情况下，膨胀作业则是单纯由机械拉力完成的。

该项目自主设计开发了我国首台实体膨胀管实物膨胀评价装备，如图18和图19所示，膨胀管实物膨胀评价系统主要参数：液压油缸行程5m；膨胀试验管材规格为外径89～340mm；机械拉伸载荷为0～1000kN，速度为0～9.15m/min，液力泵最大液压输出70MPa，排量为0～17.2m³/h，液压膨胀试验条件下，试验样管的长度不受液压缸行程约束。可进行机械膨胀、水压膨胀和机械/水压混合膨胀试验，可模拟试验各类膨胀管产品在套管井与裸眼井内的膨胀，可对膨胀管施工作业过程中遇到的各种复杂工况进行模拟试验，并完成相关试验数据的自动采集。设备关键技术指标，如试验范围、载荷能力、试验方式等方面均优于国外同类设备达到世界先进水平，填补了我国膨胀管实物膨胀性能检测评价装备领域的空白。

图18　膨胀管实物膨胀性能评价试验系统　　　　图19　正在进行膨胀试验的膨胀管

二、授权专利目录

专利号	名称	授权公告日
ZL 201810785018.5	一种埋地管道保护板及其安装方法	2020-02-14
ZL 201910110156.8	一种基于双相组织的海洋用高应变焊接钢管及其制备方法	2020-09-04
ZL201611200345.7	一种盐穴地下储气库井口水合物堵塞风险测定方法	2020-08-07
ZL201611208203.5	改进的斜Y坡口焊接裂纹敏感性试验试件及其制造方法	2020-04-10
ZL201611248328.0	地下储气库储存介质沿地层迁移泄漏半径及体积预测方法	2020-04-10
ZL201710583006.X	评价测井钢丝抗应力腐蚀开裂性能的模拟实验装置及方法	2020-01-07
ZL201710583006X	评价测井钢丝抗应力腐蚀开裂性能的模拟试验装置及方法	2020-01-07
ZL201710676044.X	一种油井管全尺寸旋转弯曲疲劳试验装置及方法	2020-07-14
ZL201710720520.3	一种基于R语言的金属拉伸试验方法	2020-01-07
ZL2017107365486	基于超声波发射法的水力压裂裂缝监测系统及方法	2020-11-06
ZL2017108614664	一种无损检测方法	2020-11-27
ZL201711039303.4	一种测试油套环空保护液长效性的试验装置和方法	2020-09-04
Zl201810002310.5	一种套管抗剪切性能评价方法	2020-08-07
ZL2018100025967	一种套管非均匀外挤能力评价方法	2020-11-06
ZL2018102453945	一种用于油气开发的高耐蚀钛合金管及其制备方法	2020-12-25

续表

专利号	名称	授权公告日
ZL2018102457804	一种用于高腐蚀性油气开发的含铁钛合金管及其制备方法	2020-12-25
ZL2018102506035	油井管接头的密封结构的设计与制造方法、密封方法	2020-09-04
ZL201810300981.X	一种超高强度钻杆用耐腐蚀钢管及其制造方法	2020-08-07
ZL201810300984.3	一种屈服强度大于1138MPa的钻杆用钢及其制造方法	2020-08-07
ZL2018103078788	一种便携式石油管内壁缺陷测量仪	2020-01-07
ZL201810312305.4	一种管道内长距离爬行检测机器人延伸装置及延伸方法	2020-06-05
ZL2018103123054	一种管道内长距离爬行检测机器人延伸装置及延伸方法	2020-06-05
ZL201810321078.1	一种高温高压气井油管气密封螺纹分析评价方法	2020-03-10
ZL2018104420838	一种管壁裂纹扩展速度测试装置的测试方法	2020-06-30
ZL201810468139.7	一种KD级表面渗铝改性抽油杆用钢及其杆体制造方法	2020-06-10
ZL2018108499374	一种用于柔性复合管拉伸性能测试的方法及试验装置	2020-11-06
ZL201810849999.5	一种环境友好型高温酸化缓蚀剂及其制备方法	2020-06-05
ZL2018108500102	一种油田地面集输管线用环保型耐氧缓蚀剂及其制备方法	2020-07-28
ZL201811022663.8	低温输气钢管承压能力及韧脆转变行为全尺寸试验方法	2020-11-20
ZL201811051810.4	一种检测低合金结构钢中残余奥氏体分布和含量的方法	2020-09-30
ZL201811056940.7	一种注气驱注入井用高温抗氧型缓蚀剂	2020-10-19
ZL201811106806.3	一种有机酸体系曼尼希碱类酸化缓蚀剂及其制备方法	2020-06-09
ZL201811149893.0	一种抗CO_2腐蚀集输管线缓蚀剂	2020-08-07
ZL201811604981.5	一种螺纹接箍	2020-09-04
ZL2019101073159	一种B型套筒的加工方法	2020-05-19
ZL2019104117568	一种输氢管道服役性能测试装置及测试方法	2020-12-11
ZL2019106831247	一种用于集输管线监测的多元传感器嵌入及保护方法	2020-11-27